ADVANCES IN MAGNETOSPHERIC PHYSICS WITH GEOS-1 AND ISEE

ADVANCES IN MAGNETOSPERIC PHYSICS WITH GEOS-1 AND ISEE

EDITED BY

K. KNOTT

and

A. DURNEY and K. OGILVIE

Reprinted from:

Space Science Reviews, Vol. 22, Nos. 4/5/6, and Vol. 23, No. 1, 1978/79

D. REIDEL PUBLISHING COMPANY

DORDRECHT : HOLLAND / BOSTON : U.S.A.

LONDON : ENGLAND

Printed in the Netherlands

TABLE OF CONTENTS

PART II: GEOS-1 PLASMA AND PARTICLE EXPERIMENTS

PART III: MAGNETOPAUSE AND MAGNETOSHEATH OBSERVATIONS

PART IV: SOLAR WIND AND BOW SHOCK OBSERVATIONS

INTRODUCTION

(Opening Address of 13th ESLAB Symposium)

With GEOS and ISEE occupying a major part of the ESA scientific programme it was clear several years ago that a 'natural' basis for the 1978 ESLAB Annual Symposium would be the early data from these two spacecraft. During the 1976 meeting of the European Geophysical Society in Amsterdam it became apparent to me that a much wider community was interested and that in particular Working Group 2 of COSPAR was considering a GEOS session during its 1978 meeting here in Innsbruck. This was of course as it ought to be because GEOS had been adopted as the reference spacecraft for the International Magnetospheric Study. After some discussions with COSPAR and with the organizers of the STP Symposium which was held here last week we felt that in the interest of customer relations and following good common sense we should give up some of our independence in 1978 and merge our Symposium with COSPAR. I would gratefully acknowledge the co-operation of COSPAR's Working Group 2 chairman and vice-chairman, the active support of the President of COSPAR and the 'gentleman's agreement' with the STP programme committee which arranged that the new ISEE and GEOS results were routed to this Symposium rather than to the STP.

I have talked about how we came to have this joint Symposium. Perhaps I should tell you too how we came close to having no Symposium. Just before dawn on 20 April last year Karl Knott and I stood on the roof of a building at NASA's Eastern Test Range watching one of the most colourful launches I have seen. At ground level it was still dark and the rocket took off illuminating the surrounding countryside in orange. Higher up dawn had arrived and the vehicle trail was beautifully illuminated in a cloudless sky. All looked perfect. A short time later, back within the control centre we learned however, that GEOS could not be placed in a geostationary orbit and indeed we were to discover that we were lucky to be in any orbit at all. Worries about radiation damage forced us to make a quick decision between the only two real alternatives which were injection into a 12-hour orbit with apogee close to the original geostationary position or injection into a 24-hour orbit with apogee in interplanetary space for much of the year. We chose the 12-hour orbit and are still in the process of being forgiven by several experimenters – especially (p)ioneers – who longed to get into the solar wind with their sophisticated new analyser.

Because of this mishap GEOS could no longer be reasonably described as the reference spacecraft for the IMS and in a formal sense had to be classified as a failure. In the next two days we are to hear of some of the excellent scientific measurements made during this 'failed' mission. The spacecraft and experiments have worked very well.

I believe the idea of ESA and NASA getting together on what is now called the ISEE mission came from Dr. Meredith of Goddard Space Flight Centre during a visit to Space Science Department at ESTEC. At that time many people were thinking in terms of flying several spacecraft together through the magnetosphere in order to sort out space/time ambiguities and NASA was considering a mission called IMP K–K′. After several changes of name we eventually arrived at 'International Sun Earth Explorer' and in October last year NASA launched the first two of the three-satellite mission – the 'daughter' spacecraft having been built by ESA. Data has been a bit slow in appearing – in particular the much discussed data pool tape – and I am looking forward to seeing the first results at this Symposium. It will be particularly interesting to me to see where the two spacecraft have made major advances over what could have been achieved by one spacecraft. This demonstration is required in order to reply to those critics who dismissed the mission as just one more trip at random into the magnetosphere.

We trust that our ESA contributions to ISEE and GEOS and our participation in the IMSCIE Boulder co-ordination centre have been useful inputs to the IMS. We hope too that this Symposium will be a significant landmark in the scientific output from the IMS. We thank you all for your interest and look forward to two days of exciting contributions and lively discussion.

D. E. PAGE
Head Space Science Department of ESA

THE GEOS-1 MISSION

K. KNOTT
SSD, ESA, Noordwijk, The Netherlands

GEOS-1 was launched on 20 April 1977 from Eastern Test Range, Florida, U.S.A. As a result of a malfunction in the separation mechanism between the 2nd and 3rd stages of the vehicle, a transfer orbit with a low apogee (12 000 km instead of the required 36 000 km) was reached. This deficiency was such that the satellite's onboard apogee boost motor could not turn this anomalous transfer orbit into anything near a geostationary orbit. The motor was, however, able to inject GEOS into an eccentric orbit with a periodicity of 12 hr. The apogee of the final orbit is at 38 000 km and the perigee at 2050 km.

One immediate consequence was that the GEOS-dedicated UHF station in Darmstadt, Germany, did not get continuous visibility of GEOS, as it would have done with a geostationary orbit. In fact, it sees only every second orbit for 4 to 5 hr before and after apogee. At the alternate apogees, the spacecraft can be seen from stations in the Pacific area. Such stations have been used to improve the total data acquisition in the period 2 June to 16 December 1977.

The scientific aim of the original (geostationary) mission of GEOS was to advance our knowledge of the dynamics of the magnetosphere, in particular to study the response of the near-Earth environment to varying conditions in interplanetary space. In order to achieve this goal, GEOS was to carry out integrated particle, wave, field and plasma measurements in the very centre of the magnetosphere, for which the geostationary orbit at 6.6 R_E offers an ideal platform. The orbit actually achieved by GEOS is shown in Figure 1. The apogee of this orbit is very near the geostationary position. Unfortunately, the spacecraft only spends a few hours in the regions for which its instruments were designed and then falls back to a perigee at ionospheric altitudes.

The 12-hr orbit is almost fixed in inertial space and therefore changes its position with respect to the Sun and within the magnetosphere. This variation is also demonstrated in Figure 1. At the time of the injection into this orbit – on 25 April 1977 – the apogee occurred at 19:30 LT and the line of apsides pointed into the early-evening sector of the magnetosphere. It has since drifted at a rate of 0.88 deg per day in the frame of reference of the magnetosphere and has crossed the whole dayside of the magnetosphere; from early 1978 onwards coverage of the nightside magnetosphere has been obtained. The apparent advantage of this orbit, namely increased radial coverage, is spoiled by the fact that the magnetometer saturates below $5R_E$ and the fact that most particle experiments become saturated or nearly so inside the radiation belts which are crossed four times per day.

The original scientific aims of the mission could therefore not be met. The

Space Science Reviews **22** (1978) 321–325.

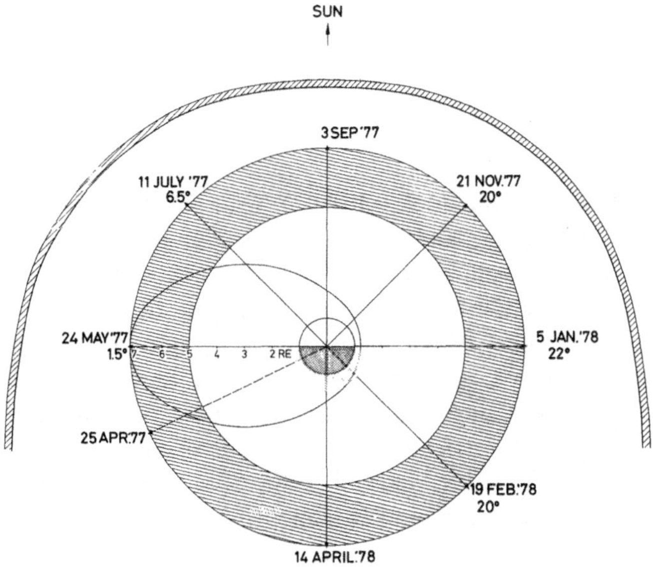

Fig. 1. Illustration of the GEOS-1 orbit in the frame of reference of the magnetosphere. The orientation of the line of obsides at various dates during 1977/78 is given. The dashed area indicates the location where most of the GEOS-1 data were collected.

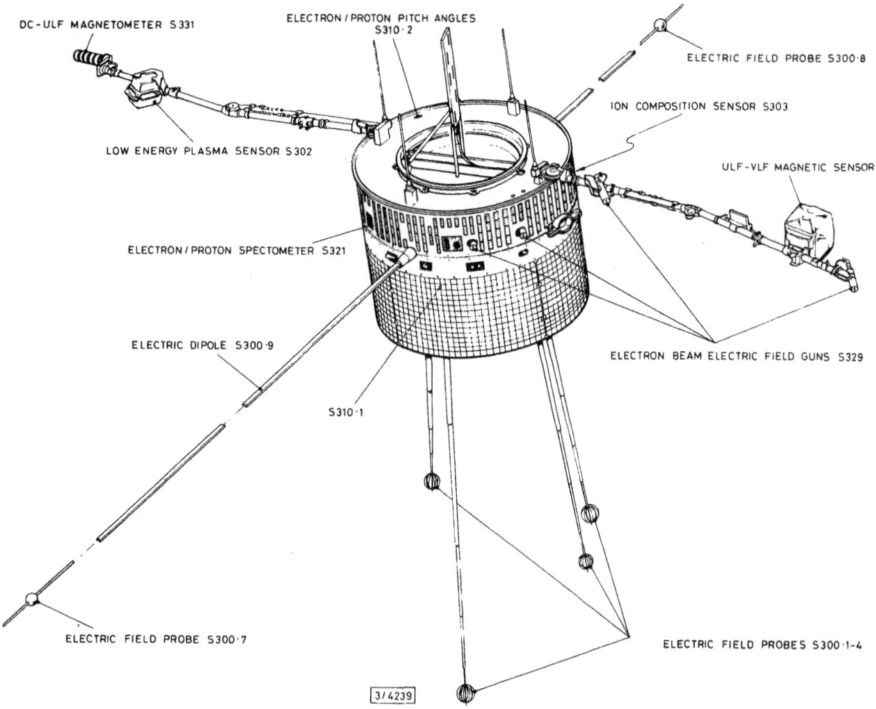

Fig. 2. Orbital configuration of the GEOS-1 satellite showing the accommodation of the various experiment sensors.

average conditions at the geostationary distance could not be established and deviations could not be measured. Consequently, instead of studying the dynamics of the magnetosphere, the GEOS-1 experimenters concentrated during 1977 on a thorough study of the dayside sector of the magnetosphere; in particular, the plasma bulge and the plasmapause have been investigated in detail. In addition, several interesting geomagnetically disturbed periods have been met when the spacecraft was around apogee. Because the apogee occurs near geosynchronous altitudes and can be changed easily in geographic longitude, it has been possible to bring GEOS-1 into geomagnetic conjugacy with existing ground-based observations in northern Scandinavia, Iceland, Syowa Base, and Alaska. Unfortunately, however, the apogee throughout 1977 occurred in the sunlit hemisphere, so optical observations were not possible from the ground at times when GEOS was in magnetic conjugacy with observations in the northern hemisphere. In early 1978 the local time of the GEOS-1 apogee drifted into the night time sector of the magnetosphere. In this situation it was possible to explore in detail that region of the magnetosphere where most of the dynamic processes develop. In addition it was possible to take full advantage of the existing ground based observation network in support of the mission.

GEOS-1 carries 7 experiments, of which 4 measure particles in various energy ranges and 3 measure fields and waves in various frequency ranges. The payload is a well-balanced mixture of instruments relying on proven techniques and of novel experiments aiming at the measurement of parameters which have so far not been measured in the outer magnetosphere. Figure 2 shows the accommodation

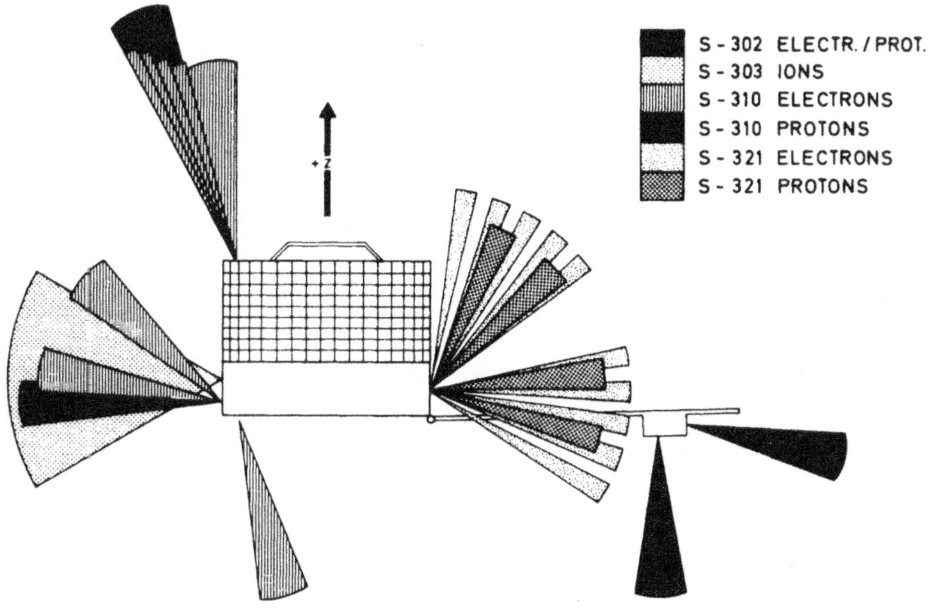

Fig. 3. Schematic illustration of the viewing angles of the four particle experiments carried onboard GEOS-1.

TABLE I
Payload of the GEOS-1 satellite

Exp. No.	Measurement	Technique	Experimenter from
S-300	AC-magnetic fields up to 30 kHz DC/AC electric fields and plasma resonances up to 80 kHz Mutual and self-impedance	Search-coil magnetometer Electric aerials Emission and reception of VLF signals	CRPE, Issy-les-Moulineaux, France Space Science Dept., ESA, ESTEC, Holland Danish Space Research Institute, Lyngby CRPE, Orléans, France
S-302	Thermal plasma up to 500 eV	2 electrostatic analysers	Mullard Space Science Lab., Dorking, UK
S-303	Composition (1–140 amu) and energy spectra of ions up to 16 keV	Combined electrostatic and magnetic analyser	University of Bern. Switzerland and Max-Planck-Institut, Garching, Germany
S-310	Pitch-angle distribution of electrons and protons in the 0.2–20 keV energy range	10 electrostatic analysers	Kiruna Geophysical Observatory, Kiruna, Sweden
S-321	Pitch-angle distribution for electrons (20–300 keV) and protons (20 keV–3 MeV)	Magnetic deflection system followed by solid-state detectors	Max-Planck-Institut, Lindau, Germany
S-329	DC electric field	Tracing of electron beam over one or more gyrations	Max-Planck-Institut, Garching, Germany
S-331	DC and ULF magnetic field	Fluxgate magnetometer	CNR, Frascati, Italy GSFC, Greenbelt, Maryland

of the various sensors employed by the field and wave experiments, and Figure 3 displays the viewing angles of the different particle spectrometers. The scientific aims, experimental techniques applied and institutes involved in the various experiments are summarised in Table I.

The rescue of the GEOS-1 mission was the most exciting and critical operation ever carried out by the European Space Agency. The initial orbit provided by the launch vehicle was not only totally uninteresting from a scientific point of view, it was also highly dangerous from a radiation damage point of view because its apogee occured inside the radiation belt of the Earth. Fortunately, the possibility existed to use the satellites built-in rocket motor to lift it into a more interesting and less dangerous orbit. In view of the continuing radiation damage - in particular to the solar cells - the necessary mission analysis had to be carried out within a few days.

The final mission and operation strategy was decided upon in a conference held at the European Space Operations Center (ESOC) in Darmstadt, Germany in the evening of 24 April 1977. From the various proposals which were presented, the one shown in the figure on the following page was finally selected. The motor was ignited by command from ESOC in the morning of 25 April, worked as predicted and placed GEOS-1 into its excentric rescue orbit with an apogee at 7 R_E, just beyond the distance of the envisaged geostationary orbit.

The GEOS design and development programme went over more than one half decade and demanded a lot of effort and devotion from everyone involved in it. The initial disappointment after the partial launch failure was therefore understandable. During the first months of operation and data acquisition it became apparent that this rescue mission would nevertheless render useful science.

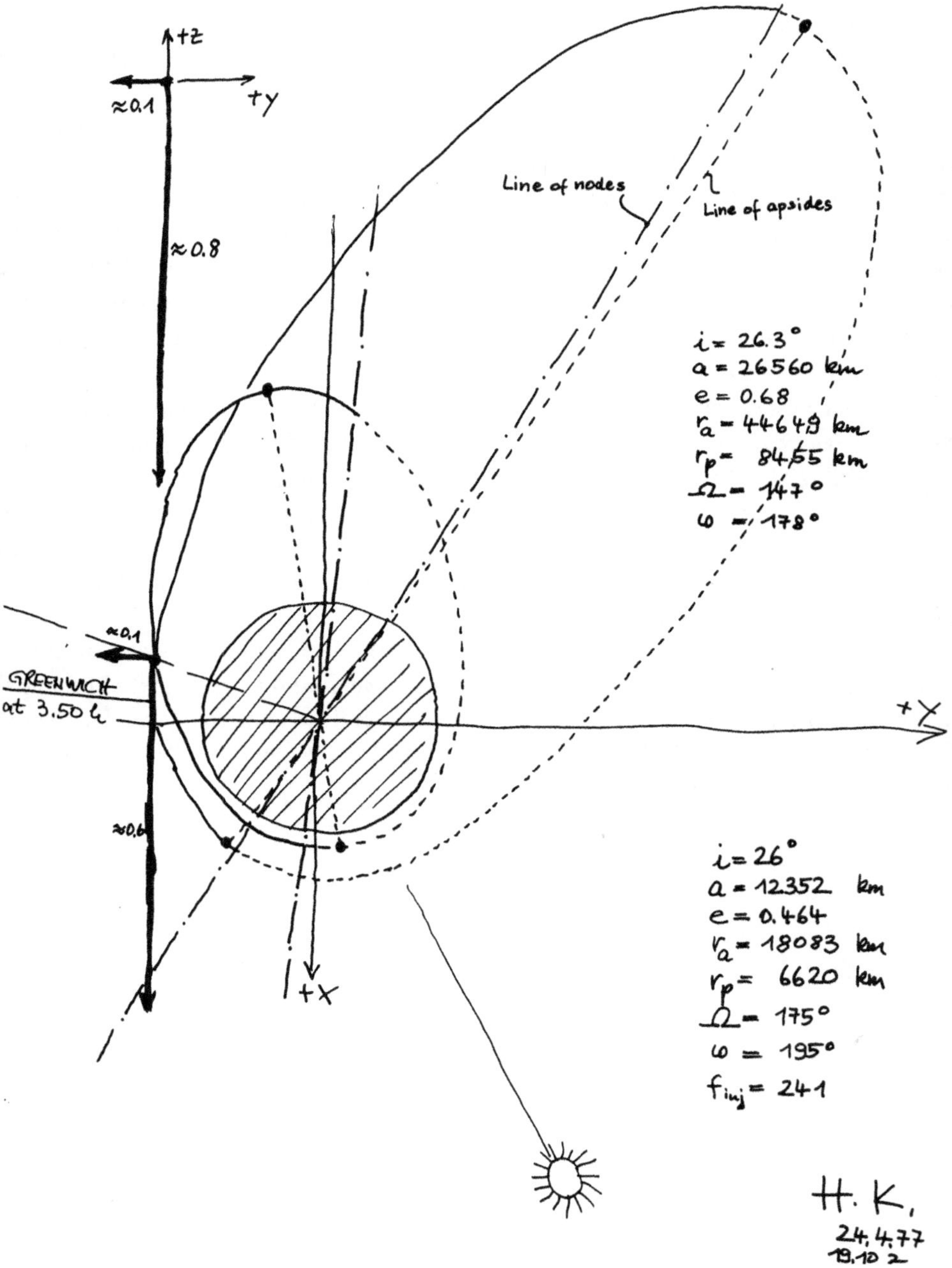

Photograph of the original scetch lining out the GEOS-1 rescue operation presented at the mission operation conference on 24 April 1977.

INTRODUCTION TO THE S-300 WAVE EXPERIMENT ONBOARD GEOS

For GEOS S-300 Experimenters

D. JONES

Space Science Dept., ESA/ESTEC, Noordwijk, The Netherlands

Abstract. The wave experiment on GEOS-1 is a cooperative enterprise involving six different European Institutes. The principle of this complex experiment is described, and the sensitivities attained for measuring electric and magnetic fields in the frequency range DC - 80 kHz are presented.

The wave/field experiment on GEOS-1 which is one of the most comprehensive wave experiments put in orbit is a cooperative enterprise involving the six European Institutes listed in Table I. The purpose of this paper is to present those technical details necessary for understanding the results presented in a number of papers to follow. A fuller description of the experiment may be obtained from S-300 Experimenters (1978).

TABLE I

List of S-300 Experimenters

C.R.P.E., Issy-les-Moulineaux (France):
J. J. Block, N. Cornilleau, J. Etcheto, B. Higel, R. Gendrin, S. Perraut, and M. Petit.

C.R.P.E., Orléans (France):
C. Beghin, P. Decreau, and F. Lefeuvre.

D.S.R.I., Lyngby (Denmark):
A. Bahnsen and E. Ungstrup.

SSD/ESA, Noordwijk (The Netherlands):
R. J. L. Grard, D. Jones, K. Knott, and A. Pedersen.

University of Sussex (England):
P. Christiansen, P. Gough, G. Martelli, and J. Hunter.

University of Sheffield (England):
W. Gibbons.

The S-300 experiment is capable of both passive and active wave measurements. The active measurements themselves are based on two techniques. One technique (S-301) requires that the plasma be pulsed over a range of frequencies and the observation of 'ringing' at certain characteristic frequencies yields information on the plasma density and magnetic field strength. The other technique (S-304) consists of transmitting continuous signals from one antenna to another and of measuring

Space Science Reviews **22** (1978) 327–332. *All Rights Reserved*

their mutual impedance at different frequencies thereby allowing a computation of plasma density and temperature. The following three sections deal briefly in turn with the S-300 sensors, the onboard analysis equipment and with the attainable sensitivities.

1. Sensors

Electric signals are measured on GEOS by means of 8 cm diameter spherical vitreous carbon spheres on radial booms of length 42 m tip to tip, denoted by E_y in Figure 1 and capable of detecting both DC and AC components of the field. In addition to E_y the DC and AC potentials of each sphere E_1 and E_2 with respect to the spacecraft are obtained. Two pairs of 10 cm diameter wire mesh spheres supported by axial booms, 1.5 m and 3 m long, also yield three mutually orthogonal components $e_{x,y,z}$ of the AC electric field.

The conductive outer layers of the long radial boom is used as a transmitting antenna for the active resonance experiment (S-301) while the sensors on the longer

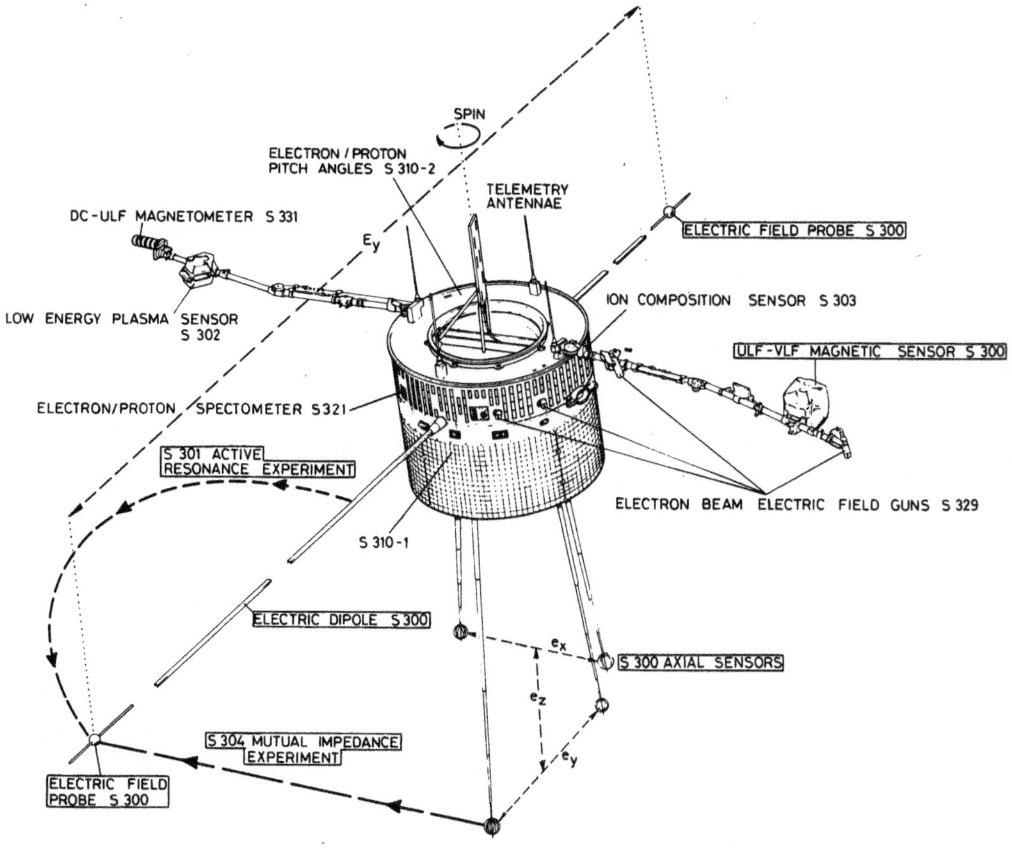

Fig. 1. General view of GEOS-1 spacecraft showing various booms and locations of experiments.

pair of axial booms (e_y) are used as transmitting elements by the mutual impedance experiment (S-304). In both cases the long radial E_y sensors are used for reception.

All S-300 magnetic sensors are situated in a single unit on one of the two short radial booms. For technological reasons, they are composed of two groups: the ULF/ELF sensors and the VLF sensors. Each group provides three mutually orthogonal components.

2. On Board Analysis

The major units within the experiment and a representation of the signal routing are shown in Figure 2. The various sensors and experiment transmitters are on the left of the Figure, the on-board analysis equipment appears in the centre and output telemetry is on the right. Certain parameters such as the electric DC-ULF signal, the filter bank output, and the magnetic ULF signals are sent to ground via the 'low-speed' (4.5 k bit/s) telemetry. Generally, the sensors whose outputs appear in this telemetry channel are permanently connected. More rapidly varying signals are

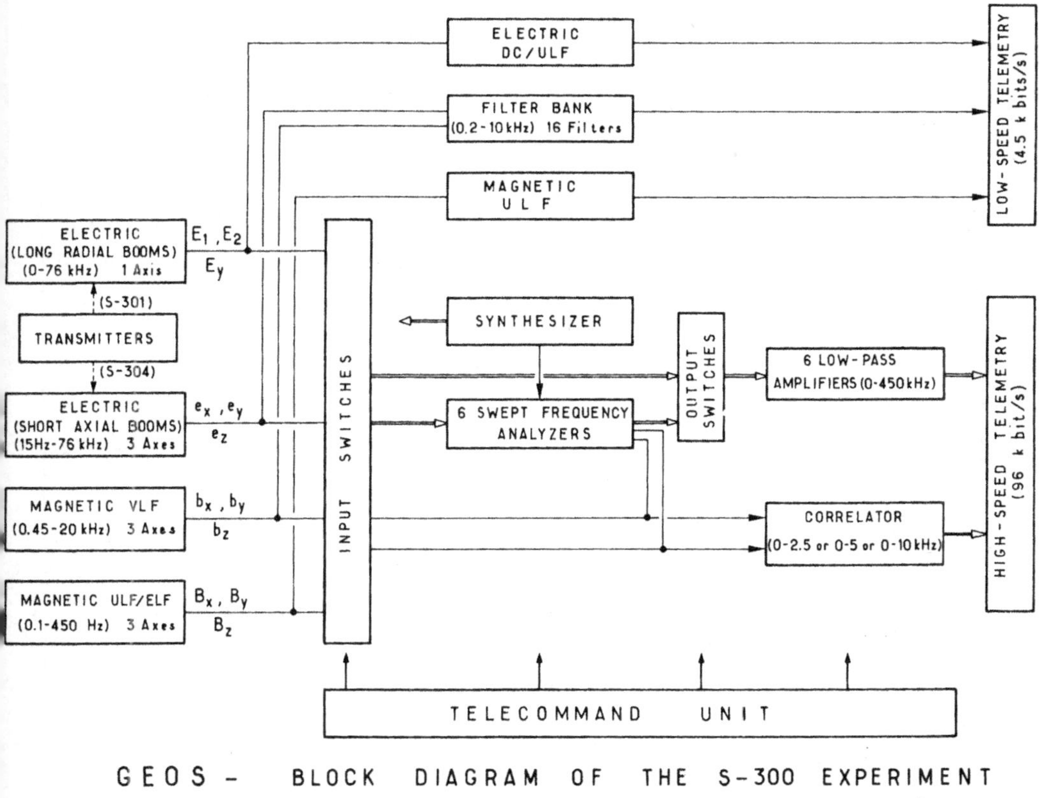

Fig. 2. Block diagram of S-300 wave experiment illustrating the manner in which signals are routed from the sensors via the onboard analysis equipment to the telemetry.

sent to ground via the 'high-speed' (96 k bit/s) link. In this case the signals required are selected by telecommanding the input and output switch boxes.

It is seen that the ELF/VLF signals are subjected to three types of on-board analyses – by the filter bank, by the step frequency analysers (SFA's) and by the correlator. The filter bank is permanently connected to two electric ($e_{y,z}$) and two magnetic ($b_{x,z}$) sensors and yields a low resolution spectrogram for signals in the frequency range 0–10 kHz for electric and 0–5 kHz for magnetic components. There are altogether 16 filters of different bandwidths between 0.3 and 4 kHz, each being sampled 5.8 times per second. The six SFA's may be connected to any sensor combination, their bandwidth being 300 Hz which may be placed or swept in parallel

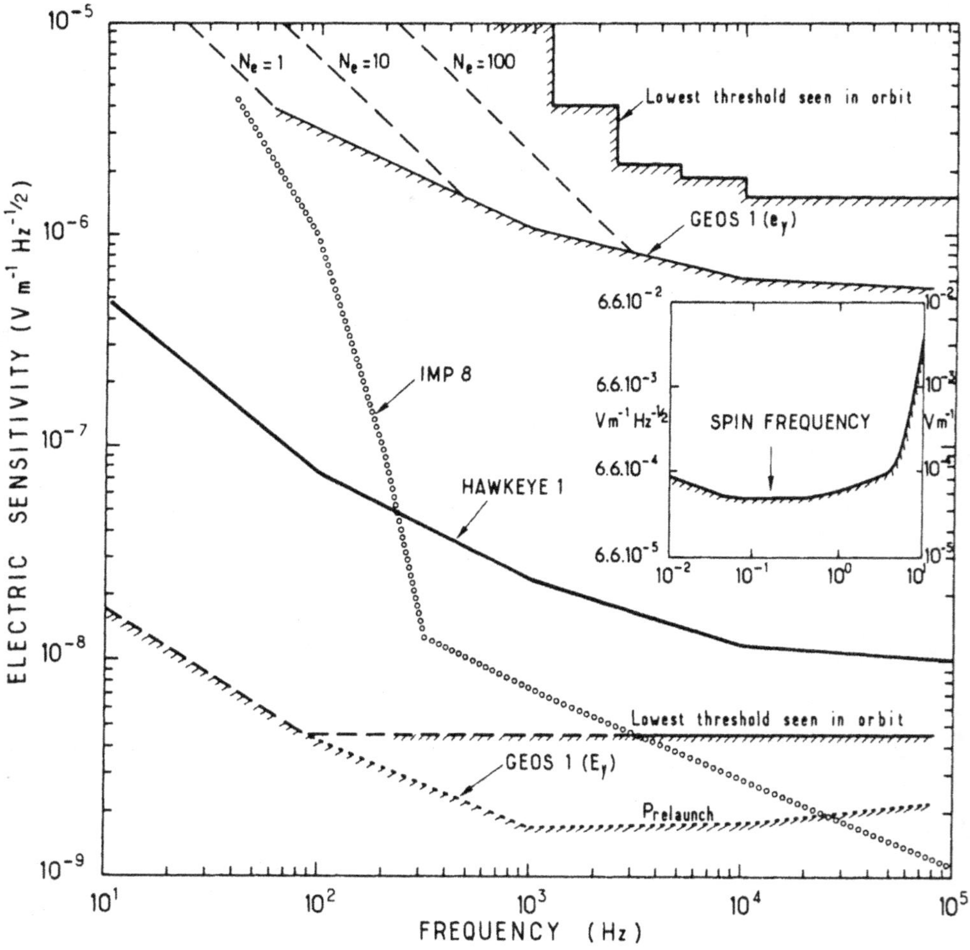

Fig. 3a Sensitivity of S-300 electric antennas in comparison with previous experiments (Gurnett *et al.*, 1976; Kintner and Gurnett, 1977). For the short electric antennas, theoretical thresholds due to shot noise have been computed for a thermal energy of 1 eV and different electron densities N_e (cm^{-3}). Sensitivity in the DC/ULF range is set by level of digitization and valid scale is on the right. Left hand scale assumes a filter bandwidth (commonly used in our analysis) of 0.022 Hz.

virtually anywhere in the range 0–77 kHz, and the six waveforms are telemetered to ground for detailed analysis. The analyser may remain on a given frequency step for various durations, the minimum duration being 86 ms.

The digital correlator can compute in real time the auto- and cross-correlation functions of signals in one of three selectable bandwidths 0–2.5, 0–5, 0–10 kHz; after Fourier transformation of correlograms on ground, high resolution (45 Hz of 2.5 kHz in 29 ms) auto- and cross-spectrograms are obtained. Alternatively, routing certain SFA output signals to the correlator allows the construction of spectrograms of bandwidth 2.5 kHz placed or swept virtually anywhere in the range 0–80 kHz.

3. Sensitivities

The sensitivities of the electric sensors are shown in Figure 3a. Also shown for comparison are the sensitivities attained by electric field experiments on other spacecraft. The long boom E_y has comparable sensitivity to the 121.8 m dipole on Imp 8. The sensitivity of antenna e_y is limited by pre-amplifier noise and shot noise; the latter is represented by dashed lines for electrons of thermal energy 1 eV and for the different electron densities marked on the Figure. The inset in Figure 3a corresponds to the DC/ULF frequency range and the reduced sensitivity in this range is a consequence of the requirement that the dynamic range be able to

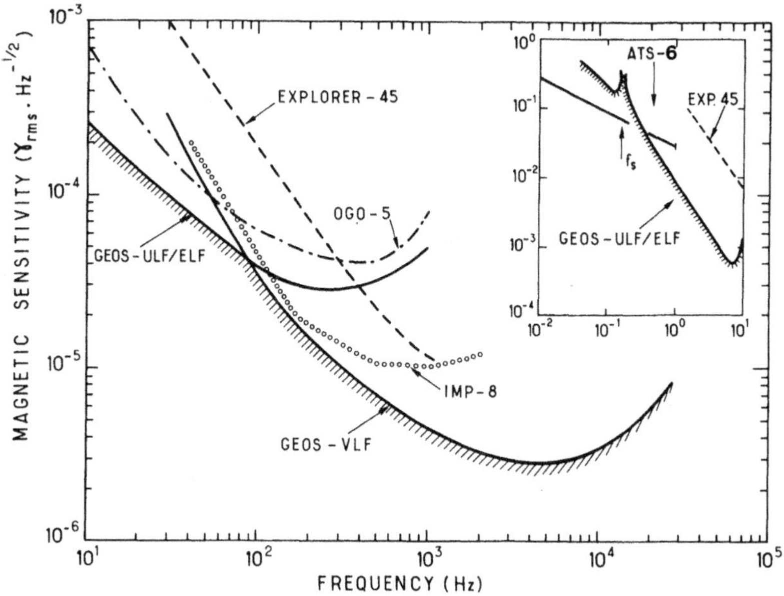

Fig. 3b. Sensitivity of S-300 magnetic antennas in comparison with previous experiments (OGO-5: Frandsen *et al.*, 1969; Explorer 45 and Hawkeye-1: Kintner and Gurnett, 1977; IMP-8: Gurnett *et al.*, 1976; ATS-1: McPherron, private communication). Sensitivity in the ULF range has a notch at the spin frequency (in the antenna reference system) and is intentionally reduced when approaching the Nyquist frequency of the low-speed telemetry.

accommodate the relatively large signal induced at the spin frequency by the DC electric field.

In Figure 3b are given the corresponding sensitivity curves for the magnetic sensors. In order to avoid a reduced sensitivity in the ULF range and because there exists another magnetometer (S-331) on GEOS to measure the DC magnetic field, a special 'despin' system on board suppresses the spin modulated component. The ULF/ELF sensors operate in the 0.1 to 450 Hz frequency range and the components are denoted by $B_{x,y,z}$. The VLF sensors cover frequencies above 450 hz and are represented by $b_{x,y,z}$.

4. Summary

Only the basic functions and sensitivities of the S-300 experiment have been presented, more detailed discussion of individual parts being reserved for papers which follow. The complete experiment has continued to perform perfectly, yielding excellent data for 14 months up to the time of GEOS-1 switch-off in anticipation of the GEOS-2 launch.

Reference

Frandsen, A. M. A., Holzer, R. E., and Smith, E. J.: 1969, *IEEE Trans. Geosci. Electron.* **GE-7**, 61.
Gurnett, D. A., Frank, L. A., and Lepping, R. P.: 1976, *J. Geophys. Res.* **81**, 6059.
Kintner, P. M. and Gurnett, D. A.: 1977, *J. Geophys. Res.* **82**, 2314.
S-300 Experimenters: 1978, *Planetary Space Sci.*, (to be published).

MEASUREMENTS OF QUASI-STATIC ELECTRIC FIELDS BETWEEN 3 AND 7 EARTH RADII ON GEOS-1

A. PEDERSEN, R. GRARD, K. KNOTT, D. JONES, and A. GONFALONE
Space Science Department of ESA, ESTEC, Noordwijk, The Netherlands

and

U. FAHLESON
Royal Institute of Technology, Stockholm, Sweden

Abstract. Quasi-static electric fields have been measured with two spherical probes supported by cable booms providing a baseline of 42 m for the measurement. The performance of the experiment is outlined to demonstrate that electric fields can be measured with accuracies of ± 0.7 mV m^{-1} and ± 1.0 mV m^{-1} in the dawn-dusk and satellite-sun directions respectively. These uncertainties can be considerably reduced under favourable plasma conditions. Examples of typical observations are described.

(a) The average electric field is always characterized by an irregular structure with time scales 0.5–5 min and with amplitudes of a few mV m^{-1}.

(b) During substorms dawn-dusk electric fields up to 20–30 mV m^{-1} have been observed over intervals of 30–60 s.

(c) Oscillating electric fields with peak-to-peak amplitudes up to 10 mV m^{-1} and periods of 3–10 min have been observed following magnetospheric disturbances.

The observations are discussed in terms of plasma motions and possible spatial scale sizes of the phenomena, standing magnetospheric wave modes and electrostatic potentials.

1. Introduction

This paper gives the first results of quasi-static electric field measurements with a spherical double probe on the GEOS-1 satellite. Due to the 12 hr elliptical orbit, measurements were carried out under different plasma conditions inside and outside the plasmasphere and thus provided good opportunities for determination of the double probe characteristics. The apogee of the orbit was initially near 18 hr LT and moved through 12 to 6 hr towards the end of 1977. This is the local time interval for data described in this paper.

The quasi-static electric field is derived from a measurement of the differential potential between two spherical sensors with built-in preamplifiers mounted at the tip of wire booms rotating with the satellite. This type of double probe has been analysed in considerable detail in various model calculations (Fahleson, 1971; Grard *et al.*, 1972). The sensors are also used for ac measurements up to 80 kHz and form part of a large cooperative wave and field experiment involving the following laboratories: CRPE, Issy-les-Moulineaux, CRPE-Orleans, DSRI-Copenhagen, SSD/ESA, University of Sussex and University of Sheffield. We will here only deal with measurements of the quasi-static electric field which is the responsibility of SSD/ESA.

The GEOS-1 satellite was launched 20 April 1977 and 22 October the same year ISEE-1 was launched with an electric field experiment using nearly identical

Space Science Reviews **22** (1978) 333–346. *All Rights Reserved*

spherical sensors to those flown on GEOS-1. The ISEE-1 instrument uses a longer baseline of 73.5 m between the sensors compared to 42 m for GEOS-1, and has additional capability to perform diagnostic measurements in a variety of plasma conditions encountered in the magnetosphere and the solar wind (Mozer *et al.*, 1978). The combined experience from the two double probe experiments has demonstrated their capability to measure quasi-static electric fields with an accuracy of better than 1 mV m^{-1} under most plasma conditions encountered.

We will here concentrate on the performance of the double probe experiment on GEOS-1 and give examples of data which are characteristic for the GEOS-1 orbit. All data discussed will increase in value by comparison with data from other GEOS experiments, by comparison with data from the quasi-static electric field experiments on ISEE-1 when satellite orbits permit and by comparison with ground data. These will be the natural next steps in future work.

2. Experiment Characteristics and Performance

The quasi-static electric field is derived from a measurement of the spin-modulated differential potential between two spherical probes with built-in preamplifiers mounted at the tips of wire booms spinning with the satellite.

Figure 1 shows that the double probe system spinning with a period of 6 s will detect a maximum signal $V_{p1} - V_{p2} = E_{sp}l$ when the double probe is aligned with the projection E_{sp} of the electric field into the spin plane. It is convenient to define the

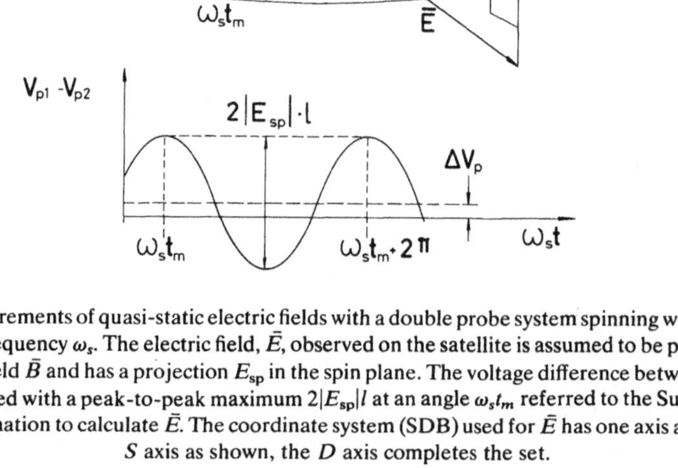

Fig. 1. Measurements of quasi-static electric fields with a double probe system spinning with the satellite, spin angular frequency ω_s. The electric field, $\bar{E}$, observed on the satellite is assumed to be perpendicular to the magnetic field $\bar{B}$ and has a projection E_{sp} in the spin plane. The voltage difference between the sensors is spin modulated with a peak-to-peak maximum $2|E_{sp}|l$ at an angle $\omega_s t_m$ referred to the Sun direction and provides information to calculate $\bar{E}$. The coordinate system (SDB) used for $\bar{E}$ has one axis along $\bar{B}$ and the S axis as shown, the D axis completes the set.

angle for this alignment as $\omega_s t_m$ with reference to the closest approach to the Sun direction.

Any long term dc offset between the two probes, ΔV_p, does not influence the measurements and in practice this offset is found to be less than 20 mV.

Since the measurement is two-dimensional we have to add an assumption about the third component to perform the data reduction. This assumption is that the electric field parallel to $\bar{B}$ is sufficiently small to permit us to write

$$\bar{E}+\bar{v}\times\bar{B}=0,$$

where $\bar{B}$ is the magnetic field, $\bar{E}$ the electric field and $\bar{v}$ the plasma drift in a given coordinate system. This seems to be a reasonable assumption for most conditions in this part of space.

With knowledge of E_{sp} and $\omega_s t_m$ it is possible to calculate $\bar{E}$ and to present it in any convenient coordinate system. In the following we will use the so-called SDB system. One axis is the $\bar{B}$ direction, the $\bar{S}$ axis is defined in Figure 1 to be in the plane defined by $\bar{B}$ and the satellite-Sun direction. The D axis completes the set and is in the general direction dawn-to-dusk. We will present data where the origin of the SDB system is fixed in inertial coordinates but allows the coordinate system to move with $\bar{B}$. The induced field due to the satellite orbital motion in this frame of reference is only a few tenths of mV m^{-1} for the parts of the orbits we will consider.

Figure 2 shows the geometry of one probe with an 8 cm diameter vitreous carbon sphere as the main sensor element connected to the input of a voltage follower amplifier. The sphere is mounted on stubs with vitreous carbon surfaces which are 'bootstrapped', i.e. they follow the potential of the sphere because they are connected to the preamplifier output. Similarly 1 m of cable braid and a connector to the main cable boom are bootstrapped.

The current-voltage characteristics for the satellite and for one probe are also shown in the same figure. The satellite and the probe will both float at an equilibrium potential with respect to the plasma potential near the probe, determined by the balance of currents from escaping photo-electrons and collected ambient electrons. Ion currents are negligible in comparison. The figure also illustrates a very important capability of the probes namely the use of a current source with an impedance which is high compared to the probe-plasma impedance in order to force a variable current to the probes. For most conditions in the magnetosphere the maximum photo-electron current from satellite or probe is larger than the ambient electron current to the satellite or probe respectively and both will be at an equilibrium potential of a few volts positive with respect to the ambient plasma. By applying a negative current bias to the probe it can be brought closer to the plasma potential and due to the increased slope of the photo-electron characteristic the probe-plasma impedance will be reduced. The satellite has a collecting area approximately 300 times that of the probes and will not change appreciably in potential. This technique is only meaningful for a satellite with conductive surfaces. GEOS-1 and ISEE-1/2 are in fact the first magnetospheric satellites with this capability.

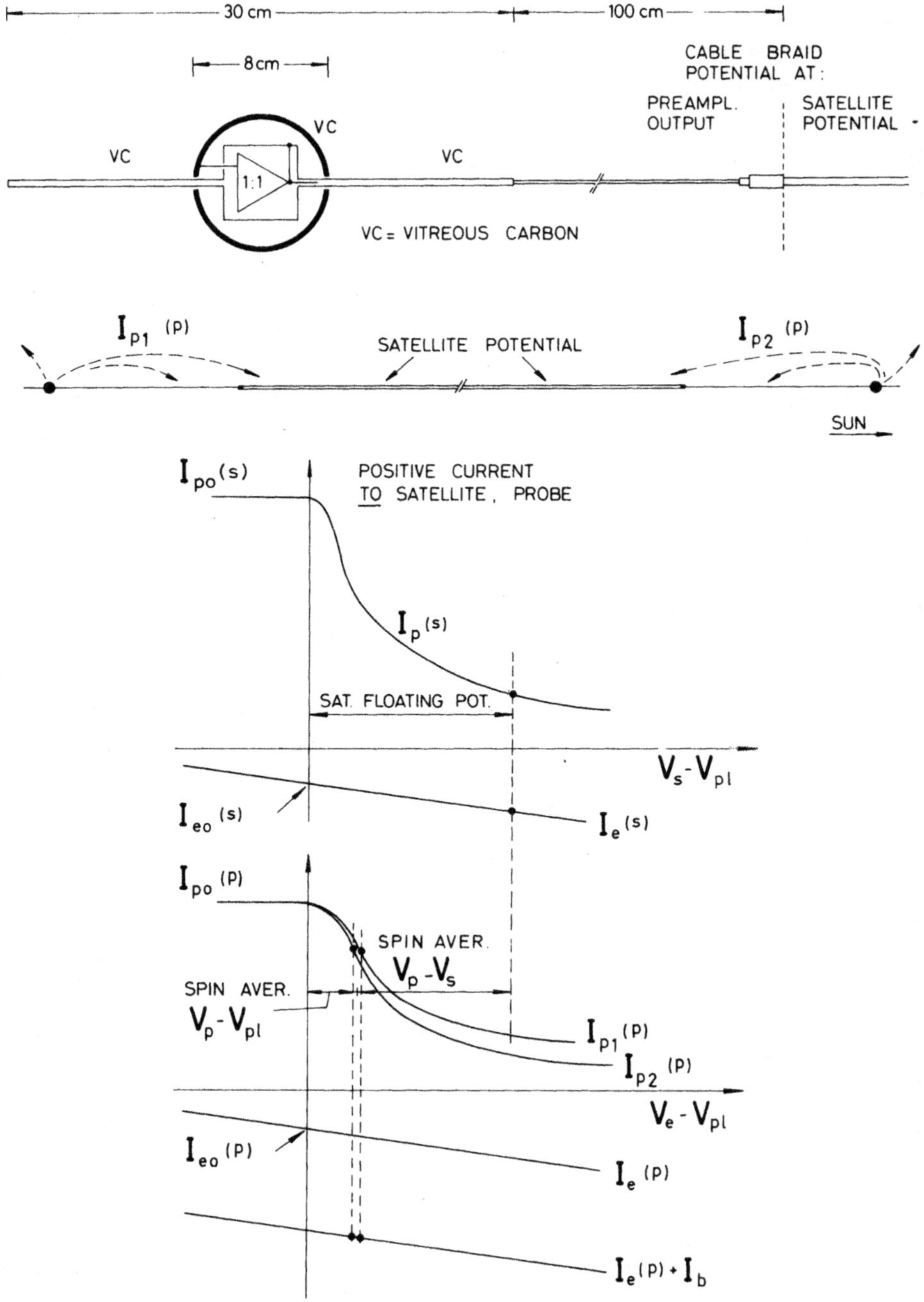

Fig. 2. Probe geometry and illustration of how the antisunward probe looses photo-electrons more easily to the boom thereby giving rise to an asymmetry and an apparent sunward electric field. The lower part shows current-voltage relationship for the satellite and a current biased probe.

In Figure 2 an effect seen both on GEOS-1 and ISEE-1 has been drawn schematically, namely the presence of an apparent electric field component $\bar{E}_a$ in the spin plane pointing sunward ($\omega_s t = 0, n2\pi$). There is a tendency for photo-electrons to escape from the probe by being collected by the bootstrapped 1 m cable as well as the long wire boom which is at satellite potential. This effect is more pronounced for the probe which is in the antisunward direction thereby driving that probe more positive and giving rise to a signal $V_{p1} - V_{p2} = E_a l \cos \omega_s t$.

During quiet magnetospheric conditions with small values of $\bar{E}$, the only signal observed is $\bar{E}_a$. When $\bar{E}$ is larger it will be observed superimposed on E_a. By studying a large number of days when $\omega_s t_m = 0$ for quiet periods when $\bar{E}$ could be assumed to be small, it was possible to establish a clear relationship between $\bar{E}_a$ and the probe-satellite potential difference, $V_p - V_s$ which is continuously measured.

This relationship is shown in Figure 3 for a bias current, $I_b = -83\ nA$ which is the normal mode for all data discussed here. $\bar{E}_a$ depends also on the solar aspect angle β which is the angle between the satellite-sun direction and the spin plane. A value of 20° for β gives better symmetry for the probes and a smaller value for $\bar{E}_a$ than $\beta = 0°$. The substraction of $\bar{E}_a$ to find the correct value $\bar{E}_{sp}$ leads to a better determination of the electric field in the dawn to dusk direction (E_D) than the sunward direction (E_s). The accuracy for E_D ($\pm 0.7\ \mathrm{mV\ m^{-1}}$) is better than that for E_S ($\pm 1.0\ \mathrm{mV\ m^{-1}}$) because E_D is usually nearly perpendicular to the disturbance field, $\bar{E}_a$, whereas E_S is nearly parallel to it and thus more directly affected, see Figure 3. These uncertainties are valid for the *absolute* values of E_s and E_D. Relative variations of $\bar{E}$ on time scales both smaller and larger than the spin period can be established with an uncertainty of a few tenths of mV m^{-1} as long as the plasma conditions as monitored by $V_p - V_s$, see below, are constant.

From Figure 2 it can be seen that the probe is biased with a negative current and is near the plasma potential. The potential difference $V_p - V_s$ can then be used to find the satellite floating potential and thereby give support to particle experiments. The use of $V_p - V_s$ for plasma diagnostics is described more in detail in Decreau *et al.* (1978).

3. Examples of Electric Field Observations

3.1. Electric Fields in the Quiet Magnetosphere

In order to illustrate the quality of the measurements the spin modulated differential signal between the probes is given in Figure 4a. The dotted cosine function is the projection of $\bar{E}_a$ corresponding to the value of $V_p - V_s$ which is also shown. The spin averaged value of the latter is constant which indicates that the plasma density and/or temperature did not change whereas the electric field increased. This can be seen near 16:39:30 UT in the treated data shown in Figure 4b. The variation of $V_p - V_s$ during one spin is mainly due to variations of satellite photoemission influencing V_s.

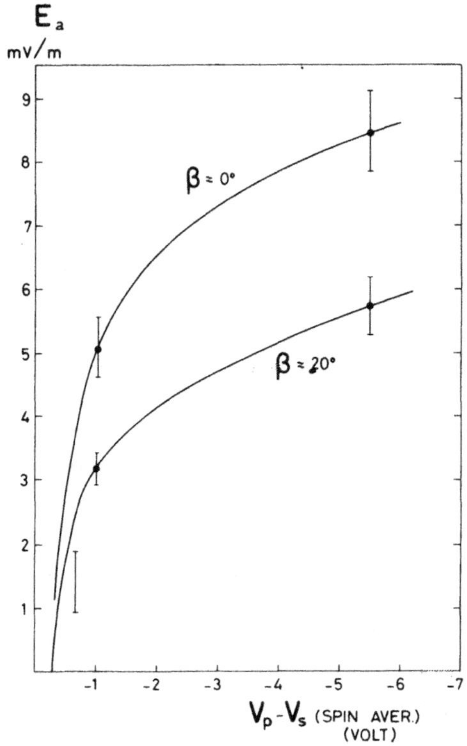

Fig. 3. Apparent sunward field, E_a, caused by asymmetric photoelectron currents between probes and their supporting booms as a function of probe-satellite potential difference $V_p - V_s$, E_a is smaller for an increased angle β between Sun direction and spin plane.

Taking into account that there are 6 s between successive determinations of $\bar{E}$ displayed as E_S and E_D and the comments made concerning the accuracies for absolute values and relative variations of $\bar{E}$, we can see that both E_S and E_D on the average are positive 2–3 mV m^{-1} and that superimposed there are significant variations with time scales 0.5–5 min.

In Figure 4 is also shown the plasma drift $\bar{v} = \bar{E} \times \bar{B}/B^2$ in the SD-plane averaged over 10 min intervals. The $\bar{v}$-vectors are placed along the trajectory, which is described as a polar plot of radial distances vs local time, in order to indicate magnetospheric positions. The plasma flow is basically sunward and slightly towards dawn for a long period near 18 hr local time in agreement with the established convection in the magnetosphere. If we take the average dawn to dusk electric field as shown in Figure 3 to be 3 mV m^{-1} and integrate this over $30 R_E$ we find a cross-magnetosphere potential which is ~ 600 kV or ten times larger than what is inferred from models and ionospheric measurements (Mozer, 1976). We must therefore conclude that we are observing a local region with electric field and corresponding plasma flow above average.

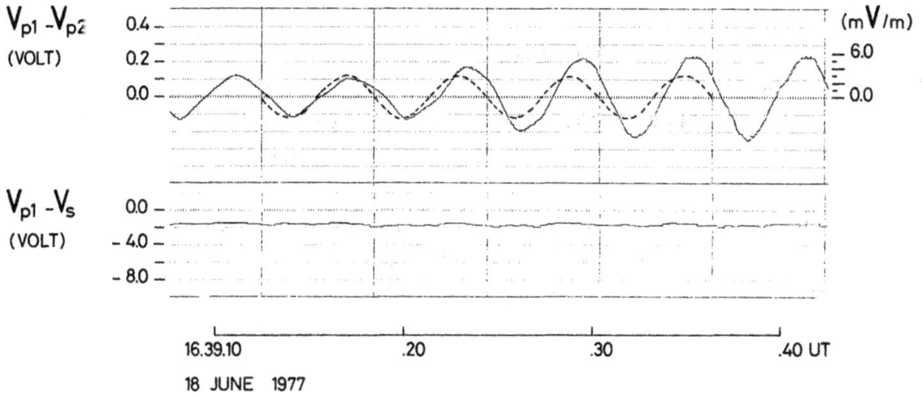

Fig. 4a. The spin modulated differential signal between probes, $V_{p1} - V_{p2}$ is shown together with the projection of E_a (dotted curve) on this probe system. Vertical lines mark each spin when the probe system is perpendicular to Sun direction. The probe-satellite potential $V_{p1} - V_s$ has spin ripples due to satellite photo-emission changing V_s. The spin averaged $V_{p1} - V_s$ indicates that plasma conditions were constant.

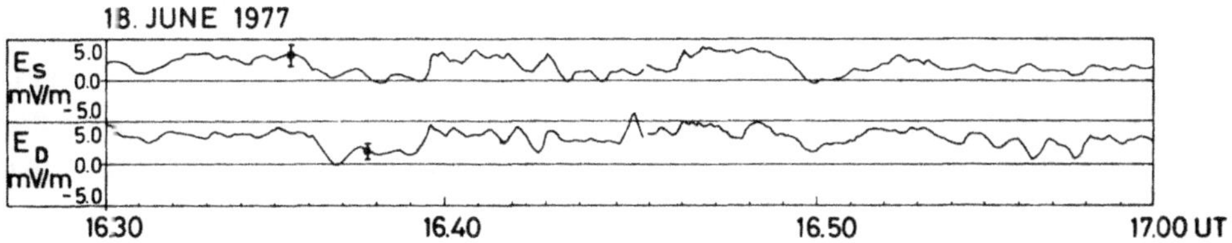

UT	16.30	17.00
LOCAL TIME	17.48	18.10
R/RE	6.02	5.60
L	6.60	6.32
MAGN. LAT.	16.74°	18.88°

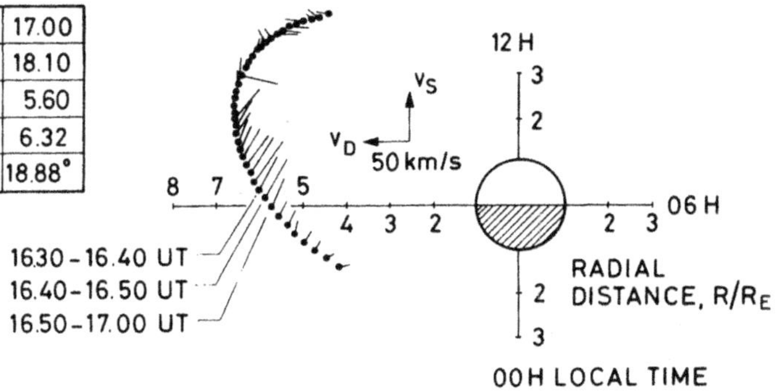

Fig. 4b. Measurements of the quasi-static electric field in the afternoon sector near $L = 6.5$ E_S and E_D are sunward and dawn to dusk fields respectively. The magnetic field was very quiet and varied from 150 γ to 200 γ over the time shown. Ten minute averages of plasma drift are shown in the SD-plane, plotted along the satellite orbit.

The whole month of June was very quiet with the period shown in Figure 4 slightly more disturbed, $K_p = 3-$. The average electric field on most quiet days in the afternoon sector is found to be smaller than those shown in Figure 4. However there are always variations of several mV m^{-1} which gives the impression of a turbulent plasma motion.

This feature is much more pronounced on apogees near noon where there are positive as well as negative deviations in E_S and E_D up to 5 mV m^{-1}. However, the average electric field over a longer period (10–20 min) is in most cases less than 1.0 mV m^{-1}. On same occasions these variations in the electric field are seen as waves with periods of several minutes, but the more pronounced waves are seen away from noon after magnetospheric disturbances as we will show later. If we assume that the observations are related to plasma drift patterns fixed to the plasma which flows sunward with an average velocity of approximately 30 km s^{-1} we find that the spatial scale corresponding to a 60 s variation is approximately 1800 km. We must however await two point measurements when GEOS-1 and ISEE-1 are on close orbits to get discrimination between temporal and spatial variations.

3.2. Substorm electric fields

An example of the electric field during a substorm is shown in Figure 5. The magnetospheric conditions can be outlined as follows; a sudden commencement was observed near 20:30 UT on December 1 and during the first hours of December 2 the ring current as measured by Dst increased from Dst = -50γ at 03:00 UT to -120γ at 06:00 UT.

The first major increase in the electric field was observed near 02:41 UT when a strong fluctuating generally sunward electric field was observed implying plasma flow dawn to dusk or earthwards considering that the local time was near 06:00 hr. A very

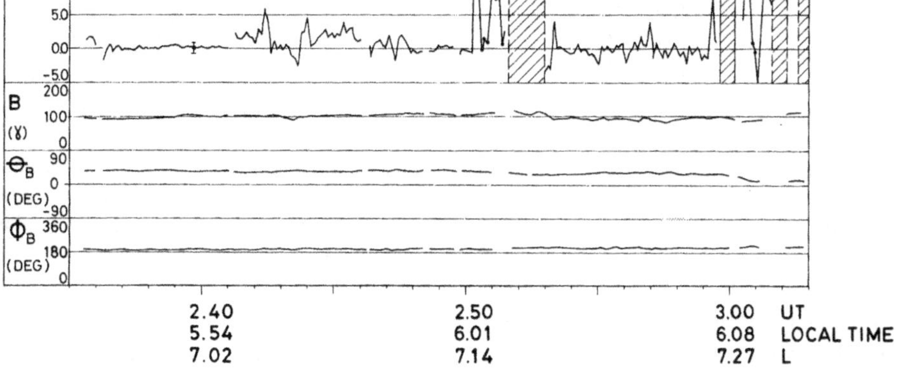

Fig. 5. The quasi-static electric field during a substorm. Large dawn to dusk components of the electric field were measured near 02:50 UT and 03:00 UT. Near these times the electric field was perturbed and often varied within a spin period (6 s) to the extent that $\bar{E}$ could not be measured (cross hatching).

strong electric field up to 20 mV m^{-1} in the dawn to dusk direction was observed near 2:50:20 UT. The field was in fact larger than expected and saturated the data channel normally used. The signals could be seen however on another channel with reduced sensitivity and time resolution. Data from this period was evaluated spin by spin and the resulting S and D components evaluated in this way are marked as black dots in Figure 5.

The duration of this strong electric field was of the order 20–30 s for the first impulsive event and the more pronounced dawn to dusk electric field demonstrates that we observed an injection of plasma from the tail of the magnetosphere in the sunward direction with an $\bar{E} \times \bar{B}/B^2$ velocity of approximately 200 km s^{-1}.

The magnetic field measured by the S-331 experiment is fairly quiet during the plasma injection event described and during subsequent smaller events. This is probably due to the fact that acceleration of the plasma and associated substorm magnetic field perturbations occur further down along the tail. After 03:00 UT when similar events with large dawn to dusk electric fields of short duration were observed the magnetic field near GEOS was very perturbed; values of B were reduced to 70–80 and tilted nearly towards the Earth.

The high speed plasma streams have other pronounced features. Superimposed on the spin modulated signals due to $\bar{E}_a + \bar{E}_{sp}$ there are periodic, nearly monochromatic, signals. For the time interval 2:50:30–2:53:00 UT their period was (0.29 ± 0.01) s, or frequency (3.45 ± 0.1) Hz, and with an amplitude in the range 1–3 mV m^{-1}. The same waves were seen with half the amplitude when observed on a channel measuring between one probe and the satellite (half effective length) as compared to the channel giving the differential signal between the two probes. This demonstrates that the wavelength is larger than the double probe (42 m tip/tip). From a spin modulation of the wave amplitude it can be concluded that the wave electric field is perpendicular to $\bar{B}$. The observed frequency is near to 2 times the local proton cyclotron frequency but we must recognize that we most likely observe a Doppler shifted frequency due to the waves being generated in the frame of reference of a fast streaming plasma. An additional effect is caused by the probe system spin (0.165 Hz) which must be added or subtracted to the wave frequency depending on wave polarization.

There are times following large impulsive plasma injections from the magneto-tail when no quasi-static electric field can be derived from the data because the electric field fluctuates with large amplitudes within a spin period, or density and temperature changes as monitored continuously by $V_p - V_s$ (see Figure 2) vary within a spin period and may cause the two probes to be at different plasma environments due to sharp gradients. These periods are crosshatched in Figure 5. The monochromatic waves of frequency 3.5 Hz have a clear onset coinciding with the impulsive plasma injections and last through the crosshatched perturbed intervals.

Aggson and Heppner (1977) have reported similar large electric fields from measurements on IMP-6 when the satellite was at L-values near 7 and at higher geomagnetic latitudes and shorter radial distances. The magnitude and duration of the events described here are in good agreement with their measurements. However in contrast to their conclusion that the direction of the electric field has quasi-stochastic nature we find a clear dominance of the dawn to dusk electric field i.e. injection of plasma from the magneto-tail. Aggson and Heppner find that their observations are well correlated with sharp reductions in the X-component of magnetograms from ground stations near the conjugated field lines. Magnetograms from Kiruna show the same effect for the December 2 event (G. Gustavsson, 1978, private communication).

3.3. Hydromagnetic wave electric fields

Figure 6 shows examples of electric field pulsations with periods of several minutes and peak to peak amplitudes of 6–8 mV m^{-1} observed after disturbed periods. In the

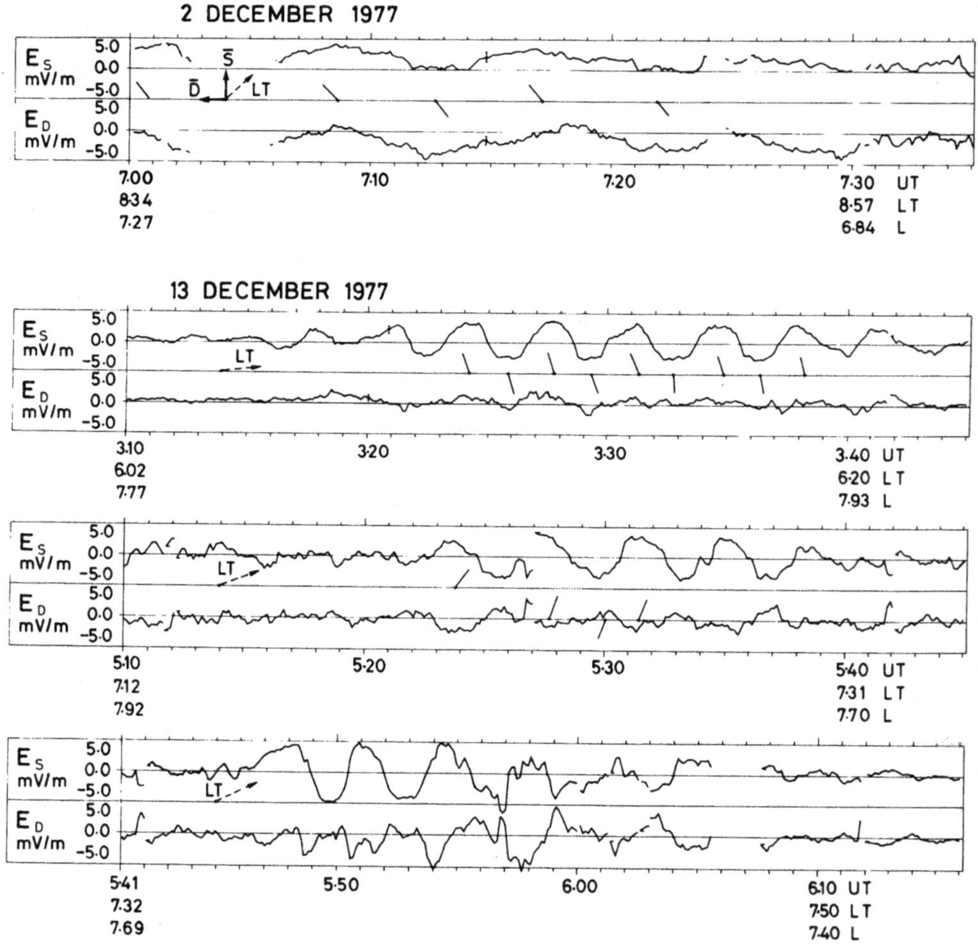

Fig. 6. Pulsating (PC-5) electric fields following disturbed periods in the magnetosphere. For the first two time intervals the wave electric field was azimuthal i.e. perpendicular to the radial or local time direction marked as 'LT' in the SD plane. Because the wave magnetic field was small we conclude that the wave plasma motion was radial.

first two time intervals from respectively December 2 and December 13 we find that the electric field is linearly polarized or strongly elliptically polarized and varies in the azimuthal direction. This is illustrated by vectors in the S-D plane of the instantaneous wave electric field. The radial direction corresponding to the local time of the observation is marked 'LT' and indicated in the S-D coordinates. It can be seen that the wave electric field is perpendicular to this direction i.e. azimuthal. In the two examples taken from later times on December 13 the wave electric field general orientation lies between the azimuthal and radial directions.

A wave magnetic field with the same periodicity as the wave electric field was observed with variations mainly in the azimuthal direction perpendicular to the background magnetic field (120–150γ) which did not show significant changes in magnitude at the wave frequency. The wave magnetic field has a 90° phase advance with respect to the wave electric field and has peak to peak azimuthal amplitudes of approximately 30γ and 12γ for the events of December 2 and December 13 respectively.

The pulsating electric field is most likely due to a standing hydromagnetic wave in the PC-5 frequency range. Such waves have been observed on many previous occasions both on satellites and simultaneously on satellites and on the ground by the use of magnetic sensors (Kokubun *et al.*, 1976; Lanzerotti *et al.*, 1976). In a recent paper by Cummings *et al.* (1978) particle detectors were used for detecting the pulsating plasma flow as well as the pulsating and background magnetic fields of hydromagnetic waves.

For the electric field $\bar{E}_w$ of a hydromagnetic wave we have the following equation:

$$\bar{E}_w + \bar{v}_w \times \bar{B} = 0\,, \tag{2}$$

where $\bar{B} = \bar{B}_0 + \bar{B}_w$ and $\bar{v}_w$ is the plasma velocity associated with the wave.

The wave magnetic field $\bar{B}_w$ is a perturbation in addition to the steady background field, $\bar{B}_0$.

The quoted magnitude of $\bar{B}_w$ does not change the direction of $\bar{B}$ by more than approximately 10° and 5° respectively, and $\bar{v}_w$ can in a first approximation be described as a motion perpendicular to $\bar{E}_w$ and $\bar{B}_0$.

We see that $\bar{v}_w$ is radial and considering that B is of the order 100γ, v_w is of the order of 30–40 km^{-1}s. It is difficult to give amplitudes of plasma motions because we do not know the exact position of observation relative to the wave. However these amplitudes must be of the order 1R_{E} for the long period wave observed on December 2.

It is interesting to try to relate the wave period to independent measurements of plasma density by the use of the electric field sensors. The method is described in Decreau *et al.* (1978). Table I gives a summary of observations for December 2 and December 13 respectively.

If we assume that the wave period is mainly determined by the Alfvén velocity near the equator we have

$$\tau_1/\tau_2 = v_{\mathrm{A}2}/v_{\mathrm{A}1} = B_2/B_1\,\sqrt{N_{e1}}/\sqrt{N_{e2}}\,. \tag{3}$$

TABLE I

	December 2 07:00 UT	December 13 03:30 UT
Background magnetic field, B	$B_1 = 120\gamma$	$B_2 = 110\gamma$
Wave period	$\tau_1 = 9$ min	$\tau_2 = 3.5$ min
Electron density, N_e	$N_{e1} = (10 \pm 1)\,\mathrm{cm}^{-3}$	$N_{e2} = (1\text{–}2)\,\mathrm{cm}^{-3}$
Alfvén velocity v_A	$v_{A1} = (0.79\text{–}0.87) \times 10^6\,\mathrm{m\,s}^{-1}$	$v_{A2} = (1.72\text{–}2.4) \times 10^6\,\mathrm{m\,s}^{-1}$

We observe $\tau_1/\tau_2 = 2.6$. From the right side of Equation (3) we find $B_2/B_1\sqrt{N_{e1}}/\sqrt{N_{e2}} = 2.0\text{–}3.0$, the uncertainty is mainly due to the poor determination of N_e on December 13.

Nevertheless, there is good agreement with Equation (3).

If we try to interpret the observations in terms of standing hydromagnetic wave modes the azimuthal $\bar{B}_w$ and the radial $\bar{v}_w$ do not fit any single mode as reviewed in papers by Lanzerotti and Fukunishi (1974) and by Orr (1973).

We suggest the following tentative explanation:

The oscillating azimuthal electric field and corresponding radial plasma flow is due to the first harmonic of an odd poloidal mode which implies radial motion of the plasma and $\bar{B}_w = 0$ near the geomagnetic equator. Observations are near 20°geomagnetic latitude, however the wave node where $\bar{B}_w = 0$ may be slightly off the theoretical geomagnetic equator.

The question is why do we observe in both cases $\bar{B}_w$ in the azimuthal direction. The only obvious explanation is that a torsional even mode implying no plasma motion near the equator but an azimuthal $\bar{B}_w$ near the equator is co-existing with the poloidal mode at the same frequency. This means that the plasma along a magnetic flux tube has a torsional motion in addition to a radially oscillating motion.

The study of such PC-5 phenomena must be pursued in more detail with more information in particular about the magnetic field variations. This description is mainly aimed at demonstrating a new possibility of detecting wave electric fields and corresponding plasma motions also for cases when the wave magnetic fields are small.

3.4. Conclusions

In this paper we have demonstrated the capability of the quasi-static electric field experiment on GEOS-1 and characteristic observations during a substorm and during periods when large amplitude electric field oscillations of several minutes period (PC-5) have been observed.

The next step will be to carry out more detailed studies on separate subjects e.g. on substorms and on the PC-5 electric fields. This will involve comparison of data from other GEOS experiments. It is particularly interesting that with a conductive satellite, low energy ion detectors can measure with confidence the spin modulation due to plasma drifts. These are often unexpectedly large and therefore also rather easy to detect. Furthermore it is essential to carry out more detailed studies of the wave magnetic field during the PC-5 events and correlate with ground observations.

The PC-5 waves reported here have in fact been observed by conjugated auroral radars in Scandinavia as an oscillating ionospheric plasma drift (R. Greenwald, 1978, private communication).

It will be very important to analyse events like those described with simultaneous data sets from ISEE-1 and GEOS-1. Such situations exist but have not been analysed so far. Future opportunities exist for continued comparison of electric field data and other data on ISEE-1 and GEOS-2 to be launched during the summer of 1978.

Acknowledgement

The development of the quasi-static electric field experiment on GEOS has involved a large number of people in the fields of science as well as technology.

We want to thank colleagues at the Royal Institute of Technology and University of California, in particular C. G. Fälthammar and F. S. Mozer for their contributions at all stages of the development. Thanks are due to E. A. Trendelenburg, the Head of SSD at the time of start of this project for his initiative and support.

We also want to thank members of the Astronomy Division of SSD, M. Anderegg, B. Feuerbacher, and B. Fitton, for giving valuable support to the electric field probe design by studying photo-emission characteristics of probe materials.

This development would not have been possible without devoted and skilled technical support; A. Butler has been responsible for the electrical design which has close interfaces with the large wave and field experiment on GEOS involving other staff; J. Eeman, D. Klinge, and M. Trischberger. Finally we want to thank R. Scheper and A. Smit responsible for mechanical design and C. Bouman and his group responsible for electrical assembly and integration, for excellent support.

We want to acknowledge the good cooperation with the S-331 experimenters (principal investigator, F. Mariani, University of Rome) for providing magnetic field data which is extensively used in our data presentations in the SDB system and for determining plasma drift.

We also acknowledge the work by both the ESTEC and the industrial project team in designing, building and testing the GEOS-1 spacecraft as well as the effort of ESOC in acquiring and distributing the data.

References

Aggson, T. L. and Heppner, J. P.:1977, *J. Geophys. Res.* **32**, 5155.
Cummings, W. D., DeForest, S. E., and McPherron, R. L.: 1978, *J. Geophys. Res.* **83**, 697.
Decreau, P., Etcheto, J., Young, D. T., Wrenn, G. L., Pedersen, A., and Knott, K.: 1978, *Space Sci. Rev.* **22**, 633.

Fahleson, U.: 1971, 'Problems of Electric Field Measurements from a Geostationary Satellite', ESRO SP-60, pp. 249–265.
Grard, R., Knott, K., and Pedersen, A.: 1972, *Photon and Particle Interactions with Surfaces in Space*, D. Reidel Publ. Co., pp. 163–192.
Kokubun, S., McPherron, R. L., and Russell, C. T.: 1976, *J. Geophys. Res.* **81**, 5141.
Lanzerotti, L. J. and Fukunishi, H.: 1974, *Rev. Geophys. Space Phys.* **12**, 724.
Lanzerotti, L. J., Fukunishi, H., and Maclennan, C. G.: 1976, *J. Geophys. Res.* **81**, 4537.
Orr, D.: 1973, *J. Atmospheric Terrest. Phys.* **35**, 1.
Mozer, F. S.: 1976, in K. Knott and B. Battrick (eds.), *The Scientific Satellite Programme during the IMS*, D. Reidel Publ. Co., pp. 101–131.
Mozer, F. S., Torbert, R. B., Fahleson, U. F., Fälthammar, C.-G., Gonfalone, A., and Pedersen, A.: 1978, *IEEE Trans. Geoscience Electronics* **16**, 258.

ULF WAVES OBSERVED WITH MAGNETIC AND ELECTRIC SENSORS ON GEOS-1

S. PERRAUT, R. GENDRIN, P. ROBERT, A. ROUX, and C. DE VILLEDARY

Département Etudes par Télédétection de l'Environnement, Centre National d'Etudes des Télécommunications, 92131 Issy-les-Moulineaux, France

and

D. JONES

Space Science Department, European Space Agency Technology Center, Noordwijk, the Netherlands

Abstract. Preliminary results on ULF waves in the range 0.2–11 Hz detected on board GEOS-1 are described. In this frequency domain one must consider two different categories of emissions: waves with frequencies below the proton gyrofrequency f_{ci} and frequencies above. Belonging to the first category are the structured ULF events called 'pearls' which are very often observed on the ground, but which were observed only twice on board GEOS-1 in seven months of operation. Most of the monochromatic emissions with a frequency below f_{ci} were detected while the satellite was near its apogee ($L \approx 7$). These waves are left hand polarized, perpendicular to the magnetic field. Other monochromatic events, whose frequency lies above the proton gyrofrequency are predominantly right hand polarized perpendicular to the magnetic field, whereas phenomena with complex, harmonically related structures, also detected above the proton gyrofrequency, are clearly polarized along the magnetic field. These waves can be probably interpreted in terms of Ion Bernstein modes.

We make here a morphological description of these different waves. We present preliminary results which relate their occurrence or characteristics to the parameters of the magnetospheric plasma. This study is mainly based upon data obtained via the magnetic detector although on some occasions, the electric component of the waves has been also identified and measured.

1. Introduction

GEOS is equipped with detection chains of magnetic and electric field fluctuations in the frequency range 0.05–11 Hz. This range covers three types of pulsations from low frequency to high frequency: Pc-3, Pc-1–Pi-1, and ELF waves. Taking into account the trajectory of the satellite (see in this issue the general introduction to GEOS where the orbital configuration is given), the proton gyrofrequency falls generally within the frequency domain detected by the sensors.

The study of Pc-3 is hampered by the fact that the spin rate of the satellite which is approximately equal to 0.18 Hz, has obliged us to introduce a filter at this frequency. The analysis of these waves is therefore more complicated than the analysis in the rest of the passband and we have not yet tried to study these waves, whose geophysical importance has however been stressed in a series of recent papers (see for instance Singer *et al.*, 1977; Cummings *et al.*, 1978; Hughes *et al.*, 1978).

Since a long time intensive studies of Pc-1–Pi-1 have been carried out using ground data, and a classification of the different types of events which can be observed in this frequency range has been made based on their spectral shapes on

Space Science Reviews **22** (1978) 347–369. *All Rights Reserved*

a three dimensional representation (Gendrin, 1970). Generally these pulsations observed on ground have frequencies below the minimum proton gyrofrequency along the field line and they are left hand polarized. One considers that these pulsations are ion cyclotron waves which can propagate along the field line at frequencies below the proton gyrofrequency.

Much was expected from direct in situ observations of these waves by means of satellites. The first measurements were carried out on board OGO-3 and -5. The trajectory of these satellites was such that they explored a large volume of the magnetosphere, leading Heppner *et al.* (1967, 1970) to find a Pc-1 distribution with a maximum near the magnetopause at the 6 hr local time meridian with high percentages extending tailwards near the magnetopause. Fredricks and Russell (1973) observed ion cyclotron waves in the polar cusp. Using data from Explorer 45, Taylor *et al.* (1975) found, on the contrary, about ten cases occurring near the plasmapause and in the proton ring current but they could not conclude that the waves were generated by the ion cyclotron instability or that the ion cyclotron waves played a dominant role in the loss of the ring current ions. A detailed comparison between the time of occurrence of these waves and the particle distribution functions led Taylor and Lyons (1976) to distinguish between two categories of waves: those which were associated with increasing fluxes of protons and those which were not. In fact the waves studied by these authors cover a wide frequency range (1–30 Hz) and no detailed spectral analysis of these waves was available. With a similar purpose in mind and with a similarly poor spectral information, Kintner and Gurnett (1977) have looked at records from Hawkeye 1. Only five events (over a period of 18 months) were detected and they were confined to the plasmasphere.

A statistical study of the occurrence of Pc-1 recorded on ATS-1 as a function of the local time and the magnetic activity was reported by Bossen *et al.* (1976a); these authors found a maximum in the afternoon sector at geostationary orbit. Comparison between ground measurements in the conjugate area and on board ATS-1 led the authors to conclude that most of the events detected in situ are IPDP's with a rather large amplitude (Bossen *et al.*, 1976b); this result is in agreement with the preferential occurrence of pulsations around 18:00 LT.

Waves with frequency above the proton gyrofrequency have been detected in the magnetosphere. These pulsations were considered to belong to the ELF range by Russell *et al.* (1970). The events observed on board OGO-3 occur only in the outer plasmasphere ($L \approx 4$–5) and are confined near the magnetic equator (Russell *et al.*, 1970). These waves are propagating perpendicular to the magnetic field. Similar phenomena were detected at $L \approx 2$–3, on board IMP-6 and Hawkeye-1 by Gurnett (1976) who shows clearly the harmonic structure of such emissions. Gurnett confirmed the strict localization of this noise near the magnetic equator.

Another type of electrostatic (or quasi electrostatic) noise attributed to plasmapause gradients has also been detected by Kintner and Gurnett (1978) in the frequency range above the proton gyrofrequency.

As appears from the above review of results obtained with the different satellites in the frequency domain covering 0.2–30 Hz, we can conclude:

(1) Few emissions were observed due to the insufficient sensitivity of the on board sensors.

(2) No precise information on the spectral characteristics of the phenomena (monochromaticity or wide spectrum for example · · ·) has been brought out due to the lack of systematic spectral analysis.

(3) No unambiguous classification (based upon the morphological properties or on the mechanism of generation) has been defined so far.

As we will see in Section 2 where our magnetic experiment is briefly described, the sensitivity of our own equipment is much better than the ones used previously. In that section, the analysing method which leads to a systematic three dimensional representation (frequency, time, intensity) is also described.

In the following sections, we will give successive examples of magnetic phenomena with a frequency below the proton gyrofrequency (Section 3), then above the proton gyrofrequency, (Section 4). Some examples of the relationships which exist between the occurrence of the characteristics of the ULF emissions and the variations of the plasma parameters are given in Section 5. The last section is devoted to examples of electric signals some associated with magnetic fluctuations and some not.

2. Description of the Magnetic Experiment

For the detection of the magnetic fluctuations, in the ULF/ELF range, there are three orthogonal magnetic sensors whereas the electric signals are detected only on one component measured with the long radial boom E_y (see Figure 1 in S-300 experimenters, 1978). The magnetic component aligned along the spin axis is called B_z (in fact there is a small angle, ≈0.8° between the z antenna and the spin axis so that this antenna is slightly affected by the rotation of the satellite, especially in the inner parts of the orbit where the DC magnetic field is strong). The two other components in the plane perpendicular to the spin axis are called B_x and B_y. The frequency sampling of these signals is ≈23 Hz, giving a Nyquist frequency ≈11.7 Hz.

Before going into the details of the treatment applied to these three signals, it is important to consider the sensitivity curves of the magnetic chain (see Figure 3b of Jones (1978)). A comparison with the sensitivity curves of ATS-6 and Explorer-45 shows that our ULF/ELF sensors operate with a better sensitivity ($\approx 2\times 10^{-3}$ γ rms $\mathrm{Hz}^{-\frac{1}{2}}$ at 2 Hz) than the others (more than a factor 10). The notch which appears at the spin frequency (in the antenna reference system) is due to the special despin system which is incorporated within the S-300 magnetic ULF unit to attenuate the spin modulated component due to the DC field which would have otherwise saturated the telemetry dynamics range (70 dB). This despin system plays simply the role of a filter whose central frequency is synchronized to the

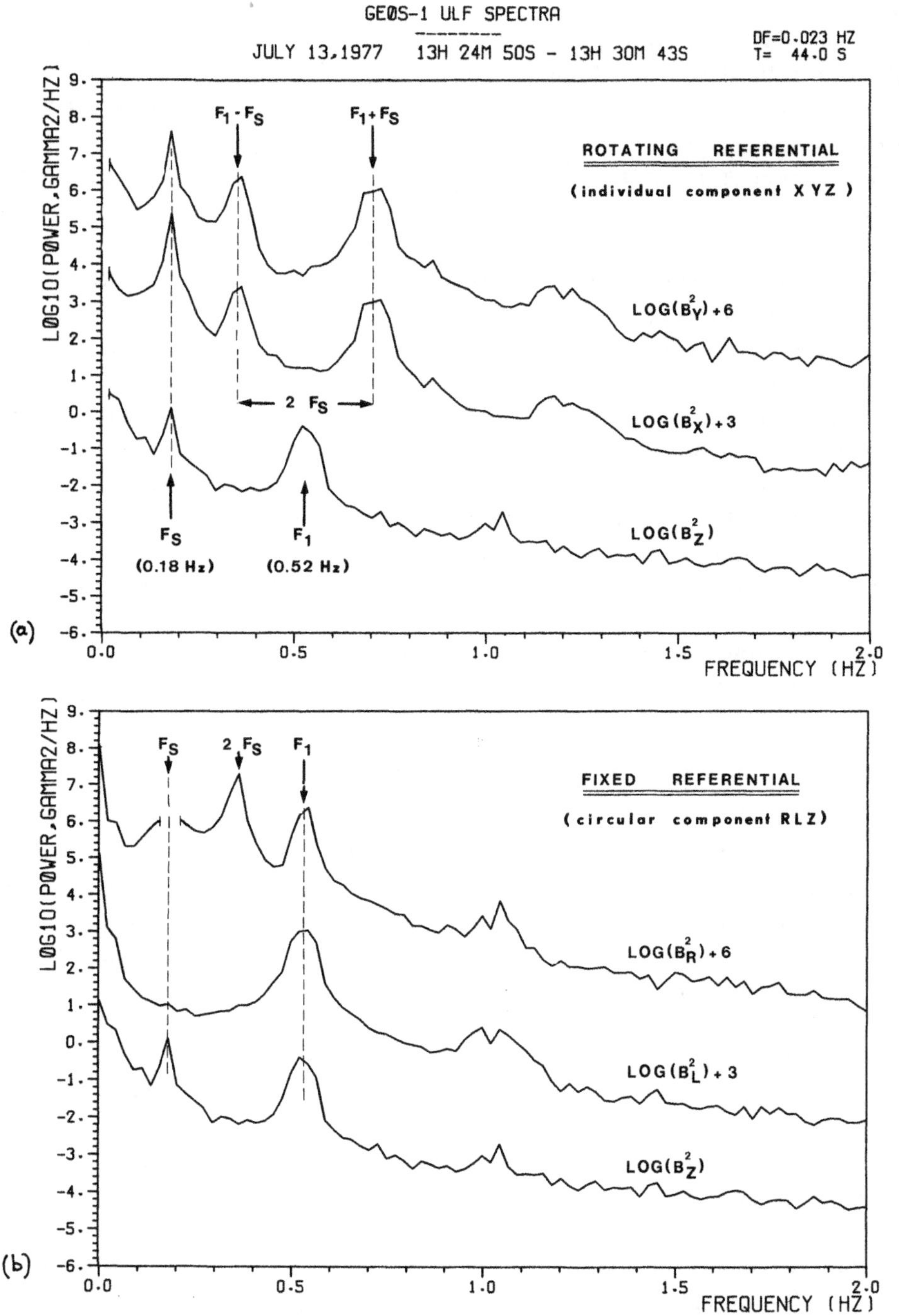

Fig. 1. Spectra of the three components of the ULF signal. (a) B_x, B_y, and B_z in the antenna reference system, B_z being aligned (to within 0.8°) along the spin axis. In spite of the despin system, the three components present a parasitic signal at the spin frequency F_s. An emission at F_1 appears on the B_z component, whereas the energy in splitted into two signals at $F_1 \pm F_s$, on B_x and B_y. (b) B_L, B_R, and B_z in a stationary frame of reference, B_L and B_R, being the circular components of opposite polarities. Now, the emission appears at the same frequency on the three components.

spacecraft spin. The attenuation is of the order of 26 dB, insufficient however to remove completely the strong influence of the DC magnetic field during the inner part of the GEOS-1 trajectory.

Signals are detected in the antenna reference system (upper part of Figure 1). Parasitic peaks at the spin frequency F_s appear on the three components in spite of the despin system. One must note on this figure that the emission during this period of time has a unique frequency F_1 on the B_z component which is almost unaffected by the satellite spin whereas the energy is split into two parts at the frequencies $F_1 \pm F_s$ on the orthogonal components B_x and B_y. It is more interesting to consider a stationary frame of reference and to distinguish between left hand (B_L) and right hand (B_R) components. This has been done applying Kodera *et al.* (1977) method with which complex modulation effects induced by the spin are easily removed by just a translation in frequency. The result obtained with such a method is represented in the lower part of Figure 1. The natural emission is seen now at the same frequency F_1 on the three components. The signal at the spin frequency has almost disappeared whereas a small component at $2F_s$ remains on B_R (which corresponds to the sense of rotation of the satellite at that time).

The advantages and difficulties of this analyzing method are explained elsewhere (Robert *et al.*, 1979). It has been systematically applied to the ULF/ELF data on a routine basis. Spectra were computed over 512 points (i.e. 22 s). The transfer function of the chain and the gain factor of the step amplifiers (which can take the values 5, 25, and 125) were also taken into account. The results are plotted as a grey-scaled representation in the range 0–1.5 Hz as it can be seen on the next pictures. We must also mention that a complementary picture is also systematically generated. It consists of integrated powers of the signal in different frequency ranges up to 11 Hz plotted as a function of time. Evaluation of the amplitude is more precise on this second view.

It must be noted since now that the electric signal being detected only along one direction E_y, it is impossible to work with circular components to eliminate the rotation influence. Conversely, knowing the frequency of the signal by comparison with the magnetic emissions, it is possible to obtain an information on the polarization of the electric waves. If the magnetic pulsations are not observed simultaneously, it is not easy to exploit directly the electric data.

The two complementary pictures described above which are called 'experimenter summaries' were generated on a routine basis, in real time by the ESA Operation Centre in Darmstadt. The French Centre National d'Etudes Spatiales also performs off-line computation on these same data and provides us with two dimensional or three-dimensional pictures covering different frequency bands. All this set of analyzed data has been used for making the study which will be presented now.

3. Magnetic Emissions with $f < f_{ci}$

Pulsations in the Pc-1 range commonly observed on the ground generally present a typical structure called 'pearls'. On board GEOS-1 this characteristic spectral shape was observed clearly only two times over seven months of 'experimenter summaries' data. However this small number may be due to the limited time and frequency resolutions used in these summaries. A study, now in progress, made on more elaborate displays seems to indicate that the number of structured events may be slightly larger. One of these examples is shown on Figure 2 (upper panel). The repetition period of the order of 150 s is larger than expected at this distance ($R/R_e \approx 4.5$) and for this frequency ($f \approx 1.2$ Hz), especially if one takes into account that a repetition period of 150 s at the level of the satellite corresponds to a repetitive period of ≈300 s on the ground.

In fact the spectral shape presented by most of the fluctuations observed on our records is given on the lower panels of the Figure 2. It consists of monochromatic emissions without clear repetitive structures. The duration of such events can be long (more than 3 hr on Aug 31, 1977). The wave frequency can be stable although the satellite is crossing various L shells. A study of the simultaneous appearance of these emissions at different ground stations is done elsewhere (Gendrin *et al.*, 1978).

Sometimes a noise with more or less wide spectrum is superimposed as appears around 10:15 UT on the Aug 31, 1977 event. The origin of the harmonic structure seen on these spectra is not yet known, but we know that it is not due to a saturation of the detector because the amplifiers were in their lowest gain.

Events presented in Figure 2 are typical of quiet magnetic activity. During periods of high magnetic activity ($K_p \approx 6$), bursts of noise are detected. They present a repetition period ≈10–20 mn (see the upper event of Figure 3).

Oscillations similar to Irregular Pulsations of Diminishing Periods (IPDP), Auroral Agitation (AA) and Short Irregular Pulsations (SIP) observed on the ground are also observed on board GEOS. An example of SIP is given on the lower panel of Figure 3.

Up to now, no precise statistics on local time of occurrence, amplitude, polarization were developed for the different types of phenomena. But a first examination of the data leads us to the following conclusions:

(1) Many ULF events have been detected in the magnetosphere (over a seven month period, we have observed ≈180 cases of Pc-1, ≈150 SIP's, ≈30 bursts events, characteristic of high magnetic activity and 4 IPDP's).

(2) For monochromatic emissions, the average amplitude varies between 0.3 and 1 γ at 0.5 Hz, explaining why similar events were scarcely observed on previous satellites.

(3) Many events were observed around apogee, i.e. outside the plasmasphere. However we must note that the cold plasma density, as measured by the active wave experiments on board GEOS-1 (see for instance Etcheto and Bloch (1978) and Decreau *et al.* (1978a)) was very high (≈10–20 cm^{-3}) at large radial distances, at least during the morning sector explored by GEOS-1 orbit during these seven months.

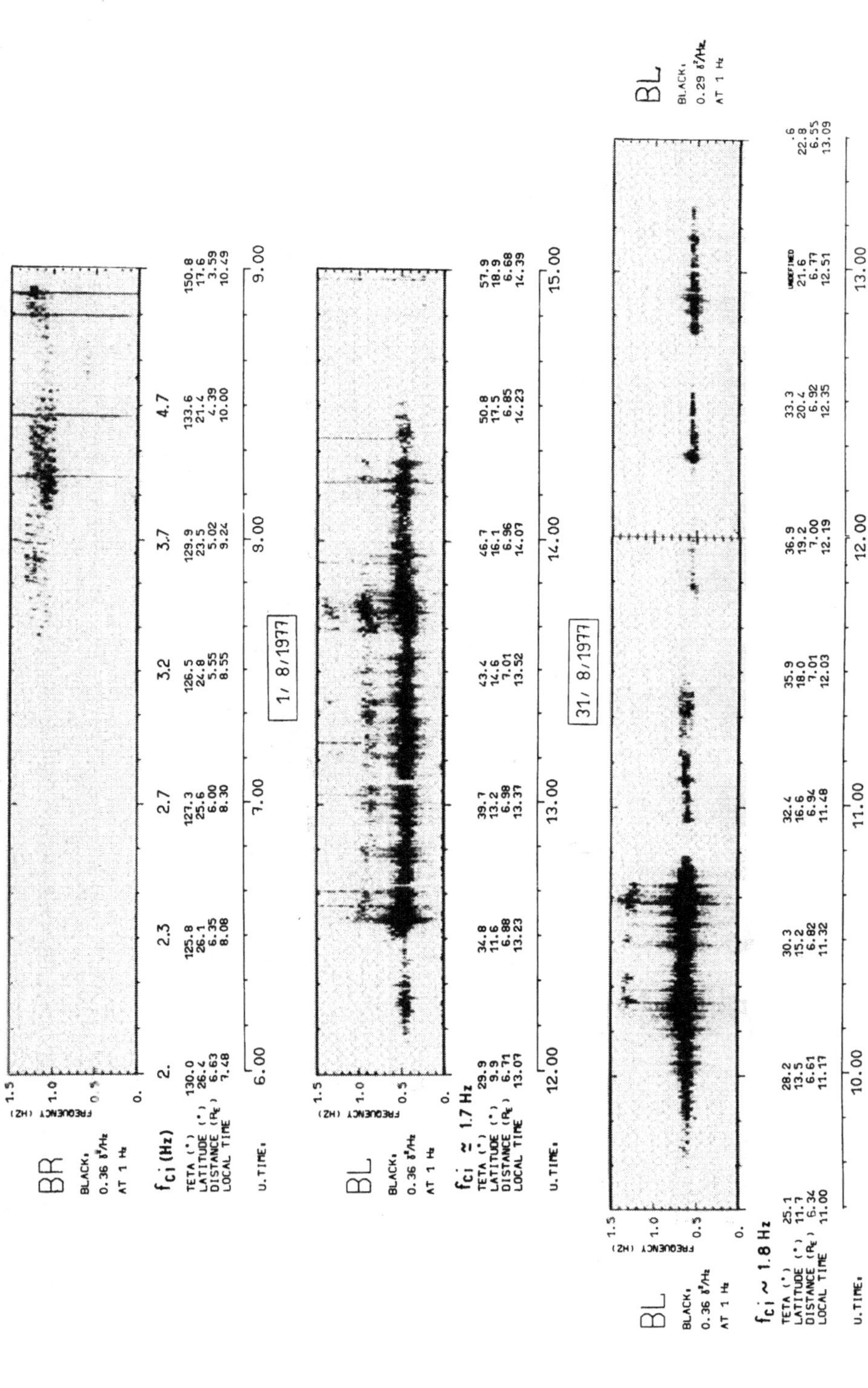

Fig. 2. Typical spectra of quasi-monochromatic ULF emissions. The upper panel represents one of the two cases we have observed up to now with a clear repetition structure (classical 'pearl' event). The other two examples are presented to illustrate both the long duration of an emission at a constant frequency and the presence of faint harmonics (not due to experimental saturation).

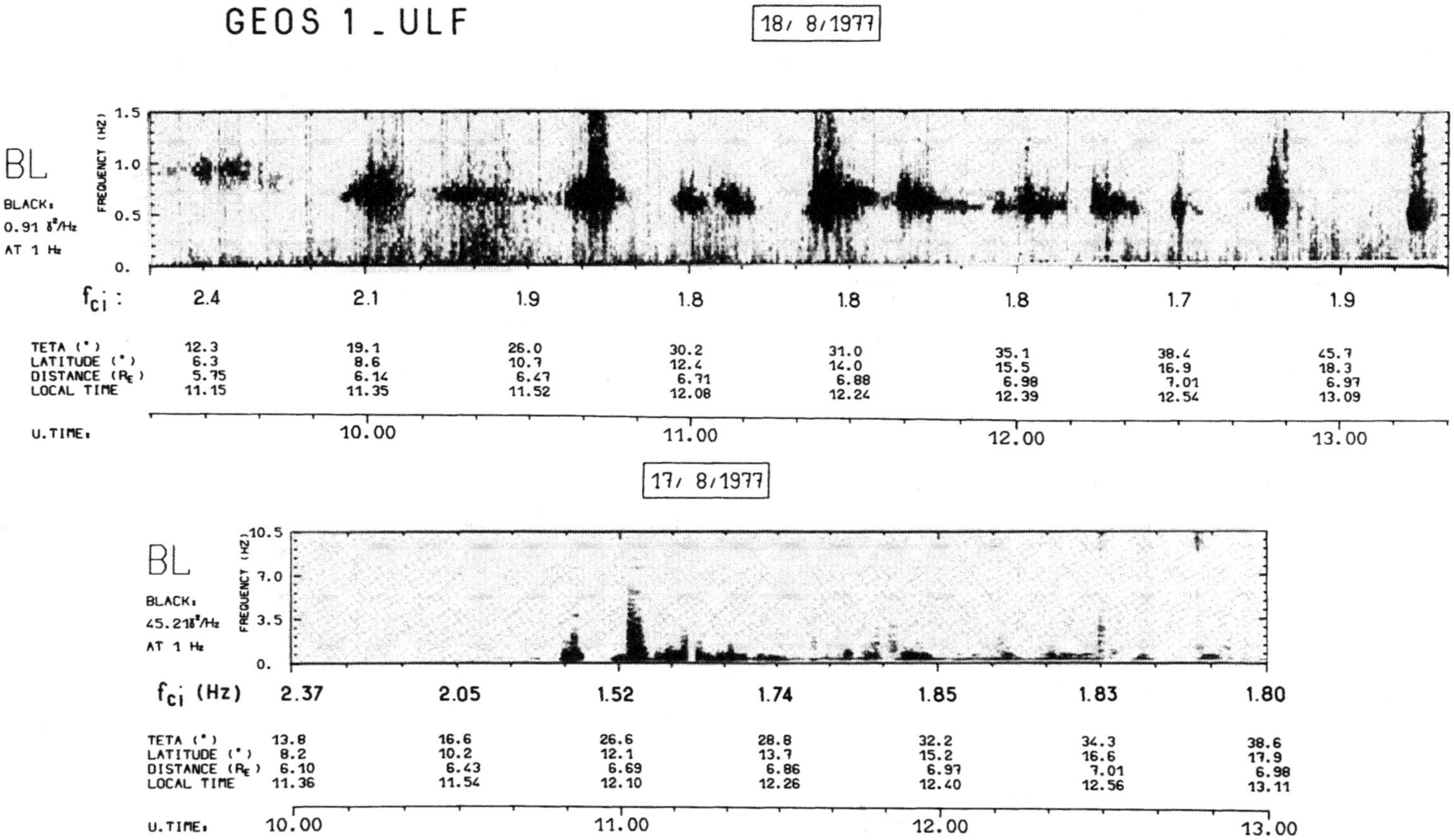

Fig. 3. Typical events occurring during periods of high magnetic activity ($K_p \approx 6$). The event presented in the upper panel consists of repetitive bursts, with a repetition period of the order of 10 mn. The main frequency is lower than f_{ci}. It corresponds to left handed ion cyclotron wave. The event presented on the lower panel looks like SIP's observed on the ground. Note that on some occasions, the spectrum extends above the local proton frequency.

(4) The mean ratio between the frequency of the monochromatic emission and the local proton gyrofrequency is of the order of 0.3.

(5) Although it is difficult to characterize the wave polarization, because the angle θ between the spin axis and the magnetic field B_0 often is large, the B_L and B_R components of these $f<f_{ci}$ emissions have a larger amplitude than the B_z component.

In this section, we have considered SIP's because they are observed on the ground and that a fraction of their spectrum is less than f_{ci}. But in some circumstances, emissions with wide spectra clearly extending above the proton gyrofrequency are detected in space as we will see in the next section which is devoted to phenomena above the proton gyrofrequency.

4. Magnetic Emissions with $f>f_{ci}$

On two occasions, the magnetopause compression was so large that GEOS-1 crossed the magnetospheric boundary. This happened on July 29, 1977 and on Dec 2, 1977. At the time during which the satellite was located in the magnetosheath, noise of large amplitude and with a continuous spectrum, extending at least up to 10.5 Hz (the Nyquist frequency of our equipment is $\approx$11.7 Hz) was observed (Figure 4). These emissions are characteristic of the very strong plasma turbulence which occurs in the magnetosheath. A quantitative spectral study of this turbulence is in progress.

Different types of magnetic fluctuations above the proton gyrofrequency have been detected on board GEOS-1. We have listed approximately 40 such events. The most simple ones consist of a single monochromatic emission similar to those observed below f_{ci} but with a mean frequency higher than f_{ci}. An example of such events is given on Figure 5. It presents a fine structure (at the end of the emission) as do classical Pc-1 pulsations. Note that the signal is stronger on the B_R component (B_L and B_R having the same black level). Therefore this kind of emission can be considered to be a right hand polarized magnetosonic wave near ion gyroresonance.

The other type of events presents a more complex structure. Generally it consists of several superimposed emissions separated or not by the proton gyrofrequency. These events look similar to the ones detected by Gurnett (1976) on board IMP-6, in a higher frequency range (30–100 Hz) but at lower L-values (2–3.5). Different cases are examined below.

In the examples presented on Figure 6, two different emissions are detected simultaneously. The emission around 0.4 Hz below f_{ci} (especially clear on the third panel from the top) is left handed polarized perpendicular to the magnetic field (B_L component) whereas, at the same time, higher frequency emissions composed of several harmonics (8 harmonics are visible around 14:50 UT) are clearly polarized along the magnetic field (large B_z). Occurrence on the B_L

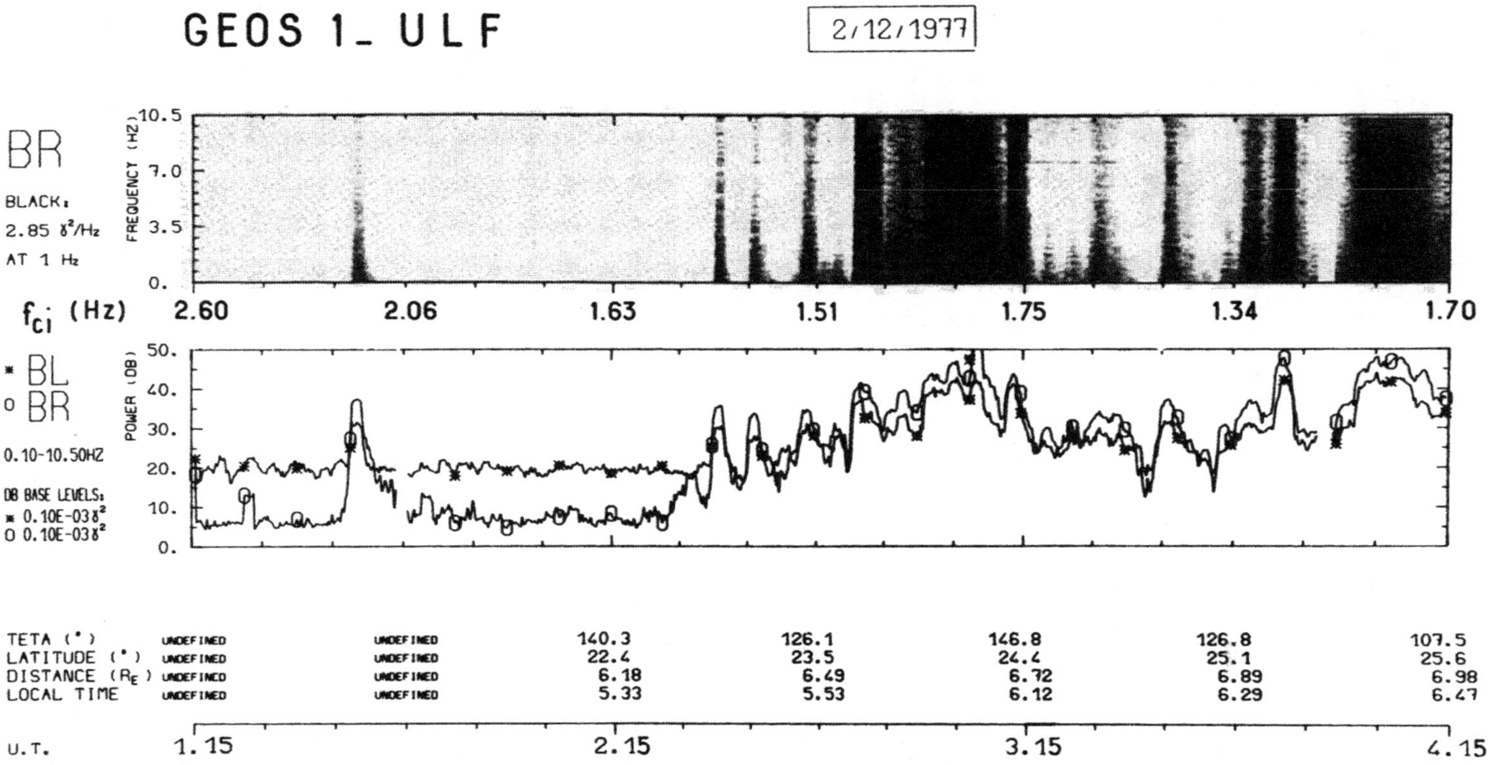

Fig. 4. The December 2, 1977 event. This noise with a continuous spectrum extending to the Nyquist frequency of our transmitting equipment is typical of magnetosheath multiple crossings. Signals detected in this region have large amplitude (see the power curves below the spectrum, which indicate a signal level as high as 10γ (near 03:00), integrated in the whole frequency band.

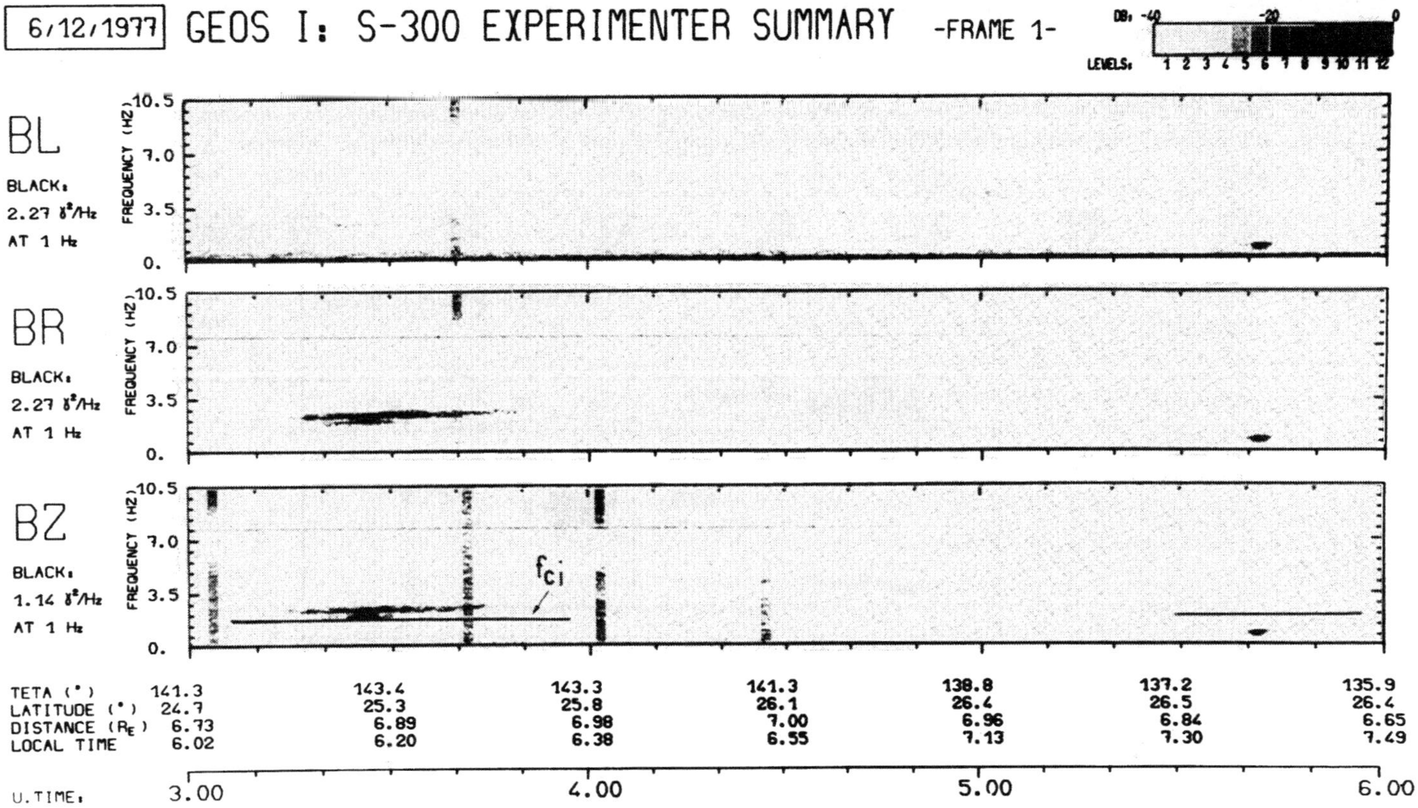

Fig. 5. Example of a monochromatic wave observed above the local proton gyrofrequency. The two emissions around 03:30 and 05:40 UT respectively above and below the proton gyrofrequency, approximately occur at the same distance. The θ angle between the spin axis and the magnetic field being almost equal for the two emissions one sees that the emission below f_{ci} (around 05:40) presents a left mode of larger amplitude than the emission above f_{ci} (around 03:30). The emission around 03:30 UT has a large right hand component.

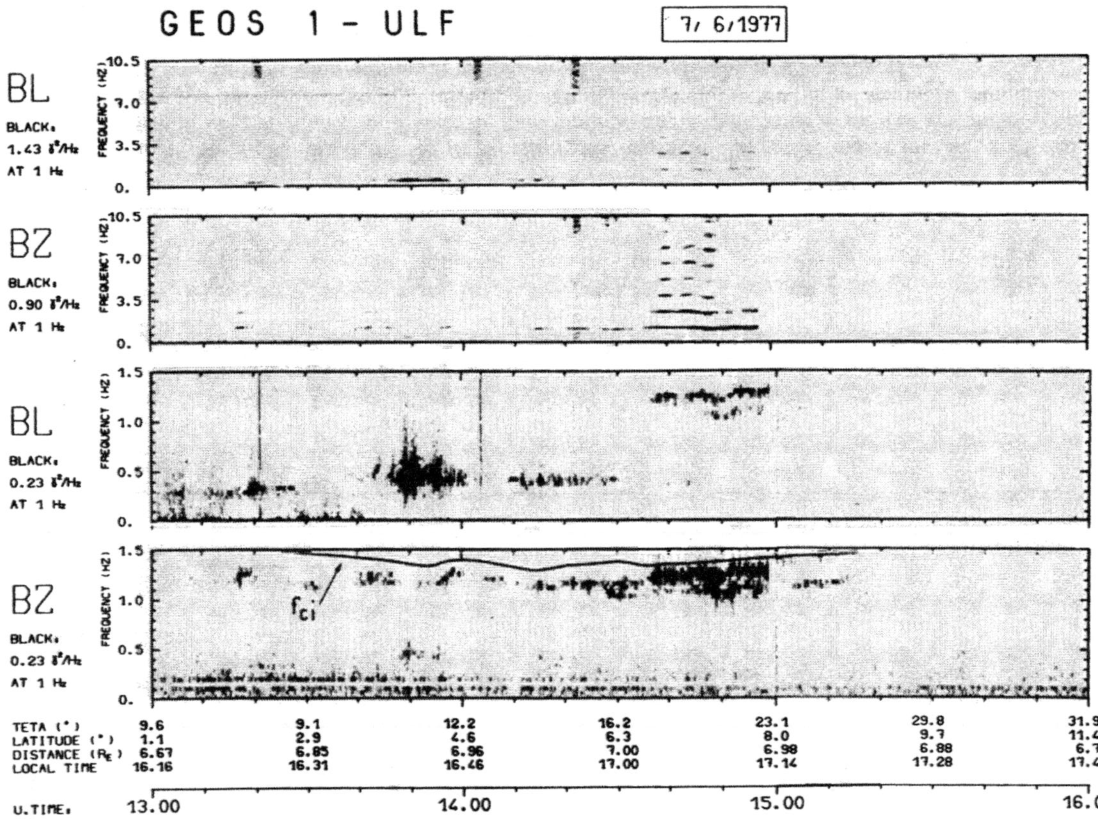

Fig. 6. Example of harmonically related emissions above the proton gyrofrequency. In this case one observes simultaneously a low frequency emission around 0.4 Hz below the proton gyrofrequency (especially clear on the third panel from the top but which is also seen on the upper panel) which is left hand polarized perpendicular to the magnetic field, and several harmonically related emissions above the proton gyrofrequency which are clearly polarized along the field line. The variations of the fundamental emission frequency seems to follow the fluctuations of the DC magnetic field and the frequency separation between the different elements is slightly lower than the proton gyrofrequency.

component after 14:30 UT is due to the fact that the θ angle (between B_z and the spin axis) increases with time. Other characteristics of this emission can be pointed out: the frequency variations of the fundamental seems to follow the fluctuations of the DC magnetic field (see the lower panel of Figure 6) and the frequency separation between the harmonics is slightly smaller than the proton gyrofrequency. Finally we must note the splitting of the emission at the end of the event. Such a splitting is visible on the two first bands. The frequency separation is of the order of $f_{ci}/8$. Very often similar fine structures are observed in these banded emissions as is shown in the next example.

Because of the complexity of its configuration, the phenomenon presented on Figure 7 is especially interesting. It consists of four distinct bands (even five around 15:10 UT). Each band itself is divided into a main emission and some fainter sideband lines. The main emissions of the different bands are harmonically related whereas the sideband lines are not. A careful examination of the structures leads to find that the frequency separations of a main emissions and its adjacent sidebands are the same for all the bands. This last observation certainly proves that the harmonic emissions have a natural origin and are not due to a saturation effect. In the previous example we had seen that the harmonically related bands were multiples of a frequency slightly different from the proton gyrofrequency. In this example, the situation is even clearer: the bands are harmonically related with a fundamental frequency of $\approx$2 Hz whereas the local proton gyrofrequency is 1.3 Hz. The frequency separation between the main emission and associated sidebands are multiples of $f_{ci}/4$.

In the last example, we show emissions harmonically related at exact multiples of the proton gyrofrequency (Figure 8). Again, we have two simultaneous events. The first one is a low frequency monochromatic emission around 1 Hz seen mainly between 11:50 and 12:00 UT and around 12:25 UT. The fact that this emission is seen on B_L (upper panel) and not on B_z (middle panel) indicates that it is mainly polarized (in the left hand mode) in the plane perpendicular to the DC magnetic field (note that the θ angle remains small throughout the event). The higher frequency part of the event (between 3.5 and 10.5 Hz), seen only on the B_z component (middle panel) consists apparently of a mixture of decreasing and increasing frequencies. However, if we consider that the stronger rising tones are due to the aliasing of emissions occurring above the Nyquist frequency of 11.7 Hz, the interpretation becomes easier. On the lower panel of Figure 8, we have reconstructed what should be the original signal, by taking the symmetrical image with respect to the Nyquist frequency. We therefore see that the picture becomes coherent; the high frequency emission consists of a series of bands, harmonically related and which, in this example and within the accuracy of the measurements, are exact multiples of the proton gyrofrequency, which is slightly decreasing from 1.85 Hz at 12:00 UT to 1.55 Hz at 12:40 UT.

Although some differences appear between the various events with $f > f_{ci}$, it must be underlined that the emissions showing a harmonic structure are clearly

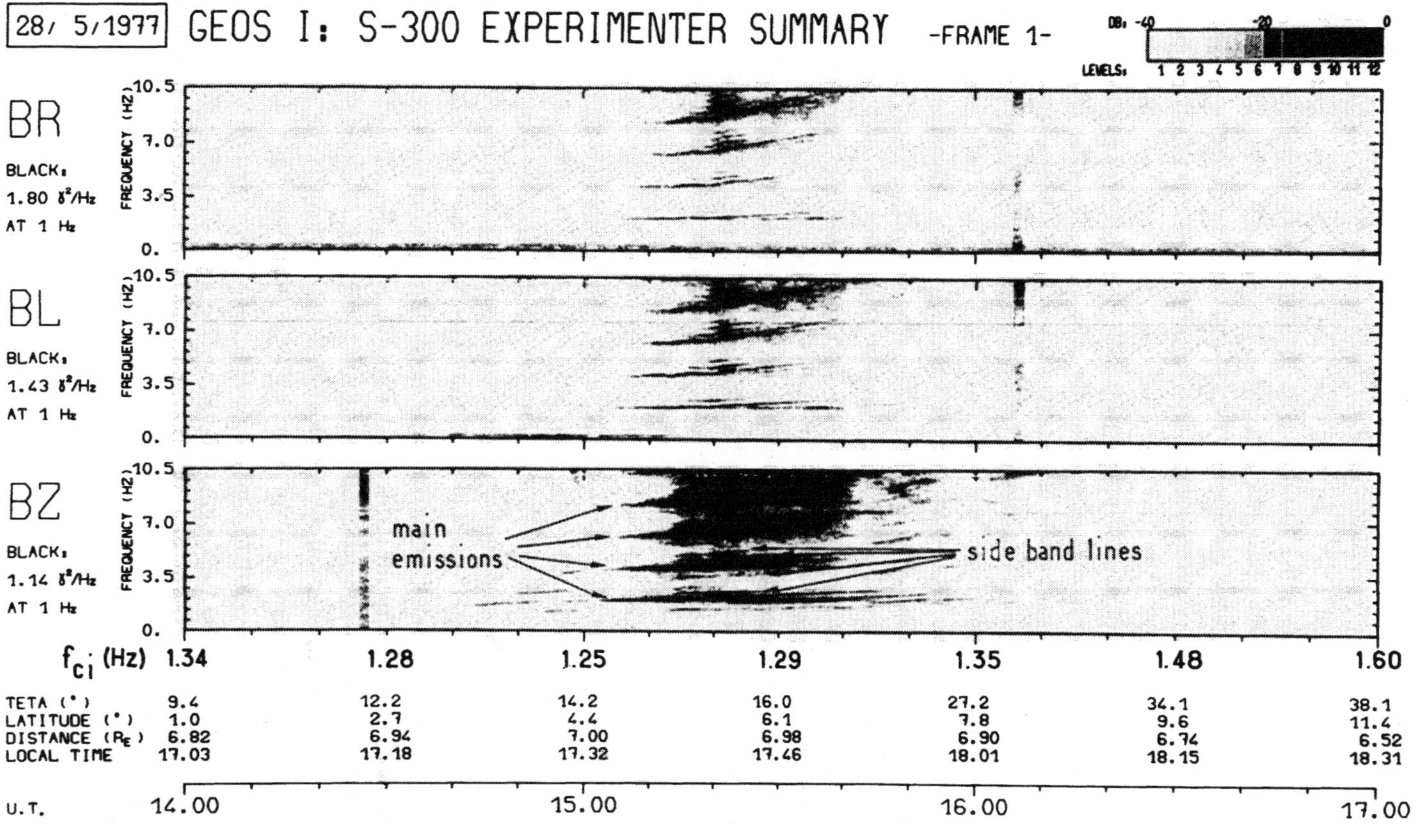

Fig. 7. Example of a more complex structure in a harmonically related emission. This event consists of several superimposed structures. Each structure is composed of a main emission and of sideband lines. The main emissions are harmonically related, whereas sideband lines are not. The frequency separation between sideband lines and the adjacent main emission is independent on the structure number. This observation proves that we are in presence of natural harmonic emissions which are not due to experimental distortions.

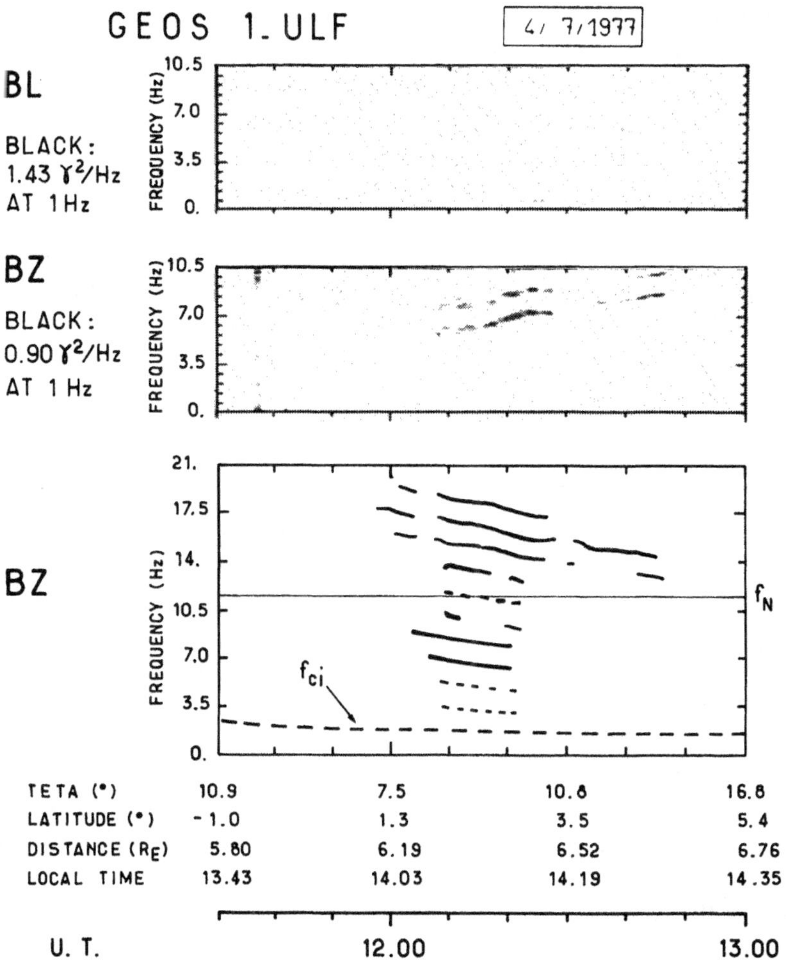

Fig. 8. Harmonically related emissions at multiples of the local proton gyrofrequency. On the upper panels are shown spectrograms of the phenomenon observed on B_L and B_z components. Again one observes simultaneously a low frequency emission below f_{ci}, which is left handed polarized perpendicular to the magnetic field (note that the θ angle is small) and higher frequency emissions polarized along the magnetic field. The experimental spectrum between 0 and 10.5 Hz is represented in the middle panel. After taking the image of the most prominent lines with respect to the Nyquist frequency, we obtain the schematic representations of the lower panel. Note that all the lines are harmonics of the fundamental frequency which follows the variation of the local proton gyrofrequency.

polarized along the magnetic field. Such waves cannot propagate along the field lines and this explains why they are so seldom recorded on the ground.

A first approach for the interpretation of these emissions is to identify them with ion Bernstein waves (Kennel *et al.*, 1970; Fredricks and Scarf, 1969; Cuperman and Gomberoff, 1977) but theoretical work has still to be done in order to understand better the different behaviour of these waves. If it were proved, by further study, that the different bands are always at exact multiples of the fundamental frequency, we should abandon the idea that these emissions are related to the so-called $(n+\frac{1}{2})f_{ci}$ instability (Ashour-Abdalla and Thorne, 1977). The strong magnetic component of these waves lead us to interpret them as magnetosonic waves, non-linearly coupled to each other. The fact that there is a difference between the frequency of the fundamental and the local proton gyrofrequency could indicate that these waves were generated on different L shells and had propagated to the satellite perpendicularly to the DC magnetic field.

On some occasions, however, it is possible to compare the characteristics of the emissions with some important local plasma parameters. This is the subject of the next section.

5. Relation with Plasma Parameters

We present here some preliminary results obtained in trying to relate the characteristics of ULF emissions to the local plasma parameters as determined by other experiments on board GEOS. Besides the local proton gyrofrequency, the plasma parameters which play an important role in the amplification of ULF/ELF waves are the cold plasma density and the high energy particle fluxes. For the cold plasma density, we use the results of the active part of the S-300 experiment (see Etcheto and Bloch, 1978, and Decreau *et al.*, 1978a). One of the drawbacks of these experiments is that they are not continuously operating, so that it is not always possible to have detailed variation of the cold plasma density. For the high energy proton population, we use the result of experiment S-321 (Korth *et al.*, 1978). In this case the results are very preliminary because the particle data have not yet been sorted out according to their true energy and pitch angle distributions.

Studying the role of the cold plasma density on the appearance of ion cyclotron instability, Cornwall *et al.* (1970) have argued that such instability should preferentially take place within the plasmasphere. However a more refined study (Perraut *et al.*, 1976) has shown that this should not always be the case if one takes into account the variation with L of the thermal energy and of the anisotropy of the hot particle population. The fact that the emissions below f_{ci} are mostly observed in the vicinity of GEOS-1 apogee is in agreement with this refined theory. Such an agreement is more easily understood if we note that the cold plasma density at these high L values ($L \approx 6$–7) is indeed much higher ($n_c \approx 5$–$20\ \mathrm{cm}^{-3}$) than the values currently admitted ($n_c \approx 1\ \mathrm{cm}^{-3}$) on the basis of measurements made with ion detectors whose results must in fact be considered cautiously (Decreau *et al.*, 1978b).

The influence of the cold plasma density on the emitted frequency of the waves has also been considered. It is well known that the larger the cold plasma density, the lower the amplified frequency, assuming that the other parameters are constant. It has been possible to study such an effect on the August 14 event (Figure 9). Knowing the magnetic field B_0 obtained from the GEOS

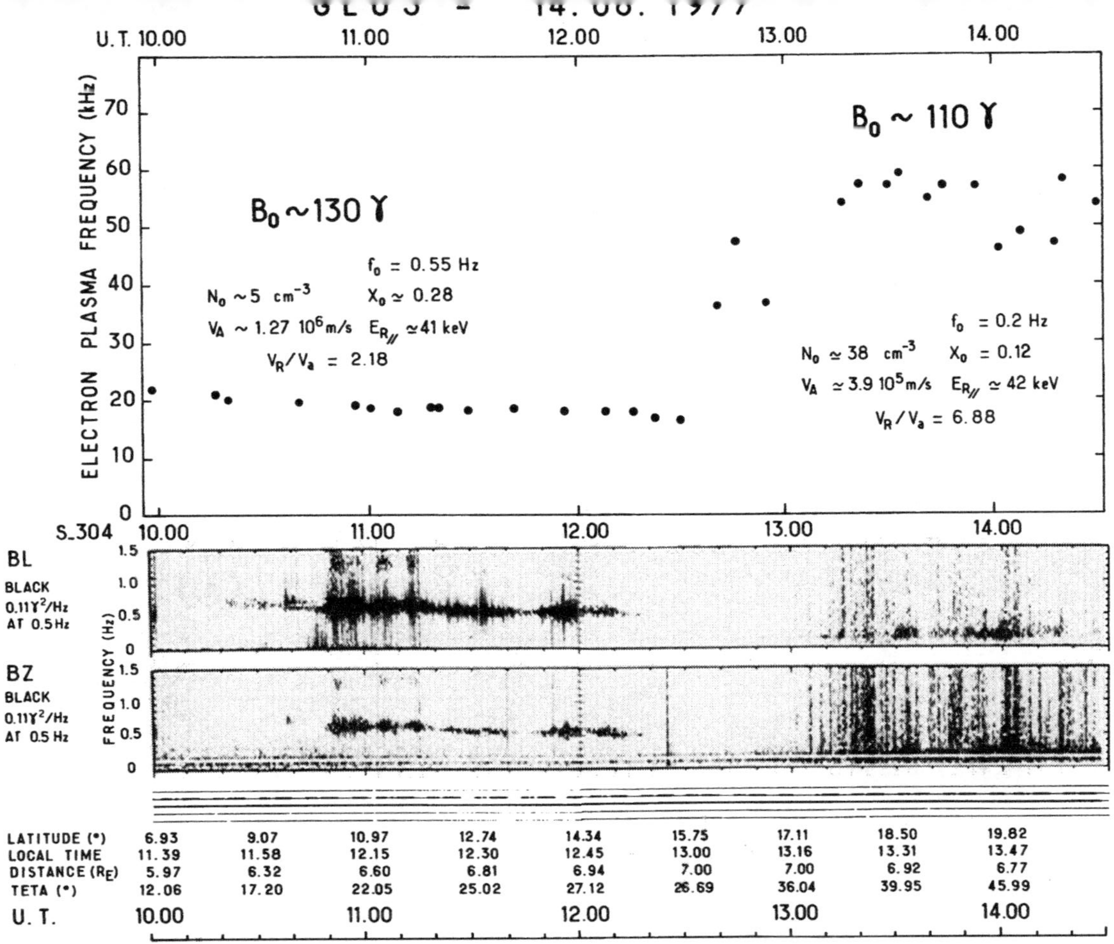

Fig. 9. Variation of the central frequency of an ion cyclotron emission with cold plasma density. The cold plasma density is deduced from the mutual impedance experiment (Decreau *et al.*, 1978a). The density profile shows a noticeable increase around 12:45 UT. The two successive ULF wave trains have different central frequencies. Taking into account both the DC magnetic field intensity and the cold plasma density variations, we can compute the parallel resonant energy $E_{R\parallel}$. The thermal energy of the interacting electrons is found to be the same in both cases.

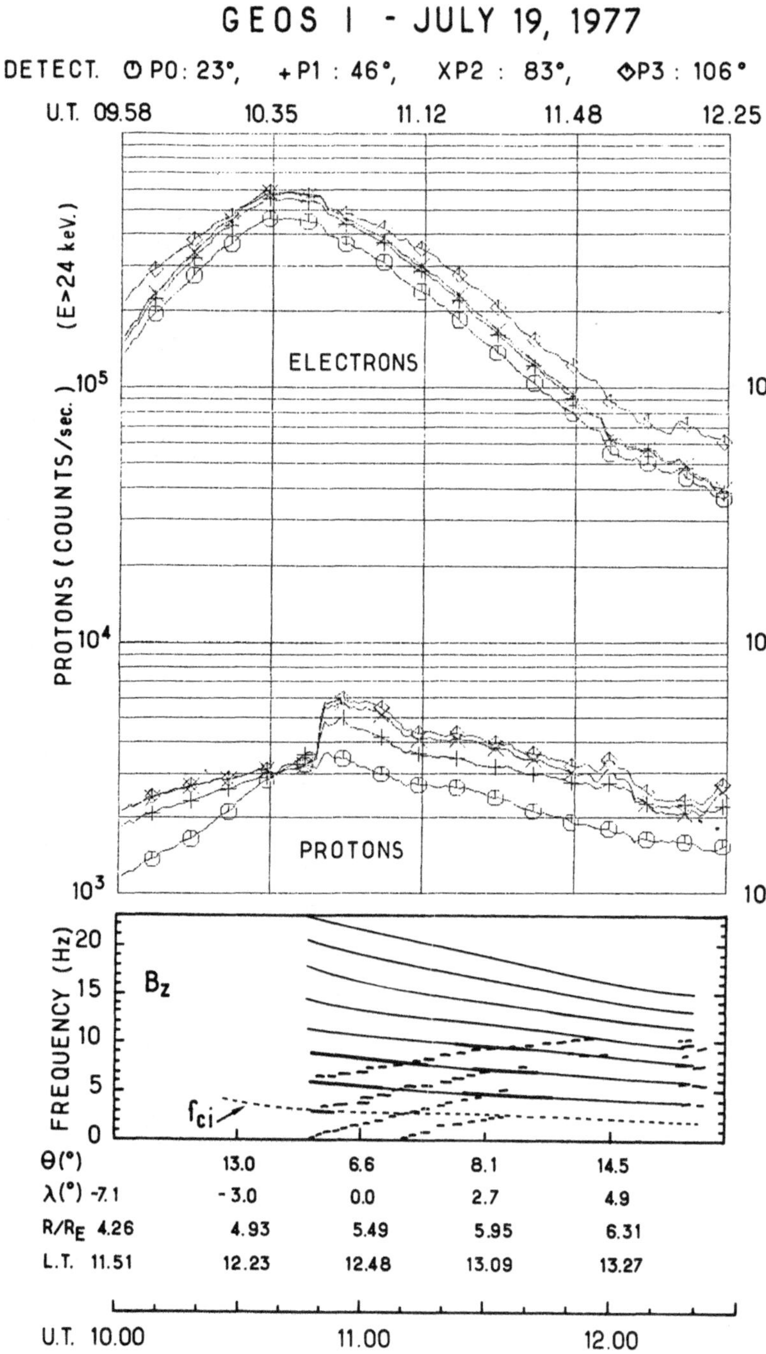

Fig. 10. Simultaneous increase of the proton flux and anisotropy and appearance of a harmonically related ULF emission. The angle θ between the spin axis and the magnetic field being small, the different angles quoted at the top of the figure for the different proton detectors correspond roughly to true pitch angles. The thick lines on the spectrogram correspond to the experimentally observed signal below the Nyquist frequency. The thin lines are obtained by a procedure similar to the one which was used in Figure 8. The emission is also polarized along the magnetic field.

fluxgate magnetometer and the cold plasma density n_e obtained from the mutual impedance experiment (Decreau *et al.*, 1978a), it is easy to deduce the Alfvén velocity V_A and the reduced frequency $x_0 = f/f_{ci}$. The parallel resonant energy is given by the following relation:

$$E_{R\parallel} = \tfrac{1}{2} m V_R^2 = \tfrac{1}{2} m V_A^2 \frac{(1-x_0)^3}{x_0^2}.$$

With the data measured on board GEOS-1 for B_0 and n_e around 11:00 and 14:00 UT, one finds an equal value for the resonant energy ($E_{R\parallel} \approx 40$ keV). In both parts of the event V_R/V_A is larger than 2. Therefore, provided that the anisotropy is larger than 1, the mean thermal velocity $U_\parallel$ of the particle distribution function is almost equal to the resonant velocity (see for instance Figure 8 of Gendrin *et al.'s* (1971) paper). The change in the mean frequency of the emission can therefore be attributed solely to the change of the cold plasma density, the high energy proton thermal energy remaining constant.

From a rapid comparison with the proton data in the energy range above 20 keV obtained from the S-321 experiments (Korth *et al.*, 1978), we have verified that there exist clear relationships between the occurrence of ULF waves and variations in the flux or anisotropy of high energy protons. A striking example is given on Figure 10. In this event high frequency, harmonically related emissions similar to those presented on Figure 8 are present. The beginning of the event is associated with a sudden increase of the integral flux in the direction perpendicular to the magnetic field, corresponding to a large increase in anisotropy of the distribution (Note that the θ angle is small, allowing us to consider the particles detected by the perpendicular detector have pitch angles near 90°). At the same time, no characteristic perturbation appears on the electron distribution. The same kind of particle fluctuations has not been observed in the other events. It may be due to the fact that we should look for lower energy protons (<20 keV) for which the data are not yet available.

6. Electric Component of ULF Fluctuations

One electric component of ULF fluctuations is measured by the quasi-DC experiment employing the sensors at the ends of the long radial boom E_y. The experiment's sensitivity to ULF signals is shown inset in Figure 3a, 'Introduction to the S-300 Wave Experiment', this volume. The reduced sensitivity in the ULF range as compared to that at higher frequencies (>10 Hz) is a consequence of the requirement that the dynamic range be able to accommodate the relatively large signal induced at the spin frequency by the DC electric field. At the high frequency end of the ULF range, the sensitivity is further reduced by a low-pass filter which was included to avoid aliasing of signals above the Nyquist frequency of the relevant telemetry channel.

ULF spectrograms are constructed by performing dynamic Fast Fourier Transforms (FFT) on the signal. The example shown in Figure 11 illustrates an event

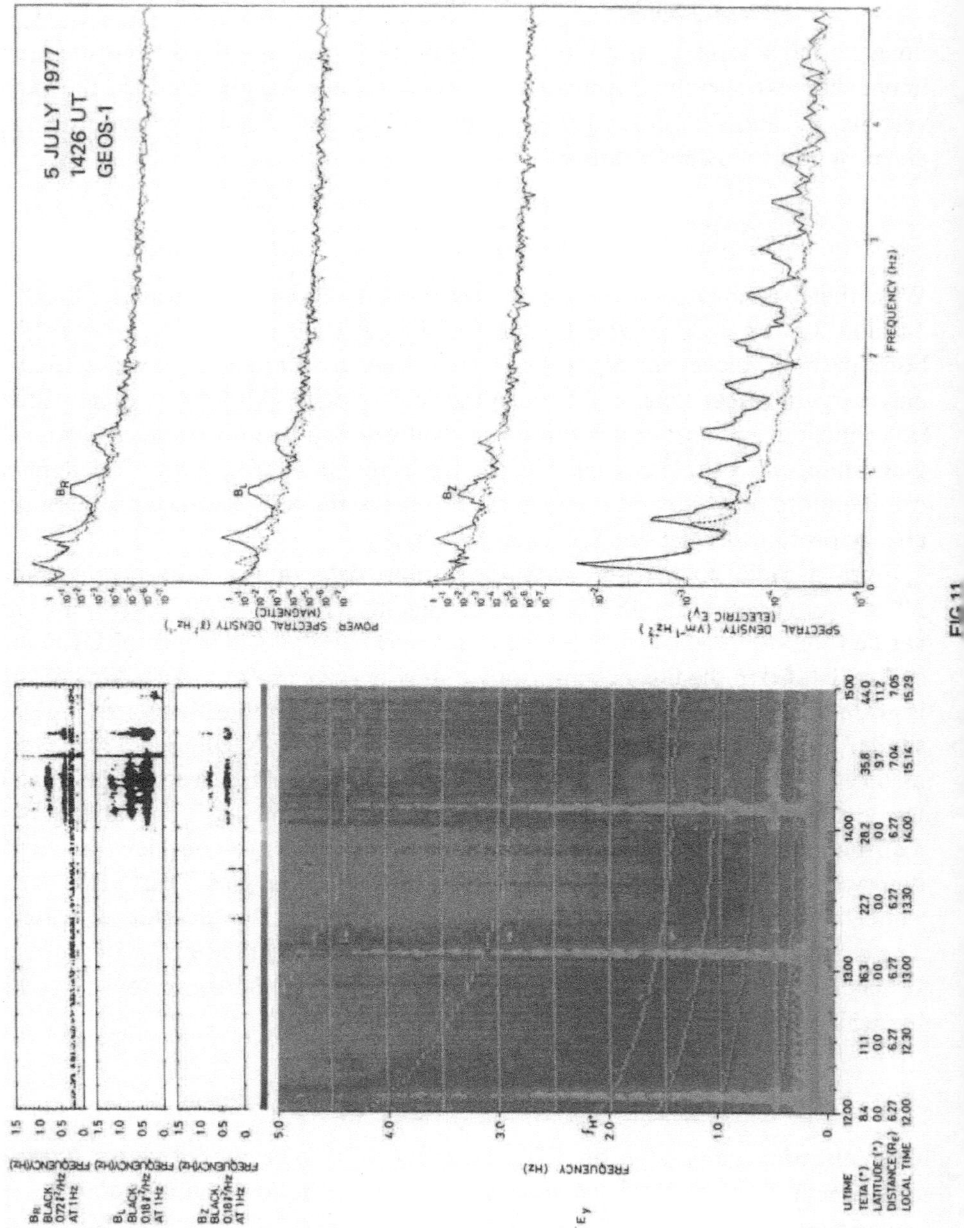

Fig. 11. ULF emissions seen in both magnetic (upper) and electric (lower) spectrograms between 14:00 and 15:00 UT, 5 July 1977. The harmonics in the magnetic channel are of natural origin, as is the harmonic structure in the electric channel. The intensity scale of the colour spectrogram is shown above, red being most intense. The dotted lines are the He^+ and H^+ gyrofrequency harmonics. The wide-band noise occurring every hour is of instrumental origin (bias sweep on spherical sensors).

Integrated spectra of B_R (right hand polarized), B_L (left hand polarized), B_Z (wave normal component) and E_y (electric signal) are shown on the right for the time indicated. The lower frequency components which appear in both magnetic and electric channels lie below the proton gyrofrequency. The higher harmonics seen only in the electric channel are probably ion-cyclotron harmonic waves.

The approximate threshold noise levels for the spectra are shown dashed.

when both electric and magnetic ULF components are detected below the proton gyrofrequency.

The frequency correspondence between the electric and magnetic components is somewhat obscured by the fact that with a single rotating electric antenna, signals are shifted in frequency by an amount equal to the spin frequency (0.17 Hz); the shift may be positive or negative depending on the sense of wave polarization. The red line at 0.17 Hz in Figure 11 represents the DC electric field which has been Doppler shifted from zero frequency.

It is seen in Figure 11 that the electric signal has a structure extending to higher frequencies, the waves being apparently located between harmonics of the helium gyrofrequency.

These are probably ion cyclotron harmonic waves which have been reported to exist at low geocentric distances by Gurnett (1976) and Kintner *et al.* (1978); a theoretical approach to these phenomena has been recently made by Ashour-Abdalla and Thorne (1977).

On other occasions particularly during disturbed conditions signals detected on both electric and magnetic sensors extend up to high frequencies with no distinct harmonic structure. Both types of phenomena are the subject of further study.

Conclusion

We have presented here results which show that indeed the magnetosphere is very active in the ULF range. A large number of events have been recorded. The systematic spectral analysis of the events, associated with a polarization analysis, has allowed us to make a preliminary classification of these events. If we except the rare wide band events which are encountered when GEOS-1 was within the magnetosheath, all the events can be classified by their simultaneous frequency and polarization characteristics. Quasi-monochromatic emissions are mainly polarized perpendicular to the DC magnetic field. Often there is a mixture of left hand and right hand waves when $f<f_{ci}$, but when f is larger than f_{ci} the wave is generally polarized in the right hand mode. Emissions which present a harmonically related structure are mainly polarized along the DC magnetic field. The frequency of the fundamental emission may be above, below or at the proton gyrofrequency. Such emissions do often present sideband lines.

Preliminary results have been obtained on the relationships which exist between the ion cyclotron emissions ($f<f_{ci}$) and the cold plasma density. It has been shown that ion cyclotron waves generally occur near the GEOS-1 apogee, that is to say outside the dayside plasmasphere, in contradiction with previous theories. However the local cold plasma density is much higher than previously assumed for these large radial distances. On one example we were able to verify that the frequency of the emission was decreasing when the cold plasma density was increasing, and a quantitative estimation allowed us to check that the observed frequency was in agreement with the one we could deduce by assuming a constant distribution of high energy protons.

The conditions of generation of the harmonically related structures have not been studied in enough details. In particular, it would be interesting to have more measurements about the angular distribution of protons at energies below 20 keV. Similarly we should verify if these emissions are restricted to a well defined region around the geomagnetic equator as evidenced by Gurnett (1976) for low L-values.

Finally, a better understanding of the regions of the magnetosphere in which the low frequency emissions ($f < f_{ci}$) take place will be obtained by a comparison between ground and satellite measurements (Gendrin *et al.*, 1978).

Acknowledgements

We are grateful to Dr P. Decreau (Centre de Recherche en Physique de l'Environnement, Issy les Moulineaux, France), to Dr A. Korth (Max Planck Institute, Lindau, Germany) and to their colleagues for having provided us respectively with cold plasma density and high energy electron flux measurements.

We also acknowledge the work of Messrs M. Avignon and D. Forcioli (Centre d'Etudes Spatiales, Toulouse) who provided us with refined summaries of our ULF data.

References

Ashour-Abdalla, M. and Thorne, R. M.: 1977, *Geophys. Res. Letters* **4**, 45.
Bossen, M., McPherron, R. L., and Russell, C. T.: 1976a, *J. Geophys. Res.* **81**, 6083.
Bossen, M., McPherron, R. L., and Russell, C. T.: 1976b, *J. Atmospheric Terrest. Phys.* **38**, 1157.
Cornwall, J. M., Coroniti, F. V., and Thorne, R. M: 1970, *J. Geophys. Res.* **75**, 4699.
Cummings, W. D., Deforest, S. E., and McPherron, R. L.: 1978, *J. Geophys. Res.* **83**, 697.
Cuperman, S. and Gomberoff, L.: 1977, *J. Plasma Phys.* **18**, 391.
Decreau, P., Beghin, C., and Parrot, M.: 1978a, *Space Sci. Rev.*, **22**, 581.
Decreau, P., Etcheto, J., Young, D. T., Wrenn, G. L., Pedersen, A., and Knott, K.: 1978b, *Space Sci. Rev.*, **22**, 633.
Etcheto, J. and Bloch, J. J.: 1978, *Space Sci. Rev.*, **22**, 597.
Fredricks, R. W. and Scarf, F. L.: 1969, in J. O. Thomas and B. J. Landmark (eds.), *Plasma Waves in Space and Laboratory*, University Press, Edinburgh. Vol. I, p. 97.
Fredricks, R. W. and Russell, C. T.: 1973, *J. Geophys. Res.* **78**, 2917.
Gendrin, R.: 1970, *Space Sci. Rev.* **11**, 54.
Gendrin, R., Lacourly, S., Roux, A., Solomon, S., Feigin, F. Z., Gokhberg, M. V., Troitskaya, V. A., and Yakimenko, V. L.: 1971, *Planetary Space Sci.* **19**, 165.
Gendrin, R., Perraut, S., Glangeaud, F., Fargetton, H., and Lacoume, J. L.: 1978, *Space Sci. Rev.* **22**, 433.
Gurnett, D. A.: 1976, *J. Geophys. Res.* **81**, 2765.
Heppner, J. P., Sugiura, M., Skillman, T. L., Ledley, B. G., and Campbell, M.: 1967, *J. Geophys. Res.* **72**, 5417.
Heppner, J. P., Ledley, B. G., Skillman, T. L., and Sugiura, M.: 1970, *Ann. Geophys.* **26**, 709.
Hughes, W. J., McPherron, R. L., and Barfield, J. N.: 1978, *J. Geophys. Res.* **83**, 1109.
Jones, D.: 1978, *Space Sci. Rev.*, **22**, 327.
Kennel, C. F., Scarf, F. L., Fredricks, R. W., McGhee, J. H., and Coroniti, F. V.: 1970, *J. Geophys. Res.* **75**, 6136.
Kintner, P. M. and Gurnett, D. A.: 1977, *J. Geophys. Res.* **82**, 2314.
Kintner, P. M. and Gurnett, D. A.: 1978, *J. Geophys. Res.* **83**, 39.

Kintner, P. M., Kelley, M. C., and Mozer, F. S.: 1978, *Geophys. Res. Letters* **5,** 139.
Kodera, K., Gendrin, R., and de Villedary, C.: 1977, *J. Geophys. Res.* **82,** 1245.
Korth, A., Kremser, G., Wilken, B., and Munch, J.: 1978, *Space Sci. Rev.*, **22,** 501.
Perraut, S., Gendrin, R., and Roux, A.: 1976, *J. Atmospheric Terrest. Phys.* **38,** 1191.
Robert, P., Gendrin, R., Kodera, K., Perraut, S., and de Villedary, C.: 1979, *Ann. Telecom.*, (to be published).
Russell, C. T., Holzer, R. E., and Smith, E. J.: 1970, *J. Geophys. Res.* **75,** 755.
S-300 Experimenters: 1978, *Planetary Space Sci.*, (to be published).
Singer, H. J., Russell, C. T., Kivelson, M. G., Greenstadt, E. W., and Olson, J. V.: 1977, *Geophys. Res. Letters* **4,** 377.
Taylor, W. W. L. and Lyons, L. R.: 1976, *J. Geophys. Res.* **81,** 6177.
Taylor, W. W. L., Parady, B. K., and Cahill, Jr., L. J.: 1975, *J. Geophys. Res.* **80,** 1271.

VLF ELECTROMAGNETIC WAVES OBSERVED ONBOARD GEOS-1

NICOLE CORNILLEAU-WEHRLIN and ROGER GENDRIN
CRPE/CNRS, Avenue de la Recherche Scientifique, 45045 Orléans-Cédex, France
FRANCOIS LEFEUVRE and MICHEL PARROT
CRPE/CNRS, Avenue de la Recherche Scientifique, 45045 Orléans-Cédex, France
REJEAN GRARD and DYFRIG JONES
SSD/ESTEC, Noordwijk, Netherlands
AXEL BAHNSEN and EIGIL UNGSTRUP
Danish Space Research Institute, Lundtoftevej 7, 2800-Lyngby, Denmark
and
WARWICK GIBBONS
Physics Department, Sheffield University, Sheffield S37 RH, U.K.

Abstract. This paper is concerned mainly with the information which can be extracted from frequency-time spectra in the VLF range. The instrument used is the correlator which has a good frequency resolution (50 Hz) and time resolution (30 ms) in one magnetic and one electric component simultaneously. By suitable computer analysis, it is possible for instance to distinguish between the two dominant electromagnetic emissions, hiss and chorus, as well as to display the complete spectra. This treatment is applied to the Survey periods, which are a fixed sequence of modes, repeated every hour on the hour in order to have reference data from GEOS analogous to many ground-based observatories. One result of this treatment obtained already is that hiss and chorus normally appear together, although one or the other may be dominating in intensity. The occurrence rate of these emissions in local time is also given.

For continuous surveillance the filterbank data are used. There are 16 frequency filters supplying magnetic and electric amplitude at few different frequencies. Using these data, a storm sudden commencement can be followed with good time resolution (1 s), and an interesting correlation has been found in a few cases between the VLF signal amplitude and the cold plasma density (as measured by the active part of the S-300 experiment).

1. Introduction

This paper deals with the electromagnetic waves observed below the electron gyrofrequency. In this frequency range, chorus and hiss are the main emissions, which have been extensively studied in the past. However, we will see how both the sensitivity of the S-300 wave experiment and its flexible data handling capability – both onboard and on the ground – can improve our knowledge of these waves. We will show also the importance of comparisons between the VLF emission occurrence and structure and the fluctuations of other plasma parameters measured simultaneously.

2. Characteristic Phenomena

In order to study in detail electromagnetic waves below the electron gyrofrequency f_{ce}, we make use of several sensor combinations and different modes of

Space Science Reviews **22** (1978) 371–382.

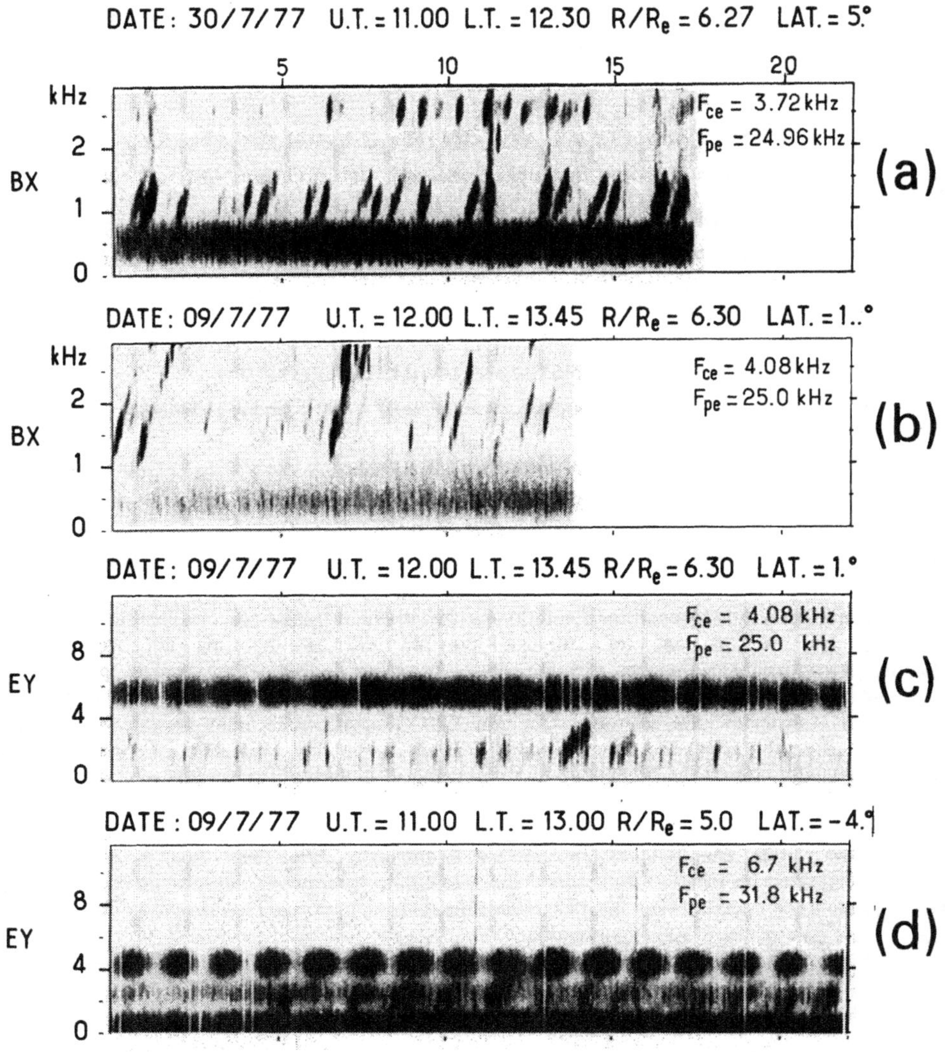

Fig. 1. Examples of spectra obtained during different subcycles of the Survey Mode. On the two upper spectrograms, the time between two tick marks is 5 s. On the two lower ones it is 10 s.

onboard data processing. (For a detailed description of the S-300 wave experiment, see Jones (1978), and S-300 Experimenters, 1978). The Swept Frequency Analyser (SFA) and the correlator data being transmitted to the ground at a very high bit rate (respectively 76 kbit s^{-1} and 20 kbit s^{-1}) cannot be analyzed on a routine basis. However, during 3 min every hour on the hour, the satellite is telecommanded with a fixed sequence of modes, the whole sequence being called the Survey Mode. This mode yields a short but extensive survey of the wave field prevailing in the magnetosphere in the whole experiment frequency range, and for the most relevant combinations of sensors and analysers. The cold plasma density is also measured during this 3 min period, by the two S-300 active experiments

(Etcheto and Bloch, 1978; Decreau *et al.*, 1978). The resulting data are systematically processed at the French Centre National d'Etudes Spatiales (CNES) in Toulouse. Some examples of the results of such a processing will be given throughout this paper.

Figure 1 for instance shows the result of a Fourier transform applied to the onboard-computed correlograms, for two different subcycles of the Survey Mode. The two upper panels give the spectrograms corresponding to a magnetic component (from 0 to ~3 kHz), and the two lower panels give the spectrograms of an electric component (from 0 to ~11 kHz). The subcycles corresponding to these two kinds of spectrograms follow each other in the sequence of the Survey Mode.

In Figure 1(a), we show an example of chorus observed simultaneously with ELF hiss. A frequency gap is observed in the chorus event, at around half the gyrofrequency (~1.8 kHz). Such gaps around $f_{ce}/2$ have been previously reported by Maeda (1976), Maeda *et al.* (1976) and Tsurutani and Smith (1977). They are a more or less general feature of the events we have recorded so far. However, on some occasions there is no gap in the chorus spectrum, as evidenced in Figure 1(b).

Figure 1(c) was obtained during the same Survey Mode as Figure 1(b), but it refers to an electric component and it covers a wider frequency range. The chorus event is clearly seen at the lower frequencies. But around $1.3f_{ce}$ we notice a strong electrostatic emission: the corresponding magnetic spectrogram (not represented here) does not show any signal at this frequency. Such an emission clearly corresponds to the $(n+\frac{1}{2})f_{ce}$ emissions reported previously (Kennel *et al.*, 1970; Fredricks and Scarf, 1973; Shaw and Gurnett, 1975) but concerning which GEOS-1 has brought up new information (Christiansen *et al.*, 1978a, b; Rönmark *et al.*, 1978).

Finally, Figure 1(d) shows an example of a multiple band emission (below f_{ce}). This emission is strongly spin modulated (the spin period being ~5.7 s) due to the large inclination of the spin axis with respect to the magnetic field. The intensity of the modulation is different for the three frequency bands.

3. Simultaneous Occurrence of Hiss and Chorus

We have already shown an example (Figure 1(a)) of a simultaneous occurrence of hiss and chorus. Contrary to what is generally thought, this is an almost permanent situation which has escaped previous experimenters probably because of a lack of sensitivity. In order to illustrate this, we present the result of a real-time analysis performed at the ESA Operations Center in Darmstadt (Figure 2).

During this period, the correlator was working continuously in its lowest frequency range. On the display, the spectrograms are not corrected for the variable gain of the correlator, which is plotted above in a reverse sense (so that a long vertical bar means a strong signal). The correlator gain can vary, by steps of ~4 dB over a total dynamic range of 80 dB. Such a representation has the

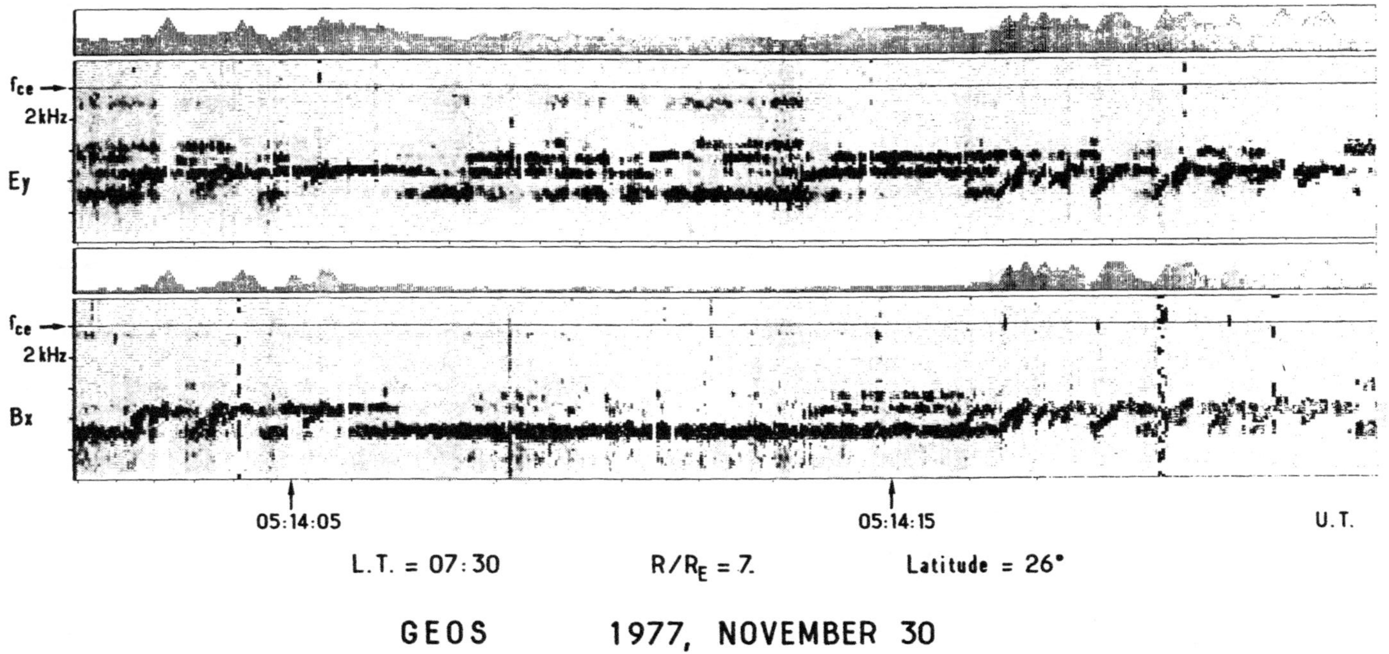

Fig. 2. Simultaneous occurrence of hiss and chorus. The grey scaling being independent of the true signal level (see text), the hiss band around 800 Hz, which occurred between 0514:06 and 0514:16 UT is visible although its amplitude is ~40 dB below the amplitude of the chorus which follows.

advantage that even weak signals can be visible on the grey scaled plot. Among the different emissions which are detected we notice, on the magnetic spectrogram mainly, a low intensity hiss around 700 Hz, out of which stronger (~40 dB) chorus seems to emerge around 0514:16. With less sensitive equipment, the hiss would not have been detected.

Using six months of data, a study of hiss and chorus occurrence has been performed on the Survey Mode, when GEOS was at radial distances larger than $6R_e$ and when K_p was less than 3^-. Its results are presented in Figures 3 and 4. In Figure 3, events were classified according to the strongest signal (chorus or hiss).

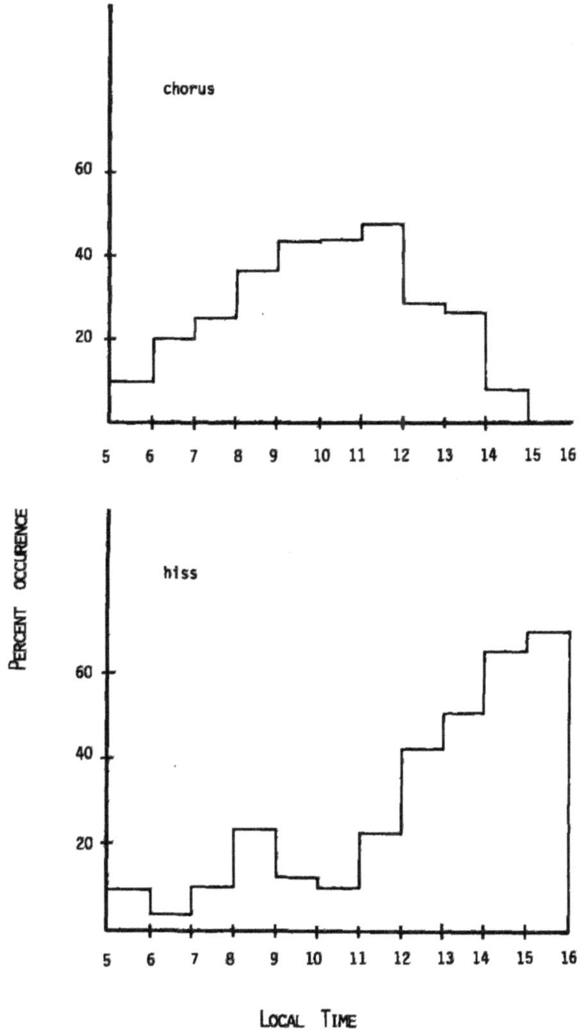

Fig. 3. Statistical occurrence of hiss and chorus (when $R/R_e \geqslant 6$ and $K_p \leqslant 2^+$), as deduced from the examination of the Survey Mode. An event is characterised as 'hiss', even when chorus is present simultaneously, if the hiss intensity is larger than the chorus intensity.

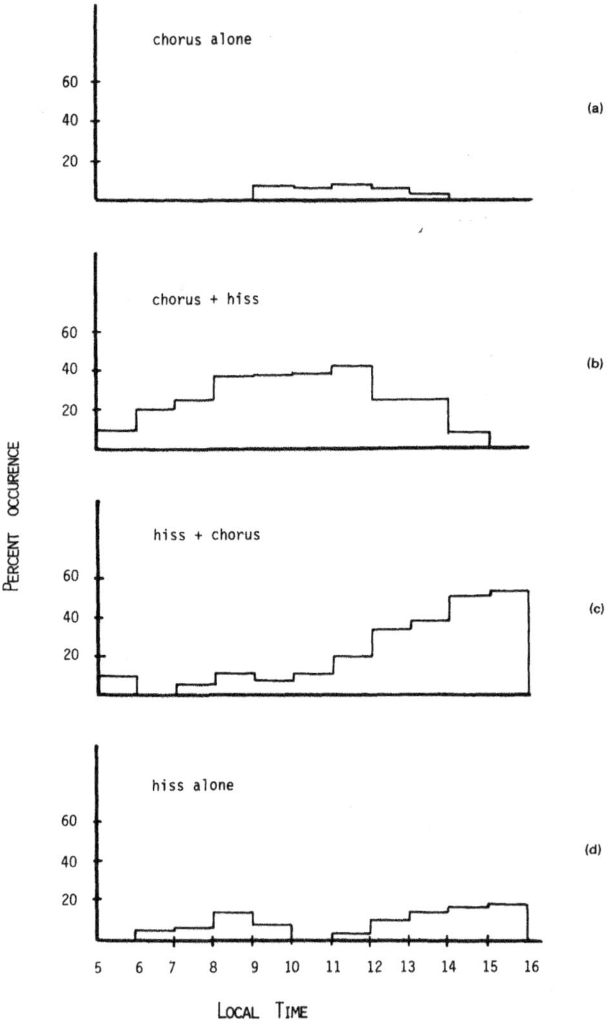

Fig. 4. Same as Figure 3, but events have been characterised differently: (a) chorus alone; (b) choru and hiss, chorus being the more intense; (c) hiss and chorus, hiss being the more intense; (d) hiss alone

The occurrence pattern (chorus occurs mainly in the morning sector and hiss in the evening sector) is in good agreement with previous results which were based on a distinction between 'burst like' and 'steady' noises (see for instance Russell *et al.*, 1972). But the striking result of our study is that nearly each time that a chorus event is observed, hiss is also present in the same frequency band (Figure 4). Apart from the cases of chorus alone or hiss alone, the events can be classified according to the stronger phenomenon, whose name is put first (chorus + hiss or hiss + chorus). Then we can recover, by adding $a+b$ and $c+d$ the 'classical' occurrence pattern represented in Figure 3.

These results have important implications with respect to the theories of chorus generation. They strongly suggest that some kind of VLF turbulence must be present before the strong chorus emission process can start.

4. Other Data Processing

Because of the huge amount of data, we have been obliged to establish procedures which enable us to recognize quickly the different types of phenomena recorded, in order to select the corresponding high speed telemetry tapes for further detailed studies. The examination of the coherence function or of the filter bank data are two of these procedures which we will now describe. A third one, the visualisation of the 'compressed high speed' data will not be discussed here. It is extensively used for characterizing resonances or harmonically related emissions (Etcheto and Bloch, 1978; Christiansen *et al.*, 1978a, b).

The coherence function, which is obtained by taking the ratio between the square of the modulus of the cross-spectrum B_xE_y and the product of the two auto-spectra B_x^2, E_y^2, is computed every 5.5 s during one subcycle of the Survey Mode. Its value can slightly exceed 1, because the three spectra are not taken

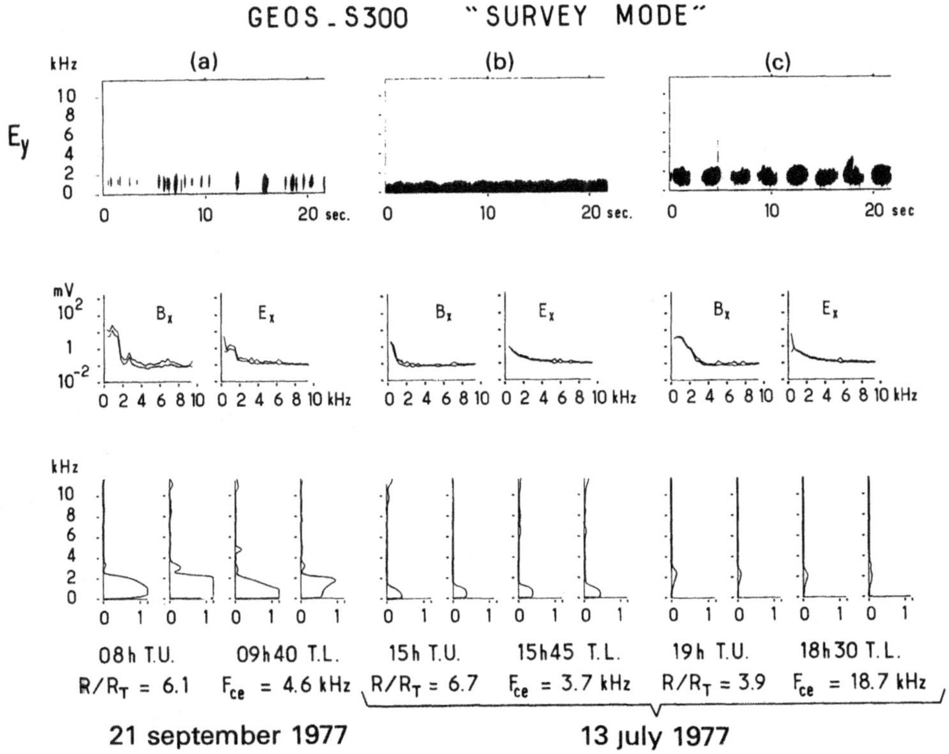

Fig. 5. Computation of the coherence function. The coherence function (lower panel) is computed over 5.5 s. In case (a), its value is large because the VLF emission is chorus, as evidenced on the upper panel. The middle panel corresponds to mean amplitudes (see text).

exactly at the same time, but some 29 ms apart. A large number of measurements has shown that this function takes large values for chorus events, which are coherent waves, whereas it never exceeds ~0.6 for hiss events, which are essentially uncoherent waves (Lefeuvre and Parrot, 1978). This is demonstrated in Figure 5, on which the coherence function is plotted as a function of frequency in the lower panel. One sees that in case (a), which clearly corresponds to a chorus event, the coherence function has a high value in the frequency range where the signal is strong. On the contrary, in cases (b) and (c), it has a low value when hiss is present.

Although it is not so efficient, one can also distinguish between chorus and hiss in computing the mean amplitude M and its variance σ_0, over 0.69 s, and in comparing the curve at $M+\sigma/2$ to the one at $M-\sigma/2$ (middle panel of Figure 5). The two curves are well separated for chorus which is non stationary in time (large variance), while they are identical for hiss which is essentially stationary in time (low variance).

Both informations (coherency and stationarity) are taken care of in a computerized selection algorithm for characterising automatically the nature of the observed signal.

The second way of quickly determining the interesting periods which are to be studied in more detail is to examine the filter bank data. These correspond to

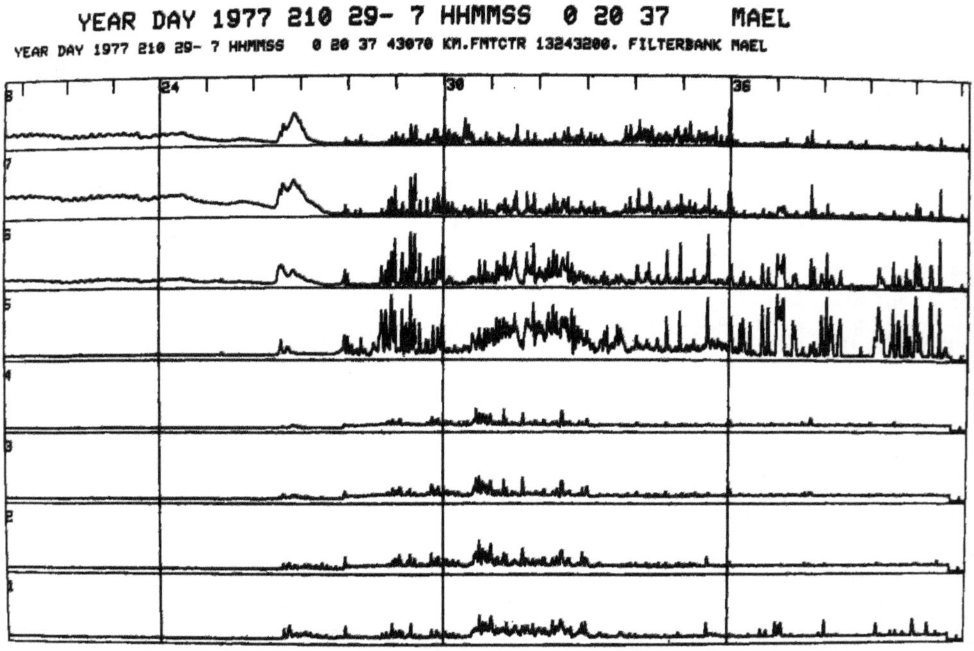

Fig. 6. Filter bank signals at the beginning of a magnetic storm. Data were obtained during a pass over Alaska, the 29th of July, 1977. An Ssc was reported to occur at 0027. The top 4 curves are magnetic, *bx*, the lower 4 curves electric, *ey*, frequencies going down are: 2.5–5, 1.25–2.5, 0.625–1.25, 0.200–0.625 kHz. The scale is logarithmic covering approx. 3 decades in amplitude.

onboard integrated signals, in 5 frequency channels, from 4 sensors, sampled 5.8 times per second. An example of some outputs of this filter bank is given in Figure 6 for an Alaskan pass of GEOS during the July 29, 1977 storm event. A sudden commencement was reported to occur at 0027 UT. A strong VLF signal, with a smooth amplitude variation, starts at 0026:30, the signal being stronger at higher frequencies. Although a precise timing of this event is still to be done, it seems that magnetospheric electrons react immediately to an Ssc, in such a way that their anisotropy exceeds the critical anisotropy (Kennel and Petschek, 1966), in a wide frequency range.

A second shock occurs at 0028:00. Here the VLF signal fluctuates strongly with a quasi-periodicity of ~6 s at the beginning; it is stronger at the lower frequencies. In view of other data from GEOS-1, it is likely that, at this time, GEOS has penetrated into the magnetosheath.

5. Cold Plasma Density and VLF Amplitude

According to hiss generation theories (see Gendrin, 1975 for a review), the parameters which play an important role are the d.c. magnetic field intensity, the

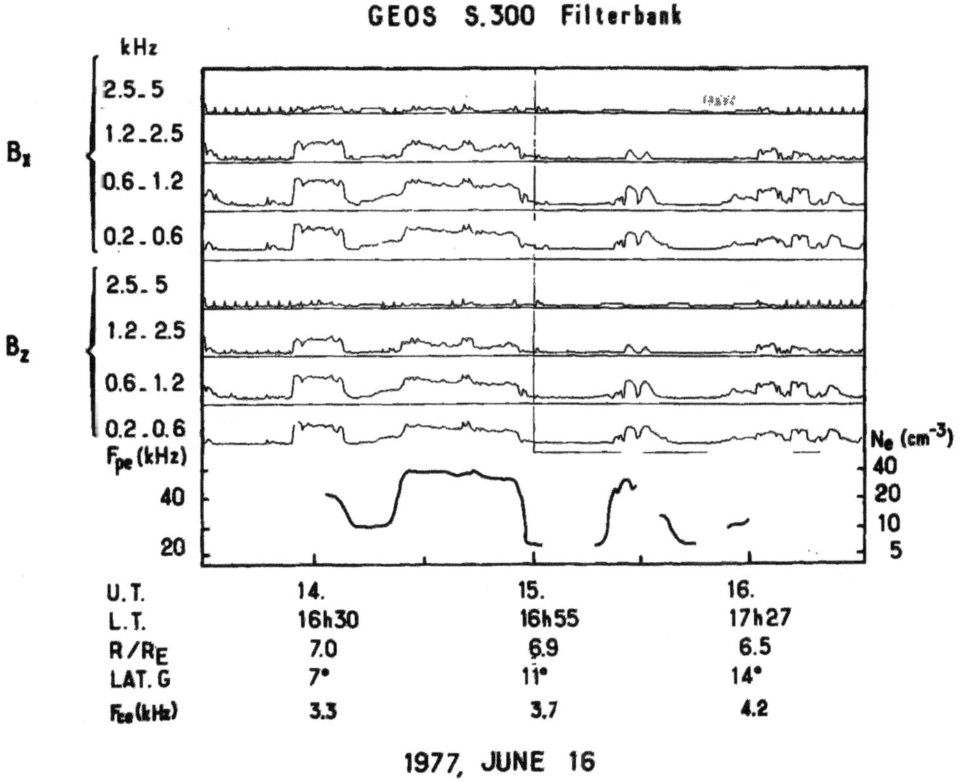

Fig. 7. Simultaneous fluctuations of the cold plasma density and of VLF amplitude in different frequency bands.

cold plasma density, the distribution function of high energy electrons and the propagation characteristics of the waves (magnetospheric ducting, ionospheric reflection). The influence of the last two parameters is studied in a separate paper where ground-satellite relationships are examined in some detail (Cornilleau-Wehrlin *et al.*, 1978). We will present here some data which show a strong correlation between the local cold plasma density and the VLF hiss amplitude. Such a comparative study can be made only on a few occasions, when there is a fluctuating VLF emission and when, simultaneously, one of the S-300 active experiments is operating, continuously or at least with a high sampling rate (of the order of 5 mn).

In Figures 7 and 8 we show two examples of such simultaneous fluctuations. We see on these examples that there is an almost one-to-one correspondence between the increase of the cold plasma density and the enhancement of the VLF hiss. The plasma density changes by up to a factor of 8. This indicates that we are more probably dealing with a spatial effect (crossing of detached plasma regions) rather than with a temporal effect. On such a short azimuthal distance (~1000 km in 5 mn near the apogee), the other parameters (high energy particle flux, ionospheric reflection coefficient) are generally not varying much, although some exceptions can be found (Cornilleau-Wehrlin *et al.* (1978)). During the December 16 event, which corresponds to a very quiet day ($Kp < 1^+$ between 1500 UT the 15th and 1200 the 16th), the particle data show no visible fluctuations. The June 16 event is more complicated, because a storm sudden commencement occurred at

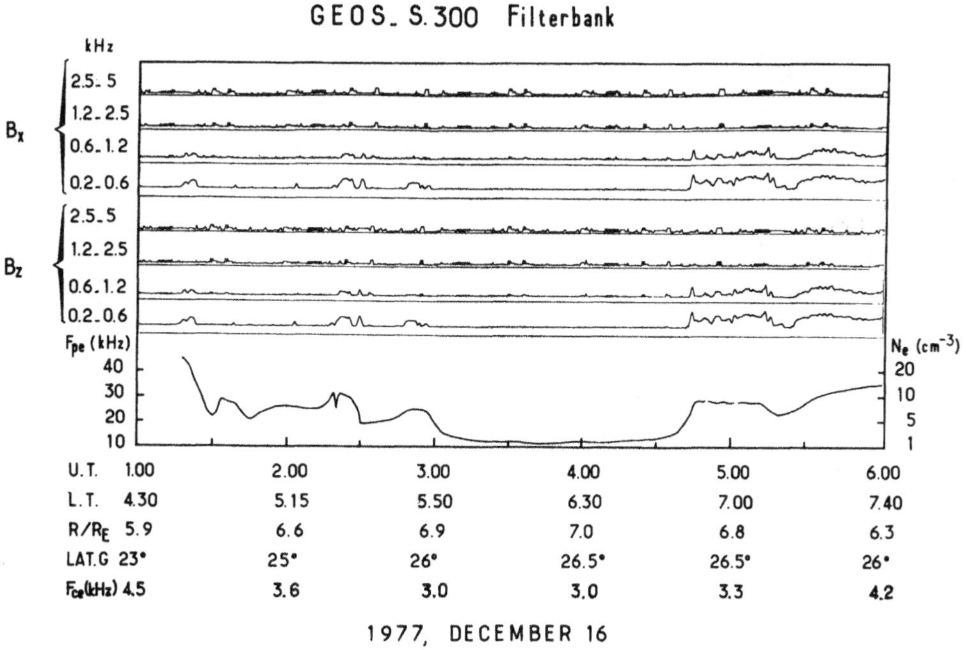

Fig. 8. Same as Figure 7, for December 16, 1977.

1250 UT and a strong depletion in the electron flux was noticed around 1400 UT. But there were no fluctuations associated with the enhancements and decreases of the VLF intensity. Consequently, we can think that the observed hiss enhancement is due to the cold plasma density increase, as expected from theoretical considerations (Etcheto *et al.*, 1973) which have been already confirmed at least partially by experiment (Chan *et al.*, 1974).

As explained by Cornilleau-Wehrlin *et al.* (1978), there seems to exist two conditions in which the hiss amplitude is suddenly varying: either an increase of the cold plasma density – with no associated fluctuation of the high energy electron flux – or a variation of the particle distribution function – not associated with a change of the cold plasma density. Further studies are needed to see if these two cases lead to noticeable differences in the characteristics of the VLF hiss: spectrum, polarization, wave normal directions, etc.

6. Conclusion

The continuous survey of VLF electromagnetic waves in the magnetosphere can still bring new information about generation mechanisms, propagation, influence on high energy particle dynamics, etc., provided that it is combined with direct measurements of other parameters, or with simultaneous ground observations. The good sensitivity of our equipment and its data handling capabilities, both onboard and on the ground, has allowed us to survey most of the electromagnetic waves below f_{ce}. In particular, the analysis of the coherence function between electric and magnetic components allows a simple distinction to be made between hiss and chorus.

We have shown that chorus is almost always associated with hiss. On two occasions, the crossing of detached plasma regions was associated with characteristic enhancements of VLF noise, thus validating presently admitted theories of the self-consistent equilibrium between cyclotron waves and electrons.

Acknowledgements

We must thank our colleagues P. Décréau and J. Etcheto for providing us with their plasma density measurements. Our gratitude goes also to Drs A. Korth and G. Kremser (Max Planck Institute, Lindau) who gave us very kindly their preliminary particle data, and to D. Barria, G. Charles and C. Huc (Centre National d'Etudes Spatiales, Toulouse) for their competent and constant assistance in the data handling.

References

Chan, K. W., Holzer, R. E., and Smith, E. J.: 1974, *J. Geophys. Res.* **79,** 1989.

Christiansen, P., Gough, P., Martielli, G., Bloch, J. J., Cornilleau, N., Etcheto, J., Gendrin, R., Jones, D., Beghin, C., and Decreau, P.: 1978a, *Nature, London* **272,** 682.

Christiansen, P., Gough, P., Martelli, G., Bloch, J. J., Cornilleau, N., Etcheto, J., Gendrin, R., Beghin, C., Decreau, P., and Jones, D.: 1978b, *Space Sci. Rev.* **22,** 383.
Cornilleau-Wehrlin, N., Gendrin, R., and Tixier, M.: 1978, *Space Sci. Rev.* **22,** 419.
Decreau, P., Beghin, C., and Parrot, M.: 1978, *Space Sci. Rev.* **22,** 581.
Etcheto, J. and Bloch, J. J.: 1978, *Space Sci. Rev.* **22,** 597.
Etcheto, J., Gendrin, R., Solomon, J., and Roux, A.: 1973, *J. Geophys. Res.* **78,** 8150.
Fredricks, R. W. and Scarf, F. L.: 1973, *J. Geophys. Res.* **78,** 310.
Gendrin, R.: 1975, *Space Sci. Rev.* **18,** 145.
Jones, D.: 1978, *Space Sci. Rev.* **22,** 327.
Kennel, C. F. and Petschek, H. E.: 1966, *J. Geophys. Res.* **71,** 1.
Kennel, C. F., Scarf, F. L., Fredricks, R. W., McGehee, J. H., and Coroniti, F. V.: 1970, *J. Geophys. Res.* **75,** 6136.
Lefeuvre, F. and Parrot, M.: 1978, *J. Atmosph. Terr. Phys.* (to be published).
Maeda, K.: 1976, *Planet. Space Sci.* **24,** 341.
Maeda, K., Smith, P. H., and Anderson, R. R.: 1976, *Nature, London* **263,** 37.
Rönmark, K., Borg, H., Reme, H., Vigo, J. M., Ungstrup, E., Christiansen, P., Gough, P., Cornilleau, N., Etcheto, J., Gendrin, R., Jones, D., and Wrenn, G.: 1978, *Space Sci. Rev.* **22,** 401.
Russell, C. T., McPherron, R. L., and Coleman, P. J., Jr.: 1972, *Space Sci. Rev.* **12,** 810.
Shaw, R. R. and Gurnett, D. A.: 1975. *J. Geophys. Res.* **80,** 4259.
S-300 Experimenters: 1978, *Planet. Space Sci.* **26,** (to be published).
Tsurutani, B. T. and Smith, E. J.: 1977, *J. Geophys. Res.* **82,** 5112.

GEOS-1 OBSERVATIONS OF ELECTROSTATIC WAVES, AND THEIR RELATIONSHIP WITH PLASMA PARAMETERS

P. J. CHRISTIANSEN, M. P. GOUGH, and G. MARTELLI

Plasma and Space Physics Group, School of Mathematical and Physical Sciences, University of Sussex, Falmer, Brighton, Sussex, U.K.

J. J. BLOCH, N. CORNILLEAU, J. ETCHETO, and R. GENDRIN

Centre de Recherches en Physique de l'Environment Terrestre et Planetaire, CNET, Issy-les-Moulineaux, France

C. BEGHIN and P. DECREAU

Centre de Recherches en Physique de l'Environment Terrestre et Planetaire, CNRS, Orleans-la-Source, France

and

D. JONES

Space Science Department, European Space Agency, ESTEC, Noordwijk, Netherlands

Abstract. In this paper we describe and discuss the occurrence of natural wave emissions detected by GEOS-1 at frequencies above the electron gyrofrequency. The bulk of the data presented comes from the first six months of satellite operation and thus concerns mainly dayside phenomena. The paper is arranged as follows:

After some general remarks, a classification of the wave phenomena is developed in Section 2, and experimental evidence and morphological information relevant to this classification are contained in Section 3. Section 4 includes some preliminary comments on nightside observations. The results are discussed in Section 5, where it is argued that they can be understood as manifestations of electron cyclotron harmonic (Bernstein) wave emission in a plasma parameter range which has only very recently received any theoretical examinations. This theme is further developed in a comparison paper (Ronnmark *et al.*, 1978).

1. The Wave Experiment

The S-300 wave experiment on GEOS-1 is a complex combination of electric and magnetic sensors, wave analysing units, and active plasma diagnostic experiments. The experiment is comprehensively described elsewhere (S-300 experimenters, 1978; Jones, 1978), as are the associated active experiments: resonance sounder S-301 (Etcheto and Petit, 1977) and mutual impedance experiment S-304 (Decreau *et al.*, 1978). Data presented here comes exclusively from the long electric antenna studied with the stepped frequency analysers, and interpreted with the help of results from the S-301 resonance sounder and S-304 mutual impedance experiment.

The natural wave experiment therefore has several features which make it more comprehensive than previously flown instruments:

(1) It has a sensitivity higher than that of previous instruments (>10 dB greater at 20 kHz).

Space Science Reviews **22** (1978) 383–400.

(2) There are related plasma diagnostics which enable the immediate identification of the gyrofrequency, f_{ce}, and plasma frequency, f_{pe}.

(3) The GEOS telecommand system is highly flexible allowing the whole wave package to be concentrated on to any given phenomenon within minutes of its existence being recognized by the ground operator.

(4) The high telemetry rate allows waveform analysis of six sensors simultaneously up to 77 kHz.

It must be stressed that all data presented in this paper is from compressed high speed tapes. Future analysis of specific events using the high speed data tapes can be expected to reveal a great deal of new information.

2. General Remarks: Electrostatic Waves Above the Electron Gyrofrequency

An important ordering parameter in the classification of possible wave modes is the ratio of the plasma frequency (f_{pe}) to the electron gyrofrequency (f_{ce}). In the plasma encountered by GEOS-1 during dayside passes, on board diagnostics show the ratio $f_{pc}/f_{cc} \sim 5$ typically, and the upper hybrid frequency $f_{uh} = (f_{pe}^2 + f_{ce}^2)^{1/2}$ is thus only a few percent greater than f_{pe}.

In the frequency band from f_{ce} to just below f_{pe}, electrostatic waves can exist as electron plasma waves near f_{pe} (which propagate primarily parallel to the ambient magnetic field), and as perpendicular propagating electron cyclotron harmonic (ECH) waves which occupy bands between harmonics of f_{ce} up to and beyond f_{uh}.

By contrast, electromagnetic waves can exist for example in the well known Whistler mode below f_{ce}, and at frequencies above the z mode cut off. Whistler mode emissions are treated in a companion paper (Cornilleau *et al.*, 1978), and while examples of the latter, continuum and kilometric radiation will be noted below, the emphasis of this paper is on electrostatic modes since they are strongly generated by local, wave particle interactions.

Figure 1 summarizes these wave mode regimes and compares the fundamental plasma frequencies with the ECH dispersion relation for perpendicular propagation. In the simplest case where the wave number k is exactly perpendicular to the ambient magnetic field B_0, these waves propagate without damping, and for a Maxwellian plasma, their dispersion relation is:

$$D(\omega, k_\perp) = 1 + \omega_p^2/k_\perp^2 v_t^2 \left[1 - \sum_{s=-\infty}^{\infty} I_s(\lambda)\, e^{-\lambda} (\omega/(\omega - s\omega_c))\right] = 0,$$

where $\lambda = kv_t/\omega_c$, v_t is the electron thermal speed, and I_s is the modified Bessel function of order s.

Also in Figure 1 are tabulated those frequencies at which natural emissions may occur and those frequencies which will exhibit resonances during the S-301 active sounding experiment. Resonances are observed during the monitoring period of the S-301 experiment when the pulsed emission contains frequencies that couple to weakly damped low-group velocity modes in the surrounding plasma. In a hot

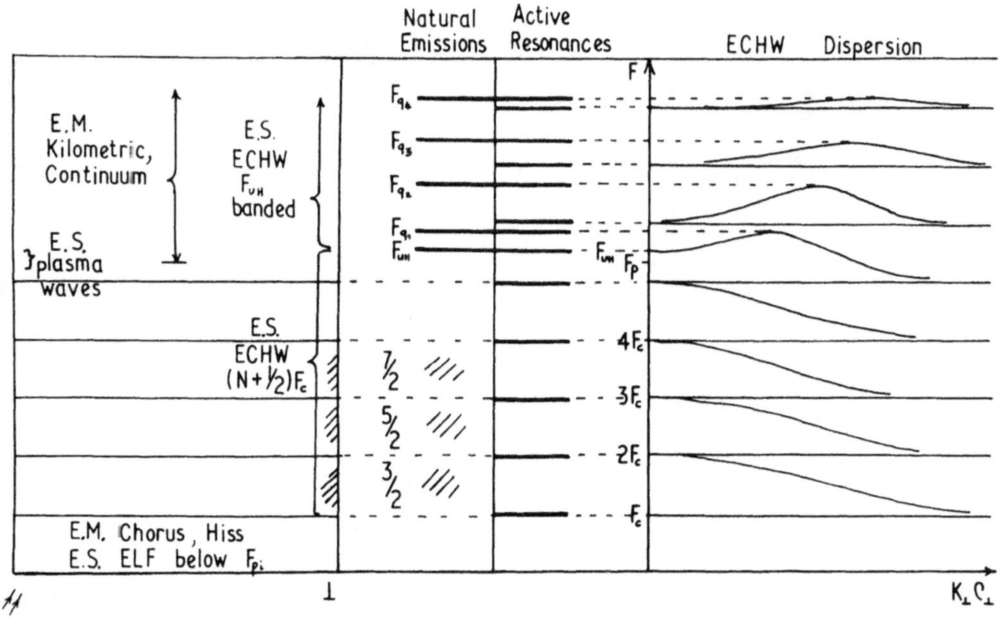

Fig. 1. A comparison of the dispersion relation of the Electron Cyclotron Harmonic (ECH) wave modes with the expected natural emissions and active resonances.

magnetized plasma the resonance frequencies include the gyrofrequency, and its harmonics (nf_{ce}), the electron plasma frequency, f_{pe}, the upper hybrid frequency, f_{uh}, and the f_q resonances. These f_q resonances are those frequencies lying between successive cyclotron harmonics above f_{uh} on the dispersion relation of Figure 1 at which the group velocity, v_g, is zero.

The wave data for 18 July 1977 when the S-301 sounder was active for the whole satellite pass has been compressed on to Figure 2 and the expected resonances can be clearly identified. Of course, natural emissions at non-resonant frequencies will also be observed during the monitoring period (after the transmitted pulse) and two examples are evident: an emission at $\frac{3}{2}f_{ce}$ (between active resonances at f_{ce} and $2f_{ce}$) around 11:30 UT; and a burst of kilometric radiation above the plasma frequency around 12:15 UT. Note that both the plasma frequency and gyrofrequency reach their minimum values around the centre of the picture, close to apogee, reflecting the minimum plasma density and magnetic field strength respectively.

In this and following spectrograms the grey scaling (black to white = 20 dB) is set to show general spectral detail, and the grey level is set for each 22 s sweep of the 0–77 kHz frequency range. The same procedure is used for 0–5 kHz correlograms when they are included.

We have adopted a classification scheme for the natural wave emissions which appears naturally from the compressed GEOS-1 data of the quiet dayside magnetosphere. The data will be discussed within the framework of this classification

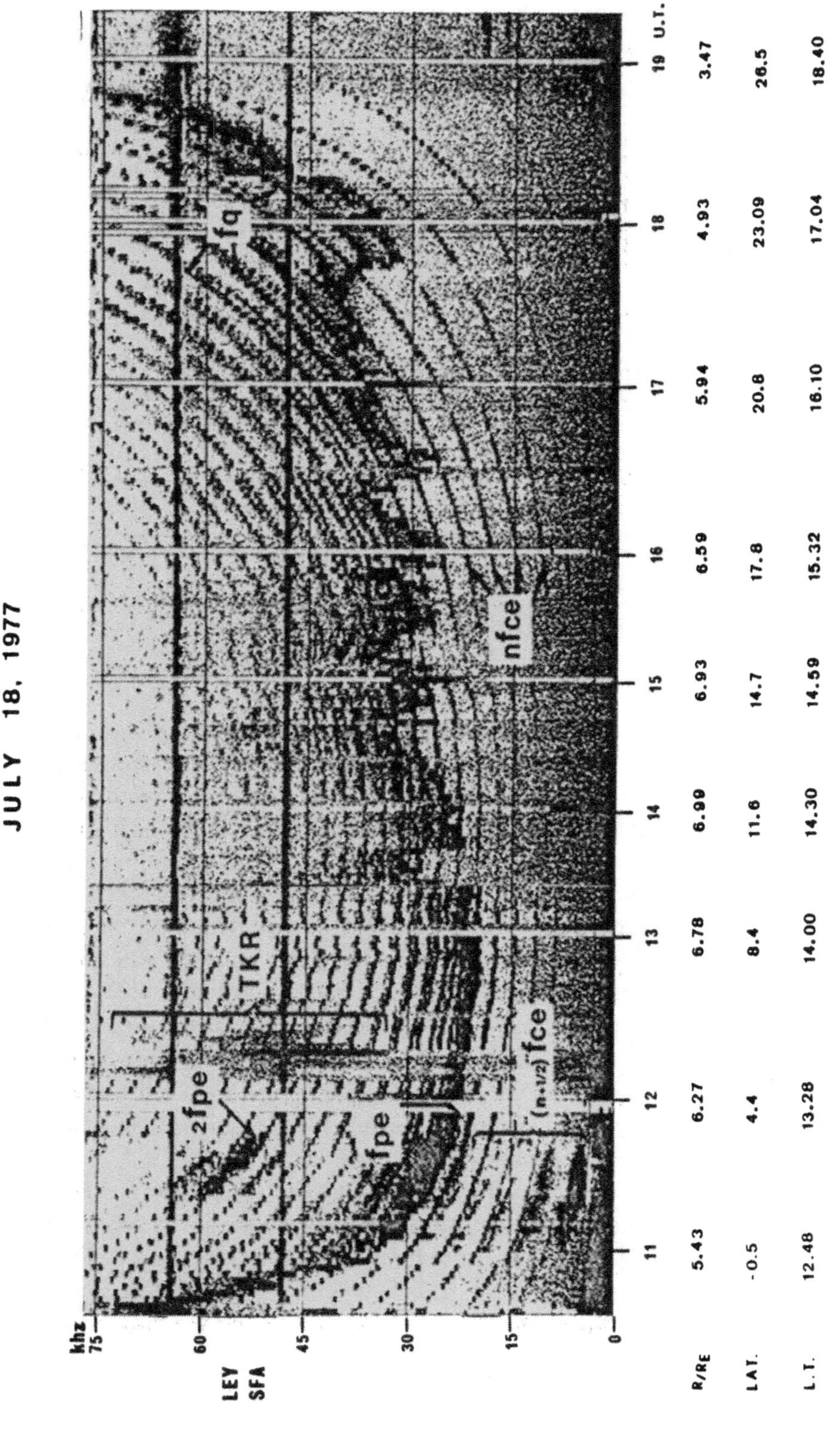

Fig. 2. The SFA spectrogram for 18 July 1977, when the S-301 active sounder was on for the majority of the day. This spectrogram is composed of many individual 22-s frequency sweeps. In the grey scaling, black to white = 20 dB. The grey level of each sweep is set to the spectral mean to show most spectral detail.

scheme and will be related to earlier work below. The emissions are classified primarily according to spectral type and are shown in order of decreasing rate of occurrence and increasing strength of the observed phenomena.

(a) Weak banded emissions around f_{uh}.

(b) Emissions between low harmonics of $f_{ce}((n+\frac{1}{2})f_{ce}$ emissions).

(c) Strong bursts around f_{pe}.

Note that the nf_{ce} resonance frequencies observed during S-301 sounding are not expected to be frequencies at which natural waves will occur because of their relatively high damping.

3. Quiet Dayside Phenomena

A. WEAK BANDED EMISSIONS AROUND f_{uh}

The most persistent of the natural electrostatic wave signals but weakest of all observed by GEOS-1 on the dayside are the weak banded emissions in the region of the local upper hybrid frequency. It appears to be a steady feature of the quiet-time dayside magnetosphere since the bands are observed essentially whenever the local upper hybrid frequency lies within the range of the wave detectors (<77 kHz). They have been observed in the region covered by GEOS-1: between $L=5$ and $L=8$; between $\lambda_m=-10°$ and $\lambda_m=+30°$; and from 6 to 19 h LT. There is no reason to believe that they are confined to this latitude range or that $L=5$ is a minimum value. The emissions have field amplitudes typically of 10^{-7} (V m^{-1}) Hz$^{-1/2}$.

It is worthwhile noting at this stage, that these weak banded emissions around f_{uh} can also be accompanied by even weaker, rather more diffuse emissions between cyclotron harmonics below f_{uh}. We group these in category (a) as we believe them to be significantly different from the stronger more narrow banded signals included in category (b).

Figure 3 shows a spectrogram of these natural emissions recorded on 14 July 1977 after the SFA had been commanded to sweep just around f_{pe} in order to optimize time resolution. At the end of the display a single full frequency S-301 active resonance sweep was made (shown here on a time expanded scale to identify the signals. The natural bands closely match the f_{pe}, f_{uh}, and f_q resonances shown on the active sweep). There are many examples from every day of GEOS-1 data for which similar matches of these natural emissions can be made to the active resonances.

Single power spectra from adjacent passive and active sweeps of the SFA on 15 June 1977 are shown in Figure 4. The natural wave spectrum is shown relative to the noise level while the strong active sounder resonances are marked as vertical lines for clarity. Note that the natural emission at the first f_q is stronger than that at f_{pe}, but the matching of passive to active frequencies is clear. This kind of variability in detail is typical of these emissions and is possibly related to the fact that significant cyclotron damping occurs when the upper hybrid frequency is close to a cyclotron harmonic.

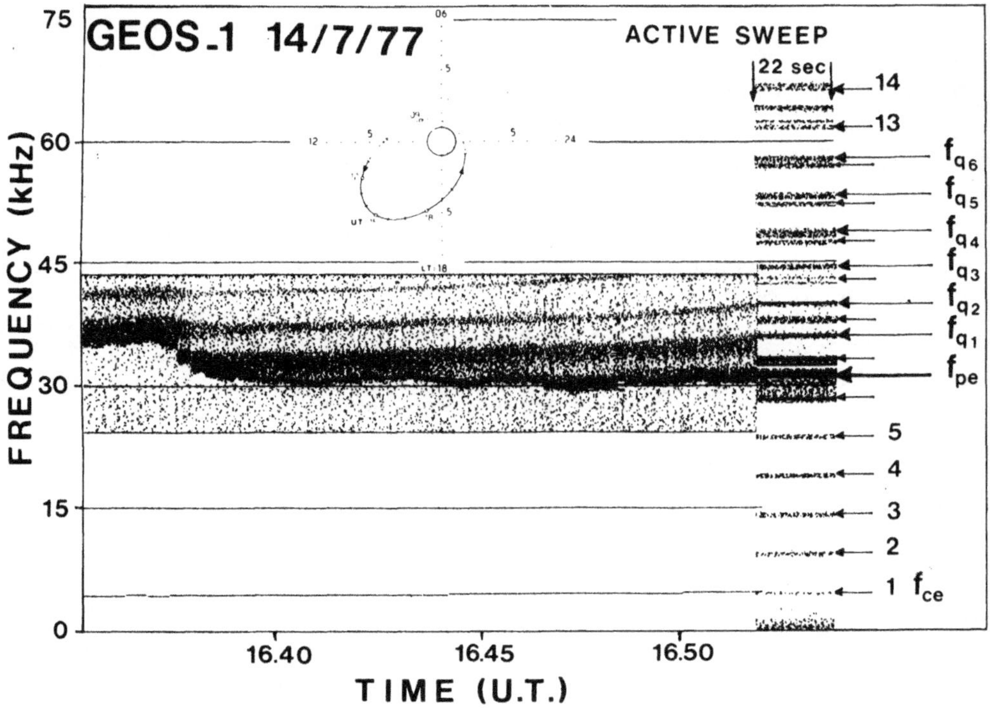

Fig. 3. Spectrogram of natural emissions recorded by the SFA sweeping around f_{pe} on 14 July 1977. (*Inset*: satellite trajectory showing variation of local time with universal time.) LT = 16:20, $L = 7.5$, $\lambda_m = 23°$.

The mechanism by which these weak banded emissions can be generated is discussed in Section 5.

B. $(n+\frac{1}{2})f_{ce}$ EMISSIONS

Signals between low harmonics of the gyrofrequency are observed regularly by GEOS-1 in the dayside magnetosphere within an hour or so of crossing the plasmapause on the outbound part of the orbit. These emissions exist from a few minutes up to an hour in duration and have wave field intensities in the range 10^{-6} to 10^{-5} (V m^{-1}) Hz$^{-1/2}$, and are accompanied by an increase in the level of the f_{uh} signal by 20–30 dB above its quiet time value.

Figure 5 illustrates a typical day (6 July 1977) with the plasma frequency descending into the range of the instrument as the satellite leaves the plasmapause, followed almost immediately by the commencement of strong $\frac{3}{2}$, $\frac{5}{2}$ and $\frac{7}{2}f_{ce}$ emissions. (f_{ce} is marked as a strip chart recording line calculated from the field strength measured by the onboard magnetometer). On this occasion, towards the end of the emission, the $\frac{3}{2}f_{ce}$ signal itself is divided into discrete bands with separations close to the value of the ion plasma frequency. This data was recorded during the operation of the mutual impedence experiment, S-304, and the increase in the background level

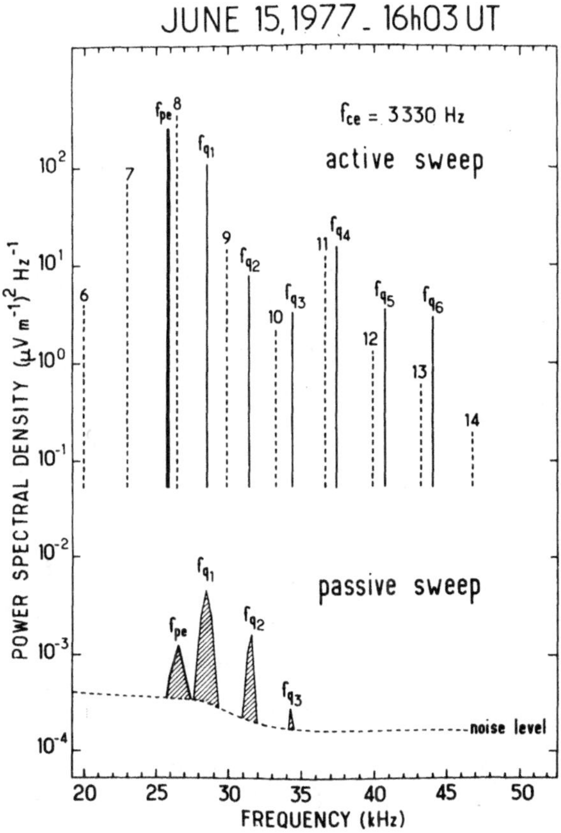

Fig. 4. Matching of f_{uh}, f_q frequencies and power levels shown in detailed comparison of a natural (lower) and active power spectrum recorded 22 s later, where $f_{ce} = 3.33$ kHz (16:03 UT, 15 June 1977). LT = 17:45, $L = 6.8$, $\lambda_m = 13°$.

at higher frequencies reflects the enhancement of the mutual impedence in the region of f_{pe} (Decreau *et al.*, 1978).

Another interesting feature of Figure 5 is the simultaneous occurrence of VLF chorus (up to $\frac{1}{2}f_{ce}$) as shown as two bands in the correlator output below. Chorus emissions and $\frac{3}{2}f_{ce}$ emissions have been observed to occur simultaneously before (Anderson and Maeda, 1977) but this correlation is by no means the rule for GEOS-1 data analysed so far.

A comment on the spectral distribution of these so called $\frac{3}{2}f_{ce}$ emissions is necessary here. The emission frequency is in fact rather variable, often showing jumps of $0.2f_{ce}$ within a given event. A study of 20 events shows that the centre frequency can lie anywhere between 1.1 and $1.9f_{ce}$, with a preference for the range 1.5–$1.6f_{ce}$.

The $\frac{5}{2}$ and $\frac{7}{2}f_{ce}$ emissions occur much less often and for shorter periods than $\frac{3}{2}f_{ce}$. When they do occur, in general they lie in the same relative position in their f_{ce} band as the accompanying $\frac{3}{2}f_{ce}$.

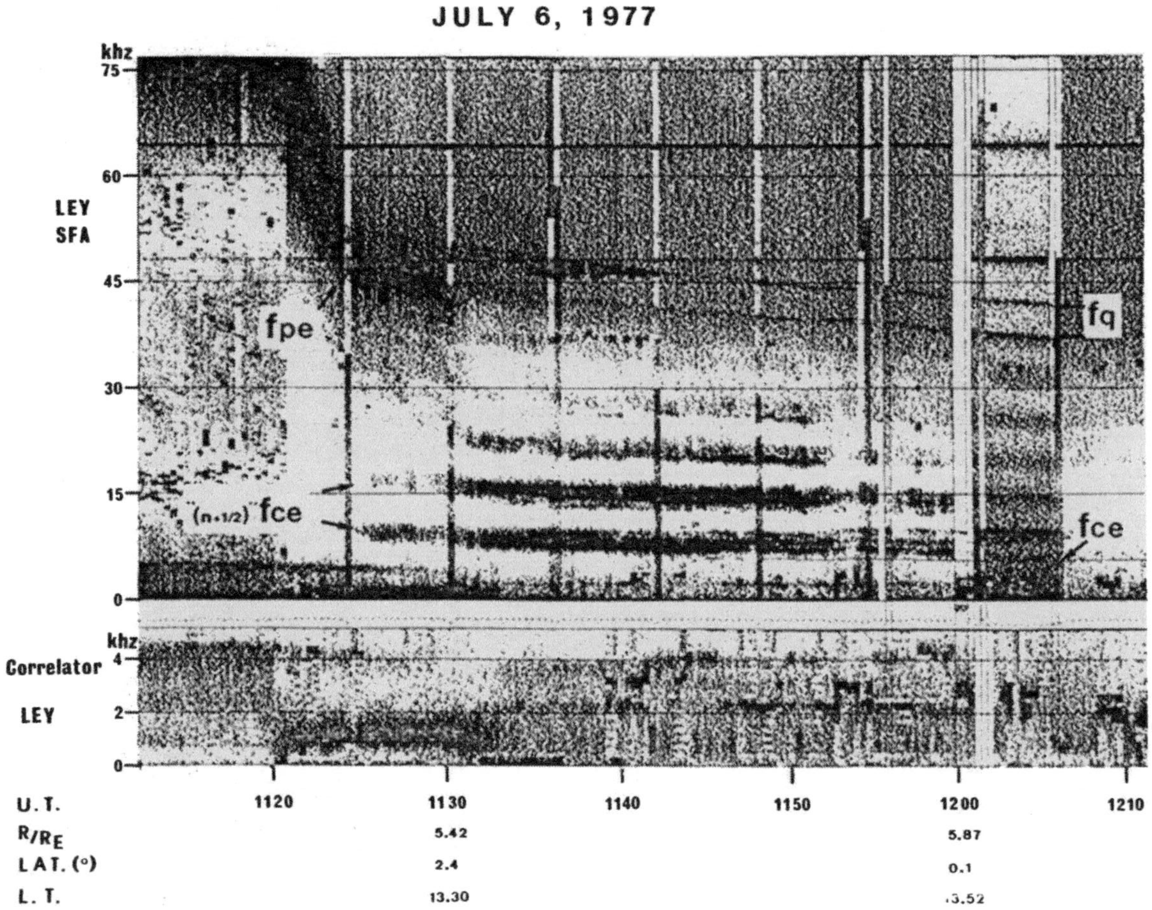

Fig. 5. Spectrogram of natural $\frac{3}{2}$, $\frac{5}{2}$, and $\frac{7}{2}$ f_{ce} emissions after crossing the plasmapause on the outbound pass of 6 July 1977.

We have studied the morphology of occurrence of $\frac{3}{2}f_{ce}$ emissions in order to attempt to understand the emission process. Figure 6 is a plot of those parts of the orbits (between days 130 and 250 of 1977) during which $\frac{3}{2}f_{ce}$ emissions were observed, plotted in the geomagnetic co-ordinates, L and local time using Barraclough internal field 75 model and Olson–Pfitzer tilt external field, (Kosik, 1978). Some plasmapause crossings are also matched and it would appear at first sight that the $\frac{3}{2}f_{ce}$ lies just outside the plasmapause with a morphology following the well known evening bulge of the plasmapause. Note however that the orbit passes through the emissions near apogees at 18 hr LT, with orbit tracks tangential to the emission region, but the orbit crosses through the emission region more normally before apogees at 12 hr LT. Now the duration of the emissions in time do not differ between the 12 hr LT and 18 hr LT regions. This observed morphology can either be explained by an emission region which is very thin at 18 hr LT and much wider at 12 hr LT, or more probably, the observations are better ordered by other parameters.

If the same observations are plotted against geomagnetic latitude data are ordered much more significantly, and the $\frac{3}{2}f_{ce}$ emissions are only found close to the geomagnetic equator. It was unfortunate that the orbit of GEOS-1 is such that the locus of

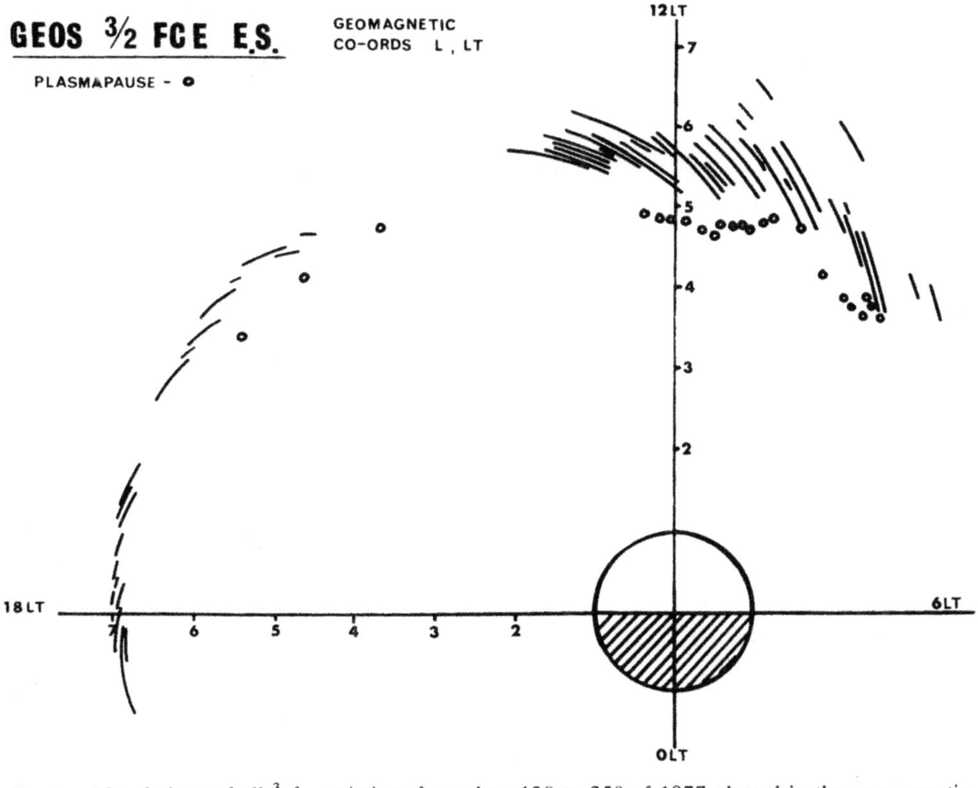

Fig. 6 Morphology of all $\frac{3}{2}$ f_{ce} emissions from days 130 to 250 of 1977 plotted in the geomagnetic co-ordinates L, and local time.

geomagnetic equator crossings should closely follow the shape of the plasmapause. This shows how necessary it is to resolve this type of time and space ambiguity with the geostationary satellite GEOS-2.

A more detailed illustration of the spread in latitudinal widths of these emissions is afforded by Figure 7(a). The most likely width of these emissions is between 3 and 4 deg in geomagnetic latitude with a 75% probability that the centre of the emission occurs within 2° of the geomagnetic equator (Figure 7(b)).

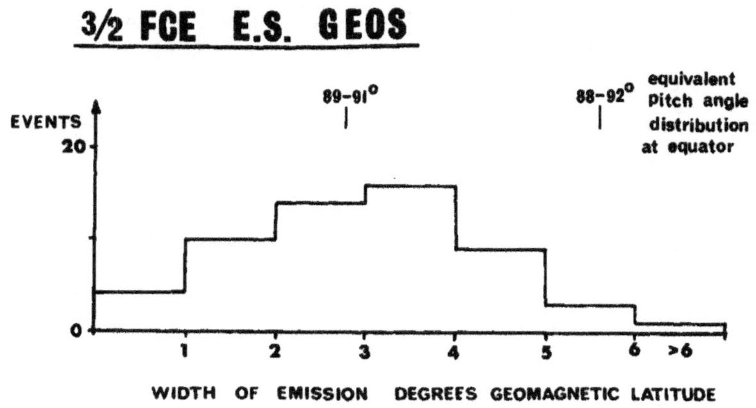

Fig. 7(a). The spread in width of $\frac{3}{2} f_{ce}$ emissions from days 130 to 250 in degrees of geomagnetic latitude.

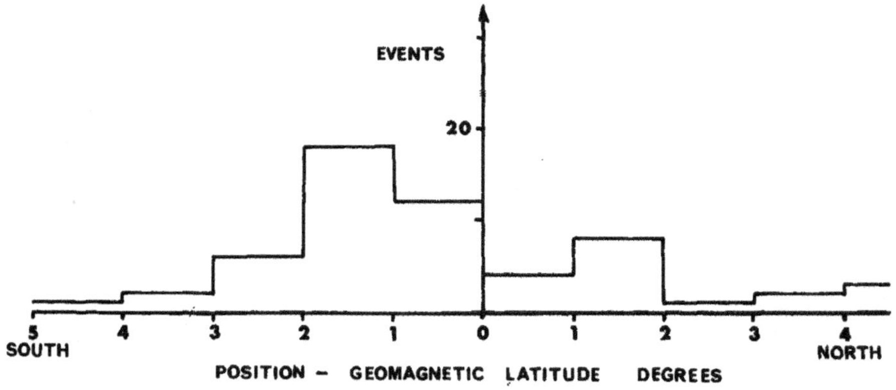

Fig. 7(b). The variation in the location of the centres (in time) of $\frac{3}{2} f_{ce}$ emissions from days 130 to 250 relative to the geomagnetic equator.

C. STRONG SIGNALS AT f_{pe}

On occasions the weak emission observed near f_{uh}/f_{pe} (described in Section 3A) becomes very strong, increasing in power by 65 dB over the quiet emission level. Figure 8 shows the spectrum of such an event on 17 June 1977. Here there is no difference in spectral levels between the active and passive sweeps which are themselves some 30 dB above that of the normal, active sweep. (The slight difference

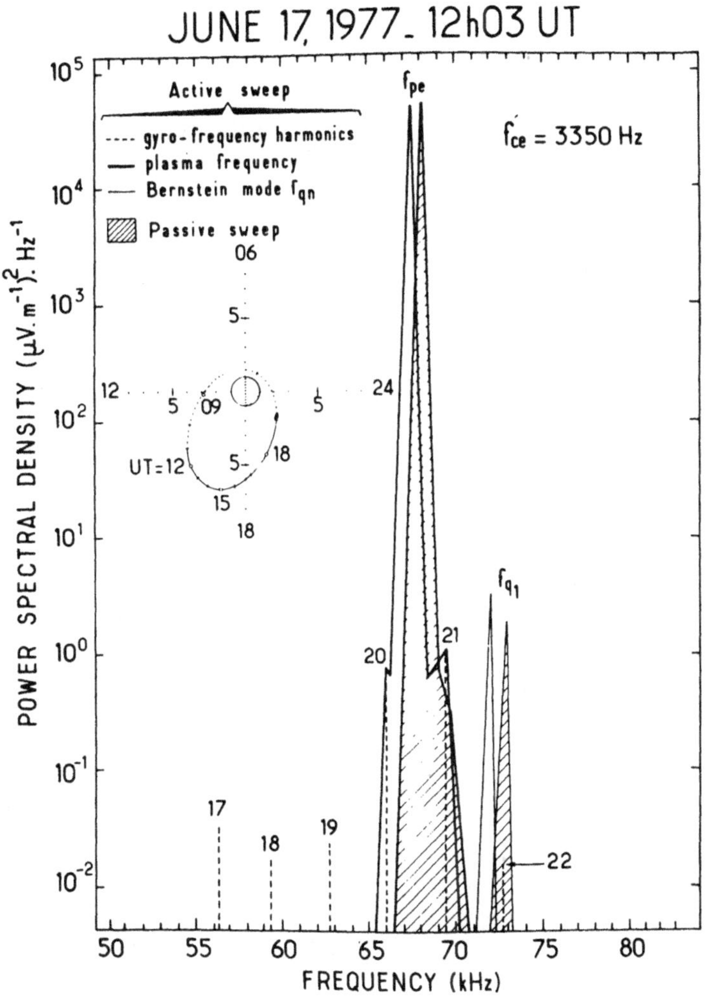

Fig. 8. A comparison of a strong burst of emission at f_{pe} with the active resonance sounder on 17 June 1977. LT = 15:00, $L = 6.5$, $\lambda_m = -6°$.

in frequency between active and passive is to be expected for measurements made 22 s apart.) The large amplitude natural emission is so strong as to obscure any resonance caused by the S-301 sounder and all nearby cyclotron resonances are depressed.

This strong natural emission lasts from a few seconds to minutes and attains wave field amplitudes of over 10^{-4} V m^{-1} Hz$^{-1/2}$ (around saturation of the long electric antenna). Of 22 events studied all except one occurred in the same narrow geomagnetic latitude range around the equator in which we find the $\frac{3}{2}f_{ce}$ emissions.

Figure 9 shows such an example, on 25 August 1977. Here the emission at f_{pe}/f_{uh} becomes very strong (seen as a widening of the black area at f_{pe} because of the grey scaling used). Three bursts occur during the $\frac{3}{2}f_{ce}$ event. Of particular interest is the

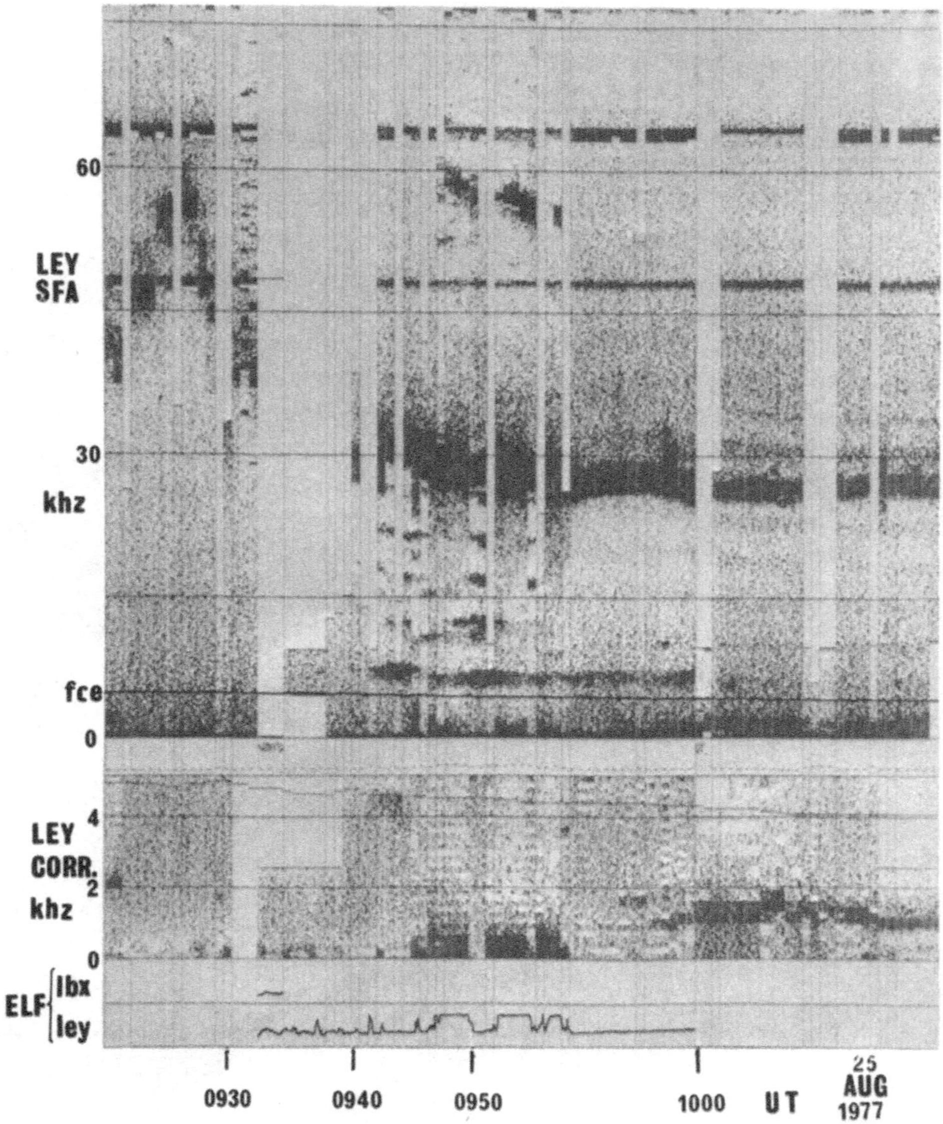

Fig. 9. An example of several bursts of strong emissions at the plasma frequency on 25 August 1977. Of particular interest are the simultaneous bursts of emissions at ELF as seen in the correlator display. Note the very strong f_{pe} signal is seen as a widening of the black area at f_{pe} because of the grey scaling used. Co-ordinates at 9:50 UT; LT = 11:20, $L = 6.6$, $\lambda_m = 10°$.

simultaneous appearance of emissions at ELF (seen both in the bottom strip chart recording and at low frequency in the correlator display) and of an emission at twice the plasma frequency.

Some care must be taken in interpreting this kind of wave event because of harmonic generation due to non-linearities in the saturated long antenna preamplifiers. There is however corroborative evidence to support the belief that the ELF bursts are real natural ELF waves. A detailed study of the SFA sweeps reveals that whenever the stepped frequency analyser steps in frequency through f_{pe} or its harmonics, the strength of the ELF at that time is always high if the f_{pe} emission is bursting, and low if it is not. As the analyser moves to frequencies far from f_{pe} and its harmonics, the ELF will remain high if the burst lasts until the next sweep and there appears to be no relation between ELF signal level and SFA step frequency as would be expected if there was any cross-talk or intermodulation of channels. Furthermore, there is evidence of a strong relationship between the wave bursts and significant changes in the low energy electron distribution measured by the S-302 experiment (J. J. Sojka, G. L. Wrenn, and J. F. Johnson, personal communication).

4. Nightside and Disturbed Times

A. DISTURBED DAYSIDE MAGNETOSPHERE

During major geomagnetic storms the frequencies f_{ce} and especially f_{pe} become very erratic, quickly changing in frequency sometimes faster than the time taken to make a full SFA sweep. Therefore the data is often difficult to interpret.

On a few very rare occasions the natural f_{uh} emissions and f_{ce} move to very low frequencies for short periods of a minute or so. All of the experiments on GEOS-1 give the same consistent explanation for these events: the satellite is moving out of the magnetosphere, crossing the magnetopause at times when the magnetosphere is severely compressed.

On moderately disturbed days the $\frac{3}{2}f_{ce}$ emission can be seen at geomagnetic latitudes far removed from the equator. These days are typified by several bursts of $\frac{3}{2}f_{ce}$ each of up to 1 hr duration spread throughout the satellite pass ($-10°$ to $+30°$ latitude). This kind of $\frac{3}{2}f_{ce}$ emission only occurs on disturbed days and it is otherwise confined to the quiet-time narrow band around the geomagnetic equator.

B. NIGHTSIDE MAGNETOSPHERE

As yet little data has been made available from this the most recent period of measurement. However we can make a few general comments emphasizing those features which must differ from the general quiet dayside magnetosphere.

The plasma frequency f_{pe} on the nightside becomes very low and is found below f_{ce} for most of the satellite pass. At such times f_{uh} lies just above f_{ce} (typically in the range 1.0–$1.2f_{ce}$) and we no longer have the natural banded emissions following the plasma frequency as we had on the dayside. A good comparison between day and night is afforded by Figure 10, where the data from 16 May 1977 is shown when the apogee was around 19 hr UT. During the pass the satellite moved from dayside to nightside type of phenomena. In the first part of the pass we see the usual dayside weak banded f_{uh} and $\frac{3}{2}f_{ce}$ emission (at the equator). On entering the nightside at 18:10 UT the f_{uh} emission appears in the correlator display at low frequency and towards the end of the pass the weak f_{uh} bands reappear rising up from f_{ce} as the

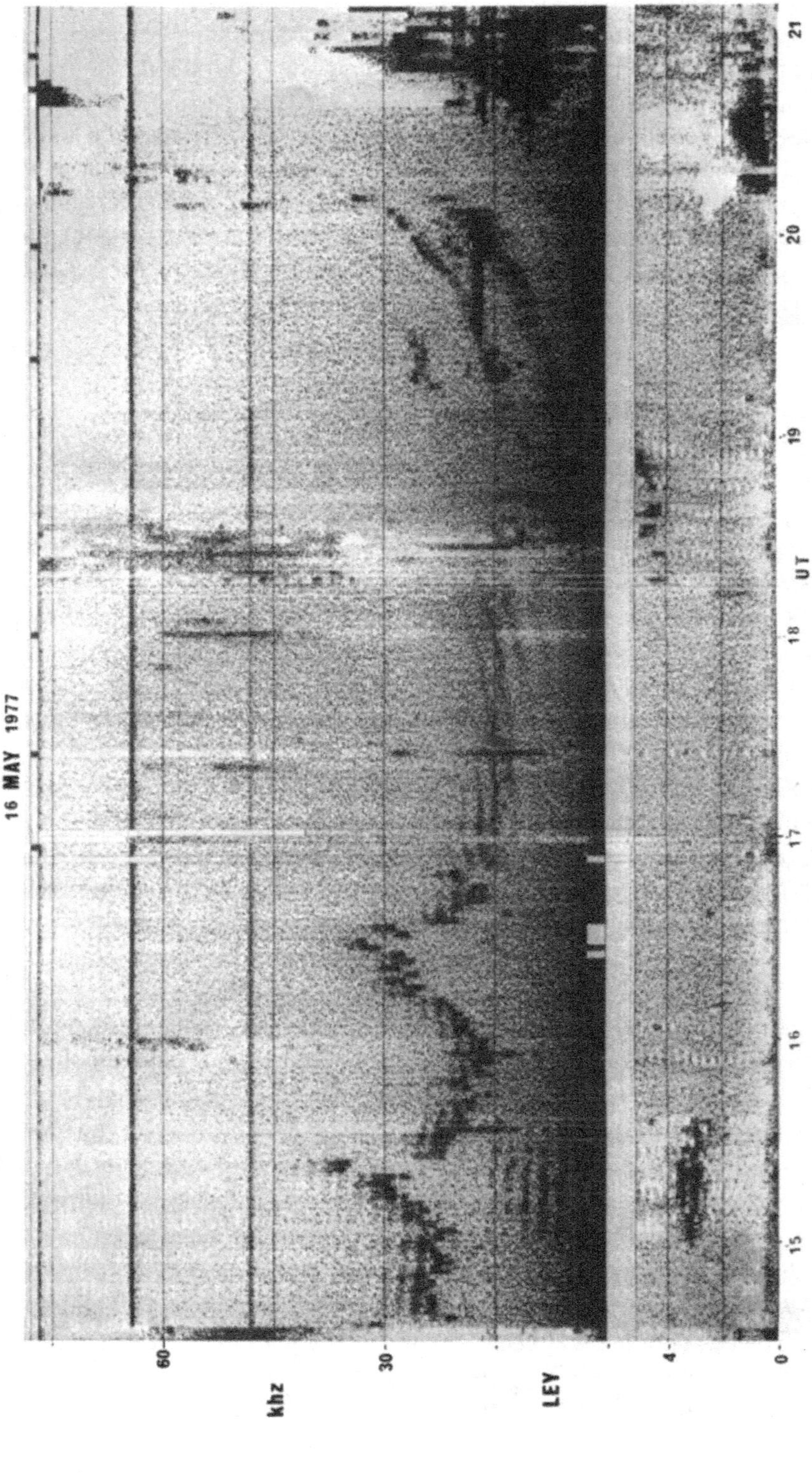

Fig. 10. Spectrograms from 16 May 1977 when the satellite moved from dayside to nightside types of phenomena during the pass. Note the sudden drop in f_{uh} frequency around 18:10 UT. At that time LT $\simeq$ 20:00, $L \simeq 6.5$, $\lambda_m \sim 5°$.

satellite moves back in towards the plasmasphere. This f_{uh} emission just above f_{ce} on the nightside is generally somewhat stronger than the average $\frac{3}{2}f_{ce}$ emission. A typical strength would appear to be 10^{-5} (V m^{-1}) Hz$^{-1/2}$.

5. Discussion

These early GEOS-1 observations of wave phenomena related to the harmonics of the electron gyrofrequency f_{ce} and local upper hybrid frequency f_{uh} complement and enrich those made from earlier satellites and, in particular, those from OGO-5 (Kennel *et al.*, 1970; Fredricks and Scarf, 1973) and from IMP-6 (Shaw and Gurnett, 1975). These satellites lacked active plasma diagnostics and thus the local plasma frequency f_{pe}, which is an important parameter controlling wave dispersion, could, at best, only be inferred. Actually, in the case of OGO-5, f_{pe} was not really known, whereas for IMP-6 it could be estimated in favourable conditions only from the low frequency cut-offs at the continuum radiation.

Nevertheless, by comparing our spectra with those observed by Shaw and Gurnett, it would appear that the emissions which these authors call 'narrow band' could almost certainly be identified with our 'upper hybrid banded' emissions, and their 'diffuse emissions' may fall into the same category. Furthermore, we would group the very strong noise bands at $\frac{3}{2}f_{ce}$ and, occasionally, at $\frac{5}{2}f_{ce}$, first detected by OGO-5, and also seen by IMP-6, with our own observations in the late evening and early morning sector.

There is now powerful experimental evidence that these waves are all propagating locally in the ECH or Bernstein mode. Certainly, for large values of f_{pe}/f_{ce}, this is the only available mode between the f_{ce} and the f_{uh}. With the aid of on-board plasma diagnostics, we have been able to determine f_{pe}/f_{ce} with considerable accuracy. Typically, on the dayside, between $L = 5$ to 7, we find this ratio to be 5 to 7. Thus we identify the $(n+\frac{1}{2})f_{ce}$ emissions described above with ECH waves. This is also true for the persistent weak banded emissions around and above the upper hybrid frequency (which occur in an enhanced form during a $(n+\frac{1}{2})f_{ce}$ event). These are discussed in detail by Christiansen *et al.* (1978), and the natural emission frequencies can be clearly identified, from Figure 3, with those corresponding to low group velocity ECH waves excited by the on-board sounder (the so called f_q frequencies).

In the many theoretical studies of magnetosphere ECH wave emissions, interest has concentrated primarily on developing linear instability theories to explain the strong nightside emission between f_{ce} and $2f_{ce}$ (Fredericks, 1971; Young *et al.*, 1973; Ashour Abdalla *et al.*; 1975, Karpman *et al.*, 1975; Ashour Abdalla and Kennel, 1976). Because of the lack of simultaneous detailed particle and density measurements, the early models out of necessity invoke plausible electron velocity distributions. These models, however, have helped in clarifying the requirements that must be met for the onset of ECH magnetospheric instabilities, and shown that the electron velocity distribution must have at least one hot component, with a loss cone, which provides the energy for the instability, and a cool component, which both destabilizes the waves and controls, to a large degree, the dispersion characteristics of unstable waves.

These early models use values of the important ratio of hot to cold number density that are more appropriate to the nightside magnetosphere ($n_c/n_h \approx 0.1$) and predict strongly unstable waves in the region of the cold plasma upper hybrid frequency. Thus, if the plasma density is sufficiently low, strong emissions occur between f_{ce} and $2f_{ce}$.

The GEOS-1 observations of daytime $(n+\frac{1}{2})f_{ce}$ emissions clearly cannot be described by these 'nightside' models since f_{uh} is measured to be many times f_{ce} and we know that the cold component is completely dominant ($n_c/n_h \approx 10$) during the emissions. Comparison is however possible with the predictions of recent calculations by Hubbard and Birmingham (1978), and Ronnmark *et al.* (1978) who have dealt with this range of parameters.

The role of the cold upper hybrid frequency (f_{uhc}) in controlling the maximum unstable frequency is also seen in these models, and they predict a series of instability bands between low gyroharmonics, terminating at or just above f_{uhc}. This agrees with our observations, in broad outline. Details of the theoretically predicted instability spectrum, such as the frequencies and growth rates of the fastest growing modes are, however, dependent on details of the electron velocity distribution and we refer the reader to the companion paper (Ronnmark *et al.*, 1978) in which a preliminary instability model is developed, based on a specific event measured by GEOS-1.

There is a further general trend in the data which confirms the broad character of the emissions as the cold upper hybrid frequency decreases towards f_{ce}. The GEOS-1 observations show that as the measured upper hybrid frequency decreases (for example on the outbound leg of a nightside pass), the highest natural emission band decreases until a single band is often observed lying just above the f_{ce}. During such a pass the ratio n_c/n_h certainly decreases and it is also reasonable to predict that the hot plasma density is also enhanced by the contribution of plasma sheet electrons. This trend is also seen in the models.

6. Conclusions

We can summarize our conclusions concerning the GEOS-1 observations of dayside phenomena as follows:

(i) *Weak banded emissions.* These are clearly related to emission on the ECH wave bands in the vicinity of the upper hybrid frequency. They are the weakest ($1\ \mu V\ m^{-1}$ broadband) and the most persistent feature of all waves observed. When seen alone the incoherent observed power levels could be explained by emission from a linearly stable hot electron component (Christiansen *et al.*, 1978), although a detailed calculation of the kind made by Trulsen (1971) is needed. When seen, at enhanced level in conjunction with $(n+\frac{1}{2})f_{ce}$ emissions, they probably arise from a weak instability, as discussed in (ii) below.

(ii) *$(n+\frac{1}{2})f_{ce}$ emissions.* These are observed at levels of 10–100 μV m^{-1} broad-band, and propagate in the ECH wave mode in bands which decrease in amplitude when their frequency increases towards the local upper hybrid frequency. The levels are such that they are the result of a linear instability of a hot loss cone component dominated by a dense, cool (but with an upper limit to n_c/n_h) component. Details of the frequency distribution in any given band are highly variable (occurring anywhere between, for the lowest band, 1.1 and $1.9f_{ce}$).

They show a particularly interesting morphology in the dayside magnetosphere, the emitting regions being confined to within a few degrees of the modelled magnetic equator on quiet days, and this equatorial confinement opens a number of avenues for conjecture:

(a) Does it result from the instability of a particular group of hot electrons convected from the plasma sheet in such a way as to produce an equatorial population with extremely large pitch angles?

(b) If convection produces the loss cone distribution necessary for weak instability in the vicinity of the magnetic equator, can the localization be explained by a filling of the loss cone, or cold plasma heating (by consequent wave particle interactions) sufficient to quench the weak instability at higher latitudes?

(c) What role is played by the highly variable refractive properties of the ECH waves as they propagate along latitudinal and radial magnetic gradients?

(iii) *Strong emission close to f_{uh}.* These are sporadic, powerful events with broad-band instability, amplitudes of 10 μV m^{-1}, their rare occurrence being associated with the weaker $(n+\frac{1}{2})f_{ce}$ emissions. Their energy density $(\bar{E}_0^2/2nkT)\times 10^{-3})$ is such that non-linear effects are to be expected, and we believe that evidence exists for their non-linear coupling to waves below the ion plasma frequency. Detailed studies of these and other possible non-linear processes are in progress.

Acknowledgements

We thank M. Petit, B. de la Porte, R. Grard, A. Pedersen, M. Debrie, F. Lefeuvre, M. Hamelin, A. Bahnsen, E. Ungstrupp, J. Johnson, J. Sojka, G. Wrenn, H. Borg, K. Ronnmark, and F. Mariani for their valuable scientific collaboration. The contributions of the European Space Operations Centre, the Centre National d'Etudes Spatiales, Toulouse, E. Gershuny and J. Hunter of Sussex University to the data handling are acknowledged, as is the work of K. Knott, Project Scientist, and the GEOS project team. This research is supported by the SRC of Great Britain, CNET, CNRS of France, and ESA.

References

Anderson, R. R. and Maeda, K.: 1977, *J. Geophys. Res.* **82**, 135.

Ashour-Abdalla, M. and Kennel, C. F.: 1976 in B. M. McCormac (ed.), *Magnetospheric Particles and Fields*, D. Reidel Publ. Co., Dordrecht, p. 181

Ashour-Abdalla, M., Chanteur, G., and Pellat, R.: 1975, *J. Geophys. Res.* **80**, 2775.
Christiansen, P., Gough, P., Martelli, G., Bloch, J. J., Cornilleau, N., Etcheto, J., Gendrin, R., Jones, D., Beghin, C., and Decreau, P.: 1978, *Nature* **272**, 682.
Cornilleau-Wehrlin, N., Gendrin, R., Lefeuvre, F., Parrot, M., Grard, R., Jones, D., Bahnsen, A., Ungstrup, E., and Gibbons, W.: *Space Science Rev.* **22,** 371.
Decreau, P., Beghin, C., and Parrott, M.: 1978, *Space Sci. Rev.* **22**, 581.
Etcheto, J. and Petit, M.: 1977, *Compt. Rend. Acad. Sci.* **285B**, 329.
Fredricks, R. W.: 1971, *J. Geophys. Res.* **76**, 5344.
Fredricks, R. W. and Scarf, F. L.: 1973, *J. Geophys. Res.* **78**, 310.
Hubbard, R. F. and Birmingham, T. J.: 1978, *J. Geophys. Res.*, to be published.
Jones, D.: 1978, *Space Sci. Rev.* **22**, 327.
Karpman, V. I., Alekhin, Ju. K., Borisov, N. D., and Rjabova, N. A.: 1975, *Plasma Phys.* **17**, 361.
Kennel, C. F., Scarf, F. L., Fredricks, R. W., McGhee, J. H., and Coronitti, F. V.: 1970, *J. Geophys. Res.* **75**, 6136.
Kosik, J.: 1978, *Space Sci. Rev.* **22**, 481.
Ronnmark, K. K., Borg, H., Christiansen, P. J., and Gough, M. P.: 1978, *Space Sci. Rev.* **22**, 401.
Ronnmark, K. K., Borg, H., Christiansen, P. J., and Gough, M. P.: 1978, this volume, p.
Shaw, R. R. and Gurnett, D. A.: 1975, *J. Geophys. Res.* **80**, 4259.
S-300 Experimenters: 1978, *Planetary Space Sci.*, (in press).
Young, T. S., Callen, J. D., and McCune, J. E.: 1973, *J. Geophys. Res.* **78**, 1082.

BANDED ELECTRON CYCLOTRON HARMONIC INSTABILITY – A FIRST COMPARISON OF THEORY AND EXPERIMENT

K. RÖNNMARK and H. BORG

Plasma Physics Department, University of Umea, and Kiruna Geophysical Institute, Sweden

P. J. CHRISTIANSEN and M. P. GOUGH

Plasma and Space Physics Group, School of Mathematical & Physical Sciences, University of Sussex, U.K.

D. JONES

Space Sciences Division, E.S.A., E.S.T.E.C., Noordwijk, Holland

Abstract. A study, which is the first of its kind, uses information derived from simultaneously measured wave spectra and particle distributions as the input to a theoretical linear instability model of an electrostatic cyclotron harmonic wave event recorded on GEOS-1. The presence of a hot loss cone component of the particle distribution is established experimentally, and the model accounts reasonably for the observed frequencies and relative strengths of the $(n+\frac{1}{2})f_c$ and upper hybrid emission features.

1. Introduction

In this paper we report results of a first attempt to compare GEOS-1 spectra of magnetospheric wave instabilities with the prediction of a linear theory of electron cyclotron (ECH) wave instability. The novel feature of this work lies in the fact that the input parameters used for the theoretical model, in particular, the electron distribution function, are derived whenever possible from measurements made concurrently with those of the wave spectra.

The understanding of magnetospheric instabilities with frequencies above the electron gyrofrequency ($f_c = \Omega/2\pi$) has been seen as an important problem since the detection of strong (electric fields >1 mV m^{-1}) electrostatic emissions between low gyroharmonics by the OGO-5 electric field experiment (Kennel *et al.*, 1970). Subsequent observations from OGO-5 (Fredricks and Scarf, 1973) and from IMP-6 by Shaw and Gurnett (1975) enriched this observational base and provided the stimulus for further theoretical effort. The literature on ECH wave instability is extensive, though much of the early work, e.g. Dory *et al.* (1965), Hall *et al.* (1965), Brambilla (1970), Tataronis and Crawford (1970a,b) is oriented to laboratory plasmas. It is with the paper by Fredricks (1971) that the problem of explaining the occurrence of the so called '$(n+\frac{1}{2})f_c$' magnetospheric emissions began to be tackled. From this and the subsequent work of Young *et al.* (1973), Karpman *et al.* (1975), Ashour-Abdalla *et al.* (1975), Ashour-Abdalla and Kennel (1976, 1978) the basic requirements for a reasonable (in the sense that plausible magnetospheric electron velocity distributions are used) linear stability theory have emerged. These are basically that a plasma containing at least two components should exist – a hot, weak loss-cone component (density n_A), and a cooler

Space Science Reviews **22** (1978) 401–417. *All Rights Reserved*

component (density n_c). Put somewhat crudely, the hot component, which contains regions of velocity space where $\partial f/\partial v_\perp > 0$, contains the free energy for the instability, while the cool component aids destabilization and controls, to a large degree, the dispersion of the waves. The role of the cool plasma upper-hybrid frequency (f_{uhc}) is particularly important in that strong instability is predicted in its vicinity.

The development of a good linear theory has more than intrinsic interest since it forms the basis for an understanding of quasi-linear and non-linear effects associated with the instability. In fact, the quasi-linear pitch angle and energy diffusion of kilovolt electrons has been discussed by Lyons (1974) in relation to the generation of diffuse aurorae, and by Ashour-Abdalla and Kennel (1978) as a wave stabilization mechanizm.

Because the earlier satellites lacked the direct methods for the measurement of important plasma parameters available on GEOS-1, the models discussed above have been constructed more or less independently of direct experimental input, other than spectral information. This in turn has emphasised emissions between low harmonics of the electron gyrofrequency – the so called $\frac{3}{2}$, and $\frac{5}{2}f_c$ instabilities. Since instability is expected close to f_{uhc}, and in gyroharmonic bands below it, the models presuppose a rather low value of the cool plasma density, and by setting the ratio $n_c/n_A \leq 0.2$ have tended to treat what might be called 'nightside' phenomena, where the cool component, presumably of ionspheric origin, is dominated by the hot component composed of electrons convected from the tail.

From the GEOS-1 observations (Christiansen *et al.*, 1978) it is known that ECH wave instability also occurs close to the geomagnetic equator, on the dayside, where the cool plasma is dense $f_{uhc}/f_c \gg 1$, and dominant, and in which signals are observed occupying the bands between gyroharmonics all the way up to f_{uhc}. This case is treated briefly by Karpman *et al.* (1975) and more recently by Hubbard and Birmingham (1978) who model various spectral types of ECH wave emissions seen in IMP-6 data and argue that the spectra can be used as a plasma diagnostic of the cool electron density n_c.

The approach adopted in this paper is rather different, since we emphasize the use of directly measured plasma parameters and thus hope to avoid the difficulty of dealing with a large number of variable input parameters which besets theoretical work in this area.

The paper is arranged as follows; in Section 2 we describe various experimental inputs which are used for the theory, which is described together with the model velocity distribution in Section 3. The predictions of this model are discussed in relation to measured wave spectra in Section 4.

2. Experimental

The data used for this study is taken from the 25th August 1977, in a short period from 09:50–09:53 UT. In order to orient the reader, a grey scaled spectrogram,

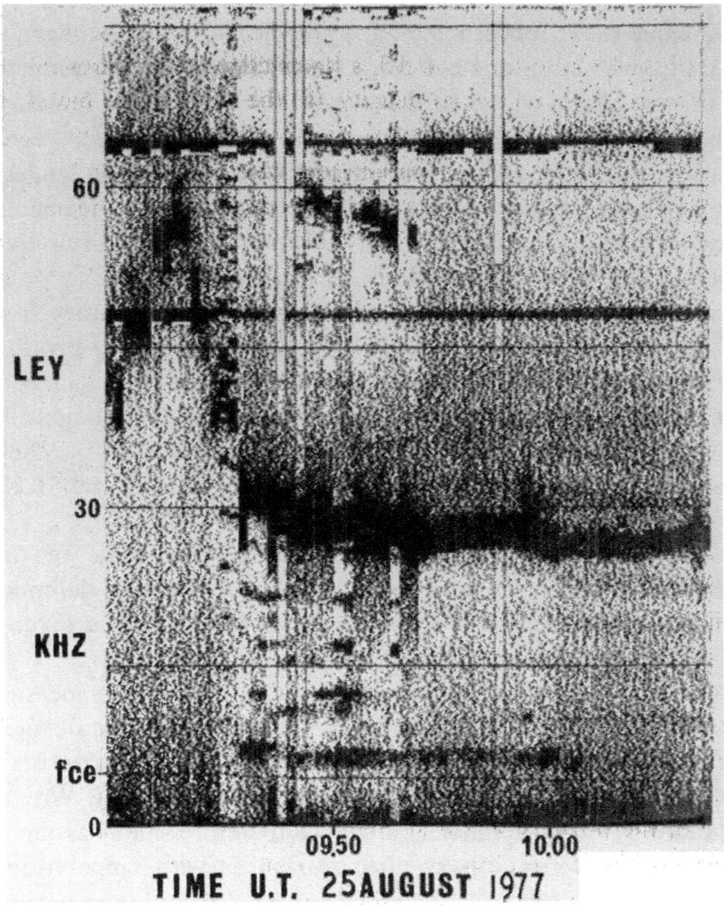

Fig. 1. Spectrogram emphasising the sporadic wave event under study. The display begins at 09:20 and up to 09:50 UT the time scale is not uniform. The model discussed here concerns wave spectra measured between 09:50 and 09:53 UT. At 09:50 UT coordinates are: LT = 11:20, $L = 6.6$, $\lambda_m = 10°$, with $Kp = 5$.

which includes this time, is shown in Figure 1. The wave spectra are derived from the large electric antenna (Jones *et al.*, 1978) and consist of sets of 0–77 kHz sweeps of the steeped frequency analyser covering the range 0–77 kHz. The display begins at 09:20 UT where the most pronounced feature is a variable high frequency emission known to be in the vicinity of the upper hybrid frequency.

At about 09:40 the onset of an $(n+\frac{1}{2})f_c$ event can be seen, when signals at $\sim\frac{3}{2}$, $\frac{5}{2}$, $\frac{7}{2}$, and $\frac{11}{2}f_c$ appear. At the same time the upper hybrid feature broadens (indicative of increased signal strength in this method of grey scaling) and significant second harmonic generation can be seen occurring periodically between 09:45 and 09:53 UT. This second harmonic generation is believed to be (at least partly) instrumental but is indicative of a very strong signal at the fundamental. After about 09:53 the $(n+\frac{1}{2})f_c$ upper hybrid emissions weaken and we are left with the persistent banded upper hybrid emission beyond about 10:00 UT.

2.1. Experimental inputs

In order to define the theoretical model, a knowledge of the following parameters is required: (i) the electron number density, (ii) the value of the ambient magnetic field, and (iii) the shape of the complete electron distribution function.

The rapidity with which these parameters can be measured, even approximately, varies somewhat, and we use values averaged over the longest measurement time scale of ~3 min. This is the time required for the medium energy particle experiment, S-310 (Borg *et al.*, 1978) to construct a full energy versus pitch angle distribution (at a satellite polar angle of 18°).

The electron number density is estimated within 5% accuracy from the frequency of the strong natural emission in the region of the upper hybrid frequency, $f_{uh}^2 = f_p^2 + f_c^2$ where f_p is the electron plasma frequency. This can be done with reasonable confidence in view of an earlier study of the relationship between emissions in this frequency range and resonances stimulated by the S-301 sounder (Christiansen *et al.*, 1978) and a corroborative check made at 09:30 UT during a short active sounding period.

The spacecraft fluxgate magnetometer provides the value of the ambient magnetic field, and thus f_c (4.3 kHz), which is also used via the definition of f_{uh} to establish the number density. This value, $n = 9.5 \times 10^6\ \text{m}^{-3}$ is taken to be the density of the cool electron component.

The low energy component of the electron distribution is measured by the S-302 experiment (Johnson *et al.*, 1978). At the time of writing detailed information is not available for this event, but it is known that it consists of several components in the 0–10 eV range (J. Johnson, J. Sojka and G. Wrenn, personal communication). For this preliminary study we therefore simply assume that it can be represented by an isotropic maxwellian distribution with temperature $T_C = 5$ eV.

The medium energy (300 eV–20 keV) particle distribution is provided by the S-310 experiment (Borg *et al.*, 1978). As will be seen in detail in the next section it is of low density (~3% of that of the cool component) and exhibits a loss-cone structure.

3. Theory and Method

In this section we describe the use of a graphical technique for establishing regions of parameter space in which ECH wave instability is expected to occur. The technique is applied to a model derived from actual measurements of the electron distribution function, and qualitative predictions about possible instabilities are discussed.

The dielectric constant $\varepsilon(\omega, \mathbf{k})$ for electrostatic waves with complex frequency $\omega = \omega_R + i\gamma$ and real wave vector $\mathbf{k}$, with components $k_\parallel$ parallel and $k_\perp$ perpendicular to the magnetic field, in an electron plasma, is

$$\varepsilon(\omega, \mathbf{k}) = 1 - \frac{\omega_p^2}{k^2} \sum_{n=-\infty}^{\infty} \int d\mathbf{v} \frac{J_n^2[(k_\perp v_\perp)/\Omega]}{k_\parallel v_\parallel - \omega - n\Omega} \left[\frac{n\Omega}{v_\perp} \frac{\partial F}{\partial v_\perp} + k_\parallel \frac{\partial F}{\partial v_\parallel} \right]. \tag{1}$$

Here, ω_p is the plasma frequency, Ω the electron gyrofrequency, $J_n[(k_\perp v_\perp)/\Omega]$ the Bessel function of order n, $v_\parallel$ and $v_\perp$ are the components of the particle velocity and F is the electron distribution function. When the growth rate is small ($\gamma \ll \omega_R$), we can calculate $\omega(\mathbf{k})$ from

$$\begin{cases} \varepsilon_R(\omega_R, \mathbf{k}) = 0, \\ \gamma = -\dfrac{\varepsilon_I}{\partial \varepsilon_R/\partial \omega}, \end{cases} \tag{2}$$

where

$$\varepsilon_R(\omega_R, \mathbf{k}) = 1 - \frac{\omega_p^2}{k^2} \sum_n P \int \mathrm{d}\mathbf{v} \, \frac{J_n^2[(k_\perp v_\perp)/\Omega]}{k_\parallel v_\parallel - \omega_R - n\Omega} \left[\frac{n\Omega}{v_\perp} \frac{\partial F}{\partial v_\perp} + k_\parallel \frac{\partial F}{\partial v_\parallel} \right] \tag{3}$$

and

$$\varepsilon_I(\omega_R, \mathbf{k}) = -\pi \frac{\omega_p^2}{k^2 |k_\parallel|} \sum_n \int 2\pi v_\perp \, \mathrm{d}v_\perp \, J_n^2\left(\frac{k_\perp v_\perp}{\Omega}\right) \times \left[\frac{n\Omega}{v_\perp} \frac{\partial F}{\partial v_\perp} + k_\parallel \frac{\partial F}{\partial v_\parallel} \right]_{v_\parallel = (\omega_R - n\Omega)/k_\parallel}.$$

The real part of the frequency will depend mainly on the first few moments of the distribution function, such as the electron density N and temperature T. Furthermore, as noticed by Young *et al.* (1973), the low energy particles alone will often determine the behaviour of ε_R. If the isotropic low energy plasma is characterized by a temperature T_C, we can thus solve the real part of the dispersion relation for $\omega_R(\mathbf{k})/\Omega$ when ω_p/Ω is known. In Figure 2 we have plotted the curves $\omega_R/\Omega = \text{constant}$ for $\Omega < \omega_R < 2\Omega$ and $\omega_p = 6.25\Omega$ in the $\mathbf{k}\rho_c$-plane, where $\mathbf{k}\rho_c = \mathbf{k}(V_c/\Omega)$ is a normalized wave vector and V_c, the thermal velocity, $(2T_c/m_e)^{\frac{1}{2}}$. A few curves of constant γ/Ω have also been included to show the cyclotron damping by low energy electrons when $(\omega_R - n\Omega)/k_\parallel \approx V_c$, and the contour $\partial\omega_R/\partial k_\parallel = 0$ is shown by a circled line. The imaginary part of $\varepsilon(\omega_R, \mathbf{k})$ is given by Equation (3). In this paper we will restrict ourselves to loss cone instabilities and thus assume $\partial F/\partial v_\parallel < 0$ at all velocities. A necessary (but not sufficient) condition for instability is then that $\partial F/\partial v_\perp > 0$ in some region of velocity space. The strength of the interaction between a wave with frequency $\omega_R = (n + \frac{1}{2})\Omega$, and a group of particles with velocity $\mathbf{v}_0$, is proportional to

$$\sum_m J_m^2\left(\frac{k_\perp v_{0\perp}}{\Omega}\right) \delta\left(v_{0\parallel} - \frac{|n + \frac{1}{2} - m|\,\Omega}{k_\parallel}\right).$$

The important terms in this sum are $m = n$ and $m = n + 1$. A strong interaction will result, if we choose the parallel wave number to be

$$k_\parallel = \tfrac{1}{2} \frac{\Omega}{v_{0\parallel}}. \tag{4}$$

The perpendicular wave number must then be chosen to maximize $J_n^2[(k_\perp v_{0\perp})/\Omega] + J_{n+1}^2[(k_\perp v_{0\perp})/\Omega]$. This can be done by defining a strongly interacting

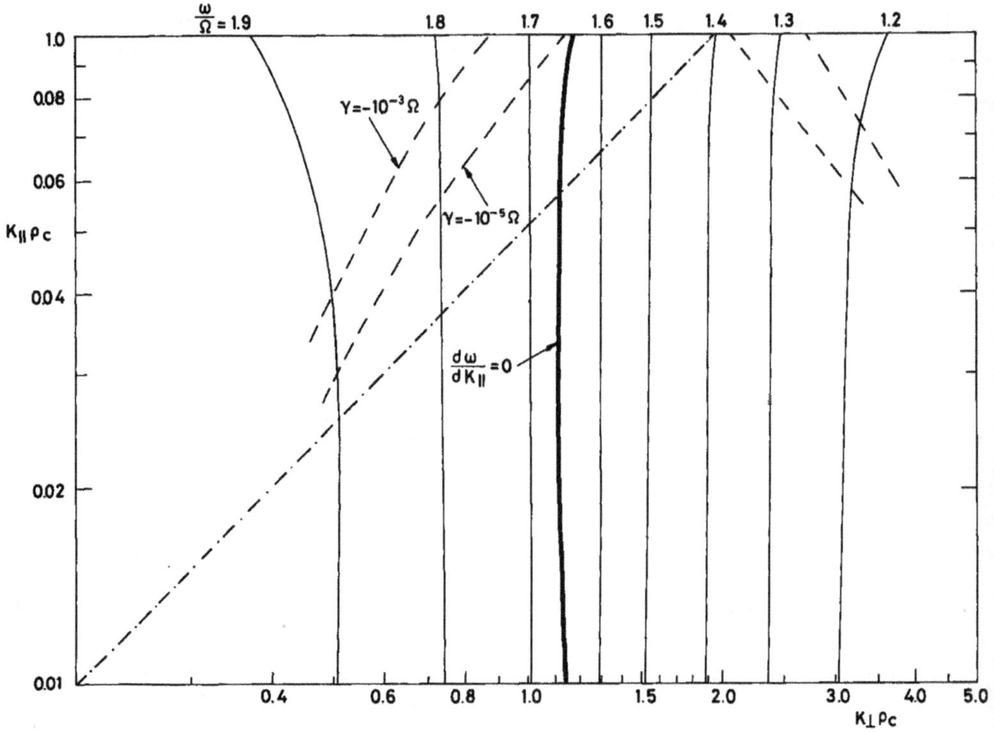

Fig. 2. The dispersion relation of a Maxwellian plasma illustrated by frequency contours in the $\mathbf{k}\rho_c$ plane. Solid curves correspond to constant $\mathrm{Re}\,\omega$, and dashed curves to constant $\mathrm{Im}\,\omega = \gamma$. The dot-dashed oblique line is to be used in conjunction with Figure 4 (see text).

$k_\perp$ as

$$k_\perp \approx \frac{\Omega}{v_{0\perp}} C_n, \tag{5}$$

where

$$C_n = \frac{j_n J_n^2(j_n) + j_{n+1} J_n^2(j_{n+1})}{J_n^2(j_n) + J_{n+1}^2(j_{n+1})}$$

and j_n is the argument corresponding to the first maximum of J_n.

These formulas are of course very rough, but they still provide a useful first estimate of possibly unstable wave numbers. The simplest way of applying them, is to use a plot of the measured distribution function to choose a $\mathbf{v}_0$ where $\partial F/\partial v_\perp |_{\mathbf{v}_0} > 0$, calculate $\mathbf{k}$ from Equations (4) and (5), and read off the corresponding ω_R in Figure 2. If the resulting ω_R and $k_\parallel$ satisfy $k_\parallel V_c < |\omega_R - n\Omega|$ and $k_\parallel V_c < |\omega_R - (n+1)\Omega|$, the instability is likely to occur. Otherwise it is quenched by the isotropic, low energy, particles. The predicted unstable frequency can then be compared with observed wave emissions.

A more systematic and quantitative comparison can be obtained by means of an analytical model of the particle distribution. We have chosen a model distribution function f of the form

$$f = n_A f_A + n_B f_B + n_C f_C. \tag{6}$$

Here, the constants n_A, n_B, and n_C include the normalization factors which are necessary to obtain f in the desired units. The cool, or low energy component f_c is the previously discussed isotropic Maxwellian with temperature T_C, while the other components have the form (with $i = A$ or B)

$$f_i = \Delta_i \exp\left(-\frac{v^2}{V_i^2}\right) + \frac{1-\Delta_i}{1-\beta_i}\left[\exp\left(-\frac{v_\perp^2}{V_i^2}\right) - \exp\left(-\frac{v_\perp^2}{\beta_i V_i^2}\right)\right]\exp\left(-\frac{v_\parallel^2}{V_i^2}\right). \tag{7}$$

The thermal velocities are denoted by V_i and the corresponding temperatures by T_i. The constants Δ_i and β_i determine the depth and width of the loss cone. Similar models have previously been used by Ashour-Abdalla and Kennel (1976).

A comparison between the measured electron distribution at 09:52 UT on August 25, 1977, and a model distribution is shown in Figure 3. The measured

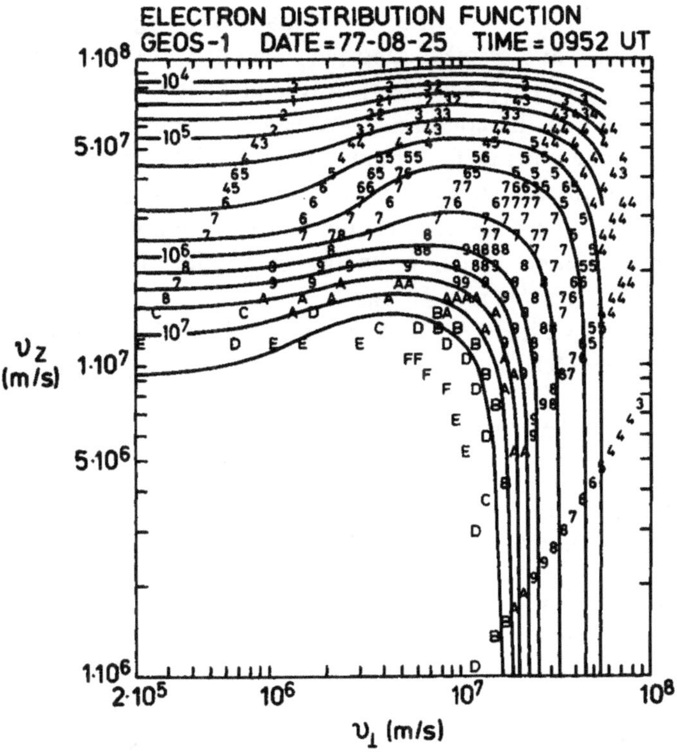

Fig. 3. Comparison between the measured electron distribution function and a model distribution. The measured fluxes are coded according to Table I, and the contours show constant values of the model distribution defined in Equations (6) and (7). Parameter values used in the model are $n_A = 1.43 \times 10^5\ \mathrm{m}^{-3}$, $n_B = 1.19 \times 10^5\ \mathrm{m}^{-3}$, $\omega_p/\Omega = 6.3$, $T_A = 0.35$ keV, $T_B = 5.0$ keV, $T_C = 5.0$ eV, $\Delta_A = \Delta_B = 0.3$, $\beta_A = 0.05$, and $\beta_B = 0.01$.

TABLE I

Symbol	$F(\text{el cm}^{-2}\,\text{s}^{-1}\,\text{sr}^{-1}\,\text{keV}^{-2})$
0	10^4
1	1.0×10^4–1.8×10^4
2	1.8×10^4–3.2×10^4
3	3.2×10^4–5.6×10^4
4	5.6×10^4–1.0×10^5
5	1.0×10^5–1.8×10^5
6	1.8×10^5–3.2×10^5
7	3.2×10^5–5.6×10^5
8	5.6×10^5–1.0×10^6
9	1.0×10^6–1.8×10^6
A	1.8×10^6–3.2×10^6
B	3.2×10^6–5.6×10^6
C	5.6×10^6–1.0×10^7
D	1.0×10^7–1.8×10^7
E	1.8×10^7–3.2×10^7
F	3.2×10^7–5.6×10^7

differential flux ϕ has been divided by the particle energy E to obtain $F=\phi/E$ in units of electrons/(cm^2 s sr keV^2), and the value of F has been coded as shown in Table I. If the model were perfect, only one kind of symbol would be found between each pair of the contours $f=$ constant, which have been calculated and plotted in the same units as F. Although this has not been achieved, this model is certainly more firmly founded in reality than any of those used in previous studies. The cold plasma density has been estimated from the observed plasma frequency. In Figure 4 the same model is shown in greater detail for $10^7\,\text{m s}^{-1}<v_\parallel<10^8\,\text{m s}^{-1}$ and $10^6\,\text{m s}^{-1}<v_\perp<10^7\,\text{m s}^{-1}$.

Since the growth rate γ is independent of T_C as long as $k_\parallel\rho_c$ is small, it is convenient to introduce a new normalization of the wave vector by

$$\mathbf{k}\rho_A=\frac{\mathbf{k}V_A}{\Omega}.$$

Recalling Equations (4) and (5), we find that particles with velocity $\mathbf{v}$ will interact strongly with a wave with frequency $\omega_R\sim(n+\frac{1}{2})\Omega$ if the components of $\mathbf{k}\rho_A$ are

$$k_\perp\rho_A\approx\frac{V_A}{v_\perp}C_n$$

and

$$k_\parallel\rho_A\approx\frac{1}{2}\frac{V_A}{v_\parallel}. \tag{8}$$

The frame of Figure 4 has been labelled with these values of $k_\parallel\rho_A$ and $k_\perp\rho_A$ (for $n=1$) as well as the corresponding velocities. The reader is recommended to make a transparent copy of Figure 4 and place it upside down upon Figure 1, with

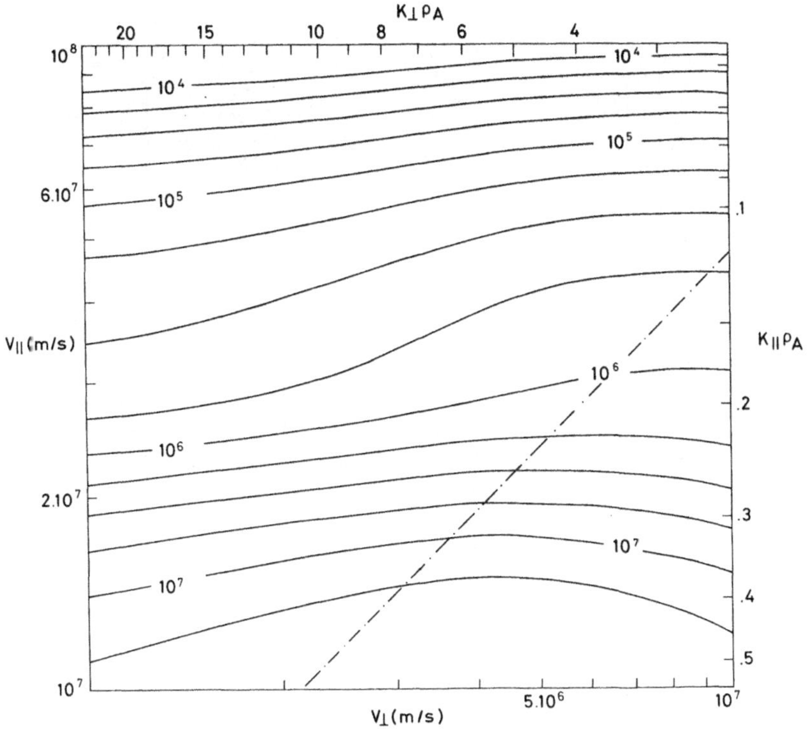

Fig. 4. Velocity space contours for the model hot electron distribution function, showing the part of velocity space containing the weak loss cone in greater detail. All parameters of the model are the same as in Figure 3.

the dashed lines covering each other. When doing so, one obtains a picture like Figure 5, where the position corresponding to $T_C = 5$ eV is shown. Since the wave numbers are plotted on logarithmic scales, different temperature ratios $T_A/T_C = (k\rho_A/k\rho_c)^2$ are illustrated by different relative positions of the figures. Keeping the dashed-dotted lines aligned ensures that $k_\perp\rho_A/k_\perp\rho_c = k_\parallel\rho_A/k_\parallel\rho_c$. By varying T_C, one finds that the instability caused by $\partial f/\partial v_\perp > 0$ in a certain region of velocity space, will occur at higher frequencies when T_C is lowered and vice versa. This leads to the somewhat surprising conclusion, that cyclotron damping by the low energy electrons may quench an instability if T_C is decreased.

The dispersion relation

$$\varepsilon(\omega, \mathbf{k}) = 0, \tag{9}$$

with ε given by Equation (1), has also been solved for $\omega = \omega_R + i\gamma$ by standard numerical methods, with real $\mathbf{k}$ and the same model distribution function f as in Figure 3. The result is shown in Figure 6. By comparing the curves $\omega_R = \text{constant}$ in Figure 6 to those in Figure 1, we conclude that calculating ω_R from the cool plasma alone is a very good approximation in this case. We also find that regions

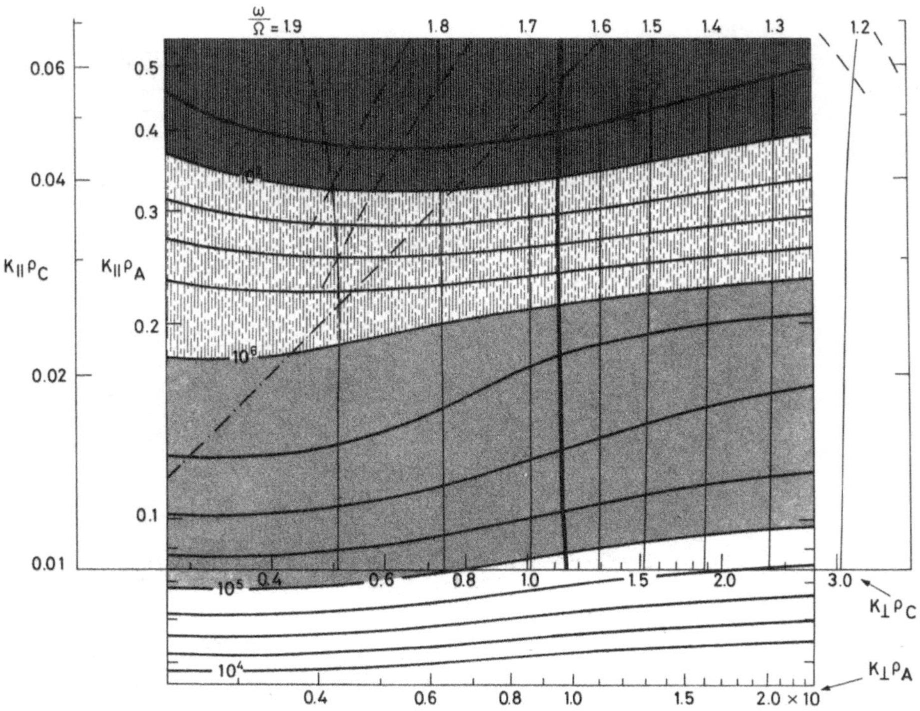

Fig. 5. A combination of Figures 4 and 1, which can be used to identify the particles interacting with a given wave, or conversely the wave interacting with a given group of particles. The velocity distribution is shaded between decade contours to emphasize the loss cone structure.

with $\partial f/\partial v_\perp > 0$ in Figure 4 correspond well to regions with $\gamma > 0$ in Figure 6. Much of the information about unstable frequencies and wave vectors can thus be obtained from Figure 5, without solving the full dispersion relation.

To summarize, the procedure outlined above consists of the following steps:

(i) Solve the dispersion relation $\varepsilon(\omega_R, \mathbf{k}) = 0$ for an isotropic plasma with given ratio ω_p/Ω and plot the result as in Figure 1.

(ii) Make a contour plot of the measured distribution function F or the model f, and introduce scales for $k_\parallel \rho_A$ and $k_\perp \rho_A$ as defined by Equation (8).

(iii) Combine the two plots as in Figure 5, with the relative position adjusted for the desired ratio of T_A/T_C.

(iv) Find potentially unstable frequencies and wave vectors by localizing regions where $\partial F/\partial v_\perp > 0$.

(v) If knowledge of growth rates is desired, solve the full dispersion relation for interesting combinations of ω and $\mathbf{k}$.

4. Discussion

Using the theoretical formulation described in the previous section it is now possible to assess the degree to which the predictions of the model agree with the

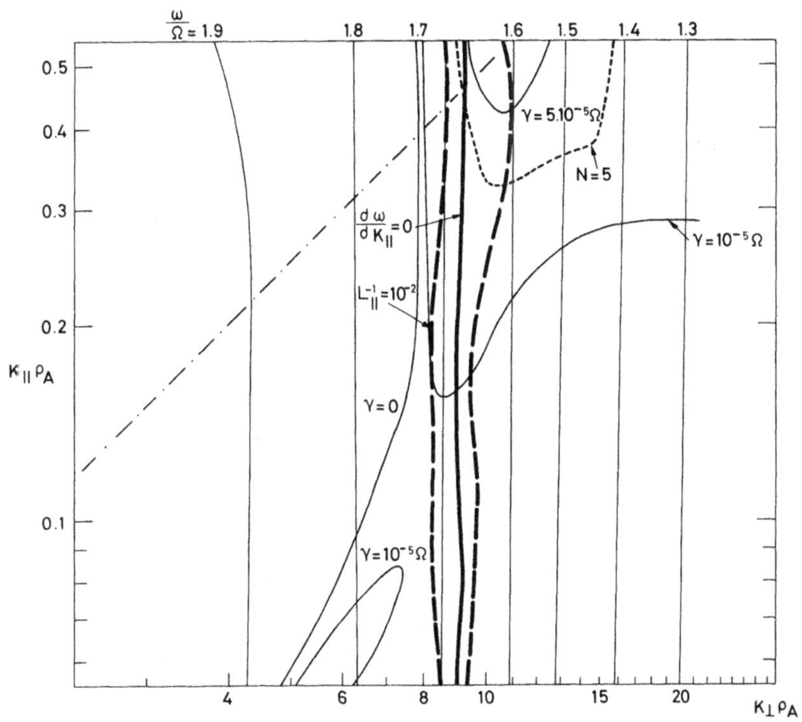

Fig. 6. Solution of the dispersion relation for the lowest instability band. Contours of constant Re ω as well as the contours of constant temporal growth rate γ are shown as thin lines. The dotted lines show the contours where the inverse amplification length parallel to B_0, $L_{\parallel}^{-1} = \gamma_A^V/(\Omega(\partial\varepsilon/\partial\omega)) = 10^{-2}$. The curve $N = 5$ is discussed in the text.

observed spectrum of the instability. To do so we examine the model dispersion relation in greater detail. The input parameters are: $\omega_{uh}/\Omega = 6.4$, $n_c/(n_A + n_B) \sim 30$, $T_C = 5$ eV, and details of the hot distribution function are included in the caption to Figure 3.

In Figure 6, the results for the lowest band ($1 < \omega/\Omega < 2$) of ECH wave propagation are shown plotted in the $\mathbf{k}\rho_A$ plane. Contours of constant temporal growth rate are shown as thin solid lines, and it can be seen, for example, that growth ($\gamma > 0$) occurs for $k_\perp\rho_A \sim 7.5$ for the whole range parallel wavenumbers and for frequencies below $\omega/\Omega \sim 1.75$. The single contour $\gamma = 5 \times 10^{-5}\Omega$ is particularly interesting as it surrounds the region of maximum growth rate, defined by the parameter range $9 < k_\perp\rho_A < 14$, $k_\parallel\rho_A > 0.5$, $1.7 \geqslant \omega/\Omega \geqslant 1.4$. However the maximum temporal growth rate is not the best characterization of the fastest growing mode, since for this frequency band, the instability is convective. By this it is meant (Sturrock, 1958; Briggs, 1964) that a localized disturbance in this frequency band amplifies as it propagates away from its source, but at a fixed point in space the amplitude eventually decays with time.

For $\gamma/\Omega \ll 1$ the spatial growth rate K_I, normalized to the hot plasma larmor radius has magnitude

$$\|K_I\| = \frac{\gamma \rho_A}{\{(\partial\omega/\partial k_\perp)^2 + (\partial\omega/\partial k_\parallel)^2\}^{\frac{1}{2}}}, \qquad \gamma > 0$$

and the amplification lengths parallel and perpendicular to the magnetic field are, respectively

$$L_\parallel = \frac{\partial\omega}{\partial k_\parallel}\frac{\Omega}{V_A \gamma}, \qquad L_\perp = \frac{\partial\omega}{\partial k_\perp}\frac{\Omega}{V_A \gamma}.$$

In order to find the modes with maximum spatial growth it is not sufficient to use the combination of γ and the group velocity v_g, which maximizes K_I; one is also restricted to directions of energy convection (ray paths) which are physically reasonable for magnetospheric wave propagation. Based on the evidence that $(n+\frac{1}{2})\, f_c$ ECH waves on the dayside are confined to within a few degrees of the geomagnetic equator (Christiansen *et al.*, 1978), we adopt for later use the suggestion of Ashour-Abdalla and Kennel (1976) that $v_{g\parallel}$ should be low enough to confine the waves to within $1R_E$ parallel to the magnetic field (B_0) while they propagate up to $3R_E$ in azimuth.

Returning to Figure 6 we note that the line of zero parallel group velocity

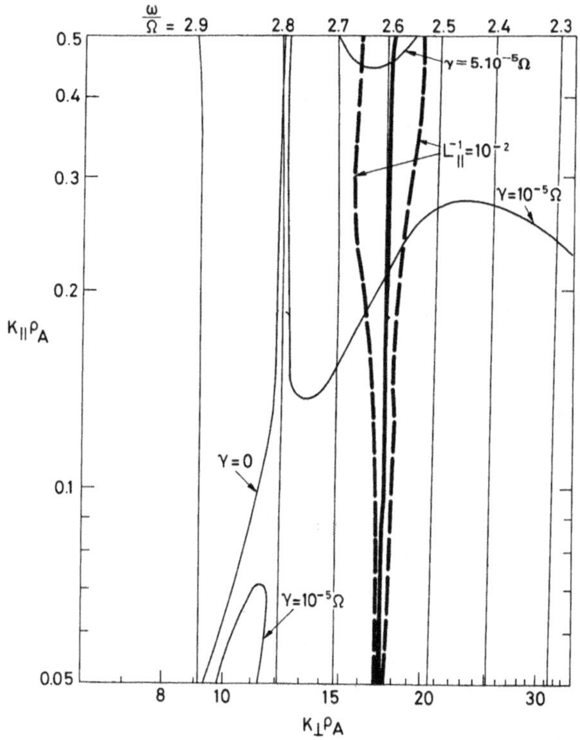

Fig. 7. Solution of the dispersion relation for the second instability band. The notation is the same as in Figure 6.

lies between $\omega/\Omega = 1.6$ and 1.7. On either side of it are contours delineating the region in which $L_{\parallel}^{-1}$ exceeds 10^{-2}, corresponding to $v_{g\parallel} \sim 5 \times 10^{-3}\ V_A$. A quick inspection of the diagram shows that in the temporal amplification region ($\omega/\Omega \sim 1.6$), $v_{g\perp} \sim 6 \times 10^{-2}\ V_A$, and that $v_{g\perp}$ decreases slightly towards lower frequencies, compensating for the effect of finite $v_{g\parallel}$ at these frequencies. The direction of energy propagation and maximum spatial growth is therefore predominantly perpendicular to the magnetic field.

To illustrate this conclusion with greater precision (while avoiding the detailed ray tracing in a model magnetosphere that will ultimately be necessary) the following model has been examined. Convective amplification is assumed to occur in a region $1R_E$ parallel and $3R_E$ perpendicular to B_0, and the frequencies of waves which have exponentiated five fold by the time they leave the region have been calculated. The frequency band so defined is shown by the line labelled $N = 5$.

As can be seen from Figure 7, a similar situation exists for the second frequency range $2 < \omega/\Omega < 3$, again with spatial growth rates approximately equal to those of the first band.

Dispersion in the harmonic band containing the upper hybrid frequency is qualitatively different as can be seen from Figure 8. The real part of the frequency

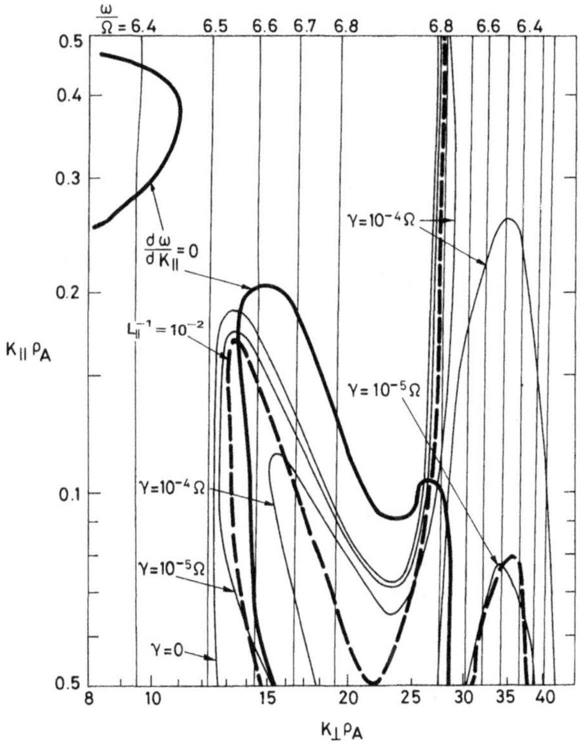

Fig. 8. Solution of the dispersion relation in the region of ω_{uh}. Note that at $\omega/\Omega \sim 6.8$, both $v_{g\parallel}$, $v_{g\perp} \approx 0$.

exhibits the well known double valued structure in $k_\perp$, and in the vicinity of $\omega/\Omega \times 6.8$, $19 < k_\perp \rho_A < 27$, *both* the parallel and perpendicular components of the group velocity are almost zero, with $\gamma > 0$. This corresponds to a region of absolute, or non-convective instability, in which the waves amplify in time at any fixed point in space. The saturation level of such an instability is determined by quasi-linear (Ashour-Abdalla and Kennel, 1978) and non linear effects which are beyond the scope of the linear theory, but it is reasonable to suppose that it will be considerably higher than that reached by convective instabilities.

In summary the predictions of the model which are relevant to an experimental measurement of wave spectra are as follows:

(a) In the two lowest harmonic bands, convective instabilities with comparable spatial growth rates exist, and are expected in the frequency ranges $1.7 > \omega/\Omega > 1.4$, $2.75 > \omega/\Omega > 2.5$. The amplification lengths of these instabilities, (a few thousands of kilometers) are such that this final amplitude could well be determined by convection out of the amplification region. Though higher bands have not yet been examined in detail there is good reason to believe that they are also convectively unstable.

(b) In the region $\omega/\Omega \sim 6.8$ just above the upper hybrid frequency, large amplitude waves are expected since there the instability is absolute, and its final amplitude will be limited by quasi-linear and non linear processes.

These results can now be compared with measured frequency spectrum, shown in Figure 9. The spectrum was recorded in the period 09:50–09:53 UT during which particle data was taken, and is the average of 8 sweeps of the full frequency range 0–77 kHz. The series of emissions between the first four gyroharmonics

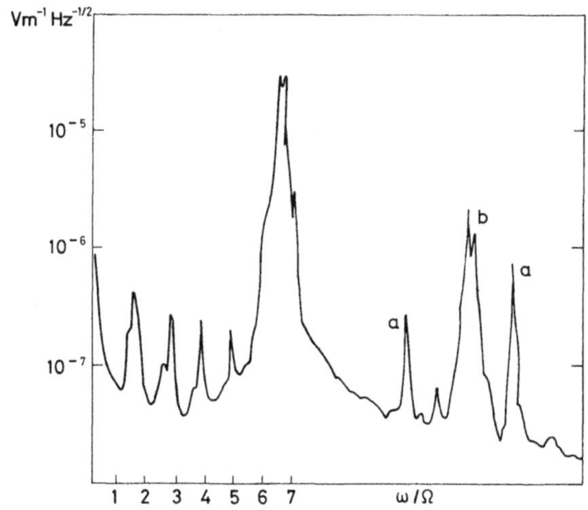

Fig. 9. Spectral amplitude against frequency averaged over 8 SFA sweeps between 09:50–09:53 UT. The spectral lines labelled *a* are interference lines, and the second harmonic of the strong instability line is labelled *b*.

have amplitudes of $\sim 3\times10^{-7}$ V m^{-1} Hz$^{-\frac{1}{2}}$ and the frequencies at the peaks are $\omega/\Omega = 1.58$, 2.8 respectively for the first two bands. The dominant spectral feature is in the vicinity of ω_{uh}, with an amplitude 40 dB greater than the lower frequency emissions and with a frequency $\omega/\Omega = 6.6$.

In several aspects this preliminary study can therefore be regarded as a success:

(a) The observed frequencies of instability are close to those predicted by the model. (The few percent discrepancy in the region of ω_{uh} is within the experimental uncertainty of the plasma density measurement made during the event).

(b) The observed amplitudes are in broad agreement with arguments based on the existence of convective and absolute instabilities in different frequency ranges.

(c) The basic theoretical requirements for instability, an electron velocity distribution which possesses a cool component and a hot, loss cone component have been experimentally verified.

(d) In the plasma region where the cold component is completely dominant, models of this kind can be expected to account for the multi-banded emissions observed from GEOS-1 in the dayside magnetosphere (Christiansen *et al.*, 1978).

We ought, however, to add a few notes of caution against over interpreting this single result.

(i) The event chosen is not typical of the class of $(n+\frac{1}{2})\,f_c$ emissions observed on the dayside by GEOS-1. For example, the wave amplitudes in the lower gyroharmonic bands are somewhat more depressed, and the upper hybrid feature stronger than is typical. It also occurred on a disturbed day, and rather further off the geomagnetic equator than the average.

(ii) The averaging of wave spectra used to produce Figure 10 smooths out considerable amplitude and frequency variation which occurs on a timescale short compared with the particle measurement period (see Figure 1). We know that these are closely related to significant variations in the distribution of low energy electrons (G. Wrenn, J. Sojka, J. Johnson, private communication) as detected by the S-302 experiment, and it is obviously necessary to include as accurate as possible a description of the thermal component in future elaborations of this model, because of its crucial role in the determination of the ECH wave dispersion.

(iii) A linear theory of the kind described is only a first step towards a description of such wave phenomena. A determination of the actual level of fluctuations requires realistic ray tracing and a consideration of quasi-linear and non-linear processes.

5. Conclusions

A theoretical model whose inputs are based firmly on experimental data has been developed together with a novel graphical technique which provides a physical insight into the process of cyclotron harmonic instability.

When applied, for the first time, to a particular wave and particle event the results are encouraging in that they successfully describe certain aspects of the natural emissions and open the way to more detailed comparisons of theory and experiment.

Acknowledgements

This work has been performed within a large GEOS-1 collaboration and our colleagues have contributed to it in many ways. We wish to thank B. Hultqvist (Kiruna Geophysical Institute); G. Martelli (University of Sussex); G. Wrenn, J. Sojka, J. Johnson (Mullard Space Science Laboratory); N. Cornilleau, J. Etcheto, R. Gendrin, B. de la Porte (CRPE, Issy-les-Moulineaux); C. Béghin, P. Décréau (C.R.P.E., Orleans); R. Grard, K. Knott (Project Scientist), A. Pedersen (S.S.D., E.S.A.), A. Bahnsen, E. Ungstrup (Danish Space Research Institute); H. Reme, J. M. Vigo (C.E.S.R., Toulouse University).

The S-310 experiment was conducted by Kiruna Geophysical Institute, in cooperation with C.E.S.R. (Toulouse), the Danish Space Research Institute (Lyngby) and Max-Planck-Institut für Aronomie (Lindau and Garching); the GEOS Low Energy Particle Cooperation - GLEPCO. The contributions by the technical staffs at KGI, DSRI and CESR to the success of this experiment, in particular those of S. Olsen and S. Laursen, are gratefully acknowledged.

The magnetometer data was provided by the S-331 experiment (principal investigator, F. Mariani, University of Rome).

We also wish to thank E. Gershuny (Sussex University), D. Barria, G. Charles, C. Huc (C.H.E.S. Toulouse) and E.S.O.C., Darmstadt, for their contributions to the data handling.

The research is supported by the U.K. Science Research Council, the Swedish Board for Space Activities, and E.S.A.

References

Ashour-Abdalla, M. and Kennel, C. F.: 1976, in B. M. McCormac (ed.), *Magnetospheric Particles and Fields*, D. Reidel Publ. Co., Dordrecht, p. 181.

Ashour-Abdalla, M. and Kennel, C. F.: 1978, *J. Geophys. Res.* **83,** 1531.

Ashour-Abdalla, M., Chanteur, G., and Pellat, R.: 1975, *J. Geophys. Res.* **80,** 2775.

Brambilla, M.: 1970, *Plasma Phys.* **12,** 903.

Briggs, R. J.: 1964, *Electron-Stream Interactions with Plasmas*, MIT Press, Cambridge, Mass.

Borg, H., Holmgren, L. A., Hultqvist, B., Cambou, F., Reme, H., Bahnsen, A., and Kremser, G.: *Space Sci. Rev.*, **22,** 511.

Christiansen, P., Gough, P., Martelli, G., Bloch, J. J., Cornilleau, N., Etcheto, J., Gendrin, R., Jones, D., Beghin, C., and Decreau, P.: 1978, *Nature* **272,** 682; and also *Space Sci. Rev.* **22,** 383.

Dory, R. A., Guest, G. E., and Harris, E. G.: 1965, *Phys. Rev. Letters* **14,** 131.

Fredericks, R. W.: 1971, *J. Geophys, Res.* **76,** 5344.

Fredericks, R. W. and Scarf, F. L.: 1973, *J. Geophys. Res.* **78,** 310.

Hall, L. S., Heckrotte, W., and Kammash, T.: 1965, *Phys. Rev.* **139A,** 1117.

Hubbard, R. F. and Birmingham, T. J.: 1978, *J. Geophys. Res.*, to be published.

Johnson, J. F. E., Sojka, J. J., and Wrenn, G. L.: *Space Sci. Rev.*, **22,** 576.

Jones, D. (ed.): *Space Sci. Rev.* **22,** 327.

Karpman, V. I., Alekhin, Ju. K., Borisov, N. D., and Rjabova, N. A.: 1975, *Plasma Phys.* **17,** 361.
Kennel, C. F., Scarf, F. L., Fredricks, R. W., McGhee, J. H., and Coroniti, F. V.: 1970, *J. Geophys. Res.* **75,** 6136.
Lyons, R. L.: 1974, *J. Geophys. Res.* **79,** 575.
Shaw, R. R. and Gurnett, D. A.: 1975, *J. Geophys. Res.* **80,** 4259.
Sturrock, P. A.: 1958, *Phys. Rev.* **112,** 1488.
Tataronis, J. A. and Crawford, F. W.: 1970a, b, *J. Plasma Phys.* **4,** 231 and 249.
Young, T. S., Callen, J. D., and McCune, J. E.: 1973, *J. Geophys. Res.* **78,** 1082.

VLF WAVES: CONJUGATED GROUND–SATELLITE RELATIONSHIPS

NICOLE CORNILLEAU-WEHRLIN and ROGER GENDRIN

Département Etudes par Télédétection de l'Environnement, Centre National d'Etudes des Télécommunications, 92131 Issy-les-Moulineaux, France

and

MICHEL TIXIER

Laboratoire de Physique de la Haute Atmosphère, Université de Poitiers, 86022, Poitiers, France

Abstract. The French mobile station for recording geophysical data has been put in operation at Husafell, Iceland (64°5′N, 20°8′W) between the 10th of July and the 22nd of September, 1977. This place was more or less conjugated with GEOS when this satellite was near its apogee. The equipments installed in the station for recording VLF and ULF phenomena have characteristics (band-pass, sampling rates) which are identical to the similar equipments installed onboard GEOS. Intercomparison between signals recorded at both points are therefore easy. We present here the results which were obtained in the VLF range.

In many occasions, VLF emissions (mainly hiss) do present identical variations in amplitude, with a very abrupt (<1 mn) and very large (>20 dB) decrease in amplitude. Because of their simultaneity at both points, such abrupt variations cannot be interpreted in terms of a sudden ionospheric absorption (associated with an enhanced particle precipitation) nor in terms of a sudden crossing of detached plasma regions. In some cases, these abrupt changes in the VLF intensity are associated with the appearance and disappearance of strong ULF emissions, in the Pc-1 frequency range. Some examples of associated onboard measurements of high energy electron fluxes or cold plasma density (when available) are given, which may help understanding these VLF conjugated relationships.

1. Introduction

One of the main purposes of GEOS was to study the ground–satellite relationships with respect to both particle precipitation and wave propagation. Would GEOS-1 have been geostationary, the first category of measurements would have led to a much larger number of experiments; but taking into account the actual orbit of GEOS-1, it was unrealistic to schedule many GEOS-associated rocket launches. In some respect, the same is true for wave propagation studies: the time during which the satellite and ground receivers remain on the same field line being short, it is difficult to study the long term fluctuations of wave intensity, frequency structure, polarization etc. which could occur due to temporal variations of the plasma properties, either at the satellite position, or along the flux tube, or in the ionosphere above the ground station. However, we know that waves propagate below the ionosphere over large distances (≤500 km for VLF and ≥1000 km for ULF waves). It is also suspected that some may propagate within the magnetosphere outside the tube of force in which they are generated, although the azimuthal or radial extension of this kind of propagation is not precisely known.

Space Science Reviews **22** (1978) 419–431.

Therefore we felt worthwhile to have a conjugated ground campaign, even in conjunction with a non-geostationary satellite: first the orbit was such that GEOS-1 had a low drift velocity near its apogee; second, some VLF rockets were scheduled to be launched from Syowa Base, Antarctica (geographic: 69.0°S, 39.6°E; geomagnetic: 69.7°S, 77.7°E). Consequently we decided to make a 3 months ground campaign at Husafell, Iceland (geographic: 64.5°N, 20.8°W; geomagnetic: 70.2°N, 74.3°E). During this period the apogee of GEOS was placed at ~6.5°E (within 1°), so that the satellite was more or less conjugated, over a large period of time, with both ground stations (Figure 1).

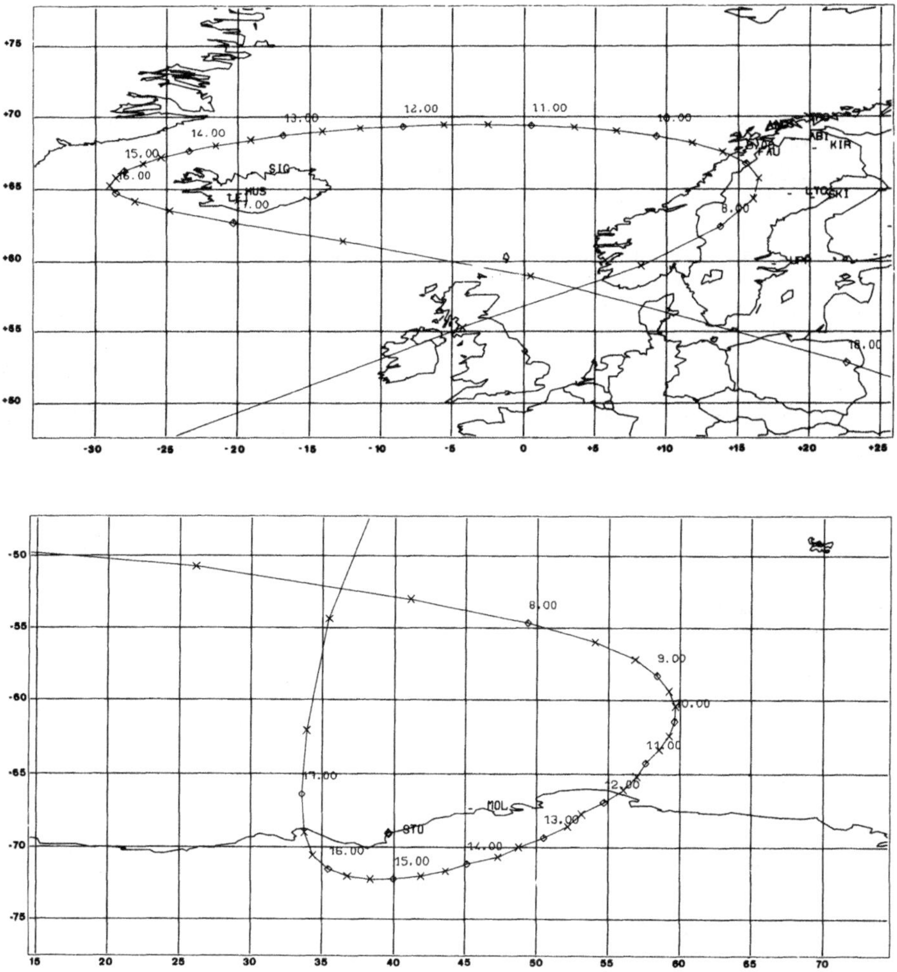

Fig. 1. Magnetic footprints of GEOS-1 in both hemispheres for August 16, 1977. Each footprint is labelled in Universal Time with a spacing between points equal to 20 min. At different dates, the footprint trajectory is not much different, but the labelling has to be modified in such a way that 30 days later, the Universal Time must be increased by approximately 90 min (computations are due to the courtesy of J. C. Kosik).

The ground campaign at Husafell took place the 10th of July and ended the 22nd of September, 1977. The French mobile station was used, whose technical characteristics are described elsewhere (Perraut *et al.*, 1978a). Simultaneous measurements were done at Husafell by the Japanese and by some British colleagues (Madden *et al.*, 1978). The results obtained in the ULF range during this conjugated campaign are described in another paper of this issue (Gendrin *et al.*, 1978). We will concentrate here on the VLF results.

2. Problems

The simultaneous measurements of VLF waves on the ground and in the conjugated equatorial region may bring answers to the following questions:

(1) Are the fluctuations in the VLF power observed on the ground the image of similar fluctuations occurring in space, or are they indicative of strong variations in the ionospheric absorption, which would prevent, from time to time, the waves from reaching the ground? In the first case we should observe similar variations onboard the satellite. In the second case, we should observe almost constant signal onboard the satellite, while the signal is fluctuating on the ground.

(2) Similarly, are the VLF fluctuations detected onboard a drifting satellite the result of a spatial or of a temporal variation? In the first case, waves should continue to be observed on the ground (because of their large extension there, due to subionospheric propagation), whereas in the second case, waves should be interrupted simultaneously on the ground and onboard the satellite.

(3) Do the local plasma parameters (cold plasma density, high energy particle flux and anisotropy) play the dominant role in determining the VLF turbulence level, or is the ionospheric reflexion coefficient an important parameter, as suggested by self-consistent theories (Kennel and Petschek, 1966; Etcheto *et al.*, 1973)? In the first case, the level of the signal detected onboard the satellite should only depend on the values of the local parameters. In the second case, it should also depend on the reflexion efficiency, which can be indirectly appreciated (a good reflexion coefficient being associated in most cases with a bad transmission coefficient).

(4) What is the exact role which is played by ULF waves associated with some characteristic VLF emissions, such as the quasi-periodic (QP) emissions (Kitamura *et al.*, 1969; Sato, 1974; Haugstad, 1976)? Do they originate in the ionosphere as a consequence of some ionospheric current instability? Do they modulate the ionsopheric reflexion coefficient, henceforth inducing a modulation of the VLF level by the process discussed above? Or do they modulate the VLF turbulence itself, in the generation region, as suggested by some theories (Coroniti and Kennel, 1970; Hagege *et al.*, 1973)?

The results which are presented here do not provide definitive answers to these questions. However we hope that we will demonstrate, on the basis of examples, that conjugated campaigns between a geostationary satellite and the ground give a unique opportunity to progress in the understanding of these complex mechanisms.

3. Results

The material which we use in this study consists of all the records (wide band and narrow band data) of the mobile station, of the VLF filter bank data from GEOS, of plasma density measurements deduced from the active part of the S-300 experiment (Etcheto and Bloch, 1978) and of some of the S-321 high energy electron data (Korth *et al.*, 1978) which were kindly communicated to us by the authors. Onboard GEOS, there are other ways than the filterbank to get VLF data (see Jones, 1978; Cornilleau-Wehrlin *et al.*, 1978), but this high resolution part of the experiment is not operating continuously in the same mode. In order to better illustrate the temporal variations of the signal over long periods of time, we found more convenient to present the results in a common format; hence we used only the filter bank for the GEOS data whereas, occasionally, some wide band spectrograms obtained in real time at the ground station are also presented.

The August 9, 1977 event is the one with which we will start our presentation, because it gives a very clear example of the similarity which may exist between VLF waves detected on the ground and onboard the satellite (Figure 2). The very abrupt disappearances of the VLF signal on the ground (and onboard GEOS) is very striking. Would the ground data be the only one available, we would have concluded that the noise disappearances was due to ionospheric absorption induced by some strong particle precipitation. The availability of simultaneous magnetospheric measurements shows that this is not the case. Therefore we have to enquire about the reasons why the VLF instability has stopped abruptly and started again, and stopped again and so on.

The simultaneous measurements of high energy electrons ($E > 21$ keV) allows us to answer, at least partially, this question. Figure 3 shows the associated fluctuation of the integral flux of electrons, above 20 keV, taken in four directions with respect to the spin axis ($\phi = 10°$, $\phi = 30°$, $\phi = 50°$ and $\phi = 90°$). We see that an abrupt decrease of this flux near 1425 UT is associated with the sudden decrease of the VLF intensity. Subsequent partial recoveries of the original flux are associated with subsequent enhancements of the VLF noise. We were fortunate this day, to have an almost regular survey (one measurement every 6 mn) of the cold plasma density, and we see that this density does not present any fluctuation associated with the VLF power.

The August 6, 1977 event gives another example of these assoicated VLF and particle oscillations (Figure 4). The noise disappearances, near 1415, 1450 and 1510 UT, which again are seen both on the ground and onboard GEOS, are clearly associated with troughs in the high energy electron flux. Near 1530 there is even a one-to-one correspondence between the electron flux and VLF intensity oscillations. The active experiment S-301 was not continuously working during this period, but indirect measurement via the DC experiment shows that the

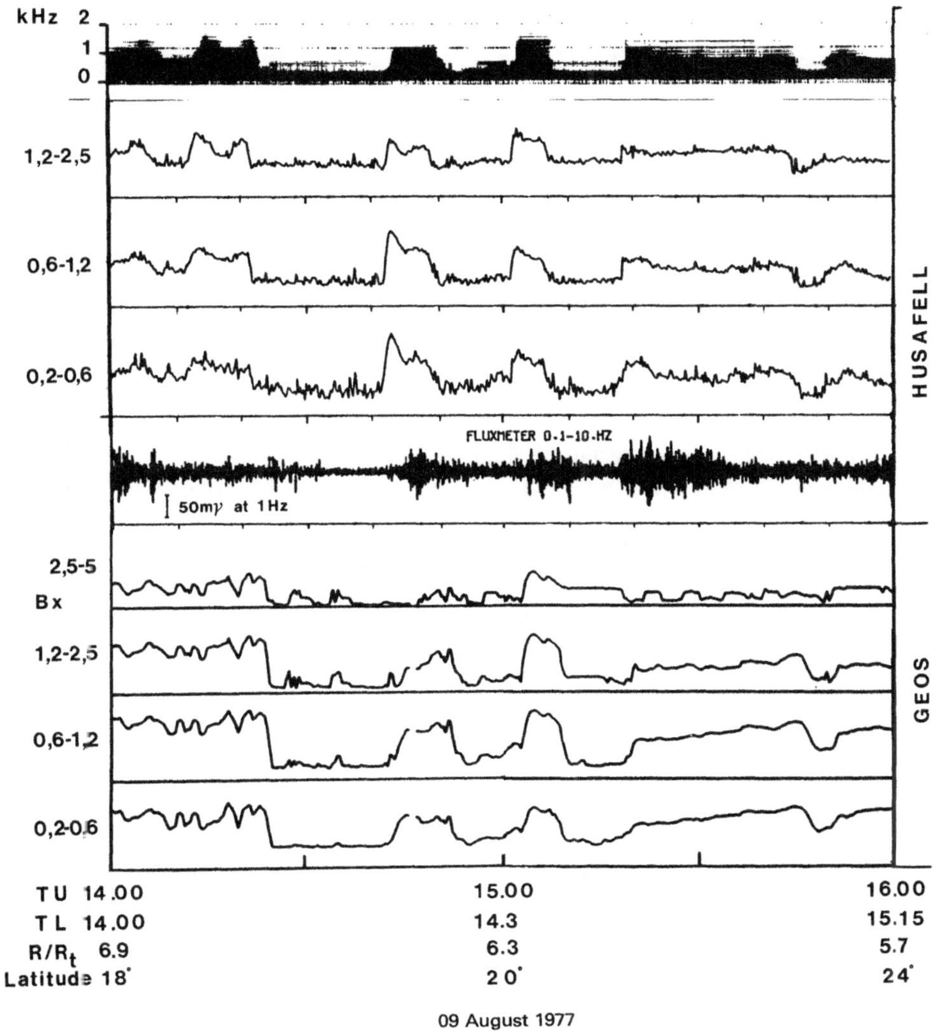

Fig. 2. Abrupt changes of VLF intensity simultaneously observed onboard GEOS and on the ground, in different frequency channels. From top to bottom: Spectrogram of the emission as observed in real time on the ground; 3 ground filter bank outputs; 1 ground ULF wave form; 4 GEOS filter bank outputs.

plasma density is not fluctuating between 15:00 and 16:00 UT (A. Pedersen, private communication). However, at about the same time, the correlation with the ground record becomes poorer, which may indicate that waves generated on the GEOS flux tube had difficulties to reach Husafell below the ionosphere (see on Figure 1 the position of the GEOS footprint at that time).

It must be mentioned that on some other occasions, similar fluctuations in the GEOS filter bank data were observed, which were not associated with high energy electron fluctuations but which were clearly associated with fluctuations of the *cold* plasma density (see Figures 7 and 8 of Cornilleau-Wehrlin *et al.*'s (1978)

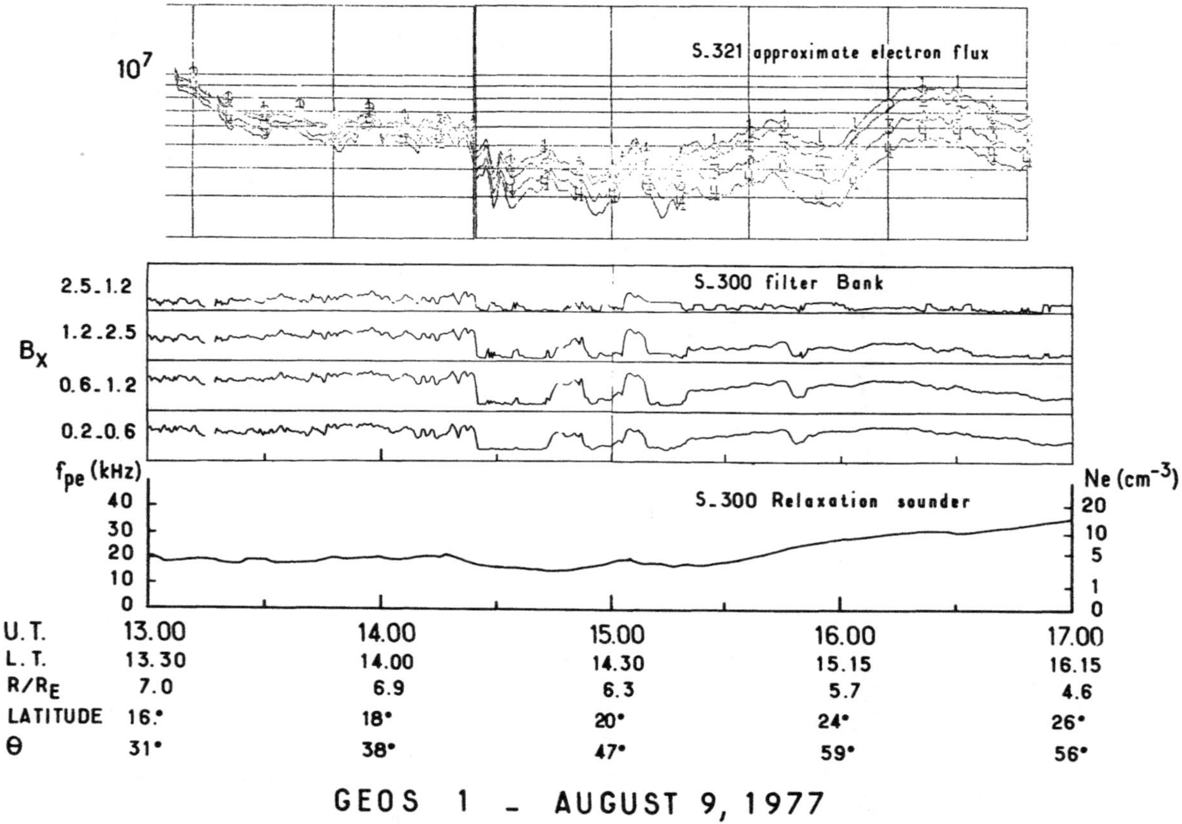

Fig. 3. Associated measurements of the high energy electron flux (upper panel), GEOS filter bank data for the magnetic component B_x (middle panels), and cold plasma density (lower panel). On this figure, as well as on the following ones, the high energy electron data were kindly provided to us by K. Korth *et al.*, and the cold plasma density by J. Etcheto *et al.*

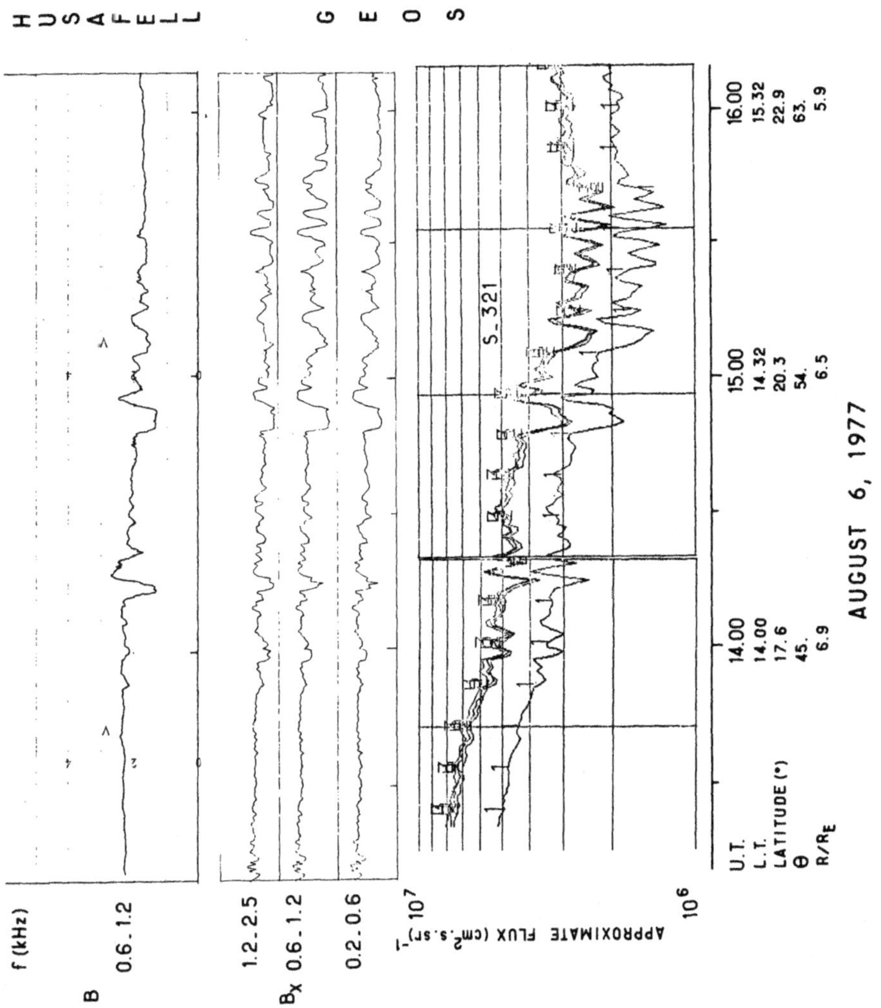

Fig. 4. Simultaneous fluctuation of VLF emissions detected on the ground (upper panel) onboard GEOS (middle panel) and of high energy electrons (lower panel). Note the almost one-to-one correspondence between the two last quantities around 1530.

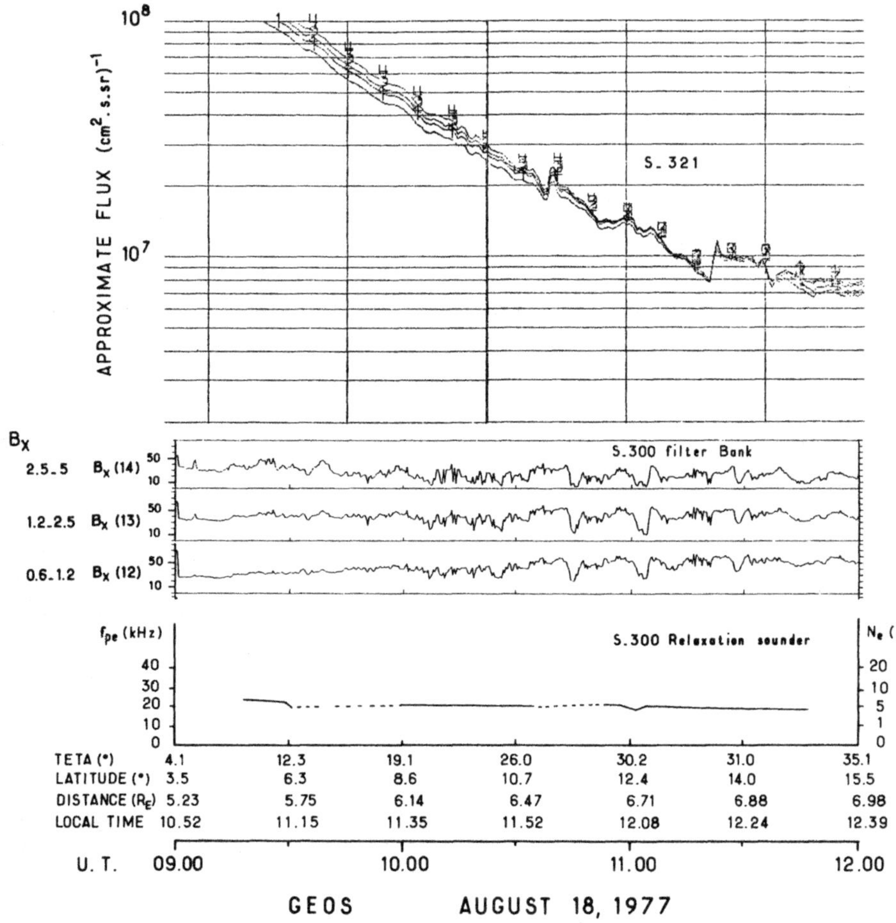

Fig. 5. The August 18, 1977 event. There is no clear relationship between the high energy particle modulation (upper panel), the GEOS filter bank data (middle panels), and even less with the cold plasma density (lower panel). On this picture, the angle θ between the spin axis and the magnetic field has a small value at the beginning of the event. Consequently detector 4 is looking almost perpendicularly to the magnetic field and we verify that the count rates are higher for this detector than for the others.

paper. These observations were intrepreted in terms of GEOS crossing regions of detached plasma. In such region, according to Etcheto *et al.*'s (1973) theory, there should be an enhancement of the VLF hiss intensity. If this is the case we should observe the enhanced hiss onboard GEOS only when it crosses such detached plasma regions, but the noise should be received continuously on the ground. It would have been interesting to see if the ground VLF intensity was varying or not during these two passes. Unfortunately, these two clear events (June, 16 and December, 16, 1977) occurred outside the period of our ground campaign.

Conversely, there are some fifteen events, which occurred during our ground campaign, in which a modulation was observed onboard GEOS with no associated

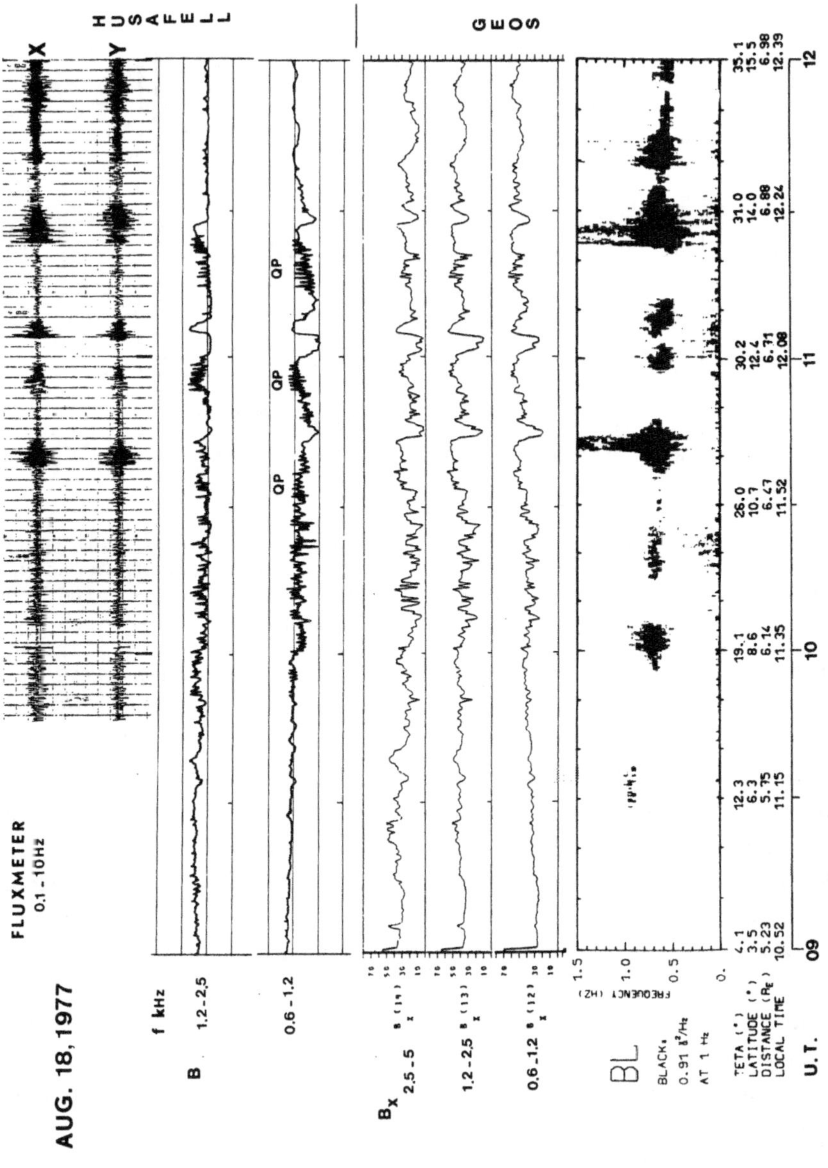

Fig. 6. Associated ULF and VLF emissions during the August 18, 1977 event. The ULF bursts seem to precede, by 3 mn or so, the large drops in the VLF intensity.

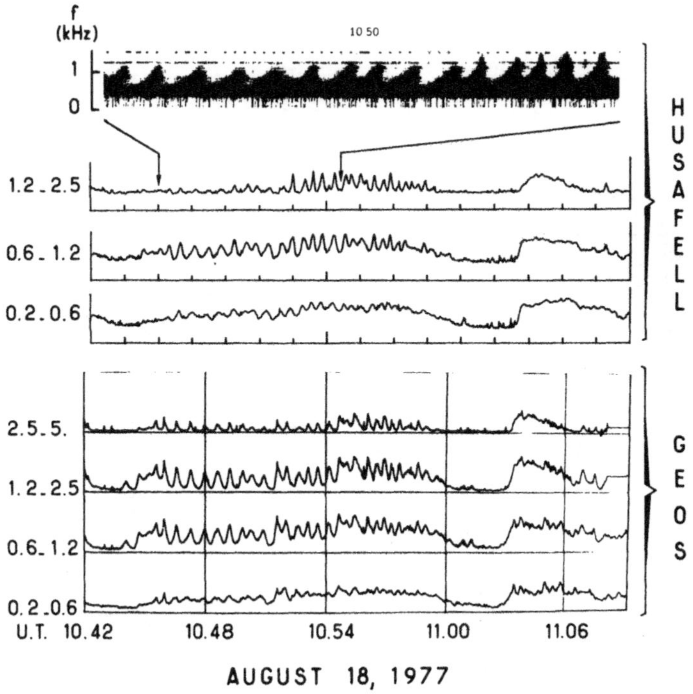

Fig. 7. Quasi-periodic emissions observed during the August 18, 1977 event. Note that the GEOS filter bank data (lower panels) show more structure on this figure than on the two previous ones. This is due to the fact that data were plotted at their maximum sampling rate ($5.8\,s^{-1}$) and were not averaged over 22 s as on the routine summaries.

ground modulation. Unfortunately, for all these cases, we had no access to cold plasma density measurements, at least with a sampling rate sufficiently short compared to the VLF modulation period. Therefore we have no formal proof, at present, for drawing the following conclusion, which we think however valid:

(a) Modulations which are observed simultaneously on the ground and onboard GEOS are associated with high energy electron flux modulations.

(b) Modulations which are observed only onboard GEOS are due to crossings of detached plasma regions.

In any case, as was said in Section 2, the cold plasma density and the high energy flux are not the sole parameters intervening in the generation of VLF hiss. This is demonstrated in Figure 5, which corresponds to the event of August, 18, 1977. Clearly there is no important fluctuation in the electron flux whereas there is a large modulation of the VLF hiss. Simultaneously, the cold plasma density (which was available by chance!) remains practically constant. Therefore, what could be the origin of this VLF modulation? When we look at what was occurring on the ground, and within different frequency bands, we see different things (Figure 6):

(a) the VLF emissions were in fact quasi-periodic emissions. We will come back to this point later.

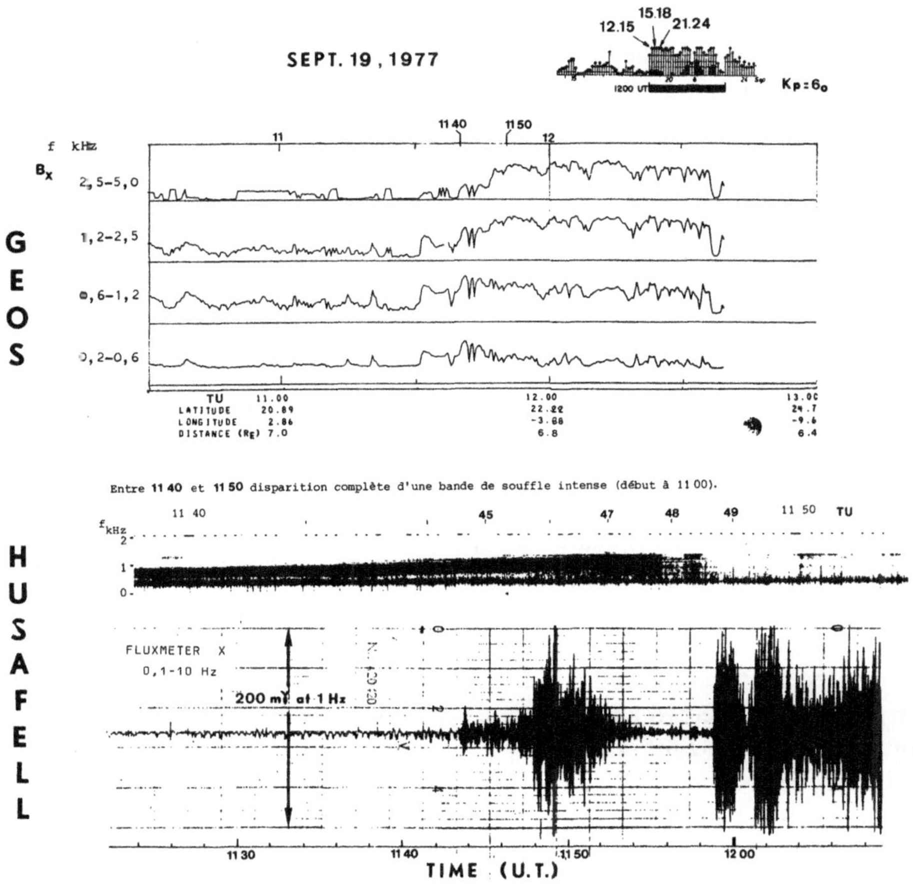

Fig. 8. A strong ionospheric VLF absorption event, associated with geomagnetic Pi's. From top to bottom: K_p indices, GEOS filter bank data, ground VLF spectrogram and ground record of geomagnetic pulsation. Around 1149 UT, there is an abrupt disappearance of the VLF signal as detected on the ground, clearly associated with the beginning of a strong Pi event. Onboard GEOS, the VLF emission is still present, but its mean frequency is increasing.

(b) There were strong ULF emissions, detected both on the ground and onboard GEOS (Perraut *et al.*, 1978b).

These ULF waves could be at the origin of the disappearances of the VLF noise in two different manners. For instance, they could quench, in the equatorial region, the VLF turbulence although most of the theoreticians suggest that ULF instabilities enhance the VLF turbulence (Coroniti and Kennel, 1970; Hagege *et al.*, 1973). However, if this hypothesis were the right one, the quenching should occur at the time of the ULF emission and not after. The second possible scenario works as follows: the ULF emissions are associated, via some mechanism with an ionospheric perturbation whose effect is to decrease the ionospheric reflexion coefficient. Consequently, the self consistent equilibrium which is needed for bringing the VLF turbulence to a sufficient level (Kennel and Petschek, 1966) is destroyed. If this is the case, there should be a time delay between the appearance of the ULF waves and the disappearance of the VLF signal, this time being of the order of the time of establishment of the equilibrium, i.e., of the order of 200 s (Etcheto *et al.*, 1973), which is approximately what we observe.

Figure 7 gives, on an extended time scale, the outputs of the filter banks and the corresponding ground spectrogram of these quasi-periodic (QP) emissions. The periodicity is visible on the GEOS filter bank in this Figure, because the data were plotted at the highest sampling rate ($5.8\ s^{-1}$), as opposed to Figure 6 where they were averaged over 22 s. This emission is a QP of Type I, according to Sato's (1974) nomenclature, because it lasts only a few tens of minutes and its periodicity is varying (from ~40 s at 1048 to ~25 s at 1058 UT). We have not yet analyzed the satellite data to see if some Pc3 activity was associated with these QP emissions. On the ground, this Pc3 activity existed but it was weak. The study of these QP emissions (Type I and Type II) and of their ground–satellite relationships is now in progress on other examples.

Finally, a typical ionospheric absorption effect is presented in Figure 8. This corresponds to the September, 19, 1977 storm (between 1200 and 1500, K_p was equal to 6). Starting around 1130, a strong VLF emission is observed which is seen both onboard GEOS and on the ground, at least until 1148. At that time a strong precipitation of particle occurs at Husafell, as evidenced by the intense Pi's, seen on the ULF record and which are typical of auroral precipitations. Examination of the GEOS ULF data shows that similar bursts, with an approximate repetition period of ~10 mn were present in the magnetosphere, with a large amplitude (~1γ) and a frequency which occupied a band between 0.7 and 1.2 Hz. These precipitations induce a complete black out on the ground, for some 5 mn, as far as the VLF emissions are concerned, whereas the level of the emission is still high at GEOS. Later on, the emission remains strong at GEOS, weaker on the ground, and its mean frequency increases.

4. Conclusion

These preliminary results show that the relationships between conjugated ground and satellite VLF emissions are rather complex. They depend on the cold plasma density, on the high energy electron distribution and on some other parameters. The role which is played hy associated ULF emissions has been evidenced, although the precise mechanism at work has not been established. More quantitative studies are needed in order to better understand the generation process of VLF emission and its interrelation with ULF turbulence, with ionospheric properties and with local plasma parameters.

Acknowledgements

We are grateful to Dr J. Etcheto (Centre de Recherches en Physique de l'Environnement, Issy-les-Moulineaux, France), to Dr Korth (Max Planck Institute, Lindau, Germany) and to their colleagues for having provided us respectively with cold plasma density and high energy electron flux measurements. We also acknowledge the work of Messrs. G. Laurent, J. P. Meunier, A. Meyer and P. Lafon and of Mrs D. Fouassier, who helped us during the operation of the mobile station at Husafell or in the data handling afterwards. The ground campaign was supported by a special contract of the Centre National de la Recherche Scientifique for the IMS.

References

Cornilleau-Wehrlin, N., Gendrin, R., Lefeuvre, F., Parrot, M., Grard, G., Jones, D., Bahnsen, A., Ungstrup, E., and Gibbons, W.: 1978, *Space Sci. Rev.* **22,** 371.

Coroniti, F. V. and Kennel, C. F.: 1970, *J. Geophys. Res.* **75,** 1279.

Etcheto, J., Gendrin, R., Solomon, J., and Roux, A.: 1973, *J. Geophys. Res.* **78,** 8150.

Etcheto, J. and Bloch, J. J.: 1978, *Space Sci. Rev.* **22,** 597.

Gendrin, R., Perraut, S., Glangeaud, F., Fargetton, H., and Lacoume, J. L.: 1978, *Space Sci. Rev.* **22,** 433.

Hagege, K., Laval, G., and Pellat, R.: 1973, *J. Geophys. Res.* **78,** 3806.

Haugstad, B. S.: 1976, *J. Atmospheric Terrest. Phys.* **38,** 781.

Jones, D.: 1978, *Space Sci. Rev.* **22,** 327.

Kennel, C. F. and Petschek, H. E.: 1966, *J. Geophys. Res.* **71,** 1.

Kitamura, T., Jacobs, J. A., Watanabe, T., and Flint, R. B., Jr.: 1969, *J. Geophys. Res.* **74,** 5652.

Korth, A., Kremser, G., Wilken, B., and Munch, J.: 1978, *Space Sci. Rev.* **22,** 501.

Madden, M. A., Rycroft, J. M., and Smith, N. E.: 1978, *Space Sci. Rev.* **22,** 465.

Perraut, S., Gendrin, R., de Villedary, C., Meunier, J. P., Fouassier, D., Lavergnat, J., Bitterly, J., Corcuff, Y., Tixier, M., Glangeaud, F., Fehrenbach, M., and Weill, G.: 1978a, in B. Battrick and M. T. Halvorsen (eds.), *European Sounding Rockets, Balloons and Related Research with Emphasis on Experiments at High Latitudes*, ESA SP/135, pp. 73–84.

Perraut, S., Gendrin, R., Robert, P., de Villedary, C., and Jones, D.: 1978b, *Space Sci. Rev.* **22,** 347.

Sato, N., Hayashi, K., Kokubun, S., Oguti, T., and Fukunishi, H.: 1974, *J. Atmospheric Terrest. Phys.* **36,** 1515.

ULF WAVES: CONJUGATED GROUND-SATELLITE RELATIONSHIPS

ROGER GENDRIN and SYLVAINE PERRAUT
Départment Etudes par Télédétection de l'Environment, Centre National d'Etudes des Télécommunications, 92131 Issy-les-Moulineaux, France

and

HERVÉ FARGETTON, FRANÇOIS GLANGEAUD, and JEAN-LOUIS LACOUME
Centre d'Etudes des Phénomènes Aléatoires et Géophysiques, Institut National Polytechnique, 38400 Saint-Martin-d'Hères, France

Abstract. We study the simultaneous occurrence of ULF waves observed on board GEOS and at two of its 'conjugated' stations: Husafell (Iceland) and Skibotn (Norway). We try to deduce some properties of the regions in which these waves are generated. The few number of simultaneous observations of 'pearl' events indicates that such structured oscillations can occur only in specific conditions which are not met generally at the geostationary altitude. We introduce a new method for measuring time delays between the satellite and the ground. We show that this time is much higher than it would be expected from a simple extrapolation of measurements done at lower latitudes on structured events.

1. Introduction

As for VLF waves (Cornilleau *et al.*, 1978), the study of ground-satellite relationships in the ULF range may bring a lot of information about wave-particle interaction mechanisms and magnetosphere-ionosphere coupling. The waves we are considering are ion cyclotron waves with frequencies between ~0.1 Hz and ~3 Hz, which roughly corresponds to the Pc 1 frequency range. At lower frequencies, the S-300 ULF magnetic detector is not sensitive enough (Jones, 1978); at higher frequencies, the structured waves which are observed onboard GEOS-1 do not propagate to the ground (Perraut *et al.*, 1978b). In the frequency range we are studying, the main problems which are still to be solved can be listed as follows:

(1) At which radial distance do most of the ULF emissions take place? What is the longitudinal extension of the generation region? Assuming that the generation region covers a wide area, is the mean frequency of the emission dependent on radial distance and/or longitude?

(2) What is the exact influence of cold plasma distribution (in the magnetosphere and in the ionosphere) on generation and propagation mechanisms? To be more explicit, are magnetospheric ducts necessary to insure a good amplification factor of unstable waves? Are the ionospheric reflexion and transmission coefficients dependent on the polarization of the wave? Is one mode therefore more easily amplified and transmitted to the ground than the other?

Clearly, these questions can be answered, at least partially, by having simultaneous ground and satellite measurements. Although GEOS-1 was not geostationary, a

Space Science Reviews **22** (1978) 433–442. *All Rights Reserved*

conjugated ground campaign was performed (between 10 July and 22 September 1977) for reasons which are explained by Cornilleau *et al.* (1978). The difference with respect to the VLF conjugated experiment described by these authors is that, besides of the two stations of Husafell, Iceland (geographic coordinates: 64.5° N, 20.8° W; geomagnetic coordinates: 70.2° N, 74.3° E) and of Syowa, Antarctica (geographic: 69.0° S, 39.6° E; geomagnetic: 69.7° S, 77.7° E), we had a third station at Skibotn, Norway (geographic: 69.4° N, 20.3° E; geomagnetic: 66.6° N, 117.6° E) where ULF data only were recorded. At Husafell, the French mobile station (Perraut *et al.*, 1978a) was fully operated.

2. Spatial Extension of ULF Sources

We give on Figures 1 and 2 two examples of spectrograms of ULF emissions detected simultaneously at Skibotn, Husafell and onboard GEOS. The magnetic foot print of GEOS, labelled in UT is plotted in the lower part of the figures, so as to enable the reader to track back the position of GEOS as a function of time.

Let us start with the August 14 event (Figure 1) and, within this event, let us consider the first part of the emission which occurred around 0.55 Hz and between 10:30 and 12:20 UT. We notice that at the beginning (from 10:30 to 11:20) the signal was strong at Skibotn and weak at Husafell. At the end (from 11:30 to 12:20) it was strong at Husafell and almost not visible at Skibotn. During the whole event, it was strong onboard GEOS. Such an effect can be interpreted by a westward displacement of the interaction region, at approximately the same velocity as GEOS. This is compatible with a westward drift of interacting protons, which has been already invoked to explain the longitudinal time shift of Pc 1 appearance at different ground stations (see Gendrin, 1970, and references therein). Two things must be mentioned:

(a) The central frequency of the emission as seen onboard GEOS is almost constant. This can be explained by the fact that GEOS was more or less following a given *L*-shell (therefore encountering high energy protons with similar distribution functions), and that the cold plasma density was almost constant during this period of time (see Figure 9 of Perraut *et al.*, 1978b).

(b) Already on the spectrograms, but more clearly on integrated spectra (see for instance Figure 5), we notice that the central frequencies of the emission detected onboard GEOS and at the ground are not exactly the same. This means that the interaction region also has a radial extension: although they are excited simultaneously, the ULF waves have not exactly equal frequencies on different *L*-shells. However the frequency difference is small and it has no evident relationship with the difference in *L*-value (at Skibotn, f is smaller and at Husafell, f is larger than f at GEOS, whereas both ground stations correspond to lower *L*-values than GEOS).

The second part of the event, whose central frequency is approximately equal to 0.2 Hz and which starts, at GEOS, around 13:00, has a different explanation. If we refer to Figure 9 of Perraut *et al.* (1978b), we see that the cold plasma density n_e,

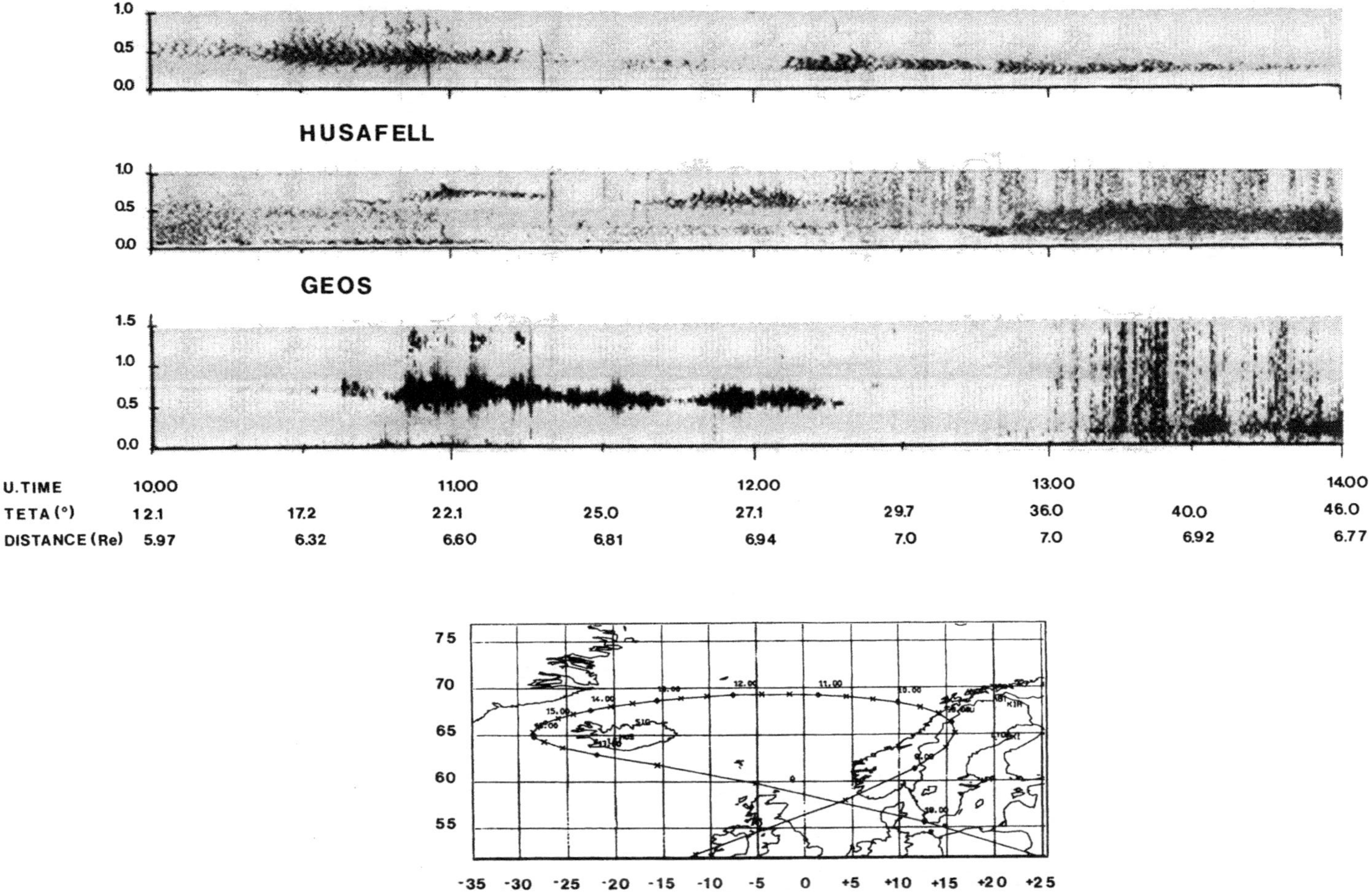

Fig. 1. Simultaneous record of ULF pulsations (14 August 1977) at Skibotn (Norway), Husafell (Iceland) and onboard GEOS-1. Two distinct emissions are seen onboard the satellite: one around 0.55 Hz from ~10:30 to 12:20 UT, the other around 0.2 Hz starting at ~13:00 UT onboard GEOS but seen well before at Husafell (see text).

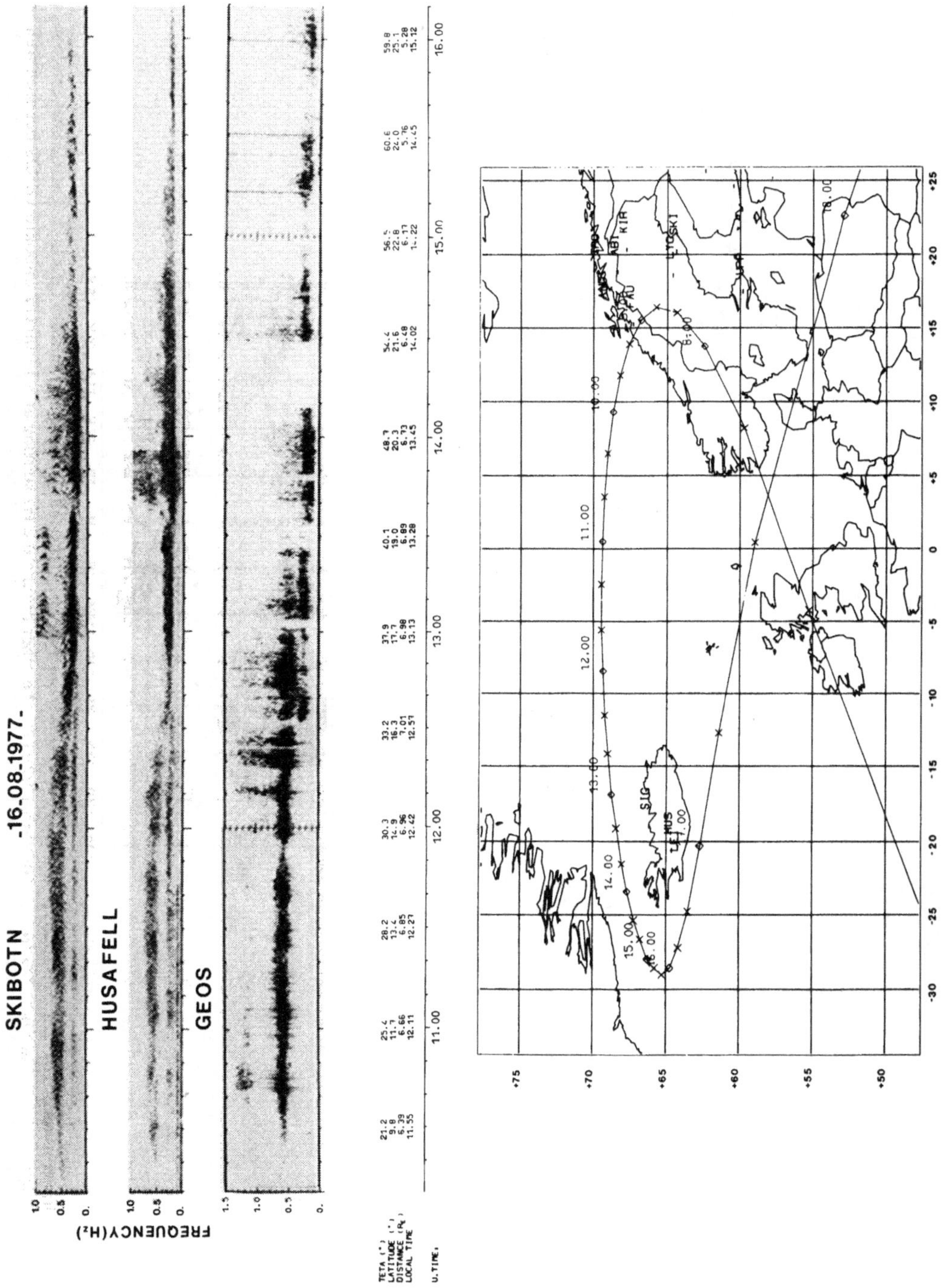
SKIBOTN
-16.08.1977-
HUSAFELL
GEOS
FREQUENCY(Hz)
TETA (°)
LATITUDE (°)
DISTANCE (R$_E$)
LOCAL TIME
U.TIME.

which had a constant value of ~5 cm^{-3} before 12:30 UT, suddenly changes and reaches high values (~40 cm^{-3} after 13:00 UT). This is probably related with the moderate storm which occurred that day (the K_p values for the whole day being equal to $1^+ 0^+ 2^+ 2^+ 4^+ 3^- 3^\circ 2^+$). Perraut *et al.* (1978b) have shown that the frequency change from 0.55 Hz to 0.2 Hz could be explained by the observed change of n_e, the resonant energy of the interacting protons remaining the same and equal to ~40 keV (this is easily understood by using Cornwall's (1965) simplified formula: $f = B^2/u\sqrt{n_e}$ in which B is the DC magnetic field intensity and u the parallel resonant velocity). It is worth noticing that the 0.2 Hz ULF noise is seen at Husafell well before it is seen onboard GEOS. We could think that the denser plasma region (in which lower frequency waves are generated) existed at the longitude of Husafell before GEOS reached such a longitude. Apparently the large magnetospheric plasma gradient prevented the waves generated on the western side of the discontinuity from propagating eastward the satellite. This is a good example of a case for which ground-satellite intercomparison enables us to distinguish between time and space variations.

The second example which we would like to comment on is presented on Figure 2. Again we notice two different emissions: one near 0.6 Hz between 10:30 and 13:30 and one near 0.3 hz between 12:30 and 16:00. However, we do not observe for the higher frequency emission the same progressive change of amplitude between Skibotn and Husafell: the intensities recorded at these two stations are very similar. The lower frequency emission also is observed with almost the same intensity at the two ground stations, and it is detected there much earlier than onboard GEOS. This day was also magnetically active after 12:00 UT (the K_p values being equal to $1^\circ 2^- 1^\circ 2^- 4^\circ 2^- 2^+ 2^\circ$), but the corresponding plasma density measurements are not yet available. If our previous explanation applies to this case, we can deduce that:

(a) An enhanced plasma region existed westward from GEOS well before 12:30 UT.

(b) Waves generated within this high density region can be observed, via subionospheric propagation over distances of the order of 2000 km, much higher than previously assumed (Greifinger and Greifinger, 1968).

(c) On the contrary, within the magnetosphere, waves cannot propagate easily through the plasma discontinuity.

A third example of ground-satellite non conjugacy, can be found in S–300 Experimenters' (1978) Figure 8.

3. Coherency Analysis and Time Delays

The best ULF events on which we can measure the propagation time between the equatorial region and the ground are the so-called 'pearl' events, which are known to propagate easily between two conjugated regions (Gendrin and Troitskaya, 1964). However, such events are seldom recorded onboard GEOS (2 cases only during 6 months, of which none occurred during the conjugated campaign). On the contrary, between the 10th of July and the 22nd of September, 10 structured events were recorded at the ground, during the 8 hr day^{-1} of GEOS operation, and they were not observed – as structured events – onboard GEOS. Such an effect may be interpreted as follows: structured events can be generated only within fluxtubes for which boundary conditions allow for a good laser effect: either the waves cannot escape from the flux tube (duct) or there is an especially good ionospheric reflexion. A weak leakage out of this flux tube should occur so that the emission is seen nevertheless onboard the satellite. But, because of multipath propagation, the ordered structure of the event is lost in the magnetospheric observation (Figure 3). Only when the flux tube has a large cross-section, can a satellite detect the Pc 1 structure. Obviously, such an interpretation has to be verified, with the help of precise wave normal direction determination and of simultaneous plasma density measurements.

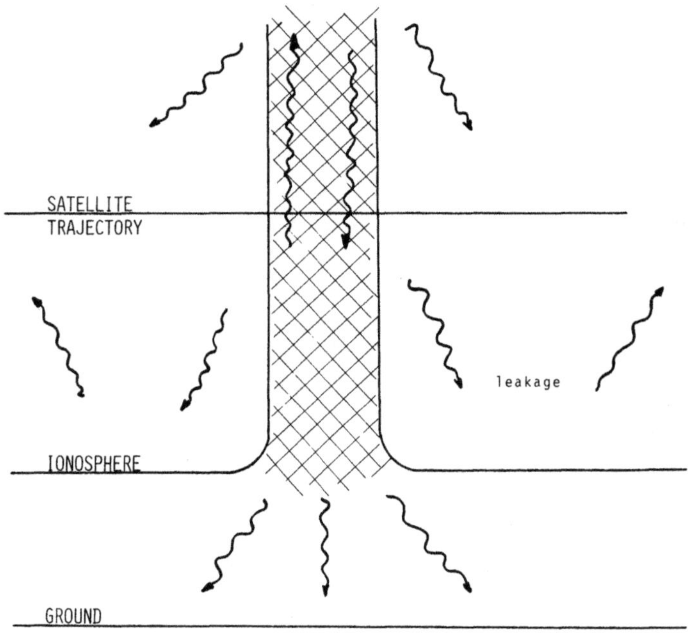

Fig. 3. Possible magnetospheric structure favouring the generation of structured ULF events.

Because of the lack of conjugated structured events, we could not make time delays measurements by looking at the group velocity of wave packets. We had to use unstructured events and to perform a less simpler analysis. The method we used consists of computing the coherence function of the two signals and of studying the frequency dependence of its phase ϕ. ϕ is a linear function of the frequency, whose slope gives the time delay between the two signals. If we artificially change the time delay, by shifting in time one record with respect to the other, we must obtain a horizontal line in the $\phi - f$ plan when the artificial increment is equal (but opposite) to the true time delay.

Such a method has been proven to be very useful for measuring ULF time delays between two conjugated stations (Glangeaud *et al.*, 1978). When applied to the GEOS-Husafell or GEOS-Skibotn data, it also gives valuable information. For instance, Figure 4 (which corresponds to the August 16 event) shows that the curve representing the variation of ϕ with frequency is horizontal, in a narrow range of frequency (0.58–0.63 Hz) for a time shift equal to 80 ± 1.5 s. At slightly lower frequencies (0.50–0.55 Hz) the time delay is smaller, as expected from the shape of the dispersion curve of ion cyclotron waves. Note that the $\phi(f)$ curves look regular only in those frequency ranges for which the modulus of the coherency function is large (~0.5). On the same figure are also plotted the autospectra for the left hand and

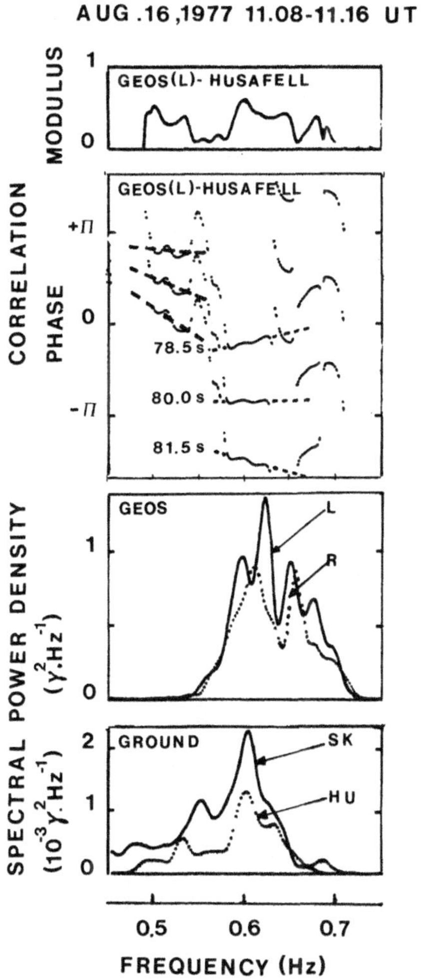

Fig. 4. Coherency analysis of ULF signals. From top to bottom: modulus of the coherency function between the GEOS left hand mode and the Husafell X component; phase of this coherency function; autospectra of GEOS left hand right hand modes; autospectra of the X components recorded at Skibotn and Husafell.

right hand modes observed onboard GEOS. The left hand mode has a higher intensity than the right hand mode, as expected. We also note that

(a) The ground spectra maximize at a lower frequency than the satellite spectra.

(b) The peak power intensity ratio *r* between the satellite and the ground is of the order of 10^3, a value much higher than the ones previously quoted ($8 < r < 100$) for Pc 1 waves observed simultaneously onboard ATS-1 and at the ground (Bossen *et al.*, 1976). It would be interesting to follow the variation of *r* with the position of the GEOS foot print; this has not yet been done. It would be interesting also to study the dependence of *r* on the different polarization modes, in order to check the theories related with ionospheric mode coupling in the ULF range, at high latitudes (see for instance Altman and Fijalkov, 1969; Gendrin, 1970, and references therein).

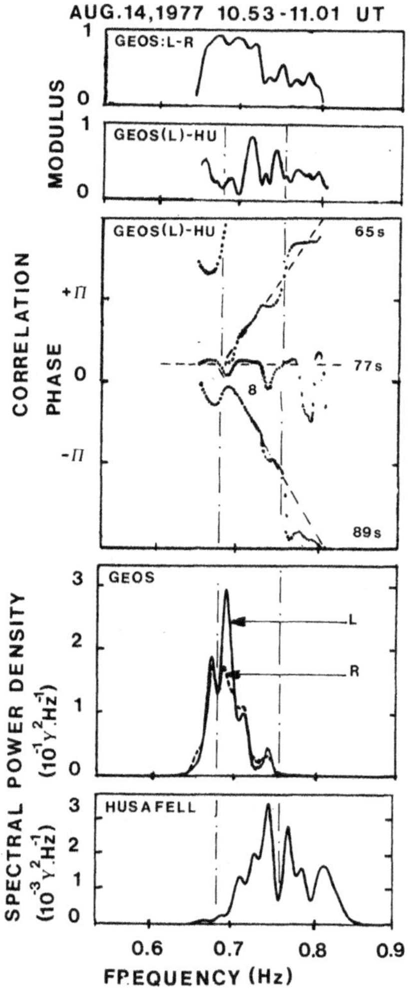

Fig. 5. Same as Figure 4 for the 14 August 1977 event, except that the magnetospheric correlation coefficient between the left- and right-hand modes has been also plotted. Note the difference between the satellite and the ground spectra.

Figure 5 gives another example of the application of the method. Again we see that $\phi(f)$ is linear only in a restricted frequency range (0.69–0.76 Hz) for which the modulus of the coherency function between GEOS and the ground is large. The reason why this modulus is not very large can be understood by looking at the two autospectra: the maximum power does not occur at the same frequency, as already mentioned in the previous section.

A striking result of this study is that the time delay between the satellite and the ground is of the order of 80 s. This means that the repetition period T which would be observed on the ground if these events were structured is of the order of 320 s, much higher than the one we would deduce from the empirical law $fT \sim 120$ (Gendrin, 1963; Tepley, 1965). This means that the group velocity is much smaller than the one we would deduce from extrapolation of results obtained at lower latitudes, where structured events are more frequent. In other words the cold plasma density is higher than the extrapolated value. This can be due to the high value of the cold plasma density at geostationary distances, during day hours, as experimentally measured with the S-300 active experiments. But one could as well imagine, on the contrary, that mid-latitude structured events propagate in regions where the Alfvén velocity is much higher than it should be, i.e. in ducts of *decreased* ionization.

4. Conclusion

This study has shown that:

(a) Some specific conditions must be met (ducts of decreased ionization?) for pearl oscillations to be generated. Either these conditions are not met at the geostationary orbit or the size of the inhomogeneity is too small to allow for detection of ULF structured events onboard drifting satellites.

(b) A given ULF event is seen for long periods of time (2–3 hr) both onboard the satellite and on the ground. It means that ULF waves can be observed at very distant position from their generation region. In some occasions, the magnetospheric azimuthal propagation of such waves is stopped by cold plasma density gradients. In some other cases, the generation region seems to move westward at the drift velocity of high energy protons.

(c) At high latitudes, the propagation time of ULF waves is much larger than previously expected.

These conclusions are preliminary. They need to be verified on more examples, by making more refined analysis (wave normal direction in particular) or by using the forthcoming GEOS-2.

Acknowledgements

We would like to thank the engineers and technicians who have operated the ground station during the Iceland campaign (Messrs P. Lafon, G. Laurent, J. P. Meunier, and A. Meyer), and Mr Kosik (Centre National d'Etudes Spatiales, Toulose) for his help in computing the magnetic foot print of GEOS.

References

Altman, C. and Fijalkow, E.: 1969, *Alta Freq.* **38**, 183.

Bossen, M., McPherron, R. L., and Russell, C. T.: 1976, *J. Atmospheric. Terrest. Phys.* **38**, 1157.

Cornilleau-Wehrlin, N., Gendrin, R., and Tixier, M.: 1978, *Space Sci. Rev.* **22**, 419.

Cornwall, J. M.: 1965, *J. Geophys. Res.* **70**, 61.

Gendrin, R.: 1963, *Ann. Géophys.* **19**, 197.

Gendrin, R.: 1970, *Space Sci. Rev.* **11**, 54.

Gendrin, R. and Troitskaya, V. A.: 1964, *Radio Sci.*,**69**, 1107.

Glangeaud, F., Lacoume, J. L., Gendrin, R., and Perraut, S.: 1978, *J. Atmospheric Terrest. Phys.*, to be published.

Greifinger, C. and Greifinger, P. S.: 1968, *J. Geophys. Res.* **73**, 7473.

Jones, D.: 1978, *Space Sci. Rev.* **22**, 327.

Perraut, S., Gendrin, R., de Villedary, C., Meunier, J. P., Fouassier, D., Lavergnat, J., Bitterly, J., Corcuff, Y., Tixier, M., Glangeaud, F., Fehrenbach, M., and Weill, G.: 1978a, in B. Battrick and M. T. Halvorsen (eds.), *European Sounding Rockets, Balloons and Related Research with Emphasis on Experiments at High latitudes*, ESA SP/135, pp. 73–84.

Perraut, S., Gendrin, R., Robert, P., Roux, A., de Villedary, C., and Jones, D.: 1978b, *Space Sci. Rev.* **22,** 347.

S-300 Experimenters: 1978, *Planet. Space Sci.* **26**, (to be published).

Tepley, L. R.: 1965, *Radio Sci.* **69D**, 1089.

RECEPTION OF THE NKL (JIM CREEK) TRANSMITTER ONBOARD GEOS-1

NICOLE CORNILLEAU-WEHRLIN, ROGER GENDRIN, and RICHARD PEREZ
Département Etudes par Télédétection de l'Environnement, Centre National d'Etudes des Télécommunications, 92131 Issy-Les-Moulineaux, France

Abstract. At the beginning of the GEOS lifetime, some attempts have been made for taking advantage of the passes over Alaska. GEOS was then commanded in a fixed mode and the corresponding telemetry data were recorded at the NASA stations. For two passes over Jim Creek (48°2′N − 121°9′W) where a powerful VLF transmitter ($f_0 = 18.6$ kHz) is located, GEOS was put in a specific mode in order to study the magnetospheric electromagnetic field in the vicinity of f_0. The results of one pass (June 11, from 0755 UT) are presented here.

During this pass, a strong enhancement of all the e.m. components at f_0 has been observed for a specific period of time, when GEOS was very near to the exact conjugacy with NKL. The distance, as measured on the ground, over which the signal was above −6 dB from the maximum is of the order of 800 km. During the corresponding period of time (0740–0750 UT), the satellite altitude varied between 8000 and 6000 km. The magnetospheric region where the signal is strong appears to be structured, as if there were many ducts.

Preliminary results concerning the polarization characteristics of the signal are presented. In the absence of precise measurements of these characteristics, the comparison between the electric and magnetic components of the received signal is not easy to interpret. An examination of the onboard computed correlograms (in the frequency range from $f_0 - 0.6$ kHz to $f_0 + 3.3$ kHz) shows that, for this pass, no VLF emissions were triggered by NKL, at the altitude of the satellite.

1. Introduction

The mechanism of generation of triggered emissions (TE) or artificially stimulated emissions (ASE) is one of the most fascinating problem of the magnetospheric wave-particle interactions. Because it may lead to the simplest 'active' or 'controlled' experiments in the magnetosphere (one needs just a ground based powerful VLF transmitter), it is really interesting to understand the mechanism by which these triggered emissions are generated. Different theories have been proposed to explain their generation: quasi-linear diffusion in an inhomogeneous medium (Ashour-Abdalla, 1972; Welti *et al.*, 1973), a phase bunching effect (Helliwell, 1967) or an equilibrium between non linear trapping and a detrapping effect due to the inhomogeneity (Roux and Pellat, 1978; Cornilleau and Gendrin, 1978). The first theories are valid for a low field intensity of the monochromatic wave ($\leqslant 1$ mγ at $L \sim 4$), the others for a higher field intensity ($\geqslant 10$ mγ). It is therefore of utmost importance to have quantitative information about the VLF field strength of monochromatic waves prevailing in the equatorial magnetosphere, whether they are due to power lines, to VLF transmitters or of natural origin (chorus).

Some indirect or direct measurements have been obtained in the past (see Helliwell, 1977 for a review) but the direct *in situ* measurements were generally obtained with wide band receivers and with a restricted number of detectors,

Space Science Reviews **22** (1978) 443–451. *All Rights Reserved*

therefore preventing from a complete and precise restitution of the full electromagnetic wave field. Consequently, we thought it would be interesting to use the flexibility of the wave experiment onboard GEOS-1 (S-300 Experimenters, 1978), associated with the fact that passes over Alaska and North-West America were available, to make a detailed study of the wave field radiated by a powerful VLF transmitter such as NKL at Jim Creek (geographic coordinates: 49.2°N–121.9°W).

The purpose of this paper is to describe some preliminary data which were obtained during the pass of June 11, 1977, around 0740 UT. After a brief description of the experimental set up, we will show the results which were obtained with a narrow band, six components system and with the wide band onboard correlator.

2. Experimental Set Up

The six components of the electromagnetic wave field in the frequency range from ~0 to 77 kHz can be received onboard GEOS and transmitted to the ground via two different systems (Jones, 1978). For the June 11 pass, the telecommand configuration was such that:

(a) The full wave form of the 3 magnetic components and of 3 electric components, within a band of 300 Hz, were transmitted to the ground (these correspond to the output of our swept frequency analyser (SFA) system). Each step had a duration of 0.69 s (8 high-speed telemetry formats) and we swept repeatedly over 4 steps from 18 360 Hz to 19 250 Hz (central frequencies). A full sweep lasted 2.75 s. The idea was to be able to detect and to study with a good frequency resolution the contingent stimulated emissions which could have been triggered just above or just below the transmitter frequencies (18 600 and 18 550 Hz). For this part of the data, the electric components which were recorded were e_x, e_z and E_y, i.e., two components detected by the axial electric antennas (less sensitive) and one component detected by the more sensitive long radial antenna (for the geometrical and electrical characteristics of these antennas, see Jones' (1978), Figures 1 and 3). Besides, we used the three magnetic antennas, b_x, b_y, b_z.

(b) The correlograms obtained over a ~3 kHz bandwidth placed just above the central frequency of the SFA's were also transmitted to the ground. The correlator was working in Mode 3, allowing for the successive transmission of the autocorrelations of b_x and E_y every 58 ms.

3. Narrow Band Analysis

For their presentation in this preliminary report, the narrow band data were analyzed as follows. Each of the 6 wave forms, for each of the 4 frequencies, was averaged over 0.69 s, in order to obtain the r.m.s. value of the signal. The gain levels of the step amplifier and the transfer functions were carefully taken into account in order that results could be plotted in physical units (mγ or mV m^{-1}).

The results of such a computation is plotted for the e_z component and for the frequency band which contains the transmitter frequency in the middle panel of Figure 1. The upper panel of this figure represents the foot prints, at the 100 km level, of the magnetic field lines passing through GEOS. This trace was computed by using a modified Olson–Pfitzer model (Kosik, 1978). It is labelled in Universal Time. The time scales of the middle and lower panels are not linear: they are such that a vertical line directly gives the correspondence with the UT time of the foot print.

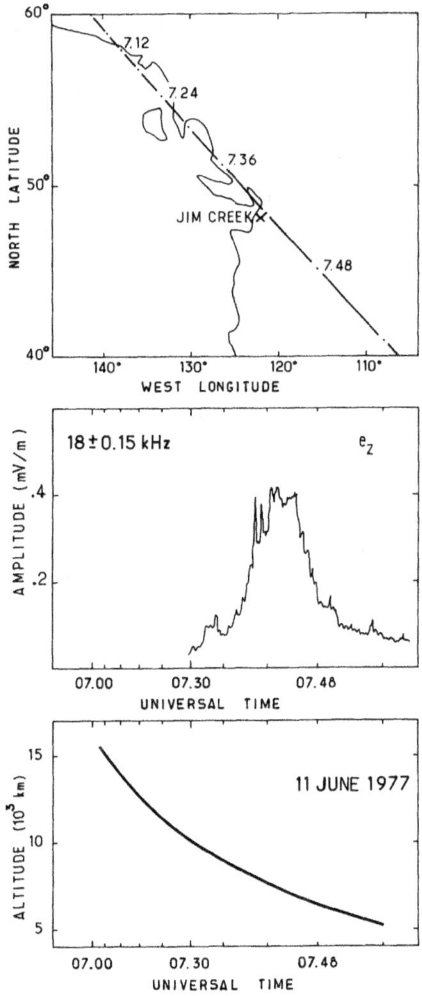

Fig. 1. Reception of NKL onboard GEOS-1 during the June 11, 1977 pass. Upper panel: magnetic foot print trajectory, labelled in UT (one point every 6 min). Middle panel: mean amplitude (averaged over 0.69 s) of the signal recorded on one small electric antenna at the transmitter frequency (one point every 2.75 s). Lower panel: altitude of the satellite above the surface of the Earth.

We immediately notice that the amplitude of the signal is large for approximately 12 mn in the vicinity of the conjugacy of GEOS with Jim Creek. However the maximum amplitude occurs approximately two minutes after the exact conjugacy. Changing the magnetic field model (within all possible reasonable limits) does not change significantly this effect (Kosik private communication). Therefore we conclude that the main entrance points of the waves into the magnetosphere were situated that day slightly South-East of the transmitter.

A more extended plot of the amplitudes of the 6 components, within the same frequency range, is given on Figures 2 and 3. Let us first examine the B components (Figure 2). We see that above the smooth variation of the intensity which is centered around 0745, there are three enhancements of the signal around 0739, 0745 and 0751. These may correspond to crossings of ducts of ionization. Unfortunately, for the Alaskan passes, it was impossible to telecommand the S-300 experiment in an active mode, so that no information about the plasma density fluctuations is available.

Examination of Figure 3 shows two things. First the mean amplitude of the e_x and E_y components are similar (around 0.5 mV m^{-1} near the maximum). This demonstrates that the calculation of the small axial antennas (which was expected to be much more difficult because of the proximity of the satellite skin and possible imbedding in the photoelectron sheath) is good. Second we notice a modulation

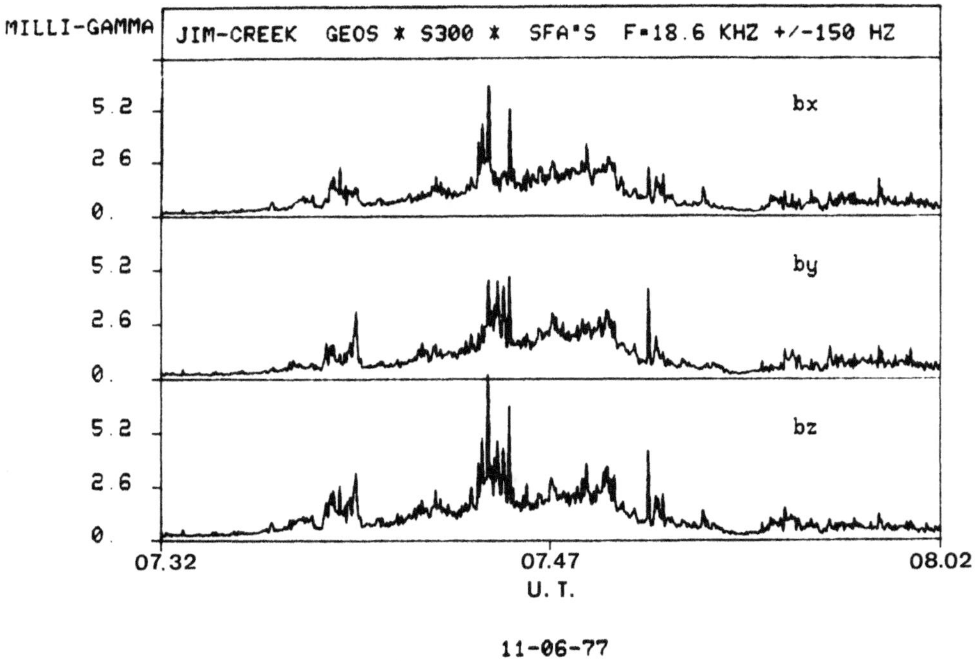

Fig. 2. Amplitude of the magnetic compounds at the transmitter frequency as a function of time. Note the three enhancements of the signal level around 0739, 0745 and 0751 UT. These could correspond to the crossings of ionization ducts.

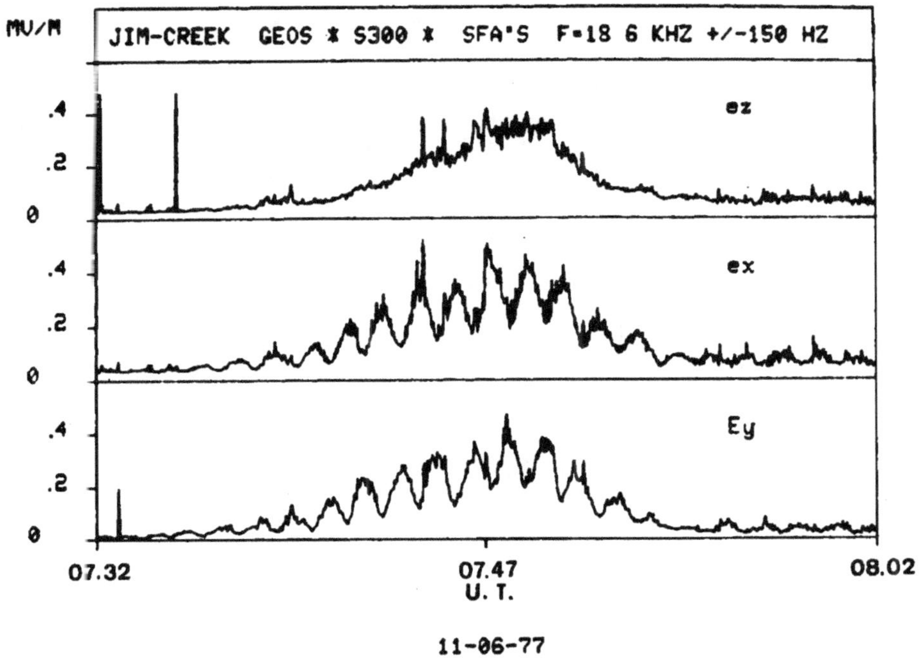

Fig. 3. Amplitude of three electric components at the transmitter frequency as a function of time. The modulation which is seen on e_x and E_y can be interpreted in terms of a large ellipticity of the projection of the electric signal onto the xy plane, which is the plane perpendicular to the spin axis.

with a ~75 s periodicity in both e_x and E_y which is in antiphase on these two antennas. This modulation can be explained if we assume that the electric component of the signal is polarized in such a way that the projection of the polarization ellipse into the x, y plane is strongly elliptical (Figure 4).

As explained above, the average amplitude of each component is computed over 0.69 s. This corresponds to approximately $\frac{1}{8}$ of a turn, the spin period T of the satellite being equal to 5.7 s. During 0.69 s, a given antenna has explored a sector of the polarization ellipse which is represented by the hatched area of Figure 4. Because the analyser sweeps in frequency over 4 steps, it comes back to the same frequency when the satellite has almost completed half a turn. Let us assume that the e_x antenna was at time $t = 0$ more or less aligned with the major axis of the projected polarization ellipse. The average power detected by this antenna will be large, and it still will be large at time $t = T/2$. On the contrary, the average power measured on the E_y antenna will remain small in both cases. But T being not exactly equal to a multiple of 2.75 s, after 5.5 s there will be a slight shift α between the sectors of the projected polarization ellipse over which the average is made: $\alpha = (1 - 5.5/T) \times 360°$. When α will be equal to 90°, the two antennas will have exchanged the sector of the projected polarization ellipse which they explore, and the signal which was maximum on one antenna will now

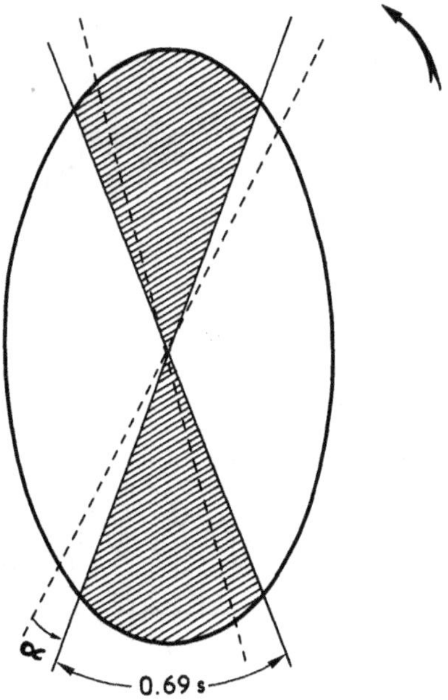

Fig. 4. Projection of the electric polarization ellipse in the xy plane. The hatched antenna corresponds to the angle over which a given antenna has rotated during the time we compute the average. Some 2.75 s later, we explore the same frequency band and almost the same portion of the ellipse. But each turn there is a small shift α because the spin period is not exactly equal to 2×2.75 s.

be minimum. This occurs after a time equal to $T/4\ (1-5.5/T)$, i.e., ~ 38 s as observed.

A similar effect was not observed on the magnetic components, which means that the projection of the magnetic polarization ellipse in the x, y plane is almost circular. Unfortunately, the d.c. magnetometer on GEOS was not working at these low L-values; therefore, it was impossible to study in detail the wave normal angle to check the polarization characteristics (both electric and magnetic) of the signal.

As a consequence of these unknown polarization characteristics, the simple computation of the ratio $|B|/|E|$ does not give any reliable information about the refractive index. However we can still compute this ratio in order to see how it varies along the satellite trajectory (Figure 5). Because of calibration problems which are not yet clarified, we must not attach any significance to the absolute value of the ratio $c\ |B|/|E|$ (where c is the velocity of light), but we see that the relative values are greatly enhanced around the three periods already mentioned and which probably correspond to ionization ducts.

4. Wide Band Analysis

The 64 points of the autocorrelograms of each component have been Fourier transformed on a PDP 11 computer and the corresponding power spectra have

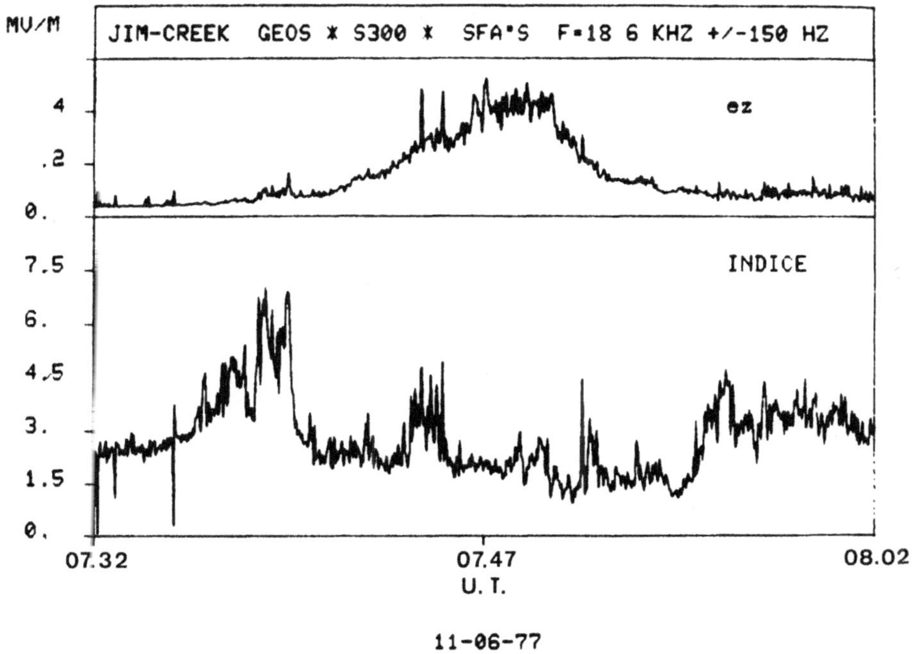

Fig. 5. Value of the ratio $c\,|B|/|E|$ as a function of time (lower panel). The absolute values are meaningless (see text) but the relative values show enhancements at the times which probably correspond to the crossings of ducts.

been plotted on a three-dimensional plotter, in two different ways. Figure 6 shows the result of these computations for a limited period near the middle of the pass. In the first representation (upper and middle panels of Figure 6) the grey scaling is adjusted so that black corresponds to the maximum power observed during one full sweep (2.75 s). Simultaneously the correlator gain (which can vary by steps of 4 dB) has been plotted above each spectrogram, so that one can see where the most intense signal occurs (this takes place during the two last steps of each sweep, when the transmitter frequency is well within the frequency range of analysis). In the second representation (lower panel of Figure 6) the grey scaling is such that black corresponds to the maximum power recorded during a single correlogram. Such a representation has the advantage of making visible even the lowest signals, but the black level does not correspond to the same power throughout a whole sweep. For instance the apparently strong line which appears on the first step of every sweep around 21.5 kHz is in fact 28 dB below the true line at the transmitter frequency (and it is due to an aliasing effect of this same signal).

From the examination of these high sensitivity spectrograms of which only a small portion is represented, we see that there was no visible triggered emission associated with the Jim Creek transmitter. If this transmitter were not emitting continuously, we could say that this is due to the fast displacement of GEOS

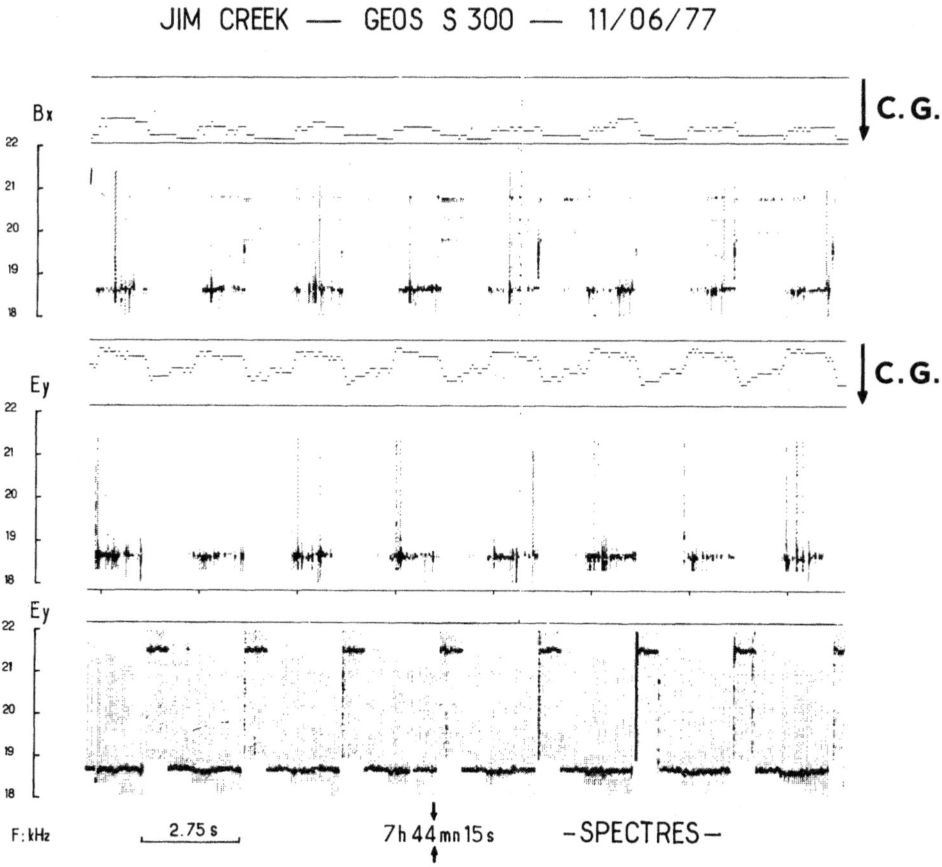

Fig. 6. A portion of the wide band analysis performed during the pass. For the upper part of the figure, the grey scalings are such that black corresponds to the maximum level recorded during a full sweep (2.75 s). The correlator gain (CG) for each correlogram is also plotted (in stepds of 4 dB). For the lower spectrogram the grey scaling is adjusted for each correlogram: the apparent line at 21.5 kHz is in fact 28 dB below the line at 18.6 kHz. No triggered emissions are visible.

across L-shells: any emission triggered in the equatorial region and propagating backwards towards the transmitter would not reach the satellite when arriving at its altitude some 2–3 s later. But the transmitter was operating continuously and we should have seen emissions triggered before if there were any. This means that the conditions for triggered emissions to be generated by NKL were not met on that day. This is probably related with the rather low level of the wave field intensity which was recorded during this pass.

5. Conclusion

We have made a preliminary analysis of the electromagnetic wave field received onboard GEOS-1 when it crossed the field lines passing through the powerful

VLF transmitter of Jim Creek. We noticed an enhancement of the field intensity, at the transmitter frequency, well above the sensitivity threshold of our instrument, and over distances corresponding approximately to 1000 km at the ground level. Field aligned ducts were probably encountered, in which the field intensity was larger, mainly on the magnetic components.

The projection of the electric polarization ellipse in the plane perpendicular to the spin axis of the satellite was strongly elliptical, in contrast to the projection of the magnetic polarization ellipse which was more circular. As a consequence, the value of the refractive index which one deduces from the simple computation of the average value of the ratio $|B|/|E|$ is erroneous. This means that one must be very cautious when trying to deduce the intensity of one component of the electromagnetic field from the measurement of the other and the use of a theoretical index of refraction. In any case the magnetic field intensity which we have measured at an altitude of 7000 km (~ 3 mγ) is smaller than the ones (~ 50–100 mγ) which these authors have reported. In the future, we will do a more refined analysis of the data (by using the full wave forms and not averaged values). Then we will be able to restitute the true polarization figure of the signal and hopefully to clarify the question of the meaning of the $|B|/|E|$ ratio.

Finally, during this experiment, we did not see any triggered emission, below or above the transmitter frequency.

Acknowledgements

We are grateful to the US National IMS Coordinator, R. L. Manka and to R. A. Helliwell (Stanford University) for their continuous support to the operation of GEOS-1 in conjunction with the Jim Creek transmitter. We are also grateful to R. Riguet, from our Institute, for his help in analyzing the data. Finally we wish to thank J. C. Kosik (Centre National d'Etudes Spatiales) for his computations of the magnetic footprints of GEOS.

References

Ashour-Abdalla, M.: 1972, *Planet. Space Sci.* **20**, 639.
Cornilleau-Wehrlin, N. and Gendrin, R.: 1978, *J. Geophys. Res.* (to be published).
Helliwell, R. A.: 1967, *J. Geophys. Res.* **72**, 4773.
Helliwell, R. A.: 1977, *Phil. Trans. Roy. Soc. London* **279**, 213.
Jones, D.: 1978, *Space Sci. Rev.* **22**, 327.
Kosik, J. C.: 1978, *Space Sci. Rev.* **22**, 481.
Roux, A. and Pellat, R.: 1978, *J. Geophys. Res.* **83**, 1433.
S-300 Experimenters: 1978, *Planet. Space Sci.* **26**, (to be published).
Welti, R., Pellat, R., and Roux, A.: 1973, *Ann. Geophys.* **29**, 263.

OBSERVATION ON GEOS-1 OF 10.2 TO 13.6 kHz GROUND BASED TRANSMITTER SIGNALS

E. UNGSTRUP, T. NEUBERT and A. BAHNSEN
Danish Space Research Institute, Lyngby, Denmark

Abstract. The Omega transmitter signals on 10.2, 11.33 and 13.6 kHz are frequently observed on GEOS-1 near the end of the pass. The signals show large fluctuations in amplitude and in *E/B* ratio on a time scale of 0.1 s. The time delay of the signals vary considerably over a period of a few minutes, and often the duration of the received pulses on the dipole antenna is longer than the duration of the transmitted pulses. The propagation aspects of these observations will be discussed on the basis of raytracing calculations from the ground to the vicinity of GEOS-1.

1. Introduction

In this paper we report observations of the GEOS-1 satellite of VLF-signals from the Omega transmitter in Norway. We will describe the characteristics of the signals as seen by the GEOS-1 correlator system, and discuss their propagation in the magnetosphere. In the discussion we shall see which conclusions can be drawn from the observations themselves, and when the observations are compared with raytracing calculations in a diffusive equilibrium model of the magnetosphere.

2. Observations

GEOS-1 is placed in an elliptic orbit with apogee at about 45 000 km distance from the centre of the Earth. The satellite traverses its orbit twice for each Earth rotation, thus keeping an almost constant position of the projected magnetic foot track. Figure 1 shows a map of the northern projection of the GEOS magnetic foot track, eastern apogee, Day 355 (December 21.) 1977, and the position of the Omega transmitter (J. I. Vette, private communication). The plot is made by use of the Olson–Pfitzer 1977 external field model, which includes the effects of tilt and the Barraclough 1975 internal field model updated to the epoch of the plot. The tick marks on the orbit track indicate universal time. The southern part of the magnetic foot track is of no importance, as it lies in the South Atlantic far from any transmitter.

The Omega station in Norway, located at 66°25′15″N, 13°09′10″E transmits three different frequencies: 10.2 kHz, 11.33 kHz and 13.6 kHz in the sequence shown in Figure 2 (Burhans, 1974). The sequence is repeated every 10 s, and starts at 4 s, 14 s, etc, past the minute.

We picked five days at random in late December 1977 to see if the transmitter was received on GEOS-1. Four of these days provided observations of the transmitter namely: Day 351, 353, 355 and 361. A selection in May 1978 gave as result that the transmitter was received two out of four randomly selected days.

Space Science Reviews **22** (1978) 453–464.

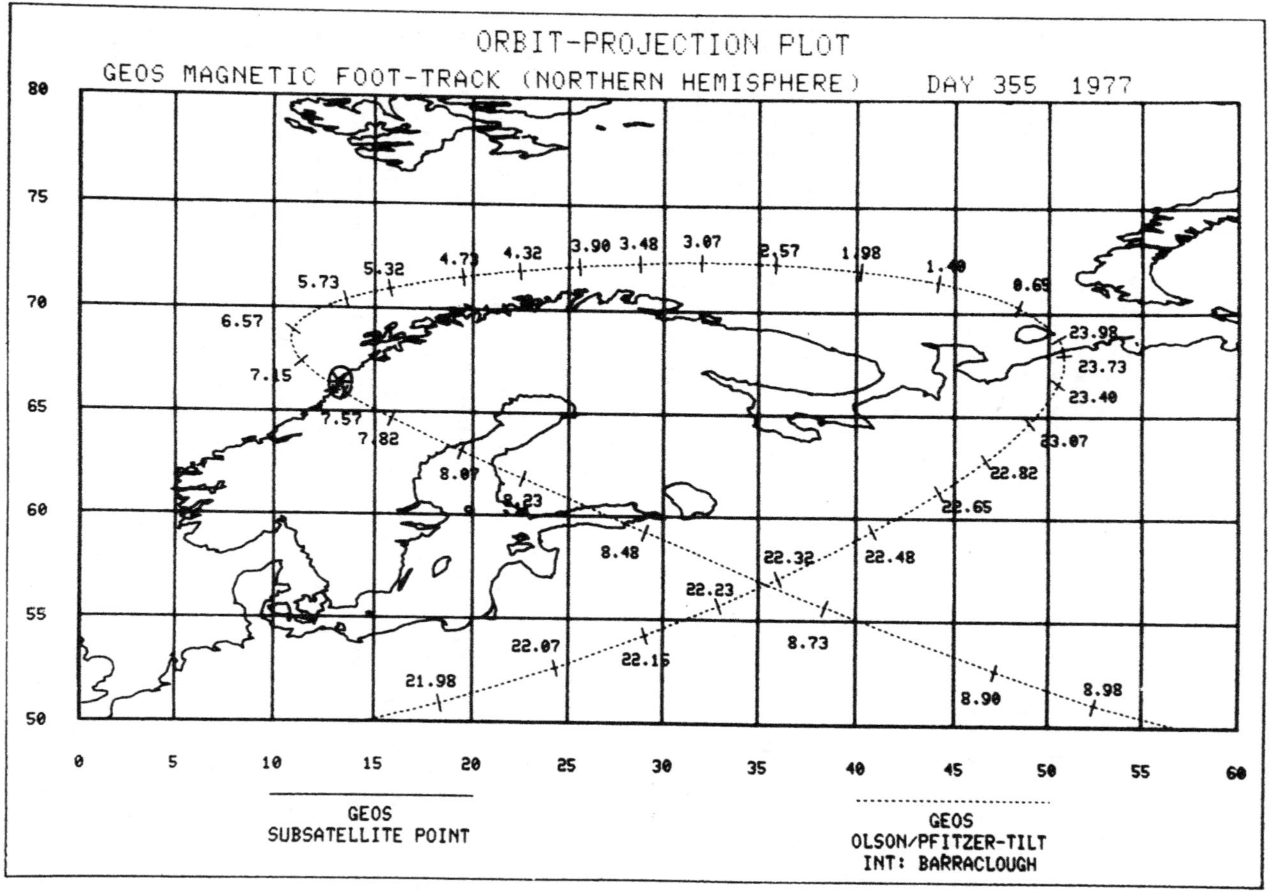

Fig. 1. Northern part of projected GEOS-1 magnetic foot track, eastern apogee, Day 355 (December 21) 1977 (J. I. Vette). Experiment was turned on from 2227–0842 UT. ⊗ = Omega transmitter.

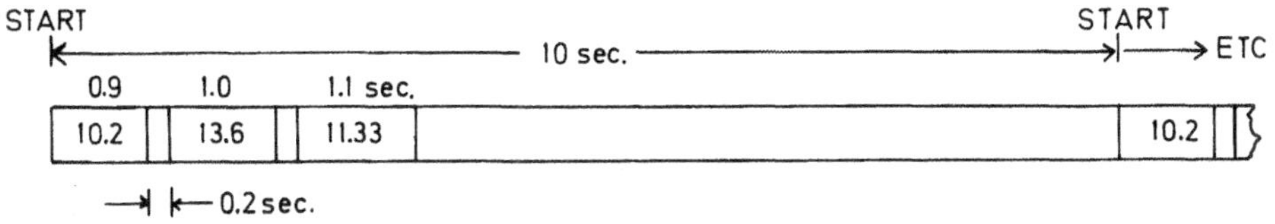

Fig. 2. Time sequence of the 10.2 kHz, 11.33 kHz and 13.6 kHz transmissions from the Omega station in Norway (R. W. Burhans). At the time of the observations, start time was 4 s, 14 s, etc. past the minute, the sequence being repeated every 10 s.

In Figure 3 we have shown the GEOS-1 track projected on the magnetic meridian plane for the two periods December 1977 and May 1978. Signals are observed from a certain point of the orbit and either until loss of contact from the telemetry station as the satellite goes beyond the horizon, as happened in May 1978, or until an unfavourable change of mode of the experiment, as happened in December 1977. Therefore Omega signals may well propagate to lower latitudes than Figure 3 suggests, in fact this seems highly probable, as the signals are very strong at cut-off time.

On board the GEOS-1 satellite a correlator system is analysing frequencies up to 11.75 kHz of one magnetic and one electric field component, when set in a proper mode. Due to aliasing, the 13.6 kHz appears at 10 kHz, although strongly attenuated by the aliasing filters (30 dB/oct.).

An example of a spectrogram displaying the 10.2 kHz and 11.33 kHz electric field components on Day 355 (December 21.) at 0805 UT is shown in Figure 4, and all three frequencies for both the electric and the magnetic field are seen on the spectrogram from 0814 UT, same day, shown in Figure 5.

At the time of the observations GEOS-1 is moving towards the earth into areas of increasing electron gyro frequency, f_{ce}. As VLF-whistler mode waves propagate

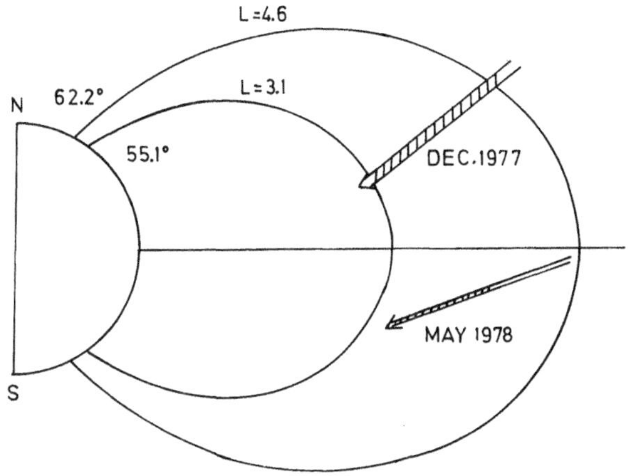

Fig. 3. The last part of the GEOS-1 track in the magnetic meridian plane for two periods: Days 351–361, 1977 (upper tracks) and Day 136–139, 1978 (lower tracks). Hatched areas indicate the position of the satellite, when detecting the Omega transmitter. At $L = 4.6$ the first Omega signal was detected, at $L = 3.1$ the experiment was turned off. The L-shells are calculated from a simple dipole model.

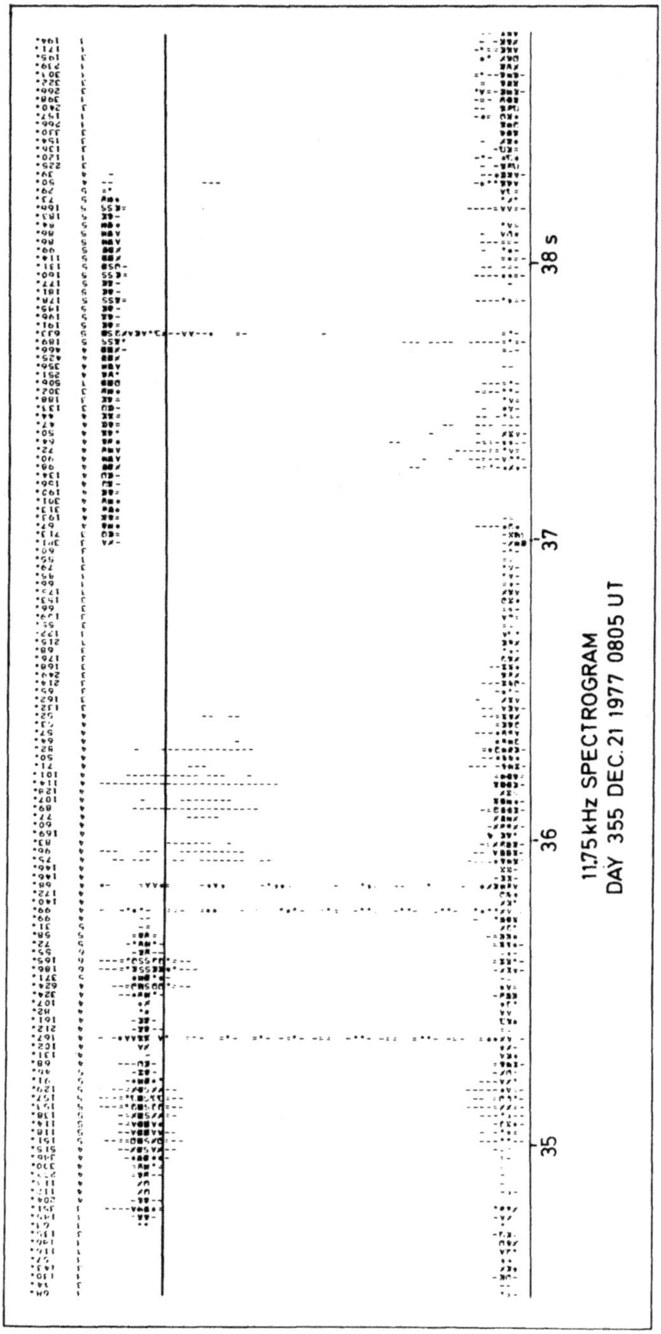

Fig. 4. 11.75 kHz spectrogram of the electric field component for Day 355, (December 21.) 1977, 0805 UT, displaying the 10.2 kHz and 11.33 kHz frequencies of the Omega transmitter. Horizontal line shows position of 10.0 kHz.

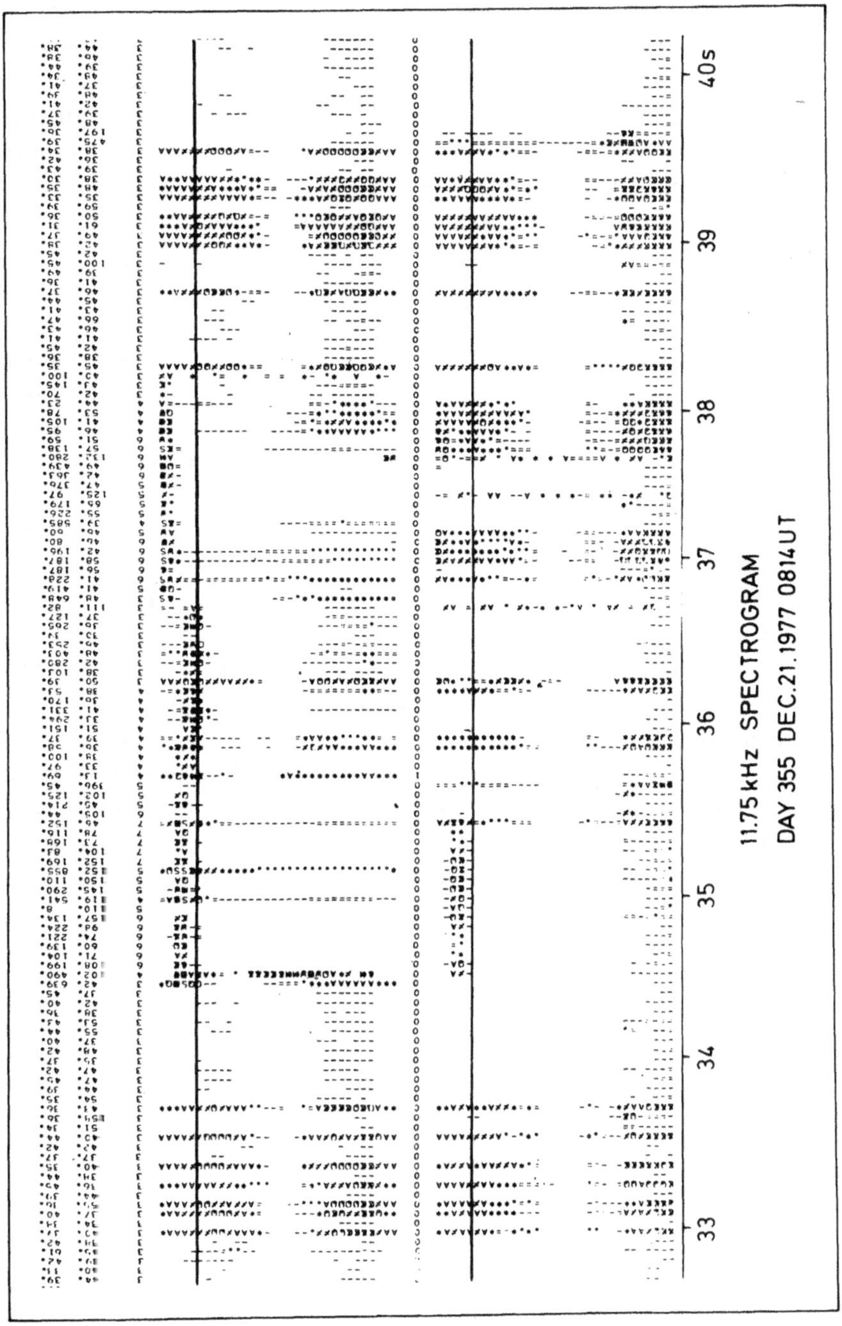

Fig. 5. 11.75 kHz spectrogram of the electric (top) and magnetic (bottom) field components for Day 355 (December 21.) 1977, 0814 UT, displaying the 10.2 kHz, 13.6 kHz and 11.33 kHz frequencies of the Omega transmitter. Horizontal line shows position of 10.0 kHz. Because of aliasing, 13.6 kHz shows up at 10 kHz.

at frequencies below the local f_{ce}, we should expect to see the lowest frequencies first. That this is so, is seen from Figure 6, which shows the times at which the signals were detected by the correlator for Day 355 and 361. In Figure 6 we see the low frequencies appearing first, as predicted by theory. The strong damping by the aliasing filters mentioned above could be affecting the first appearance of the 13.6 kHz signals, but would not significantly disturb the 11.33 kHz. Furthermore, as shown in Figure 6, the reception of the Omega transmitter signals is intermittent on a timescale of minutes. This seems to indicate that the signals are ducted (Angerami, 1970). If so, the dimensions of the ducts are ~100 km when projected along the Earths magnetic field to the ionosphere.

Since early December 1977 there has been a malfunction in the magnetometer on GEOS-1, and hence the direct measurement of f_{ce} is uncertain. To get an idea of the value of the normalized Omega frequency f/f_{ce}, we have used model calculations for the magnetic field of the Earth (J. O. Kosik) to calculate f_{ce} at the position of the satellite, and find $f/f_{ce} \leqslant 0.7$. This is a little less than the values reported by Dunckel and Helliwell (1977). They find $f/f_{ce} \leqslant 0.9$ analysing signals from ground-based transmitters in the 60–100 kHz range observed by the OGO-1 satellite.

The start time and length of the pulses were obtained from plots of signals vs. time in selected channels centered around the frequencies in question. Figures 7 and 8 show typical examples of such plots. The two upper curves on each plot represent the electric component of the signal at 10.2 kHz and 11.33 kHz measured by the 40 m long electric dipole antenna, while the two lower curves represent the magnetic component measured by the magnetic VLF antenna. The curves show the logarithm of the mean power of the signals in arbitrary units in 1.1 kHz bands around 10.2 kHz and 11.33 kHz. The plot in Figure 9 illustrates the delay and duration of the signals. It shows 10.2 kHz and 11.33 kHz signals in a time scale relative to the transmitted pulses. Time of transmission is shown at

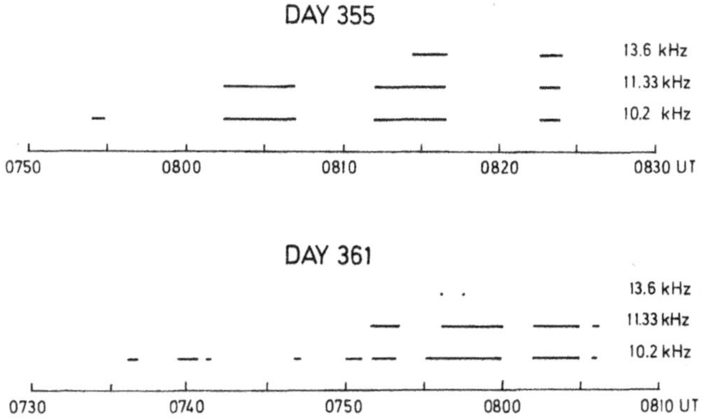

Fig. 6. Time of reception of the 10.2 kHz, 11.33 kHz and 13.6 kHz Omega signals on Day 355 and 361, 1977. From 0730–0750 UT on Day 361, strong electrostatic turbulence rendered detection of Omega signals impossible.

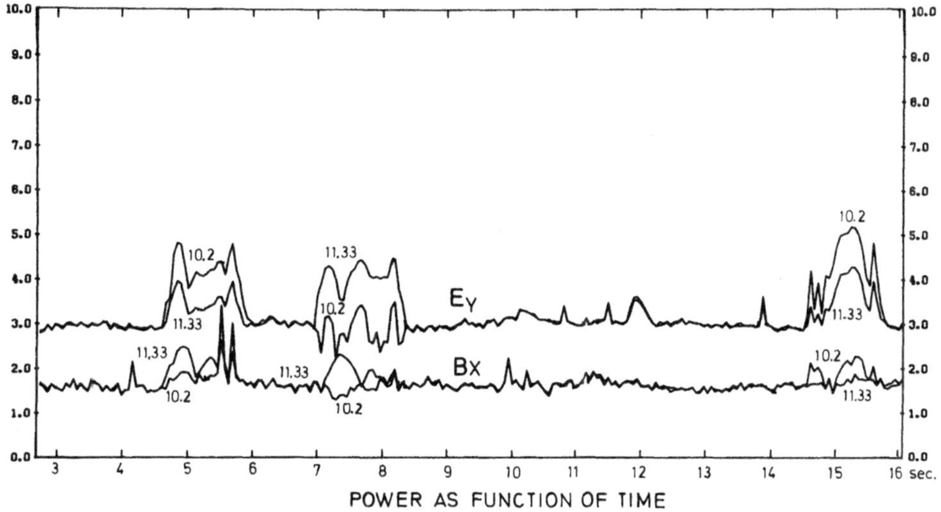

Fig. 7. Electric and magnetic field components of 10.2 kHz and 11.33 kHz showing the Omega signals on Day 355, 0806 UT. The ordinate is logarithm of power in arbitrary units.

the bottom. From the 4 min. interval on Day 355, 0802–0806 UT shown in Figure 9 we have selected the pulses with well defined lengths only. The average characteristics are listed in Table I. We note that, since the transmitter cycle is repeated every 10 s, the measured delays could in principle be larger by 10 or even 20 s. Table I shows that the delay of the 11.33 kHz signal is approximately 10% larger than that of the 10.2 kHz signal, and that the prolongation of the 10.2 kHz signal is greater than that of the 11.33 kHz signal. A more detailed analysis of the pulses has revealed, however, a dependence of prolongation on the

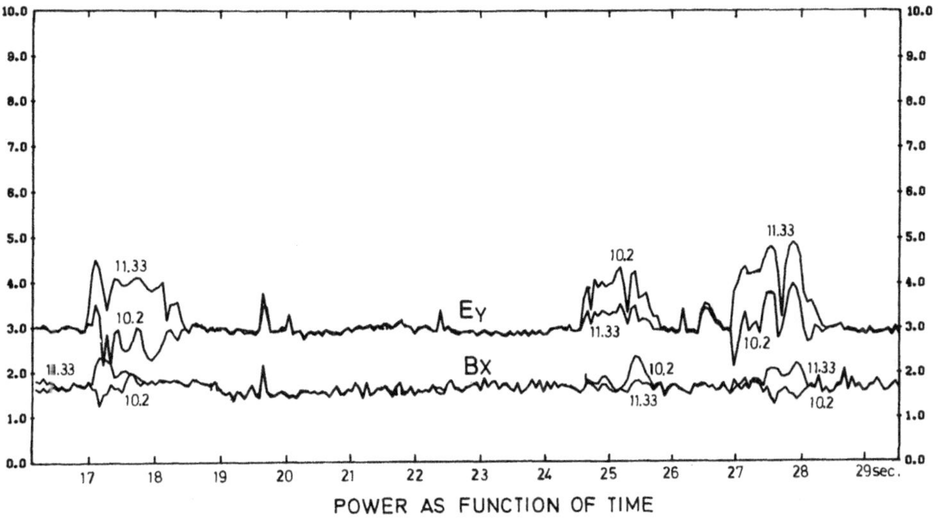

Fig. 8. Electric and magnetic field components of 10.2 kHz and 11.33 kHz showing the Omega signals on Day 355, 0806 UT. The ordinate is logarithm of power in arbitrary units.

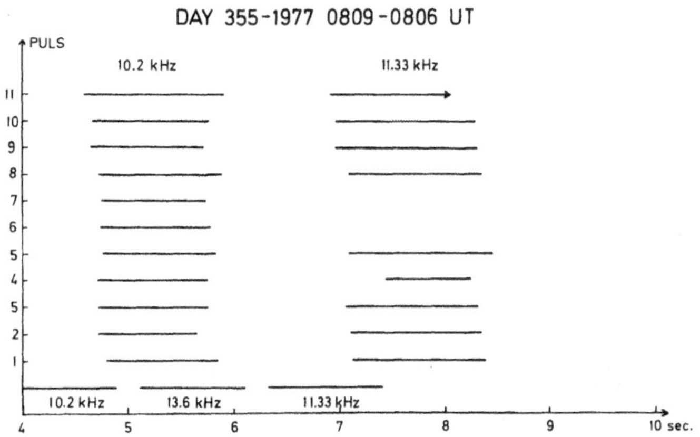

Fig. 9. Pulse length and start time relative to transmitted signal (bottom) for 10.2 kHz and 11.33 kHz Omega pulses received on Day 355, 0802–0806 UT. Only well defined pulses are shown.

amplitude of the signals, i.e., a stronger signal gives a greater prolongation. For the signals on Figure 9 and in Table I the 10.2 kHz pulses were usually much stronger than the 11.33 kHz pulses. An example of equal amplitude pulses gives a prolongation of 0.113 s at 10.2 kHz and 0.184 s at 11.33 kHz.

It is interesting to notice that the magnetic amplitude in the beginning of the pulses grows more slowly than the electric amplitude, furthermore there is strong amplitude modulation of the pulses (~ 10 dB) on a time scale of 0.1 s, often with the electric and magnetic components in antiphase giving rise to large E/B variations. To illustrate this effect, we have plotted the E/B ratio corresponding to Figures 7 and 8 as function of time in Figures 10 and 11, showing E/B variations up to a factor 5. Figures 7, 8, 10 and 11 are plots based on 1.1 kHz bandwidth.

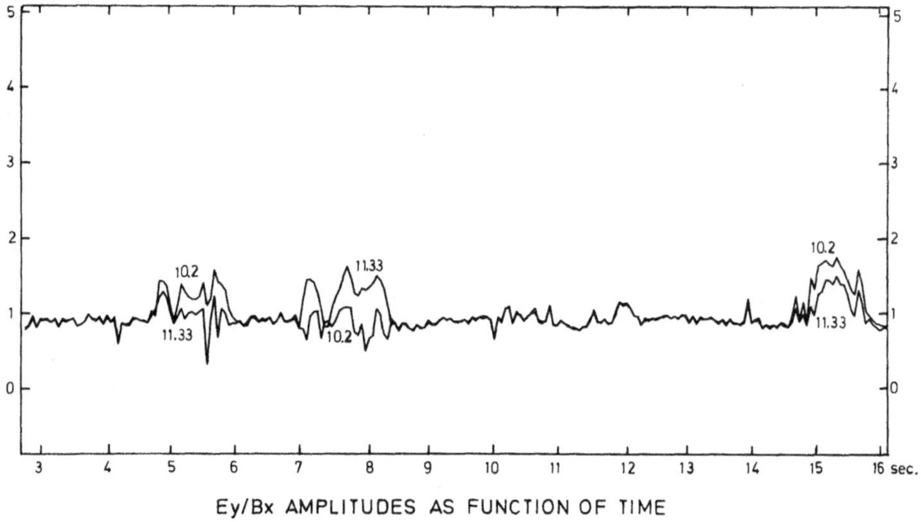

Fig. 10. *Ey*/*Bx* ratio on Day 355, 0806 UT of 10.2 kHz and 11.33 kHz showing large fluctuations in the Omega pulses. The ordinate is logarithm of amplitude in mV/m/nT (preliminary values).

TABLE I
Average properties of the pulses shown in Figure 9

	10.2 kHz	11.33 kHz
Delay	0.705 s	0.770 s
Duration	1.091 s	1.248 s
Prolongation	0.191 s	0.148 s

Similar plots for a single channel, (0.359 kHz bandwidth) have shown even stronger amplitude modulation.

So far only a few comments have been made about the 13.6 kHz signals. Because of aliasing 13.6 kHz appears on the spectrogram as 10 kHz, as seen in Figure 5. This is so close to 10.2 kHz, that it becomes impossible to make plots showing discrete, well defined pulses like Figures 7, 8, 10, and 11.

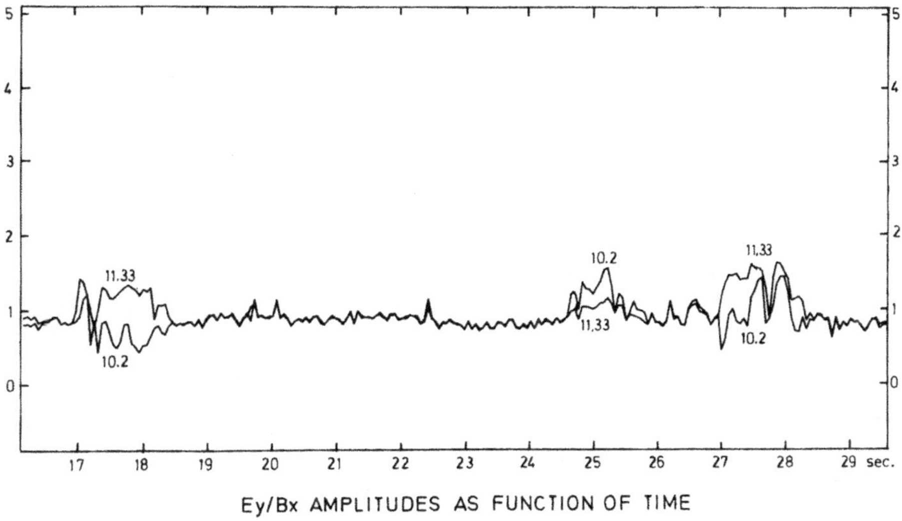

Fig. 11. *Ey/Bx* ratio on Day 355, 0806 UT of 10.2 kHz and 11.33 kHz showing large fluctuations in the Omega pulses. The ordinate is logarithm of amplitude in mV/m/nT (preliminary values).

3. Raytracing Calculations

We have compared the observations with raytracings based on the Hazelgrove equations. A simple diffusive equilibrium model was assumed, neglecting the plasmapause and ducting effects. Raytracing was carried out from the ionosphere (400 km) to the vicinity of GEOS-1. Figure 12 shows the result of the calculations for 10.2 kHz. Rays were started at magnetic latitudes of 66°, 64°, 62°, 60° and 58° with vertical wave normals. Tick marks indicate successive 0.5 s delays of the ray

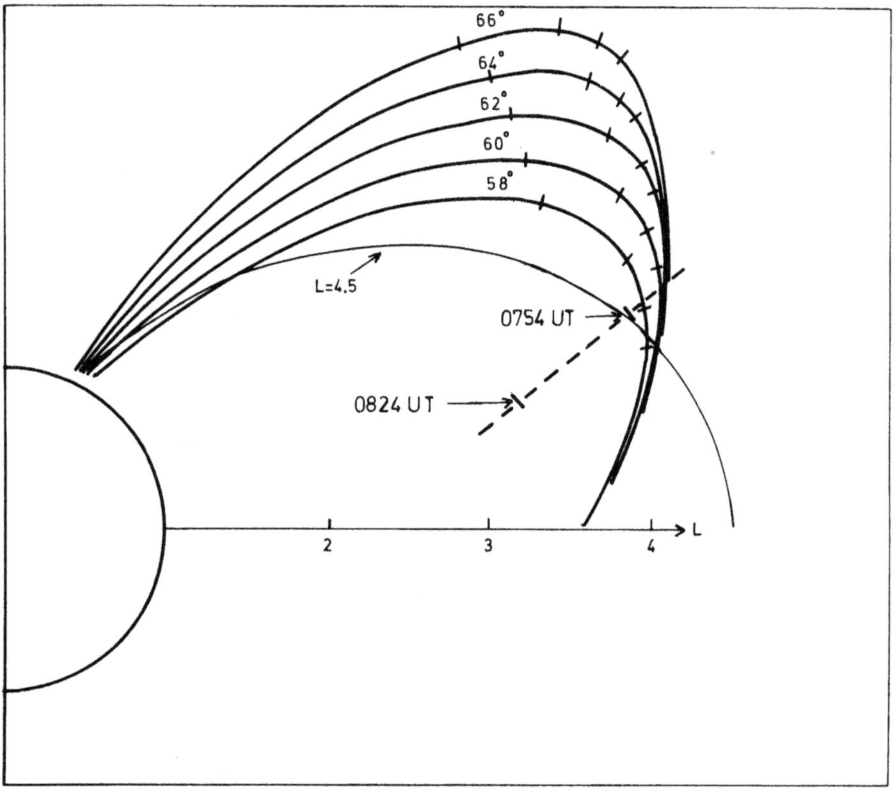

Fig. 12. Raytracing at 10.2 kHz. Rays commence at 400 km altitude at magnetic latitudes of 66°, 64°, 62°, 60° and 58° with vertical wave normals. Tick marks indicate successive 0.5 s delays of the ray for the first two seconds of propagation. Dotted line shows the GEOS-1 track in the magnetic meridian plane. From 0754–0824 UT the Omega transmitter in Norway was received on the satellite.

for the first two seconds of propagation. The dotted line shows the GEOS-1 satellite track in the magnetic meridian plane on Day 355. From 0754 UT to 0824 UT the Omega transmitter in Norway was received on GEOS. It is clear from Figure 12 that the signals are observed at lower altitudes and with a smaller propagation delay than predicted by the calculations.

We have tried the effect of varying the wave normal direction at 400 km. If we consider the lower boundary of the ionosphere as a sharp boundary, the refractive index will jump typically from 1 to 30. A ray with a large angle of incidence will then after refraction at the lower boundary propagate at an angle ~2° to be vertical. In Figure 13 we show ray paths for 10.2 kHz rays commencing at 400 km altitude at 62° magnetic latitude with wave normals having angles of 0°, 2° and 4° to the vertical and pointing South. Again the dotted line shows the GEOS-1 track in the magnetic meridian plane. It is evident that a variation of the wave normal direction does not give a better fit to the data, we should still observe the signals earlier according to the calculations.

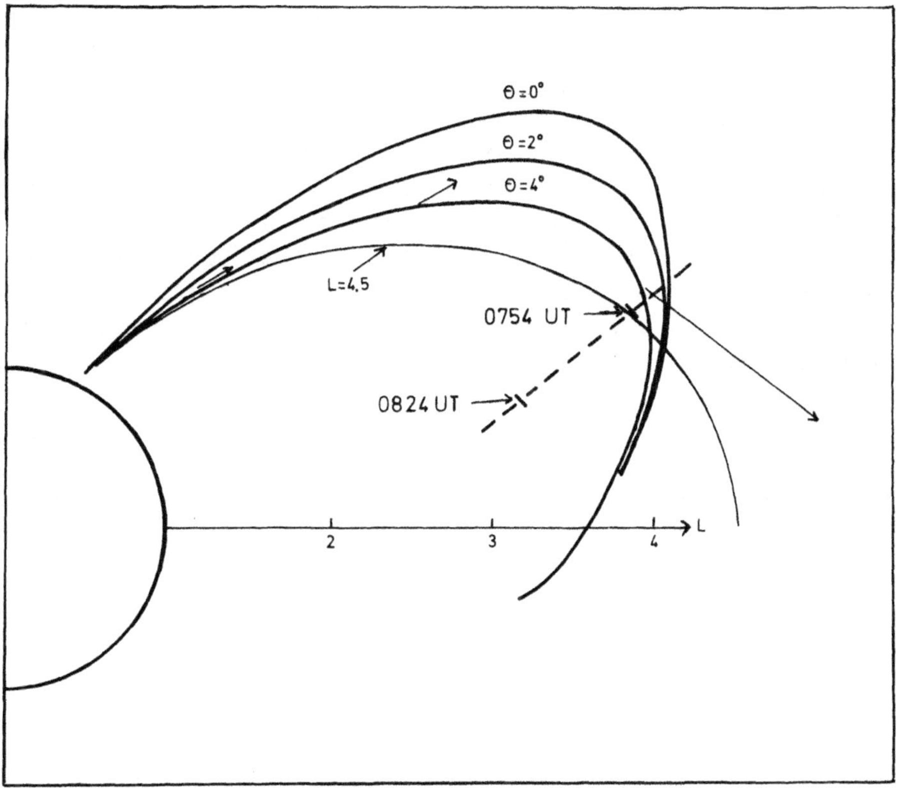

Fig. 13. Raytracing at 10.2 kHz. Rays commence at 400 km altitude at 62° magnetic latitude with wave normals having angles of 0°, 2° and 4° to the vertical, pointing southward. Wave normal directions are shown by arrows for the ray 4°, dotted line shows the GEOS-1 track in the magnetic meridian plane. From 0754–0824 UT the Omega transmitter was received on the satellite.

4. Conclusion

The 10.2 kHz, 11.33 kHz and 13.6 kHz signals transmitted by the Omega station situated at 63.5° magnetic latitude in Norway are received on board GEOS-1. The lowest frequencies come in at the highest L-values as expected, at a ratio of $f/f_{ce} \leqslant 0.7$. In a simple dipole model magnetosphere the Omega signals are detected for values of $L = 4.6$–3.1, corresponding to magnetic latitudes 62.2°–55.1°. Only experimental restrictions (change of mode) prevented observations of the signal below $L = 3.1$.

The magnetic amplitude of the pulses grows more slowly than the electric, and both components show strong amplitude modulation, often in antiphase, which gives rise to E/B variations of up to a factor of 5. The average signal delay is approximately 0.7 s (with an ambiguity of $n \times 10$ s). Strong pulses show greater prolongation than weak pulses, and for equal amplitudes, the transmitter pulse at

11.33 kHz (1.1 s long) is extended by 0.184 s when received at GEOS-1, while the 10.2 kHz pulse (0.9 s long) is extended by 0.113 s.

The raytracing calculations presented in this paper are based on a simple diffusive equilibrium model magnetosphere. As a first approximation these calculations are in reasonable agreement with observations, as they show that rays commencing at 400 km altitude in the ionosphere propagate to the vicinity of GEOS-1. In a diffusive equilibrium model magnetosphere the Omega transmitter signals should, however, have been observed earlier in the GEOS-1 passes. This discrepancy indicates that the waves are somehow guided to lower latitudes, thus again, as in connection with the intermittent observations in Figure 6, we come to the conclusion that the waves are ducted.

Acknowledgements

The work was supported in part by the Danish Natural Science Research Council.

References

Angerami, J. J.: 1970, *J. Geophys. Res.* **75,** 6115.
Burhans, R. W.: 1974, *Electronics* **47,** No. 18, 90.
Dunckel, N. and Helliwell, R. A.: 1974, *Radio Science* **12,** 821.

GROUND-BASED ELF/VLF OBSERVATIONS AT HIGH LATITUDES DURING PASSES OF GEOS-1 AND ISEE-1 AND -2

M. A. MADDEN, M. J. RYCROFT, and N. E. SMITH
Dept. of Physics, The University, Southampton S09 5NH, U.K.

Abstract. Recordings of ELF/VLF radio signals were made, as a contribution to the International Magnetospheric Study, in Iceland (17 August to 5 September 1977) and Norway (21 October to 15 December 1977; and 11 January to 27 February 1978) by the Space Radio Physics group. The equipment used at each of three sites was a goniometer (direction finding) receiver. As an example of the results obtained, recordings of risers, occurring at a rate up to $\sim 10\ \text{min}^{-1}$ and with frequencies ($\sim$1.0 to 1.5 kHz) just greater than those of simultaneous hissy chorus signals, made between 10:20 and 11:00 UT on 31 August 1977, are discussed. These risers (downcoming whistler mode signals) are shown to have well defined exit points from the ionosphere which are located, to within an uncertainty of typically $\pm$40 km, by triangulation. The observations are broadly consistent with there being a single exit point which, on this occasion, happens to be almost on the flux tube through the geomagnetic observatory at Leirvogur. Simultaneous ground-based magnetometer observations, and also wave and energetic charged particle observations made aboard GEOS-1, have been studied. The electron spectra and pitch angle distributions are as required for the operation of the electron cyclotron instability in which whistler mode signals are amplified.

1. Introduction

It is well known that whistlers constitute an important diagnostic tool for studying the distribution of thermal plasma in the magnetosphere. Discrete ELF/VLF radio emissions are a manifestation of an interaction between energetic electrons trapped in the magnetosphere's radiation belts and whistler mode radiation, which require further study before they can – by 'remote sensing' – provide detailed information on parameters of the Van Allen belts.

As part of a research programme centred on the International Magnetospheric Study, the Space Radio Physics group at Southampton University has made ELF/VLF observations at auroral latitudes. Signals in the band 0.5 to 16 kHz have been recorded using goniometer (direction finding) receivers. At each remote recording site, the equipment used consisted of an aerial (two electrostatically screened vertical loops, of area 60 m^3, mounted in the N–S and E–W planes), a preamplifier and a goniometer receiver. The signals from the two loops are combined electronically to simulate the output from a single loop rotating about a vertical axis with a frequency $\sim$27.5 Hz, and recorded on one channel of a stereo magnetic tape recorder. A stationary source gives a tape recorded signal with two maxima and two minima per simulated rotation. The phase difference between this modulation of the input signal and a calibration pulse occurring once per rotation period at a known orientation, say N, gives the azimuthal bearing of the source, with an ambiguity of 180°. This ambiguity is removed by triangulation from two or more stations. The

Space Science Reviews **22** (1978) 465–479. *All Rights Reserved*

region in space where a particular downcoming whistler mode signal emerges into the earth-ionosphere waveguide (termed its exit point) is thus located within a quadrilateral of uncertainty (see also Rycroft *et al.*, 1975).

2. Chorus and Risers Observed on 31 August 1977, and Discussion

Recordings made in Iceland on 31 August 1977 were subjected to detailed study. Selected tapes were studied at Sheffield University using their Semi-Automatic Whistler Analyser (Bullough *et al.*, 1975). Having calibrated the systems using signals from a small radiating coil mounted at a known angle at the centre of the large loop aerials, the azimuthal bearings of interesting naturally occurring signals could be determined.

As an example of the types of signals received at high latitudes, and of the procedure and the results obtained, Figure 1 shows the dynamic spectra of signals received at all three stations at 10:49:23 UT during a particularly active period. At the time of these signals, geomagnetic activity had been steady ($K_p \sim 2$) for 24 hr previously. Evident are two rising frequency emissions, between 0 and 1 s and between 3 and 4 s, with frequencies increasing from ~1.0 to ~1.5 kHz. These signals, with a sweep rate ~100 Hz s^{-1}, emerge from a band of hissy chorus near its upper frequency limit.

The right hand side of Figure 1 shows spectra obtained with the narrow band (22.5 Hz) filter of a Sonagraph. The bandwidth of the emission is well resolved to be ~200 Hz. By contrast, the left hand side of Figure 1 shows spectra of identical signals obtained with the wide band (150 Hz) filter in which the temporal variations are well resolved. Below 0.5 kHz the Sonagrams show the spectra of the calibration pulses. Just evident in the left hand part of Figure 1 are two maxima and two minima of the received signal per interpulse period, not only for the risers, but also for the hissy chorus.

Figure 2 shows this goniometer modulation much more clearly; it is the dynamic spectrum of the earlier riser in Figure 1. The improved temporal resolution necessitates a worsening of the frequency resolution, the bandwidth of the wideband setting on this 0 to 16 kHz range being 600 Hz. Particularly in the upper (Husafell) and lower (Siglufjördur) Sonagrams, the modulation of the riser is evident. Also evident in Figure 2 are strong atmospherics at ~0.2 s, by which the identification of the same signal at the three stations is aided. For this purpose also, signals from VLF transmitters are helpful; these include signals of ~1 s duration from the Omega long range navigation system at 10.2, $11\frac{1}{3}$, 13.6 kHz and other frequencies. Also shown here are signals of ~0.4 s duration at ~12.6 kHz, presumed to be from the U.S.S.R. version of Omega (Klass, 1971).

Returning to the risers, these occurred at a rate not exceeding 10 min^{-1} from 10:20 to 11:00 UT. Rather than perform triangulations on each individual riser, mean bearings from both Husafell and Skaftafell were calculated for all risers observed in 5 min intervals. In each case, an estimate of the standard error of the

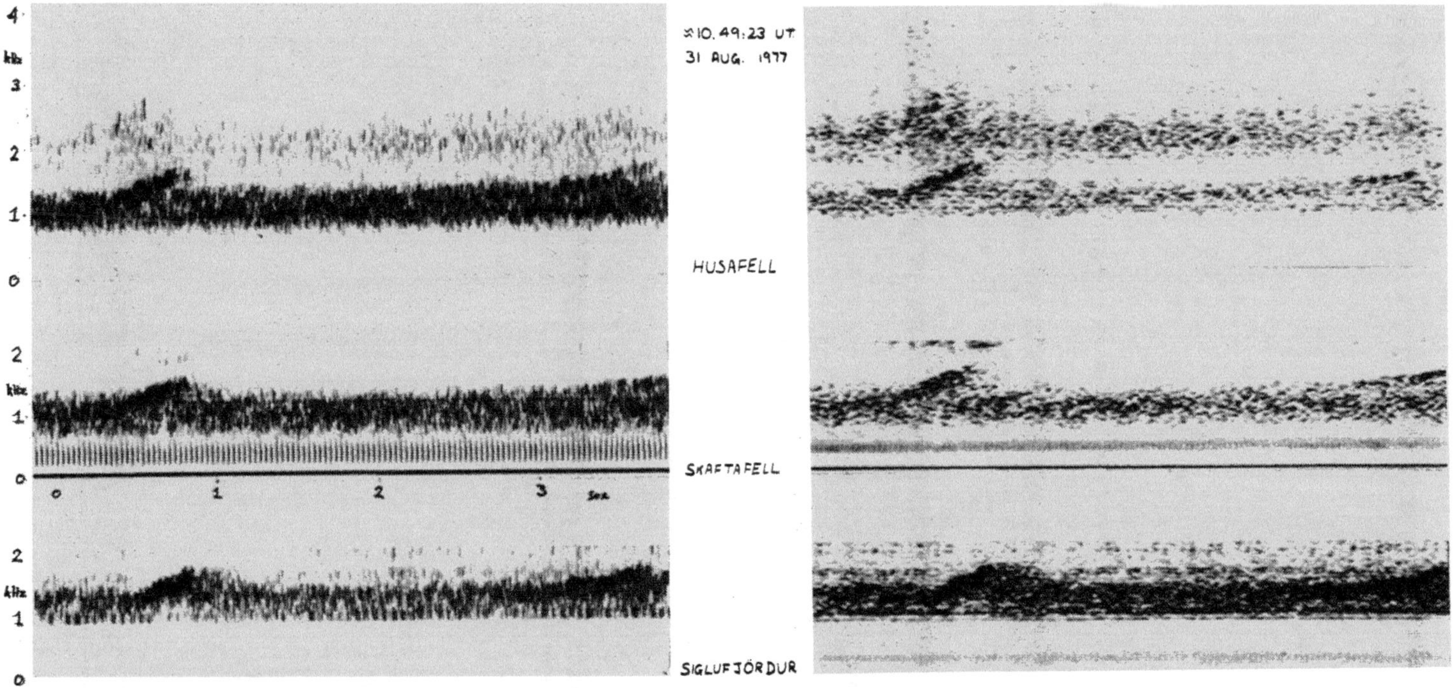

Fig. 1. Wideband (left) and narrowband (right) 0–2 kHz or 0–4 kHz Sonagrams of 4 s sample of risers and hissy chorus received at 10:49:23 UT on 31 August 1977 at Husafell (top), Skaftafell (centre) and Siglufjördur (bottom), all stations being in Iceland.

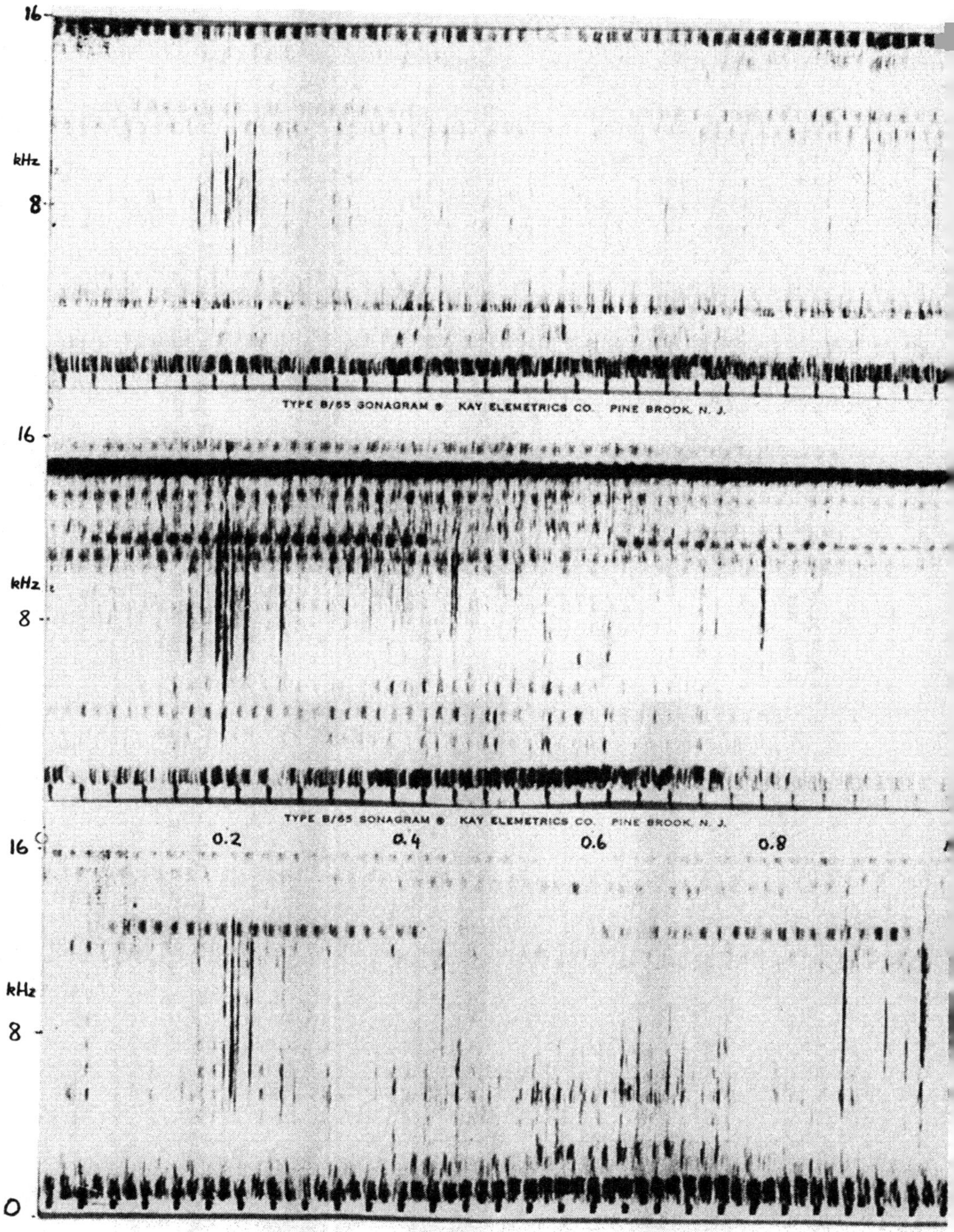

Fig. 2. Wideband 0–16 kHz Sonagrams of a 1 s sample showing a riser observed at 10:49:23 UT on 31 August 1977 at Husafell (top), Skaftafell (centre) and Siglufjördur (bottom); for discussion see text.

mean was also made, and found to be typically ±10° from Husafell and ±4° from Skaftafell. (Recordings made at Siglufjördur could not be analysed by the Semi-Automatic Whistler Analyser due to the presence of a time signal with the calibration pulses on the pulse channel of the stereo tape recorder).

Figure 3 shows, for example, a map of Iceland and triangulations from Husafell and Skaftafell to find the mean exit point for risers received between 10:45 and 10:50 UT on 31 August 1977. The full lines give the mean azimuthal bearings; the broken lines give the bearings of ±the standard deviation of the mean which define the quadrilaterial of uncertainty in the position of the exit point determined.

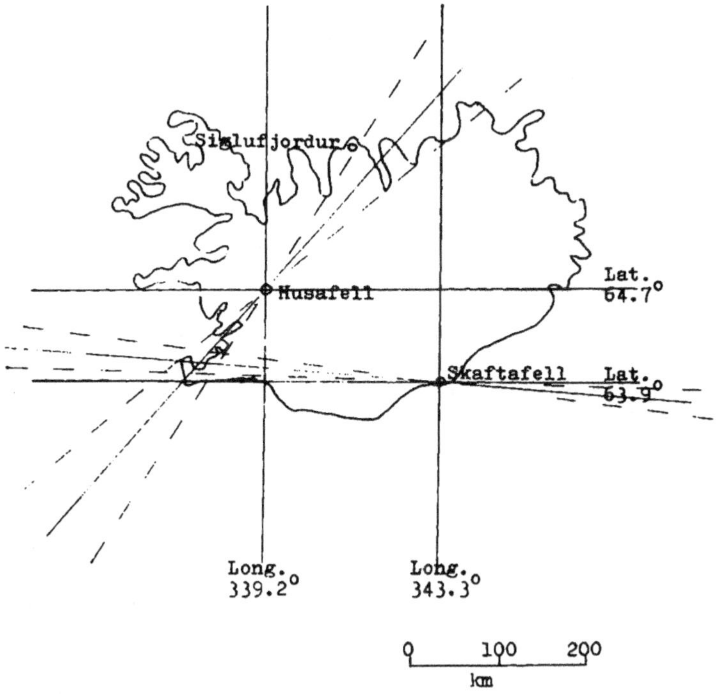

Fig. 3. Map of Iceland showing mean bearings, and standard errors of these, for the triangulation of risers observed between 10:45 and 10:50 UT on 31 August 1977.

In Figure 4, *A* marks the exit point for those risers observed between 10:40 and 10:45 UT, and *B* that for those between 10:45 and 10:50 UT (Figure 3). A preliminary analysis of the risers recorded at Siglufjördur at these times shows that their bearings are 220° ± 20°. Their propagation, from SW to NE, is consistent with the results shown in Figures 3 and 4. For risers observed between 10:35 and 10:40 UT, and also between 10:50 and 10:55 UT, the quadrilateral of uncertainty is open to the West of Iceland because the bearings from the two stations are rather

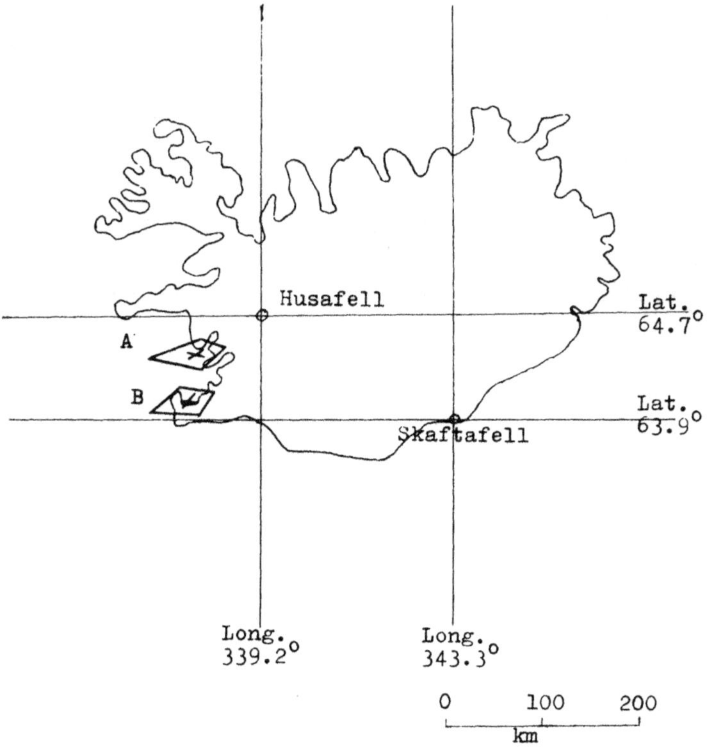

Fig. 4. Map of Iceland showing the mean exit points and their quadrilaterals of uncertainty, for downcoming risers observed between 10:40 and 10;45 UT (A), and between 10:45 and 10:50 UT (B).

similar. It can be stated, however, that exit points over the Southern of the two bays to the West of Iceland would fit these results.

Overall, the following six results are apparent.

(a) Sensible bearings are obtained by goniometer receivers when the distance of propagation of signals at 1.2 ± 0.2 kHz in the Earth-ionosphere waveguide is $\gtrsim 100$ km.

(b) Risers have well defined exit points, as do whistlers (Rycroft *et al.*, 1975).

(c) The error in the determination of the position of the exit point is typically ± 40 km, as it is for whistlers (Jarvis, 1976).

(d) The standard error of the mean bearing is, as expected, smaller when the distance travelled in the Earth-ionosphere waveguide is larger.

(e) *Either* the risers emerge from a few closely neighbouring exit points, with different proportions emerging from each during the different 5 min intervals, *or* they emerge from one exit point which may drift ~100 km in ~5 min.

(f) On this occasion, the exit point happens to be almost on the flux tube through the geomagnetic observatory at Leirvogur.

The conclusions can be strengthened by more detailed analyses, and by checking the exit points using bearings from the third station at Siglufjördur.

3. Simultaneous Observations aboard GEOS-1, and of Geomagnetic Pulsations, on 31 August 1977

The University of Lancaster group operate six magnetometers in Iceland as an IMS programme. With data recorded digitally every 10 s, Figure 5 gives an example of the magnetograms, that recorded at Siglufjördur, between 09:00 and 15:00 UT on 31 August 1977. It is evident that one of the most noticeable features of the records between 10:00 and 11:00 UT was from 10:36 to 10:50 UT. The Y component, i.e. D, the E–W horizontal component of the geomagnetic field, decreases by 15γ then. A rather similar decrease of Y, by 10γ, occurred from 10:17 to 10:22 UT. In both of these time intervals, the discrete risers mentioned earlier were particularly evident; then they were more frequent and more intense than at other times.

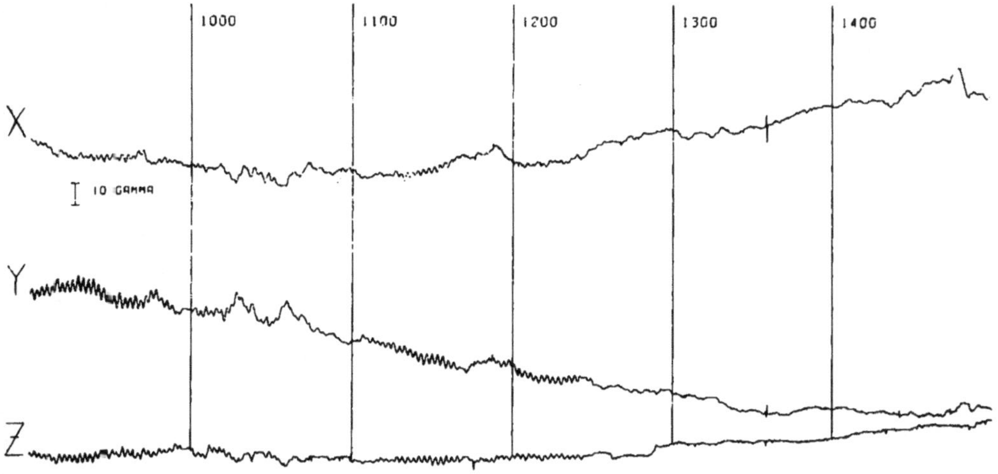

Fig. 5 X(N–S), Y(E–W), and Z (vertical) components of the geomagnetic field recorded at Siglufjördur between 09:00 and 15:00 UT on 31 August 1977.

The magnetograms also show quasi-monochromatic oscillations, with a period of ~2 min and an amplitude of ~2 or 3γ typically, from 09:00 to 09:55, from 10:03 to 10:15, from 11:07 to 11:39 and from 11:55 to 12:25 UT. In the future, we shall examine our ground-based VLF recordings in detail at these times; we request GEOS experimenters to study their observations at times between 09:00 and 12:00 UT on 31 August.

Electric and magnetic wave fields, with frequencies between 0.01 Hz and 77 kHz, are observed aboard GEOS-1 (S-300 experimenters, 1978). At the lowest frequencies, the electric field observations show energy at the proton cyclotron frequency (~4 Hz). From 10:00 to 10:40 UT there is a strong, apparently natural signal at 0.4 Hz, and also at harmonics up to the sixth.

Before discussing signals observed in the VLF band, it is necessary to consider the orbit of the satellite. Figure 6 shows a map of the projection, onto the ground, of the

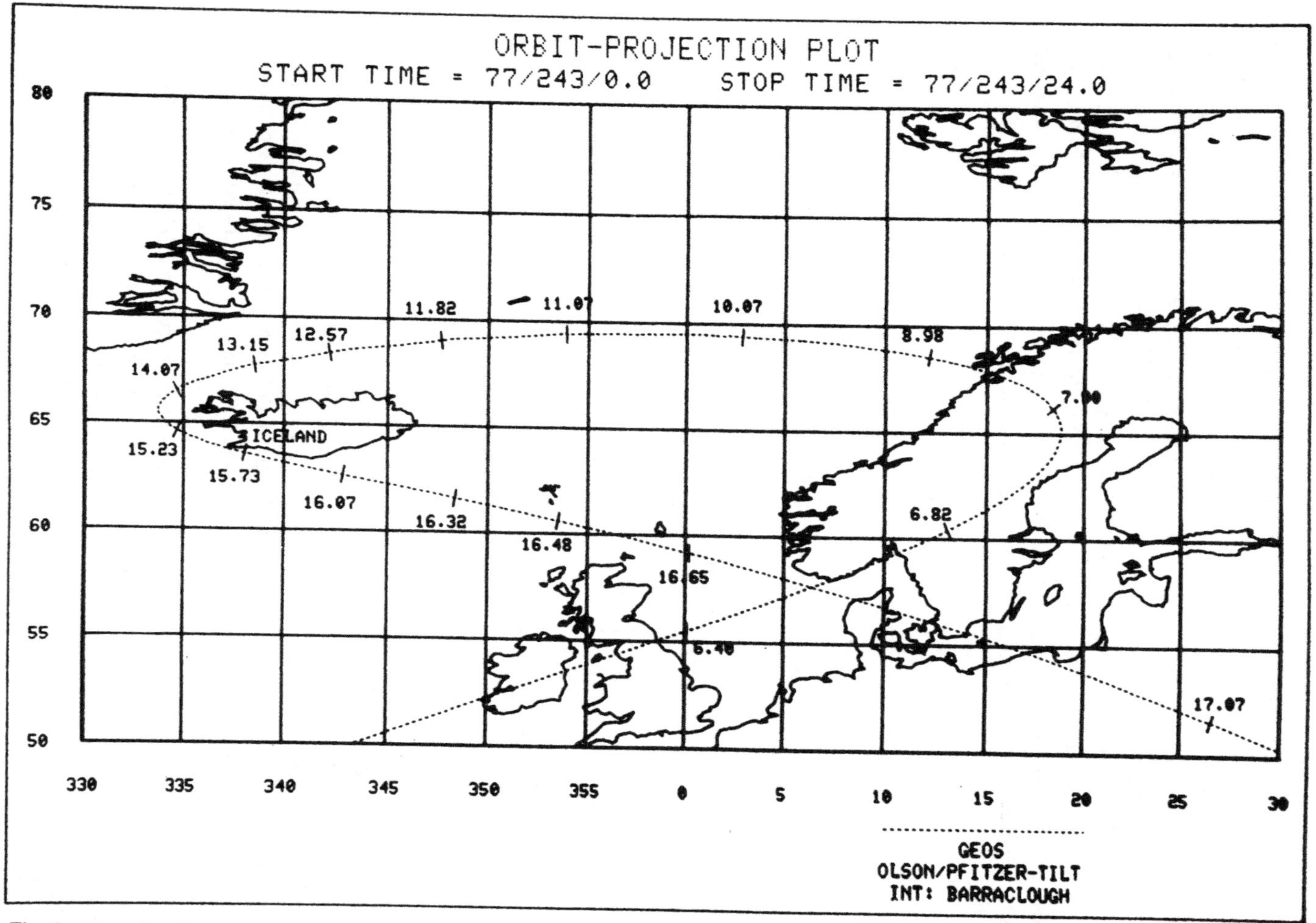

Fig. 6. Map showing the footprint of the geomagnetic flux tube through GEOS-1 at different times (in hours UT and hundredths of hours) on 31 August 1977. The Barraclough internal field model and the Olson-Pfitzer 1977 external field model (which includes the effect of the tilt of th

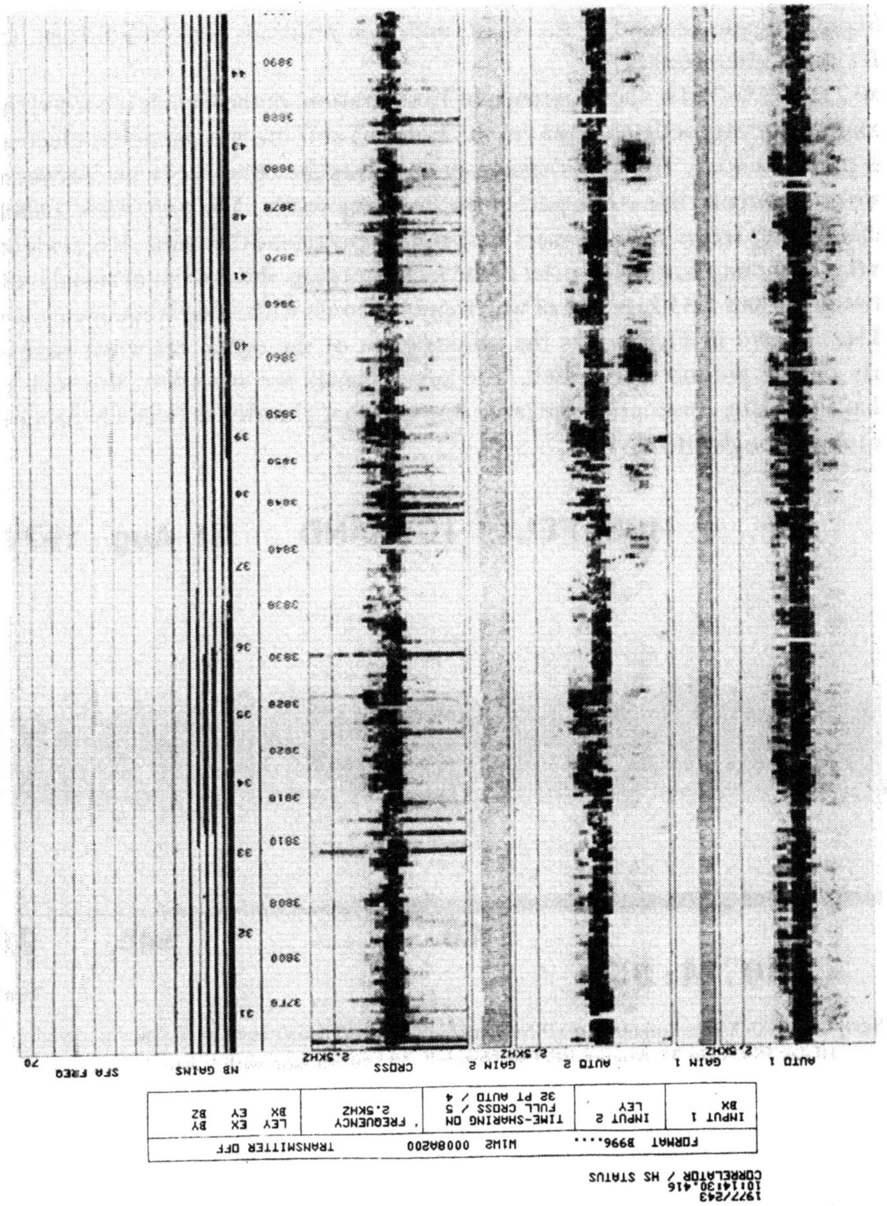

Fig. 7. GEOS-1 0–2.5 kHz spectrograms, from 10:14:30 to 10:14:45 UT on 31 August 1977. The lower panel is for the x component, magnetic antenna, the centre panel is for the y component, electric antenna, and the upper panel is the cross-spectrum between the two.

foot of the geomagnetic flux tube through GEOS-1. At 09:00 UT this is just to the West of Northern Norway, is within 5° of the Greenwich (0°) meridian between 10:00 and 11:00 UT, and passes to the North of Iceland between 12:00 and 14:00 UT. Apogee is reached at 11:43 UT (12:10 local time). The foot of the flux tube circumnavigates Iceland to the West, and then proceeds East South East, to lower L values, after 16:00 UT.

Figure 7 shows 0–3 kHz spectrograms, of 15 s duration, of the signals recorded by the x component magnetic antenna (at the bottom) and the y component electric antenna (in the centre). The upper spectrogram shows the cross-spectrum between these two components; this is one part of the Poynting vector. The uppermost panel shows that the step frequency analysers were not in operation at the time. It is evident that, on this outbound part of the orbit in the local morning, the dominant signal is of hiss between 1.0 and 1.35 kHz, out of which emerge tones with rising frequencies, up to 1.6 kHz. Shown in Figure 8 is the spectrogram of the signal recorded simultaneously on the ground at Husafell. The same signals are recorded; this fact is somewhat surprising considering the large distance that Husafell is from the foot of the flux tube through GEOS-1.

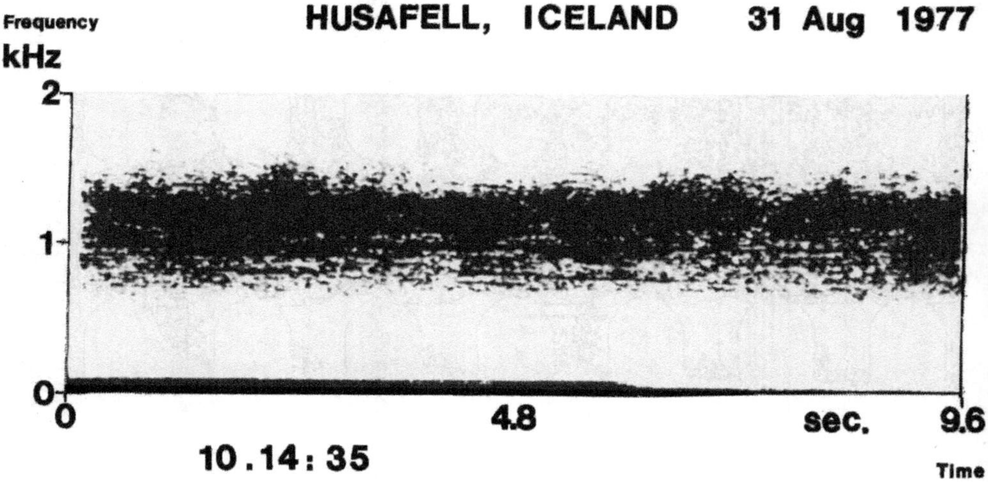

Fig. 8. Narrowband 0–2 kHz Sonagram of 9.6 s sample of hiss and rising frequency tones received at 10:14:35 UT on 31 August 1977 at Husafell; for comparison with figure 7.

Figures 9 and 10 show, similarly, simultaneous GEOS-1 and Husafell spectrograms of signals recorded at 15:18:45 UT on 31 August 1977. On GEOS-1, strong, rising frequency tones (from 1.1 to 1.4 kHz) are evident; these are not recorded on the ground. Also observed on the satellite are signals between 0.35 and 1.0 kHz. The components of these signals above 0.7 kHz are clearly also seen on the ground, the lower frequency components being removed by the filtering needed to remove power line radiation and its harmonics. The reason for this peculiar

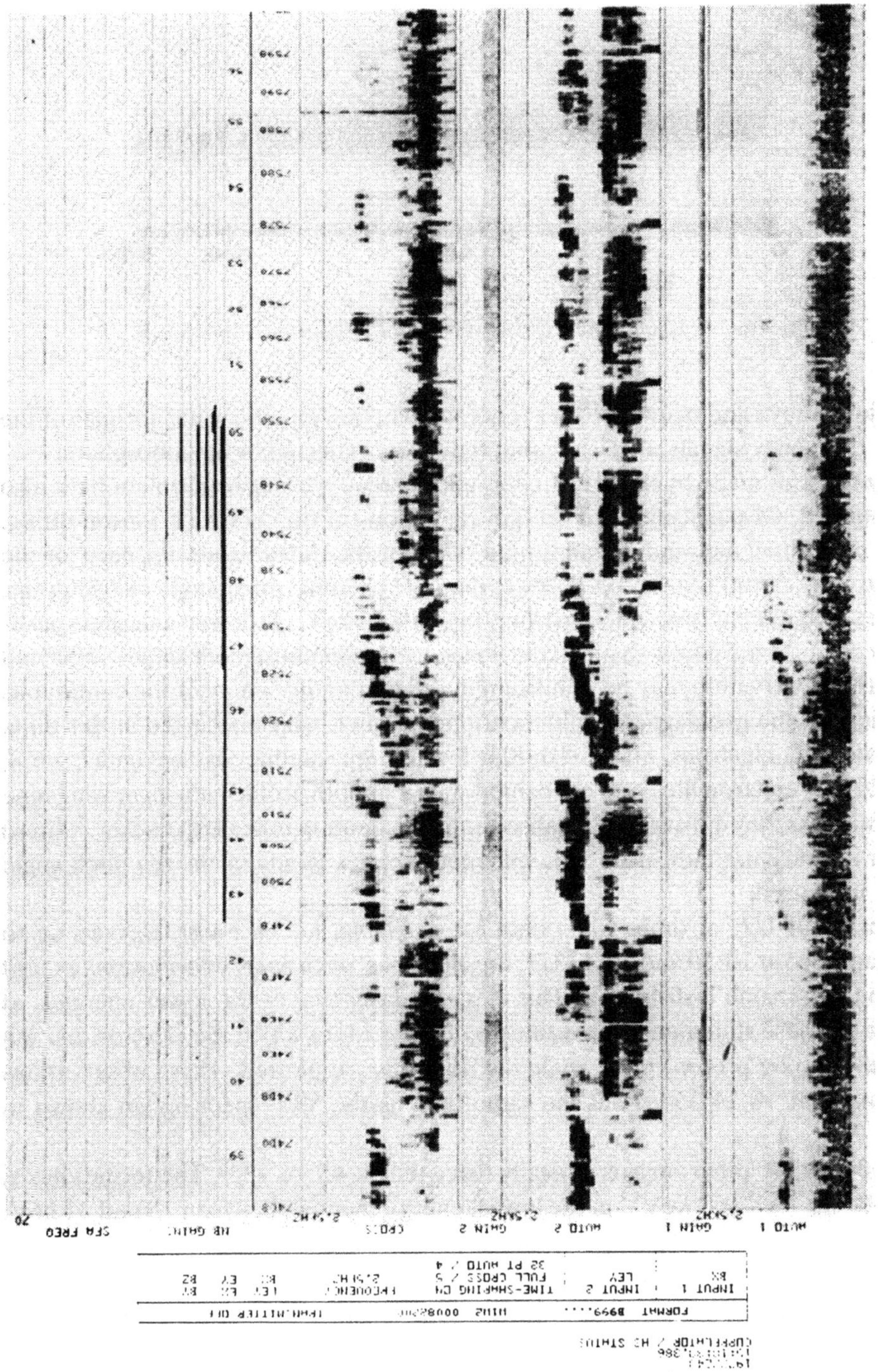

Fig. 9. As Figure 7, but from 15:18:38 to 15:18:57 UT.

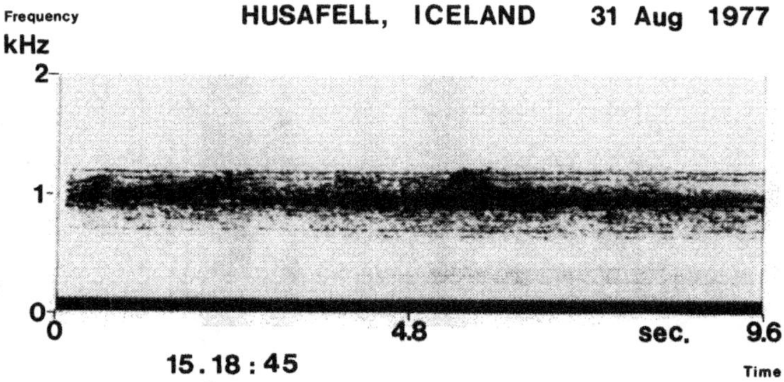

Fig. 10. As Figure 8, but at 15:18:45 UT; for comparison with Figure 9.

behaviour above and below 1.1 kHz is not known; possibly the wave normals of the higher frequency signals were such that they were reflected in the ionosphere.

Observations made by the S-321 energetic charged particle experiment have also been studied. Of particular interest are oscillations of the >24 keV proton fluxes, with a period ~2 min and an amplitude ~3% of the mean value, for parts of the interval from 09:00 to 09:55 UT. Regarding the protons' pitch angle distributions, these exhibit simple loss cone distributions at >43 keV. At lower energies, particularly in the 24 to 33 keV channel, the fluxes are maximum at pitch angles of 60° and 120°. This observation may be significant in relation to the source of the continuous pulsations of the geomagnetic field mentioned earlier, which occurred at this time.

Considering electrons, after ~10:00 UT when the satellite was beyond $L = 6.8$, the >20 keV electron fluxes do not exhibit such a smooth profile with increasing time (or L value) as they did earlier. Localised enhancements in time (with a scale ~3 min) and/or in space are indicated. This statement applies to any of the ten pitch angle ranges measured.

After 10:00 UT, at times when data are available, i.e. at 2 min intervals up to 10:16 and also at 10:30 and 10:43 UT, the electrons' pitch angle distributions exhibit a definite loss cone distribution. This is particularly true at the lower energies, at ≤50 keV. These statements are verified by Figure 11, in which the electron and ion (presumed to be proton) pitch angle distributions are plotted. These observations were made at 10:14:35 UT, at the same time as the VLF spectrogram shown in Figure 7.

The flux at 90° pitch angle is typically twice that at 45° or 135°. The former flux is $\sim 3 \times 10^5$ cm^{-2} s^{-1} sr^{-1} keV^{-1} in the lowest energy channel, between 15 and 21 keV.

Examination of the electron spectra (see Figure 12 for pitch angles near 90° at 10:14:35 UT) shows that there is a kink in the spectrum at energies ~50 keV. This is particularly marked at pitch angles ≤45° and ≥135°. This kink, or knee in the spectrum, is such that the fluxes of higher energy (>50 keV) electrons is less than would be expected from the observed electron fluxes at lower energies (<50 keV). It is concluded that the appropriate conditions are set, in the presence of an electron

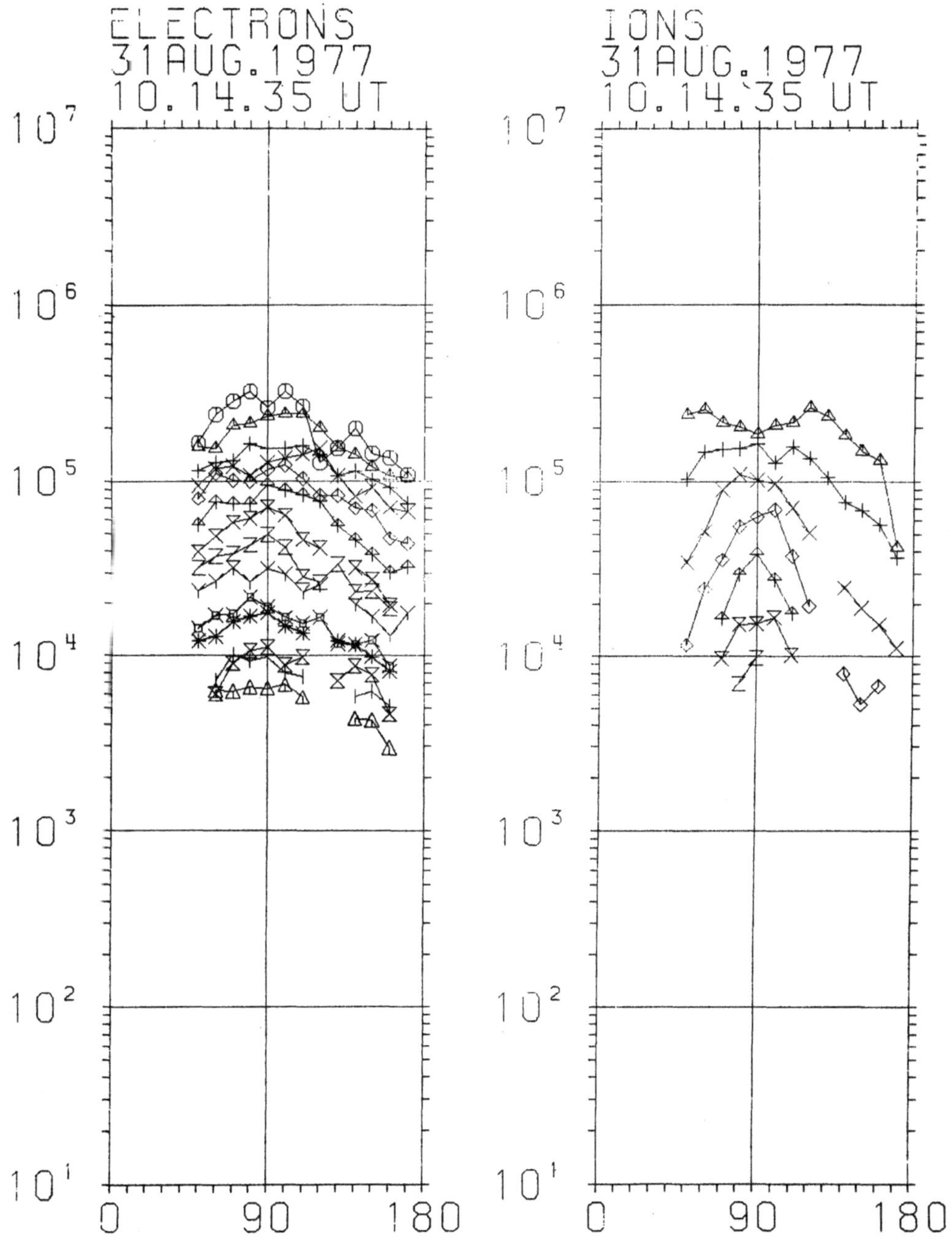

Fig. 11. Charged particle pitch angle distributions (0° to 180°) observed at 10:14:35 UT on 31 August 1977 aboard GEOS-1, in units cm^{-2} s^{-1} sr^{-1} keV^{-1}, at 15–21 keV (○), 21–27 keV (△), 27–35 keV (+), 35–41 keV (×), 41–48 keV (◇), 48–57 keV (⩚), 57–66 keV (⊼), 66–78 keV (Z), 78–90 keV (Y), 90–108 keV (⊗), 108–127 keV (✱), 127–151 keV (⧗), 151–174 keV (1), and 174–206 keV (△) for electrons, and at 17–24 keV (△), 24–33 keV (+), 33–43 keV (×), 43–56 keV (◇), 56–73 keV (⩚), 73–96 keV (⊼), and 96–125 keV (Z) for ions.

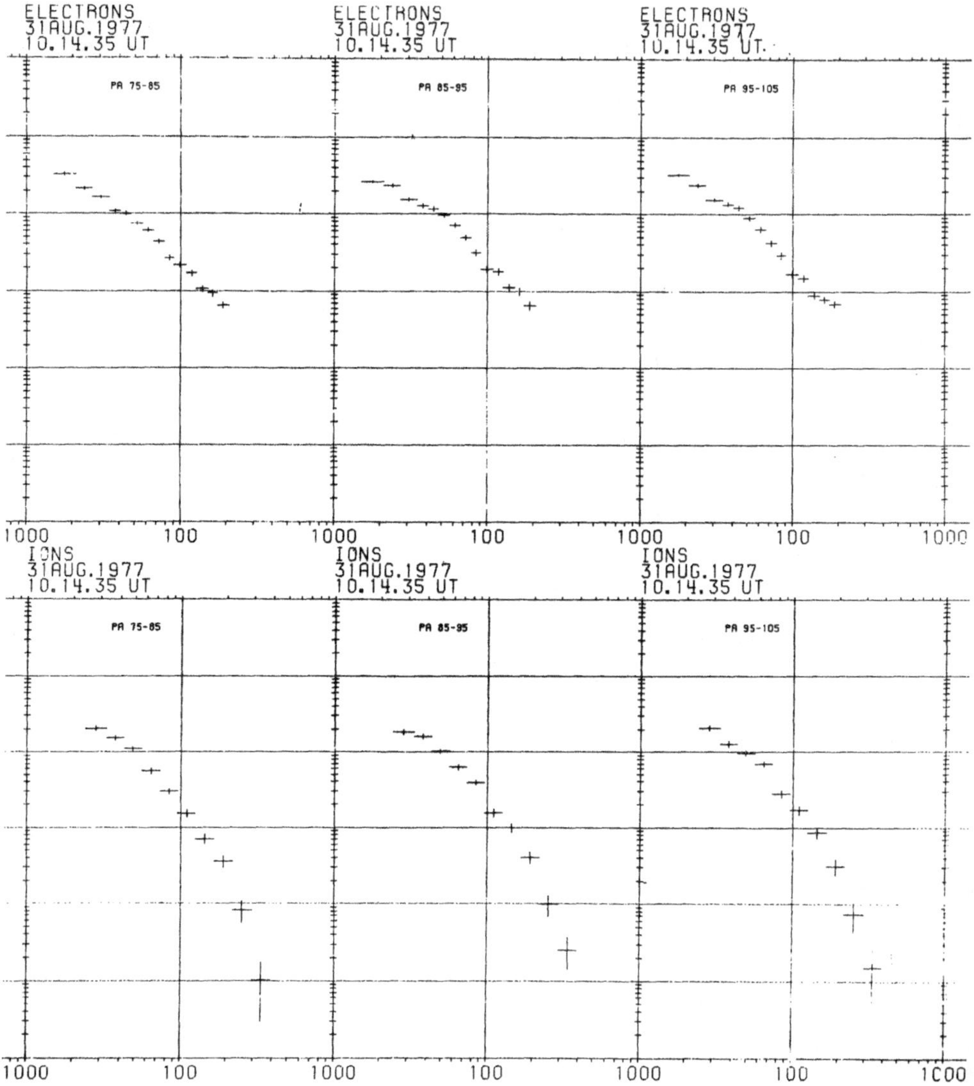

Fig. 12. Energetic electron (upper) and ion (lower) spectra from 1000 (10 keV) through 100 keV to 1000 keV, for three pitch angle ranges (75–85° left, 85–95° centre, and 95–105° right); horizontal lines occur at 10, 10^2, 10^3, 10^4, 10^5, 10^6, and 10^7 cm^{-2} s^{-1} sr^{-1} keV^{-1}.

loss cone distribution, for the operation of the electron cyclotron instability (see Kennel and Petschek, 1966). In this, whistler mode radiation grows in intensity, to a value above the observational threshold in space and on the ground.

Analyses similar to those reported here can be conducted at times of the reception of other interesting signals. It would be particularly valuable to do this when appropriate observations are being made aboard GEOS-1, ISEE-1, and ISEE-2 and when they are situated on flux tubes which cut the Earth's surface within a few hundred kilometers of the recording station.

Acknowledgements

The authors are most grateful to the Science Research Council for funding the research discussed, to many volunteers and to scientific colleagues, especially at the Universities of Oslo and Tromsø, who helped arrange suitable field sites, to all students who participated in these recording campaigns, and to colleagues at Sheffield University for the use of the Whistler Analyser. They also thank Prof. A. N. Hunter and Mr G. R. Moody of the University of Lancaster for providing Iceland magnetometer data, Dr D. Jones of ESTEC for sending GEOS-1 electric field ULF spectra, Dr F. Lefeuvre of CRPE, Orleans, for sending GEOS-1 VLF spectra, and Dr A. Korth of MPI, Lindau, for communicating GEOS-1 energetic charged particle results, all for 31 August 1977. They are indebted to Dr J. I. Vette and Dr M. Teague of the IMS SSC, Maryland, for providing Figure 6.

References

Bullough, K., Smith, A. J., Garside, G., and Deeley, A. M.: 1975, 'The Semi-Automatic Whistler Analyser', Paper presented at URSI General Assembly, Lima, Peru.

Jarvis, M. J.: 1976, 'An Investigation of the Morphology of Whistler Ducts by Triangulation using VLF Goniometer Receivers', Ph.D. thesis, University of Southampton.

Kennel, C. F. and Petschek, H. E.: 1966, *J. Geophys. Res.* **71,** 1.

Klass, P. J. 1971, *Soviets Using Omega-like Navaid System*, Aviation Week and Space Technology, p. 22, November 22.

Rycroft, M. J., Jarvis, M. J., and Strangeways, H. J.: 1975, 'Determination by Triangulation of the Regions where Whistlers Emerge from the Ionosphere', ESRO Special Publication No. 107, pp. 225–232.

S-300 Experimenters: 1978, *Planetary Space Sci.* **26,** (to be published).

THE USES OF PAST AND PRESENT MAGNETOSPHERIC FIELD MODELS FOR MAPPING FLUXES AND CALCULATING CONJUGATE POINTS

J. C. KOSIK

Division Mathematiques et traitement, Centre National d'Etudes Spatiales, Toulouse, France

Abstract. Two different utilisations of magnetospheric models are described in this paper. In the first one very crude analytical expressions are used to describe the motion of energetic particles and the evolution of fluxes along a given orbit. In the second one, the exact internal and external fields are used for field line tracing and the determination of conjugate points. Applications to the calculation of synchronous fluxes and the Geos conjugate points are given.

1. Introduction

In the present work we examine two important applications of the magnetospheric models: classification and identification of particle fluxes – calculation of magnetic conjugate points. In the first part we give some results obtained with the well-known Mead 64 magnetospheric model in absence or presence of electric fields. We also draw some consequences that can be expected from the use of the Olson–Pfitzer 74 model by comparing the field line topology of these two models. In the second part we give some results obtained in the determination of magnetic conjugate points. The origin of the diurnal and seasonal motion is clearly stressed with analytical results obtained with the Choe–Beard model. Then, GEOS conjugate locations obtained with the Olson–Pfitzer tilt model are presented.

2. Flux Mapping

Since 1960 classification and mapping of fluxes was performed using the McIlwain (B, L) coordinate system. In the outer zone this coordinate system suffers from shell splitting effects due to the magnetic field asymmetry and the presence of convection and corrotation electric fields. Different suggestions were made by Roederer (1970). We would like to examine here the possibility offered by the use of the approximate analytical topologies of the magnetospheric models. For this purpose, the Mead 64 model is an adequate tool as it provides very simple analytical solutions. Improvements can be obtained by adding the electric fields or by using a more up to date model like the Olson-Pfitzer 74 model.

2.1. Fluxes in absence of electric fields

In the synchronous region ($r < 8R_e$), the outer magnetospheric field can be considered as a perturbation of the dipole field. Thus an approximate analytical descrip-

Space Science Reviews **22** (1978) 481–497.

tion of the field line topology of the Mead magnetosphere can be obtained easily (Kosik, 1971):

$$r = L_0 \sin^2\theta \left\{ 1 - \frac{1}{2}\frac{\bar{g}_1^0}{g_1^0} L_0^3 \sin^6\theta + \frac{3}{4}\left(\frac{\bar{g}_1^0}{g_1^0}\right)^2 L^6 \sin^{12}\theta + \right.$$

$$\left. + 2\sqrt{3}\frac{\bar{g}_2^1}{g_1^0} L_0^4 \left(\frac{\sin^7\theta}{7} - \frac{\sin^9\theta}{3}\right)\cos\varphi_0 \right\}, \tag{1}$$

$$\varphi = \varphi_0 + \frac{\sqrt{3}}{7}\frac{\bar{g}_2^1}{g_1^0} L_0^4 \sin^7\theta \sin\varphi_0 .$$

In these equations L_0 is the field line label at longitude φ_0 ($L_0 = 1/\sin^2\theta_0$, where θ_0 is the colatitude of the field line foot). $\bar{g}_1^0$, $\bar{g}_2^1$ are the geomagnetic potentials of the Mead model, $\bar{g}_1^0 = 0.25/r_b^3$ G, $\bar{g}_2^1 = 0.12/r_b^4$ G, where r_b is the subsolar distance expressed in Earth radii). Longitude is counted from noon. From these equations it is possible to obtain the magnitude of the magnetic field along a dipole field line of given L_0, φ_0:

$$B = \frac{|g_1^0|(1+3\cos^2\theta)^{1/2}}{L^3\sin^6\theta} \left\{ 1 + \frac{1}{2}\frac{\bar{g}_1^0}{g_1^0} L_0^3 \sin^6\theta \left(\frac{5+3\cos^2\theta}{1+3\cos^2\theta}\right) - \right.$$

$$- \frac{3}{4}\left(\frac{\bar{g}_1^0}{g_1^0}\right)^2 L_0^6 \sin^{12}\theta \frac{1+15\cos^2\theta}{(1+3\cos^2\theta)^2} +$$

$$\left. + \frac{3\sqrt{3}}{7}\frac{\bar{g}_2^1}{g_1^0} L_0^4 \sin^7\theta \left(\frac{5-13\cos^2\theta}{1+3\cos^2\theta}\right)\cos\varphi_0 \right\}. \tag{2}$$

Using Equations (1) and (2) it is possible to obtain analytical expressions of I and B_m. These two invariants will depend on three parameters L, φ, and θ_m which is the mirror point colatitude. This last parameter can be replaced also by the equatorial pitch-angle α_e. The invariants I and B_m can be split in internal field and external field components $I_d + I_e$ and $B_{md} + B_{me}$. Perturbation technique is applied a second time: the label L is found to be constant on first order but dependent on the pitch-angle to the next order:

$$L = L_0 \left[1 + \frac{\bar{g}_2^1}{g_1^0} L_0^4 q(\alpha_e)(\cos\varphi - \cos\varphi_0) \right]. \tag{3}$$

This equation gives the value of the label L of the field line where is to be found a particle of initial location L_0 and equatorial pitch-angle α_e after the drift from φ_0 to φ. The change in L depends on the value of the initial pitch-angle α_e, through the function $q(\alpha_e)$ drawn on Figure 1. One notices that $q(\alpha_e)$ is maximum for equatorial particles. The change in pitch-angle as the particle drifts can be deduced from the

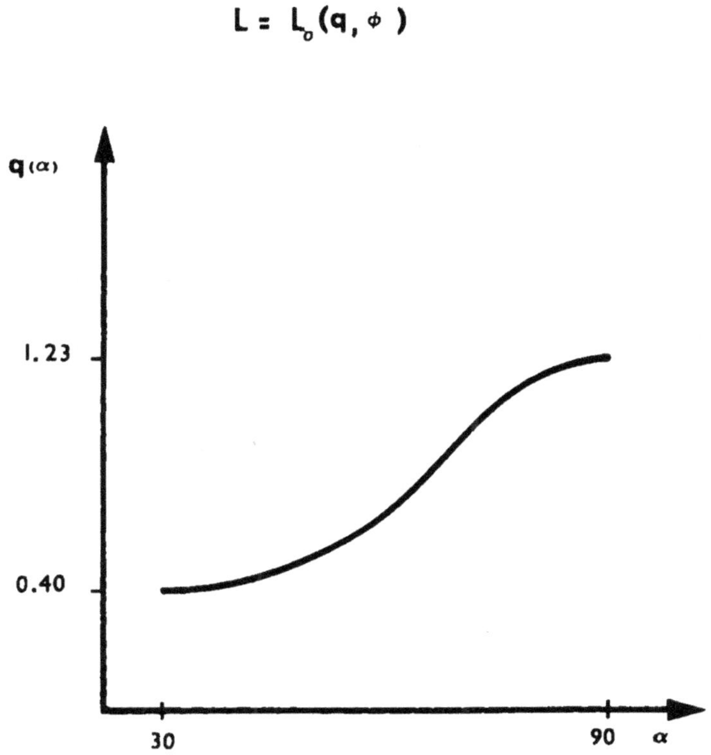

Fig. 1. Function $q(\varphi_c)$ gives the shell-splitting dependence on pitch-angle α_e.

conservation of the first adiabatic invariant μ and Equation (4):

$$\sin \alpha_e = \sin \alpha_{e0} \times \exp\left[3 \frac{\bar{g}_2^1}{g_1^0} L_0^4 (q(\pi/2) - q(\alpha_c))(\cos \varphi - \cos \varphi_0)\right]. \tag{4}$$

This equation shows simply that the equatorial pitch-angle of a particle decreases when the particle drifts from noon to midnight. We can use the last equations and a map of directional fluxes at noon to built fluxes at any other longitude and L value. Let us give three examples of this, in the case of synchronous satellites ATS-1 and OV2-5. One first notices that the shells sampled in one day by a synchronous satellite is given by:

$$L = L_0\left[1 + \frac{8\sqrt{3}}{21} \frac{\bar{g}_2^1}{g_1^0} L_0^4 (\cos \varphi - 1),\right] \tag{5}$$

where L_0 is the field line label at noon.

ATS-1 has a spin axis which is approximately perpendicular to the plane of the orbit. We choose the following distribution of directional fluxes

$$j = A e^{-\gamma \cos^2 \alpha_e - \beta (L - L_s)}. \tag{6}$$

We then use Equations (7) and (5, 6):

$$j = j_0 \exp\left[-\frac{\bar{g}_2^1}{g_1^0} L_s^4 (\cos\varphi - 1)\, \beta L \left(\frac{8\sqrt{3}}{21} - q(\alpha_e)\right)\right] \times$$
$$\times \left[1 - 3\frac{\bar{g}_2^1}{g_1^0} L_s^4 (\cos\varphi - 1)\ \gamma\{q(\pi/2) - q(\alpha_2)\}\right].$$

This equation gives the directional fluxes observed for any longitude and L value at arbitrary pitch-angle α_e. Usually experimenters neglect the effect of the pitch-angle, considering only equatorial particles. Figure 2 shows the flux variation obtained with Equation (9). However, in some cases, the influence of the pitch-angle is of importance. This is particularly true in the proton fluxes detected by the satellite OV2-5: its spin axis is located in the plane of the orbit and the detector samples particles of various pitch-angles and with guiding centers located on a shell different from the satellite shell. If we adopt the proton distribution given by Stevens *et al.* (1970).

$$j = j_0 \sin^7 \alpha' \, e^{-\beta(L' - L_s)} \tag{8}$$

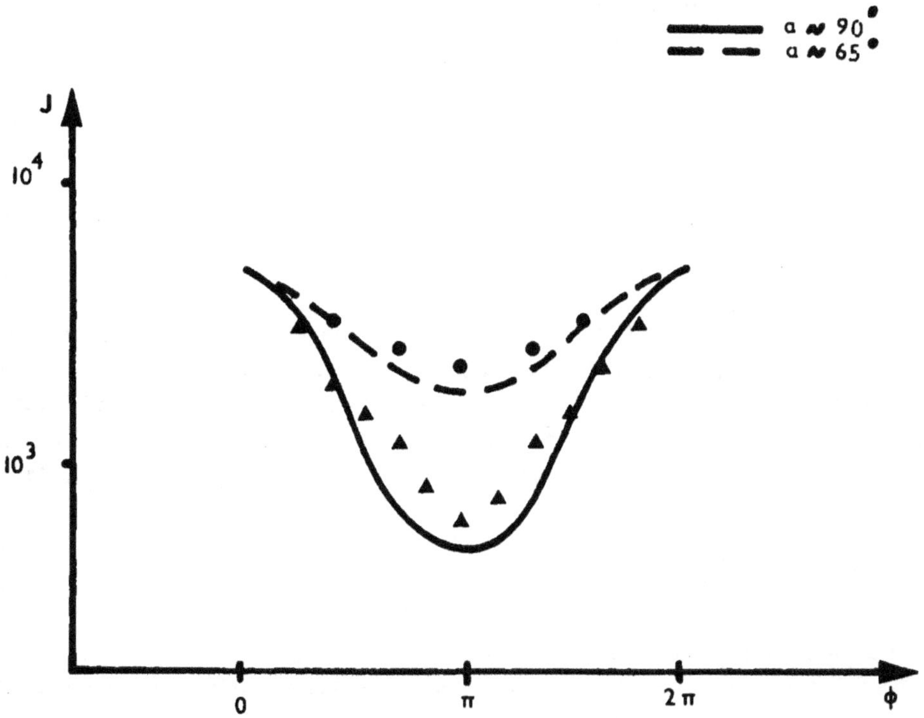

Fig. 2. The continuous and dotted curves show the diurnal variation of directional electron fluxes ($Fe \geq 150$ keV) observed by ATS-I detectors at pitch-angles $\alpha_e = 65°$ and $\alpha_c = 90°$ respectively.

we obtain the following flux distribution at longitude φ (Kosik, 1971);

$$j = j_0 \sin^7 \alpha \left\{ 1 + \tfrac{7}{2} K(1 - \cos \varphi) \right\} e^{-\beta \Delta L_s} \times$$
$$\times \exp\left[-\beta \frac{\bar{g}_2^1}{g_1^0} L_s^5 \left(\frac{8\sqrt{3}}{21} - q(\alpha_c) \right) (\cos \varphi - 1) \right]. \tag{9}$$

In this equation ΔL_s is the gyro-radius contribution which is not negligible for protons. The corresponding fluxes observed at noon and midnight are shown on Figures 3 and 4 respectively. One notices that the maxima are greater when the

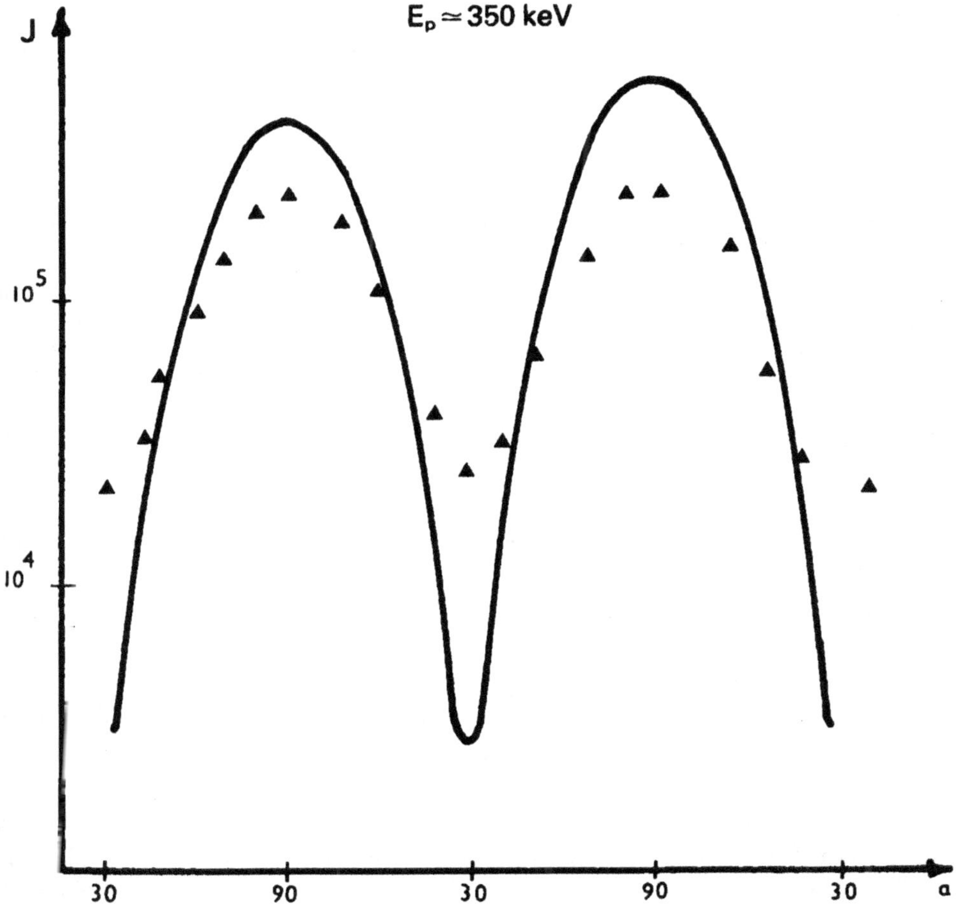

Fig. 3. The OV2-5 spacecraft is located at noon with its spin axis in the plane of the orbit. During a complete revolutions protons of 25° to 90° pitch-angles are measured. The maxima correspond to equatorial particles. The right shoulder correspond to detector looking eastward and counting protons located on an inner shell (relative to satellite) and the flux is higher.

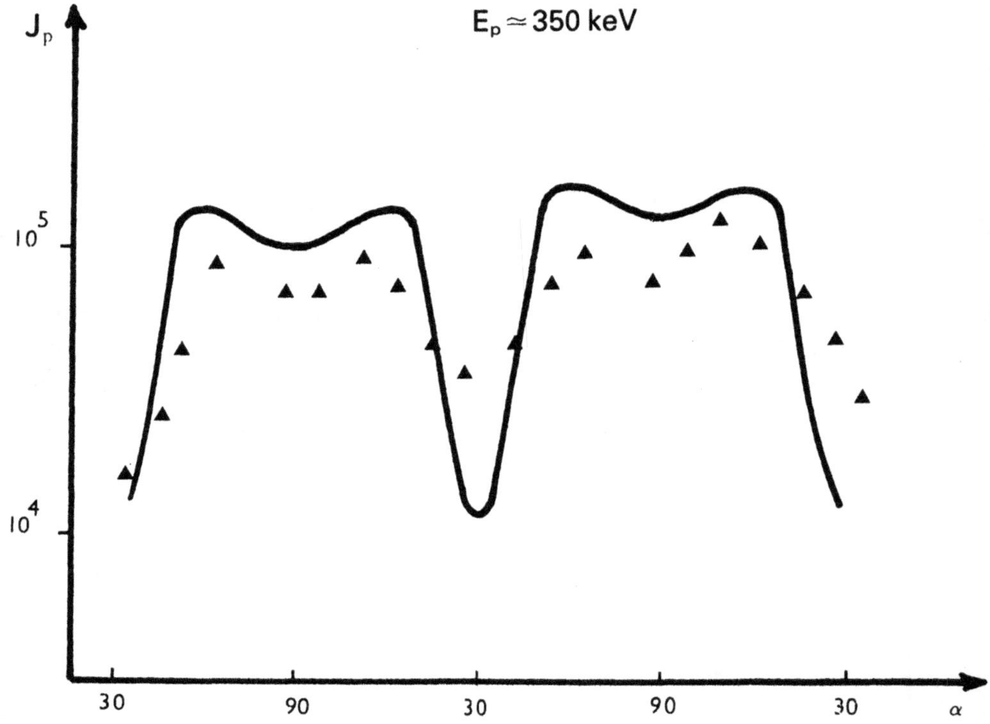

Fig. 4. The OV2-5 spacecraft is located at midnight. As the satellite rotates it first measures protons of low pitch-angles with guiding centers located on an outer shell (relative to the satellite location). Then as the pitch-angle increases the detector measures higher fluxes, but more external shells. A maximum is obtained resulting of the balance between shell-splitting effect pitch-angle flux dependance. As the rotation continues, the shell-splitting effect is predominant and measured fluxes decrease: a small minimum is observed. As in Figure 3, the right shoulder corresponds to guiding centers located on an inner shell(relative to satellite location).

detector looks eastwards: the protons detected have their guiding centers on an inner shell and fluxes are higher.

2.2. Fluxes in presence of electric fields

In presence of electric fields three cases can be considered:

- Below 1 keV, the asymmetry of the magnetic field can be neglected compared to the asymmetry induced by electric fields: the motion of particles can be studied using a dipole field only as was done by different authors (Chen, 1970; Stern, 1975; Roderer and Schulz, 1971; Kivelson and Southwood, 1975). It is then possible to explain some features observed by Explorer 45 (Cowley and Abdallah, 1975, 1976).

– Between 1 keV and 50 keV the magnetic and electric asymmetries are of equal importance and only purely numerical solutions can be obtained.

– Above 50 keV, the electric field effects play a minor role and can be considered as perturbations of the strictly asymmetric magnetospheric magnetic field. Analytical solutions are possible and will be presented here.

We adopt also the Mead magnetosphere to which are added convection and corotation electric fields given by:

$$V_1 = -AL \sin \varphi_0, \qquad V_2 = \frac{B}{L}. \tag{10}$$

The particle motion verifies now the three invariants μ, J, and W, where W is the total energy of the particle:

$$W = E + q(V_1 + V_2). \tag{11}$$

Using the field line topology given by Equations (1) it is still possible to solve the corresponding equations:

$$\mu = C^{st}_{-}, \qquad J = C^{st}_{-}, \qquad W = C^{st}_{-}. \tag{12}$$

The field line label L is now found to be dependent on both the magnetic field asymmetry and on the electric field:

$$L = L_0\left[1 + M\frac{\bar{g}_2^1}{g_1^0} L_0^4(\cos(\varphi - U) - \cos(\varphi_0 - U))\right], \tag{13}$$

where $$M = \sqrt{p^2(\theta_m) + \frac{A^2 q^2}{E^2}\left|\frac{g_1^0}{\bar{g}_2^1}\right|^2 \frac{g^2(\theta_m)}{L^6}} \tag{17}$$

and $$\tan U = \frac{Aq}{E}\frac{1}{L^3}\frac{g_1^0}{\bar{g}_2^1}\frac{g(\theta_m)}{p((\theta_m)}. \tag{15}$$

In these equations E is the kinetic energy, A the convection electric field potential, $g(\theta_m)$ and $p(\theta_m)$ are given in Table I. This equation shows that the plane of symmetry of the shell is tilted by some angle U towards the noon–midnight meridian, U being positive for electrons and negative for protons. When a particle drifts in longitude, it samples different field lines given by Equation (13), the value of L being minimum at longitude $\varphi = U + \pi$ and maximum at longitude $\varphi = U$. The value of angle U

TABLE I

$p(\theta_m)$ gives the dependance of magnetic shell-splitting on θ_m. $g(\theta_m)$ gives the dependance of electric shell-splitting on θ_m.

θ_m	$p(\theta_m)$	$g(\theta_m)$
30°	0.42	0.47
40°	0.43	0.46
50°	0.47	0.43
60°	0.57	0.40
70°	0.76	0.37
80°	1.06	0.345
85°	1.16	0.34
90°	1.23	0.33

depends greatly on the value of kinetic energy of the particle and on the strength of the convection electric field and the mirror point colatitude θ_m (Figure 5).

We can associate also to this change of field line a simultaneous change in pitch-angle α_e and in energy. The change in energy is given by:

$$\Delta E = qAL_0(\sin \varphi - \sin \varphi_0) + \\ + qAL_0^5 \frac{\bar{g}_2^1}{g_1^0} M \{\cos(\varphi - U) - \cos(\varphi_0 - U)\} \sin \varphi - \\ - qBL_0^3 \frac{\bar{g}_2^1}{g_1^0} M \{\cos(\varphi - U) - \cos(\varphi_0 - U)\} . \quad (16)$$

This equation shows that the kinetic energy increases when the particle drifts to the post midnight sector. The energy gain can reach 20 keV for particles located on shells of $L \sim 6$. As a consequence the detector of a synchronous satellite will measure near

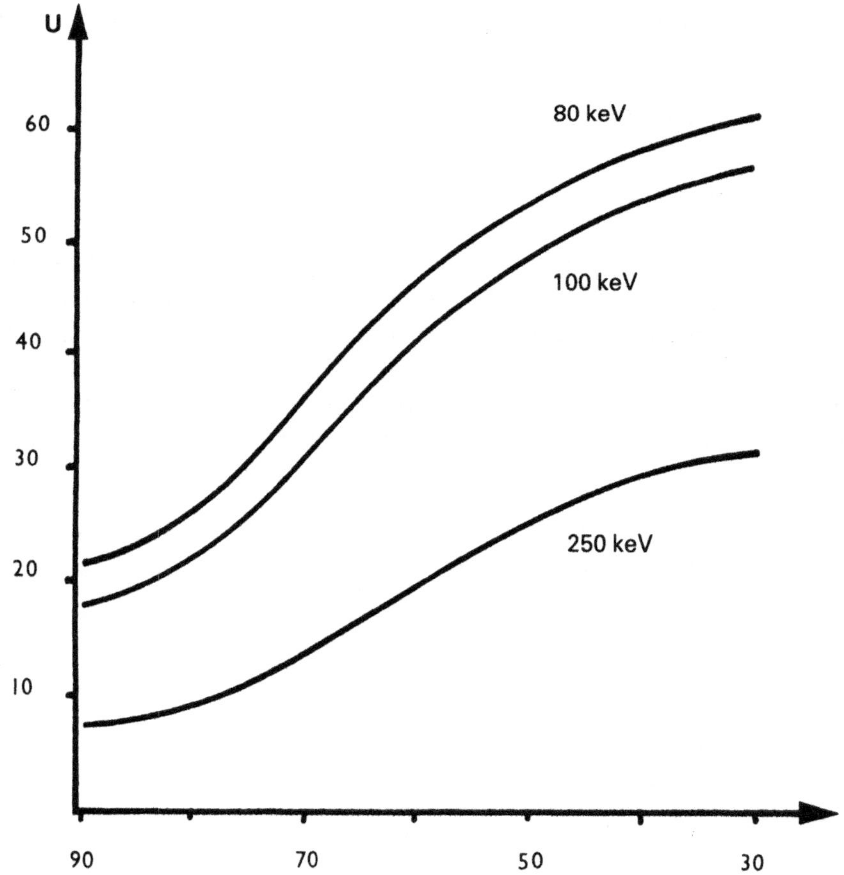

Fig. 5. Three curves which correspond to electrons of 250 keV, 100 keV, 80 keV show the dependance of angle U with the mirror point colatitude.

midnight particles that had initially lower energies at noon. If one assumes that the fluxes increase with the decrease of the energy, this effect will tend to increase observed fluxes in the midnight sector, at the opposite of the magnetic shell-splitting which produces a dip (Figure 2). As a consequence, during quiet periods, the dip observed by ATS-I is quite small compared to the one predicted assuming only a magnetic field asymmetry (Figure 6).

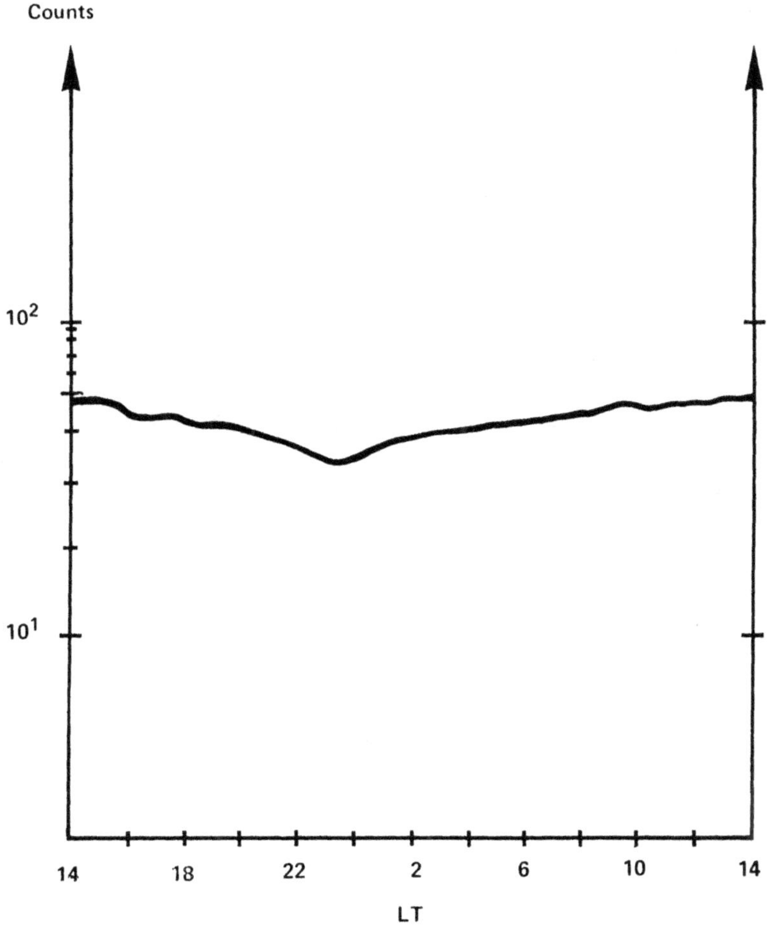

Fig. 6. Observed fluxes by a detector aboard ATS-1 (Lezniak *et al.*, 1968).

2.3. Contribution of more recent models

Since 1964 various models have been obtained which have brought various improvements. Among these, the Olson–Pfitzer 74 model adds the depression due to the ring current. Unfortunately, the three components B_x, B_y, B_z lead to rather complex field line equations:

$$r = L\sin^2\theta\left\{1+\frac{L^3}{g_1^0}\left[-1.25\times\sin^6\theta+0.89\sin^8\theta+L^2\sin^{10}\theta,\right.\right.$$

$$(0.76-3.23\sin^2\theta)+L^3\sin^{12}\theta(-0.15+0.617\sin^2\theta)+L^4\sin^{14}\theta,$$

$$\left.(0.0075-0.03\sin^2\theta)\right]+\frac{L^3}{g_1^0}\cos\varphi\left[L\sin^7\theta(-0.978+3.07\sin^2\theta)+\right.$$

$$+L^2\sin^9\theta(0.127-0.419\sin^2\theta)+$$

$$\left.+L^3\sin^{11}\theta(-0.0057+0.0197\sin^2\theta)\right]+$$

$$+\frac{L^3}{g_0^1}\cos 2\varphi[\ldots], \tag{17}$$

$$\varphi = \varphi_0+\frac{L^3}{g_1^0}[-0.3802C\sin^9\theta\sin\varphi+0.0518C^2\sin^{11}\theta\sin\varphi+$$

$$+0.0292C^3\sin^{14}\theta\sin\varphi+\ldots]+\ldots.$$

Figure 7 shows clearly the differences in the topologies of the Mead magnetosphere and the Olson–Pfitzer magnetosphere: in this new magnetospheric model a flaring of the field lines is observed even in the day side up to $L=8$, the day side compression appearing only for field lines of higher L value. On the opposite, all the field lines of the Mead magnetosphere are subject to compression on the day side. In the

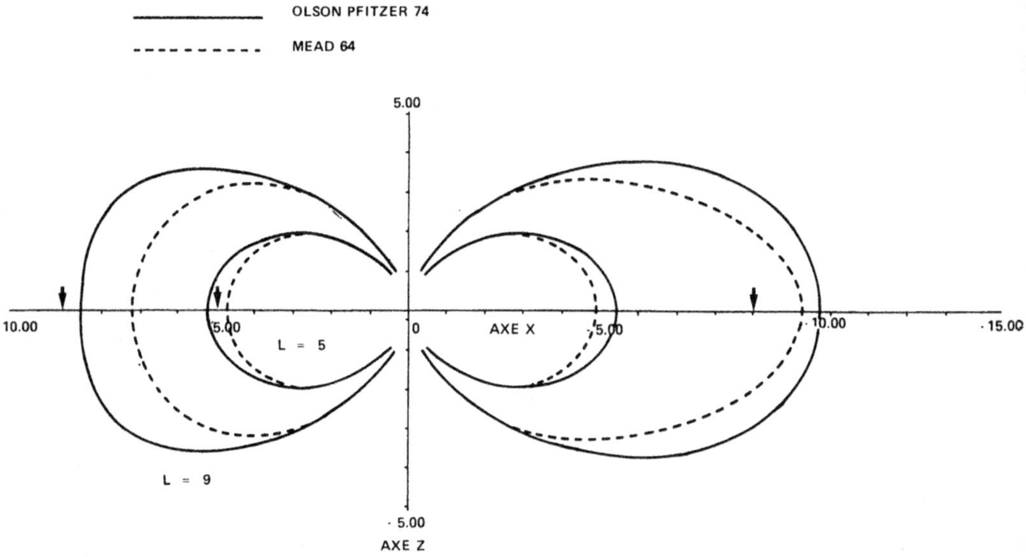

Fig. 7. The discontinuous curves represent the Mead 64 field lines defined by Equations (1), the continuous curves correspond to the Olson–Pfitzer 74 model with Equations (17).

synchronous region the drift trajectories of equatorial particles stress the dipole-like structure of the Olson model (Figure 8). The determination of the trajectories of non-equatorial particles will require further calculations. Significant changes are to be expected due to the different topology of the Olson–Pfitzer model, with more complex equations than those obtained for the Mead model.

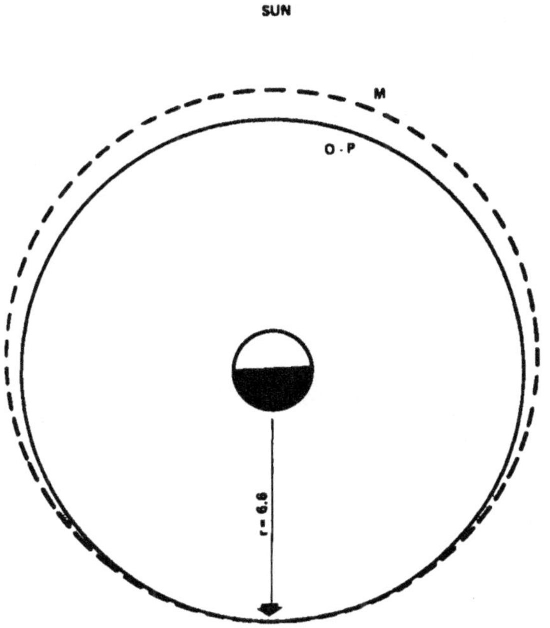

Fig. 8. Drift trajectories of an equatorial particles detected at midnight at $r = 6.6R_e$ in the Mead 64 and Olson–Pfitzer 74 models.

Very recently, Olson and Pfitzer (1977) have extended their model to include the dipole tilt effects. The use of this last model for particle studies must be discarded until an abbreviated version is published. On the opposite, this tilt model is particularly suited to conjugate point calculations which are now presented in the following paragraphs.

3. Conjugate Point Calculations

Up to 1975, all the magnetospheric models were built on the hypothesis of a perpendicular flow of the solar wind against the Earth dipole. The three components of the external magnetic field B_r, B_θ, B_φ verified the classical conditions of symmetry with respect to the geomagnetic equator (Mead, 1964). Thus, the analytical Equations (1) of the field lines had only constant coefficients and even trigonometric expressions. Since 1975, new magnetospheric models have arisen, each of them built on a given set of hypotheses, the most important one being the non-perpendicular flow of the solar wind against the dipole (Mead and Fairfield, 1975; Choe and Beard, 1974; Olson and Pfitzer, 1977; Voigt, 1972).

Based on purely physical conditions, the boundary currents and the shape of the magnetopause were calculated for the whole range (−35°, +35°) of the dipole tilt which is the combination of the inclination of the dipole and the inclination of the ecliptic about the Earth geographic axis. As a result the external magnetic field is no more symmetric with respect to the geomagnetic equator. We will now examine the consequences for the calculation of the magnetic conjugate point locations.

3.1. Seasonnal and Diurnal Motions of the Magnetic Conjugate of an Earth Station

Let us consider the three components of the external field in a frame of reference where the Z-axis is the dipole. It is always possible to split the three components in symmetric and antisymmetric parts:

$$B_r = B_{rs} + B_{ras}, \qquad B_\theta' = B_{\theta s} + B_{\theta as}, \qquad B_\varphi = B_{\varphi s} + B_{\varphi as} \tag{18}$$

At the station location we can define an elementary flux tube associated with the field line passing through it: the magnetic flux through this tube at the Earth Station location (θ_0, φ_0) is defined as:

$$\Delta\Phi = B_r \Delta S, \tag{19}$$

where ΔS is the cross section. At the conjugate point, we should have the same magnetic flux $\Delta\Phi$. But we have in general $B_{ras}(\theta_0) \neq B_{ras}(\pi - \theta_0)$ and this condition will not be verified at the point $(\pi - \theta_0, \varphi_0)$ but at some neighbouring point (θ_1, φ_1) such that: $\theta_1 = \pi - \theta_0 - \delta\theta_0$, $\varphi_1 = \varphi_0 + \delta\varphi_0$, $\delta\theta_0$ and $\delta\varphi_0$ are the longitude and latitude displacements of the real field line relative to a pure dipole field line which passes at points (θ_0, φ_0) and $(\pi - \theta_0, \varphi_0)$. From Equation (19) one notices that the relative displacement depends on the strength of the asymmetric part of B_r. In the case of the Choe-Beard model we have found in a first approximation (Kosik, 1977):

$$\begin{cases} \delta\theta_0 = \dfrac{3\pi}{32} R_{11} \tan\theta_1 \cos\varphi_1, \\[2ex] \delta\varphi_0 = \dfrac{3\pi}{8} R_{11} \sin\varphi_1, \end{cases} \tag{20}$$

where

$$R_{11} = \frac{\bar{g}_1^1}{g_1^0}\left(\frac{L}{2_b}\right)^3 \quad \text{with} \quad \bar{g}_1^1 = -0.06822\lambda + 0.04967\lambda^3,$$

λ being the Sun geomagnetic latitude.

3.2. Seasonal and diurnal motions of the conjugate point of a synchronous satellite

The seasonal and diurnal motions of the conjugate points have a greater amplitude in this case. Furthermore, this motion would appear also in a magnetosphere having the geomagnetic equatorial symmetry. The only reason of this phenomenon is the noon–midnight asymmetry of the magnetosphere: in one day a synchronous satellite samples different field lines, their labels being given by Equation (5) in the Mead model. At midnight we have $L < L_0$, where L_0 is the field line label at noon. One must also take into account the fact that for the same L value the day side field lines are less elongated than the night side field lines. The colatitude of the conjugate point is approximately given by:

$$\sin\theta = \sin\theta_0\left[1 - \frac{4\sqrt{3}}{21}\frac{\bar{g}_2^1}{g_1^0}L_0^4(\cos\varphi - 1)\right] \tag{21}$$

in the Mead model. This equation shows simply that the colatitude of the conjugate point is smaller on the day side. In this simple model where the orbit is perpendicular to the dipole, the conjugate point performs only a vertical oscillation. If we choose an orbit tilted by 11° 4 about the geomagnetic equator, the motion of the conjugate would be already elliptical. It is possible to give a more detailed analytic formulation, for example with the Choe Beard model (Kosik, 1977). We get in this case:

$$\begin{aligned}\delta\theta_1 = {} & -\tfrac{1}{2}\tan\theta_1[R_{10}\{G(\theta_1) - G(\theta_s)\} + R_{10}^2\{G(\theta_1) - G(\theta_s)\} + \\ & + R_{11}\{H(\theta_1) - H(\theta_s)\}\cos\varphi_s + R_{20}\{K(\theta_1) - K(\theta_s)\} + \\ & + R_{21}\{L(\theta_1) - L(\theta_s)\}\cos\varphi_s],\\ \delta\varphi_1 = {} & R_{11}\{M(\theta_1) - M(\theta_s)\}\sin\varphi_s + R_{21}\{N(\theta_1) - N(\theta_s)\}\sin\varphi_s,\end{aligned} \tag{22}$$

where the R_{ij} involve the external coefficients of the magnetospheric model, G, H, K, L, M, N are trigonometric expressions in θ, θ_1 is the colatitude of the field line foot on the Earth θ_s the colatitude of the synchronous satellite. Figure 9 shows the diurnal motion of a synchronous satellite located at 10° East longitude.

3.3. Numerical calculations of GEOS-1 conjugate points

The precise calculation of the location of conjugate points requires the numerical integration of the differential equations which give the field line topology. At such distance as synchronous altitude it is necessary to add the complete external field to the complete internal field (Cain, 1968). We had the choice between three external field models which took into account the dipole tilt: Choe-Beard, Mead-Fairfield, Olson–Pfitzer 77). The first two were discarded as their formulation did not take into account the ring current field depression. The choice of the internal field was less evident as three models were of same quality: POGO 8/69, IGRF 75, BARRACLOUGH 75 (Barraclough *et al.*, 1975). Following the advice of J. Vette and M. Teague we have adopted this last model. Our computer code was set to treat 24 hr

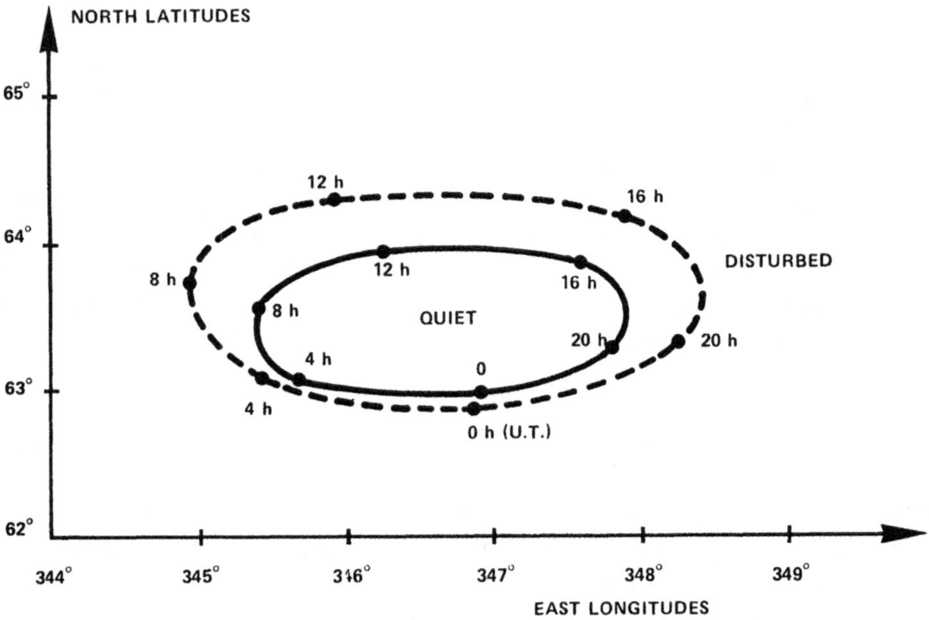

Fig. 9. The diurnal motions of the conjugate point of a synchronous satellite located at 10° E longitude is represented for quiet and disturbed periods.

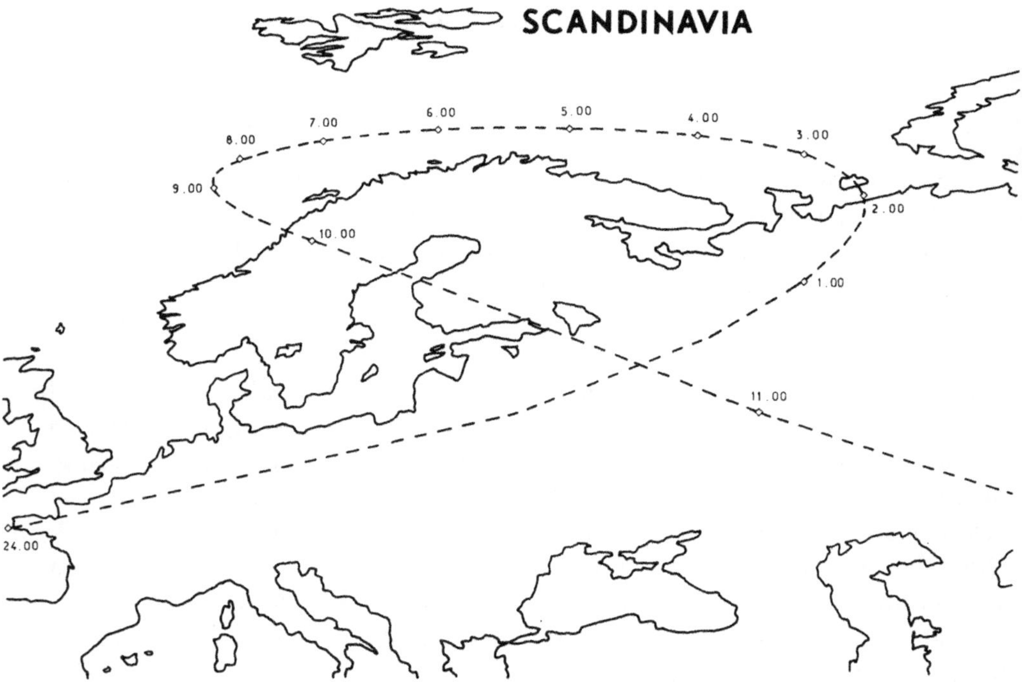

Fig. 10. Plot of the conjugate points of GEOS-1 over Scandinavia.

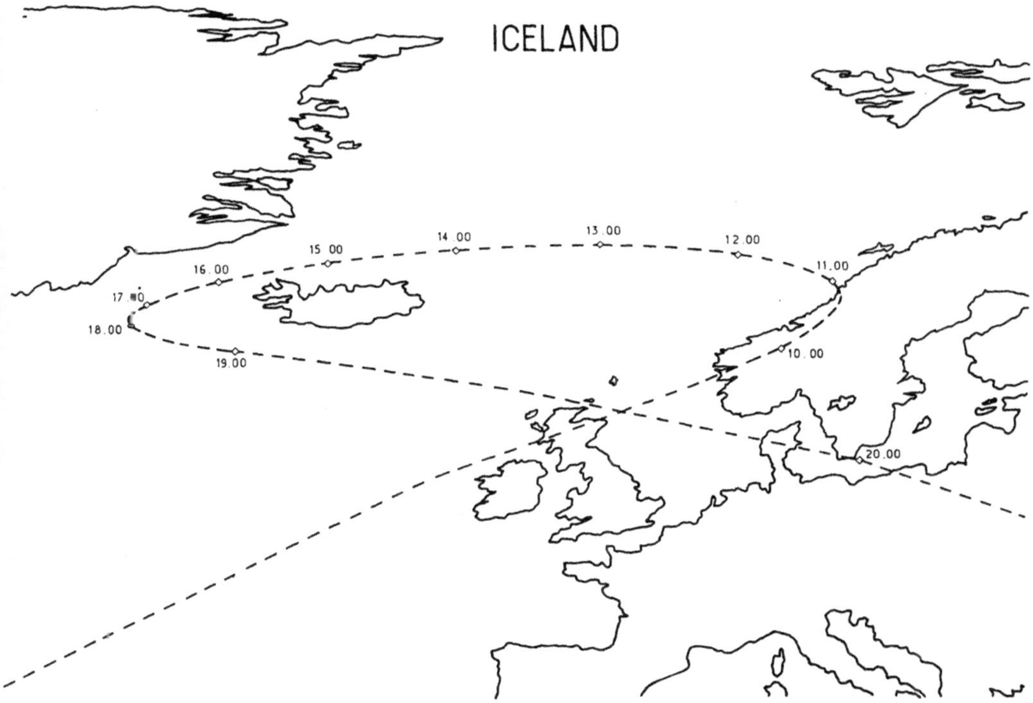

Fig. 11. Plot of the conjugate points of GEOS-1 over Iceland.

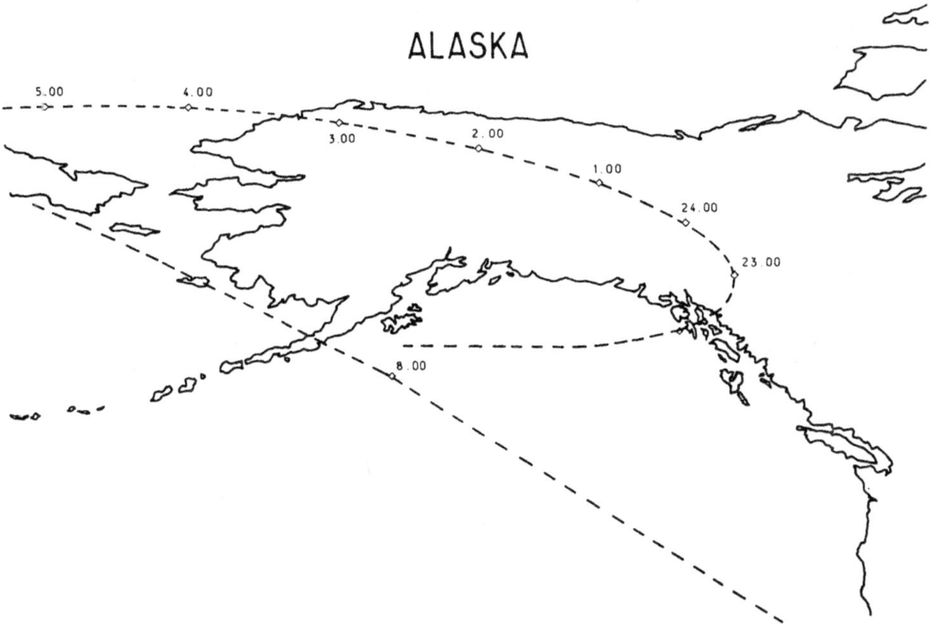

Fig. 12. Plot of the conjugate points of GEOS-1 over Alaska.

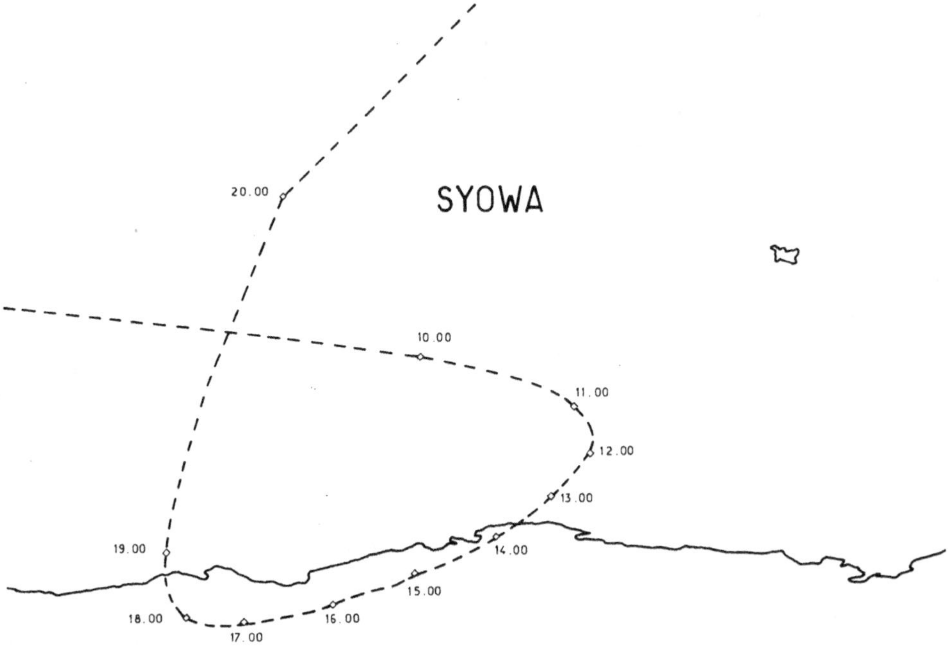

Fig. 13. Plot of the conjugate points of GEOS-1 over Syowa.

orbit with 20 min time intervals, involving northern and southern conjugates. We also included auxilliary data like: L value, geomagnetic local time, Universal time, Local time, ... World maps and particular enlargements are also produced on a routine basis. The total computer time is 16 s on CDC-7600. Figures 10, 11, 12, and 13 give some samples of the information *t* provided to GEOS S-300 group and CRPE in particular. Inspection of these plots shows that when GEOS is near the apogee, the conjugate points move quite slowly, and conjugate observations are possible during 6 hr approximately. Furthermore, near apogee, the angular motion of the satellite along the orbit is slower than the Earth angular motion: the field line foot moves westwards during 6 hr. Outside the apogee the field line foot moves quite fast and this enhances the difficulty of ground-satellite observations. Nevertheless, this has been possible during some GEOS passes over the Jim Creek transmitter (Cornilleau *et al.*, 1978).

4. Conclusion

In this work we have shown two possible applications of the magnetospheric models, the first one uses the magnetospheric model as a tool to map fluxes, the second one uses the magnetospheric model to calculate conjugate points. These applications show that a magnetospheric model can be used on two different ways:

– An approximate and explanatory one for the physics of magnetospheric phenomena (fluxes, motion of conjugate points, motion of particles, . . .).

– A precise way through numerical calculations with powerful computers (calculation of exact conjugates, exact field lines, mission analysis, . . .).

These two ways are complementary, the relative inaccuracy of the first one leading to far better understanding of what occurs and being a first step, the second permitting accuracy when it is needed.

Acknowledgements

We gratefully acknowledge Roger Gendrin for his support and help in GEOS-1 mission analysis. This was made possible through the help of W. P. Olson, J. Vette, and M. Teague who provided the external field coefficients, and P. Lemke who provided the orbit predictions. The calculations were performed on CDC-7600 in CNES.

References

Barraclough, D. R., Harwood, J. M., Leaton, B. R., and Malin, S. R. C.: 1975, *Geophys. J. Roy. Astron. Soc.* **43**, 645.
Cain, J. C.: 1968, *Radio Sci.* **3**, 7.
Choe, J. Y. and Beard, D. B.: 1974, *Planetary Space Sci.* **22**, 595.
Cornilleau, P. and Gendrin, R.: 1978, *Space Sci. Rev.* **22**, 443.
Cowley, S. W. H. and Ashour Abdallah, M.: 1975, *Planetary Space Sci.* **24**, 1527.
Cowley, S. W. H. and Ashour Abdallah, M.: 1976, *Planetary Space Sci.* **24**, 805.
Kivelson, M. G. and Southwood, D. J.: 1975, *J. Geophys. Res.* **80**, 56.
Kosik, J. Cl.: 1970, *Ann. Geophys.* **27**, 11.
Kosik, J. Cl.: 1971, *Ann. Geophys.* **27**, 2.
Kosik, J. Cl.: 1977, *Planetary Space Sci.* **25**, 457.
Lezniak, T. N., Arnoldy, F. L., Parks, G. K., and Winckler, J. R.: 1968, *Radio Sci.* **3**, 7.
McIlwain: 1961, *J. Geophys. Res.* **66**, 3681.
Mead, G. D. and Fairfield, D. H. 1975, *J. Geophys. Res.* **80**, 523.
Olson, W. P. and Pfitzer, K. A.: 1974, *J. Geophys. Res.* **79**, 25.
Olson, W. P. and Pfitzer, K. A.: 1977, Airforce Report No. AFOSR-77-0156TR.
Roederer, J. G.: 1970, *Dynamics of Geomagnetically Trapped Radiation*, Springer-Verlag, New York.
Roederer, J. G. and Schulz, M., 1971, *J. Geophys. Res.* **76**, 1055.
Stevens, J., Martina, F., and Stephen White, R.: 1970, *J. Geophys. Res.* **75**, 28, 5375.
Voigt, H.: 1972, *Z. Geophys.* **38**, 319.

GEOS-model in orbital configuration with artist impression of magnetosphere as background.

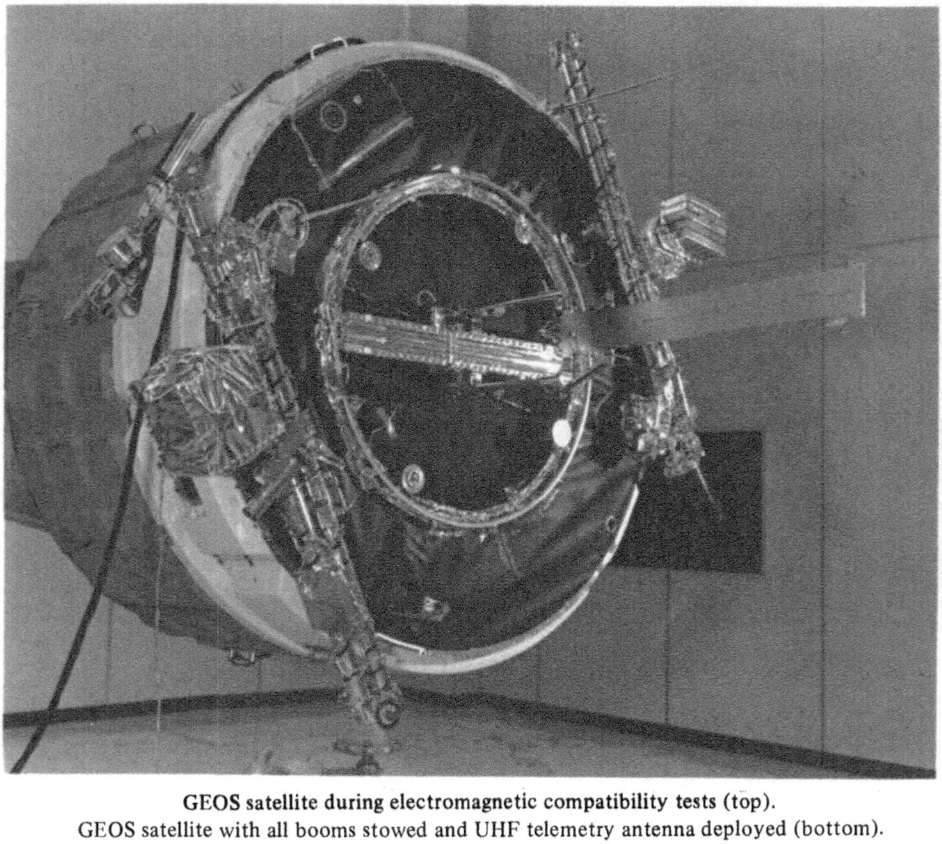

GEOS satellite during electromagnetic compatibility tests (top).
GEOS satellite with all booms stowed and UHF telemetry antenna deployed (bottom).

OBSERVATIONS OF SUBSTORM-ASSOCIATED PARTICLE-FLUX VARIATIONS AT $6 \leq L \leq 8$ WITH GEOS-1

A. KORTH, G. KREMSER, and B. WILKEN

Max-Planck-Institut für Aeronomie, D-3411 Katlenburg-Lindau 3, F.R.G.

Abstract. Measurements made with the charged particle spectrometer of the Max-Planck-Institut für Aeronomie onboard GEOS-1 were used to investigate the behaviour of energetic electrons and ions in the dusk sector of the magnetosphere. During substorms the integral ion flux ($\gtrsim 24$ keV) increased whereas the integral electron flux ($\gtrsim 20$ keV) first decreased and later on also increased. The dependences of these flux variations on pitch-angle and particle energy are described and discussed in terms of particle drift in the geomagnetic and geoelectric fields and adiabatic energy variations. The results also provide some information on the source region of the drifting particles.

1. Introduction

Magnetospheric substorms strongly affect the distribution of charged particles in the magnetosphere. The drift of freshly injected particles in the geomagnetic and geoelectric fields is an important process at energies above about 20 keV (e.g. Bogott and Mozer, 1974; Williams *et al.*, 1974; Walker and Kivelson, 1975; Barfield *et al.*, 1977). The drifting particles are related to several local-time dependent phenomena. In the morning sector electrons may increase wave-particle interactions and thus lead to a loss of electrons via precipitation into the ionosphere (e.g. Kremser *et al.*, 1973; Kangas *et al.*, 1975). At dusk protons may form the asymmetric ring current, that inflates the magnetic field and leads to betatron deceleration of electrons (Bogott and Mozer, 1973; Erickson and Winckler, 1973; Barfield *et al.*, 1975).

Until now the source of the freshly injected drifting electrons is not well known. It seems very difficult to transport energetic particles from the tail deep enough into the magnetosphere without extremely strong dawn–dusk electric fields (Barfield *et al.*, 1977). McIlwain (1974) introduced the concept of an injection boundary that was inferred from tracing back to substorm onset particles observed with geostationary satellites. It was postulated that at substorm onset particles are energized and injected along or outside this boundary. Konradi *et al.* (1975) e.g. successfully used this concept to interpret measurements made by Explorer 45. Still the relation of this boundary to the dynamical processes in the magnetosphere during substorms is not fully understood.

We have further investigated the problems related to the appearence of the drifting particles using data from the charged particle spectrometer of the Max-Planck-Institut für Aeronomie (MPAE) onboard GEOS-1. The orbit of this satellite enables measurements at L-values up to $L \approx 8$, i.e. well outside the geostationary orbit. Around apogee local time and L-value change very slowly. The instrument provides measurements with high energy and pitch-angle resolution over most of the energy range in which particle drift phenomena occur. In the present paper we report on first results obtained from observations at local times near dusk.

Space Science Reviews **22** (1978) 501–509. *All Rights Reserved*

2. Experiment

The MPAE charged particle spectrometer provides particle measurements at energies above about 15 keV. It consists of two identical units that both together cover an angular range of 5–125° with respect to the spin axis of the satellite (see Wilken *et al.*, 1977; Korth and Wilken, 1978, for more detailed descriptions).

Each unit contains a magnetic deflection system and an array of solid state detectors (Figure 1). The deflection system serves to separate electrons and ions. In addition, the magnetic field geometry was designed such that a parallel beam of electrons entering the instrument at a fixed angle with respect to the axis of the aperture is focussed to a single point, irrespective of the electron energy The focussing can be achieved for electron energies between 15 and 300 keV. The deflection system defines a focal curve on which each point corresponds to a given angle of incidence to the spectrometer. Along this focal curve five rectangular solid state detectors are mounted which define five angular intervals within the 60°-aperture of each unit. The electron energy is determined from pulse height analysis for each detector.

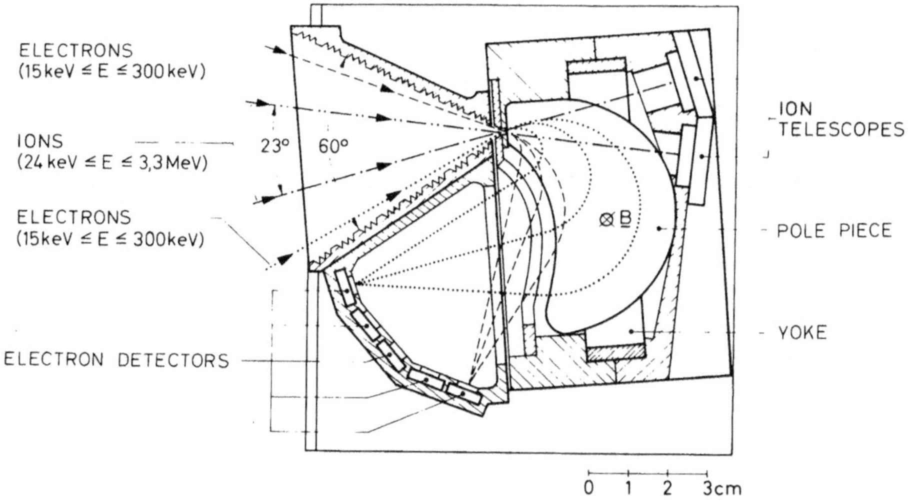

Fig. 1. View of one of the detector units of the MPAE charged particle spectrometer (experiment S-321 on GEOS; after Wilken *et al.*, 1977).

The deflection of ions with energies above 20 keV is small. They propagate on almost straight lines and are detected in two solid state detector telescopes.

The data from the spectrometer can be transmitted in three modes. In the normal mode integral and differential data are provided from all 14 detectors. A full energy and pitch-angle distribution is obtained every 5.5 s with a sampling time of 0.344 s, whereas the integral data have a time resolution of 0.172 s. In the other two modes data from only four detectors are transmitted, but with higher time resolution, reaching to 0.043 s for integral data and to 0.688 s for differential data with a sampling time of 0.172 s. The energy ranges and the angular coverage of the experiment are summarized in Table I.

TABLE I
Energy ranges and angular coverage of the MPAE charged particle spectrometer

	Electrons	Ions
Energy range	15–300 keV	24–3300 keV
Energy channels	>20	>24
	15–21	24–33
	21–27	33–43
	27–35	43–56
	35–41	56–73
	41–48	73–96
	48–57	96–125
	57–66	125–165
	66–78	165–219
	78–90	219–294
	90–108	294–390
	108–127	390–3300
	127–151	
	151–174	
	174–206	
	206–300	
Look directions of detectors with respect to the spin axis[a]	10° (E0); 20° (E1); 30° (E2); 40° (E3); 50° (E4); 80° (E5); 90° (E6); 100° (E7); 110° (E8); 120° (E9)	23° (P0); 46° (P1); 83° (P2); 106° (P3)
Angular resolution		
in elevation	±3°	±5°
in azimuth	±2°	±2°

[a] Detector number in brackets.

3. Variations of the Integral Particle Flux

Characteristic particle flux variations in the dusk sector are presented in Figures 2 and 3, where count rates of integral energy channels are displayed for two of the electron and ion detectors, whose axes of aperture differ by about 80°. During these events the ion fluxes increased whereas the electron fluxes first decreased and later on also increased. The count rates varied by a factor of 2–3 for the ions and by almost 10 for the electrons. The whole events lasted for 2–3 hr.

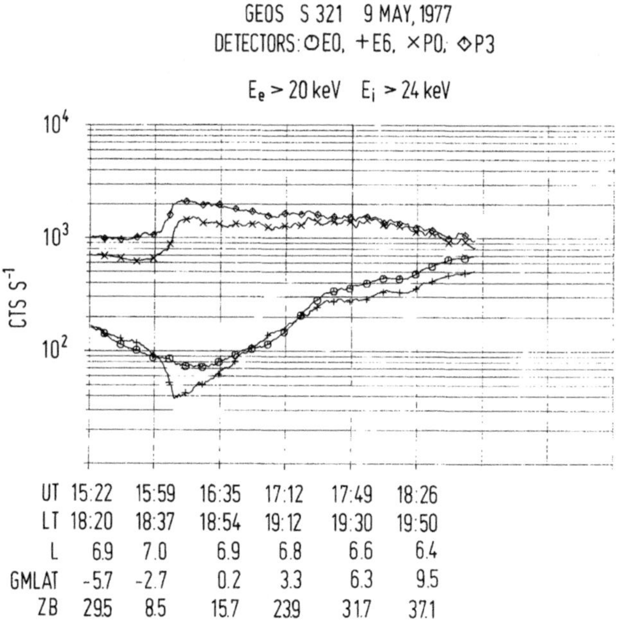

Fig. 2. Variations of the integral count rates of two electron (E0 and E6) and two ion (P0 and P3) detectors during the event of 9 May 1977. The look directions with respect to the spin-axis of the satellite are: E0:10°, E6:90°, P0:23°, P3:106°. GMLAT = geomagnetic latitude, ZB = angle between geomagnetic field and spin axis, E_e = electron energy, E_i = ion energy.

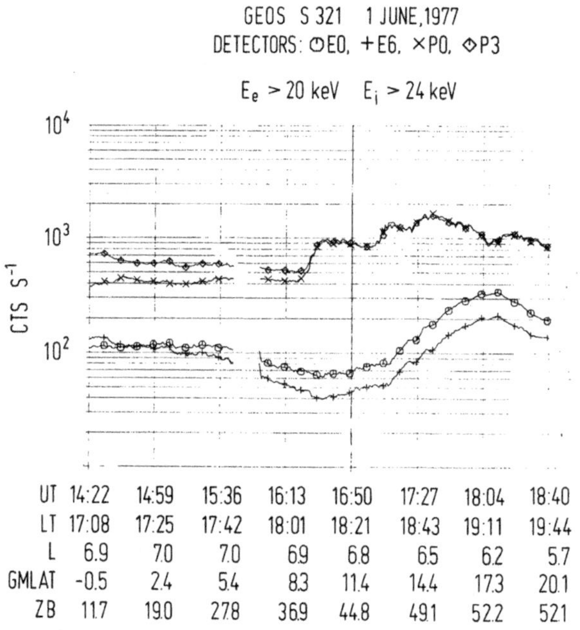

Fig. 3. Same as Fig. 2 for the event of 24 May 1977.

We observed a total of 12 events of this kind in May and June 1977, when the apogee of GEOS-1 was near dusk. An analysis of simultaneous magnetograms from Chelyuskin ($L = 9.6$, local time during events appr. 22–24 hr) and Dixon ($L = 7$, local time appr. 20–22 hr), the only stations with local time near midnight for which magnetograms were available, showed that 10 of these events occurred during polar magnetic substorms. During the two other events no substorm was recorded at these stations. Four further substorms were recorded during which no particle flux variation of this kind was found. In spite of the lack of a one-to-one correlation we regard it as certain that all events were related to magnetospheric substorms.

4. Energy Spectra

More details of the particle flux variations can be revealed from the energy spectra of electrons and ions in Figures 4 and 5. The data presented here were obtained during successive 10-min intervals at pitch-angles near 90°. The hatched area shows the variation of the energy spectrum as compared to the spectrum just before the onset of the event. On 9 May 1977 the electron flux first decreased at all energies, but the decrease was strongest at low energies. On 1 June 1977 only the low energy electron fluxes decreased. This decrease lasted for more than 1 hr. About 1 hr after onset electron flux maxima appeared first at high energies and then at continuously decreasing energies. These maxima moved down to about 30 keV, an energy well above the lower threshold of the instrument. The ion spectra showed no initial decrease, but one or several maxima moving from high to low energies. In the later phase of the events the electron and ion fluxes increased in a broad energy range above about 30–50 keV.

A striking feature during these events are the flux maxima moving with time from high to low energies. In most cases these particle flux maxima can be interpreted as the arrival of freshly injected particles that were moving on drift orbits. In the example of 1 June 1977 with the very clear energy dependent drift effects the occurrence times and energies of maximum electron fluxes corresponded almost exactly to what one expects if the electrons started to drift eastward from the midnight meridian at 16:30 UT. The calculated drift onset times for the electrons in the different energy channels ranged between 16:25 and 16:32 UT. It should be noted here that at about 16:30 UT a strong geomagnetic field variation was recorded at Dixon during a substorm.

5. Pitch-Angle Distributions

Pitch-angle distributions for one of these events are shown in Figure 6 for selected energy channels. This figure indicates that the decrease of the electron flux at low energies (21–27 keV) first occurred at pitch-angles around 90° and was less

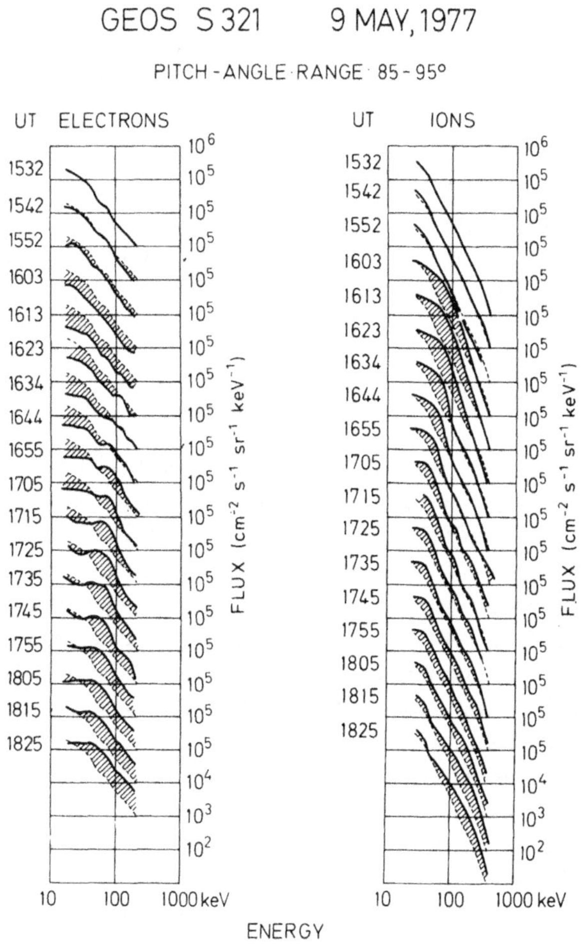

Fig. 4. Energy spectra of electrons and ions with pitch-angles near 90° for the event of 9 May 1977. Data obtained during consecutive 10-min intervals are displayed from top to bottom. The spectrum from a time interval prior to the onset of the event (uppermost curve) is repeated as dashed curve for comparison. The hatched area indicates the deviations from conditions before the event. The spectra are separated by a distance corresponding to one decade of the particle flux.

important at pitch-angles near 180°. Thus a temporary flux minimum at 90° occurred in the early phase of the event.

At higher energies (66–78 keV) the flux decrease was quickly compensated by freshly injected drifting electrons that were seen first at pitch-angles around 90° and later on also at pitch-angles near 180°. There is some indication of a similar pitch-angle dependence for the arrival of the ions. These pitch-angle dependent arrival times are a result of the pitch-angle dependence of the drift velocity which is greatest for pitch-angles of 90° and thus essentially confirm the results that were obtained in the preceding section from the interpretation of the energy spectra.

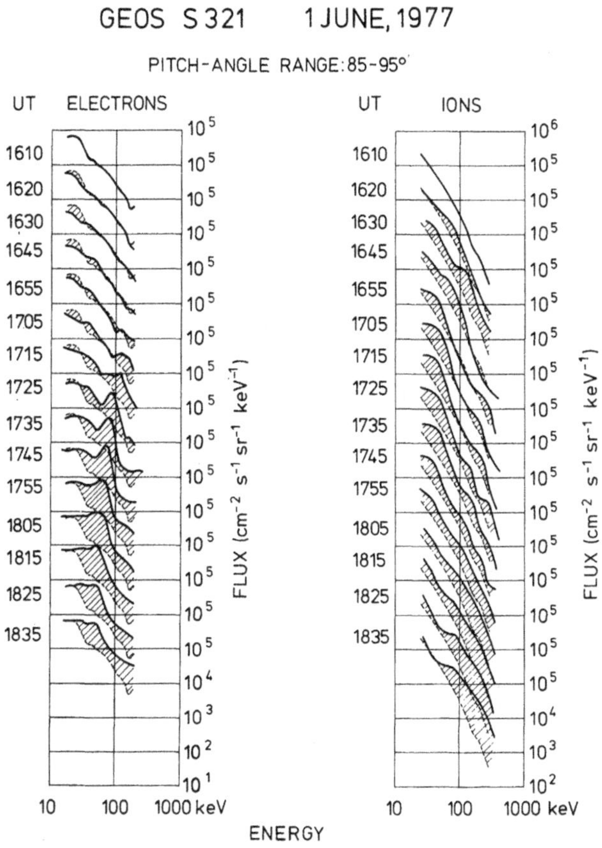

Fig. 5. Same as Figure 4 for the event of 1 June 1977.

6. Discussion

Our observations of energy spectra and pitch-angle distributions confirm the interpretation of data obtained previously at distances at and inside the geostationary orbit (e.g. Bogott and Mozer, 1973; Erickson and Winckler, 1973; Barfield *et al.*, 1975).

Taking also into account these earlier interpretations we can summarize the main sequence of processes that most probably led to the observed particle flux variations in the dusk sector as follows:

(1) Ions and electrons were injected on drift shells in a region near the midnight meridian.

(2) The ions drifted westward and arrived first at the dusk meridian. They were observed as energy-dependent ion flux increases and contributed to the formation of the 'asymmetric ring current'.

(3) The 'asymmetric ring current' led to an inflation of the magnetic field and thereby to betatron deceleration of the electrons. This effect was seen as a flux decrease in a fixed energy range.

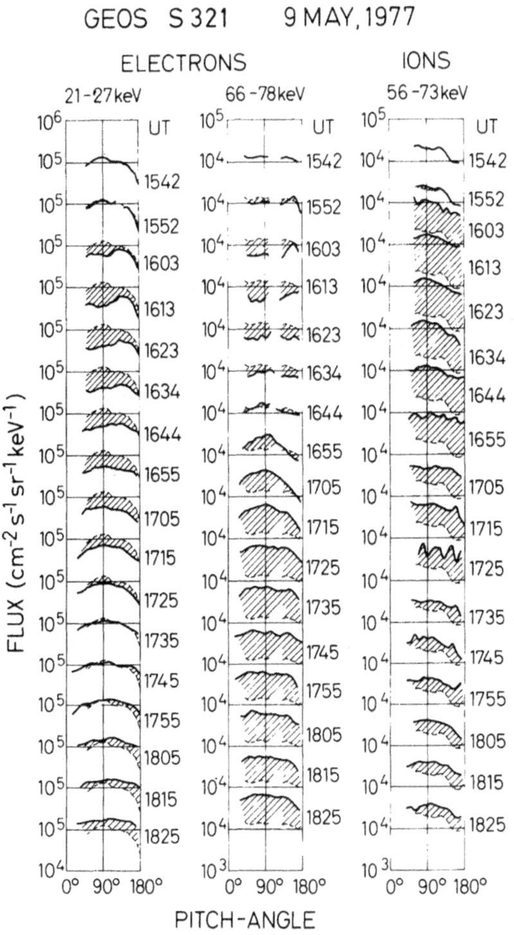

Fig. 6. Pitch-angle distributions of the electron and ion flux for selected energy channels obtained during the event of 9 May 1977. Data obtained during consecutive 10-min intervals are displayed from top to bottom. The distribution from a time interval prior to the onset of the event (uppermost curve) is repeated as dashed curve for comparison. The hatched area indicates the deviations from conditions before the event. The distributions are separated by a distance corresponding to one decade of the particle flux.

(4) The electron flux recovery in the later phase of the events was caused mainly by the arrival of electrons that had drifted eastward from midnight.

In addition to this main sequence the results obtained in the present study suggest several more detailed specifications:

(a) The initial electron flux decrease was strongest at low energies. It may have been caused by drift shell splitting that acted in addition to the betatron deceleration.

(b) Freshly injected drifting electrons were observed only at energies above about 30 keV. This limit may have been caused by drift shell splitting, but it may also be characteristic for the injection process. At present we cannot decide between these two possibilities.

(c) Near the end of the events (see e.g. Figure 5) a decrease of the low-energy ion flux was observed. First comparisons of our data with recordings of ULF-magnetic waves on GEOS-1 indicated that this decrease may have been caused by wave particle interaction (S. Perraut and A. Roux, private communication).

The substorm-related particle flux variations in the dusk sector obviously are a part of the magnetospheric substorm. Any attempt to explain them has to take into account a great number of interrelated physical processes. Still these first results indicate that further investigations of the data obtained with the same satellite at other local times than dusk may well lead to an explanation of details of the drift process and moreover provide information on the injection of charged particles onto drift shells.

Acknowledgements

The development and fabrication of the particle spectrometer was accomplished at the Max-Planck-Institut für Aeronomie. We want to acknowledge the assistance of G. Flentje, W. Grill, B. Wand, Ch. Wasshausen, and H. P. Winterhoff. F. Both. H.-J. Müller, J. W. Münch, and W. Stüdemann participated in the data reduction. We thank Prof. F. Mariani, who provided on a routine basis the information on the magnetic field that we needed for the calculation of pitch-angles. Dr J. C. Kosik, CNES, Toulouse calculated geomagnetic coordinates of the GEOS orbits. The World Data Center in Copenhagen provided copies of magnetograms.

The GEOS-Project office at ESA very effectively managed the preparation and performance of the whole project.

This work was supported by the Max-Planck-Gesellschaft zur Förderung der Wissenschaften and by the Ministerium für Forschung und Technologie through the DFVLR-BPT under contract no. RV14-B12/73 (WRK 243)-SF21.

References

Barfield, J. N., Burch, J. L., and Williams, D. J.: 1975, *J. Geophys. Res.*, **80**, 47.
Barfield, J. N., DeForest, S. E., and Williams, D. J.: 1977, *J. Geophys. Res.* **82**, 531.
Bogott, F. H. and Mozer, F. S.: 1973, *J. Geophys. Res.* **78**, 8113.
Bogott, F. H. and Mozer, F. S.: 1974, *J. Geophys. Res.* **79**, 1825.
Erickson, K. N. and Winckler, J. R.: 1973, *J. Geophys. Res.* **78**, 8373.
Kangas, J., Lukkari, L., Tanskanen, P., Trefall, H., Stadsnes, J., Kremser, G., and Riedler, W.: 1975, *J. Atmospheric Terrest. Phys.* **37**, 1289.
Konradi, A., Semar, C. L., and Fritz, T. A.: 1975, *J. Geophys. Res.* **80**, 543.
Korth, A. and Wilken, B.: 1978, *Rev. Sci. Instr.* **49**, 1435.
Kremser, G., Wilhelm, K., Riedler, W., Brønstad, K., Trefall, H., Ullaland, S. L., Legrand, J. P., Kangas, J., and Tanskanen, P.: 1973, *J. Atmospheric Terrest. Phys.* **35**, 713.
McIlwain, C. E.: 1974, in B. M. McCormac (ed.), *Magnetospheric Physics*, D. Reidel, Publ. Co., Dordrecht, Holland, pp. 143–154.
Walker, R. J. and Kivelson, M. G.: 1975, *J. Geophys. Res.*, **80**, 2074.
Wilken, B., Korth, A., and Keppler, E.: 1977, *Z. Flugwiss. Weltraumforsch.* **1**, 298.
Williams, D. J., Barfield, J. N., and Fritz, T. A.: 1974, *J. Geophys. Res.* **79**, 554.

SOME EARLY RESULTS OF THE keV PLASMA EXPERIMENT ON GEOS-1

GEOS Low Energy Particle Cooperation (GLEPCO)

H. BORG, L.-A. HOLMGREN, and B. HULTQVIST
Kiruna Geophysical Institute, S-981 01 Kiruna, Sweden

F. CAMBOU and H. RÈME
C.E.S.R., Toulouse University, B.P. 4346, Toulouse, France

A. BAHNSEN
Danish Space Research Institute, DK-2800 Lyngby, Denmark

G. KREMSER
Max-Planck-Institut für Aeronomie, D-3411 Katlenburg-Lindau, Germany

Abstract. Some preliminary new results are presented of the keV plasma experiment on GEOS-1. Electrons and ions have been observed to stream along the magnetic field lines in the dayside magnetosphere from the ionospheric side of the satellite towards the equatorial plane during magnetic storms, with streaming velocities corresponding to a kinetic energy of the order of a keV. The opposite streaming velocity has also been seen, but primarily in the ions only and with a smaller flux ratio for the two opposite directions along the field lines. The transition between the two opposite streaming directions, as seen by the satellite, has been found to occur even in a fraction of a second.

In magnetic storm conditions azimuthal asymmetries in ion fluxes corresponding to electric field intensities of several tens of millivolts per meter have been observed.

1. Introduction

Observations of the magnetospheric hot plasma near the equatorial plane at L values greater than three was first made by means of spacecraft with highly eccentric orbits making rapid traversals of the magnetospheric regions outside $3R_e$ (Gringauz, 1961; Frank, 1967a,b,c, 1970; Vasyliunas, 1968, and others), and later from circularly orbiting spacecraft of large orbital radii (the Vela spacecraft; Bame *et al.* (1967) and a large number of later reports) from geosynchronous satellites (Freeman and Maguire, 1967; DeForest and McIlwain, 1971; McIlwain, 1975, and others) and with Explorer-45 which had an elliptical equatorial orbit with an apogee of about $5R_e$ (e.g. Williams, 1975). The most detailed measurements of the magnetospheric plasma distribution function have been made by means of the UCSD experiment on ATS-6 at $6.6R_e$ (McIlwain, 1975) and the Explorer-45 within $5R_e$ (see e.g. Williams, 1975).

The ATS-6 geostationary satellite contained, besides plasma experiments, magnetometers but neither electric field and wave experiments nor ion mass analysing instruments. The GEOS spacecraft provides the most complete fields and particles measurements made in the inner magnetosphere hitherto. It was launched on 20 April 1977 and was aimed to be the first geostationary satellite with a

Space Science Reviews **22** (1978) 511–535

very complete magnetospheric payload. Because of a malfunction in the third stage of the Thor-Delta launcher, it was placed into an eccentric orbit with 12 hr period and apogee at $7R_e$ from the Earth's center. The orbit thus became more similar to that of Explorer-45 than to a geostationary one. The apogee visible from Europe was near dusk (in eccentric dipole time) at launch. It moved into the dayside, passed noon in the latter half of August and reached dawn (magnetic time) at the end of the year 1977. The inclination of the orbit is about 26.5° from the equatorial plane and the spin period is 5.7 s after an initial period of a few weeks. The angle between the satellite spin axis and the magnetic field lines varies in the range from a few degrees near the equatorial plane up towards 90 at the highest magnetic latitudes that are reached.

The keV plasma experiment on board GEOS-1 covers the energy range 0.3-20 keV. It has the ESA number S-310. There are two more particle experiments on GEOS, one (S-302) for lower energies (from thermal to 500 eV) and one (S-321) for higher energies (15-250 keV electrons and 20-2000 keV protons). Some data samples taken with the S-310 experiment in various parts of the dayside magnetosphere are shown and some interesting new observations are presented and discussed in this report.

2. Some examples of Distribution Functions Observed in the Dayside Magnetosphere

The KGI experiment on GEOS provides 576 values of the distribution in the energy range 0.3-20 keV when the temporal variations during 3 min are not significant: 32 energy levels for each of 18 pitch angle intervals. When the distribution function varies, fewer points are taken in a shorter time interval in order to obtain meaningful data. In mode B only 72 points are available for the definition of the distribution function. These points are determined within a spin period of 5.7 s.

Measurements in the loss cone are of particular interest. The routine magnetic field data provided by ESOC, giving the angle between the satellite spin axis and the magnetic field lines, may not be accurate enough for detailed studies of the loss cone particles. Such investigations may have to be carried out in collaboration with the magnetic field experimenter. At this stage we can only report that no narrow beams of the angular width of the loss cone, or narrower, have been seen in the limited amounts of data investigated. We have, however, to look a lot more. The field aligned electron and ion populations that have been seen hitherto in the GEOS data are fairly wide, as was the case for the source cone particles reported by McIlwain (1975).

The keV plasma experiment measures the differential number flux as function of energy, E, for different pitch angles, α, so the energy-pitch angle space is the natural one to work in. The phase space density, f, can easily be obtained from

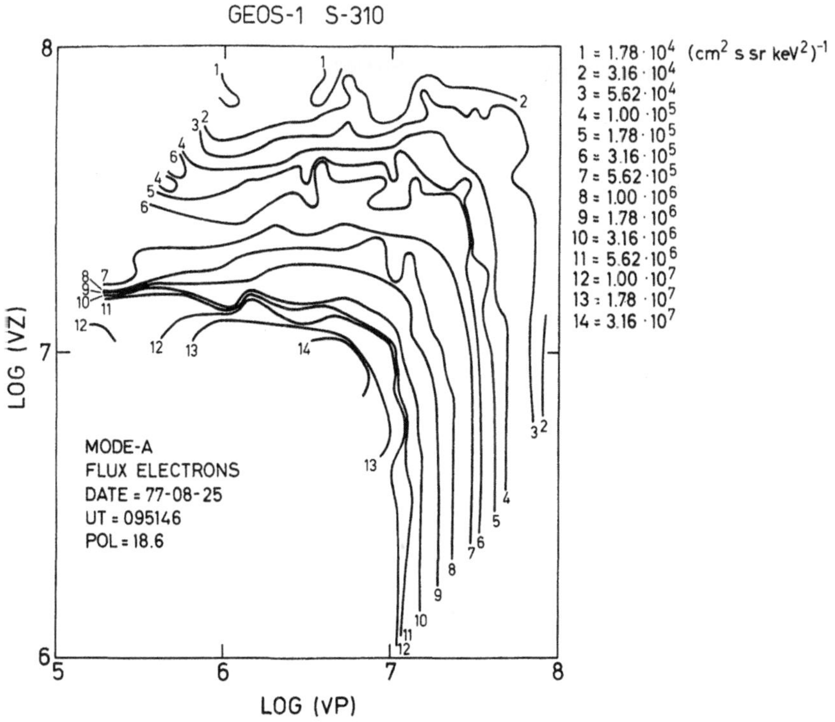

Fig. 1. Part of the hot plasma velocity distribution determined by means of the S-310 experiment.

the differential number flux, J, according to the relation

$$f(v_{\parallel}, v_{\perp})[\mathrm{s^3\,km^{-6}}] = 1.62 \times 10^{-7} \frac{J(E, \alpha)}{E} [\mathrm{cm^{-2}\,s^{-1}\,sr^{-1}\,keV^{-2}}].$$

An example of a distribution function in $v_{\parallel}$, $v_{\perp}$ space is shown in Figure 1. The distribution of the phase space density in velocity space is of particular interest in studies of wave-particle interactions. The distribution of f as function of $v_{\perp}$ for constant $v_{\parallel}$ is what enters in instability calculations. The example shown in Figure 1 has been plotted in connection with such an investigation. Some results of this kind, based on S-310 plasma data and wave data from GEOS-1, are presented at the ESLAB-symposium by Rönnmark *et al.* (1978). In this report we show the S-310 data as differential number flux in the energy-pitch angle space.

2.1. *Examples of partial hot plasma distribution functions in the dayside magnetosphere*

We present below, as an introduction, some examples of pitch angle distributions and energy spectra in the range 0.3–20 keV, obtained near apogee ($7R_e$ geocentric distance) on the dayside. We believe the examples are fairly representative for quiet and disturbed conditions in this part of the magnetosphere. However, we

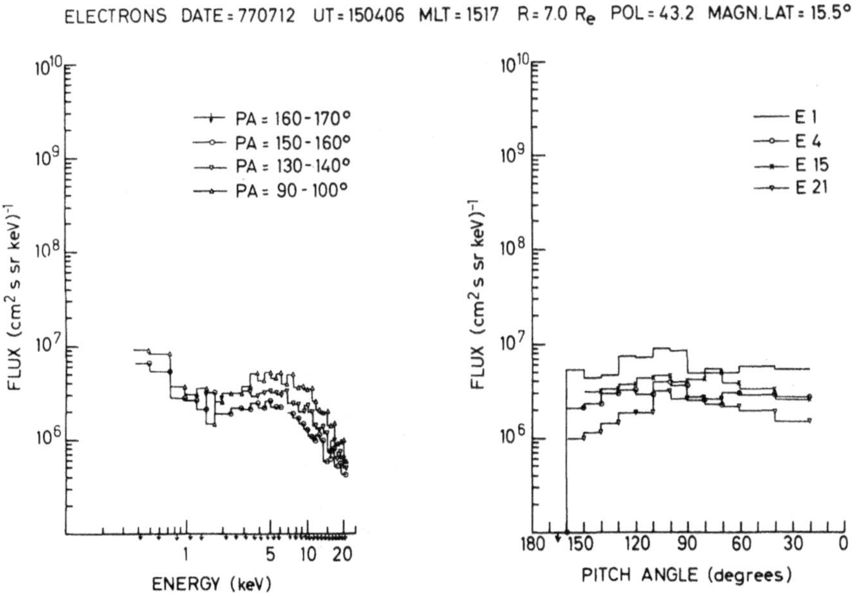

Fig. 2. Energy spectra and pitch angle distributions in the dayside magnetosphere during very quiet conditions.

have not yet been in a position to carry out a full study and do, therefore, not claim that the comments we make and the conclusions we draw below are applicable to 'typical' situations.

July 12, 1977 was a quiet day ($\sum K_p = 8-$, Dst $= -8$ to $-32\ \gamma$) and also the preceding day was quiet. GEOS-1 was at $7R_e$ distance between ~14:00 and 15:00 UT. Its magnetic latitude was 0° at about 10:45 UT and it increased continuously to 29.7° at 19:00 UT. The magnetic local time is less than one hour different from universal time except at the very end of the passage.

Figure 2 shows the energy spectrum and pitch angle distribution for the keV electrons at UT 15:04 (MLT 15:17). The pitch angle distribution shows flux maximum near the perpendicular direction. There are in this case slightly larger fluxes in the southern hemisphere than in the northern one, but this may not be a common situation. What is of particular interest is the wide loss cone. Out to 20° pitch angle there is no measurable electron flux. The atmospheric (100 km) loss cone is only about 3° wide.

The energy spectrum is peaked at about 5 keV for all pitch angles. The flux values are very low. Only at the lowest energies measured do they reach $10^7\ (\mathrm{cm}^2\ \mathrm{s\ sr\ keV})^{-1}$.

Figure 3 too represents fairly quiet conditions but there was some disturbance present. The data in the figure were taken at $6.1R_e$ geocentric distance on June 13, 1977 at 11:43 UT (MLT 14:35). $\sum K_p$ was 120 and the K_p value for the three-hour period of interest was 1−. Dst had values $\geqslant -30\ \gamma$ during that day. The pitch angle distributions in Figure 3 differ from those in Figure 2 mainly in

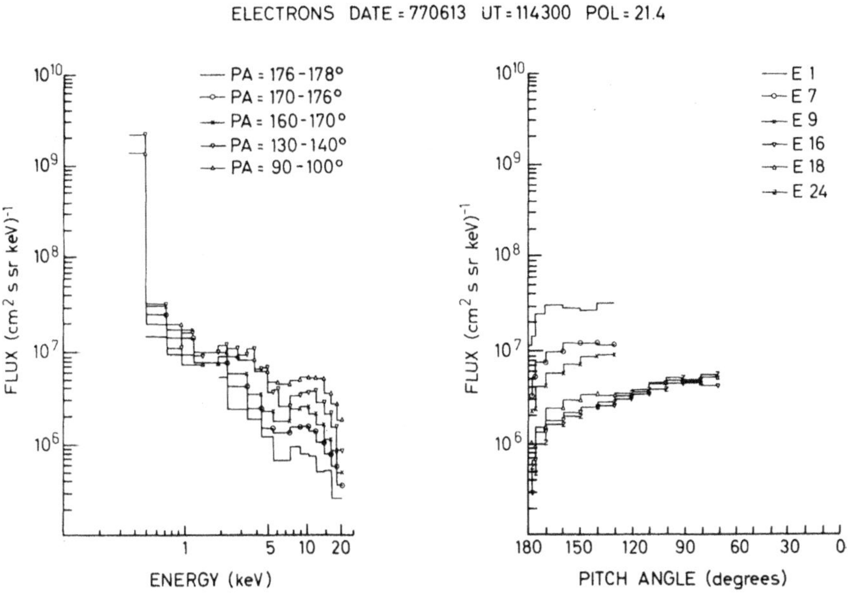

Fig. 3. Energy spectra and pitch angle distributions in the dayside magnetosphere during fairly quiet conditions.

the flux decrease near the loss cone being much less. The energy spectra are peaked also in this case, but the peak is located between 10 and 20 keV. The fluxes are higher than in Figure 2.

Storm conditions are represented by Figure 4, the data of which were obtained on July 29, 1977, during the first magnetic storm that occurred after the launch of GEOS-1. On this day the satellite was at $7R_e$ distance approximately between 13:00 and 14:00 UT (13:20–14:05 MLT). The magnetic latitude of GEOS-1 was 12° at the time shown in Figure 4. $\sum K_p$ was 40- and at the time of interest $K_p = 7$-. Dst had a secondary negative peak value of -80γ for the hour 12:00–13:00 UT. The primary peak (-100γ) occurred 6 hr earlier. Figure 4 shows a (weak) source cone situation. The flux peaks near the magnetic field line (the 100 km loss cone was not reached). There appears again to be some asymmetry between the hemispheres with the peak flux of trapped electrons not at 90° but between 70 and 80°.

The energy spectra shown in the figure show some small scale structure the significance of which is questionable, but on the whole the spectra are monotonically decreasing with energy. The fluxes observed near noon at $K_p = 7$- are the lowest of any of those in Figures 2–4. The lowering during disturbed conditions of electron fluxes in limited regions of the dayside magnetosphere has been seen in GEOS-data during several disturbed periods. We do not present here any more data showing this effect but will return to it in future papers.

Some two hours later than the time in Figure 4 the electron fluxes had

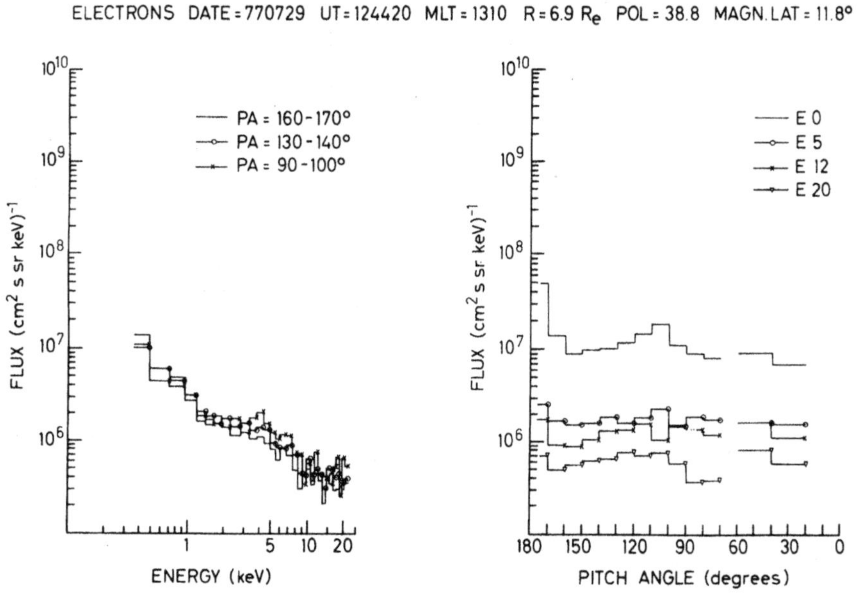

Fig. 4. Energy spectra and pitch angle distributions for disturbed conditions ($K_p = 7$-, Dst $= -80\ \gamma$) for electrons in a region of the dayside magnetosphere with very low fluxes.

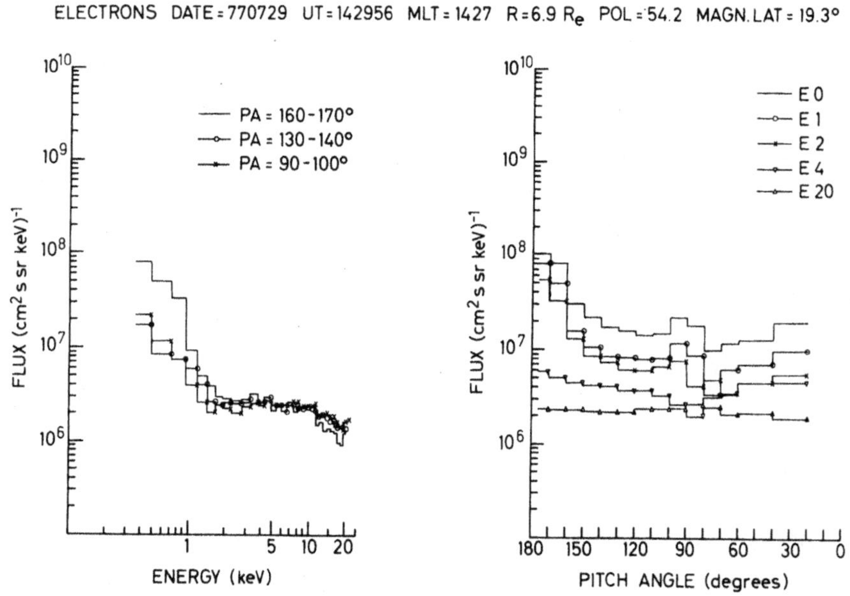

Fig. 5. Energy spectra and pitch angle distributions for electrons in disturbed conditions, obtained one hour and three quaters later than those in Figure 4.

increased, as shown in Figure 5, and the source cone in the pitch angle distribution is wider. For energies above 10 keV the pitch angle distribution is isotropic.

From the above we may think that a source cone distribution characterizes disturbed conditions, in contrast to undisturbed conditions in which the loss cones are frequently depleted of electrons. That this is not generally true is demonstrated by Figure 6 which shows pitch angle distributions and energy spectra obtained in the same magnetic storm, only one hour before the data in Figure 4 were taken. Figure 6 shows strongly anisotropic distributions peaking at 90°. The fluxes were at this time almost one order of magnitude higher than at 12:44 UT. So there is in the dayside magnetosphere no clear relation between flux and anisotropy for the entire distribution of the kind seen for the loss cone distribution in the nightside upper ionosphere, namely that the higher the flux the closer to isotropic is the pitch angle distribution.

The keV ion population, observed at about the same time as the electrons shown in Figures 2 to 6, is much less variable that the electron population. This is illustrated by Figures 7 to 9 which contain ion data from July 12, 1977 (quiet day) and from the storm day July 29, 1977, before and after the electron flux decrease at the satellite. All the pitch angle distributions are quite similar to each other with peak flux near 90° and a depleted loss cone. A comparison of the three figures indicates that the depletion of the loss cone flux is greater on the quiet day than on the storm day, but the significance of this relation has not yet been demonstrated. The shapes of the ion energy spectra in Figures 7 to 9 are not very

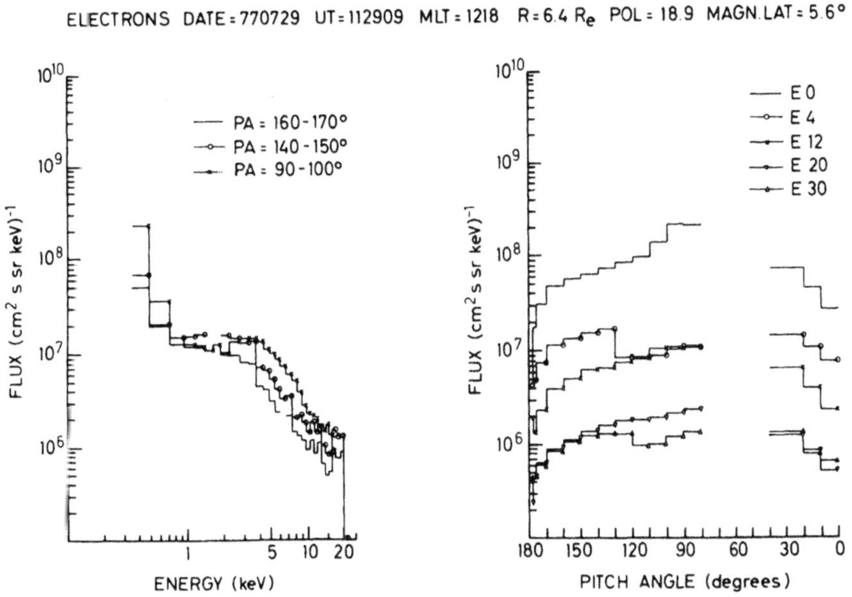

Fig. 6. Energy spectra and pitch angle distributions for electrons in disturbed conditions, obtained one hour and a quater before those in Figure 4.

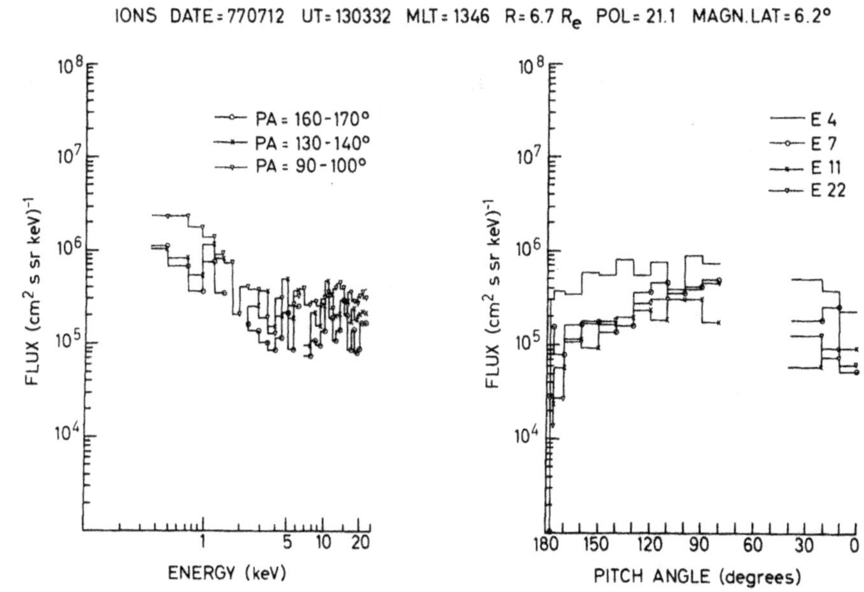

Fig. 7. Energy spectra and pitch angle distributions for ions in the dayside magnetosphere during very quiet conditions.

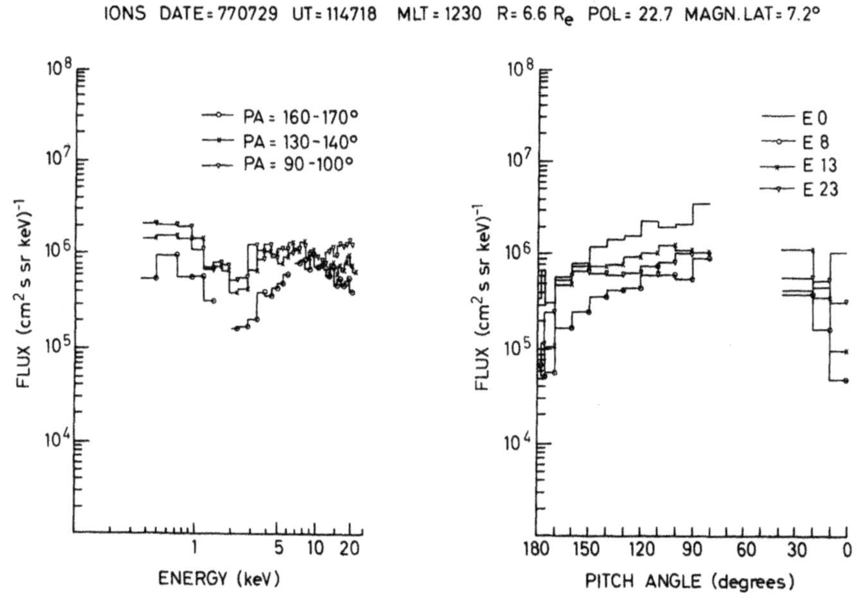

Fig. 8. Energy spectra and pitch angle distributions for ions in the dayside magnetosphere obtained on the storm day July 29, 1977 before the electron flux decrease was seen (cf. Figures 4 and 6).

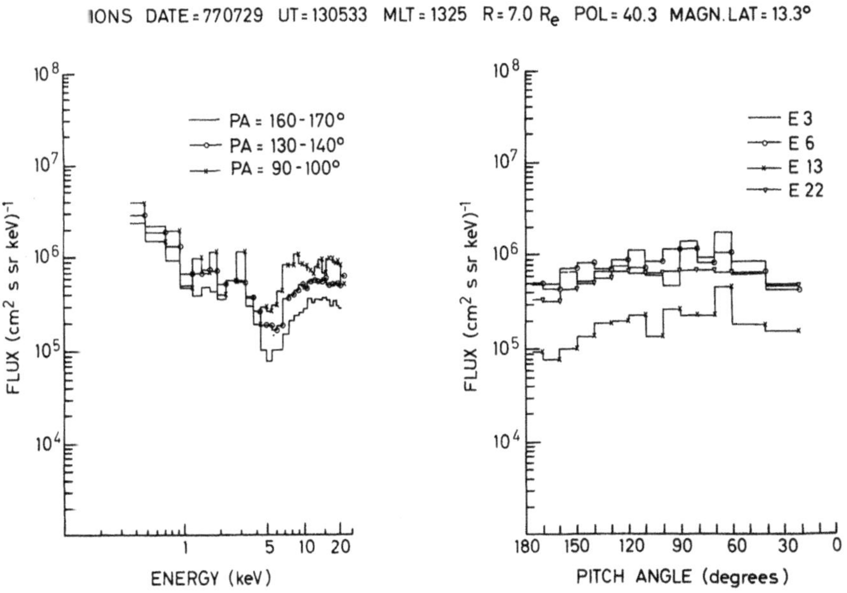

Fig. 9. Energy spectra and pitch angle distributions for ions in the dayside magnetosphere obtained during the 29 July, 1977 storm in the period with low electron fluxes (cf. Figure 4).

different and the fluxes are well within the same order of magnitude in all three figures.

With the examples shown in this section as a background we describe and discuss below some interesting distributions which deviate strongly from those presented above.

2.2. *Observation during magnetic storms of keV electrons and ions with extreme pitch angle distributions*

The December 1977 magnetic storm was associated with very large temporal and/or spatial variations in the magnetospheric hot plasma. The storm commenced in Scandinavia at about 16:00 UT on December 1, 1977. The Dst diagram for December is shown in Figure 10. The 2nd of December was the most disturbed day in the Scandinavian sector, where GEOS-1 was located. The auroral zone magnetic disturbance is represented by the Kiruna magnetogram in Figure 11. An outline of the hot plasma observations around UT 04:20–04:35 is given in Figure 12, which shows the output from the 10 detectors of the S-310 experiment on GEOS-1. The satellite was in that period near its apogee ($7.0 R_e$ geocentric distance) at a magnetic local time of about 07 hr and at a magnetic latitude of approximately 20°. Its velocity was slightly less than the corotation velocity (about 3 km s^{-1} in the solar-magnetospheric coordinate system). The positive spin axis was directed towards south and the majority of the S-310 detectors were mainly observing particles on their way towards the northern

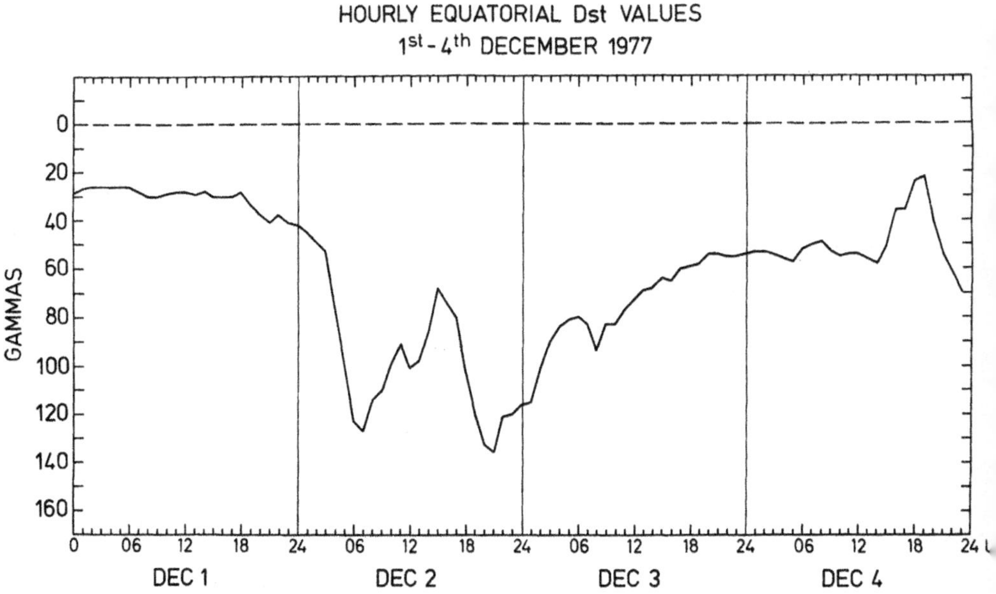

Fig. 10. Dst values in the period December 1–4, 1977.

hemisphere. However, in the time interval shown in Figure 12 the angle between the spin axis and the magnetic field lines was only slightly above 90°, so detectors 7 (electrons) and 10 (ions) were scanning practically the entire pitch angle range of 0–180°.

During a period of some twenty minutes before the dramatic event to be discussed here, the electron and ion fluxes decreased continuously, the electrons to the extent that they disappeared practically completely for the period 04:20 to 04:27 UT (see Figure 12). At 04:27:30 UT the electron flux above 0.3 keV increased by at least three orders of magnitude in about one minute of time. The increase was not monotonic but showed strong variations. The detectors D2–8 were in electron stepping mode when the electron flux increase started and remained in that mode until the end of the energy cycle, which was completed at UT 04:30:06. From this moment on D2–8 were in sweep mode. The ion detector D9 was in the sweep mode during the entire period of interest and the other ion detector, D10, was going through a 'proton convection mode' measuring at two pairs of nearby energy levels, until 04:28 when it too was switched into sweep mode.

Already in the period of absence of significant keV electron fluxes, preceding the injection of plasma, the keV ions showed a strongly field aligned population together with another component having a narrow distribution around 90° pitch angle. This is illustrated in Figure 13. The field aligned ions were streaming away from the earth within an angle of a few tens of degrees from the magnetic field lines.

The electron detectors, D2–D8, were at energy level 23 (~13 keV) in an energy

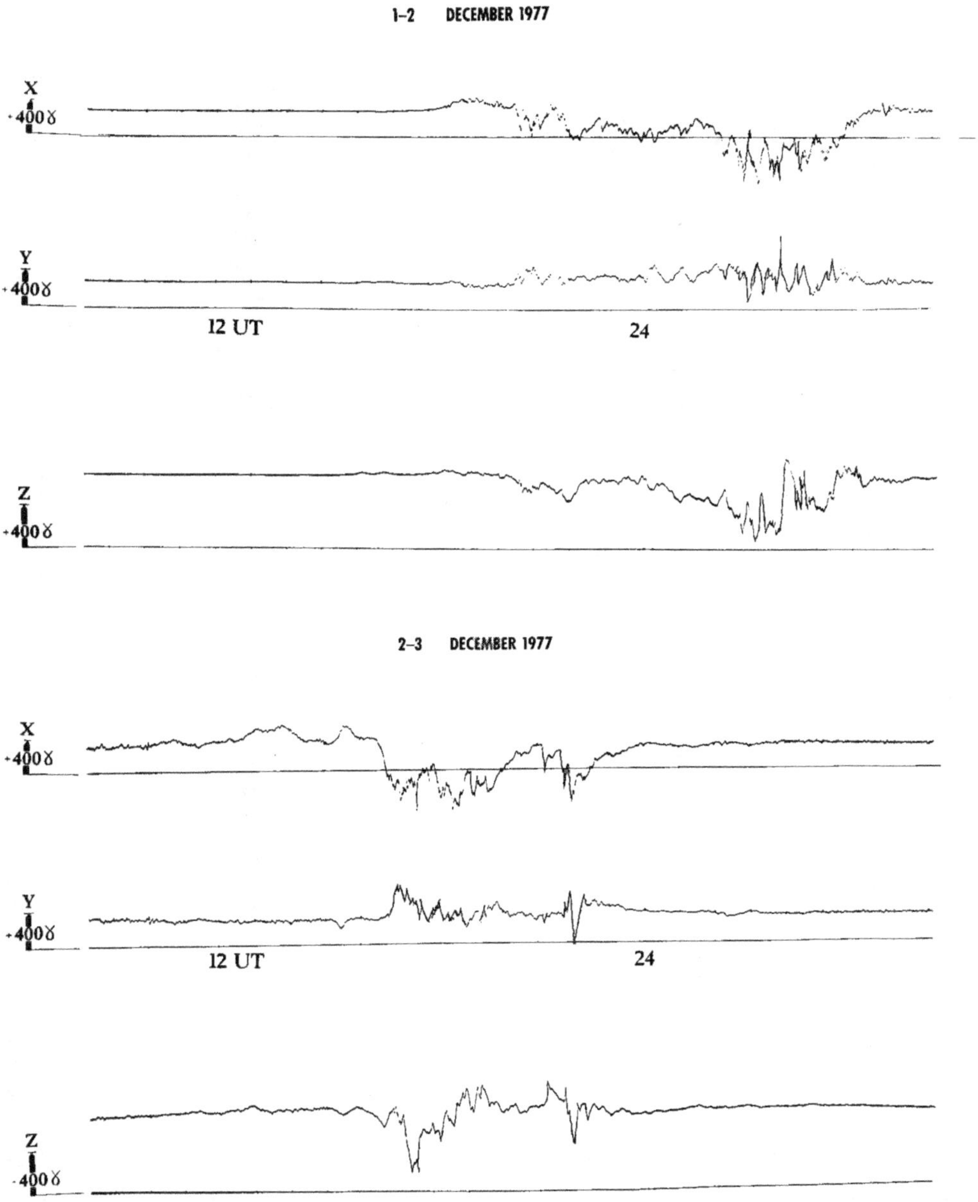

Fig. 11. Magnetograms from Kiruna for the period December 1–3, 1977.

cycle going downward from level 31 to level 0 when the first measureable electron fluxes appeared. These first observed electrons were also field aligned in the direction from the Earth towards the equatorial plane. Some eleven seconds later the pitch angle distributions for electrons and ions shown in Figure 14 were observed. As can be seen there, the ions still have a narrow component around 90° pitch angle. The degree of field alignment is about the same for electrons and

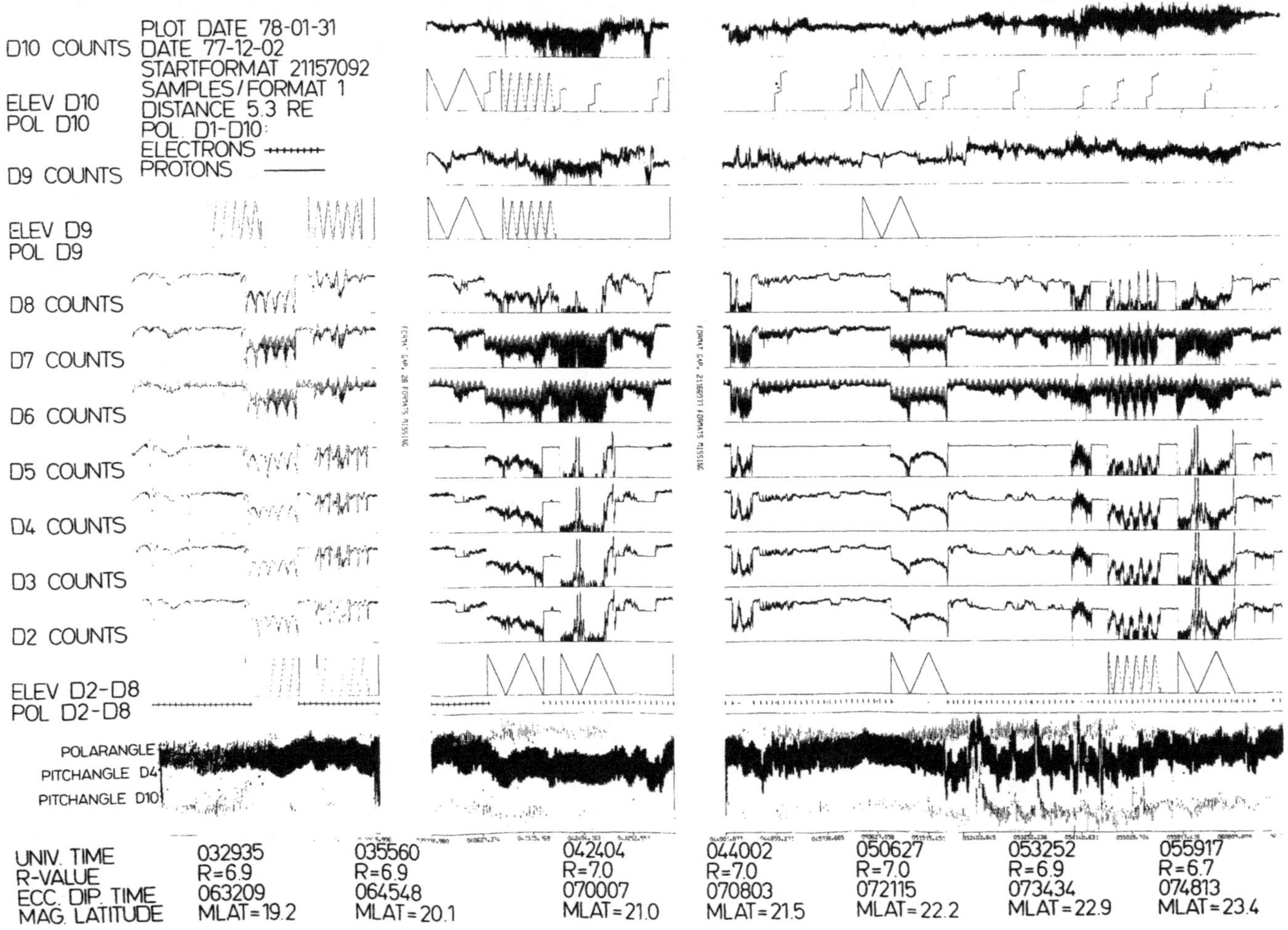

Fig. 12. Count rates for detectors 2–10 in the period 03:20–06:17 UT on December 2, 1977. ELEV means energy level. When nothing is shown for ELEV the detectors are in sweep mode. POL gives the particle species (electrons or ions). When nothing can be seen along the POL line (or only a very weak line) the detectors measure ions. The lowest frame shows the pitch angle scans for D4 and D10. The angle

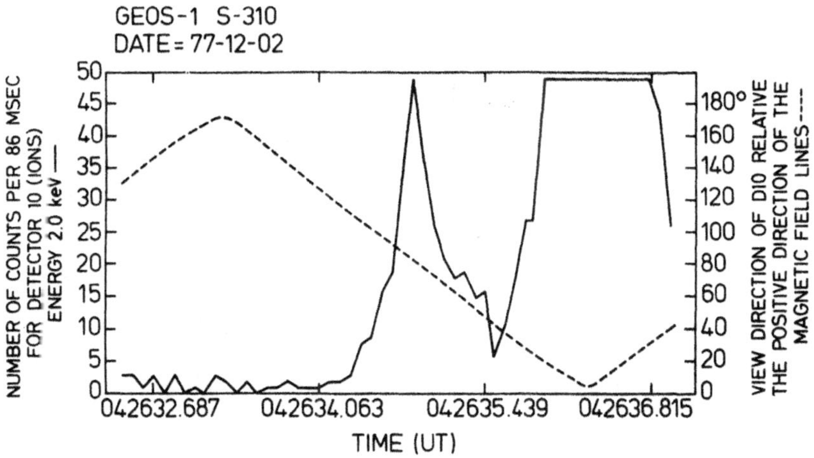

Fig. 13. Count rate of D10, measuring ions, as a function of pitch angle in a period during the December 2, 1977 storm when this detector scanned almost the entire pitch angle range of 0–180°. A strongly field aligned and a trapped component can be seen.

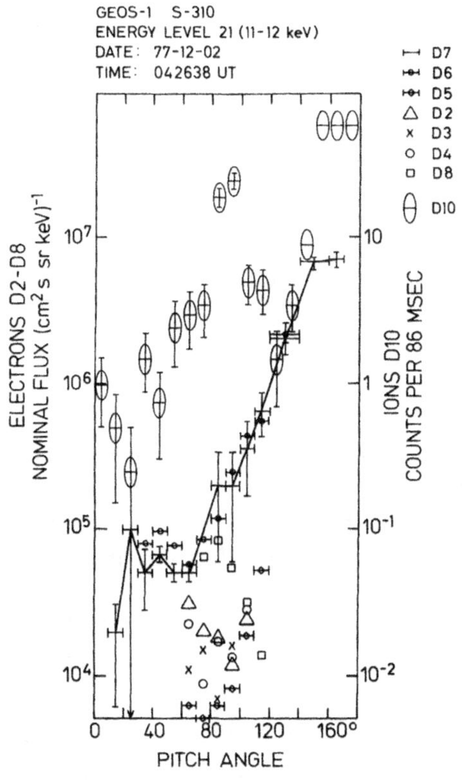

Fig. 14. Pitch angle distribution for electrons and ions of 11–12 keV energy some few seconds after the electron flux started to increase at 04:26 on December 2, 1977. The trapped ion component is still visible.

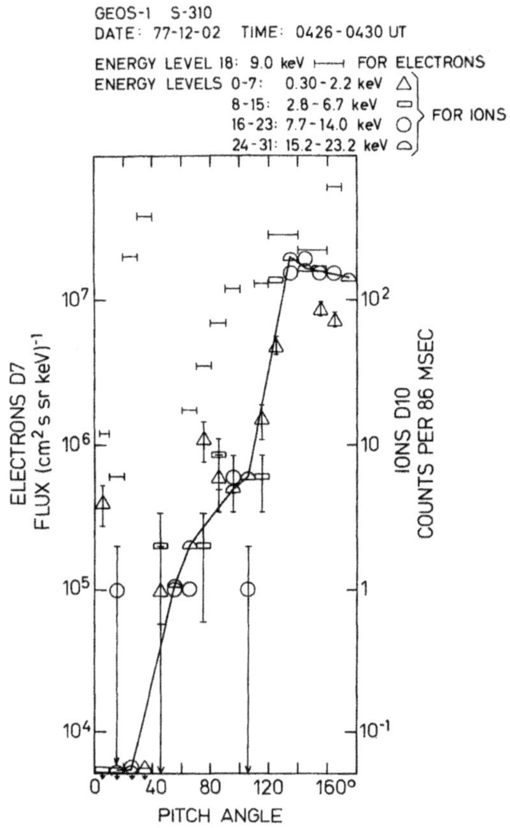

Fig. 15. Pitch angle distributions for electrons of 9 keV energy and ions in four energy bands, obtained 17 s after the distributions shown in Figure 14. The trapped ion component has disappeared. The electron peak at low pitch angle is likely to be due to a rapid time variation.

ions of equal energy; there is about two orders of magnitude higher flux in the direction from the earth than in the opposite direction.

The electron and ion pitch angle distributions stayed field-aligned through several more satellite spins. About 17 s after the situation shown in Figure 14, the one in Figure 15 was observed. At this time D10 was operating in sweep mode. Again, the fluxes are about two orders of magnitude higher in the direction from the Earth than in the opposite direction for both electrons and ions. The ions have flux maximum not at, or near, 180° but at 130–140°.

During energy level 17 (~8 keV) the flux increase reached a maximum seen in Figure 12, where it stayed for some seconds, after which it decreased. The pitch angle distributions during this satellite spin are shown in Figure 16. The electrons are much less field-aligned than in the previous spins, and the ions have as large a flux along the field lines in the direction towards the Earth as in the direction

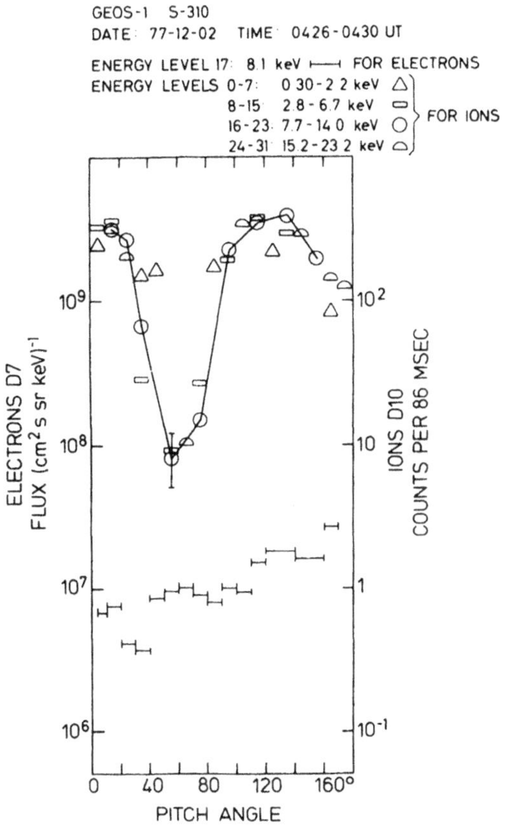

Fig. 16. Pitch angle distributions for electrons of 8 keV energy and ions in four energy bands obtained at the peak of a rapid flux increase, 6 s after those shown in Figure 15.

away from it. In the pitch angle range 50–140° there is still a strong asymmetry between the hemispheres.

During the next satellite revolution, when the electron detectors were at energy level 16 (~7.3 keV) the situation was very similar to that preceding energy level 17: about two orders of magnitude higher flux from the Earth than towards it, for both electrons and ions. This is shown in Figure 17. The situation remained very similar to that in Figure 17 down to and including electron energy level 13 (5.3 keV; see Figure 18). From energy level 12 on, more isotropic distributions were observed. The ions showed even slightly higher fluxes in the opposite direction, i.e. towards the Earth (see Figure 19). Ion observations during the spin period in which D2–8 were measuring electrons at energy level 11 (4.2 keV) are also shown in Figure 20. It stayed similar to this for some 40 more seconds. During energy level three the electron and ion pitch angle distributions again became field aligned in the direction from the Earth for a few seconds. This repeated itself for a somewhat longer period during energy level zero for the electrons, which is illustrated by the raw data output of detector 7 in Figure 21.

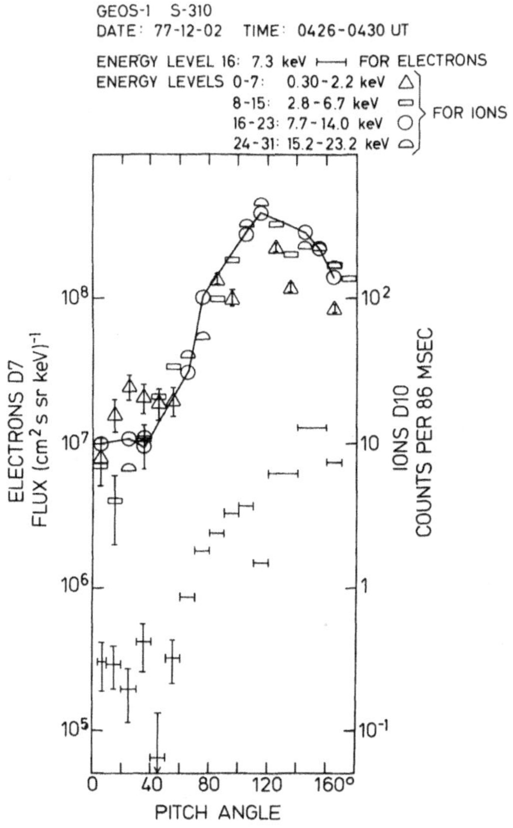

Fig. 17. Pitch angle distributions for electrons at 7 keV energy and ions in four energy bands obtained in a period of decreasing fluxes, 6 s after those shown in Figure 16.

In a short time interval in which D10, sweeping over most of the pitch angle range, showed ion flow in the direction from the Earth, D9 which scanned some 30° around 90° pitch angle indicated a slight flux maximum in the direction towards the Earth (see the left part of Figure 22). Then, suddenly, the flow direction along field lines seen with D9 switched (see at 04:30:13 UT in Figure 22). This change took place in a fraction of a second.

Another injection event, with electrons and ions streaming from the Earth towards the equatorial plane, was observed during the September magnetic storm, namely on 21 September at about 10:00 MLT. This event was very limited in time (a few minutes), of less intensity than the one on December 2 and the degree of field alignment was lower. It occurred near the central part of the dayside magnetosphere and at a geocentric distance of only $4.4R_e$.

3.3. *Observation of strong azimuthal asymmetry in keV ion distribution*

Figure 23 illustrates a situation of strongly asymmetric azimuthal distribution of keV ions which persisted for about one minute, starting some two minutes after

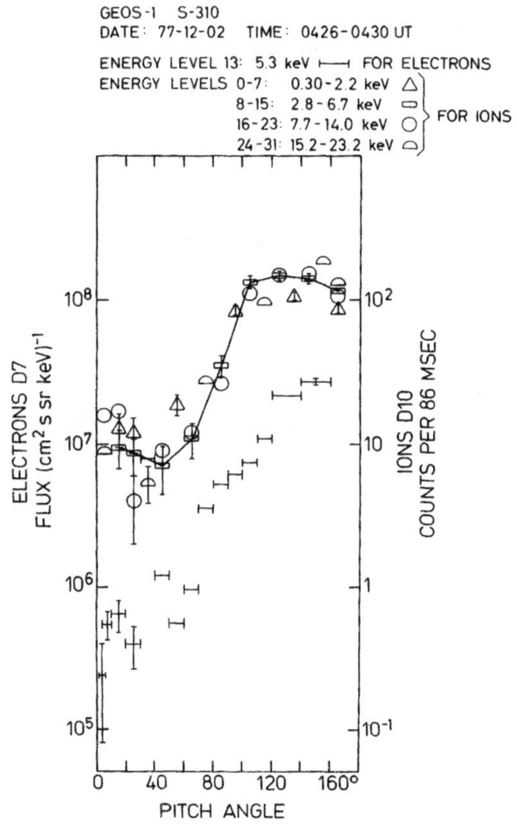

Fig. 18. Pitch angle distributions for electrons at 5 keV energy and ions in four energy bands obtained 17 sec after those in Figure 17, in a period of flux increase.

the situation shown in Figure 22. The highest peak of the event outlined in Figure 12 at 04:26–04:34 was passed. The magnetic field still behaved fair regularly. The amount of asymmetry is very large.

During most of the one minute period in which azimuthal asymmetry in the ions was observed, no corresponding effect could be seen in the electron fluxes. However, in the time interval of Figure 23 there was some tendency to azimuthal asymmetry also in the electrons. The effect was, however, much less than in the ions and its experimental significance may even be questioned.

It should be emphasized that the basis for the conclusion that the data shown in Figure 23 are azimuthally asymmetric is based on the assumption that the error of the magnetic field direction provided by ESOC is possibly of the order of a few degrees but certainly not tens of degrees. As the angle between the satellite spin axis and the magnetic field lines was near 90° in the period of interest, the azimuthal error of the magnetic field direction would have to be more than 50° for the situation shown in Figure 23 to appear for an azimuthally symmetric distribution. This is considered to be extremely unlikely. An inspection of the time series

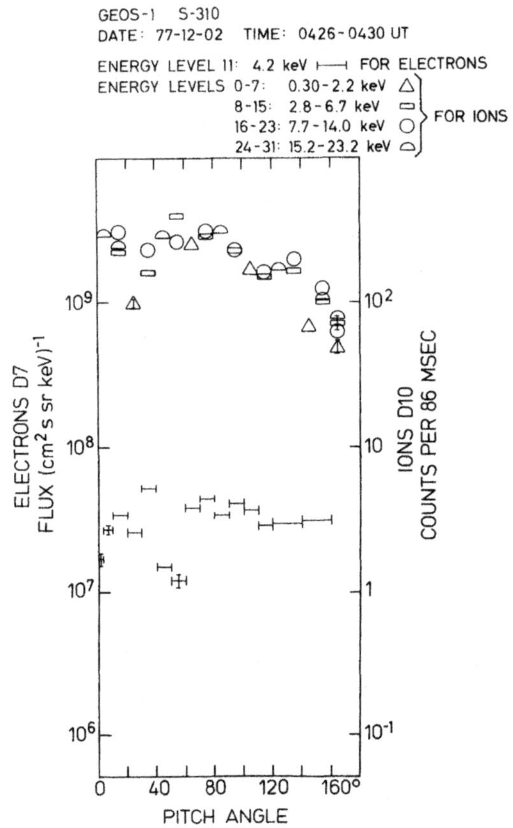

Fig. 19. Pitch angle distributions for electrons of 4 keV energy and ions in four energy bands obtained at the end of the flux increase, some 11 s after those in Figure 18.

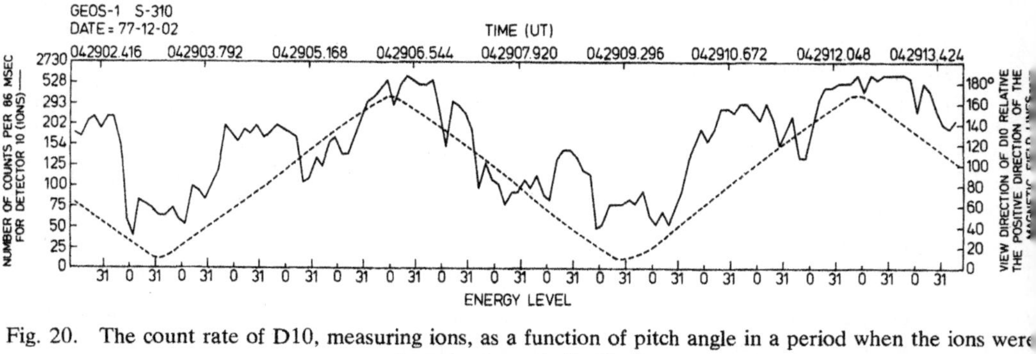

Fig. 20. The count rate of D10, measuring ions, as a function of pitch angle in a period when the ions were streaming towards the Earth.

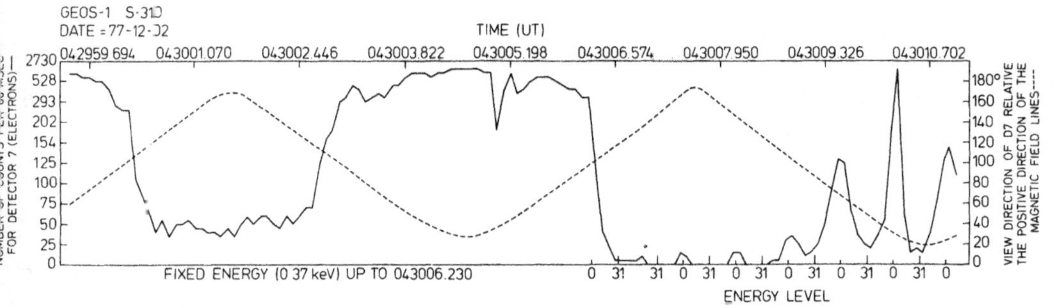

Fig. 21. The count rate of D7, measuring electrons, as a function of pitch angle during energy level 0 (0.37 keV), up to 04:30:06.230, and in sweep mode. The electrons are streaming away from the Earth.

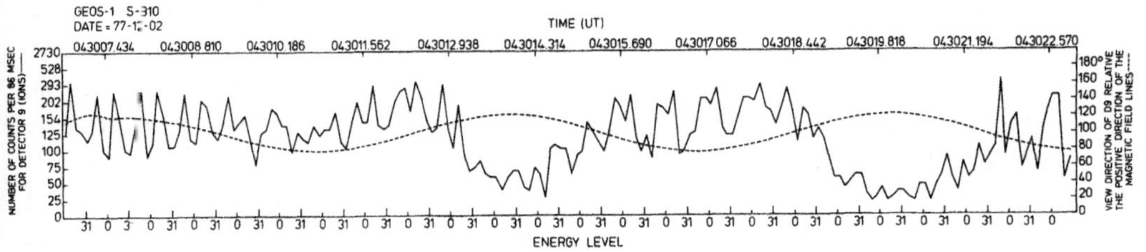

Fig. 22. The count rate of D9, measuring ions, as function of pitch angle in an interval around 90°. Before 04:30:13 the upper envelope of the count rate curve showed a weak spin modulation which was in phase with the dotted curve. After 04:30:13 the spin modulation was much larger and ~180° out of phase with the dotted curve. The transition occurred in a fraction of a second.

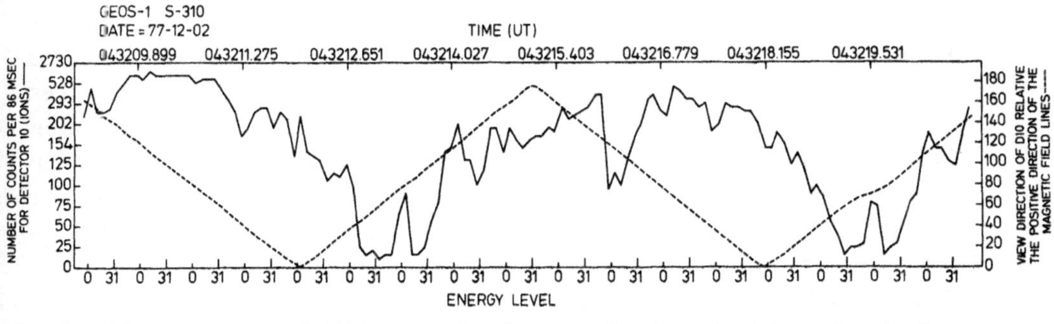

Fig. 23. The count rate of D10, measuring ions, as function of pitch angle, showing a pronounced azimuthal asymmetry.

of particle and magnetic field data over several minutes before the interval shown in Figure 23 shows continuous and consistent magnetic data with no sign of any malfunction of the magnetometer. On that basis we conclude that the azimuthal asymmetry shown in Figure 23 is a real one.

4. Discussion

Detailed measurements of pitch angle distributions for the hot magnetospheric plasma as function of energy have not been published before for the geocentric distances greater than 5 R_E. Explorer-45 made measurements within this geocentric distance (see e.g. Williams, 1975, Lyons, 1976) and McIlwain (1975) has published some limited amount of pitch angle distribution data obtained from ATS-6 in the geostationary orbit. No distribution function data at all from these distances in the dayside magnetosphere seem to have been presented before.

One preliminary result of the GEOS-1 measurements is that no very narrow ion beams, corresponding to those observed by the S3-3 satellite (Shelley *et al.*, 1976; Fennel *et al.*, 1977; Ghielmetti *et al.*, 1977, 1978b; Mizera and Fennel, 1977; and Mizera *et al.*, 1977) have been seen in the data investigated hitherto. An ion on an $L = 7$ field line, which has a pitch angle of 20° at an altitude of one Earth radius, will have a pitch angle of only 2–3 deg within the magnetic latitude range of GEOS-1. We have not yet looked very extensively, but it at least appears fairly safe to state that the ion beams are not at all so frequently seen in the vicinity of the equatorial plane as they are reported to be between 4000 and 8000 km altitude. If this is right, the obvious conclusion is that the ion beams are scattered over a fairly wide angular range on their way to the equatorial plane.

The source cone type of distribution reported for keV electrons by McIlwain (1975) has been observed by means of GEOS-1 in the dayside magnetosphere during disturbed conditions. Examples are shown in Figures 4 and 5. These distributions are much wider than the loss cone.

The lowest keV electron fluxes of any reported above for the dayside magnetosphere have been found during magnetic storms. On December 2, 1977 the keV electrons practically disappeared for a period of 7 min (cf Figure 12). On the storm day of July 29, 1977 the flux decreased by an order of magnitude during a little more than 1 hr and stayed low for more than 1 hr, after which it increased again. A similar electron flux decrease in the dayside magnetosphere has been seen in the first or second day of most magnetic storms which have occurred after GEOS-1 was launched and also during some substorms which have been investigated. Most likely these flux decreases correspond at high altitudes to the decrease, or even disappearence of electrons in and close to the loss cone, that has been observed during substorms in the upper ionosphere by means of the ESRO-1 satellite (Hultqvist *et al.*, 1974; Hultqvist, 1974, 1975a,b). It remains to be determined whether this electron flux decrease is caused by the plasma convecting out of certain regions of the dayside magnetosphere during storms and substorms or whether the electrons are scattered into and lost in the atmosphere much more efficiently during disturbed conditions.

The observation that the fairly general relation, for keV electrons on the nightside in and near the loss cone, between increasing flux and decreasing anisotropy appears not to be generally valid for the whole electron population in the field tube on the dayside (see Figures 4 and 6) illustrates that increased auroral electron density is not generally accompanied by increased pitch angle diffusion in the dayside magnetosphere. It remains to be determined how frequent this behaviour is.

The observation during the December 2, 1977 magnetic storm of both electrons and ions streaming along the magnetic field lines from the northern hemisphere of the Earth towards the magnetic equatorial plane at 7 o'clock a.m. magnetic local time appears to be the first one of this kind anywhere in the magnetosphere. If we assume that the distribution functions are drifting Maxwellians, the electron observations may be represented by a drift velocity of $\sim 10^4$ km s^{-1} and a temperature of 1.4 keV. The drift velocity alone corresponds to an energy of 0.34 keV for these electrons, of which there was more than one order of magnitude higher flux in the directions from the Earth than in the opposite direction (see e.g. Figures 15 and 17).

Whereas the electrons tended to have their highest flux at the highest pitch angles observed, the ions had their maximum at an intermediate angle between 90 and 180°. From the pitch angle of the flux maximum and the ratio of the maximum flux to the 90° flux, the streaming velocity and the ratio of the perpendicular to parallel temperatures may be evaluated, assuming displaced Maxwellian distributions. For the case shown in Figure 17, assuming the ions to be protons, a streaming velocity of ~ 400 km s^{-1} corresponding to a kinetic energy of ~ 1 keV is found. The $T_\perp/T_\parallel$-values is 4.3. If the ions are assumed to be O^+ the drift velocity is found to be ~ 100 km s^{-1}. The corresponding kinetic energy as well as the $T_\perp/T_\parallel$-value remain the same as in the proton case.

The drift velocities for the electrons and ions are thus quite different but still the amount of field alignment is almost the same. The acceleration process working along the magnetic field lines obviously does not add the same velocity value to electrons and ions, but rather the same velocity expressed in a characteristic velocity.

The fact that the field aligned flow was observed over periods much longer than the bounce period for the electrons indicates that a loss process is working more or less continuously, or that the spatial dimensions of the region with field aligned particle motion are so small that the mirroring electrons are not seen by GEOS-1. Which of these alternatives is the right one has not been possible to determine, but the latter seems more unlikely than the first.

The cause of the simultaneous flow along field lines of electrons and ions in the direction away from the Earth is not known. There do not seem to be too many wellknown processes to choose among. It appears that the acceleration takes place during magnetic storms at fairly high latitudes on the dayside of the Earth along auroral zone field lines ($L \sim 7$) as well as along field lines in the subauroral region ($L \sim 4$–5). The relative velocity increase in terms of the characteristic velocity appears to be of the same order of magnitude for electrons and ions. Obviously, no mechanisms involving a simple potential drop along the magnetic field lines can be responsible for the observed distributions. Conversion of magnetic field energy into kinetic plasma energy appears to be out of question too.

In connection with the 2nd of December event reported above we observed a decrease of the keV plasma density during some 20 min before the sudden plasma injection leading to almost complete absence of measurable keV electron fluxes during the last 7 min before the event. In this period the ion fluxes were already strongly field aligned (and the pitch angle distribution had also a maximum at 90° pitch angle).

In conclusion we can at present only state that some violent process fairly near the ionosphere appears to occur sometimes during storms and that this process increases the energy in the direction of the magnetic field lines by the same order of magnitude for keV electrons and ions.

The observations show that distributions streaming away from and towards the Earth may replace each other at the satellite within less than a second (see Figure 22). As the satellite was practically corotating with the Earth in the period described above and had a velocity of only about 3 km s^{-1} relative the solar-magnetosphere coordinate system, these observations demonstrate the existence of very sharp structures in time or space. Considering the inductance of the magnetospheric current system (Boström, 1972) it does not appear very likely that the structure is temporal. On the other hand, the proton gyro radius is probably of the order of 100 km, so it appears equally difficult to understand the observation as a spatial structure, unless the plasma convects past the satellite at fairly high speed.

The situations with field alignment in the two opposite directions show important differences. In the direction from the earth the degree of asymmetry between hemispheres is generally higher and about as high for electrons as for ions. In the opposite direction the field alignment is seen mainly in the ions and only to a small degree and in limited periods in the keV electrons.

Field aligned ions streaming towards the Earth in the upper ionosphere have been reported by Hultqvist *et al.* (1971) on the basis of ESRO-1 measurements, by Rème and Bosqued (1971) from sounding rocket measurements, and others. They show a high degree of field-alignment below 1000 km altitude and are not likely to be generated in the same way as those reported about here (see Hultqvist and Borg, 1978), the fraction of which that will reach 1000 km will show practically isotropic pitch angle distribution there.

Downward streaming field-aligned ion fluxes have recently been observed also at intermediate altitudes of 4000 to 8000 km by means of the S3-3 satellite

(Fennell *et al.*, 1978; Ghielmetti *et al.*, 1978). They may possibly be caused by the same process that gives rise to those field aligned ions streaming towards the ionosphere which have been reported above. Because of the long bounce period of these ions the observed fluxes may be the returning fluxes of ions, ejected field aligned from the opposite hemisphere but not visible at the satellite at the same time as the returning ones.

The azimuthal asymmetry observed in the ion angular distribution during the December 1977, magnetic storm (see Figure 23) existed for some eleven spin periods with a variable degree of asymmetry. There was frequently about ten times higher ion flux in that pitch angle scan in which the detectors looked away from the Sun that there was in that half spin in which the Sun was seen. At the same time the pitch angle distribution was field aligned in the direction towards the Earth. The most obvious way of interpreting this azimuthal asymmetry is as an effect of convection of the plasma perpendicularly to the magnetic field lines. Assuming a displaced Maxwellian distribution of protons with a typical characteristic energy of 5 keV, one finds that an asymmetry of a factor of ten for protons of the characteristic energy corresponds to a convection speed of $\sim$670 km s^{-1}. For this to be caused by an $E \times B$ drift the electric field component perpendicular to the magnetic field lines at $7R_e$ distance, B being of the order 100 γ, has to be $\sim$67 mV m^{-1}. If the ions are O^+ the convection speed and the electric field intensity values are only a quarter of the values given.

The mentioned electric fields are quite high. They are well comparable with those seen much closer to the Earth on auroral zone field lines (see e.g. Föppl *et al.*, 1968; Mozer, 1970; Cauffman and Gurnett, 1971; Lüst, 1971; Haerendel, 1972; Heppner *et al.*, 1971; Fahleson, 1972; Mozer *et al.*, 1977). The direction of the electric field was roughly dawn-dusk, giving a convection direction at 07:00 MLT approximately towards the Sun.

5. Concluding Remarks

The data from the GEOS-1 keV experiment presented in this paper show some new features of the complex plasma physics in the magnetosphere. Some of these new observational results are difficult to interpret on the basis of the plasma data alone. We look forward to a detailed study of them in collaboration with the other GEOS-experimenters.

Acknowledgements

The contributions by the technical staffs at KGI, DSRI and CESR to the success of this experiment, in particular those of S. Olsen and S. Laursen, are gratefully acknowledged. The magnetic field data on which the pitch angle calculations are based have kindly been provided by the S-331 experimenters. Prof. Mariani and his coworkers.

The Swedish part has been financed partly by means of grants from the Swedish Board for Space Activities. The CESR contribution has been made possible by grants from CNES. GEOS-1 has been developed, launched and operated by the European Space Agency.

References

Bame, S., Asbridge, J. R., and Strong, H. E. I. B.: 1967, *J. Geophys. Res.* **72,** 113.

Boström, R.: 1972, in E. R. Dyer (ed.), *Critical Problems of Magnetospheric Physics*, p. 139, Nat. Acad. Sci., Washington D.C.

Cauffman, D. P. and Gurnett, D. A.: 1971, *J. Geophys. Res.* **76,** 6014.

DeForest, S. E. and McIlwain, C. E.: 1971, *J. Geophys. Res.* **76,** 3587.

Fahleson, U. V.: 1972, in B. M. McCormac (ed.), *Earth's Magnetospheric Processes*, p. 223, D. Reidel, Publ. Co., Dordrecht, Holland.

Fennell, J. P., Mizera, P. F. and Croley D. R.: 1977, Paper GA-124 presented at the IAGA/IAMAP Joint Assembly in Seattle, August 22–September 3.

Fennell, J. F., Mizera, P. F. and Croley, D. R.: 1978, Paper SM-6 presented at the AGU Spring Meeting, 1978, *EOS* **59,** 348.

Frank, L. A.:1967a, *J. Geophys. Res.* **72,** 3753.

Frank, L. A.: 1967b, *J. Geophys. Res.* **72,** 185.

Frank, L. A.: 1967c, *J. Geophys. Res.* **72,** 1905.

Freeman, J. W., Jr. and Maquire, J. J.: 1967, *J. Geophys. Res.* **72,** 5257.

Föppl, H., Haerendel, G., Haser, L., Lüst, R., Melzner, F., Meyer, B., Neuss, H., Rabben, H. H., Rieger, E., Stöcker, J. and Stoffregen, W.: 1968, *J. Geophys. Res.* **73,** 21.

Ghielmetti, A., Shelley, E. G., Johnson, R. G. and Sharp, R. D.: 1977, Paper GA-126 presented at the IAGA/IAMAP Joint Assembly in Seattle, August 22–September 3.

Ghielmetti, A. G., Shelley, E. G., Sharp, R. D. and Johnson, R. G.: 1978a, Paper SM5 presented at the AGU Spring Meeting, 1978, *EOS* **59,** 348.

Ghielmetti, A. G., Johnson, R. G., Sharp, R. D. and Shelley, E. G.: 1978b, *Geophys. Res. Letters* **5,** 59.

Gringauz, K. I.: 1961, *Space Res.* **2,** 574.

Haerendel, G.: 1972, in McCormac (ed.), *Earth's Magnetospheric Processes*, p. 246, D. Reidel Publ. Co., Dordrecht, Holland.

Haerendel, G. and Lüst, R.: 1970, in B. M. McCormac (ed.), *Particles and Field in the Magnetosphere*, p. 213, D. Reidel Publ. Co., Dordrecht, Holland.

Heppner, J. P., Stolarik, G. D. and Wescott, E. M.: 1971, *J. Geophys. Res.* **76,** 6028.

Hultqvist, B.: 1974, *Ann. Geophys.* **30,** 223.

Hultqvist, B.: 1975a, in V. Formisano (ed.), *The Magnetospheres of the Earth and Jupiter*, p. 77, D. Reidel Publ. Co., Dordrecht, Holland.

Hultqvist, B.: 1975b, in B. Hultqvist and L. Stenflo (eds.), *Physics of the Hot Plasma in the Magnetosphere*, p. 291, Plenum Press, New York.

Hultqvist, B. and Borg, H.: 1978, *Planetary Space Sci.* **26,** 673.

Hultqvist, B., Borg, H., Riedler, W. and Christophersen, P.: 1971, *Planetary Space Sci.* **19,** 279.

Hultqvist, B., Borg, H., Christophersen, P., Riedler, W. and Bernstein, W.: 1974, NOAA Techn. Rept. ERL 305-SEL 29.

Lyons, L. R.: 1976, in D. J. Williams (ed.), *Physics of Solar Planetary Environments*, p. 701, Am. Geophys. Union, Washington D.C.

McIlwain, C. E.: 1975, in B. Hultqvist and L. Stenflo (eds.) *Physics of the Hot Plasma in the Magnetosphere*, p. 91, Plenum Press, New York.

Mizera, P. F. and Fennell, J. F.: 1977, *Geophys. Res. Letters* **4,** 311.

Mizera, P. F., Fennell, J. F. and Vampola, A. L.: 1977, Paper GA-125 presented at the IAGA/IAMAP Joint Assembly in Seattle, August 22–September 3.

Mozer, F. S.: 1970, *Planetary Space Sci.* **18,** 259.

Mozer, F. S., Carlson, C. W., Hudson, M. K., Torbert, R. B., Prady, B., Yatteau, J. and Kelley, M. C.: 1977, *Phys. Rev. Letters* **38,** 292.
Rème, M. and Bosqued, J. M.: 1971, *J. Geophys. Res.* **76,** 7683.
Shelley, E. G., Sharp, R. D. and Johnson, R. G.: 1976, *Geophys. Res. Letters* **3,** 654.
Vasyliunas, V. M.: 1968, *J. Geophys. Res.* **73,** 2839.
Williams, D. J.: 1975, in B. Hultqvist and L. Stenflo (eds.), *Physics of the Hot Plasma in the Magnetosphere*, p. 159, Plenum Press, New York.

DYNAMICS OF MAGNETOSPHERIC ION COMPOSITION AS OBSERVED BY THE GEOS MASS SPECTROMETER

J. GEISS, H. BALSIGER, P. EBERHARDT,
H. P. WALKER, L. WEBER and D. T. YOUNG
Physikalisches Institut, University of Bern, 3012 Bern, Switzerland

and

H. ROSENBAUER
Max-Planck-Institut für Aeronomie, 4311 Katlenburg-Lindau, Germany

Abstract. After one year of operation the GEOS-1 Ion Composition Experiment has surveyed plasma composition at all local times in the L range $3 \sim 8$ and the energy per charge range from thermal to 16 keV/e. From measurements made in the keV range during eleven magnetic storms we find that the percentage of heavy ($M/Q > 1$) ions present in the outer magnetosphere increases by a factor of 3 to 10 during disturbances. We conclude that two independent sources (solar wind, characterized by $^4He^{2+}$, and ionosphere, characterized by O^+) give on the average comparable contributions to injected populations, although in a single event one or the other source may dominate. However, in magnetically quiet periods protons are the dominant species with a few percent of heavy ions. With the help of special satellite manoeuvres magnetic field aligned fluxes of ≈ 0.05–3 keV/e H^+, He^+, O^+ with traces of O^{2+} have been observed which may be related to ion beams found previously at lower altitudes in the auroral zone. At still lower energies (~ 1 eV/e) the thermal plasma population is found to be made up of six ion species, three of which, D^+, He^{2+} and O^{2+}, were unknown in the magnetosphere prior to the GEOS-1 measurements. We present here a study of the evolution of doubly charged ions and their parent populations over four consecutive days. Various production mechanisms for doubly charged ions are discussed. We argue that ionization of singly charged ions by UV and energetic electrons and protons is the dominant process for plasmasphere production. Furthermore, the observed high concentrations of O^{2+} at high altitudes are a result of production in the upper ionosphere and plasmasphere combined with upward transport by thermal diffusion.

Throughout the 1 year lifetime of GEOS-1 the ICE functioned perfectly and, because of its novel design, a short review of technical performance is included here.

1. Introduction

The ion composition experiment (ICE) flown on GEOS-1 is a mass spectrometer particularly designed for determining the abundances and distribution functions of several magnetospheric ion species in the energy per charge range 0 to 16 000 eV/e. Our intention was of course to investigate the composition of the ambient plasma near the geostationary orbit ($L = 6.6$) covering all local times every 24 h. The eccentric 12 h orbit of GEOS-1 has added radial distance and latitude to the independent parameters mass per charge, energy per charge, and angular distribution sampled by the instrument. On the one hand we have benefitted from this orbit in being able to obtain radial profiles of ion composition between $L \approx 3$–8. On the other hand, one full year was needed to obtain a complete sampling at all local times and, together with the limited operations and coverage available on the western apogee orbits, it has been difficult to sort our

Space Science Reviews **22** (1978) 537–566.

systematic and relevant trends in the observed ion composition. Nevertheless, during the unexpectedly long one year lifetime of GEOS-1, valuable and in a large part novel data were obtained. With these data we have studied the main sources of magnetospheric plasma as a function of location and geomagnetic activity.

The ICE results from GEOS-1 confirm that magnetospheric plasma in the keV energy range contains ions of solar wind origin as indicated by He^{2+} (Reasoner, 1973) and ^3He (Bühler *et al.*, 1976; Geiss *et al.*, 1974) as well as ions from the atmosphere as indicated by O^+ (Shelley *et al.*, 1976) and He^+ (Johnson *et al.*, 1974). In fact on GEOS-1 we quite commonly observe a mixture of these two sources, but the relative contribution from one source or the other varies strongly in the eleven storms sampled thus far. Some results in the keV energy range are summarized in Section 4.

The eccentric orbit has provided an unexpected opportunity to investigate the composition of the plasmasphere and plasmapause. In spite of the fast motion of the spacecraft in the plasmasphere and high background due to Van Allen zone electrons at $L = 2$–4, new data have been obtained on the composition of the plasmasphere. In addition to the ions H^+, He^+ and O^+ reported earlier by Harris *et al.* (1970), we have found thermal and low energy O^{2+} and He^{2+} (Young *et al.*, 1977) and now have tentatively identified D^+ in the thermal plasma. In Section 6 of this paper further data on the spatial and energy distribution of O^{2+} are presented. A discussion of origins and possible O^{2+} production and transport mechanisms is given in Section 7.

Because of the 26 degree orbit inclination, GEOS-1 reaches magnetic latitudes of $\pm 30°$ and this introduces a correspondingly large range in pitch angles covered by the ICE. Moreover, during 3 special periods, lasting a total of 17 days, the GEOS-1 spin axis was tilted by 90° to allow the ICE to obtain complete pitch angle scans near apogee. Field aligned fluxes of H^+, He^+, O^+ and O^{2+} were observed on several occasions, one of which is presented in Section 5.

2. Instrument

The instrument has been described by Balsiger *et al.* (1976). Its basic components are shown in Figure 1. After passing through a retarding potential analyzer (RPA) ions of charge Q are accelerated by a fixed potential difference $U_a = 2.7$ kV. With their original kinetic energy E augmented by QU_a, ions are then filtered by a cylindrical electrostatic analyzer (EA). About 10% of the ions are intercepted by a channeltron which gives a 32 step energy per charge spectrum of the total ion flux. Since the energy resolution is $\pm 2.5\%$ of $E + QU_a$ it is poor at low energies ($E \ll QU_a$), necessitating use of the RPA below 100 eV/e. The major part of the ion beam enters the mass analyzer (MA) where a combination of cylindrical electric field and perpendicular magnetic field provides mass separation. By

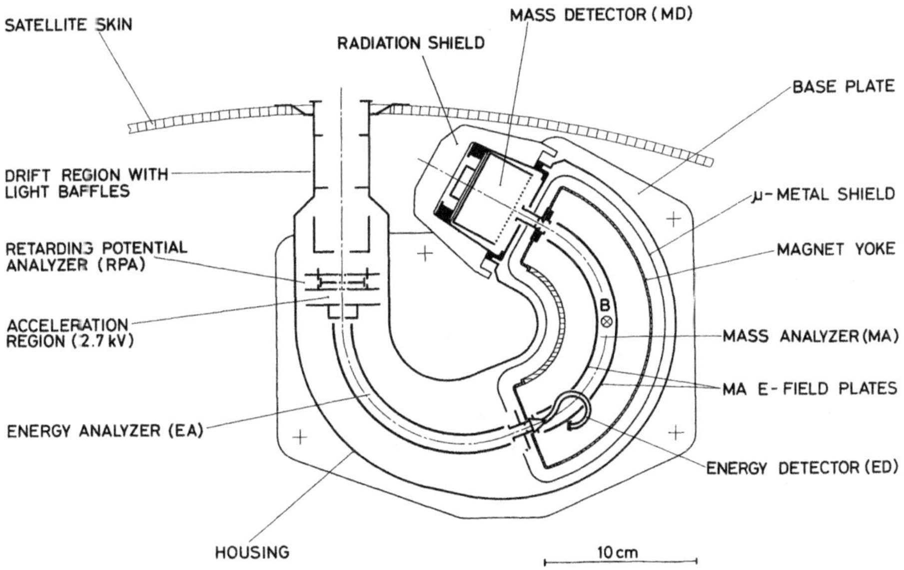

Fig. 1. Schematic of ICE sensor without electronics (see text for description).

varying the MA electric field over 64 steps the mass detector (MD) registers a mass per charge spectrum for the energy per charge window selected by the EA.

There is always a trade-off between mass resolution, energy range and sensitivity. Since the main ions (natural and artificial) expected in the magnetosphere have rather widely separated mass per charge ratios, we developed an instrument with low mass resolution. The resolution $M/\Delta M$ of this type instrument depends on the mass and energy of the ion. Resolution at 1% of peak height is $\geqslant 4$ for $E/Q \leqslant 3$ keV/e and $M/Q \leqslant 4$. For higher energies and/or masses $M/\Delta M$ decreases, but is still sufficient to separate O^+ from Ba^+ for $E/Q \leqslant 3$ keV/e. The sensitivity of the instrument depends slightly on mass and energy and is approximately 1 count per second for a flux of 1 ion/cm^2 sec ster eV in the L mode energy range (25 to 16 000 eV/e, see below).

Energy per charge analysis of incoming ions is carried out in two overlapping energy ranges each requiring a separate operating mode. The thermal (T) mode covers the 0 to 110 eV/e range with 32 RPA voltage steps (Figure 1). The low energy (L) mode fixes the RPA at +25 V and uses the EA to cover the 25 to 16 000 eV/e range with 32 steps logarithmic in $E + QU_a$. In the simplest operating program a 64 step mass per charge scan is carried out at each of the 32 energy steps (cf. Figure 2). While this energy-mass scan is sufficient for most purposes, particularly in the L mode where fluxes tend to be roughly isotropic within the detector's field of view, it must occasionally be augmented by angular scanning programs when flux anisotropies are present due to electric field drifts, satellite motion, or pitch angle anisotropies. In both T and L modes therefore, high

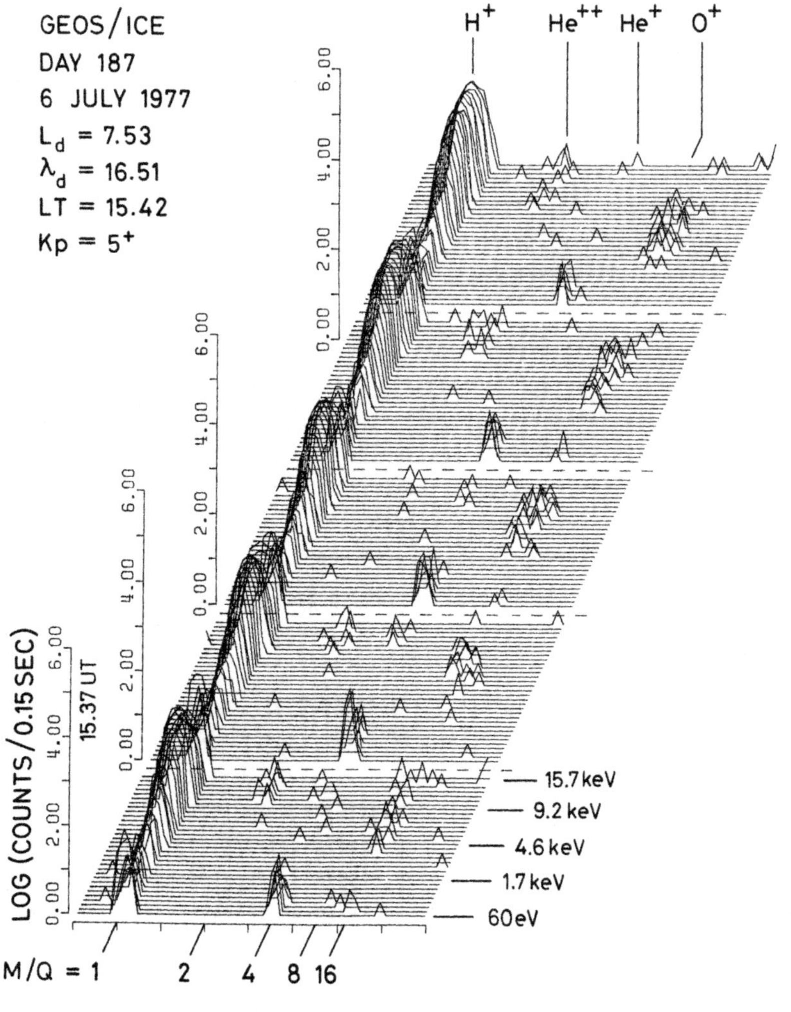

Fig. 2. Three-dimensional L mode spectra for 6 July 1977 showing 30 minutes of ion composition data. Dashed lines separate five complete mass-energy scans each lasting 5.9 min. Zero and 1 count per sample (0.15 s) are plotted as zero, but the data (from the fixed discrimination level mass detector) are otherwise shown directly as they appear from telemetry. The abscissa gives mass per charge (M/Q) with divisions every 8 mass steps, the ordinate is logarithm of counts per sample and the third axis gives time and energy. L_d and λ_d are the dipole approximations for L and magnetic latitude, LT is local time; all are referenced to the start of the first spectrum (1537 UT). Kp refers to the 3 h period in which the spectrum was taken. Energies given are the center values of steps 0, 8, 16, 24 and 31 in the L mode.

resolution ($\Delta\varphi \cong 10.5°$) and low resolution ($\Delta\varphi \cong 42°$) azimuthal scans were often carried out using experiment logic to divide each satellite revolution into ≈ 34 samples referenced to a fixed point in space (cf. Figure 12). In order to increase the signal to noise ratio of rarer ion species and provide better temporal resolution, the ICE can also be programmed to remain on a fixed mass peak while scanning energy and angle. The different modes and angular scans are programmed on a regular basis through the satellite control center at ESOC in Darmstadt.

The ICE views perpendicular to the satellite spin axis with an acceptance angle in the plane of the spin axis of $\pm 7.5°$ for $E > QU_a$. This angle increases gradually with decreasing ion energy, approaching $\pm 25°$ for $E \ll QU_a$. The azimuthal acceptance angle varies only from ± 5.3 to $\pm 5.7°$ over the full energy range. The large vertical acceptance angle combined with 26° orbit inclination has led to rather wide pitch angle coverage over parts of the orbit although pitch angles outside the range $90° \pm 45°$ are seen occasionally.

3. Instrument Performance

The instrument worked perfectly during the first 12 months of spacecraft operation with ~60 000 commands successfully executed. In-flight calibration of plate voltages (which go as high as 5500 V) was performed routinely and confirmed that the average differential voltage drifts were smaller than 0.3%. Thus, the absolute mass scale is extremely stable as are peak shapes, and spectra can be summed over long periods without any corrections. After one year of operation no gain loss in the two multipliers has been detected.

An important goal of the ICE design was to suppress detector background from several possible sources (scattered UV, penetrating electrons, electromagnetic interference and internal contamination). During thermal vacuum tests with the integrated satellite a background of 5×10^{-2} counts s^{-1} was measured at the normal discrimination level 2 (equivalent to a charge pulse of 1×10^6 electrons) for both ED and MD. In flight the detector background is higher due to penetrating radiation and depends on location and geomagnetic activity. Typical background in the MD is approximately 1 count s^{-1} at $L = 8$, 5 counts s^{-1} at $L = 6.6$, 100 counts s^{-1} at $L = 4.5$ and 3 counts s^{-1} at $L = 3$. (We have been able to increase the original 335 g of MD shielding by an additional 40 g for GEOS-2 to reduce background still further.) No interference from solar UV can be detected. Even with the ICE pointing within $\pm 5°$ of the Sun the upper limit of Sun induced background is 0.1 counts s^{-1}.

At various times on GEOS-1 we measured small fluxes of ions from satellite outgassing but only when we suspect that the satellite potential was negative, i.e., only in the high density plasmasphere. Occasionally we find a relatively broad peak of heavy background ions in mass channels 52 to 62 ($M/Q \gtrsim 30$) with $E/Q < 110$ eV/e. The occurrence of these ions is correlated with fluxes of O^+ ions with energies ~10^4 eV. Measurements in the calibration chamber have confirmed that the same heavy mass distribution with similar energies is produced with an efficiency of ~10^{-4} by a Ne^+ beam at 10^4 eV. Furthermore, the mass distribution is consistent with Cu^+ ions plus a smaller contribution from CuS^+ and S^+ ions. We therefore conclude that the major part of these heavy ions could be due to sputtering, mainly by O^+ ions, from CuS surfaces at the entrance of the instrument. However, in some instances heavy ions were observed when the O^+ and He^+ fluxes above 1 keV were relatively low. Further investigation is needed to show whether natural ions heavier than O^+ (e.g., N_2^+, Fe^+) are occasionally present in the magnetosphere.

4. Energetic Ions

Ion fluxes measured with the ICE often show the presence of three energetically distinct ion populations which we define as *thermal* ($E \sim$ eV), *superthermal* ($E \sim 0.1$ keV) and *energetic* ($E \approx 1$–16 keV). Superthermal fluxes tend to be found beyond the plasmapause during quiet periods and extend to lower L during magnetic storms. They will be the subject of a later study (for examples of superthermal ions, see H^+ and He^+ fluxes in the first 6–8 energy steps of Figures 2 and 4).

In this section we present representative data covering the keV energy range of the instrument. A more extensive account of these results will be given by Balsiger *et al.* (1978). Data presented here have been taken mainly from the part of the orbit above $L \approx 5$ where the keV ion fluxes are easily distinguished above the penetrating radiation background. However, during magnetic storms ring

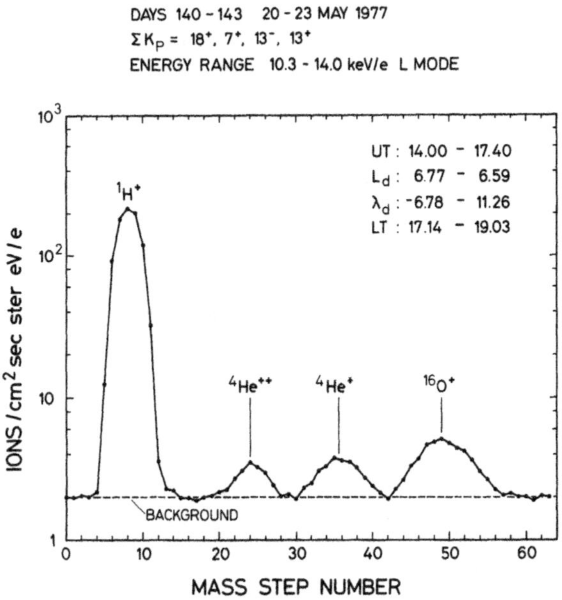

Fig. 3. Integrated L mode mass spectrum for approximately 2 h UT either side of apogee for 4 passes on 20–23 May 1977. Abscissa is mass per charge, the ordinate is differential flux per unit charge (to avoid adjusting the $^4He^{2+}$). Integration is made over energy steps 26–29 covering the range 10.4–14 keV/e. Each data point is accumulated over 272 s and a flux of 1 ion $(cm^2\,s\,ster\,eV/e)^{-1}$ corresponds to 1810 accumulated counts.

current profiles (in the range 1–16 keV/e) can be obtained down to $L \leqslant 3$, because ion fluxes are then more intense and background is lower.

The 3-D plot in Figure 2 is a mass–energy spectrum illustrating characteristic dominance of H^+ fluxes relative to He^{2+}, He^+, and O^+ during magnetic quiet. A more quantitative example of quiet time data is given in Figure 3 where the energy range 10.3–14.0 keV/e has been integrated over 4 consecutive quiet days at $L_d = 6.8$ to 8.0. The sensitivity of the instrument as well as its ability to resolve the main magnetospheric constituents even at high energies is quite apparent here.

In general, as a magnetic storm main phase begins, the heavy ion fluxes intensify by factors ≈3–10. Figure 4 is an example of data from near local noon

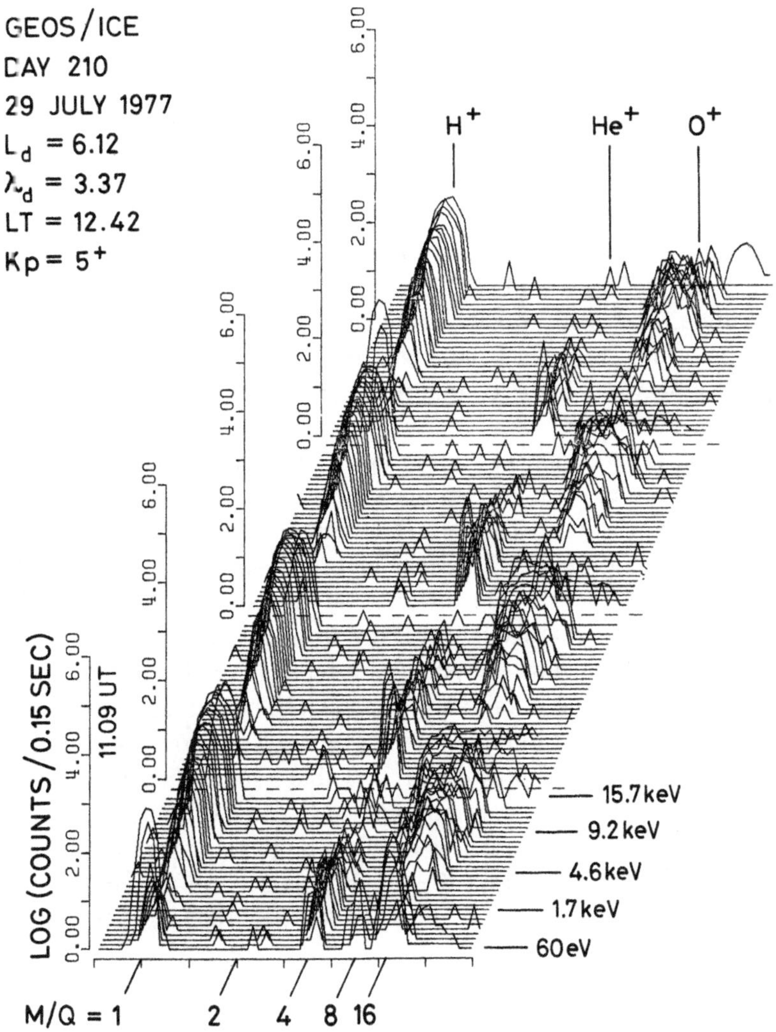

Fig. 4. Three-dimensional L mode spectra for 29 July 1977. For description see Figure 2 and text.

taken during the main phase of a storm (Dst reached a maximum -100γ at 0700 UT and was -57γ at 1200 UT). The plasma composition found in this storm (at the time and place of the GEOS-1 pass) is rich in O^+ and low in He^{2+}. Between $L_d = 5.8$ and 7.8 (1045 UT−1445 UT) the He^{2+}/H^+ density ratio is 0.007 whereas the O^+/H^+ ratio is 0.6. An interesting feature of Figure 4 is the 'superthermal' He^+ in the lowest energy channels. (A similar but smaller feature also occurs in Figure 2, possibly as the result of slightly increased magnetic activity following an SSC at 1344 UT on 6 July.) Similar distributions of H^+ and O^+ ions are sometimes observed as well, even at L values as low as 4, particularly during magnetically active periods.

In contrast to the storm fluxes low in He^{2+} in Figure 4, the 3-D plot of Figure 5 shows plasma containing ions of both solar and terrestrial origin with each source contributing comparable amounts (density ratios at this time in the storm are $He^{2+}/H^+ = 0.02$ and $O^+/H^+ = 0.34$). Data in Figure 5 were taken during the storm recovery phase (Dst reached -125γ at 1200 UT on 11 December and was -80γ at 0500 UT on 12 December). The He^{2+} ions first appeared at $\approx$0430 UT and at the same time the H^+ and O^+ fluxes intensified by a factor of 2–3. The ion fluxes decreased at $\approx$0630 UT in such a way that the composition remained fairly constant. The overall behavior is suggestive of drifting clouds of storm injected plasma similar to what has been observed previously in this region of the magnetosphere (DeForest and McIlwain, 1971). These observations, carried out at high altitudes and near-equatorial pitch angles, demonstrate the injection of solar wind plasma into the equatorial ring current and as such provide an important boundary condition for theories of ion radial diffusion (Cornwall, 1972; Spjeldvik and Fritz, 1978).

We have analyzed a data set made up of data taken from 2 h near apogee ($L_d = 6.9$ to 8.2) for 11 magnetic storms between May 1977 and March 1978. The results are summarized in Table I. The 'night side storm composition' was obtained by averaging data taken between 1800 and 0600 local time during the main or recovery phases of 6 storms (12 h of data total). Similarly the 'day side storm composition' represents the average from 5 storms (0600 to 1800 local time, total 10 h of data). For comparison, a typical 'quiet time composition' was obtained by averaging the data from 6 passes (12 h of data with $Kp = 0°$ to 2^-) spread over local time. Although there are large differences when individual storms are compared (cf. Figures 4 and 5), Table I shows that we have not so far found a systematic difference between night side and day side storm composition. The absolute and relative abundances of O^+ and/or He^{2+} always increase during a storm. However, in all cases both the number density and energy density of H^+ were still higher than the corresponding O^+ densities (at $L_d = 6.9$–8.2). Based on the averages in Table I we conclude that He^+ ions are mainly terrestrial in origin since their average energy is at least a factor of 2 too low for them to be the result of the charge exchange of He^{2+}.

For the purpose of illustrating the contributions of the two sources, the energy

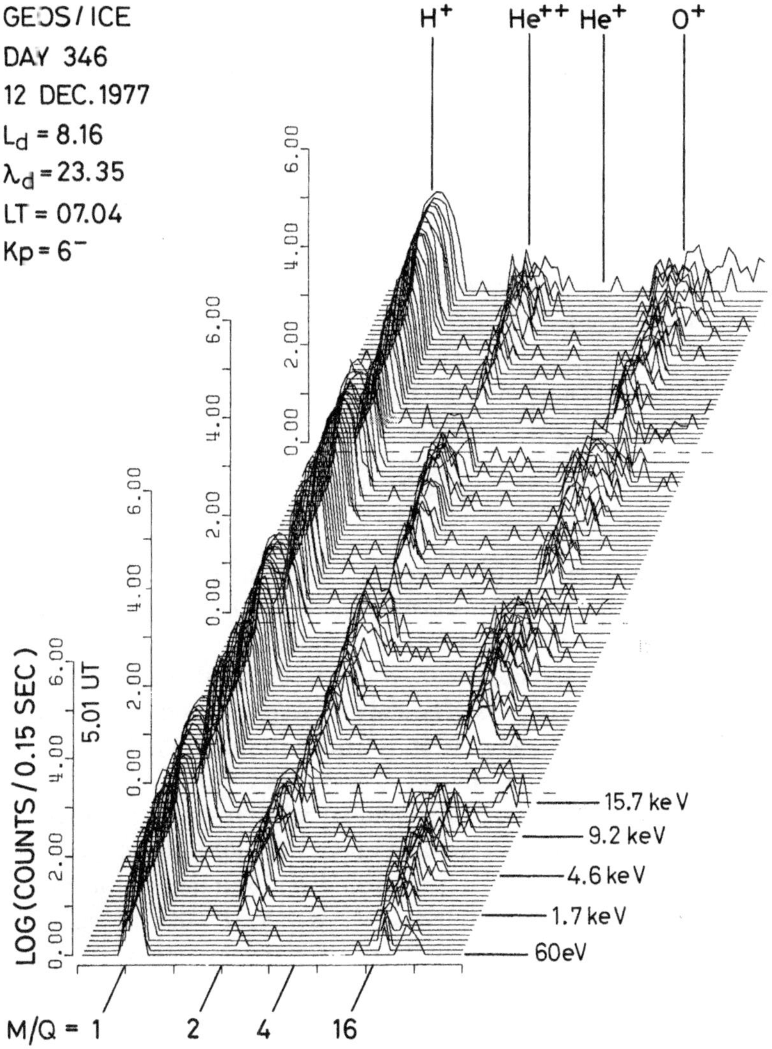

Fig. 5. Three-dimensional L mode spectra for 12 December 1977. For description see Figure 2 and text.

densities of injected solar wind plasma and injected plasma of terrestrial origin given in Table I were estimated in the following way: The average quiet time number densities of the individual ions were subtracted from the respective storm time number densities. For the remaining gas it was assumed that all He^{2+} plus a fraction of H^+, corresponding to $H^+/He^{2+} = 22$, is of solar wind origin. The remainder of H^+ as well as He^+ and O^+ are assumed to be terrestrial. This simplifying assumption neglects H^+ and He^{2+} charge exchange with neutral hydrogen, the possible presence of He^{2+} rich piston driver gas from the solar wind and possible differences in the average energy of solar and terrestrial H^+.

TABLE I

Average characteristics of ion composition in the region $L_d = 6.9$–8.2 and energy per charge range 1–16 keV/e. Average total energy is weighted by number density

		Number density (10^{-2} cm^{-3})				Average total energy (keV)				Contribution to excess energy density (percent)	
		H^+	He^{2+}	He^+	O^+	H^+	He^{2+}	He^+	O^+	Solar plasma	Terrestrial plasma
Storm day-side	Average	57	1.1	0.8	15	6.0	11.8	5.1	4.7	45	55
	Range	(20–120)	(0.2–3.6)	(0.4–1.2)	(2.1–30)	(5.0–8.5)	(10.2–18.8)	(3.9–7.3)	(4.1–6.1)		
Storm night-side	Average	59	1.0	0.7	11	6.2	14.5	4.2	4.3	40	60
	Range	(25–135)	(0.6–1.5)	(0.3–1.8)	(3.0–22)	(3.6–9.6)	(10.2–20.2)	(2.8–5.0)	(3.9–4.9)		
Magnetic quiet	Average	15	0.1	0.4	0.9	6.8	13.9	5.8	5.1	—	—
	Range	(8.2–25)	(0.02–0.2)	(0.2–0.6)	(0.1–1.6)	(5.4–8.6)	(9.0–19.6)	(3.8–8.6)	(3.8–7.2)		

Our upper energy limit is 16 keV/*e*, but data in Table I is concentrated at $L = 7.5$–8.1 so it is reasonable to assume most of the plasma is included in this sample. Despite these reservations, Table I indicates that on the average, solar wind and terrestrial plasma contribute roughly comparable amounts to the energy density of storm injected plasma. The ranges of density and mean energy in Table I suggest considerable variability from storm to storm and indeed, in the extreme cases of the individual storms studied, one was found to have $\geqslant 90\%$ terrestrial ions and one $\geqslant 90\%$ solar ions. It should be kept in mind that these results are based on 2 h averages taken at random epochs of storm development and hence do not characterize the entire storm. However, examination of more complete data sets (in the 3-D format) for these storms suggests that at least a few of the injected plasma clouds from individual substorms contain mainly terrestrial (O^+/H^+ high) ions with little or no solar component. Further analysis, including evaluation of M/Q and E/Q dependent losses, is necessary to determine whether a given storm or substorm possesses a compositional signature that is 'mainly solar' or 'mainly terrestrial'. Since there is no *a priori* reason to assume that an individual substorm would inject equal amounts of terrestrial and solar plasma, it will be of interest to investigate whether the balance of solar and terrestrial sources suggested by Table I is true for individual storms or is only true for an average over many storms.

Concerning the charge exchange process $He^{2+} + H \rightarrow He^+ + H^+$, it should be noted that the energy per charge ratio is approximately doubled and consequently He^{2+} above 8 keV/*e* is removed from the energy range covered by ICE. We often find, however, that ions within this range may be fitted quite well by a Maxwellian distribution. In these cases the graph of phase space density may be extrapolated outside the ICE energy range so that the mean energy and total density may be estimated with some confidence.

Some controversy has arisen over the composition of ions contributing to the storm time ring current (Sharp *et al.*, 1977; Johnson *et al.*, 1977; Lyons and Evans, 1976; Lyons, 1977; Tinsley, 1976). Lyons and Tinsley have presented evidence for He^+ as the dominant ion during the recovery phase of storms whereas satellite observations reported by Sharp and co-workers suggest that He^+ is a minor constituent of the ring current after H^+ and O^+. Table I shows that in the 11 storms studied so far, He^+ in the 1–16 keV range is clearly a minor constituent outside the ring current at $L \approx 7$–8. In Figure 7 we show a radial profile of ion composition in the ring current for one pass during the 29 July storm at a time when Dst was $\approx -60\gamma$. As remarked earlier for the data at higher L values, the July storm is low in He^{2+} and this trend continues down to $L = 3$.

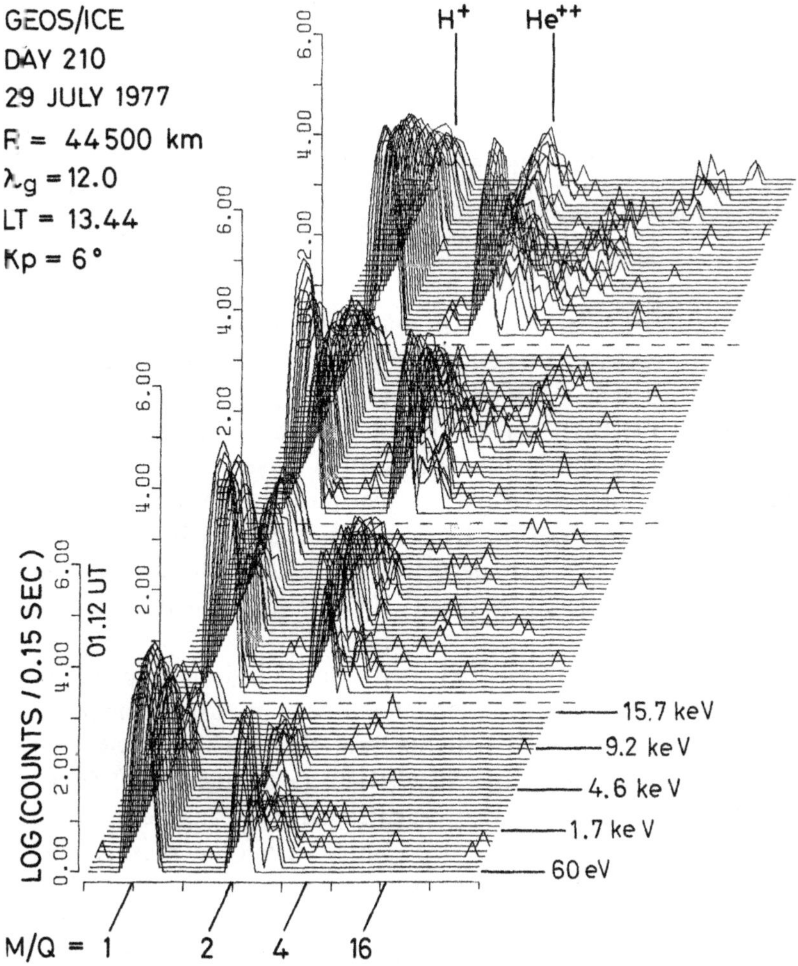

Fig. 6. Three-dimensional L mode spectra taken in magnetosheath during NASA pass on 29 July 1977. For description see Figure 2 and text. The small cluster of counts at $M/Q = 16$ in the upper right hand corner of the last spectrum is not solar wind, but rather O^+ of magnetospheric origin.

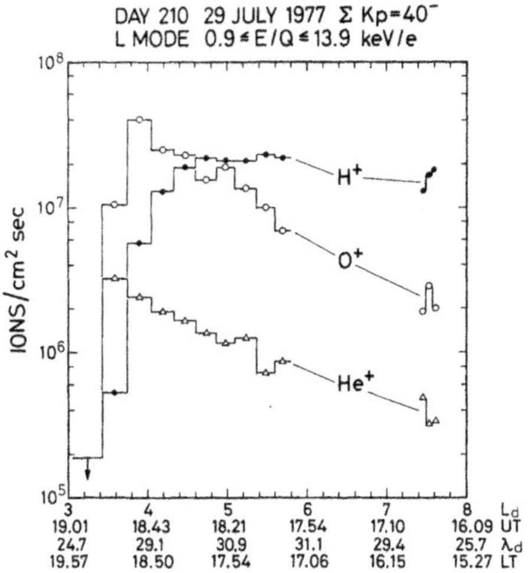

Fig. 7. Radial profile of composition during the 29 July 1977 magnetic storm. Each set of data at a given L_d is obtained from a single L mode scan by summing over the 25 energy steps between 0.9 and 13.9 keV/e. In converting to total flux differential fluxes are assumed to be isotropic. DST at 1600, 1700, 1800 and 1900 UT was -67γ, -64γ, -60γ and -57γ respectively. Data between L_d = 5.8–7.4 were missing from experiment tapes. The upper limit at $L_d = 3.0$–3.4 applies to each of the three ions.

Similar profiles from other storms do not necessarily show the clear and simple trends apparent in Figure 7 but two generalizations may be made: (a) in agreement with our data at higher L values, H^+ is generally the dominant ion but where this is not the case then O^+ is dominant and (b) He^+ fluxes are on the average a small fraction of the total ion flux and in no case have we observed He^+ to be the dominant ion. We emphasize that these conclusions apply only below the ICE energy limit of 16 keV/*e*. Our measurements may be somewhat weighted in favor of terrestrial ions which, particularly at lower L values, could be expected to have lower energies than solar wind He^{2+} or the resultant charge exchanged He^+.

One interesting feature of the July storm was that during the western apogee (NASA pass), GEOS-1 was near local noon when a strong compression of the magnetosphere occurred. ICE data indicate that GEOS-1 was in the magnetosheath (this is confirmed by GEOS-1 magnetometer data) from 0026 UT until at least 0149 UT when data coverage ended. Several spectra from this event are given in Figure 6. Although strong spin modulation is present, the 3-D plot nonetheless shows the high M/Q component of the solar wind which extends diagonally away from the $^4He^{2+}$ peak toward higher mass and energy. Several individual spectra show counts due to the $^3He^{2+}$ ion (between the large H^+ and He^{2+} peaks). With further analysis of these data we have been able to distinguish in addition to H^+, $^3He^{2+}$, and $^4He^{2+}$, the oxygen group (mainly O^{6+}), the silicon group (around $M/Q = 3.1$ to 3.5) and the iron group.

5. Field Aligned Ions

In order to avoid electromagnetic interference caused by shadowing of the solar cells by the radial booms, GEOS-1 is 'inverted' (i.e., the spin axis is flipped by 180°) at each equinox. At our request (and with cooperation of the other experimenters) two equinox and one special inversion were carried out in 2 steps, with an intermediate position where the spin axis remained oriented in the geomagnetic equatorial plane for periods up to 8 days. In this position the ICE scanned to within a few degrees of the local magnetic field direction, obtaining a pitch angle scan every 3 s during the ~1 h per pass when the spin axis was approximately perpendicular to the local magnetic meridian plane. The inversion periods were 9–10 September 1977, 14–21 October 1977, and 4–10 April 1978 with local times of apogee respectively 1135, 0930, and 0016 LT.

Figures 8 and 9 are examples of the pitch angle-composition distribution during a period of geomagnetic activity (Kp was 5^- at this time and Dst had a value of

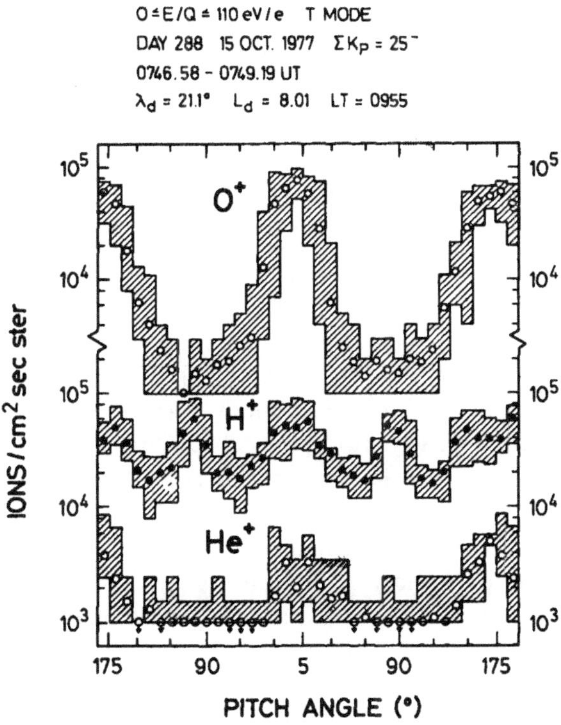

Fig. 8. Pitch angle distribution of magnetic field aligned ion fluxes on 15 October 1977 during a GEOS-1 inversion manoeuvre. Data were obtained using the high resolution angular scan mode (i.e., one spin scan per energy step) with an 8 energy step RPA scan between 0 and 22 eV/e. The shaded regions represent the extreme range of all 8 measurements, demonstrating both the stability of the ion fluxes and the fact that most of the ions are above 22 eV/e. The data have been corrected for background. Symbols show average value. Small divisions on the abscissa correspond to pitch angles of 30, 60, 120, and 150 degrees. For the averaged points a flux of 10^3 ions $(cm^2 \sec ster)^{-1}$ corresponds to 1 count.

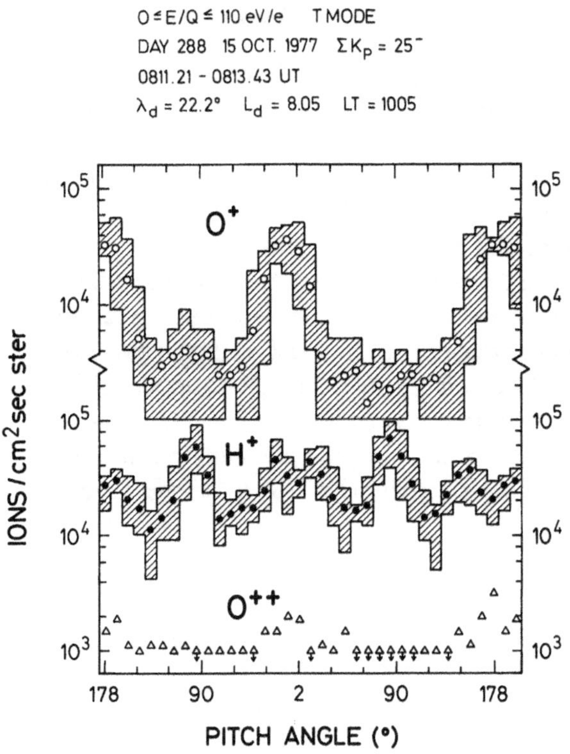

Fig. 9. Pitch angle distribution of ions on 15 October 1977. Data were obtained 22 minutes after that shown in Figure 8. Range data for O^{2+} is not statistically significant thus only averages are plotted. For further description see Figure 8.

-48γ, but there was no clear storm in progress). These examples of field aligned fluxes are more typical of disturbed periods whereas during quiet times fluxes were weaker and more nearly isotropic. All ion fluxes in this event have mean energies of ~ 50 eV with few ions present above ~ 1 keV.

Figures 8 and 9 show that O^+, He^+ and probably O^{2+} fluxes were field aligned whereas H^+ exhibited both field aligned and perpendicular components. This condition was steady over most of the 1 h observing period, but short intervals with H^+ dominant were also seen. Field aligned number density ratios were $O^+ : H^+ : He^+ : O^{2+} \approx 20:4:1:1$. The field alignment of these fluxes together with their composition can be taken as a signature of recently injected ionospheric plasma which appears to originate near or below ~ 1000 km altitudes where O^+ is the dominant ion. The distinct H^+ peak at 90° pitch angles must then be an older population, possibly of plasmaspheric origin since this component is low in O^+. Fluxes from 0° and 180° pitch angles are nearly equal, suggesting that either injection occurred from both hemispheres equally or that injection from one hemisphere was followed by several bounces. This supports a similar result obtained by McIlwain (1976) who used electrostatic analyzer observations of bouncing ion clouds to infer the existence of heavy ion bunches at geostationary orbit.

We have in addition found field aligned O^+ fluxes with energies up to 4 keV and intensities of ~10^6 ions/cm^2 sec ster during preliminary analysis of data from a magnetic storm during the third inversion period (4 April 1978, 2100 UT–2300 UT, $Kp = 6+$). Less intense He^+ and H^+ fluxes are present as well at this time. Although we should not place too much weight on a single event, these fluxes may well be the manifestation near the equatorial plane of intense beams of 0.5–16 keV H^+ and O^+ ions observed by the Lockheed group at altitudes of ~5000 km (Shelley *et al.*, 1976; Ghielmetti *et al.*, 1978) and at lower energies (~100 eV) by Wahlen *et al.* (1978). Analysis of the GEOS-1 field aligned observing periods is not yet complete and other similar events may be found. However our data base is limited to ≈15 h scanning within the ≈5° wide source/loss cone and more than half of this total was obtained in the dawn local time sector where Ghielmetti *et al.* report a sharp minimum in occurrence frequency of field aligned events.

6. Composition of Thermal Ions

An example of thermal ion composition inside the plasmasphere at $L = 2.88$ is given in Figure 10 (a sharp plasmapause crossing occurred between $L = 5.0$ and

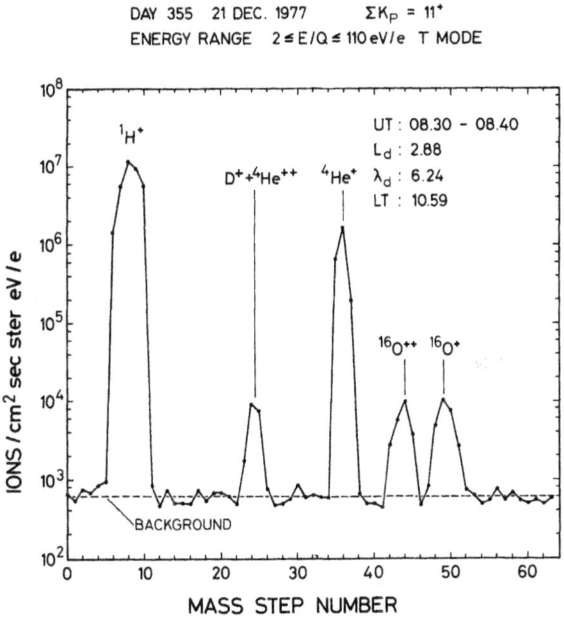

Fig. 10. Integral *T* mode mass spectrum taken with a low resolution angular scan and the RPA fixed at the 2 eV/e step. Each point is an average over 3 spin periods and 61 accumulated counts corresponds to a flux of 10^3 ions (cm^2 sec ster)$^{-1}$. The top of the $^1H^+$ peak is distorted due to saturation of detector electronics.

5.2). The most interesting features in Figure 10 are undoubtedly the ion peaks at $M/Q = 2$ and 8 observed in the high altitude magnetosphere for the first time by GEOS-1 (Young *et al.*, 1977). In Young *et al.* we argued it is plausible to identify the $M/Q = 8$ peak as O^{2+}. However the $M/Q = 2$ peak could be deuterium (D^+) or helium ($^4He^{2+}$). The relatively high O^{2+}/O^+ ratio shown in Figure 10 is not unusual for this location and history of magnetic activity. The existence of these doubly charged ions, apparently throughout that part of the plasmasphere accessible to GEOS-1 (within $\pm 30°$ of the equator above $L \approx 2.8$), has led us to investigate in some detail possible production, loss and transport processes which we discuss in the next section. The $H^+/(M/Q = 2)$ ratio of $\sim 10^3$ is rather typical.

In Figure 10 we have plotted the ion flux above 2 eV/e integrated over a low resolution angular scan. The integral above 2 eV/e eliminates background in mass channels above step 50 which, as mentioned in Section 3, is observed when the satellite potential is negative (at this time the potential is ≈ -0.5 to -1.0 V as determined by the S-300 DC experiment). The local pitch angle range covered is estimated to be $45° \leq \alpha \leq 135°$ at this time with no indication of ion flows within the detector field of view. The last complete high angular resolution RPA scan made 15 min before the spectrum in Figure 10 was taken, gives H^+ and He^+ temperatures of 0.8 ± 0.1 eV and densities of 350 cm^{-3} and 50 cm^{-3} at $L \approx 3.5$. At $L \approx 2.9$ the partially completed RPA scan gives temperatures of 0.5 ± 0.1 eV and densities of 690 cm^{-3} and ≈ 30 cm^{-3} respectively. For minor ions the RPA scan at $L \approx 2.9$ is incomplete. Temperatures and densities of the minor ions can only be measured at the earlier time (i.e., at $L \approx 3.5$), giving 0.7 ± 0.1 eV for O^+ and O^{2+} and densities of 0.15 cm^{-3} and 0.3 cm^{-3} respectively. If the $M/Q = 2$ channel is assumed to be $^4He^{2+}$ then we obtain 1.7 eV and 0.3 cm^{-3} whereas an assumption of D^+ gives 0.8 eV and 0.15 cm^{-3}. The D^+ identification of $M/Q = 2$ gives in this case the more reasonable temperature since for these densities the ion–ion collision time is short (~ 1 h), leading to the approximate thermal equilibrium observed for the other four ions. Thus we tentatively identify the major part of the $M/Q = 2$ peak in this particular case as D^+. We plan to study and apply further this identification scheme in order to test its general validity.

Figure 11 illustrates the variation in composition of the thermal plasma during a period of relative magnetic quiet. At present we do not calculate densities on a routine basis for large amounts of data, hence Figure 11 (the bottom two panels) is given in terms of counting rate ratios. Assuming equal ion temperatures (which seems generally to be the case for T mode spectra) and neglecting plasma drift velocity, then the singly charged ion ratios O^+/H^+ and He^+/H^+ are unaffected by satellite potential and may be converted to density ratios by multiplication by the factors 4 and 2 respectively. According to information from the S-300 experiment, roughly 90% of the data considered here were obtained under conditions of positive satellite potential. Since positive satellite potential can only discriminate against doubly charged ions, the doubly changed ion ratios are lower limits. The remaining $\sim 10\%$ of data were obtained at lower altitudes where the satellite

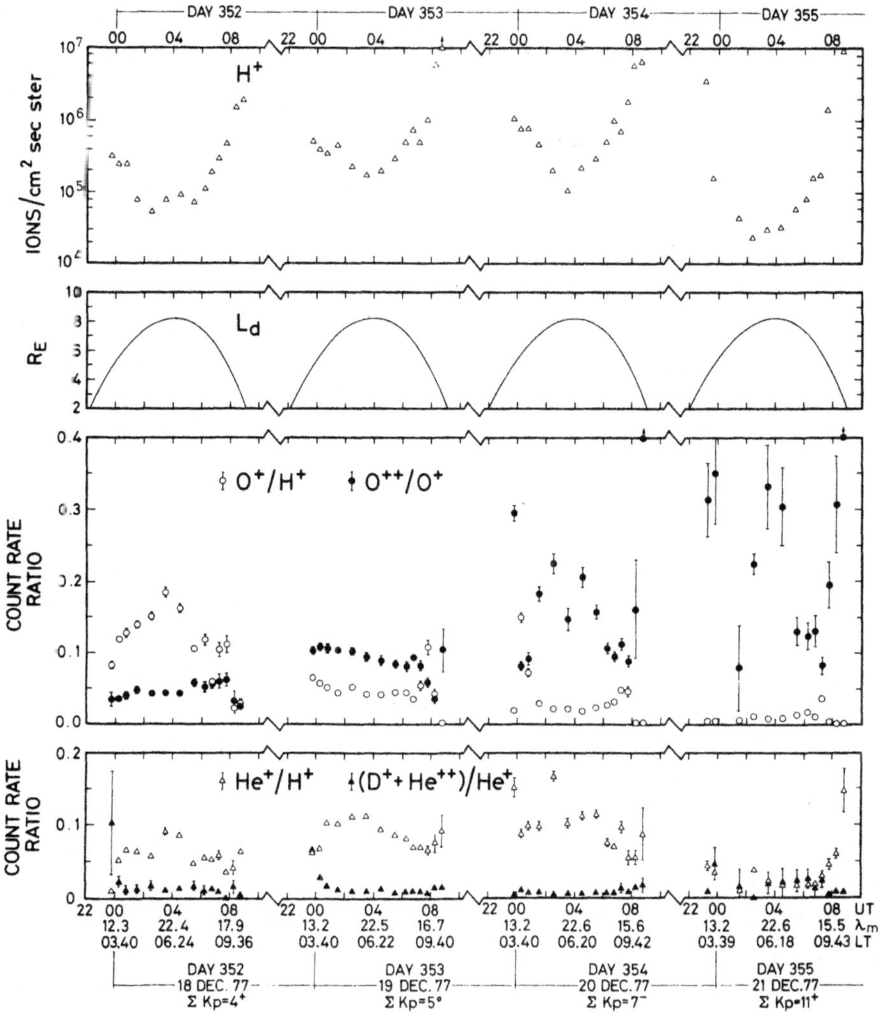

Fig. 11. Averaged T mode data for 4 consecutive GEOS-1 passes in December 1977. Each point is an average over 45 to 600 samples depending on type of angular scan mode (i.e., low or high resolution) and whether a point is a 1/2 or 1 h average. The upper panel is the H^+ flux, approximately correct to within a factor of 2. The second panel is the orbit in terms of L_d. The bottom two panels are ratios of the average counting rates of the respective ions. Ion counting rates have been corrected for background and error bars give ± 1 standard deviation of the mean, including the background error. No correction has been applied to ratios for effects of satellite potential or velocity (see text).

potential is zero or slightly ($\lesssim 1$ V) negative and the satellite velocity comparable to the thermal speed of the heavier ions. Under these conditions the counting rate ratio of the doubly charged ions is approximately equal to the density ratio.

Prior to the magnetically quiet period shown in Figure 11, Kp reached 5° at 0300–0600 UT on Day 351 and the $\sum Kp$ for that day was 16+. For the periods shown in Figure 11, on the first 3 days Kp was between 0+ and 1−, whereas the fourth day was slightly more active with Kp of 1+, 0+, 1+ between 00 and 09 hours UT. During the 3 quiet days the H^+ flux increases systematically at a given L, as expected for plasmasphere filling following a disturbance. The O^+/H^+ ratio decreases and the O^{2+}/O^+ ratio increases, both in systematic fashion such that the O^{2+}/H^+ ratio tends to remain relatively constant at $\approx 5\times10^{-3}$. Linear regression analysis of data for days 352–357 gives a statistically significant correlation coefficient $r = -0.74$ for 82 O^{2+}/O^+ vs. O^+/H^+ data points. There is a less pronounced trend in the $M/Q=2$ ratios: He^+/H^+ tends to increase and $(D^+ + He^{2+})/He^+$ to decrease. On Day 355 the effect of the small magnetic disturbance is clearly to reduce the H^+ density, creating a plasmapause at $L_d \approx 5$ on both inbound and outbound parts of the orbit. However, for oxygen ions the trend of the previous 3 days seems to continue whereas for $M/Q=2$ it is broken off.

We can try to understand these observations using the following picture. The disturbance on Day 351 has emptied the region of the plasmasphere seen by GEOS-1 and filling takes place over the next 3 days of magnetic quiet. As the H^+ density increases the O^+/H^+ ratio decreases in keeping with models of ion transport in the topside ionosphere (Murphy *et al.*, 1976). Although the O^+ density may be expected to have a large diurnal variation (Hagen and Hsu, 1974) on the average it does not build up as H^+ does. On the other hand, the combination of O^{2+} production and upward transport keeps the O^{2+} density growing over the quiet period (see next section).

In order to calculate ion density, drift velocity and temperature it is necessary to determine the distribution functions of the different ion species in the plasma rest frame. This in turn requires unfolding the effects of satellite motion and potential from the measured fluxes using a suitable theoretical treatment. Although we have developed the necessary theoretical formulation (Young, 1978) it has not yet been implemented in routine data analysis. An example of its application is given in Figure 12. Here we have plotted the flux of H^+ and He^+ as a function of the angle γ between the ICE view direction and the satellite velocity vector $\mathbf{V}_0$ in a coordinate system corotating with the Earth. At the time these high resolution angular scans were made the satellite had a large velocity component perpendicular to the ICE scan plane, hence the detector never pointed nearer than 30° to the ram direction. Differential energy spectra obtained at this time for H^+ and He^+, together with knowledge of the approximate satellite potential and $\mathbf{V}_0$, allows calculation of the ion distribution functions. From this, N and T were obtained and a theoretical curve of flux modulation was calculated as a function of γ. The theoretical curve for H^+ was then found to fit best if shifted 10° to the left of center in the figure whereas a shift of 5° fit He^+ the best. In the case of H^+ this corresponds to an ion drift velocity $\mathbf{V}_D$ of $\approx 0.5\ \mathrm{km\,s^{-1}}$ perpendicular to the satellite velocity $\mathbf{V}_0$ of $2.9\ \mathrm{km\,s^{-1}}$ (and half this value for He^+ which was measured ≈ 500 km deeper inside the plasmasphere than H^+). The shape of the curves was then recomputed using a drift velocity given by $\mathbf{V} = \mathbf{V}_D + \mathbf{V}_0$ and the result plotted in Figure 12.

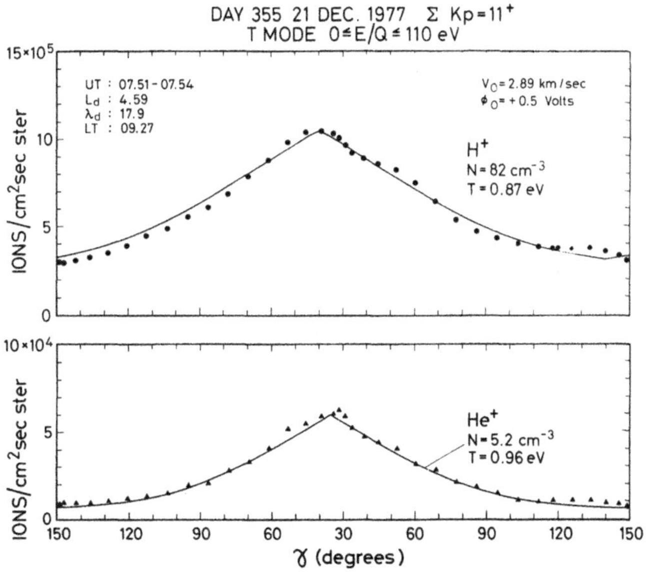

Fig. 12. Thermal mode high resolution angular scan of H^+ and He^+ fluxes at the 0 eV/e RPA step. Data are single 0.15 s samples and a flux of 10^3 ions $(cm^2\,s\,ster)^{-1}$ corresponds to 1.46 counts. The solid line is a theoretical curve based on measured *N*, *T* and *V* and an estimate of satellite potential (see text). γ is the angle between satellite velocity in the corotating system, $\mathbf{V}_0$, and the detector view direction.

In terms of an $\mathbf{E}\times\mathbf{B}$ drift, the 10° shift in the flux maximum of H^+ in Figure 12 corresponds to an electric field of $\approx 0.2\,\mathrm{mV\,m^{-1}}$ (using a calculated value of $B=450\gamma$) radially outwards from the Earth. In the rest frame of plasma corotating with the Earth this represents a 'slippage' of corotation by $\approx 25\%$ (or $\approx 12\%$ at the He^+ measurement point). This result demonstrates the use of a combined mass–energy-angle measurement to determine independently small electric fields. The advantage of using heavier ions for this purpose is seen in Figure 12 where the He^+ flux modulation is 3 times greater than the H^+ modulation and can provide proportionately greater sensitivity.

7. Production, Loss and Transport Processes

When Young *et al.* (1977) first reported thermal O^{2+} and He^{2+} ions in the high altitude equatorial magnetosphere, they proposed that these ions are produced from O^+ and He^+ by UV and energetic proton and electron radiation. We shall in this section discuss radiation and other possible production mechanisms in the plasmasphere and compare them with losses by charge exchange. Abundances of doubly charged ions in the *F*-region and the topside ionosphere have been measured by Hoffman *et al.* (1969) and Breig *et al.* (1977). The latter authors have shown that the O^{2+} abundances observed below 400 km can be understood

TABLE II

Production rates ν_1 of O^{2+} and He^{2+} near the equatorial plasmasphere by UV and energetic electrons and protons. UV spectrum ("high average solar UV flux") and cross sections after Banks and Kockarts (1973). Proton and electron fluxes for $L = 3.5$ from Chan *et al.* (1977). For $L = 2$ only electrons with $E \geqslant 40$ keV (Teague and Vette, 1972) and protons with $E \geqslant$ 100 keV (Sawyer and Vette, 1976) were included. Electron cross sections after Kieffer and Dunn (1966). The proton cross sections were scaled from the electron figures (cf. Banks and Kockarts, 1973)

L	$\nu_1(O^+ \to O^{2+})$ (s^{-1})	$\nu_1(He^+ \to He^{2+})$ (s^{-1})	$\nu_2(O^+)$ (s^{-1})
3.5	5.2×10^{-7}	7.5×10^{-8}	4.8×10^{-7}
2.0	1.5×10^{-7}	9.0×10^{-9}	3.2×10^{-6}

in terms of chemical equilibrium. The behaviour of doubly charged ions in the topside ionosphere has been discussed in theoretical work by Schunk and Walker (1969, 1970). We shall extend this work and discuss production, loss and transport processes in the plasmasphere.

7.1. Production Mechanisms

7.1.1. *Ionization of* O^+ *and* He^+ *by* UV, *protons and electrons*

In Table II we give our estimates of the production rates ν_1 for O^{2+} and He^{2+}. Whereas at $L = 2$ production is mainly due to UV, at $L = 3.5$ trapped protons, electrons and UV contribute comparably.

TABLE III

Production of doubly charged ions from singly charged ions by low energy electrons. Observed electron fluxes and energy spectra in different magnetospheric regions are taken as examples. Cross sections after Kieffer and Dunn (1966). (Measured for He^+, estimated for O^+.) Plasmasheet and magnetotail fluxes after Chan *et al.* (1977), auroral fluxes after Lui *et al.* (1977), cleft fluxes after Winningham *et al.* (1977)

Region	$\nu(O^+ \to O^{2+})$ (s^{-1})	$\nu(He^+ \to He^{2+})$ (s^{-1})
Plasmasheet (high estimate)	3×10^{-7}	3×10^{-8}
Plasmasheet (low estimate)	1×10^{-8}	2×10^{-9}
Aurora (substorm)	8×10^{-8}	1×10^{-8}
Cleft	6×10^{-8}	5×10^{-9}
Magnetotail (high latitude)	2×10^{-10}	2×10^{-11}

We estimate that the production rates in a large volume of the equatorial magnetosphere ($L \approx 3$–6; $|\lambda| \lesssim 25°$) are similar to those given for $L = 3.5$. Elsewhere production is mainly by UV ($\nu_1 = 1.6 \times 10^{-7}\,s^{-1}$ for $O^+ \rightarrow O^{2+}$ and $\nu_1 = 0.095 \times 10^{-7}\,s^{-1}$ for $He^+ \rightarrow He^{2+}$) during sunlit periods (for 'maximum average' UV spectrum, Banks and Kockarts, 1973). Lower energy (~keV) electrons are widely distributed in the outer magnetosphere. In Table III production rates are given for several examples of observed fluxes.

7.1.2. *Production by stripping of singly charged ions*

The question arises whether during the acceleration of field aligned ions from the ionosphere (cf. Section 5), or in subsequent bounces, O^{2+} is produced by stripping O^+, thus accounting for the O^{2+} observed in the field aligned superthermal plasma discussed in Section 5. Cross sections σ_{12} for the reaction $O^+ + O \rightarrow O^{2+} + O + e^-$ have been measured (Lo *et al.*, 1971) for $E > 30$ keV. Applying the systematics outlined by Hasted (1964) we have extrapolated the cross-section to lower energies and obtain $\sigma_{12} < 10^{-20}\,cm^2$ for $E = 300$ eV and $\sigma_{12} \sim 3 \times 10^{-20}\,cm^2$ for $E = 1$ keV. These cross-sections are so low that production of appreciable amounts of O^{2+} will not occur before the O^+ ions have lost their kinetic energy by collision with the neutral gas.

7.1.3. *Direct double ionization of atoms by electrons*

In electron and other particle impact phenomena, singly and doubly charged ions are simultaneously produced from neutrals. For electron energies above 3 times the threshold for double ionization, the production ratios are relatively constant. For helium $\sigma_{02}/\sigma_{01} \approx 0.01$ has been measured (cf. Kieffer and Dunn, 1966). From data on other elements reviewed by these authors, we estimate $\sigma_{02}/\sigma_{01} \approx 0.1$ for oxygen. Thus, these production rates are similar to the relative abundances of He^{2+} and O^{2+} often observed in the thermal and superthermal plasma. However, the cross-sections are low ($\sigma_{02}(\max) = 5 \times 10^{-18}\,cm^2$ at $E = 180$ eV for He; $\sigma_{02}(\max) = 4 \times 10^{-17}\,cm^2$ at $E = 140$ eV for O). Thus, this mechanism can compete locally with production by UV if $\langle \sigma_{02} f_e \rangle n_n \geqslant \nu_1 n_+$ (f_e is the electron flux, n_n and n_+ are the respective neutral and ion number densities). For instance, with an average $\sigma_{02} = 10^{-17}\,cm^2$ and $O^+/O = 10^{-3}$ (at ~500 km), we obtain equal production of O^{2+} by UV and by electrons for $f_e = 1.6 \times 10^7\,cm^{-2}\,s^{-1}$. Below ~ 400 km O^{2+} would be efficiently reduced by charge exchange with O (Breig *et al.*, 1977).

7.1.4. *Auger-effect induced by solar X-rays*

Breig *et al.* (1977) concluded that during quiet solar conditions this mechanism is significant only below 220 km, and thus not relevant for the magnetospheric ion population.

From this brief survey of ionization processes, it is concluded that ionization of singly charged ions by solar UV augmented by energetic electrons and protons is the most important source of doubly charged ions found in the topside ionosphere, plasmasphere and magnetosphere.

7.2. Local production of doubly charged ions and charge exchange processes in the plasmasphere

The neutral gas in the high altitude plasmasphere is mainly H, and thus we have to consider the following reactions for the thermal (~1 eV) oxygen ions:

$$O^+ + (UV, e, p) \rightarrow O^{2+} \qquad (\nu_1), \qquad (R_1)$$

$$O^+ + H \rightarrow O + H^+ \qquad (\nu_2 = \gamma_2 n_H), \qquad (R_2)$$

$$O^{2+} + H \rightarrow O^+ + H^+ \qquad (\nu_3 = \gamma_3 n_H). \qquad (R_3)$$

The reaction rates ν_j (s^{-1}) and rate constant γ_j ($cm^3\,s^{-1}$) are defined in parenthesis. Atomic oxygen is rapidly removed from the volume considered and can be neglected as a reaction partner. The reaction (R_2) is fast. We adopt $\gamma_2 = 1 \times 10^{-9}\,cm^3\,s^{-1}$ for $T_i = 10^4$ K and $T_H = 10^3$ K (cf. Banks and Kockarts, 1973; Ferguson, 1975). The corresponding ν_2 are given in Table II. It can be assumed that $\gamma_3 \ll \gamma_2$ (Ferguson, 1978) and thus for $\nu_3 t \ll 1$ we can neglect reaction (R_3). Disregarding for the moment ion transport, we obtain with initial conditions $O^+ = 1$, $O^{2+} = 0$,

$$O^+ = e^{-(\nu_1+\nu_2)t}, \qquad (1)$$

$$\frac{O^{2+}}{O^+} = \frac{\nu_1}{\nu_1+\nu_2}\left(e^{(\nu_1+\nu_2)t} - 1\right). \qquad (2)$$

Figure 13 shows the build-up of O^{2+} and the simultaneous neutralization of O^+.

The rate constants for the charge exchange at ~1 eV of both He^+ and He^{2+} with H should be smaller than $10^{-12}\,cm^3\,s^{-1}$ (Dalgarno, 1978), and these reactions can be neglected. We thus obtain (Figure 13)

$$\frac{He^{2+}}{He^+} = \nu_1 t. \qquad (3)$$

Figure 13 shows that 4–12 days are needed for locally building up O^{2+}/O^+ ratios of 0.2 to 1 which are often observed in the plasmasphere (cf. Figures 10 and 11). Undoubtedly transport cannot be neglected for such long periods. In fact, the investigations by Murphy *et al.* (1976) suggest that O^+ in the plasmasphere is replenished on a faster time scale.

7.3. Transport

The effect of transport will not only be to replenish O^+ but also to supply O^{2+} produced at lower altitude. In fact, as we shall see, thermal diffusion is strong relative to convection and tends to increase the O^{2+}/O^+ ratio strongly in the high temperature plasmasphere, because for ions the thermal diffusion coefficient is negative and roughly proportional to Q^2 (Chapman, 1958).

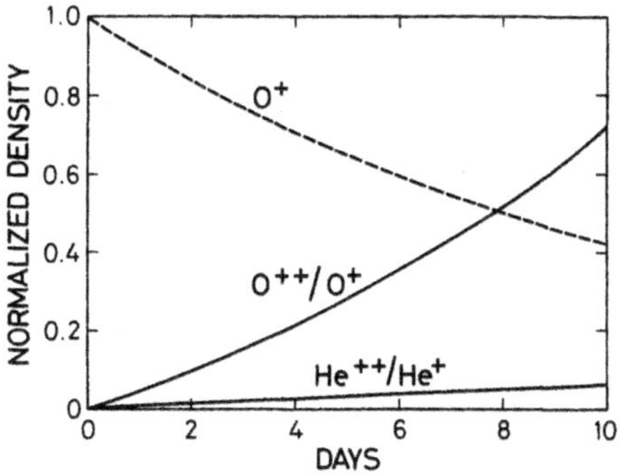

Fig. 13. Local production and charge exchange in the plasmasphere at $L = 3.5$ near the equator showing the build-up of the O^{2+}/O^+ and He^{2+}/H^+ ratios through ionization of O^+ by UV, e, p and simultaneous loss of O^+ (through charge exchange with neutral H).

In order to assess the importance of diffusion, we initially disregard production and loss as well as convection and consider only diffusive transport along a flux tube in the s-direction. Furthermore, we assume we have a major ion species with mass $M_0 m_0$, charge e, density n_0, bulk velocity u_0 (in the s-direction), and minor ion species with mass $M_i m_0$, charge Qe, density n_i and bulk velocity u_i. Since we consider the topside ionosphere and the plasmasphere, the major ion is either O^+ or H^-. At mid-latitudes, the cross-over between O^+ and H^+ dominance occurs at altitudes of about 1000 km. However, this altitude is highly variable (cf. Hagen and Hsu, 1974; Murphy *et al.*, 1976). Assuming that the temperature T of ions and electrons are the same, and assuming

$$|u_i - u_0| \ll (3KT/M_i m_0)^{1/2} \tag{4}$$

we can apply the diffusion equation (cf. Burgers, 1969; St. Maurice and Schunk, 1976; Murphy *et al.*, 1976)

$$u_i - u_0 = D_i \left\{ -\frac{\mathrm{d}}{\mathrm{d}s} \ln n_i + (\alpha_i - 1) \frac{\mathrm{d}}{\mathrm{d}s} \ln T - \left(M_i - \frac{QM_0}{2} \right) \frac{m_0}{kT} g_s \right\}. \tag{5}$$

g_s is the s-component of the gravitational acceleration. To second order the diffusion constant is given by (Banks and Kockarts, 1973; St. Maurice and Schunk, 1976)

$$D_i = \frac{kT}{M_i m_0 \nu_i} \frac{1}{1 - \Delta_i} \tag{6}$$

with the collision frequency

$$\nu_i = 1.27 \frac{\mu_A^{1/2}}{M_i} \frac{n_0}{T^{3/2}} Q^2 \quad (\mathrm{s}^{-1}) \tag{7}$$

(T in degrees K; n_0 in cm^{-3}; μ_A is the reduced mass in a.m.u.).

Values for the second order correction parameter Δ_i estimated after St. Maurice and Schunk (1976) and for the thermal diffusion coefficients $-\alpha_i$,

TABLE IV
Thermal diffusion coefficients $-\alpha_i$ and correction factor Δ_i for trace ions in fully ionized plasmas with H^+ or O^+ as the main ion components

	H^+-Plasma		O^+-Plasma	
Ion	α_i	Δ_i	α_i	Δ_i
H^+	0	0.15	−1.2	0.64
D^+	0.9	0.05	−1.2	0.6
He^+	1.7	0.02	−1.1	0.48
He^{2+}	8.7	0.02	0.2	0.48
O^+	2.4	0	0	0.15
O^{2+}	11.2	0	3.1	0.15

calculated after Burgers (1969), are given in Table IV. (The values for H^+, He^+, O^+ and O^{2+} are in agreement with those given by Schunk and Walker, 1969.) Equation (5) holds as long as ion–ion collisions dominate (collisions with neutrals can be neglected above 400 km).

In the (quiet) plasmasphere we may also assume that the effective collision frequency of ULF waves with ions is much smaller than ν_i. However, this assumption is not necessarily justified outside the plasmasphere, where densities are lower, but temperatures and wave intensities are higher. Another condition for Equation (5) to be applicable is that the ion–ion collision frequency ν_i is large compared to the bounce frequency

$$\nu_b = \frac{1}{4LR_E}\left(\frac{3kT}{M_i m_0}\right)^{1/2} \tag{8}$$

(R_E = Earth radius). For $T = 10^4$ and $L = 3$, comparison of Equations (7) and (8) gives $n_H \gg 170\ \text{cm}^{-3}$ in the case of O^{2+} and $n_H \gg 90\ \text{cm}^{-3}$ in the case of He^{2+}. Thus, for the temperature and densities found at $L = 3$ on day 355 (cf. Section 6 and Figure 10) mirroring is only beginning to slow down the diffusion of the doubly charged ions near the equator. However, for the greater part of the flux tube Equation (5) should be applicable without correction, because temperatures are lower and densities are higher than near the equator.

In order to illustrate the effect of thermal diffusion, we shall first consider a static atmosphere, i.e., $u_i = u_0 = 0$ in Equation (5). Integrating from a reference level s_0 to s, we obtain

$$N_i = \frac{n_i}{n_{i0}} = \left(\frac{T}{T_0}\right)^{\alpha_i - 1} J^{M_i - Q/2} \quad \text{for an } H^+\text{-plasma} \tag{9}$$

$$N_i = \frac{n_i}{n_{i0}} = \left(\frac{T}{T_0}\right)^{\alpha_i - 1} J^{M_i - 8Q} \quad \text{for an } O^+\text{-plasma} \tag{10}$$

with

$$J = \exp - \int_{s_0}^{s} \frac{m_0 g_s}{kT} \, ds. \tag{11}$$

Using the α_i values in Table IV and eliminating J, we can compare the variations in ion abundances above the reference level. For an H^+-plasma we obtain

$$\frac{N_{O^{2+}}}{N_{O^+}} = (N_{O^+})^{-0.03} \left(\frac{T}{T_0}\right)^{8.9} \tag{12}$$

$$\frac{N_{He^{2+}}}{N_{He^+}} = (N_{He^+})^{-0.14} \left(\frac{T}{T_0}\right)^{7.1} \tag{13}$$

and for an O^+-plasma

$$\frac{N_{O^{2+}}}{N_{O^+}} = (N_{O^+})^{-1} \left(\frac{T}{T_0}\right)^{2.1}. \tag{14}$$

Hoffman *et al.* (1969) found $O^{2+}/O^+ = 3 \times 10^{-3}$ between 700 and 1100 km. Applying Equation (12) we would therefore expect O^{2+}/O^+ ratios in the high altitude plasmasphere which are 2 to 3 orders of magnitude higher than those observed. This result illustrates that thermal diffusion can very efficiently concentrate O^{2+} and also He^{2+} (Equation (13)) in the high temperature plasmasphere.

We now turn to the question of D^+ at high altitudes. From Equation (9) we obtain

$$\frac{N_{D^+}}{N_{H^+}} = \left(\frac{T}{T_0}\right)^{0.9} J. \tag{15}$$

For estimating J we assume $T = 3000$ K at 2000 km, the reference level, and $T = 10^4$ K for equatorial $L = 3$. The vertical temperature gradient is assumed to be 1 K km^{-1} above 2000 km until T reaches 10^4 K. We obtain $J = 0.57$ and $N_{D^+}/N_{H^+} = 1.7$. Hoffman *et al.* (1969) determined $D^+/H^+ \approx 1.6 \times 10^{-4}$ at 2000 km. This ratio may have to be considered an upper limit since some He^{2+} might be present. Thus, according to this static model D^+/H^+ could only rise to $\sim 3 \times 10^{-4}$ in the high altitude plasmasphere. This figure does not change very much if another temperature profile is chosen, since T/T_0 and J are inversely related.

The charge exchange reaction $H^+ + D \rightleftarrows H + D^+$ could in principle increase the D^+/H^+ ratio above 2000 km. Deuterium is probably enriched relative to H below 1000 km, but it should decrease at high altitude towards D/H ratios even lower than 1.6×10^{-4} (Banks and Kockarts, 1973). Since the D/H profile above 2000 km is not known, it is difficult to estimate quantitatively the net effect of charge exchange on D^+/H^+ above 2000 km.

The considerations presented here thus indicate that D^+/H^+ can only be raised by a modest factor in the high altitude plasmasphere, supporting the conclusion of Young *et al.* (1977) that the $M/Q = 2$ peak is mainly He^{2+} as long as its abundance is $> 10^{-3}$ of the H^+ abundance.

We have seen above that O^{2+}/O^{+} and He^{2+}/He^{+} can be strongly enriched in the high altitude plasmasphere, unless diffusive separation is offset by convection. In order to estimate the diffusion velocities and the time scales involved in the separation of doubly charged from singly charged ions, we apply Equation (5) to a situation where the singly charged ions move with the major component ($u^{+}=u_0$) and where the relative density gradients of the corresponding doubly and singly charged ions are initially the same ($\mathrm{d}\ln n^{2+}/\mathrm{d}s=\mathrm{d}\ln n^{+}/\mathrm{d}s$). The diffusion equation for the doubly charged ions is then

$$u_{2+}-u_0=D_{2+}\left\{(\alpha_{2+}-\alpha_{+})\frac{\mathrm{d}}{\mathrm{d}s}\ln T+(M_0-\tfrac{1}{2})\frac{m_0 g_s}{kT}\right\}. \tag{16}$$

The resultant velocities are given in Table V for a proton dominated atmosphere at an altitude of 1500 km and $L=3$ (dip angle 67.3°) with $n_{H^+}=5\times10^3\ \mathrm{cm}^{-3}$, $T=3000$ K and a vertical temperature gradient of $1\ \mathrm{K\,km}^{-1}$. (The diffusion velocity of D^+ was calculated by assuming the relative density gradients in H^+ and D^+ to be the same.) The high velocities obtained are mainly due to thermal diffusion.

TABLE V
Diffusion velocities u_i of doubly charged ions after Equation (16) under conditions discussed in the text. Also given are the scale heights H_s (Equation (18)) for the singly charged ions and the separation times τ_{SEP} (Equation (17)).

Ion	u_i ($\mathrm{m\,s^{-1}}$)	H_s (km)	τ_{SEP} (s)
He^{2+}	420	1660 (He^+)	4×10^3
O^{2+}	470	310 (O^+)	7×10^2
D^+	−70	2200 (H^+)	3×10^4

The second term in Equation (16) which is due to the separation electric field, contributes only a small fraction. Thus, these velocities would not greatly change if we were to introduce differences in T_i and T_e, as thermal diffusion depends mainly on the ion temperature gradient.

Hagen and Hsu (1974) have determined ion drift velocities at altitudes between 666 and 1314 km over Aricibo. They observed strong diurnal variations, ranging typically from $+100\ \mathrm{m\,s^{-1}}$ to $-150\ \mathrm{m\,s^{-1}}$ at 1314 km. Vickrey *et al.* (1976) have shown that there is sometimes counter-streaming of H^+ and O^+. A model covering altitudes up to 4500 km for O^+ and H^+ including transport, production and loss has been developed by Murphy *et al.* (1976) and Bailey *et al.* (1977). They obtain characteristic diurnal variations of H^+ bulk velocities ranging from $+200\ \mathrm{m\,s^{-1}}$ to $-250\ \mathrm{m\,s^{-1}}$ at 1000 km and similar values at 4500 km. Comparison of these H^+ velocities (u_0) with the model diffusion velocities of the doubly charged ions in Table V shows that at 1500 km diffusion of doubly charged ions is important relative to convection. The diffusion velocities decrease strongly with decreasing altitude, because densities are higher, temperatures are lower and the plasma composition changes from H^+ to O^+ (cf. Equation (6)). Thus, at lower levels the effects of diffusion will be reduced by convection. However, for altitudes $\gtrsim 1500$ km diffusion should rather effectively separate doubly charged from singly charged ions.

For assessing the effect of diffusion on the rate of variation in the O^{2+}/O^+ and He^{2+}/He^+ ratios, we calculate a separation time

$$\tau_{SEP} = u_{2+}/H_s \tag{17}$$

where H_s is the local scale height (along s) of the respective singly charged ion:

$$H_s^{-1} = (M_i - \tfrac{1}{2}) \frac{m_o}{kT} g_s - (\alpha_+ - 1) \frac{d}{ds} \ln T. \tag{18}$$

The τ_{SEP} are given in Table V. Since the scale height of O^+ is smaller than that of He^+, the separation time is shorter for O^{2+} than for He^{2+}. This calculation, though simplified, implies that diffusion will enrich O^{2+}/O^+ faster than He^{2+}/He^+ high up field lines in the plasmasphere.

We have seen in the foregoing paragraphs that ion production and loss as well as transport by thermal diffusion are important in determining the doubly charged ion ratios observed inside the plasmasphere by GEOS-1. In order to show how these effects combine to build up the O^{2+}/O^+ ratio we introduce $N^+(z)$ and $N^{2+}(z)$, the integral numbers of O^+ and O^{2+} above altitude z contained in a flux tube of unit area at the equator. The change of the O^{2+}/O^+ ratio is determined by production, charge exchange, diffusion and convection according to

$$\frac{\partial}{\partial t} \ln \frac{N^{2+}}{N^+} = \nu_{PROD} + \nu_{CH-EX} + \nu_{DIFF} + \nu_{CONV}. \tag{19}$$

We consider a model in which the velocities of the main ions H^+ and O^+ are negligible, i.e., $\nu_{CONV} = 0$. Furthermore, we assume that charge exchange reactions other than (R_2) and its reverse can be neglected. Then we obtain

$$\nu_{PROD} = \left(\frac{1}{N^{2+}} + \frac{1}{N^+}\right) \int_\lambda^0 \nu_1 n^+ \frac{dV}{d\lambda} d\lambda \tag{20}$$

$$\nu_{CH-EX} = \frac{1}{N^+} \int_\lambda^0 \nu_2 n^+ \frac{dV}{d\lambda} d\lambda \tag{21}$$

$$\nu_{DIFF} = \frac{1}{N^{2+}} \frac{B(0)}{B(\lambda)} u_{2+}(\lambda) n^{2+}(\lambda). \tag{22}$$

n^+ and n^{2+} are the number densities of O^+ and O^{2+}, u_{2+} is given by Equation (16), dV is unit volume and λ is magnetic latitude. Daytime UV and proton and electron flux distributions according to Chan *et al.* (1977) were used in calculating ν_1, (cf. Section 7.1.). The charge exchange rate ν_2 is (Banks and Kockarts, 1973)

$$\nu_2 = 2.5 \times 10^{-11} \left\{ T^{\frac{1}{2}} n_{\mathrm{H}} - \frac{8}{9} T^{\frac{1}{2}} n_{\mathrm{O}} \frac{n_{\mathrm{H}^+}}{n^+} \right\}. \tag{23}$$

We now consider a flux tube at $L = 3$ and take the reference level $z_1 = 1000$ km. For the densities of the major constituents we assume $n_{\mathrm{H}} = 3 \times 10^4\ \mathrm{cm}^{-3}$; $n_{\mathrm{O}} = 10^4\ \mathrm{cm}^{-3}$; $n_{\mathrm{H}^+} = 2 \times 10^4\ \mathrm{cm}^{-3}$; $n^+ = n_{\mathrm{O}^+} = 2 \times 10^3\ \mathrm{cm}^{-3}$ (cf. Hagen and Hsu, 1974; Murphy *et al.*, 1976; Banks and Kockarts, 1973). The neutral temperature was assumed to be constant $T_n = 1000$ K, the ion temperature to be rising according to $T = T_\infty - (T_\infty - T_1) r_1 / r$, with $T_\infty = 1.4 \times 10^4$ K and $T_1 = 2400$ K, resulting in $T = 9526$ K and $\mathrm{H}^+ = 3.4 \times 10^3\ \mathrm{cm}^{-3}$ at the equator. This relatively large $n_{\mathrm{H}+}$ value was taken deliberately to insure that the diffusion Equation (5) is applicable throughout the flux tube. The height distribution of O^+ is given by Equation (9). An initial ratio $n^{2+}/n^+ = 3 \times 10^{-3}$ was assumed throughout the flux tube (this value was observed by Breig *et al.*, 1977 between 700 and 1100 km). For $z \leqslant 1500$ km the gradual transition from an H^+ to an O^+ plasma was taken into account in determining u_{2+}. The H^+ and O^+ densities below 1000 km were estimated from the Day-8 model of Murphy *et al.* (1976).

Figure 14 shows the initial diffusion velocity u_{2+} and the initial rates ν_{PROD} and ν_{DIFF}. For $z \geqslant 1000$ km charge exchange rates $\nu_{\mathrm{CH-EX}} \leqslant 10^{-5}\ \mathrm{s}^{-1}$ were calculated, i.e., initially charge exchange is a minor effect at high altitudes (cf. Equation (2) and Figure 13). The diffusion velocity u_{2+} rises to about 25% of the thermal velocity of O^{2+}, or about 6% that of H^+. Thus, the approximate collision theory from which D_{2+} and α_{2+} are derived is still valid. Also the $u\,du/ds$ term neglected in Equation (5) is indeed small relative to the other terms. Thus, the diffusion equation should be a fair approximation even at the high u_{2+} velocities obtained here.

The importance at high altitude of diffusion is evident from Figure 14. As the

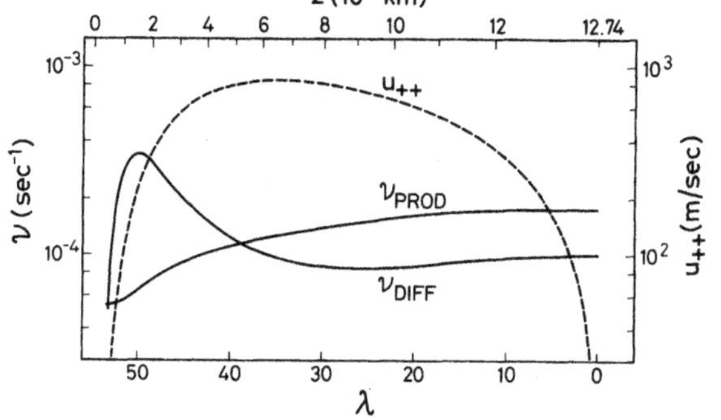

Fig. 14. Illustrative model (see text) showing the influence of production and transport by diffusion on the $\mathrm{O}^{2+}/\mathrm{O}^+$ ratio along a flux tube at $L = 3$. u_{2+} is the O^{2+} diffusion velocity as a function of magnetic latitude λ or altitude z. ν_{PROD} and ν_{DIFF} are volume integrated rates defined in Equations (19), (20) and (22).

O^{2+}/O^+ increase with time ν_{PROD} decreases proportionally and the ν_{DIFF} term dominates even at high altitudes. For lower H^+ densities u_{2+} is higher (cf. Equation (7)) and the diffusion term becomes even more important (cf. Equation (22)). At low altitude (<1000 km) convection (cf. Bailey *et al.*, 1977) and charge exchange are important and will limit the build up of O^{2+}/O^+.

In summary we find that whereas the O^{2+} and He^{2+} ions can be produced locally in the high plasmasphere, diffusive transport along the field line from lower levels is likely to be more effective on short (1–3 days) time scales. For altitudes $\gtrsim 2000$ km only O^+ is lost by charge exchange with an appreciable rate and this process further increases the O^{2+}/O^+ ratio. During long, quiet periods O^{2+}/O^+ may be raised to even higher levels than yet observed. Lastly, temperature estimates allow the identification of the $M/Q = 2$ peak. So far the data and the model calculations together indicate that both D^+ and He^{2+} are present with D^+ dominating if the abundance of the $M/Q = 2$ ions is $\lesssim 5 \times 10^{-4}$ of H^+, and He^{2+} dominating if this abundance is $> 10^{-3}$ of H^+.

Acknowledgements

The development and calibration of the ICE instrument and related GEOS interfaces represents more than 7 years effort, so it is appropriate to acknowledge here the support given by both participating institutions and their contractors as well as GEOS project personnel at ESTEC, ESOC and British Aerospace.

We are grateful to Dr A. Ghielmetti, Mr J. Fischer and Mrs R. Bänninger of the University of Bern and Messrs H. Loidl and P. Obermayer of the Max-Planck-Institut, Garching for technical assistance from development through calibration. We thank the principal ICE contractors Contraves A. G., Zurich, and Messerschmitt-Bölkow-Blohm GmbH, Munich for providing a high quality instrument.

We wish to thank the GEOS project scientist, Dr K. Knott and the ESTEC and ESOC project managers Messrs D. E. Mullinger and P. B. Lemke as representatives of the entire GEOS team. We are also grateful to the individual team members with whom we have worked, in particular Mr J. Kingdon, experiment officer, and Messrs G. Mersch, W. Smith and G. Tomaschek of ESTEC together with Messrs W. Wimmer, D. Heger and H. Nimmo of ESOC. The British Aerospace integration and test teams were most helpful and we thank Messrs D. Link and T. Harris on their behalf.

We acknowledge data provided to us in advance of publication by Drs A. Pedersen and F. Mariani and helpful discussions with Drs E. E. Ferguson, A. Dalgarno, R. F. Stebbings, E. Flückiger and F. Bühler.

Construction of the ICE and subsequent data analysis were supported by the Swiss National Science Foundation (grants 2.705.72, 2.672.76 and 2.886.77) and the State of Bern and by the German Bundesministerium für Forschung und Technologie (grant RV 14-B 44/73-SF 20) and the Max-Planck-Society.

References

Balsiger, H., Eberhardt, P., Geiss, J., Ghielmetti, A., Walker, H.-P., Young, D. T., Loidl, H., and Rosenbauer, H.: 1976, *Space Sci. Instrum.* **2,** 499.

Balsiger, H., Eberhardt, P., Geiss, J., and Young, D. T.: 1978 *J. Geophys. Res.* (to be submitted).

Bailey, G. J., Moffet, R. J., and Murphy, J. A.:1977, *Planet. Space Sci.* **25,** 967.

Banks, P. N. and Kockarts, G.: 1973, *Aeronomy*, Academic Press, New York.

Breig, E. L., Torr, M. R., Torr, D. G., Hanson, W. B., Hoffman, J. H., Walker, J. C. G., and Nier, A. O.: 1977, *J. Geophys. Res.* **82,** 1008.

Bühler, F., Axford, W. I., Chivers, H. J. A., and Marti, K.: 1976, *J. Geophys. Res.* **81,** 111.

Burgers, J. M.: 1969, *Flow Equations for Composite Gases*, Academic Press, New York.

Chan, K. W., Sawyer, D. M., and Vette, J. I.: 1977, National Space Science Data Center publ. WDG-A-R&S 77-01.

Chapman, S.: 1958, Thermal diffusion in ionized gases, *Proc. Phys. Soc. London* **72,** 353.

Cornwall, J. M.: 1972, *J. Geophys. Res.* **77,** 1756.

Dalgarno, A.: 1978, private communication.

De Forest, S. E., and McIlwain, C. E.: 1971, *J. Geophys. Res.* **76,** 3587.

Ferguson, E. E.: 1975, in B. M. McCormac (ed.), *Atmospheres of Earth and the Planets*, Reidel, Dordrecht, p. 197.

Ferguson, E. E.: 1978, private communication.

Geiss, J., Lind, D. L., and Stettler, W.: 1974, *Advances in the Astronautical Sciences*, **31,** 849.

Ghielmetti, A. G., Johnson, R. G., Sharp, R. D., and Shelley, E. G.: 1978, *Geophys. Res. Lett.* **5,** 59.

Hagen, J. B. and Hsu, P. Y.: 1974, *J. Geophys. Res.* **79,** 4269.

Harris, K. K., Sharp, G. W., and Chappell, C. R.: 1970, *J. Geophys. Res.* **75,** 219.

Hasted, J. B.: 1964, *Physics of Atomic Collisions*, Butterworth, London, p. 384 ff.

Hoffman, J. H., Johnson, C. Y., Holmes, J. C., and Young, J. M.: 1969, *J. Geophys. Res.* **74,** 6281.

Johnson, R. G., Sharp, R. D., and Shelley, E. G.: 1974, *J. Geophys. Res.* **79,** 3135.

Johnson, R. G., Sharp, R. D., and Shelley, E. G.: 1977, *Geophys. Res. Lett.* **4,** 403.

Kieffer, L. J. and Dunn, G. H.: 1966, *Rev. Mod. Phys.* **38,** 1.

Lo, H. H., Kurzweg, L., Brackman, R. T., and Fite, W. C.: 1971, *Phys. Rev.* **A4,** 1462.

Lui, A. T. Y., Venkatesan, D., Auger, C. D., Akasofu, S.-I., Heikkila, W. J., Winningham, J. D., and Burrows, J. R.: 1977, *J. Geophys. Res.* **82,** 2210.

Lyons, J. R.: 1977, *J. Geophys. Res.* **82,** 2367.

Lyons, L. R., and Evans, D. S.: 1976, *J. Geophys. Res.* **81,** 6197.

McIlwain, C. E.: 1976, *EOS, Trans. Am. Geophys. Union* **57,** 307.

Murphy, J. A., Bailey, G. J., and Moffet, R. J.: 1976, *J. Atmos. and Terr. Phys.* **38,** 351.

Reasoner, D. L.: 1973, *Rev. Geophys. Space Phys.* **11,** 169.

Sawyer, D. M. and Vette, J. I.: 1976, *National Space Science Data Center Publ.*, WDC-A-R8S, 76-06.

Schunk, R. and Walker, C. G.: 1969, *Planet Space Sci.* **17,** 853.

Schunk, R. W. and Walker, C. G.: 1970, *Planet. Space Sci.* **18,** 535.

Sharp, R. D., Shelley, E. G., and Johnson, R. G.: 1977, *J. Geophys. Res.* **82,** 2361.

Shelley, E. G., Sharp, R. D., and Johnson, R. G.: 1976, *Geophys. Res. Lett.* **3,** 654.

Spjeldvik, W. N. and Fritz, T. A.: 1978, *J. Geophys. Res.* **83,** 654.

St. Maurice, J.-P. and Schunk R. W.: 1976, *Planet. Space Sci.* **25,** 907.

Teague, M. J. and Vette, J. I.: 1972, *The Inner Zone Electron Model AE-5*, NSSDC-72-10, NASA GSFC, Greenbelt, Md.

Tinsley, B. A.: 1976, *J. Geophys. Res.* **81,** 6193.

Vickrey, J. F., Swartz, W. E., and Farley, D. T.: 1976, *Geophys. Res. Lett.* **3,** 217.

Wahlen, B. A., Bernstein, W., and Daly, P. W.: 1978, *Geophys. Res. Lett.* **5,** 55.

Winningham, J. D., Speiser, T. W., Hones, E. W., Jeffries, R. A., Roach, W. H., Evans, D. S., and Stenbaek-Nielsen, H. C.: 1977, *J. Geophys. Res.* **82,** 1876.

Young, D. T.: 1978, Note on the analysis of Ion Composition Experiment data from the GEOS satellite, University of Bern internal report, January 1978.

Young, D. T., Geiss, J., Balsiger, H., Eberhardt, P., Ghielmetti, A., and Rosenbauer, H.: 1977, *Geophys. Res. Lett.* **4,** 561.

THERMAL/SUPRATHERMAL PLASMAS OBSERVED BY THE S-302 EXPERIMENT ON GEOS-1

J. F. E. JOHNSON, J. J. SOJKA, and G. L. WRENN
University College London, Mullard Space Science Laboratory, Holmbury St. Mary, Dorking, Surrey, England

Abstract. The S-302 experiment has benefited more than most from the non-geostationary nature of the GEOS-1 orbit in so far as additional regions of quite different thermal and suprathermal plasmas were made accessible. Electron and positive ion spectra from three regions, the plasmasphere, plasmatrough and boundary layer, are described in order to highlight the variety of thermal plasmas observed.

We show how, even in the presence of the local photoelectron sheath and its associated electric potential, the thermal characteristics can be derived. The success of this technique during active periods is demonstrated by the observation of both a heating of the thermal population and the appearance of a second field aligned thermal component during a period of intense wave activity.

The detailed structure of the boundary layer adjacent to the magnetopause as observed on the 2nd December, 1977 shows only slow systematic development over a period of more than an hour. Where the thermal plasma density peaks the suprathermal positive ions show considerable drift motion as well as non-Maxwellian characteristics. This region is identified from the thermal particle data as being very close to the magnetopause, however, whether in fact the magnetopause was crossed requires corroborative data from the magnetometer and other experiments.

1. Introduction

Since the late 1950's the progress of ionospheric physics has owed much to the direct measurement of charged particle fluxes using rocket and satellite instruments. Langmuir probes and retarding potential analysers of one form or another have provided a wealth of data on both electron and ion densities and temperatures and also ionic abundances [1]. The use of such current integrating sensors generally required an assumption of thermal velocity distribution functions but it became clear that non-Maxwellian, anisotropic or suprathermal components could be very significant in many cases. Local heating due to photoelectron transport, the current carrying of low energy particles on auroral field lines, and magnetosphere–ionosphere coupling due to ion streaming at high latitudes are just three examples of study areas where a more complete description of the full distribution function for particles with energies from 0 to 100 eV is essential. In the last decade differential flux measurements have been extended down to thermal energies [2, 3] as a number of practical problems have been overcome in order to permit electrostatic analysers to bridge the suprathermal gap. At low altitudes the Earth's magnetic field tends to be a limiting constraint but attempts to shield analysers on rockets and low orbiting satellites have met with some success. It is on satellites passing to the outer plasmasphere and the magnetosphere beyond that the real challenge arises. The realisation that wave particle interactions could exert a controlling influence on magnetospheric dynamics and that any thermal/suprathermal plasma would play a vital role in these processes

Space Science Reviews **22** (1978) 567–580. *All Rights Reserved*

[4, 5] prompted S-302 to be selected for the GEOS payload with the acknowledgement that such an instrument would certainly have its limitations. The fundamental problem of all *in situ* space research is the question of whether the presence of the spacecraft modifies the medium to an extent that the measurement is seriously perturbed.

Experience gained with missions such as OGO-5 and ATS-5 left no doubt as to the magnitude of this problem with respect to the study of low energy plasma in the magnetosphere [6–8]. Secondary electrons emitted from spacecraft surfaces have energies of a few eV and can dominate ambient fluxes. Low number densities mean large Debye lengths and sheath dimensions of many metres, charging of external surfaces and resulting potentials of several volts or even kilovolts clearly will modify analyser acceptance characteristics. In sunlight the current due to photoemission from the satellite will generally exceed that due to charged particle influx, the resulting 'floating potential' can be expected to be several volts positive and the photoelectrons, unable to escape, will form a space charge cloud.

The design of the GEOS S-302 Suprathermal Plasma Analysers (SPA) include four basic features which are novel and provide some hope of overcoming the inherent difficulties which have been alluded. They are:

(1) The area of grounded conducting surface of GEOS has been maximised and sources of electrostatic charging have been eliminated where ever possible.

(2) The SPA's are boom mounted so that they are displaced from the satellite surface by more than 1 satellite diameter.

(3) The sensor package is isolated from satellite ground so that the bias potential of aperture and analyser reference planes can be controlled.

(4) The GEOS real time facilities, with considerable command flexibility, enables optimisation of analyser operation.

The decision to locate the analyser on a boom did introduce design constraints, the displacement along the boom had to be a compromise choice with regard to available weight. The consequent need for a light weight construction then imposed restriction on the number of analysers and the sophistication of the design.

S-302 is then an 'experiment' in the true sense of the word, this paper sets out to give a brief description of the instrument and typical data, outline the type of analysis procedure being employed, and establish an understanding of the limitations which are expected. Two particular studies are then presented to demonstrate that the data are indeed valid and crucial to the GEOS mission objectives.

2. S-302, The Instrument

A detailed description of the design, calibration and performance of the instrument is currently being prepared by Wrenn *et al.* [9]. We give here only an overview of the salient features and summarise the characteristics in Table I.

TABLE I
Summary of the S302 detectors

	Detector A	Detector B
Viewing direction	Parallel to spin axis	80° to spin axis
Acceptance angles	±9° × ±9°	±15° × ±4°
Geometrical factor (cm^2 ster)	0.0012	0.0033
Energy resolution	0.11	0.11
Energy range (eV)	0.5 to 510	0.5 to 510
Sampling time (m s)	43	43

The instrument consists of two hemispherical electrostatic analysers of 11% energy resolution, which use 'funnel' channel electron multiplier detectors. These analysers are mounted in a rectangular package whose exterior shell is conductive but isolated from spacecraft ground. This sensor unit is mounted on the X-axis, short radial boom of GEOS-1, at 1.8 m from the main body of the spacecraft.

One detector (denoted Detector A) views along a direction parallel to the spin axis, while the other (Detector B), views at 80° to the spin axis, and outwards at 50° to the radius vector. Detector A has a square aperture of ±9° about its viewing direction while B has a rectangular aperture of ±15° by ±4°, the plane of the larger angle lying at 10° to the spin plane.

The two detectors are operated in tandem, at any instant both sampling the same energy, with a high degree of flexibility in mode sequence. Commands can be sent to set up a programme of energy stepping with a selected bias on the sensor, and in either electron or positive ion mode.

A full energy sequence consists of 64 steps at logarithmically equal intervals between 0.5 and 500 eV. Commanding enables sequences of arbitrary start, increment and end to be selected from the set of 64 steps, including operation at a fixed energy. In addition, the package exterior and ground plane potential may be selected in the range −28 to +34 V relative to the satellite main body, to a resolution of 0.1 V, and with the capability of defining a sequence of such potentials.

3. S-302, The Data

In the first year of GEOS-1 operations the satellite has made a complete circuit in local time over an altitude range from 3 to 7 Re. The satellite's orbital inclination of 26° has resulted in data being acquired over a band of magnetic latitudes from 30° to −30°. This has given us a comprehensive data set on the variation of thermal and suprathermal plasma components in the plasmasphere, plasmapause, plasmatrough, inner edge of the plasma sheet, plasma sheet, boundary layer, magnetopause and magnetosheath. Observation in the latter three regions is restricted to storm periods with the satellite on the dayside.

Figure 1 shows electron and positive ion spectra taken from three of the above regions. The upper row of three sets of spectra correspond to the electron data for both the A and B detectors, while the lower row corresponds to the equivalent ion spectra. In each graph the logarithm of the detector count rate is plotted against energy level, i.e., logarithmic energy. Because detector B has twice the sensitivity of A, the B spectra all appear above the A spectra. In the plasmasphere, Figures 1(a) and 1(b), the thermal plasma has densities greater than $10\ \mathrm{cm}^{-3}$ and temperatures typically below 1 eV. Here the ion spectrum shows a distribution which can be fitted by a Maxwellian distribution accelerated through

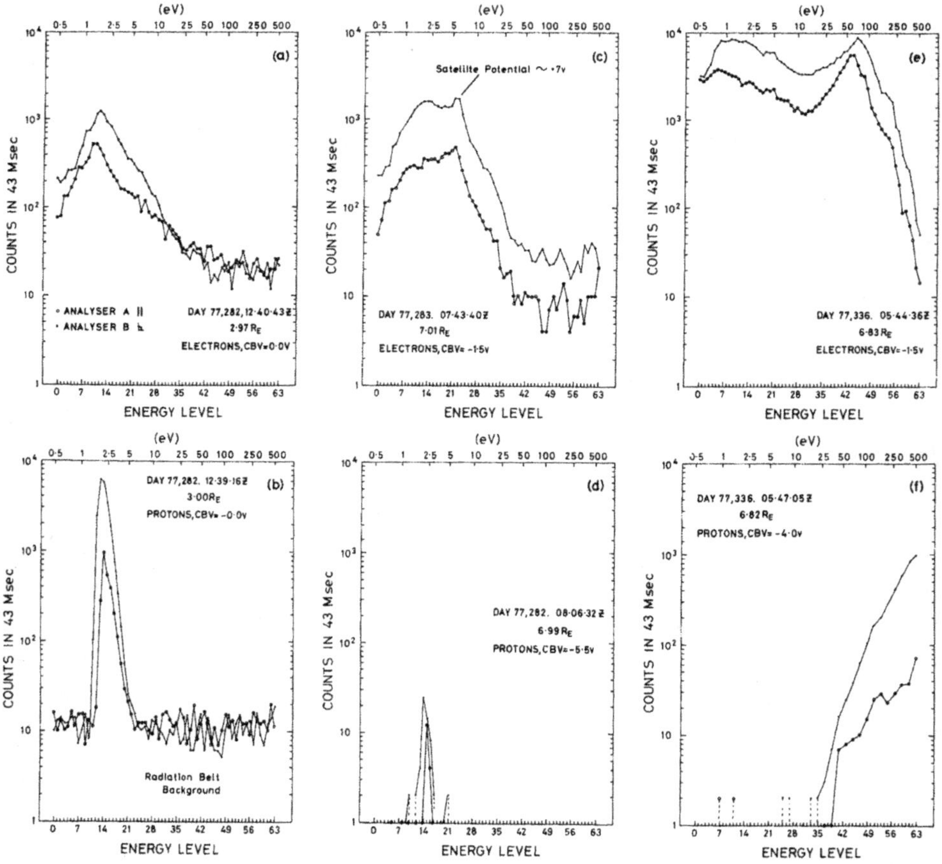

Fig. 1. S-302 raw data spectra; upper row are electron data and lower row are ion data while the columns left to right are data from the plasmasphere, plasmatrough and boundary layer.

a spacecraft potential of between 0 and −1 V and drifting relative to the satellite with the satellite velocity w.r.t. corotation. Due to this slightly negative potential the thermal electron distribution is not clearly resolved since the few electron volt tail population is superimposed upon the sheath population.

On moving out of the plasmaphere the density falls to about 1 cm^{-3} and the spacecraft potential begins to go positive. In this region both electron and ion thermal distributions can be resolved. Also in this region a population of suprathermal electrons is observed to show strong anisotropies in temperature with pitch angle, the temperatures being over the range 20 to 300 eV.

Figures 1(c) and (d) are spectra from outside the plasmapause, with the spacecraft potential now being between 5 and 10 V thermal electrons are accelerated into the detectors and are easily resolved. Because of the much lower densities and the retarding potential, positive thermal ions would not be observed, however by suitably biasing the detector package relative to the spacecraft ions can be accelerated and are in fact observed. This biasing procedure does not operate with similar success in the electron mode since there is a large flux of photoelectrons and secondary electrons produced on the spacecraft which, when accelerated by a bias potential, obscure the thermal electrons. A cold-thermal population is found to be typically 1 eV and 1 cm^{-3} but in regions of strong wave activity this population can be significantly modified.

The third region, Figures 1(e) and (f), is the boundary layer on the inside of the magnetopause. For the electrons a characteristic temperature would be about 20 eV with a density of 20 cm^{-3}. The ions are drifting with velocities up to 100 km s^{-1} and, assuming H^+, the density is comparable with that for the electrons although the characteristic temperatures are a factor of two greater.

4. Interpretation of Thermal Plasma Spectra

Before a count spectrum can be converted to part of the distribution function a background must be subtracted. Normally this count is small (<0.1 Hz) but transitions of the outer radiation belt produce count rates of up to 1250 Hz due to penetration of the energetic radiation particles through the light weight detector package. (See Figure 1(b)). The background level is readily identified from the ion spectra and can consequently be removed from both the ion and electron data. In addition a Sun pulse must be eliminated from Analyser B data.

In the electron mode, the photoelectron sheath population is clearly visible and shows a characteristic sunward orientation, see Figure 2. Here the thermal component, occurring at about 5 eV, exhibits no spin modulation. On magnetically quiet and even moderately active days this sheath population retains a particularly constant spectrum and angular distribution. The overall intensity is observed to vary systematically as a function of plasma density. This has enabled us to empirically model the sheath population and subtract it from the observed spectra. The process is most successful when the thermal electrons are accelerated

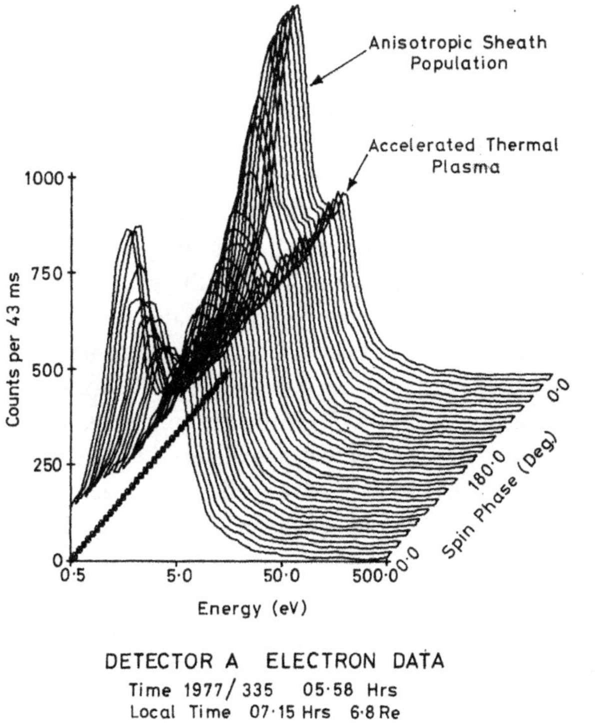

Fig. 2. A 3-D plot of Detector A count as a function of logarithmic energy and Sun phase angle; Sun phase being zero when Detector B points to Sun.

by a spacecraft potential greater than 1.5 V. Figure 3 is a composite graph showing the analysis procedure in converting from an electron count spectrum to a phase density distribution. The sheath spectrum, Figure 3(a), is subtracted from the observed count spectrum, Figure 3(b), to give the thermal electron spectrum, Figure 3(c). To make the final step the thermal electron population is assumed to be an isotropic Maxwellian which has undergone an acceleration equal to the spacecraft potential. This then enables the three parameters, temperature, density and spacecraft potential to be computed, see Figure 3(d).

An example of the thermal data acquired in the positive ion mode is used to illustrate the ion analysis procedure. Two analyser 'A' proton spectra (see Figure 4) are presented as 43 ms counts and plotted logarithmically as phase space density ($cm^{-6}\,s^3$). Data taken with the sensor bias (denoted by CB) of zero volts yields a well defined phase space density slope, with a temperature of 0.31 ± 0.1 eV. Data taken six minutes later with CB $= -4$ V shows the improved sensitivity (an instrumental ghost appears below the main peak) but the poorer energy resolution for the accelerated protons gives a temperature of 0.34 ± 0.035 eV, the cut-off at 4.12 eV is defined as the point where the count falls to half of the Maxwellian fit indicating a satellite potential of -0.12 V: the intercept then yields the plasma concentration as shown. Clearly the correct estimation of satellite

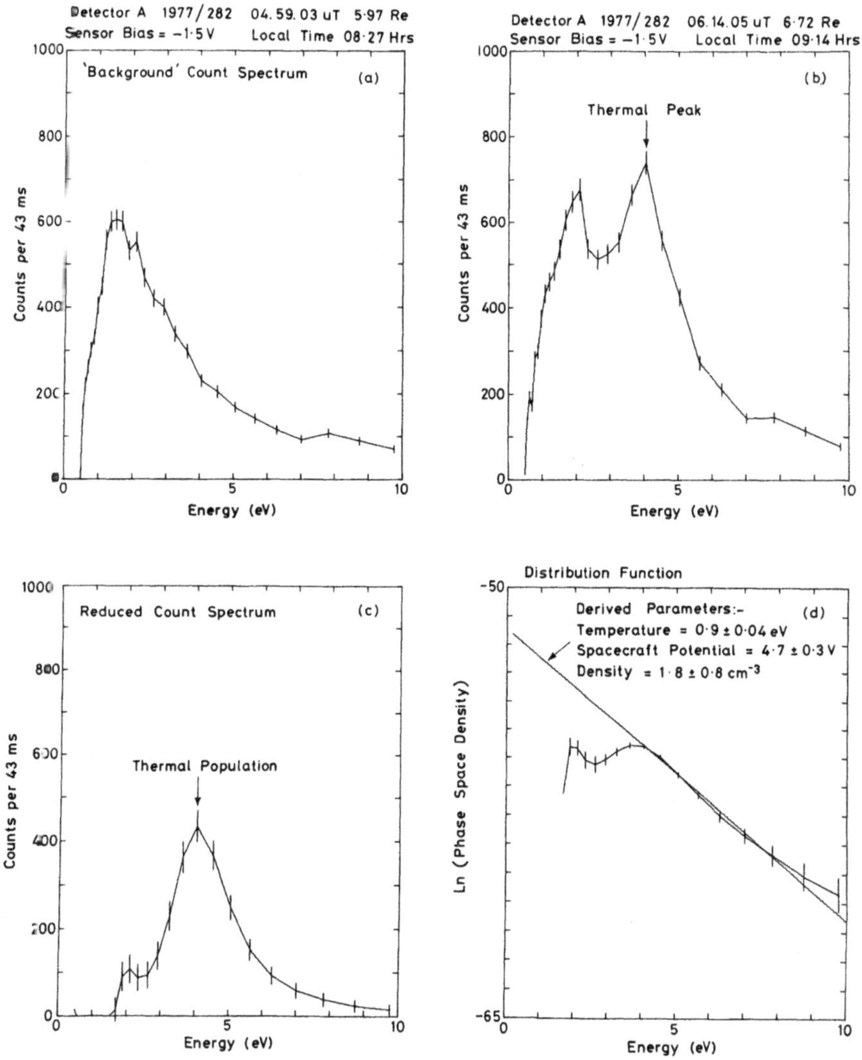

Fig. 3. A sequence of electron spectra to illustrate the sheath subtraction technique. The spectra are (a) the sheath (b) raw data (c) residual and (d) residual transformed to phase space density.

potential is crucial for an accurate determination of concentration, this is highlighted by the analysis of the CB = 0 V spectrum. Here there is no cut-off since the accelerating potential is less than 0.5 V, assumption of either constant satellite potential or constant phase space density leads to a factor of two uncertainty in concentration; in fact, increased concentration and more negative satellite potential are appropriate to the earlier time. The analysis can be complicated by the effect of ion drift; the satellite motion represents only a small fraction of an electron volt and is negligible if the sensor bias provides several volts of acceleration but large electric field drifts have been observed on occasions.

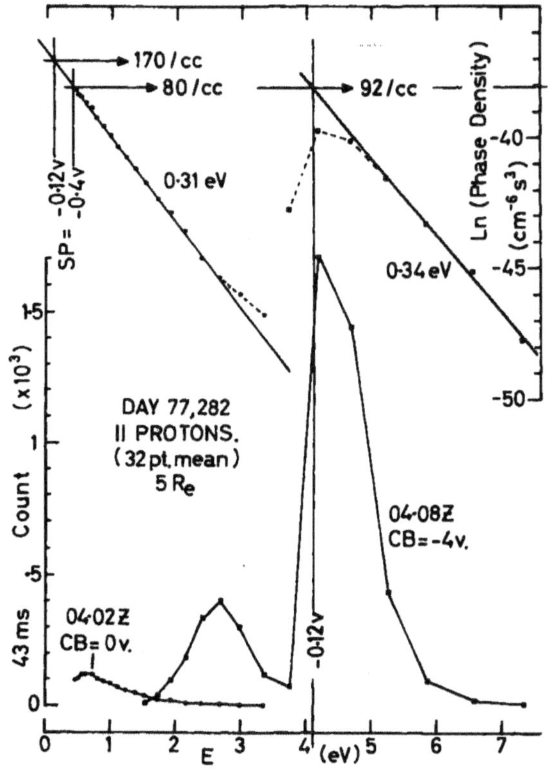

Fig. 4. A composite plot of ion data, the lower plots are count spectra for two different sensor bias values while the upper are the respective phase space density plots.

Using data from both the ion and electron modes it is possible to deduce the thermal plasma parameters (temperature, density and particle drift velocity) as well as spacecraft potential over most of the GEOS-1 orbit. An example of this analysis for day 282, 1977 is presented by Décréau *et al.* in this volume [10]. The ability of the S-302 experiment in coping with the variety of thermal plasmas it encounters is summarised in Table II. Two detailed examples follow to demonstrate the success of the analysis. The parameters which appear in Table II are defined with the aid of equation (1) which describes an isotropic Maxwellian distribution which experiences both drift motion and acceleration relative to the satellite.

$$f(E) = An \exp\left\{-\frac{1}{T}(E - E_{sp} + E_D - 2\cos\theta\sqrt{(E-E_{sp})E_D})\right\} \qquad (1)$$

where E is the energy, $f(E)$ is the energy distribution function, E_{sp} is the combination of spacecraft potential and detector package bias, E_D is the particle drift energy, θ is the angle between viewing direction and drift direction, T is the plasma temperature (eV), A is a constant dependent on the particle mass and plasma temperature, and n is the plasma density.

TABLE II
Derived thermal parameters from S302 data

	Parameters derived	Constraints on parameters	Assumptions
Electron mode	T, N, E_{sp} and angular distribution	$E_{sp} > 1.5$ V	E_D is negligible
Ion mode	T, N, E_D, E_{sp} and angular distribution	$N > 3\ cm^{-3}$ $E_{sp} < 2$ V for E_D require $E_D > E_{sp}$	H^+ is the dominant ion species

5. A Two-Component Cold Plasma

On crossing the plasmapause, on the 25th August, 1977, S-300 observed considerable wave activity despite the fact that this was a period of only moderate magnetic activity, ($Ap = 27.1$). During a period from 0930 to 1000 UT intense emissions of the $(n+\frac{1}{2})f_{ce}$ type together with strong emissions at the plasma frequency and ELF noise were observed [11]. These emissions were not all observed simultaneously, the $(n+\frac{1}{2})f_{ce}$ occurred over most of the half hour period, being most intense between 0940 and 0955. The enhanced plasma frequency and ELF noise did occur simultaneously over two periods of around two minutes between 0940 and 0955. It was during this period that the thermal particle population undergoes substantial modification. The modified thermal distribution function is no longer a single component isotropic Maxwellian but becomes a heated two component non-isotropic distribution.

During this observation the satellite was positioned within 30 minutes of the noon meridian at distances of 6.1 to 6.5 Re just outside the plasmasphere. The orientation of the satellite is such that the A detector views at 18° to the magnetic field and B covers the range of pitch angles 62° to 98°. For most of the 0930 to 1000 period the detectors are operating in the electron mode sweeping through the whole energy spectrum every 2.75 s.

Figure 5(a) shows the thermal distribution as observed by both detectors at 0949 h, after the strongest wave emission. The spacecraft potential is observed to be 1.9 V and would normally be seen to increase systematically as the density decreases. On this day in the regions of strong wave emissions the thermal density and temperature fluctuate and this causes the spacecraft potential to vary by ±0.1 V. The plasma density is around 7.5 cm^{-3} and has a temperature of 6500 K, these values being quite typical of the plasmapause region. When the distribution becomes heated this isotropic component increases both in temperature and density and a second component at a few electron volts appears. This second component is highly anisotropic and is only seen by the A detector at pitch angles around 18°. The total density of this second component is uncertain since a full

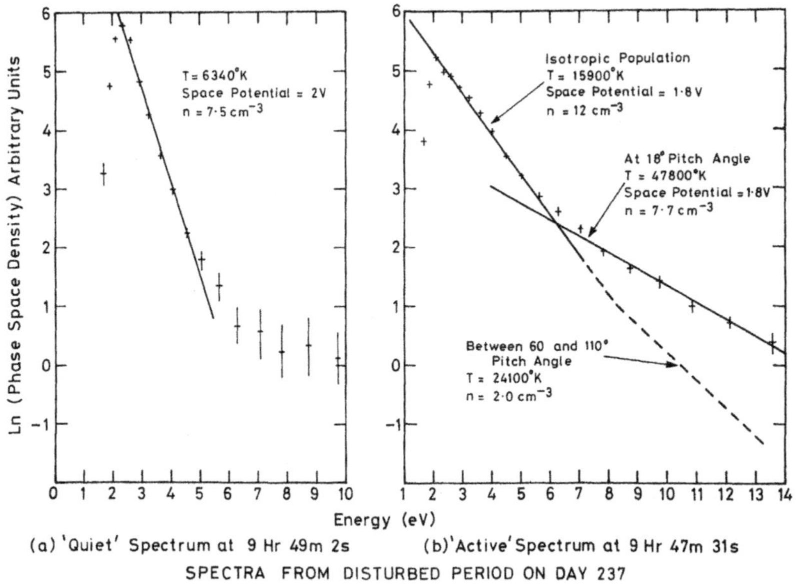

Fig. 5. Shows two electron phase space spectra from a quiet and active period during day 237.

pitch angle distribution is not available. However, assuming that over the angular range of the A detector the distribution is isotropic, the equivalent density for the observed part of the second component would be 7.7 cm^{-3}, see Figure 5(b).

Extending the above analysis procedure to all the spectra in the 0930 to 1000 period results in a high time resolution view of how the two component distribution develops. The A and B analysis results for the isotropic component are shown as a single parameter in Figure 6 together with the field aligned hot component. With the time resolution available the evolution from the isotropic distribution to the extreme case of the two-component distribution is clearly seen. Whether the evolution is due to the satellite moving through regions of quite different properties, or that a temporal development is being observed is uncertain.

6. Suprathermal Plasma in the Boundary Layer

The magnetic storm on the 2nd December 1977 ($Ap = 69.2$) was large enough to compress the magnetosphere such that the magnetopause was situated very close to the GEOS-1 apogee at a local time of 8 h and a magnetic latitude of 24°. For a period of $1\frac{1}{2}$ h the S-302 experiment observed a dense boundary layer type plasma. The density gradually increased to a maximum of 20 cm^{-3} while the electron temperature decreased from 100's of electron volts to 17 eV. Maximum density was observed at 6 h UT: around this time the ions were observed to be drifting anti-Sunward but also tangential to the Earth. The ion temperature was a factor of two hotter than the electron temperature and, assuming the dominant

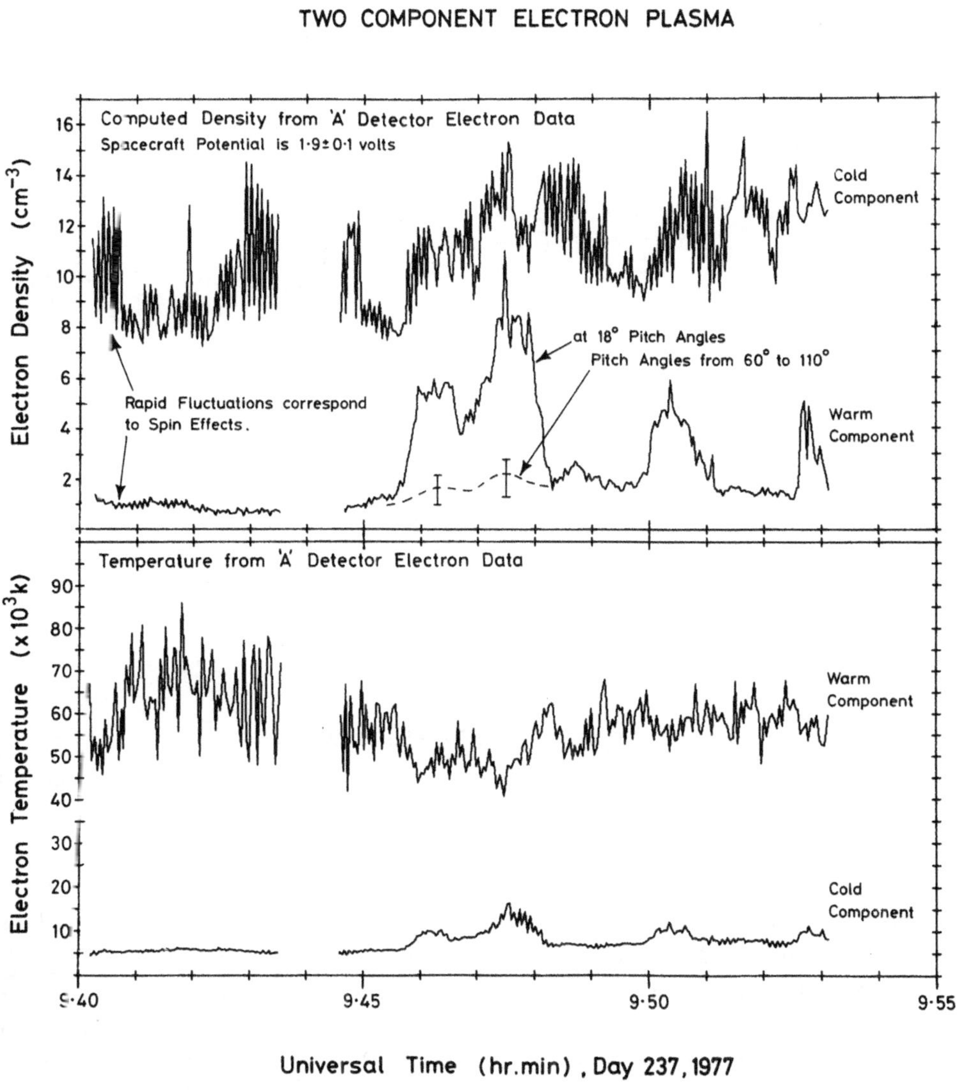

Fig. 6. Shows the development of both electron temperature and density during the period when intense wave activity was observed by S-300.

species to be hydrogen, a drift velocity of around 50 km s^{-1} would best fit the data. From 6 h UT onwards the density began to decrease with a corresponding increase in the electron temperature, by 7 h UT the suprathermal distribution returned to being magnetospheric.

The detailed analysis of the suprathermal boundary layer electron population is given in Figure 7. In the region from 0530 to 7 h UT the population is clearly defined, the distribution has a factor of two angular anisotropy but in general appears Maxwellian. Interpretation of the angular distribution is confused because the magnetic field, although weak, fluctuated sufficiently to make the pitch angle

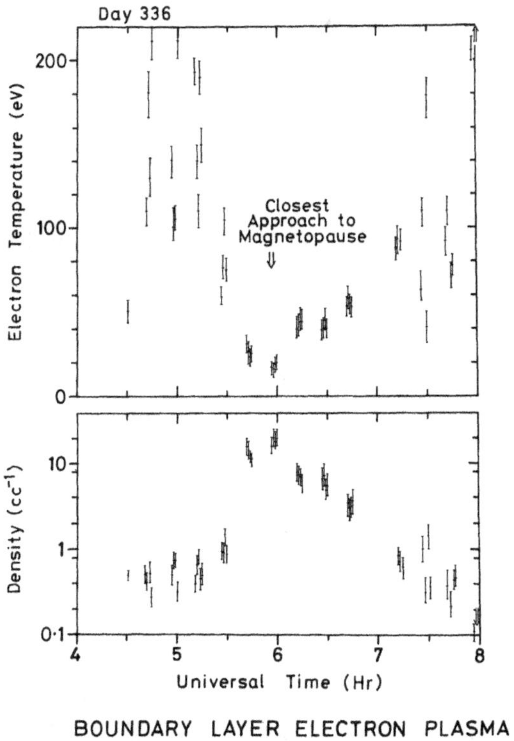

Fig. 7. The variation in electron temperature and density during a period on day 336 when GEOS-1 was near the magnetopause.

analysis extremely difficult. At times prior to 0530 and after 7 h the density falls to around 1 cm^{-3} and the electron temperature becomes ill defined. The distribution function over the 20 to 500 eV range seems to have two components, or alternatively, one component which has a strong anisotropy of temperature with pitch angle. In this region the magnetic field is even more turbulent, on a time scale that making comparisons from sweep to sweep at different angles is impossible. It is the region of plasma between 0530 and 7 h which bears strong resemblance to a boundary layer plasma adjacent to the magnetopause [12].

The ion data is more difficult to interpret since the counting rates are at least a factor of 40 less than for the electrons. Also the ions were drifting sufficiently strongly to introduce an anisotropy factor of 100 in detector B count rate. In general it was found that the distribution was not a simple drifting Maxwellian, there seemed to be a noticeable degree of pitch angle anisotropy present.

The best plasma parameter fit to this population gives an ion temperature of about twice that of the corresponding electron temperature and a drift velocity of 50 km s^{-1} for hydrogen ions. This analysis is only successful for about 20 min around 6 h UT. Outside this time interval other effects, a higher energy ion

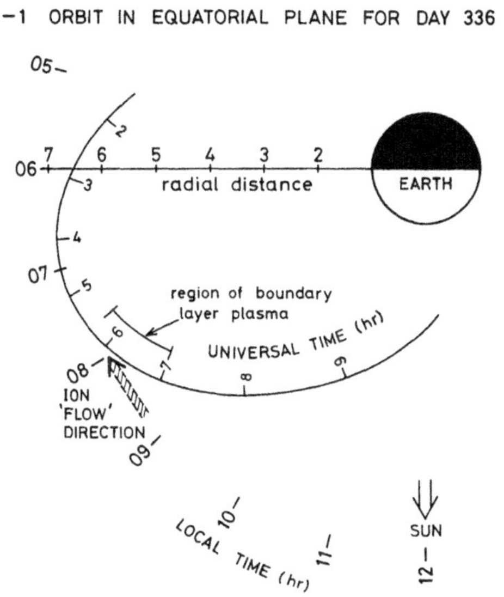

Fig. 8. A schematic indicating ion flow direction at 6 h UT on day 336.

population, low count rates and rapid temporal variations, make the analysis ambiguous.

Figure 8 schematically shows both the GEOS-1 orbit on the 2 December, 1977 and the main features of the boundary layer as detected from the suprathermal plasma data. The ion flow direction has been transformed to the Sun–Earth orbit plane. Whether there is a significant component out of this plane is uncertain since the ion population is not sufficiently well fitted by a drifting Maxwellian population to allow a full vector velocity to be deduced from the joint A and B detector data.

7. Conclusion

The interaction between the spacecraft and the ambient plasma produces many complex spectral and angular features in the particle distribution function measured by the low energy particle experiment S-302. We have presented a selection of data to illustrate the wide variety of conditions which have been encountered. The real-time control, made possible by the powerful groundbased computer, has proved to be extremely valuable in helping to distinguish vehicle-produced effects from direct measurements of the ambient plasma.

We are now able to trace the variation of the cold plasma throughout most of the orbit; the typical plasma density of one particle per cubic centimetre in the plasmatrough is seen to vary by up to an order of magnitude from one orbit to the next, while the temperature is more consistently around one electron–volt.

More complicated conditions, such as that presented on day 237, need to be analysed individually, but the resolution of two thermal populations both of under five electron-volts energy, would have been impossible with any other instrumentation. Such data will prove invaluable as we attempt to tackle the subject of wave particle interactions known to occur throughout the region traversed by the GEOS satellite.

Acknowledgements

The authors would like to thank Professor R. L. F. Boyd, CBE, FRS, for his interest and encouragement, and also a number of people at this laboratory whose specialised technical support is gratefully acknowledged. The excellent quality of the data is a direct reflection of the success of the British Aerospace and STAR in making GEOS-1 a uniquely 'clean' satellite, and without the capability of the real-time control made possible by the European Space Operations Centre at Darmstadt, Germany, the extremely complex low energy data would have been exceptionally difficult to interpret. The Mullard Space Science Laboratory's participation is funded by the United Kingdom Science Research Council, while the overall project was sponsored by the European Space Agency.

References

[1] Sayers, J.: 1970, *J. Atmos. Terrest. Phys.* **32,** 663.
[2] DeForest, S. E. and McIlwain, C. E.: 1971, *J. Geophys. Res.* **76,** 3587.
[3] Chappell, C. R.: 1972, *Rev. Geophys. Space Phys.* **10,** 951.
[4] Gringauz, K. I.: 1976, in D. J. Williams (ed.), *The Cold Plasma Throughout the Magnetosphere, in Physics of Solar Planetary Environments*, p. 672, A.G.U.
[5] Southwood, D. J.: 1978, *Nature* **271,** 309.
[6] Harris, K. K., Sharp, G. W., and Chappell, C. R.: 1967, *J. Geophys. Res.* **72,** 5939.
[7] Serbu, G. P. and Maier, E. J. R.: 1970, *J. Geophys. Res.* **75,** 6102.
[8] DeForest, S. E.: 1973, in R. J. L. Grard (ed.), *Electrostatic Potentials Developed by ATS-5 in Photon and Particle Interactions with Surfaces in Space*, D. Reidel, Dordrecht, p. 263.
[9] Wrenn, G. L., Johnson, J. F. E., and Sojka, J. J.: *Suprathermal Plasma Analysers aboard the ESA GEOS Satellites* (in preparation).
[10] Decreau, P., Etcheto, J., Knott, K., Pedersen, A., Wrenn, G. L., and Young, D. T.: 1978, *Space Sci. Rev.* **22,** 633.
[11] Christiansen, P., Gough, P., Martelli, G., Bloch, J. J., Cornilleau, N., Etcheto, J., Gendrin, R., Jones, D., Beghin, C., and Decreau, P.: 1978, *Nature* **272,** 682.
[12] Eastman, T. E. and Hones, E. W. Jr.: 1978, *Characteristics of the Magnetospheric Boundary Layer and Magnetopause Layer in High Time Resolution.* Paper presented to ISSTP, Innsbruck.

ELECTRON DENSITY AND TEMPERATURE, AS MEASURED BY THE MUTUAL IMPEDANCE EXPERIMENT ON BOARD GEOS-1

P. M. E. DECREAU

Université d'Orléans et Centre de Recherche en Physique de l'Environnement, 45045 Orleans Cédex, France

and

C. BEGHIN and M. PARROT

Centre de Recherche en Physique de l'Environnement, 45045 Orleans Cédex, France

Abstract. The mutual impedance experiment on GEOS-1 provides an original diagnostic of the thermal electron population. The electron density N_e, and temperature T_e, are derived from the plasma frequency and Debye length, the values of which determine the shape of the frequency dependent mutual impedance curves. The existing limits of the method are pointed out. They may be instrumental or arise from a lack of theoretical development, for instance when the steady magnetic field or the drift velocity of the plasma cannot be neglected. Nevertheless, first geophysical results have been derived, using measurements obtained on the dayside of the equatorial magnetosphere where most of the data enter within the above limits. In particular, we have drawn a map of the dayside magnetosphere, in terms of densities, Debye lengths, temperatures, at geocentric distances of 4 to 7 Earth radii. The conventional shape of the plasmasphere is recognized, but the temperatures obtained are lower than expected ($\sim$2 eV at apogee, outside the plasmasphere). The influence of the magnetic activity on apogee measurements is reported: N_e values and A_m indices are shown to be correlated, but it is not the case for T_e and A_m. Finally, detailed T_e and N_e profiles are shown, and the presence of a plasmapause boundary is discussed.

1. Introduction

The density and temperature of the cold plasma are key parameters in the study of the magnetospheric medium, and of a number of phenomena related to it (wave propagation, dynamics of the magnetosphere, transfer of energy, etc. . . .). The precise and reliable measurement of those parameters on board GEOS was essential on such a magnetospheric observatory, which was conceived to be as complete and sensitive as possible. This is the reason why two new techniques for such measurements, based on wave propagation, were implemented on GEOS, in the frame of S-300, the complete plasma wave experiment. Those techniques have proved to be successful, as may be inferred for instance from the comparative study of density and temperature measurements made by 5 different experiments on board GEOS (Decreau *et al.*, 1978).

The purpose of this paper is to present, with some details, the data obtained by one technique: the mutual impedance experiment. The next section treats the method used, its principle, and, in addition to satisfactory results, will show more difficult cases, illustrating the limitation of the technique. Section 3 presents some geophysical results obtained with this experiment, during the first seven months of

Space Science Reviews **22** (1978) 581–595.

the satellite operation. In particular, thanks to the non-geostationary orbit of GEOS-1, we have been able to draw a first 'map' of the dayside equatorial magnetosphere, in term of densities and temperatures, between 4 to 7 Earth radii. Section 4 is devoted to complementary and concluding remarks.

2. The Method of Measurement

2.1. Principle of the Experiment

The principle of this experiment, illustrated in Figure 1, is that of the quadrupolar probe (Storey *et al.*, 1969; Renard and Debrie, 1970; Rooy *et al.*, 1972; Chasseriaux *et al.*, 1972; Decreau and Debrie, 1977). The mutual impedance between two double-sphere dipoles is measured in the vicinity of the plasma frequency. The results are drawn as frequency dependent curves, showing the modulus and the argument of the mutual impedance Z_m, normalized to its value in vacuum Z_v. The shape of those curves is a function of the parameters of the medium, and allows a diagnostic of the plasma. In particular, when the plasma may be assumed to be homogeneous, Maxwellian and isotropic, the mutual impedance is a function of two parameters, the plasma frequency f_p and the Debye length λ_D. The impedance vs. frequency curves exhibit a resonance, the position of which is related to f_p, and the sharpness of which to λ_D. Hence, when we are able to find a good, two-parameter fit between an experimental impedance

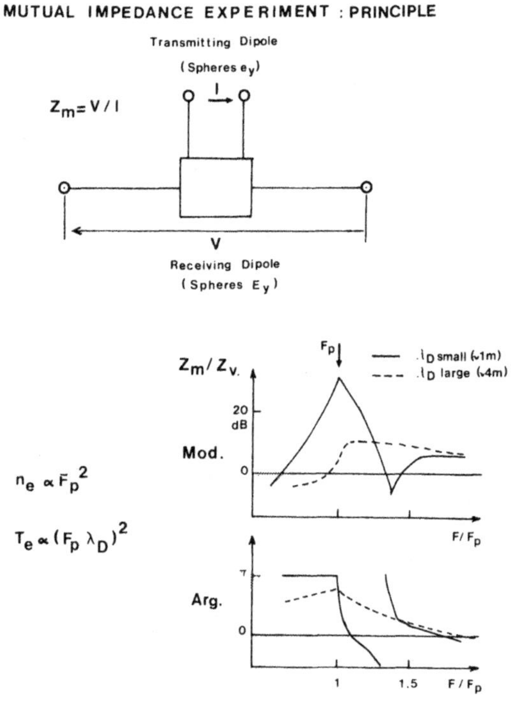

Fig. 1. Principle of the experiment.

curve, and a theoretical curve obtained with our simple hypothesis of a homogeneous, isotropic, Maxwellian plasma, we infer that this hypothesis is valid to a first approximation. Afterwards, knowing f_p and λ_D, we get a measurement of the electron density, N_e, and of the electron temperature T_e.

The realization of this experiment on a high-altitude satellite, and its implementation within the complete set of measurements used by the general S-300 experiment, is described elsewhere (S-300 Experimenters, 1978; Auger, 1974). We will only discuss a few features, limiting the domain of validity of the experiment. To get an efficient diagnostic through mutual impedance measurements, the distance between one transmitting sphere and one receiving sphere must be distinctly larger than the Debye length (Pottelette, 1976). The values expected for λ_D were of the order of, or larger than, the dimensions of the satellite which led us to use, as receiving dipole, the spheres placed at the tip of the long booms, deployed 20 m on each side of the satellite. The transmitting dipole, of length 1.4 m, is placed 3 m from the satellite body in a direction parallel to the receiving dipole. Under those conditions, the larger values of λ_D that can be measured are of the order of 6 m. The upper working frequency is ~76 kHz, corresponding to a density $N_e = 70\ \text{cm}^{-3}$. Finally, a sheath of photoelectrons surrounds the body of the satellite, as well as the antennas. The positions of the spherical sensors, placed away from the satellite body, and the meshed structure of the transmitting spheres, were chosen in order to minimize the influence of that sheath. In fact, measurements done during the transition between eclipse and non eclipse conditions have proved that the photoelectron perturbation is negligible for typical density values $N_e \geqslant 10\ \text{cm}^{-3}$ (Decreau *et al.*, 1978). Nevertheless, such measurements are not available for very low plasma densities, and we must remain cautious in this respect, when interpreting low-density results.

2.2. Typical Good Quality Results

During the first months of operation, when the orbit of the satellite stayed on the dayside, the mutual impedance experiment revealed itself as well adapted to the diagnostic of the plasmas encountered. The Debye lengths measured there, range between 0.5 and 6 m, and the electron gyrofrequency f_H stays usually well below the plasma frequency. This last condition is necessary to permit the neglect of the magnetic field, and to consider the plasma as isotropic. Figure 2 shows typical curves obtained for these conditions: the normalized mutual impedance is shown vs. frequency. Those results are obtained from compressed data, which are always available, and give the r.m.s. values of the currents and potentials measured at each frequency step; they are perturbed, at low frequencies, by the presence of the natural noise. The f_p and λ_D values are evaluated by a best fit to the experiment, of curves obtained using a hot isotropic plasma theory, which approximates the distribution function, assumed as Maxwellian, by a multiple water bag model (Navet and Bertrand, 1971). The agreement between theory and experiment over a large frequency range is remarkable, better than that usually

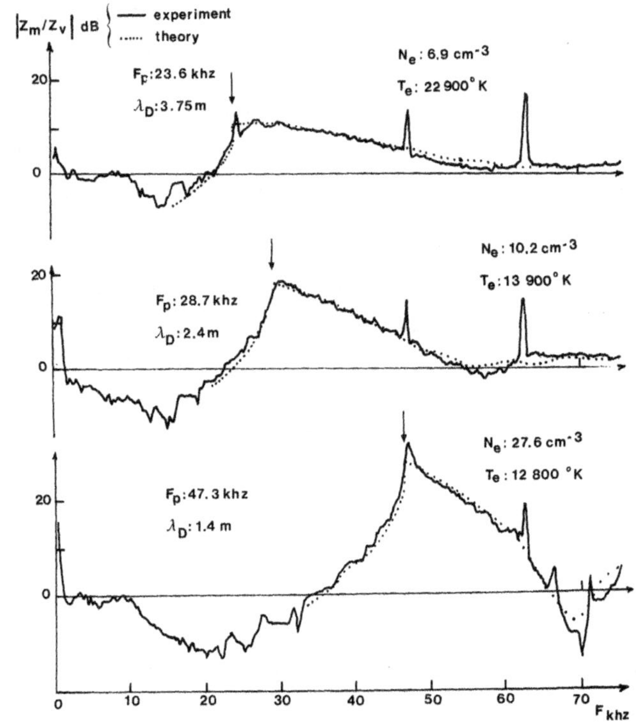

Fig. 2. Mutual impedance variation with frequency, in three different cases where the results are good. The peaks at 47 kHz and 63 kHz are due to hardware interference.

obtained in the laboratory (Decreau and Debrie, 1977; Pottelette, 1976). This is possibly due to the small dimensions of the spheres with respect to the characteristic dimensions of the antennas, as well as to the good homogeneity of the plasma encountered, in a very large volume.

Those results are confirmed when one considers the wave form of the signal, at a given frequency step, instead of a single r.m.s. value. In that case, we are able to derive both the modulus of the quantity Z_m/Z_v, which is then usually free from the natural noise of the plasma, and the argument of the same quantity. Figure 3 shows those two quantities vs. frequency, together with the corresponding theoretical curves. The presence of a harmonic structure is linked to the gyroharmonics, but is neglected to a first approximation, in our approach which considers the plasma as isotropic.

2.3. Typical difficult cases

For various reasons, we are sometimes unable to make a good diagnostic of the plasma.

(a) One reason is when the quantities to be measured are outside the range of the instrument. This is true sometimes for the plasma frequency, which may be higher than the upper measurable value, inside the plasmasphere for instance. It

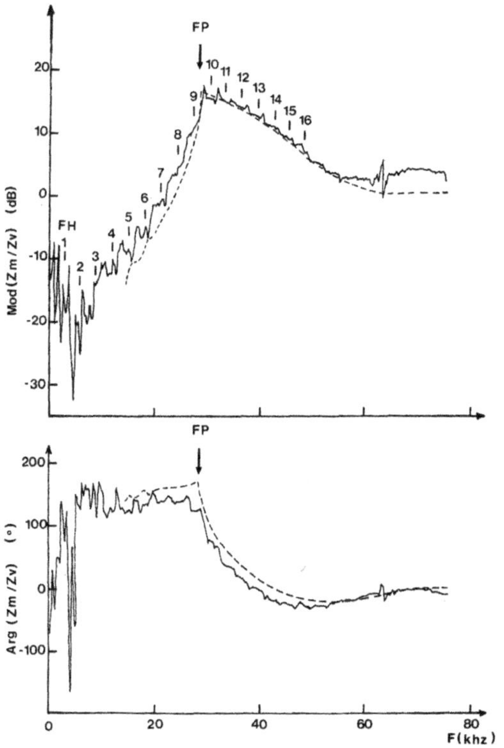

Fig. 3. Mutual impedance vs. frequency. Comparison of theoretical (– – – –) and experimental (——) results. The experimental values are derived, using the recorded wave form, through a least square best fit to a sine wave. The theoretical curves correspond to $f_p = 28.7$ kHz and $\lambda_D = 2.9$ m. The positions of the gyroharmonics (FH_i) are indicated.

happens also for the Debye length, which may rise above 10 m, especially in the morning and night sector; this leads to an almost flat impedance versus frequency response, as shown in Figure 4(a).

(b) Another reason is if the plasma departs from the simple model we have considered: homogeneous, Maxwellian, isotropic. Each deviation from those conditions is reflected in the shape of the experimental curves we record, but it is not always easy to understand what causes the misfit between experience and theory. An obvious departure from isotropy happens when the plasma frequency is low, and no longer high with respect to the gyrofrequency. Such an example is given in Figure 5(b). Here, even if we have a rough idea of the plasma frequency value, we are unable to derive a correct Debye length value. To make a valuable analysis, we have to take into account two additional parameters, the amplitude and the direction of the magnetic field, which makes the diagnostic much more complicated. In fact, further theoretical work, which is in progress, has still to be completed, before we will be able to interpret the measurements in presence of a magnetic field.

Another case is presented in Figure 4(b), where the misfit between theory and

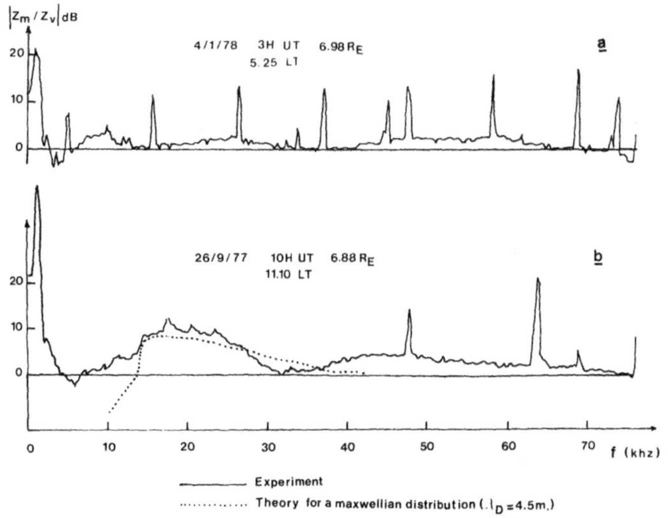

Fig. 4. Examples of not exploitable responses in preliminary analysis: (a) The Debye length being too large, we get a flat, vacuum type mutual impedance variation vs. frequency. The narrow peaks are spurious pulses due to shadow effects. (b) The plasma cannot be assumed to be steady, isotropic and Maxwellian.

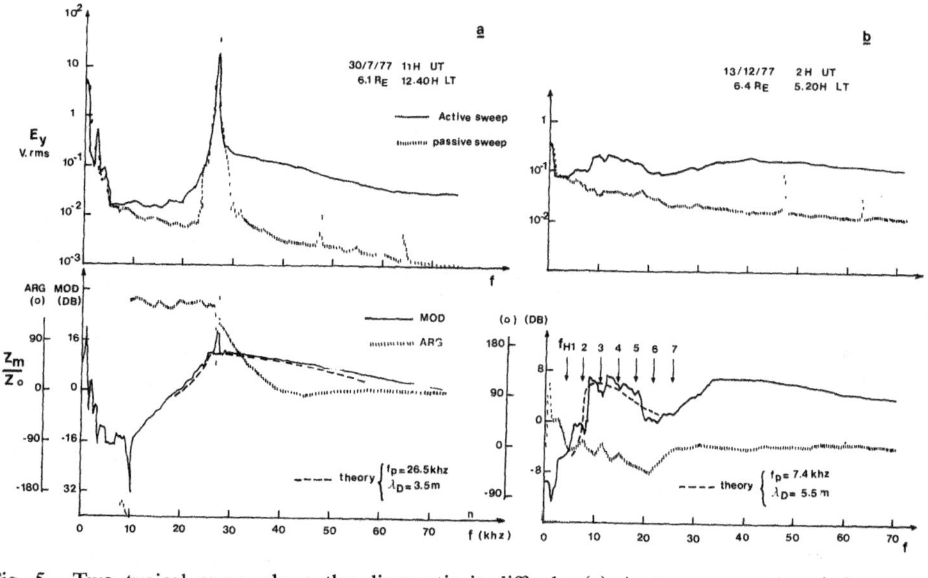

Fig. 5. Two typical cases where the diagnostic is difficult: (a) A strong natural emission occurs around f_p. (b) f_p/f_H is not large enough to assume isotropic medium.

experiment is not mainly due to the magnetic field, as for same f_p/f_H values, we usually get a better fit. Here, we think that the plasma is not Maxwellian. A rough theoretical study shows that such an experimental behaviour may be observed either in the case of two Maxwellian populations, with temperature ratios of the order of 10, or in the case of a drift velocity of the order of 0.2 times the thermal velocity. Here again, further theoretical work has to be done to achieve a better interpretation.

(c) Finally, the diagnostic may be perturbed by external phenomena, mainly by strong natural emissions around the plasma frequency (Christiansen *et al.*, 1978a). Such a case is illustrated in Figure 5(a). The r.m.s. values of the signal received (top curves, active sweep), show a strong and narrow peak at the plasma frequency, due to natural noise (passive sweep). Usually, those values (from the compressed data) are the only ones available, and the measurement of λ_D is impossible, or questionable. But, by treating the complete wave form data, a useful signal may sometimes be extracted from the noise, and a correct diagnostic recovered (bottom curves).

The review of the different types of results encountered in the analysis of the mutual impedance experiment data, allows us to define the domain of validity of the method. This domain is shown on Figure 6, as a λ_D vs. N_e frame. We evaluate the accuracy of measurements obtained in the inner frame to ~2 to 8% for N_e, and to ~6 to 12% for λ_D, which leads to ~15 to 30% for T_e. The outer frame contains rougher estimations, until accuracies of ~100% for T_e. The upper limits chosen for N_e and λ_D are instrumental; they have already been discussed. The lower limits chosen for N_e correspond to f_p values which are not much higher than usual f_H values. The lower limits chosen for λ_D correspond to sharp resonances in the mutual impedance curve, where it is difficult to distinguish between a strong

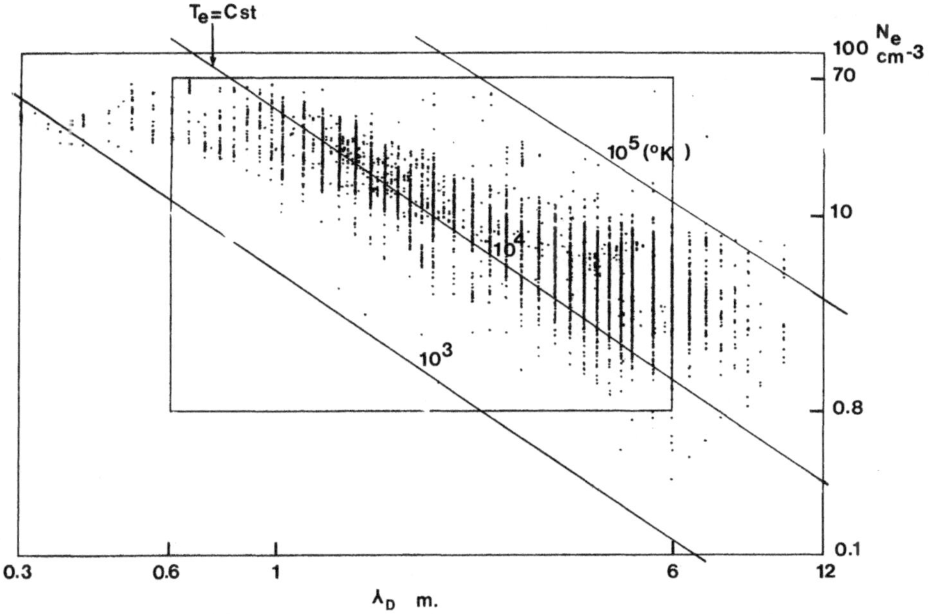

Fig. 6. Domain of validity of the mutual impedance experiment on board GEOS-1. The points drawn correspond to the measurements obtained from May to December 1977. The apparent line structure is due to the method used to fit an experimental curve to a theoretical one, involving a choice between discretized values of λ_D.

natural emission and a significant result. The points drawn on the figure correspond to all the measurements obtained from May to December 1977 (mostly dayside measurements). Their position in the plane leads directly to geophysical result, which we discuss now.

3. Geophysical Results

To present the measurements obtained from a geophysical point of view, we will first come back to Figure 6, which presents the data globally, and then proceed to an increasing level of detail. The remarkable result in Figure 6 is that the order of magnitude of the temperature observed is rather constant, and rather low: ~ 1 eV. We notice a tendency, at high densities, for the temperatures to get larger. This was to be expected from measurements made in the vicinity of the plasmapause: the plasma inside the plasmasphere is known to be denser and colder than outside the plasmasphere. Nevertheless, there is a difference between the order of magnitude of the temperatures we have measured, with respect to previous measurements of the thermal plasma made in similar regions of the magnetosphere. We may mention for instance the ion temperatures obtained by Serbu and Maier (1970), or by Bezrukikh and Gringaus (1976). The temperatures they recorded are generally higher than those we have measured for the same densities or L values, although a direct comparison is not possible, because of the differences in the orbits, dates and ranges of validity of the instruments.

In Figure 7, we have shown the behaviour of the quantities measured in terms of altitude and local time. In each of the three parts of the figure, one quantity, N_e, λ_D or T_e is mapped in the equatorial plane, in a colored scale going from low (green) to high (red) values. One elemental square corresponds to one measurement, or to the average of several measurements. The morning sector is less complete, partly because the visibility of the satellite was reduced in some periods, partly because some data is still to be processed, but also because the percentage of data which cannot be analysed (the 'difficult' cases) goes larger as the local time of the apogee goes nearer to the night sector. Nevertheless, we notice interesting features in the figure: the altitude and local time dependence of the density are qualitatively in agreement with previous measurements (Chappell *et al.*, 1970; Gurnett and Frank, 1974; Gringaus and Bezrukikh, 1976). There is a marked difference between values obtained at lower altitudes ($R < \sim 5$ RE), which we assume to be taken inside the plasmasphere, and values obtained at higher altitudes. The plasmasphere is asymmetric along the dawn to dusk line, extending sometimes higher than the apogee near 18 HR LT. We observe a similar behaviour for λ_D, which is larger at higher altitudes and in the morning sector. For the temperature, the contrast between inner and outer plasmasphere is not so clear: the values obtained outside the plasmasphere are not the highest, they are scarcely higher than 2 eV near apogee. Here the dawn-to-dusk asymmetry is not noticeable.

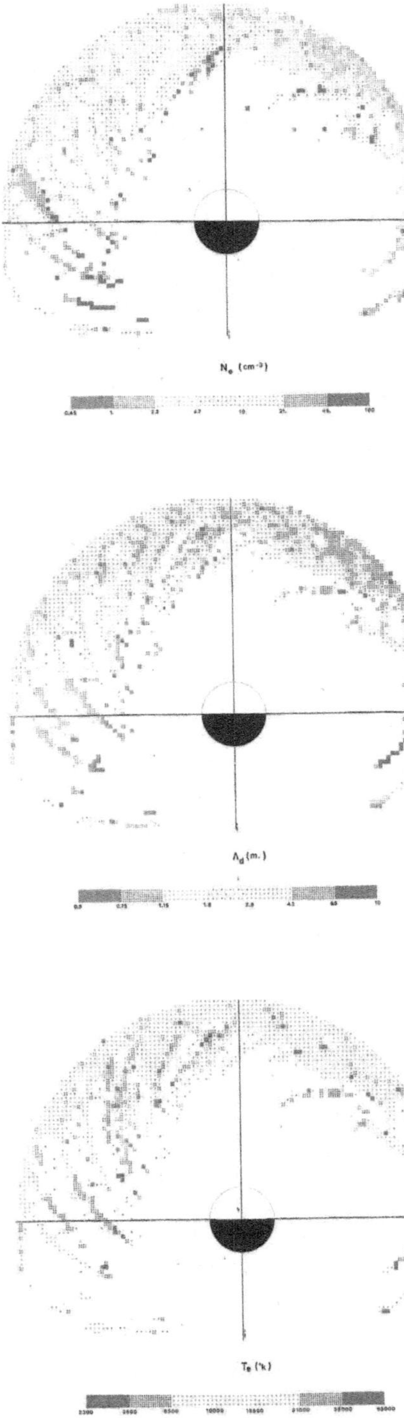

Fig. 7. Plasma parameters values, obtained from May to December 1977, mapped in the equatorial plane.

Another topic which may be studied using plasma diagnostics made *in situ*, is the dynamic of the plasmasphere, and its relation to the magnetic activity. To obtain a first idea of the possible link between our measurements and magnetic activity, we have undertaken a statistical study. For each day, we have chosen one set of values characterizing the plasma: the density and temperature obtained at apogee, or very near to the apogee, and one parameter characterizing quantitatively the activity: the A_m value for the day (Mayaud, 1968). Figure 8 shows these three quantities as well as the local time of apogee, displayed vs. the day number. An obvious feature in this figure, is the relation between the local time and the density: the mean density follows the decrease in the local time with day number. This appears already in Figure 7.

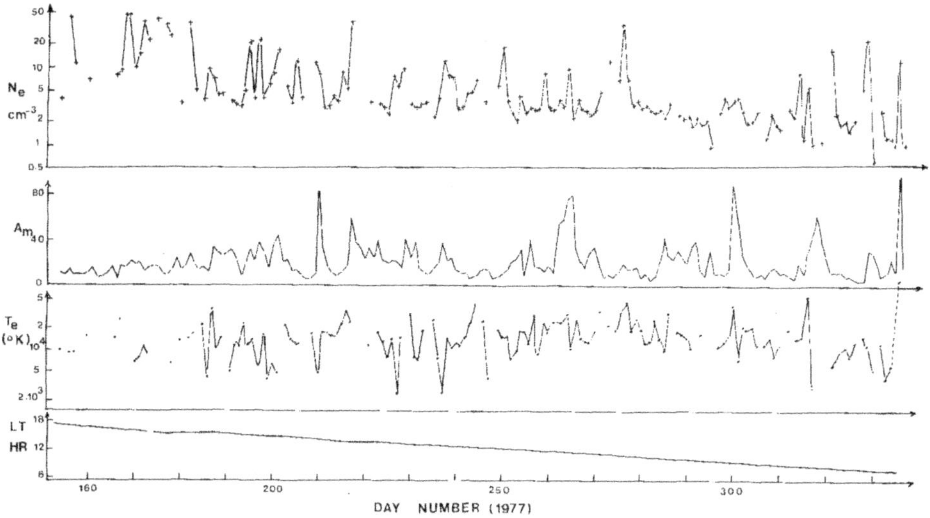

Fig. 8. Measurements of N_e and T_e at apogee, compared to the day to day variation of the index A_m and of the local time at apogee, over a six months period.

To evaluate the relation between the A_m and the plasma parameters, we have calculated a normalized linear correlation coefficient r:

$$r = \frac{\sum (x_i - \bar{x})(y_i - \bar{y})}{\sqrt{[\sum (x_i - \bar{x})^2 \sum (y_i - \bar{y})^2]}}$$

where x_i and y_i are the respective values of the two quantities A_m and $\log N_e$ (or $\log T_e$) for the same day, $\bar{x}$ and $\bar{y}$ being their average value over the whole set of data. Calculating similar quantities after having introduced a delay of l days in one set of data, we obtain a function $r(l)$, which characterizes the degree of correlation between the A_m and the densities (or temperatures) measured l days later. The results are plotted in Figure 9. In addition, we have calculated the probability P that any random sample of N uncorrelated experimental data points, would yield an experimental linear correlation coefficient as large as or larger than $|r|$,

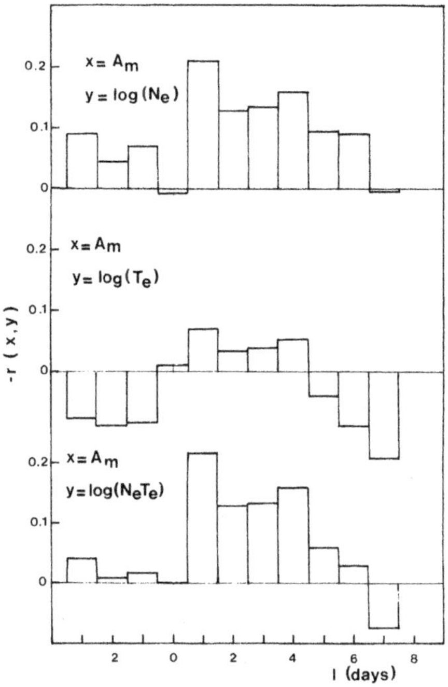

Fig. 9. Variation of the normalized linear cross correlation coefficient r with the delay l in days. The two quantities correlated are the index A_m and one of the three parameters measured at apogee: density, temperature, thermal energy.

where N is the number of data we used in our calculation (see Bevington, 1969). the values obtained are $P \sim 1.5\%$ for $|r| = 0.2$; $P \sim 7\%$ for $|r| = 0.15$; $P \sim 25\%$ for $|r| = 0.1$ and $P \sim 55\%$ for $|r| = 0.05$. Hence, the probability that a value $|r| = 0.2$ is significant, is 98.5%, whereas it is only 75% for $|r| = 0.1$, etc., ... In two cases, the curves of Figure 9 point out a relation between the data. The density N_e, and the thermal energy N_eT_e are correlated with magnetic activity, strong activity being followed by low N_e and N_eT_e values, and the correlation being highest for a delay of one day. This behaviour is already known (Chappell, 1972). What is more surprising is that the temperatures measured seem not to be significantly correlated to the magnetic activity.

To end this preliminary analysis of the geophysical results obtained by our experiment, we will present a few examples of detailed density and temperature profiles obtained on GEOS. In Figure 10, three typical profiles are drawn vs. the geocentric distance, R, which is given on a linear scale. For all three of them, the N_e profiles taken in the inbound pass resemble a $1/R^4$ function, which corresponds to diffusive equilibrium (Chappell, 1972). The measurements obtained the 22nd of June show higher densities and lower temperatures than those obtained the 3rd of September; they are taken deeper in the plasmasphere, as the local time is nearer to dusk. The profile obtained the 4th of September is seldom

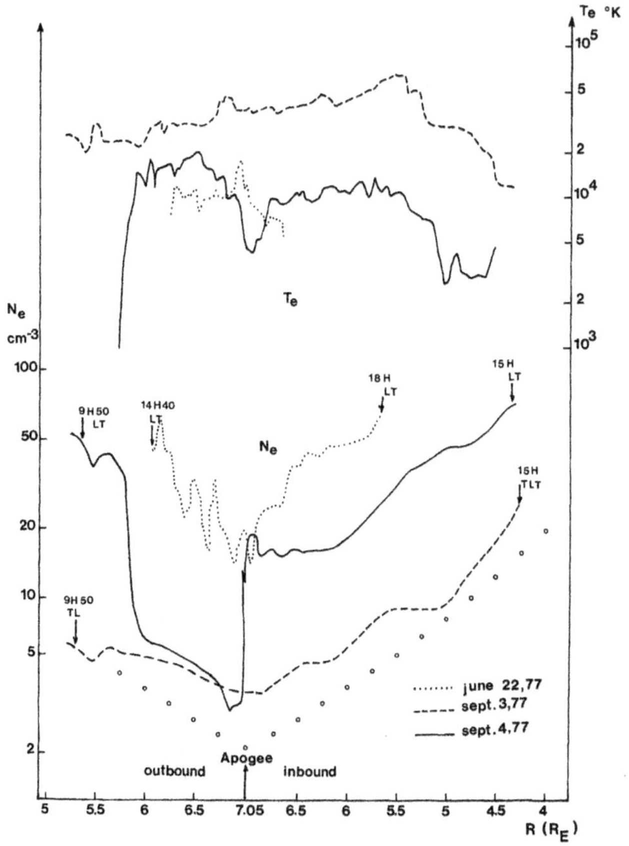

Fig. 10. Typical profiles of the density and temperature for three different days, versus the geocentric distance R in Earth radii. The magnetic index $\sum K_p$ for the preceding 24 hr had the following values: 12^- the 22nd of June; 16^+ the 3rd of September; 9^+ the 4th of September. The curve with open circles corresponds to a $1/R^4$ profile.

seen, but is not a unique example. The density profile shows two sharp plasmapause crossings, the satellite being outside the plasmasphere only between $R = 6$ and 7 Earth radii, on the outbound pass. The temperature profiles does not show the same behaviour. The first plasmapause crossing is clearly seen, with a sharp increase in temperature, whereas the second one is associated with a hole in the temperature profile: the temperature value is correlated to the gradient in density, more than to the density value. Another decrease in T_e is seen near $R = 5$, without any associated change in density. Those details show that a conventional plasmaspause boundary, which separates dense and cold plasma on the inner side, and tenuous and hot plasma on the outer side, is only a rough model. This is confirmed by the two profiles presented in Figure 11, both obtained in rather quiet magnetic conditions. Densities and temperatures are drawn vs. the time given on a linear scale. The 28th of September, the N_e profile is very regular,

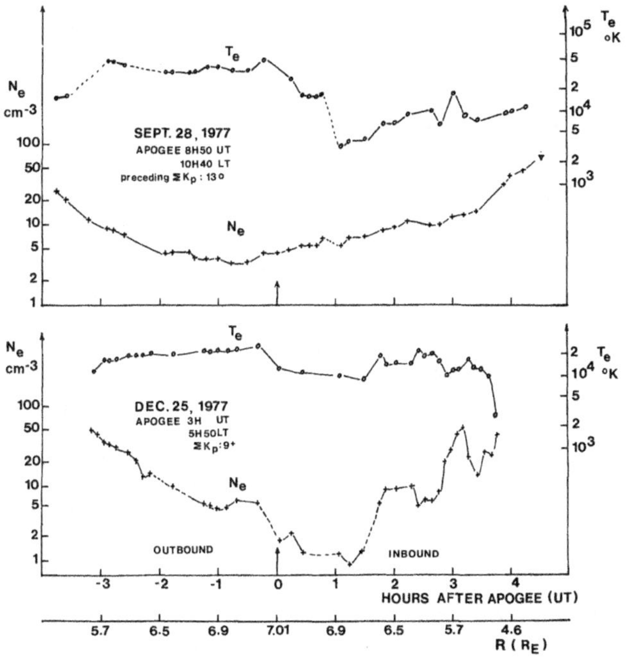

Fig. 11 Profiles of the density and temperature for two different days vs. the time around apogee. The geocentric distance R in Earth radii is indicated.

showing low values, whereas a sharp decrease is seen for T_e, near the apogee. On the other hand, the T_e profile obtained the 25th of December is fairly regular, showing only a decrease at the end of the pass, whereas the inbound density profile shows several successive increases and decreases. Here, it is difficult to distinguish between the crossing of detached plasma regions, and the entering of the plasmasphere.

4. Concluding Remarks

The mutual impedance experiment on board GEOS-1 produced promising results from the very beginning of data analysis. The results obtained as the study proceeded confirmed this promise. The principal outcome of the experiment is that the diagnostic of the density and temperature is generally easy to carry out and gives good results in the dayside of the equatorial plane. The density behaviour, vs. altitude, local time and magnetic activity, is consistent with features already reported in previous observations. Moreover, the temperature measurements provide new material for geophysical studies. It appears that the cold part of the plasma outside the plasmasphere is properly revealed for the first time, and that detailed joint analysis of the N_e and T_e data should help to improve the known models of plasmasphere dynamics.

The instrumental limits prevent the study of the inner plasmasphere, where the density is too high, as well as the study of the hot tenuous plasmas which are encountered sometimes outside the plasmasphere, where the Debye length is too large. However, the majority of cases which are uninterpretable through the simple analysis currently used, do contain material for a useful diagnostic of the plasma. Even on the night side of the equatorial plane, the large majority of the recorded experimental curves of mutual impedance against frequency are not of the type seen for λ_D too large. They show resonances, associated often with a sharp anti-resonance above f_p, which cannot usually be interpreted in term of isotropic, Maxwellian and steady plasma. Those data require a rather complex analysis which can be undertaken only after further studies in the field of plasma physics have been completed. These studies involve mainly theoretical work, but may also include plasma experiments in laboratory. The first step of the study, will be to take into account the presence of the magnetic field. This will allow the diagnostic of N_e and T_e when the plasma frequency is not very large with respect to the gyrofrequency; it will also lead to the quantitative interpretation of the harmonic structure observed above the plasma frequency, which was recognized as the f_q resonances (Christiansen *et al.*, 1978b). By so doing, we aim to measure independently the electron temperature parallel and perpendicular to the magnetic field. Thus, we hope to have access to an eventual anisotropy of the plasma temperature, which we suspect to be present when we observe severe fluctuations of the f_q's amplitudes. The second step of the study will be to introduce a drift velocity, with the hope of measuring that quantity when it is high enough to modify significantly the shape of the impedance curves, which usually seems to be the case on the nightside. Another step in the study would be to take into account a non-Maxwellian velocity distribution of the plasma. But, apart perhaps from the case of a two Maxwellian distribution, our experiment is not adapted to the study of non-Maxwellian plasmas.

In conclusion, we think the data given by the mutual impedance experiment on board GEOS are significant. We may reasonably hope that after some improvement of the theoretical model of the plasma used for interpreting the experimental response, we will have an original powerful tool to study dynamics of the plasmasphere.

Acknowledgements

The authors associate themselves with the other experimenters to acknowledge all the individuals and organizations who contributed to the success of GEOS in an amicable and cooperative spirit.

We are particularly indebted to those who contributed to the mutual impedance experiment data handling, especially to C. Huc and D. Barria of the Centre National D'Etudes Spatiales (Toulouse, France). We are grateful to M. Navet, who provided computer programs on the multiple water bag model, to V. Corbin and Y. Legendre for their participation in the analysis of the experimental curves.

We are glad to thank A. and J. J. Berthelier, J. J. Bloch, R. Debrie, J. Etcheto, I. Freeston, M. Hamelin and J. P. Lebreton, for valuable discussions about data evaluation.

References

Auger, J.: 1974, Technical note S-300 No. 10, ed. by CNET, Issy-les-Moulineaux, France.

Bevington, P. R.: 1969, *Data Reduction and Error Analysis for the Physical Sciences*, McGraw Hill Book Company, New York.

Bezrukikh, V. V. and Gringaus, K. I.: 1976, *J. Atmosph. Terrest. Phys.* **38,** 1085.

Chappell, C. R., Harris, K. K., and Shapp, G. W.: 1970, *J. Geophys. Res.* **75,** 50, 219, 348.

Chappell, C. R.: 1972, *Rev. Geophys. Space Phys.* **7,** 951.

Chasseriaux, J. M., Debrie, R., and Renard, C.: 1972, *J. Plasma Phys.* **8,** 231.

Christiansen, P. J., Gough, M. P., Martelli, G., Bloch, J. J., Cornilleau, N., Etcheto, J., Gendrin, R., Beghin, C., Decreau, P. M. E., and Jones, D.: 1978a, *Space Sci. Rev.* **22,** 383.

Christiansen, P., Gough, P., Martelli, G., Bloch, J. J., Cornilleau, N., Etcheto, J., Gendrin, R., Jones, D., Beghin, C., and Decreau, P. M. E.: 1978b, *Nature* **272,** 682.

Decreau, P. M. E. and Debrie, R.: 1977, *Rev. Phys. Appliquée* **12,** 1043.

Decreau, P. M. E., Etcheto, J., Knott, K., Pedersen, A., Wrenn, G. L., and Young, D. T.: 1978, *Space Sci. Rev.* **22** 633.

Gringauz, K. I. and Bezrukikh, V. V.: 1976, *J. Atmosph. Terrest. Phys.* **38,** 1071.

Gurnett, D. A. and Frank, L. A.: 1974, *J. Geophys. Res.* **79,** 2355.

Mayaud, P. N.: 1968, *Indices K_n, K_s and K_m 1964–1967*, Ed. CNRS, Paris.

Navet, M. and Bertrand, P.: 1971, *Phys. Letters* **34A,** 117.

Pottelette, R.: 1976, Thèse de Doctorat d'Etat, Paris VI.

Renard, C. and Debrie, R.: 1970, *C. R. Hebd. Sean. Acad. Sci.* **B270,** 649.

Rooy, B., Feix, M. R., and Storey, L. R. O.: 1972, *Plasma Phys.* **14,** 275.

Storey, L. R. O., Aubry, M. P., and Meyer, P.: 1969, in J. O. Thomas and B. J. Landmark (eds.), *Plasma Waves in Space and in the Laboratory*, Edinburgh Univ. Press **1,** 303.

S-300 Experimenters: 1978, *Planet Space Sci.* **26,** (to be published).

PLASMA DENSITY MEASUREMENTS FROM THE GEOS-1 RELAXATION SOUNDER

J. ETCHETO and J. J. BLOCH
C.R.P.E., CNET/ETE, 92131 – Issy-les-Moulineaux, France

Abstract. The relaxation sounder uses the characteristics of the propagation of radiowaves to sound the plasma surrounding the spacecraft. It determines, in particular, the plasma frequency, which gives the electron density.

We have now measurements over the whole dayside of the magnetosphere, from the evening to the night sectors.

The behaviour of the plasma resonance depends on local time, the nightime echoes being generally weaker.

Density measurements thus obtained are shown and discussed in the context of what is presently known about the plasma distribution in the magnetosphere. In particular, the density around apogee is studied as a function of magnetic activity. On the dayside, it appears to vary between a few and a few tens of electrons per cubic centimeter. The evolution of the density profile for several consecutive days is studied and interpreted tracing back the drift of the particles.

1. Introduction

The relaxation sounder experiment which was flown on board GEOS constitutes an entirely original method for measuring electron densities as low as $1\ \text{cm}^{-3}$. Plasma frequency and electron gyrofrequency are indeed measured both accurately and reliably. The effect of the photoelectron cloud surrounding the spacecraft is proved to be negligible (see also Lockwood, 1963; Calvert and Goe, 1963; Etcheto, 1976; Decreau *et al.*, 1978). Basically the experiment is an extrapolation of the resonance spikes observed with the ionospheric topside sounders to plasmas 10^5 times less dense than the ionospheric one (Calvert and McAfee, 1969; Heikkila *et al.*, 1967; Higel *et al.*, 1972; McAfee *et al.*, 1972).

The results obtained are independent of the spacecraft potential. The general behaviour of the plasma density is in agreement with earlier data and with the predictions of a crude theoretical model.

2. Description of Measurements

We shall now describe the measurements that we get and how we can use them for magnetospheric studies.

The resonance sounder is part of the S-300 experiment and uses the hardware described earlier (Jones, 1978). The technical features are those of an ionosonde: a short radio frequency pulse – 3.3 ms long – is transmitted via the shield of the long radial boom, then the sounder monitors the returning wave, received on the spheres during 83 ms.

The working frequency is swept in 256 steps, 300 Hz apart through the range 0 to 77 kHz. In our usual mode of operation, 22 s are necessary to achieve a

Space Science Reviews **22** (1978) 597–610. *All Rights Reserved*

complete sweep, however the time resolution can be increased (Higel, 1978). The pulse length is chosen in order to transmit a 300 Hz (4 dB) wide spectrum, consistent with the receiver bandwidth. In the general case, the transmitted wave propagates away from the satellite and no signal is received on board. However, if the frequency is close to a resonance frequency of the medium (low group velocity) a signal is received during up to hundreds of milliseconds. The medium exhibits a very sharp frequency response and the 3 dB bandwidth of the received signal is never larger than some 50 Hz. These signals have first been interpreted as waves moving along with the satellite, their group velocities being equal to the satellite's one with respect to the ambient plasma (Fejer and Calvert, 1964; Dougherty and Monaghan, 1965; Deering and Fejer, 1965; Shkarovsky, 1968; Dougherty, 1969). McAfee (1968, 1969) first stressed the importance of the non uniformity of the plasma: he showed that a small density gradient modifies completely the group path for waves close to the plasma frequency, so that this path, instead of being a straight line, is bent over and can even form a closed loop. He called this phenomenon 'oblique echoes'. This is due to the fact that, in an anisotropic plasma the group velocity of a wave close to a resonance frequency is low in a given direction with respect to the magnetic field, leading to a very strong refraction, whenever the plasma is slightly inhomogenous. Most of the subsequent ttheoretical work on resonances made use of the oblique echoes concept (Aubry *et al.*, 1970; Bitoun *et al.*, 1970; Fejer and Yu, 1970; Graff, 1971; McAfee, 1973; Bitoun, 1974; Bitoun *et al.*, 1975).

3. Observed Resonances

On Figure 1 is shown an example of the resonance signals usually observed on board GEOS-1. On this ionogramme the frequency at which the sounder operates is displayed in abscissa while the vertical scale displays the time elapsed since the transmitter frequency was changed (increasing downwards). The amplitude of the received signal is shown by grey scaling. Reception is inhibited during transmission, which takes place within the 20 ms following the frequency stepping.

A reference passive sweep is performed before the active sweep in order to discriminate easily between the signals associated with the transmission and those which were already present (in particular, two fixed satellite generated interference lines are present at 47.6 kHz and 63.5 kHz). The resonances can be easily identified as transient signals appearing after transmission and rapidly decaying with time. While only a few resonances are usually observed in the ionosphere (Calvert and McAfee, 1969), along the GEOS orbit we get many of them. They can be sorted out in three classes:

(a) The electron gyrofrequency and its harmonics ranging from one to the upper frequency limit (76 kHz). (In the case of Figure 1 the gyrofrequency is 3.0 kHz corresponding to a magnetic field of 107γ and there are 25 harmonics visible). They correspond to waves propagating in the Bernstein's modes with

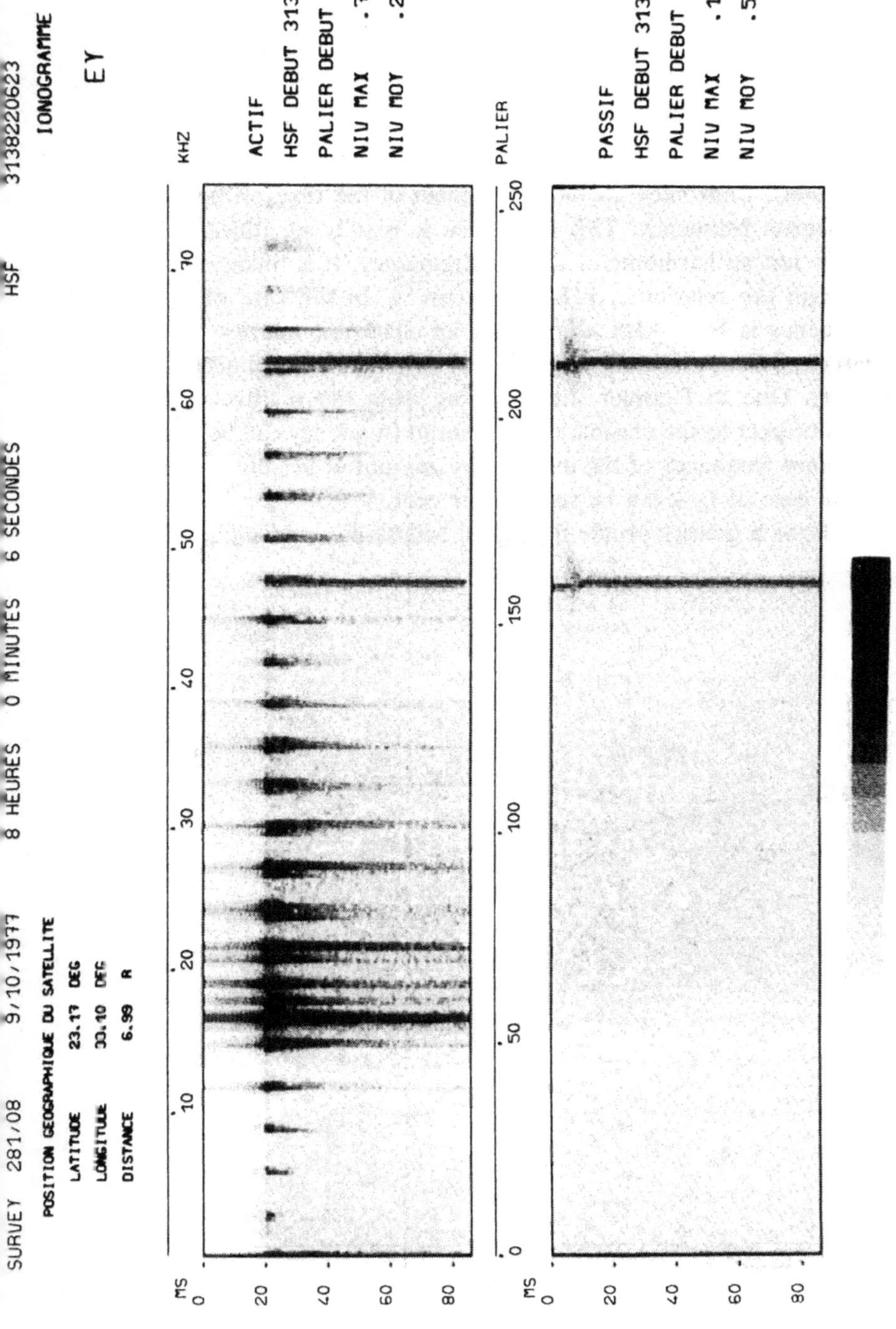

Fig. 1. Ionogramme recorded on October 9, 1977, at 08:00 UT. Top frame illustrates the behaviour of the plasma when the transmitter is on, bottom frame, recorded just before the latter shows the signals observed when the transmitter is off (the two interference lines at 47.6 kHz and 63.5 kHz are easily identified).

wave vectors nearly perpendicular to the magnetic field (Bitoun, 1974). We usually observe resonances at the harmonics of the gyrofrequency through the whole frequency range of the sounder.

(b) The f_Q frequencies. These resonances are observed above the upper hybrid frequency, each one between two successive harmonics of the gyrofrequency; as the frequency increases the frequency shift from the harmonic just below decreases. These waves correspond to another zero in the group velocity of the Bernstein modes, appearing on each branch above the upper hybrid frequency (Crawford, 1965). They have been sometimes observed in the ionsphere (Warren and Hagg, 1968), while they are observed most of the time on board GEOS-1.

(c) The plasma frequency. This resonance is readily identified as the lowest trace which is not an harmonic of the gyrofrequency. It is linked to the electron density through the relation: $f(\text{kHz}) = 9\sqrt{n(\text{cm}^{-3})}$. In the case of Figure 1, the plasma frequency is 16.66 kHz which gives an electronic density of 3.43 cm^{-3}.

The frequency of the resonances can be determined with an accuracy of a few tens of Hertz. Due to Doppler shifts coming from the relative velocity of the satellite with respect to the plasma, the measured frequency can be away from the actual resonance frequency of the medium by amount of usually less than 1% but which, in the case of f_Q's can be several per cent.

Figure 2 shows a density profile during an eclipse pass; no significant variation

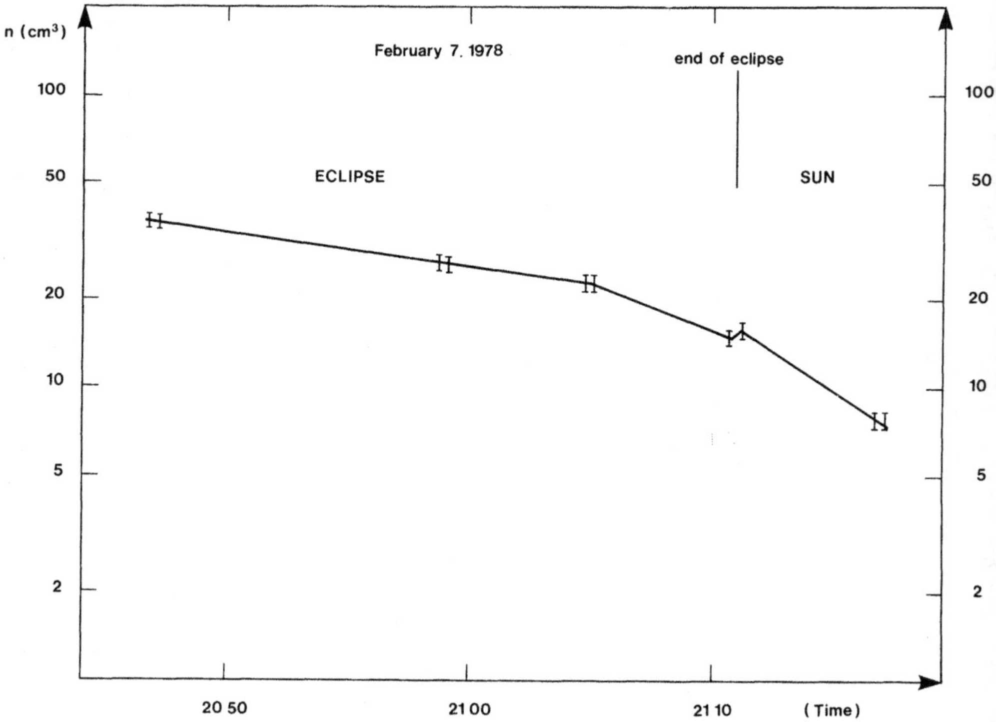

Fig. 2. Density profile for February 7, 1978. The satellite stayed in the shade until 21:12 UT then moved out of the eclipse zone. No significant change occurs at the crossing zone. (The slight peak in the profile at 21:12 UT remains within the measurement incertainty values).

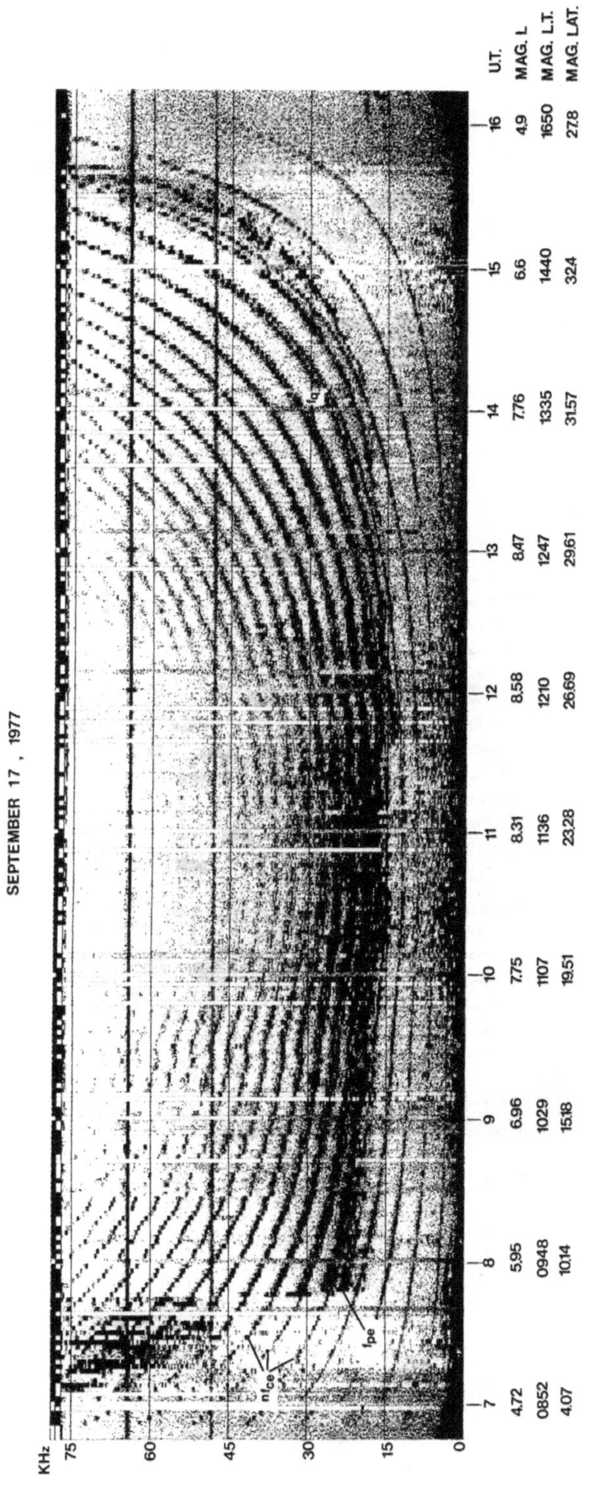

Fig. 3. A whole pass as seen with sounder permanently active. The plasma frequency f_p is identified as the inferior limit of the dark region of the figure. Note the sharp f_p gradient during the outbound pass at 07:50 UT and the smooth gradient during the inbound period.

is observed when the satellite goes out of the eclipse zone. This is a proof that the photo-electrons have an effect on our density measurements which is smaller than the experimental error.

Figure 3 shows the evolution of the received signals along a whole GEOS pass. The amplitude of the received signal is plotted using grey-scaling in a frequency versus time frame. This type of display does not exploit the high frequency resolution of the experiment but gives a global picture of what is going on during

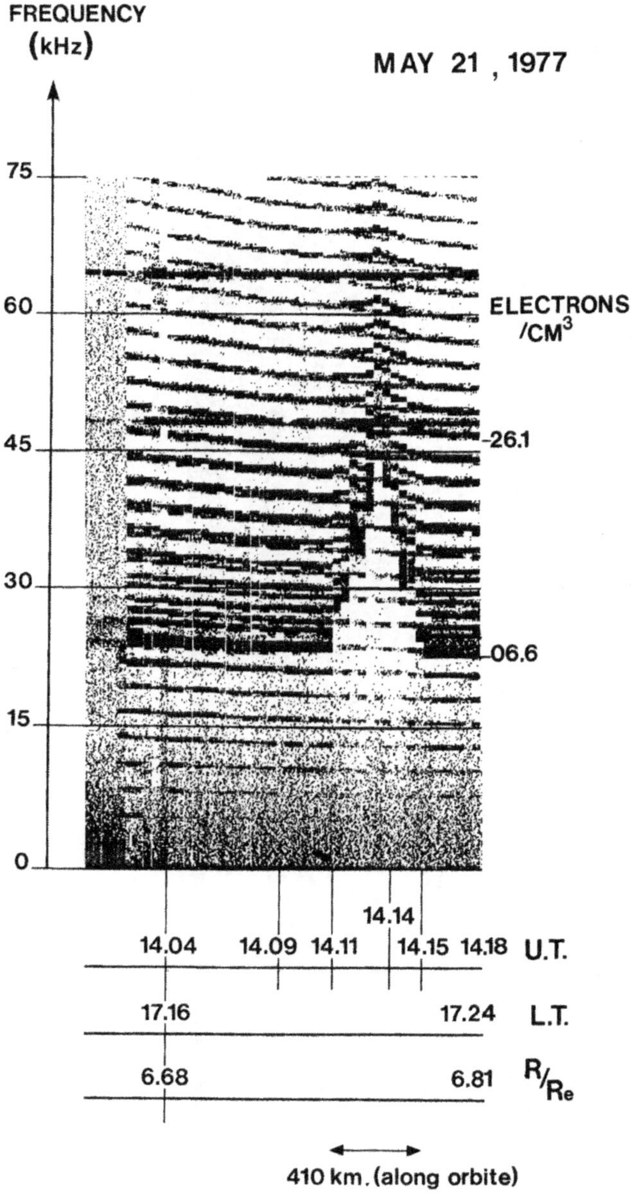

Fig. 4. A detached plasma zone observed on May 21, 1977.

a pass. The regularly spaced resonances at the harmonics of the gyrofrequency are clearly visible. The f_Q resonances are seen between the harmonics of the gyrofrequency at the ends of the pass but seem merged with them at apogee, while the inferior limit of the dark area gives an estimate of the plasma frequency. The large-scale density fluctuations at the edge of the plasmasphere on the morning side visible on Figure 3, are often observed. They correspond to either an oscillation of the plasmapause position with time or a spatial structure through which the satellite is passing. So far we have no clue to choose between these two interpretations.

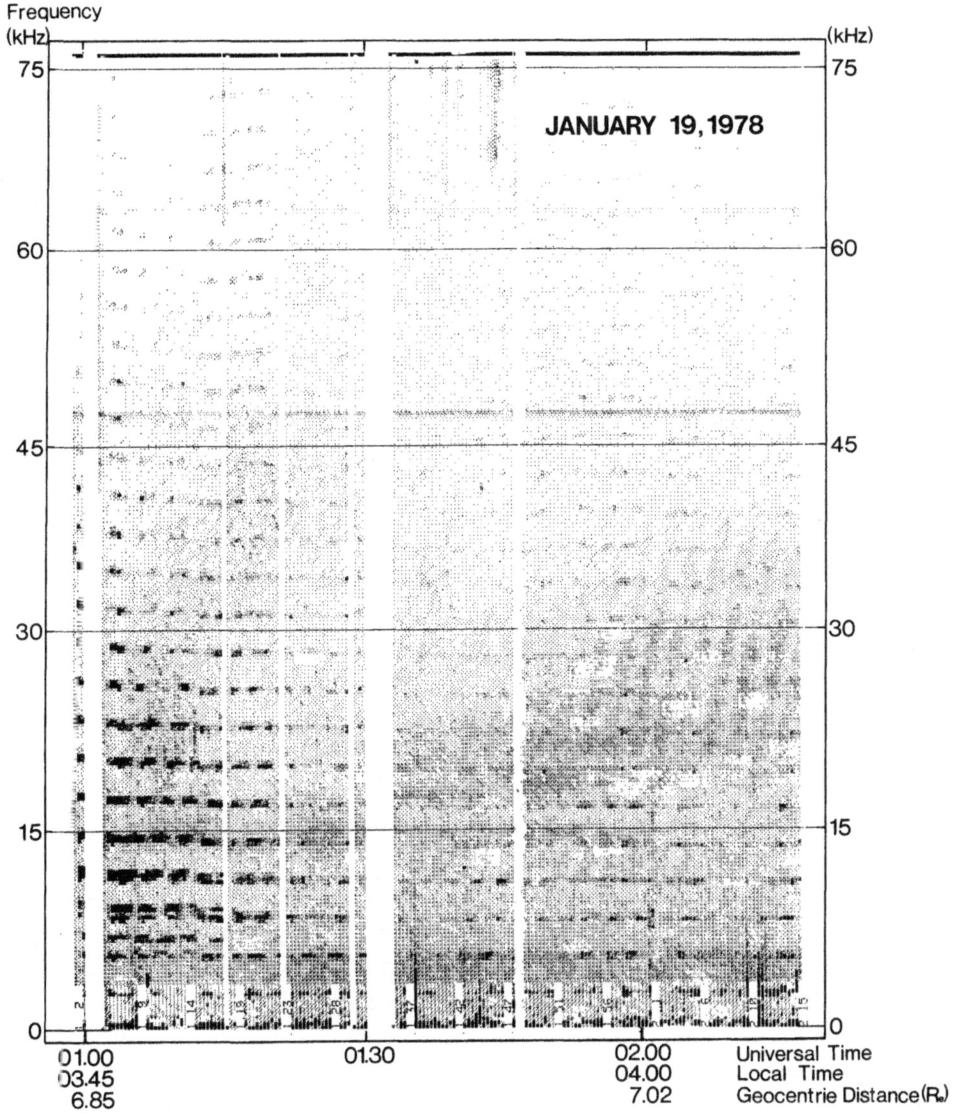

Fig. 5. Fading out of the received echoes during a local night-time pass on January 19, 1978.

Figure 4 shows the same type of display as Figure 3, with an enlarged time scale for a period during which the satellite crossed a detached plasma region, from 14:11 till 14:15 UT the density increased sharply from 6.6 cm^3 to 26.1 cm^{-3} and returned to 6.6 cm^3. The size of this plasma island was about 400 km along the GEOS orbit. This is in agreement with previous observations of detached plasma (Chappell, 1974). This pass occurred during a recovery following an increase in the magnetic activity during the previous night (where Kp was up to 4). Note that the intensity of the resonance signals at the harmonics of the gyrofrequency is maximum near the plasma frequency and decreases away from it. In the midnight sector the usual pattern of resonances is often reduced to low frequency signals only; this is shown on Figure 5. This feature is usually associated with low plasma density (here 0.5 cm^{-3} at 01:10 UT) and large Debye length, (here greater than 5.5 m at 01:00 UT (Decreau, private communication)).

It is known (Deering and Fejer, 1965; Fejer and Yu, 1970) that the amplitude of the plasma resonance should decrease very quickly when the density decreases and the temperature increases, this concerns the part of the signal accompanying the satellite which is received during the early receiving period, and also the oblique echoes which are received later. The wave field decreases as $f_{pe}^{9/2}$ for the accompanying signal and as $f_{pe}^{7/2}$ for the oblique echoes.

4. Density Profiles

We shall now look at some density profiles and see how they fit previous observations and convection theory. Figure 6 shows a typical profile measured on the dayside for a moderately disturbed day (previous Kp of the order of 2). On the morning side the plasmapause is very steep around $L = 5.3$, the density decreasing from 50 to 6 cm^{-3} within a few hundred kilometers while in the afternoon the gradient is much smoother. The density varying from 3 to 70 cm^{-3} in two earth radii. The density outside the plasmasphere is usually a few electrons per cubic centimeter.

Figure 7 shows some typical profiles for a magnetically quiet period at the top (Kp less than 2 during three days) and for a magnetically disturbed period at the bottom (Kp around 6 for 12 h). Note that the two profiles have different density scales. During quiet periods the density remains high (for the example displayed greater than 25 cm^{-3}). During disturbed periods the density stays low and the satellite is lost by the tracking station before encountering the plasmapause. These observations are consistent with previous measurements (see reviews by Chappell, 1972; Carpenter and Park, 1973; Maynard and Grebowsky, 1977) showing that, during magnetically disturbed period, the plasmapause is found closer to the Earth. Our measurements however seem to locate the plasmapause further away from the Earth by about one earth radius than reported by Maynard and Grebowsky (1977) for similar magnetic conditions and even more compared to Carpenter (1966). The measurements of Gringauz and Bezrukikh (1975) are in

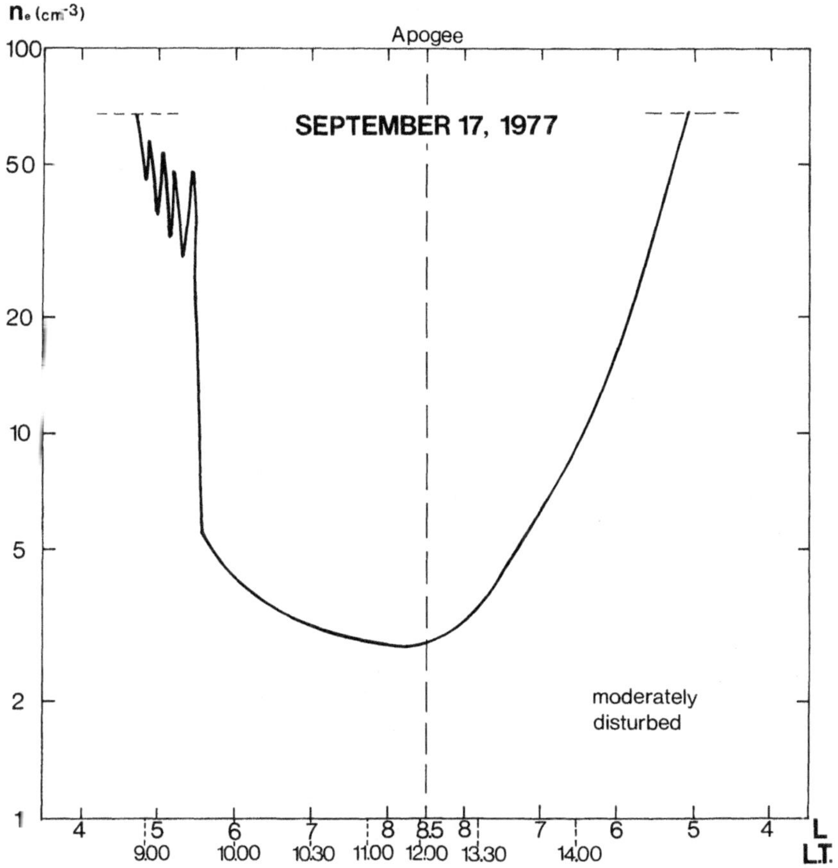

Fig. 6. A typical density profile (correspondihg to the same day as Figure 3). Note the large scale fluctuations at the edge of the plasmasphere around 09:00 LT.

better agreement with our observations. This could be due to the different techniques used to determine the plasmapause, the first mentioned determinations being more indirect. A statistical study is necessary to be able to compare Chappel *et al.* (1971) results.

We shall now study in more details a series of profiles measured during consecutive days in July 1977.

On Figure 8 are plotted density profiles measured between July 25 and July 29 during a period of very low magnetic activity until July 29 at 00:27 UT when a storm began. At the top of the figure are plotted the Kp and D_{st} indexes for that period. At that time the satellite was monitored in the dayside of the magnetosphere (see the orbit on Figure 9). Notice that the universal time is not much different from the local time at the satellite. To get a quick evaluation of the phenomena we took a very simple convection model using a uniform electric field super-imposed on a corotation field (Kavanagh *et al.*, 1968) and we computed the

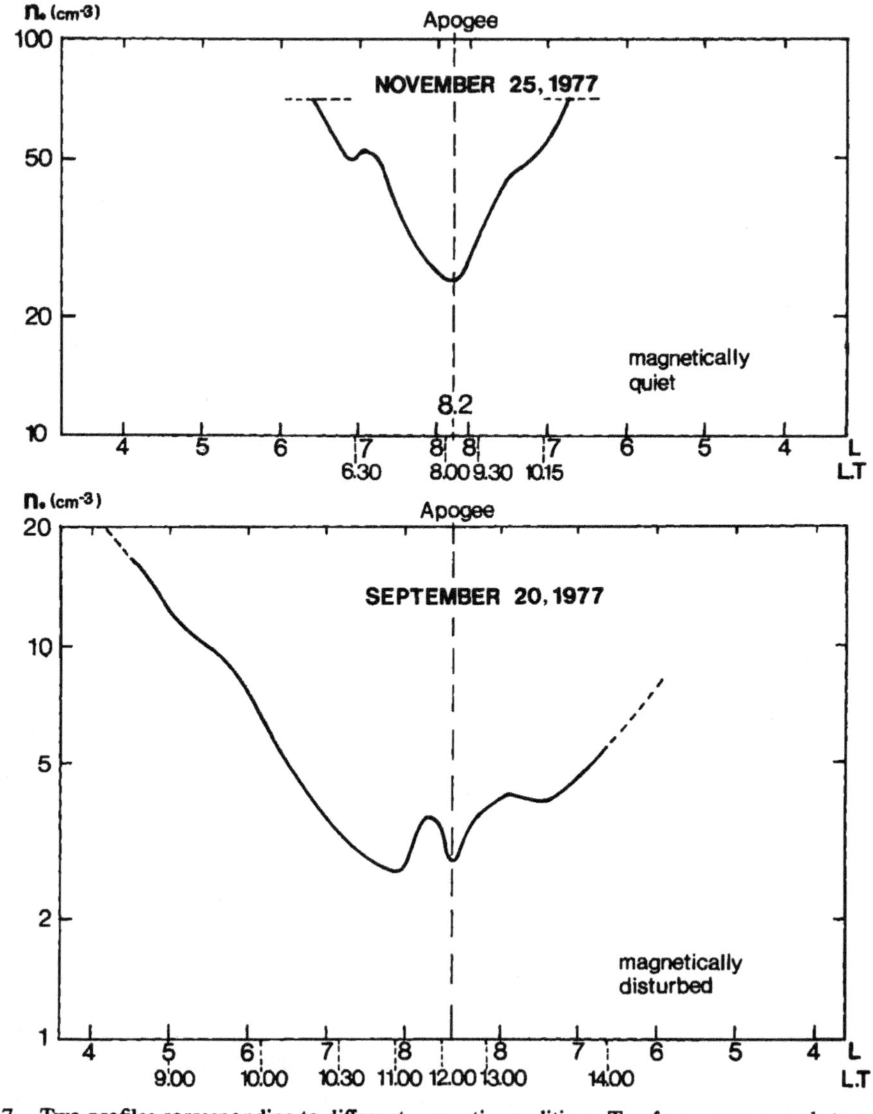

Fig. 7. Two profiles corresponding to different magnetic conditions. Top frame corresponds to a very quiet period, lowest density is about 30 cm^{-3}. Bottom frame corresponds to a disturbed period, lowest density is about 3 cm^{-3} and it keeps to low value during the whole pass.

equipotentials of the electric field along which the cold plasma should drift. To determine the value of the uniform electric field we used the empirical relation between the electric field and the *Kp* index given by Kivelson (1976):

$$E(\text{kV/Re}) = 0.44/[1 - 0.097\,Kp]^2.$$

It appears that the predictions of this model agree fairly well with the observed position of the plasmapause.

During steadily quiet conditions ($Kp < 1$) the closed equipotentials extend over

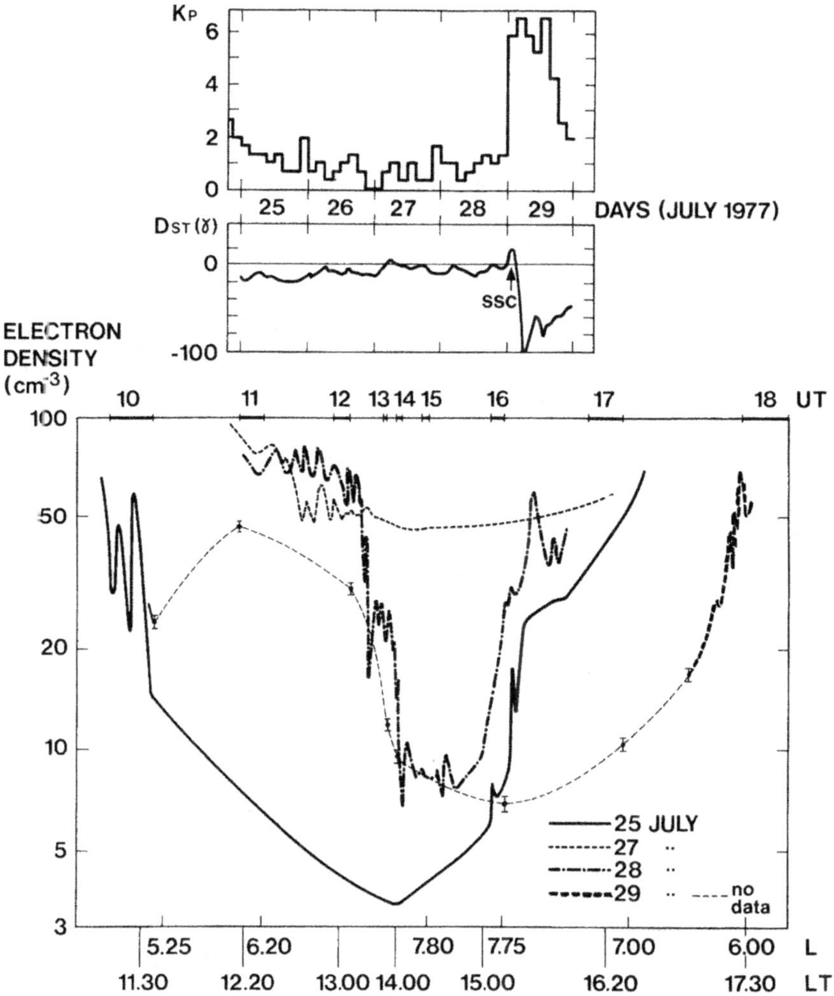

Fig. 8. The top frames represent the Kp and D_{st} index for the last days of July 1977. At the bottom of the figure, the density profiles corresponding to that same period are drawn. On July 29, from 10:00 UT to 17:30 UT we have only measurement every hour, a smooth interpolation curve between those points has been drawn. The error bars for that day are also adapted for the other profiles.

all the GEOS-1 orbit, (cf. Figure 9) so that we should not see any plasmapause under those conditions. This is indeed the case on July 27, 1977 (the Kp index remained under 1+ during July 26, 1977 and July 27, 1977). Looking at the density profile represented on Figure 8 we see that the density remains above 20 cm^{-3} all day long and no plasmapause is readily seen. On the other hand during disturbed conditions the pattern of the equipotential lines changes (Figure 9) and the last closed equipotential (corresponding theoretically to the plasmapause) is much closer to the earth. Then GEOS, along its orbit from around 10 UT till 18 UT, should stay outside the plasmasphere. If we consider the profile

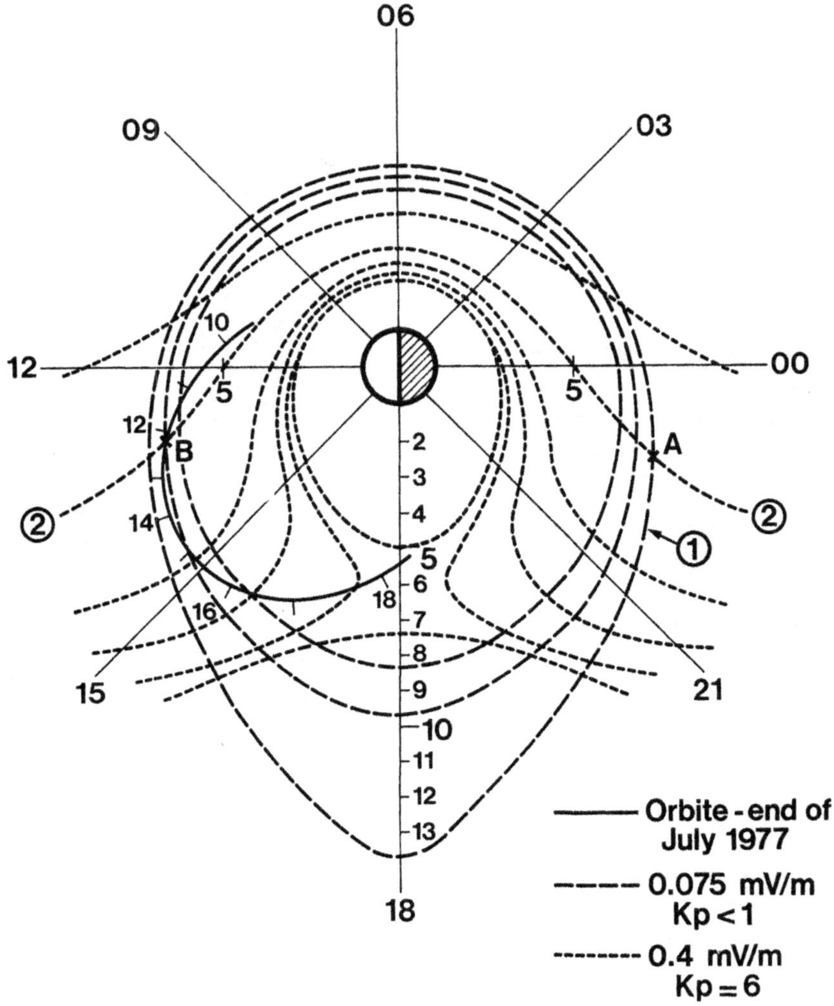

Fig. 9. Equipotentials of the global magnetospheric electric field following Kavanagh-Kivelson model (see text) for quiet conditions (main dashed lines) and disturbed conditions (thin dashed lines). The orbite of GEOS-1 corresponding to the end of July 1977 is also drawn (solid line) the figures along the orbit represent universal time.

of July 29 on Figure 8 we still see relatively dense zone of plasma up to 12:30 UT. Around apogee the density is roughly 7 times less than on day 27, then around 18:00 UT the plasmapause gradient appears. We are now going to present a tentative explanation for the high plasma density observed before apogee. The tubes of force concerned are located at point *B* of Figure 9 at 12:30 UT on day 29, before that, they drifted along line 2 of Figure 9 from point *A* which is located at 00:27 UT, time of the SSC. At that point before the SSC the plasma was drifting along line 1 (corresponding to quiet conditions) the sudden increase in the convection field after the SSC forced the plasma to drift along line 2. That is why we still see much plasma at 12:30 UT on day 29, it is a remnant of the past magnetic conditions. If the conditions remain steadily disturbed, then the whole pass of the satellite remains in a plasma-depleted zone as it is the case for the profile shown at the bottom of Figure 7, concerning September 20, 1977. Then the *Kp* index had been above 4 for about 24 hr.

On July 29, after 12:30 UT the profile exhibits a sharp plasma depletion which was already present the day before July 28, due to some magnetic activity during the night 27–28 July.

The density profile of July 25 is, like the one displayed on Figure 6 concerning September 17, 1977, typical of unsettled magnetic conditions. It is surprising to get such a good agreement between the data and the crude theoretical model we used, which can account for our observations on the dayside even though it is usually considered that this part of the plasmapause is irregular and difficult to relate precisely to *Kp* (Maynard and Grebowsky, 1977). It is probably because a detailed study is needed, the situation being too complex for systematic studies.

5. Conclusion

The sounder experiment provides very accurate and reliable measurement of the electron density and magnetic field strength. It thus permits detailed studies of wave propagation and amplification mechanisms, also it can be used for studies of magnetospheric plasmas. Our results allow us to confirm that the night side sector responds immediately to increased magnetic activity, the enchancement of the convection field digs out the plasmasphere so that the plasmapause gets much closer to the Earth (below $L = 4$). The behaviour of the dayside sector is delayed in time with respect to the onset of the perturbation. Any analysis of this sector must be linked to the conditions existing in the night sector. Most profile features are best understood by following the drift back in time. No statistical studies seem adequate. We intend to continue this kind of analysis, for GEOS-1 and GEOS-2. Also the filling of plasma from the ionosphere during the local morning hours is presently under study.

Acknowledgements

We wish to thank the Sussex University group for their help in providing Figures 3 and 4. We are also grateful to the CNES Division Mathematiques for their precious assistance in data handling. This work was performed under CNES contract No. 214.

References

Aubry, M., Bitoun, J., and Graff, Ph.: 1970, *Radio Sci.* **5**, 635.
Bitoun, J., Graff, Ph., Aubry, M.: 1970, *Radio Sci.* **5**, 1341.
Bitoun, J.: 1974, *Radio Sci.* **9**, 17.

Bitoun, J., Fleury, L., and Higel, B.: 1975, *Radio Sci.* **10,** 875.
Calvert, W. and Goe, G. B.: 1963, *J. Geophys. Res.*, **68,** 6113.
Calvert, W. and McAfee, J. R.: 1969, *Proc. Inst. Elec. Electron. Engrs.* **57,** 1089.
Carpenter, D. L.: 1966, *J. Geophys. Res.* **71,** 693.
Carpenter, D. L. and Park, C. G.: 1973, *Rev. Geophys. Space Phys.* **11,** 133.
Chappell, C. R., Harris, K. K., and Sharp, G. W.: 1971, *J. Geophys. Res.* **76,** 7632.
Chappell, C. R.: 1972, *Rev. Geophys. Space Phys.* **10,** 951.
Chappell, C. R.: 1974, *J. Geophys. Res.* **79,** 1861.
Crawford, F. W.: 1965, *Radio Sci.* **69D,** 789.
Decreau, P., Etcheto, J., Knott, K., Pedersen, A., Wrenn, G. L., and Young, D. T.: 1978, *Space Sci. Rev.* **22,** 633.
Deering, W. D. and Fejer, J. A.: 1965, *Phys. Fluids* **8,** 2066.
Dougherty, J. P. and Monaghan, J. J.: 1965, *Proc. Roy. Soc. London* **A289,** 214.
Dougherty, J. P.: 1969, in J. O. Thomas and B. J. Landmark (eds.), *Plasma Waves in Space and Laboratory*, Vol. 1, Edinburgh University Press, Edinburgh, PR 83–95.
Etcheto, J.: 1976, in K. Knott and B. Battrick (eds.), *Scientific Satellite Program during the International Magnetospheric Study*, D. Reidel Publ, Co., Dordrecht, pp. 87–97.
Etcheto, J. and Petit, M.: 1977, *Compt. Rend. Acad. Sci.* **285,** 329.
Fejer, J. A. and Calvert, W.: 1964, *J. Geophys. Res.* **69,** 5049.
Fejer, J. A. and Yu, W. M.: 1970, *J. Geophys. Res.* **75,** 1919.
Graff, Ph.: 1971, *J. Geophys. Res.* **76,** 1060.
Gringauz, K. I. and Bezrukikh, V. V.: 1975, Rep. D217, Space Res. Inst., Acad. Sci. of the U.S.S.R., Moscow.
Heikkila, W. J., Fejer, J. A., Hugill, J., and Calvert, W.: 1967, *Space Res.* **VII,** 395.
Higel, B., Bitoun, J., and Petit, M.: 1972, *J. Geophys. Res.* **77,** 6254.
Higel, B.: 1978, *Space Sci. Rev.* **22,** 611.
Jones, D.: 1978, *Space Sci. Rev.* **22,** 327.
Kavanagh, L. D. Jr., Freeman, J. W. Jr., and Chen, A. J.: 1960, *J. Geophys. Res.* 5511.
Kivelson, M. G.: 1976, *Rev. Geophys. Space Phys.* **14,** 189.
Lockwood, G. E. K.: 1963, *Can. J. Phys.* **41,** 190.
McAfee, J. R.: 1960, *J. Geophys. Res.* **73,** 5577.
McAfee, J. R.: 1969, *J. Geophys. Res.* **74,** 802.
McAfee, J. R., Thompson, T. L., Calvert, W., and Warnock, J. M.: 1972, *J. Geophys. Res.* **77,** 5542.
McAfee, J. R.: 1973, *Fund. Cosmic Phys.* **1,** 71.
Maynard, N. C. and Grebowski, J. M.: 1977, *J. Geophys. Res.* **82,** 1591.
Shkarofsky, I. P.: 1960, *J. Geophys. Res.* **73,** 4859.
Warren, E. S. and Hagg, E. L.: 1960, *Nature* **220,** 466.

SMALL SCALE STRUCTURE OF MAGNETOSPHERIC ELECTRON DENSITY THROUGH ON-LINE TRACKING OF PLASMA RESONANCES

B. HIGEL

CNET/CRPE ETE, Issy-les-Moulineaux, France

Abstract. The plasma resonance phenomena observed at f_{pe}, nf_{ce}, and f_{qn} by the GEOS-1 S-301 relaxation sounder are identified through a pattern recognition software process implemented in a mini-computer which receives on-line the compressed data. First, this processing system distributes in real time f_{pe} and f_{ce} measurements to the ground media. Second, it drives and controls automatically the S-301 on-board experiment by sending appropriate telecommands: the tracking of resonances is performed by shortening the frequency sweeps to a narrow range centered on the resonance location. Examples of such tracking sequences are presented, exhibiting sampling rates of the electron density measurements from once every 22 s (slowest rate) to once every 86 ms (highest rate available). The results give evidence of the existence of very small scale structures in the magnetospheric density, having characteristic sizes of the order of a few 10^2 m or/and a few 10^{-1} s. The relative amplitude of these density fluctuations is typically 1%. Because of satellite spinning, fixed frequency sounding sequences allow to measure in a few seconds the directivity features of the plasma resonance signals. Examples of directional patterns in the plane perpendicular to the geomagnetic field are presented: the electrostatic nature of the waves received at f_{pe}, nf_{ce}, and f_{qn} being consistent with these patterns, the corresponding **k** vector orientations become available. The Bernstein modes properties are used to interpret the nf_{ce} and f_{qn} results.

1. Introduction

The basic objective of a relaxation sounding experiment is to observe the waves excited in space plasma by an HF pulse transmitted in-situ. At some frequencies (f_{pe}, f_{ce}, f_{uh}, nf_{ce}, f_{qn}) occurs a resonance phenomenon: it gives rise to the reception of a relaxation signal, the record of which allows to perform a diagnosis of the plasma and to study the propagation of the related waves. This paper presents the first results obtained from the application to the GEOS satellites of a new philosophy of relaxation sounding. This consists of remote driving the on-board sounder from a ground based computer: the computer processes on-line the telemetered relaxation data, and feeds back to the sounder by building up telecommands to optimalize the observation of the resonance phenomena.

Section 2 describes the set-up of this entirely automatized, closed loop system, the time constant of which is a few seconds only. Section 3 shows examples of supplying, as first output, instantaneous measurements of the magnetospheric electron density N_e and of the magnetic field strength B_0. Section 4 presents the results obtained from the tracking of the resonance phenomena: first, the existence of very small scale fluctuations of the electron density is pointed out, and secondly the directivity features of the electrostatic waves relevant to the Bernstein modes of propagation are measured.

Preliminary, it is useful to review briefly the performance of previous relaxation

Space Science Reviews **22** (1978) 611–631.

sounding experiments, whose achievement has led to the present one. The plasma resonance phenomena have been discovered by the ionospheric topside sounding experiments on board the Alouette, ISIS and Explorer-20 satellites (see PIEEE, 1969). The resonances occur mainly near the characteristic frequencies of the plasma (electron plasma frequency f_{pe}, electrongyrofrequency f_{ce} and its integer harmonics nf_{ce}, upper hybrid frequency f_{uh}) as relaxation, long duration signals (many milliseconds) which are received immediately after the transmission of a short (a few 10 μs) RF sounding pulse. Able to provide with in-situ measurements of the electron density N_e through f_{pe} and of the Earth's magnetic field intensity B_0 through nf_{ce} (Lockwood, 1963; Calvert and Goe, 1963), the plasma resonances were also an useful phenomenon for wave propagation studies. The experimental and theoretical works performed during the first age of relaxation sounding have been reviewed by Calvert (1969), Calvert and McAfee (1969), Dougherty and Watson (1971), Graff (1971), Muldrew (1972), McAfee (1974). However, the topside sounders, designed as radar devices, presented some limitations in recording the plasma resonance signals: the dynamical range was not matched to the resonance levels and the spectral information carried by the signals was not preservated. These lacks have been filled by a second generation of so-called relaxation sounding rocket-borne experiments (Higel *et al.*, 1972; McAfee *et al.*, 1972): these sounders were stepwise frequency swept devices, outputing the waveform of the resonance signals in a large dynamical range ($\simeq$80 dB). Such improvements allowed some progress in plasma resonances study: the coherence property of the resonance signals was proofed (Higel, 1975), the nf_{ce} ($n \geqslant 2$) resonances were interpreted as Bernstein waves driven by the magnetic field gradient (Higel *et al.*, 1972; Bitoun *et al.*, 1975), the first study of the f_{ce} special case was performed (Higel and de Feraudy, 1977). Moreover, the diagnosis capabilities were improved by measuring through the f_{pe} resonances the small scale irregularities (or fluctuations) of the electron density (Higel, 1978).

However the fixed mode of sounding which was used in these experiments (repetitive sweeps of the whole frequency range, constant listening time) did not allow to answer, among others, the following questions: what is the actual spectrum of the electron density fluctuations, possibly undersampled by the constant, arbitrarily fixed sweep rate? How to profit by the whole information carried by a resonance signal, the duration of which can be longer as the fixed listening time? How to determine experimentally the influence on the resonance phenomenon of a given parameter, such as for example the antenna orientation, since the recording of a data set distributed over 360° requires so much times that other involved parameters have varied? Generally speaking, how to match the relaxation sounding tool to the various plasma conditions encountered in the space plasma?

An attempt has been made on the GEOS satellites to answer these questions, by designing a *robotic* conception of relaxation sounding. Its set-up will be now described, and then its first results.

2. Experimental Set-Up

The design of the GEOS relaxation sounder (experiment S-301, part of the S-300 wave-experiment described by Jones (1978) and by S-300 Experimenters (1978)) allows to select by telecommand all the sounding parameters: frequency range, sweep rate, listening time, transmitted pulse power level, dynamical range of the receiver. Then, a ground based computer has been dedicated to S-301 in order to process on-line the data and to feed back by carrying out various telecommand strategies accorded to the set of scientific objectives. The computer interacts permanently with the on-board experiment and performs a real-time control.

Figure 1 displays a schematic diagram of the system: a chain of computers is set in closed loop together with the satellite. This loop is covered along time successive segments numbered from 1 to 10. The relaxation sounding data flow through the high speed (HS) telemetry channel (segments Nr. 1–2). In routine mode, one sounding step takes 86 ms (3 for transmission, 83 for reception). One set of relaxation data consists of the 4 × 128 numerical samples picked up on the waveform output from the four gain stepped swept frequency analysers (SFA). After decommutation in HS1, the data are transmitted to HS2 (segment Nr. 3) which performs a two-step processing. First, the best level-matched SFA is selected through a non-saturation criterion. Secondly, the data of the selected SFA are compressed through the computation of 6 quantities describing the relaxation signal carried: the mean level from each of three successive time slices, the frequency, power and bandwidth of the peak issued from a FFT-processing performed on the whole received signal. Every 5.5 s, the HS2 computer sends

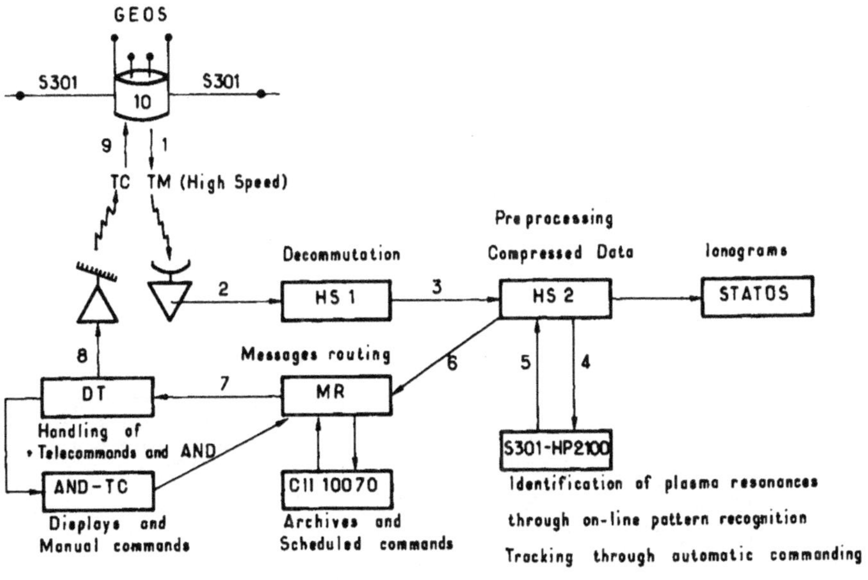

Fig. 1. Real-time control of the S-301 relaxation sounding experiment. Block-diagram of the closed loop system. The various computers are labelled by their mnemotechnic ESOC-code.

(segment 4) to the dedicated S-301 computer (labelled S-301-HP2100) a message containing 64 sets of 6 compressed data (i.e. the data from the last 64 sounding steps) plus housekeeping and experiment-status data. The S-301 computer answers immediately (segment 5) by a message containing f_{pe} and f_{ce} measurements plus telecommands for the on-board sounder. Of course, this message results from the processing performed on the previous messages sent by HS2. The segments Nr. 6–7–8 show the way for this feedback. The $f_{pe} - f_{ce}$ measurements are passed to the ground-based system for various purposes: on-line display, archiving together with the compressed data, input parameters for other running experiments.

The telecommands are sent to the on-board S-301 experiment (segment Nr. 9); their execution (segment Nr. 10) modifies the sounding parameters for the further data, and so on. The minimum time delay between an on-board event and the again on-board execution of the related feedback (closed loop delay) is 6.96 s, i.e. a little bit higher than one spin period ($\simeq 6$ s).

Figure 2 shows how the computer, by driving the experiment, allows the automatic tracking of plasma resonances. Indeed, one of the main purposes is to obtain an optimal sampling of the fast fluctuations of the magnetospheric plasma (Higel, 1978). The sounding sequence is schematized (Figure 2, top) by its frequency sweeping (in actual fact not a continuous but a stepwise one). The time unit of the horizontal axis corresponds to the rate at which messages are exchanged between the experiment and the S-301 computer (once per 5.5 s). Together with the sounding run is written versus time (Figure 2, below) the

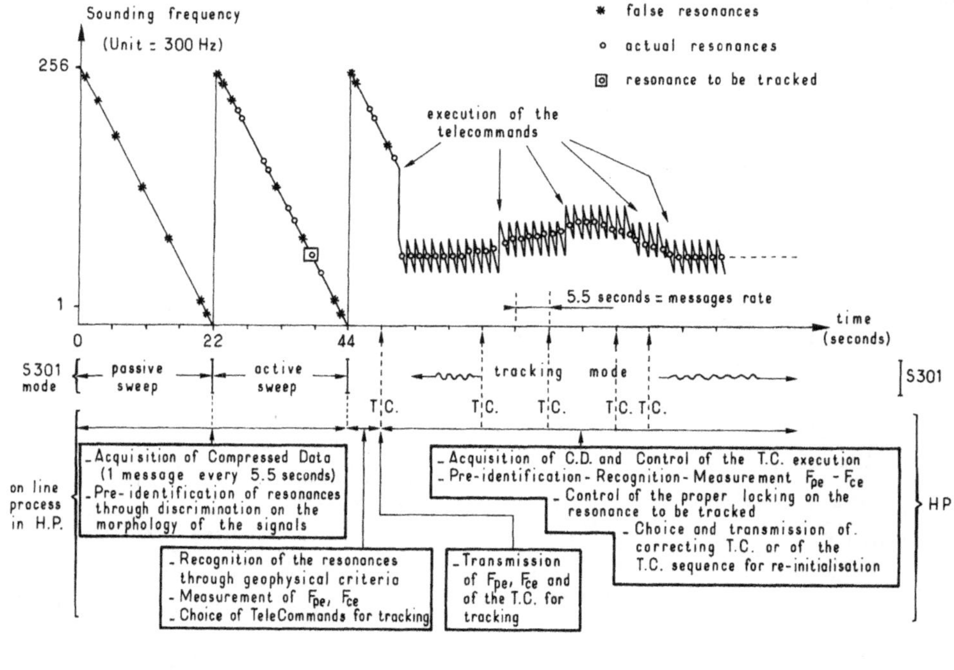

Fig. 2. Schematical display of the realisation of a tracking sequence versus time.

outline of the pattern recognition process performed in the computer. This is a numerical process which is grounded on two independent, time successive methods: first a discriminant analysis is performed to select among the observed events a lot of assumed resonances (Higel, 1975), some of the decision criteria being built up through a self learning process; then the selected resonance events are subjected to a bank of geophysical criteria for identification purpose: each submitted event is either rejected or labelled as f_{pe}, nf_{ce}, f_{uh}, f_{qn} resonance. This applies to the routine mode (whole passive and then active sweeps) set by the external prescheduled commands (see Figure 1). When the resonance to be tracked becomes identified, the S-301 computer takes the lead on the experiment by sending the tracking appropriate telecommands, and keeps it hereafter to control the proper locking. If the locking is lost, the computer reconfigurates itself the experiment in routine mode through the suitable commands, and tries again to track. A more detailed description of the computational strategy would lead too far from the present aim. One may however mention that the soft-ware program is implemented in a Hewlett-Packard computer (HP2100S, core 32k − 16 bit words) and that it uses about 3000 FORTRAN IV instructions without requiring manual intervention for 24 hr per day operations.

Figure 3 shows a real-time ionogram-display of the plasma resonance phenomena observed at the beginning of a tracking sequence. This display is to be compared with Figure 2: in addition to the frequency sweeping versus time (Figure 3, below), is plotted at each sounding frequency the intensity (as darkness) of the relaxation signal versus listening time (vertical axis). Figures 2 and 3 show how the sampling of the plasma characteristics is improved by the tracking performance: the observing rate of a given resonance, which is once every 22s with a sweep range Δf equal to 256 steps as in routine mode, can be reduced to faster values according to the smaller sweep range telecommanded. The values of Δf can be 256, 128, 64, 32, 16, 8, 4, 2, or 1 step, this last value (fixed frequency sounding) providing with a sampling rate as fast as once per 0.086 s. The intermediate values allow for example either to study the large fluctuations of the frequency location of a unique resonance, or to focus on a set of neighbouring resonances with nearly simultaneous observations at a high sampling rate. Thus, correlative studies can be done on several resonances, for instance in the case of small fluctuations. A great variety of scientific strategies can be implemented in the driving software, as suggested by the range of sounding parameters given in Table I. The results presented below have been supplied by this computer-aided relaxation sounding experiment on GEOS-1.

3. Magnetospheric Plasma Diagnosis in Real Time

The first output of the on-line control is a real-time diagnosis of the magnetospheric cold plasma density N_e and of the static magnetic field strength B_0 through the recognition of the f_{pe} and nf_{ce} resonances. As the pattern recognition process

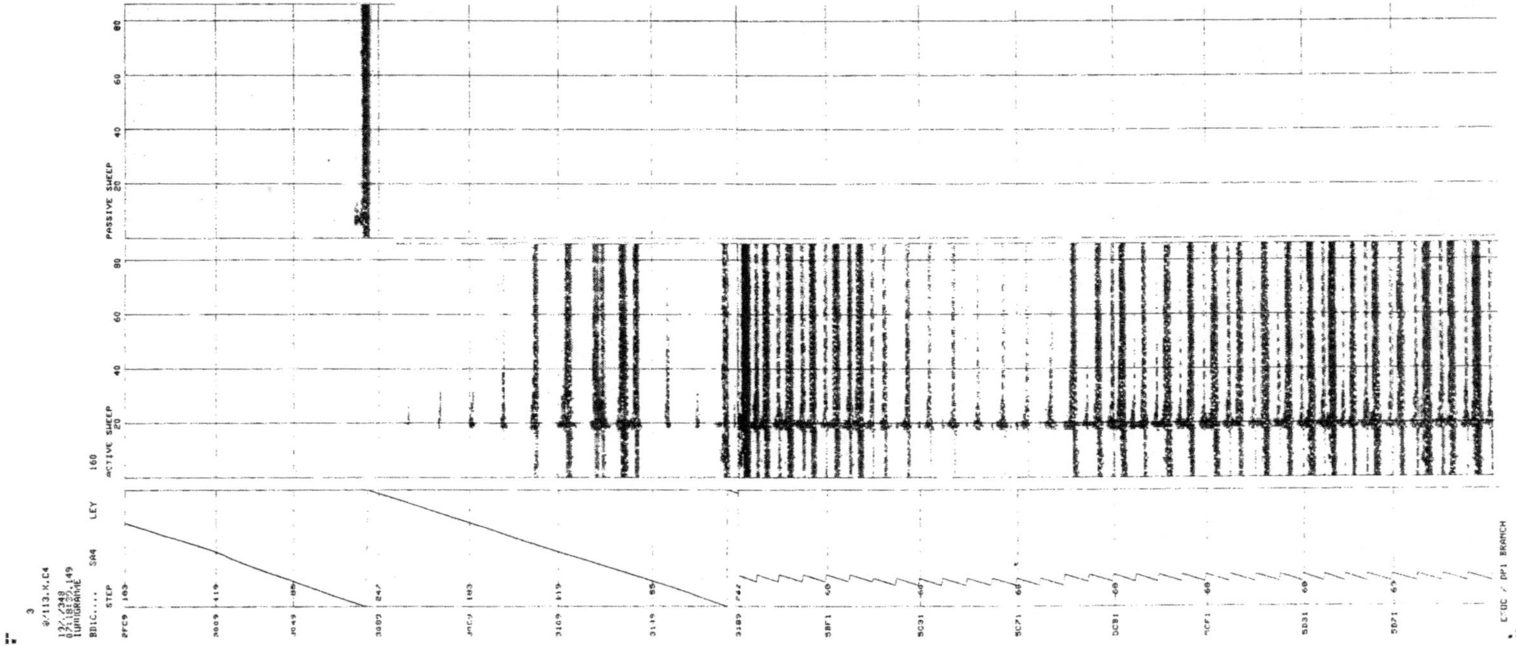

Fig. 3. Ionogram display performed during the realisation of a tracking sequence, to be compared with Figure 2. The frequency sweeping versus time (long range time) is displayed in the bottom part. During the second half, a tracking is performed. The top and the medium parts show the resonance signals received versus listening time (short range time, vertical axis) respectively in the passive and active modes.

TABLE I

Sounding parameter	Label	Unit	Values available by telecommand	Remarks
Sweep extent	ΔF	step	1 − 2 − 4 − 8 − 16 − 32 − 64 − 128 − 256	1 step ≃ 300 Hz
Lower frequency	FMIN	step	1 ≤ FMIN ≤ 256	1 step ≃ 300 Hz
Listening time	CLOC	HSF	1 − 8 − 32 − + SPIN PERIOD (≃ 6 s)	1 HSF = 86×10^{-3} s
Transmitter	EM	–	ON − OFF	pulse duration: 3×10^{-3} s
Transmitted level	LEVEL	V rms	HIGH = 100 − MED = 30 − LOW = 10	–

is generally performed in less than 5.5 s after the end of a whole sweep, this plasma diagnosis is available 27.5 s at the maximum after its actual achievement. Figure 4 displays for a GEOS-1 orbit a plot of N_e and f_{ce} measured in that manner. As the sounder remains always in routine mode (whole sweeps), the diagnosis updating rate is once every 22 s. During tracking mode (not performed here), this rate can become equal to the message rate, i.e. once every 5.5 s. The N_e and f_{ce} profiles shown by Figure 4 are typical of a quiet situation. The f_{ce} values differ from those given by the magnetometer (experiment S-331) which was unfortunately in failure during the second half of the GEOS-1 lifetime. The density measurement exhibits at first the ability of the relaxation sounding technique: values from 1 to 100 cm^{-3} are obtained here without lack. This completes in the lower range the previous diagnosis performed earlier for the higher range (10^3–10^6 cm^{-3}) in ionospheric plasmas (Higel, 1978). Secondly the results of Figure 4 demonstrate the reliability of the real time measurement process: an only very low percentage of the dots can be regarded as erroneous measurements. Most of them are a consequence of the high variability of the resonance levels, as shown in the next section: this variability is contrary to the statistical part of the pattern recognition, which may not be too permissive in selecting the events. The accuracy of one nf_{ce} location is about 30 Hz: the higher the n order, the higher the accuracy on B_0. As all the nf_{ce} resonances take part in the f_{ce} measurement, one may estimate roughly the accuracy on B_0 to a few $10^{-1}\gamma$.

Figure 5 shows an example of real-time measurements during the inbound crossing of a sharp density variation, which could be the plasma-pause or one of its stages. The density increases by a factor 3 in about 2 min, corresponding to 200 km along the orbit. Density oscillations of smaller amplitude with an apparent periodicity of about 3–4 min are superimposed on both sides of the transition region. This is a typical example of the medium scale density fluctuations observable in real time through the sounding experiment. Such density measurements are made with an accuracy better than 1% without fail. The occurrence of this kind of density perturbations, the interpretation of which is not undertaken here, has to be compared with similar events observed through a different technique (Kikuchi, 1976; Kivelson, 1976).

When tracking sequences are performed, the actual sampling rate of the

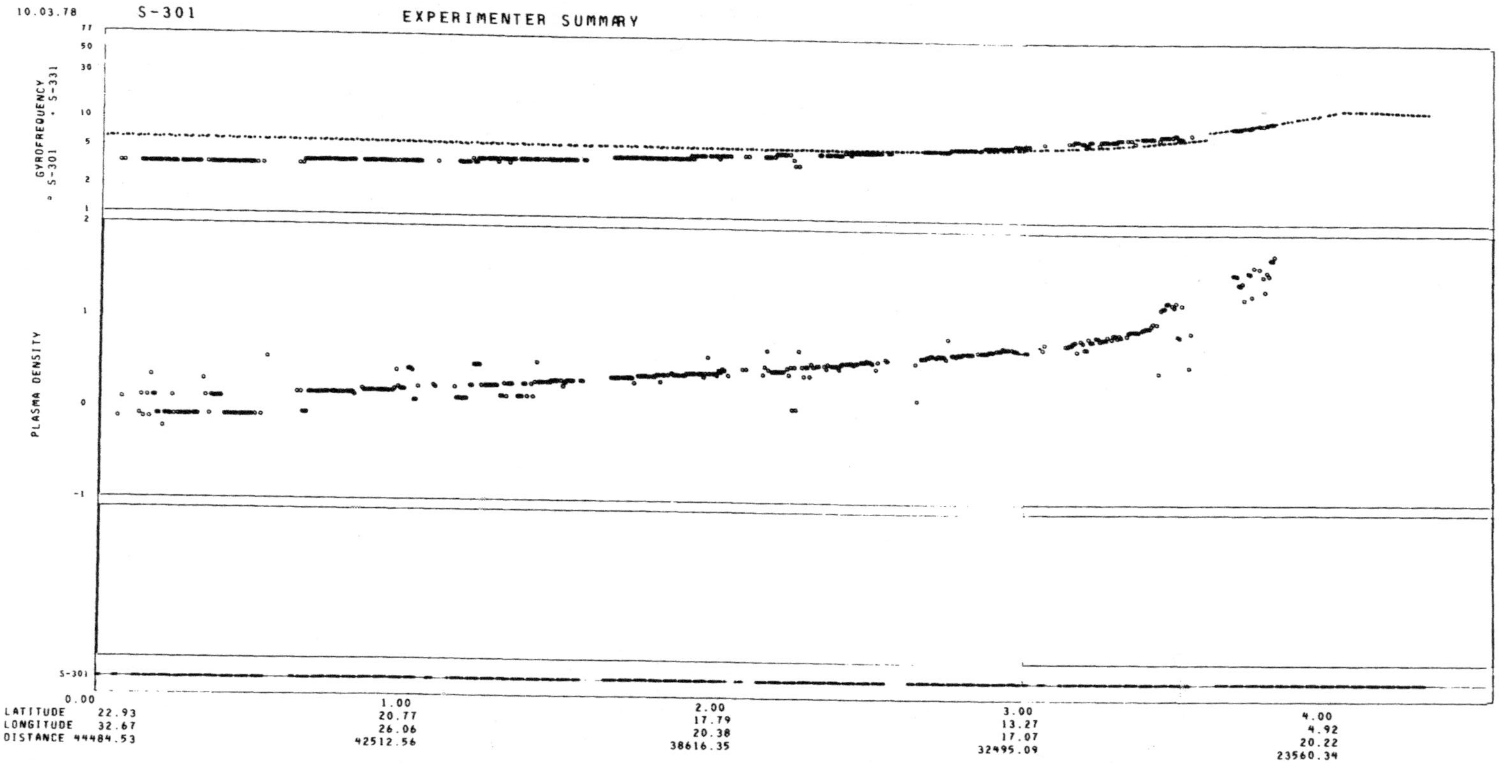

Fig. 4. The plasma density and gyrofrequency measurements performed in real-time are plotted versus Universal Time during half a GEOS-1 orbit (inbound). S-301 running time is given by the darkened line below. The density scale is a $\log_{10}$ one. The smaller dots of the gyrofrequency measurements are those given by the S-331 magnetometer (in failure at this period). The positions versus time are geographical ones.

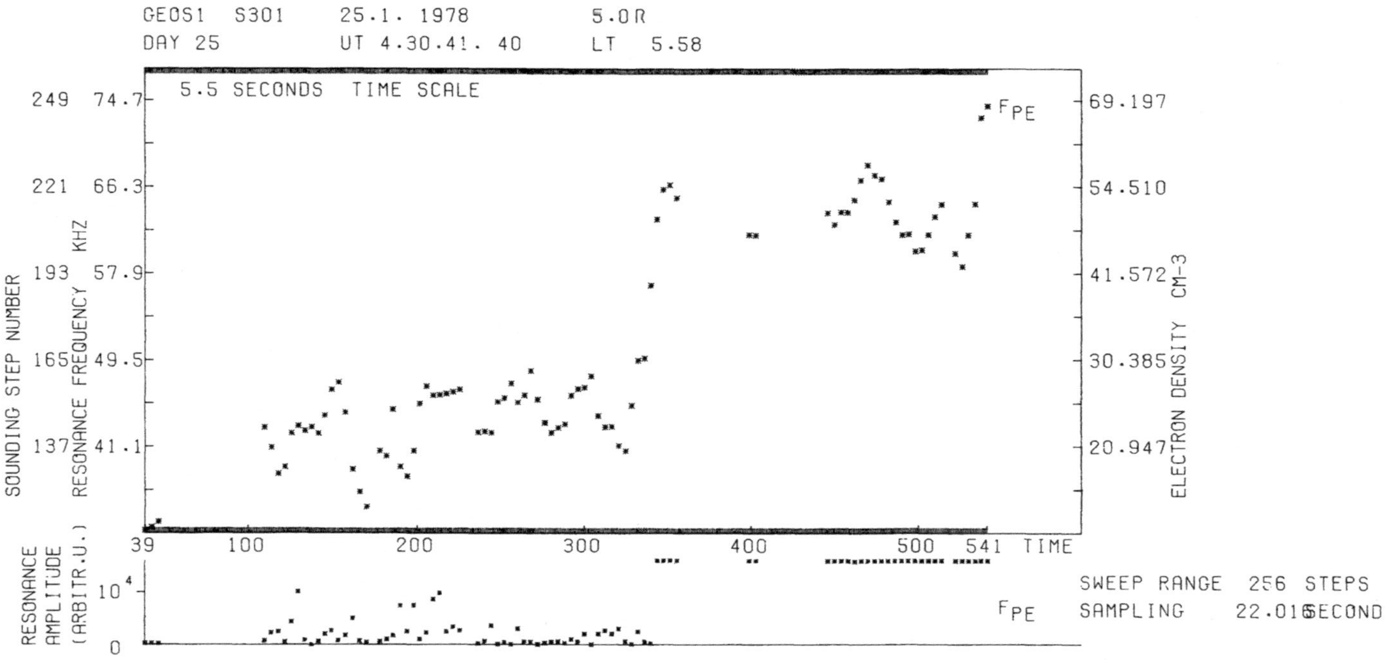

Fig. 5. The plasma frequency measured in real-time is plotted versus time during a plasmapause crossing. The time unit is the one of the messages rate, i.e. 5.5 s. The universal time (UT) given on the top is relevant to the first time unit plotted on the display (i.e. 39). The resonance amplitude plotted below with a linear scale is the spectral amplitude deduced from the compressed data (see text). The reference level for the arbitrary units is the lowest absolute spectral level measurable by the compressed data, i.e. the level 1 from the most sensitive SFA. The display shows resonance amplitudes which saturate the less sensitive SFA after the sharp increase of the density.

diagnosis exceeds the on-line facilities for displaying. Hence an off-line reprocessing of the tracking data is needed, leading to displays such as Figure 6. This tracking with a 8 steps frequency range allows to sample each one of the three resonances f_{pe}, $3f_{ce}$, and f_{q3} once every 0.688 s. A very smooth increase of a very low density can be observed in that case, with the maximum of resolution in frequency ($\simeq 12$ Hz) available from the GEOS relaxation sounder. This set of f_{pe} results demonstrates that density variations of about 0.003 cm^{-3} are detected in a

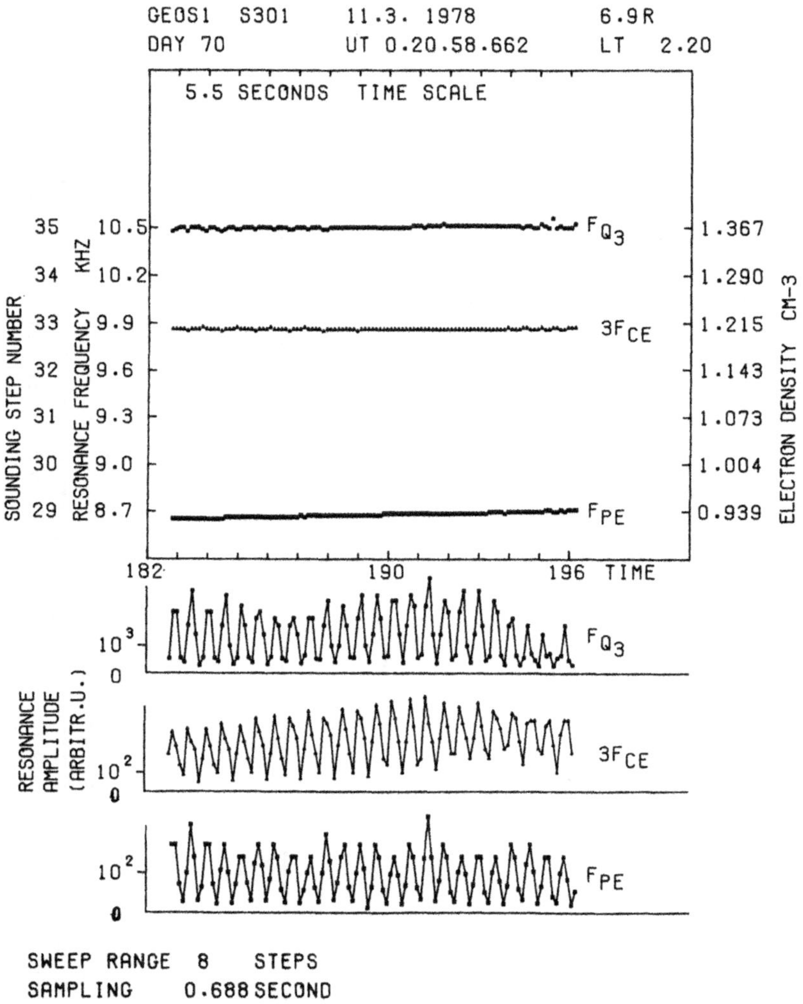

Fig. 6. The frequency location (top) and the amplitudes (below) of the resonances observed during a tracking sequence are plotted versus time through the off-line reprocessing of the tracking data. The sensitivity of the frequency scale equals 12 Hz. The amplitude scales are linear ones, homogeneous together and referenced to the lowest level available from the sounder (ref. to the commentary of Figure 5). The time unit is equal to 5.5 s (messages rate). The universal time (UT) given in reference (top) is relevant to the first time unit plotted on the display (i.e. 182 here). Local time (LT) and geocentric distance (R = Earth radii) are given too.

medium containing less than 1 electron per cm^3. This steady state plasma situation allows to display obviously the satellite spin effect on the resonance signals (Figure 6, lower part). The amplitude levels obtained from the Fourier spectra exhibit a modulation at half the spin period ($\simeq 3$ s). Each one of the three different resonances is spin modulated, with a different phase (see the first and last dots of the plot) according to the spin phase along the frequency sweep. This directional pattern will be emphasized in the next section. Another spin effect appears in Figure 6 on the frequency dots of the $3f_{ce}$ and f_{q3} resonances: a very slightly shift of the frequency occurs at half the spin period, by about 10 Hz up and down in alternance for $3f_{ce}$, and a little bit more (10 to 20 Hz) always down for f_{q3}. The perceptibility of this effect, the interpretation of which is probably relevant to the wave propagation (see below), proofs again the improvement brought to the relaxation sounding by an automatized control.

4. Plasma Resonances Tracking in the Magnetosphere

The results presented now deal with two main objectives of the experiment: the measurement of the small scale fluctuations of the electron density, and the study of the directional patterns of plasma resonances.

4.1. *Small scale fluctuations of the magnetospheric density*

Figure 7 is a first example to show what can be discovered when the routine density measurement performed once every 22 s is replaced by a tracking mode. Underlying variations of the density are revealed by the f_{pe} plot: their relative intensity lies between 1 and 3% with time characteristics of the order of 10 s. The f_{q3} fluctuations in frequency location are a consequential effect of the f_{pe} ones: indeed, Crawford's (1965) results concerning the Bernstein modes involved in the f_q resonance mechanism point out the strong sensitivity of the dispersion curves to the ratio f_{pe}/f_{ce}. An unquestionable correlation between both the sets of fluctuations appears even in the small details, taking account of the small phase difference between both the sets of samples. Doppler shifts due to plasma drifting fluctuations cannot give rise to so large a frequency jitter, since rms amplitudes greater than 10 km s^{-1} would be required according to the small **k** vector intensities (less than unity) of the waves. No relationship with the spin rate appears on the frequency versus time plot of Figure 7. Moreover the very monochromatic spin modulation of the resonance level stated from Figure 6 under stable conditions appears here very much disturbed in conjunction with the density fluctuations (see Figure 7, below). Hence any study of some directional patterns should be joint with a careful examination of the plasma density conditions.

More faster fluctuations of the density are shown in Figure 8. The maximum sampling rate is performed through fixed frequency sounding at f_{pe}, i.e. once every

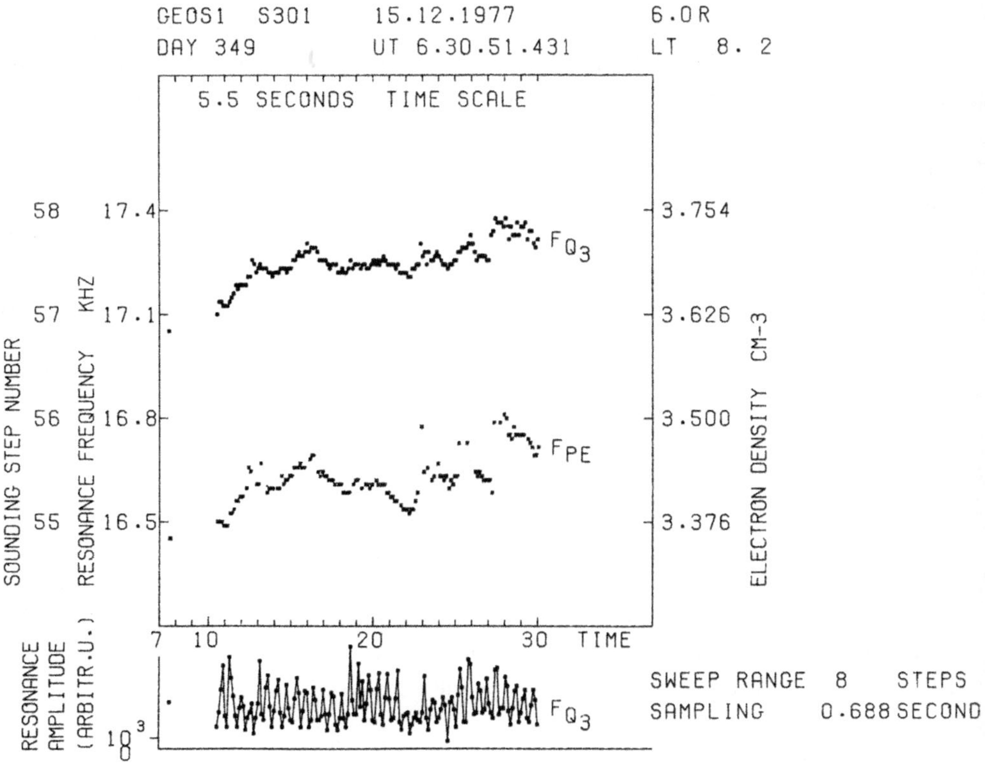

Fig. 7. Idem Figure 6. Typical example of correlated density fluctuations on f_{pe} and f_{q3} resonances. The initial dots near time = 7 are the measurements performed in routine mode. During the time gap until the next dots, the routine mode is finished off and the tracking is set up (see Figures 2 and 3).

86 ms (or every 150 m). The density varies of 1 to 2% within 1 or 2 s. The variations are continuous rather than breakwise ones. The Fourier spectra of the resonance signals, not displayed here, do not exhibit specially large values of their 3 dB-bandwidth, always less than 6×10^2 Hz. Relatively, such a bandwidth is lower than for similar kinds of observations in the ionosphere (Higel, 1978). This might indicate the existence of a stronger damping of the waves which have to travel farther (or longer) in order to be reflected back by higher electron densities. Density fluctuations like those presented here often occur on GEOS-1 data, in alternance with very stronger variations or with steady situations as displayed in Section 3. No result of a systematical study of occurrence versus magnetospheric parameters is now available. Such a work will be much more significant with the geostationary GEO-2. It must be reminded that any detached measurement of the density is sightless on such a phenomenon. The same is true about the amplitude resonance level (Figure 8, below), which exhibits, superimposed to the spin modulation, some strong disturbances linked to the density fluctuations.

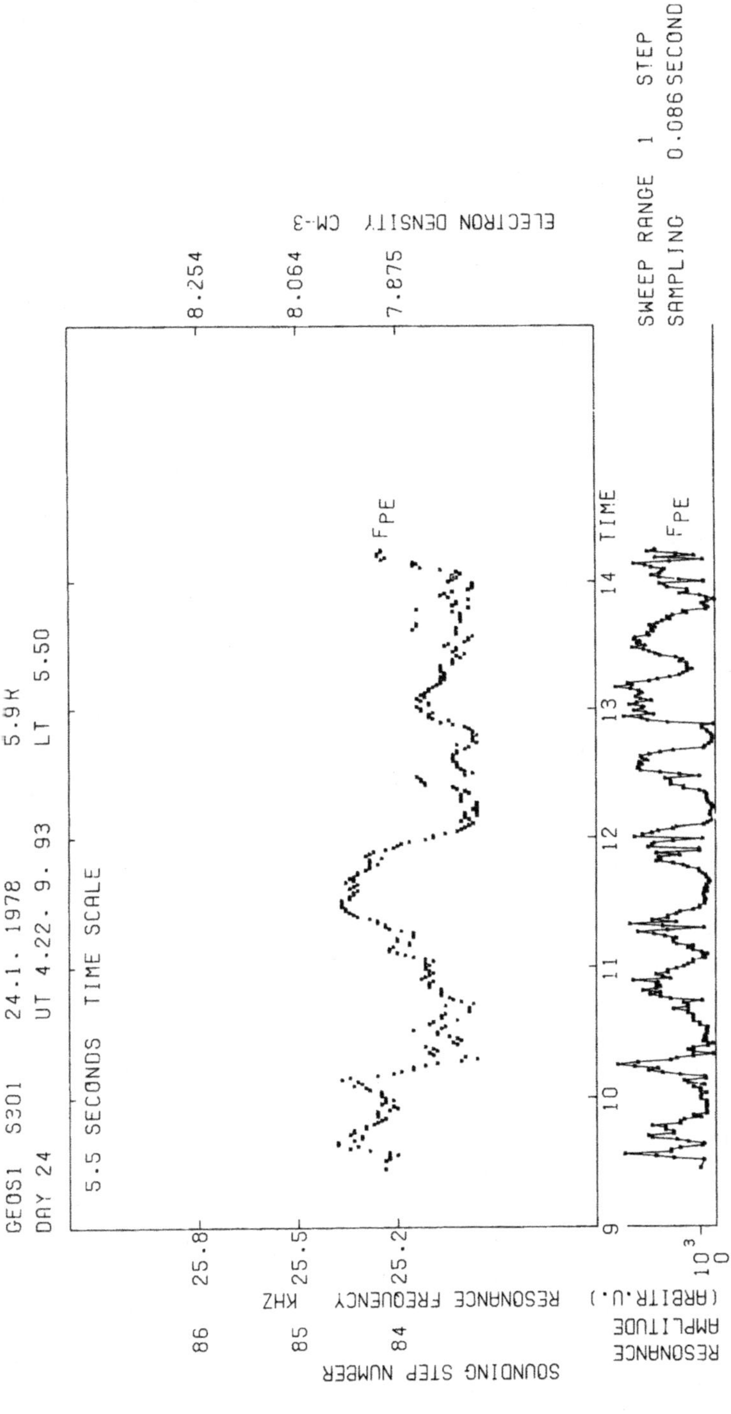

Fig. 8. Idem Figure 6. Typical set of fixed frequency tracking data exhibiting the fastest density fluctuations. The satellite velocity is 1.5 km s^{-1}.

4.2. Directivity features of the resonance signals

Steady state plasma conditions have been selected to study the influence of the antenna orientation on the resonance levels. Figure 9 and 10 display two sets of fixed frequency tracking data at f_{q3} and $3f_{ce}$ with a gap of 40 s between both the sets. In Figure 9, the only parameter which varies versus time is the antenna orientation. The sudden increase of the level which occurs during the sequence corresponds to a shift up of the sounding frequency from step 45 to step 46, as requested by the on-line computer to obtain the better level of signal. Indeed the actual receiving bandwidth location with respect to the tuning frequency is [−300, +450] Hz, but a filtering is performed out of the [−150, +150] Hz band. On the right hand side of Figure 9, the rotation plan of the antenna is defined by the two cartesian axis. This plan is perpendicular to the satellite spin axis, and the spin axis is at this time parallel to the static Earth's magnetic field $\mathbf{B}_0$. A polar plot of the resonance level has been performed in this plan, the polar angle being equal to the angular phase of the rotating antenna. The projections of the Earth and Sun directions have been plotted too, allowing to display the geographical configuration. The directivity figure thus obtained provides with a determination of the **k** vector orientation of the f_{q3} waves. Indeed, if the propagation of Bernstein modes is assumed (see check hereafter), electrostatic waves having a **k** vector perpendicular to $\mathbf{B}_0$ are involved (Crawford, 1965).

The colinear AC electric field is measured by the receiving antenna (two pin-point captors 40 m away), and the preferential direction of reception provides with the azimutal orientation of the **k** vector. This direction forms here an angle of about 25° with the projected Earth's direction. The same kind of directivity measurement is displayed in Figure 10 at $3f_{ce}$. Some level discontinuities seem to occur in the level plot: it is an artefact due to the integer quantification of the compressed data used, joint with a low level of signal in the selected SFA. This situation has some influence on the apparent frequency stability of the $3f_{ce}$ resonance, because of the poor signal over noise ratio. A valuable directivity measurement is nevertheless available, which shows a **k** vector very close to the Earth's direction.

The earlier interpretation of the nf_{ce} resonances (Bitoun *et al.*, 1975) has shown that the **k** vector is perpendicular to $\mathbf{B}_0$ and lies in the geomagnetic local meridian plane. Then one may conclude from Figure 10 that it is still true in this magnetospheric $3f_{ce}$ case, and from Figure 9 that the **k** vectors of f_q waves may have another direction. Various hypotheses can be taken to start with an interpretation. One may add an extra-one, suggested by the strong coupling shown in Figure 7 between the density fluctuations and the f_q frequency location: the orientation of the density gradient, which is not necessarily colinear to the magnetic field one, can take a noticeable part to select the **k** vectors which give rise to the f_q resonance phenomenon. This assumption is to be examined in connection with Crawford's formulation. The angular selectivity of the plasma resonance can lead to dynamical variations up to 20 dB, as in Figure 8. This could

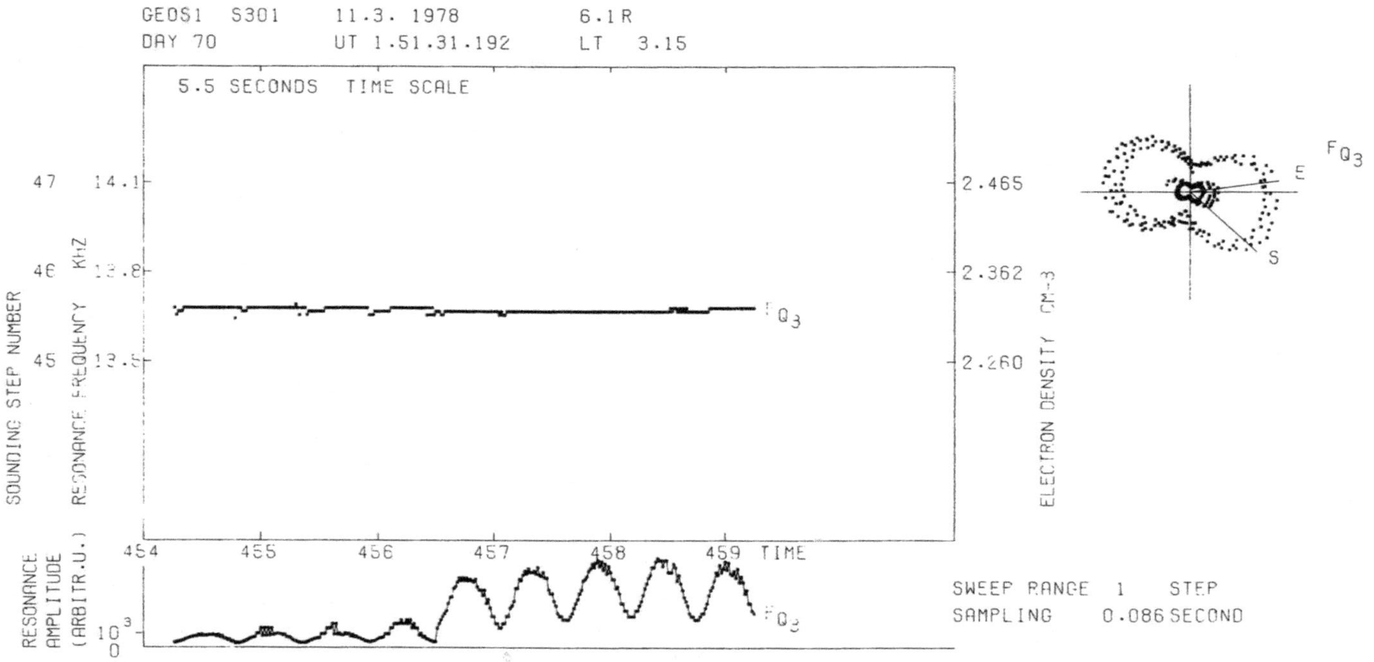

Fig. 9. Idem Figure 6. Fixed frequency tracking in steady state plasma conditions. The sudden increase of the resonance level results from the frequency matching correction order sent by the computer (see text). The polar plot of the resonance level (right hand side) results from the restitution of the dynamics of the antenna in the plan perpendicualr to the spin axis. *E* and S indicates respectively the Earth and Sun directions projected in this plan.

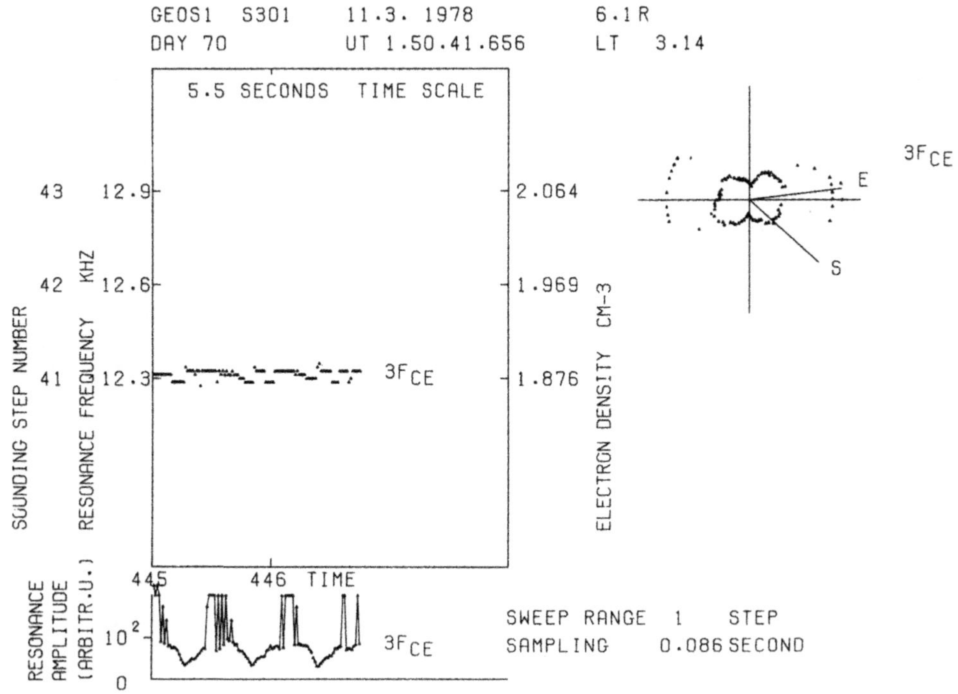

Fig. 10. Idem Figure 9. The discontinuities on the resonance level versus time result from the integer quantification of the compressed data in a case for which the level is on the lower part of the dynamical range of the relevant SFA. Same remark for the polar plot.

explain some apparent variability of signals, even natural ones, received on-board GEOS, as for example the variability pointed out by Christiansen *et al.* (1978) (see Figures 3a and 3b of Christiansen's paper).

Last, one hypothesis made above remains to be checked, concerning the electrostatic nature of the waves received. For that purpose, the small frequency shift, spin modulated, seen in Figures 6 and 9 will be used in a self consistent manner. If electrostatic waves are received, they are Doppler shifted, and the frequency shift is equal to $-(\mathbf{k} \cdot \mathbf{V}_s)/2\pi$ where $\mathbf{V}_s$ is the satellite velocity vector with respect to the plasma. This shift appears on the Fourier spectra of the signals only by antenna orientations at which sufficiently strong and long duration signals are received. By unfavourable antenna orientations (i.e. at the minima of the signal level) the received signal is of course not due to the same waves. Hence the frequency shift observed between the two different angular situation may be an approach to the amplitude of the Doppler shift, and its sign may indicate the respective orientations of $\mathbf{k}$ and $\mathbf{V}_s$. The shift is about 20–30 Hz for f_{q3} (see Figures 6 and 9). Disregarding plasma drift, a satellite velocity V_s of 1.5 km s^{-1} leads roughly to $|\mathbf{k}|$ values of about 10^{-1} m^{-1}. This result agrees well with the theoretical value deduced from numerical calculations of the Bernstein's f_q waves with the assumption of an 1 eV cold plasma (de Feraudy and Hamelin,

1978). The corresponding wavelength (50 to 100 m) explains well the shape of the directivity polar plot in Figure 9, since the antenna length is 40 m tip to tip.

4.3. *Multi-variabilities of the magnetospheric plasma resonance phenomena*

The results presented above have shown that the study of the plasma resonance phenomena in the magnetosphere should be made by paying attention to the lot of parameters which may affect the observation of a unique phenomenon, since some of these parameters exhibit very fast variations, as shown on the density results. The purpose is now to illustrate this statement by an example of resonance observation during which a high variability is observed for a lot of parameters or of events.

Figure 11 displays a fairly long tracking sequence (about 12 min) which can be regarded as typical of the variability of plasma resonance phenomena. Three resonances are simultaneously tracked: f_{pe}, and, thanks to a well splitting effect, f_{q3} and $f_{q'3}$. Each one encounters its own history versus time and is subject to strong modifications of its characteristics. F_{pe} increases at first monotonically and shows a wavelike pattern which becomes evanescent. At this place, there are some antenna orientations for which the signal level becomes suddenly stronger. Near the time 90, the increase stops. Some periodic fast oscillations occur for 30 s, and disappears, together with a decrease of the signal level. F_{q3} displays with some damping the f_{pe} fluctuations, but its level, 30 dB higher than the f_{pe} one, decreases slowly versus time. The $f_{q'3}$, which is one of the interests of this sequence, is very low and sometimes not observed at the beginning. Its frequency location displays the f_{pe} fluctuations but more damped than f_{q3}. The gap in frequency between f_{q3} and $f_{q'3}$ remains constant until the time 90, while the $f_{q'3}$ level begins to increase well before f_{q3} and $f_{q'3}$ become closer. After the time 110, $f_{q'3}$ fluctuates very fast in frequency and in level. The directivity diagrams provide with an original result concerning $f_{q'}$: whereas f_{q3} and f_{pe} have here the same preferential direction, the direction of $f_{q'3}$, at least for the strong signals, is exactly the perpendicular one. Figure 12 is the enlargement of a part of Figure 11, in order to show better the smooth wavy fluctuation of f_{pe}, the fluctuation of the f_{q3} frequency, and the respective orientations of the f_{q3} and $f_{q'3}$ directional features. Low $f_{q'3}$ signals reveal also a slight preference for a direction perpendicular to the f_{q3} one. Further investigations are of course required to interpret this directivity feature of the $f_{q'}$ resonance.

5. Conclusion

The achievement of an automatized real-time control of the S-301 relaxation sounding experiment on-board GEOS-1 has greatly improved the efficiency of the plasma resonance study in the magnetosphere. The spatial or temporal structures of the cold plasma electron density and of the magnetic field strength

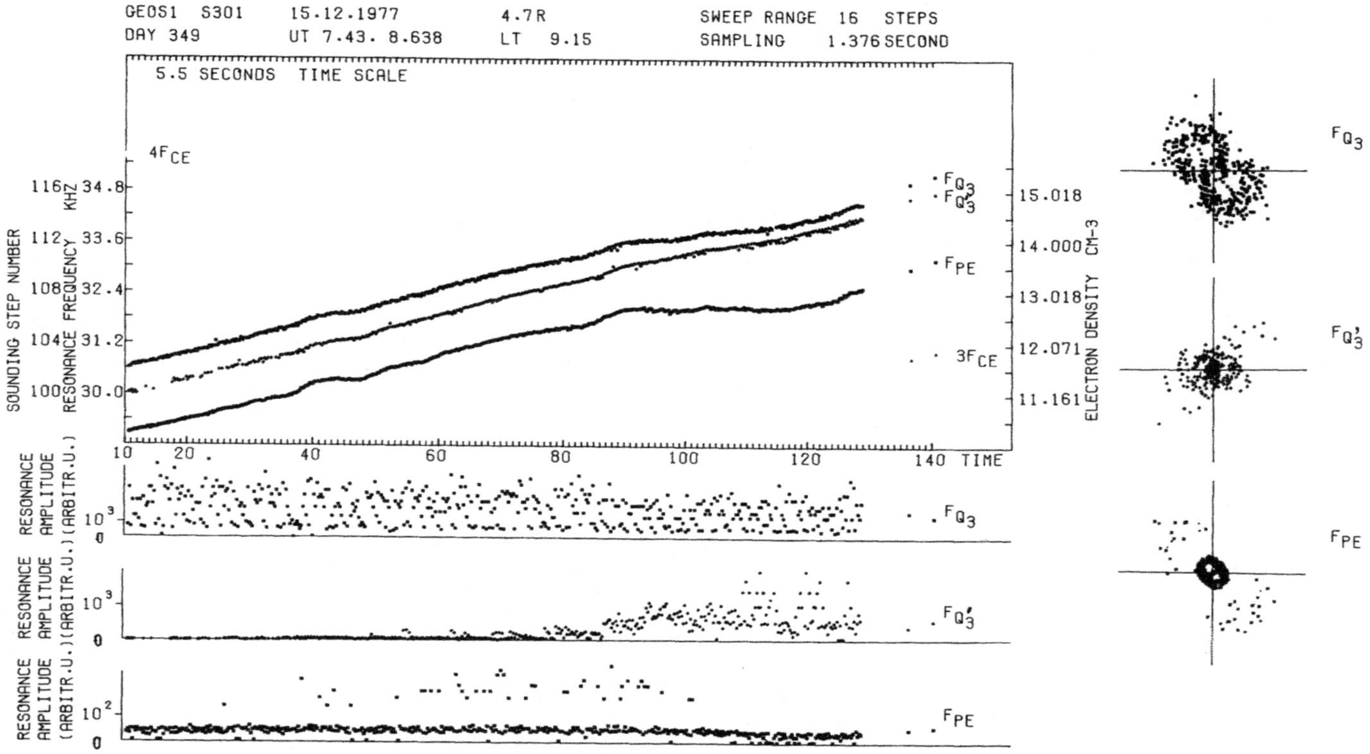

Fig. 11. Idem Figure 6. Long duration tracking sequence to illustrate the variability of the plasma resonance phenomena. The $f_q{}'$ resonance is to be pointed out.

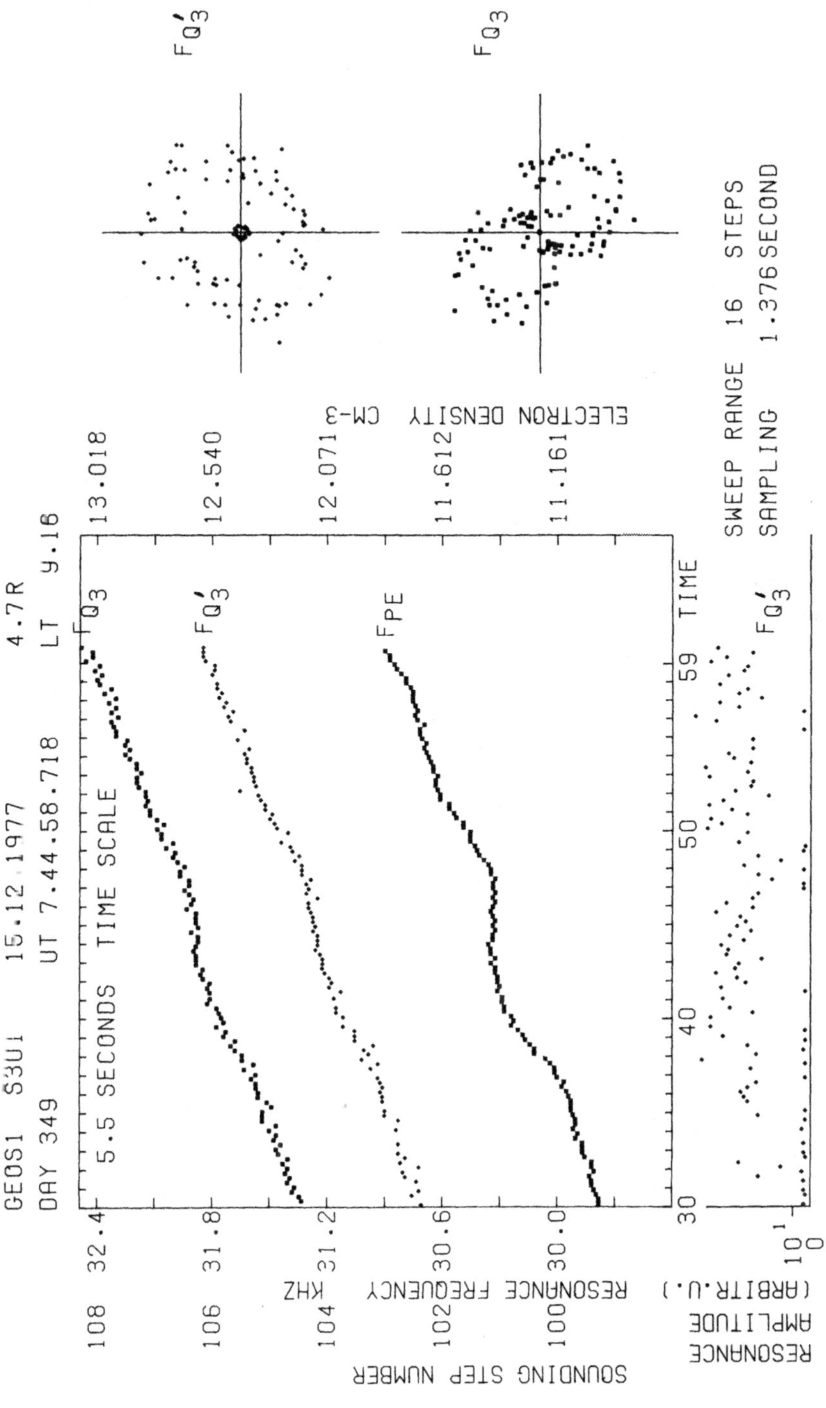

Fig. 12. Idem Figure 6. Enlargement of a part of Figure 11.

can be explored from the larger scale up to the smallest one. In addition, a reliable and precise diagnosis of the magnetospheric plasma is furnished in real time. The available range of measurement is as wide as needful in the magnetosphere. The ability to measure very small scale structures of the electron density (sensitivity of the density measurement reaching a few 10^{-3} cm^{-3}, maximum sampling rate equal to once every 86 ms or 150 m) and the given proof of the existence of such structures requires to perform their systematical study. The directivity patterns of the resonance waves are well defined, easily available; their interpretation is relevant to the wave propagation study (interpretation of the f_q Bernstein modes) and to the geophysical one (in particular the hypothetical approach to the density gradient measurement is to be checked). Hence, further detailed investigations are required on the GEOS-1 tracking data; careful measurements will be performed on GEOS-2, especially during the antenna deployment phase since variable lengths of the receiving antenna shall be available.

Acknowledgements

The author appreciates the strenuous work performed by J. P. Irlande, E. Pawela, and J. Bilbille who helped him to materialize what was at first just a scientist's idea. M. Petit, co-author of the idea, is to be acknowledged too: his weight has been a convincing support for the acceptance of this new experiment by ESA. ESOC's efforts have been determinant for interfacing the hard and soft-wares.

References

Bitoun, J., Fleury, L., and Higel, B.: 1975, *Radio Sci.* **10,** 875.

Calvert, W.: 1969, in J. O. Thomas and B. J. Landmark (eds.), *Plasma Waves in Space and in the Laboratory*, Vol. 1, pp. 41–54, Edinburgh University Press, Edinburgh, Scotland.

Calvert, W. and Goe, G. B.: 1963, *J. Geophys. Res.* **68,** 6113.

Calvert, W. and McAfee, J. R.: 1969, *Proc. Inst. Elec. Electron. Engrs.* **57,** 1089.

Christiansen, P., Gough, P., Martelli, G., Bloch, J. J., Cornilleau, N., Etcheto, J., Gendrin, R., Jones, D., Beghin, C., and Décréau, P.: 1978, *Nature* **272,** 682.

Crawford, F. W.: 1965, *Radio Sci.* **69D,** 789.

Dougherty, J. P. and Watson, S. R.: 1971, in A. Simon and W. B. Thompson, (eds.), *Advances in Plasma Physics*, Vol. 4, pp. 1–41, John Wiley, New-York.

de Feraudy, H. and Hamelin, M.: 1978, 'Analysis and Theoretical Approach of the f_q Magnetospheric Plasma Resonances Excited by the GEOS Spacecraft Relaxation Sounder', Communication at URSI XIXth General Assembly, Commission H Symposium, Helsinki.

Graff, Ph.: 1971, 'Echos ionosphériques aux fréquences de plasma et hybride haute', Thèse de Doctorat d'Etat, Paris.

Higel, B.: 1975, *Ann. Telecom.* **30,** 239.

Higel, B.: 1978, *Radio Sci.* **13,** 901.

Higel, B. and de Feraudy, H.: 1977, *Radio Sci.* **12,** 879.

Higel, B., Bitoun, J., and Petit, M.: 1972, *J. Geophys. Res.* **77,** 6254.

Jones, D.: 1978, *Space Sci. Rev.*, **22,** 327.

Lockwood, G. E. K.: 1963, *Can. J. Phys.* **41,** 190.

Kikuchi, H.: 1976, *J. Atmospheric Terr. Phys.* **38,** 1055.

Kivelson, M. G.: 1976, *J. Atmospheric Terr. Phys.* **38,** 1115.
McAfee, J. R.: 1974, *Fundam. Cosmic Phys.* **1,** 71.
McAfee, J. R., Thomson, T. L., Calvert, W., and Warnock, J. M.: 1972, *J. Geophys. Res.* **77,** 5542.
Muldrew, D. B.: 1972, *Radio Sci.* **7,** 779.
P.I.E.E.E.: 1969, *Proc. Inst. Elec. Electron. Engrs.* **57,** 861.
S-300 Experimenters: 1978, *Planetary Space Sci.*, **26,** (to be published).

MULTI-EXPERIMENT DETERMINATION OF PLASMA DENSITY AND TEMPERATURE

P. M. E. DECREAU
CRPE, Orleans, France

J. ETCHETO
CRPE, Issy-les-Moulineaux, France

K. KNOTT and A. PEDERSEN
Space Science Department, ESTEC, Noordwijk, Holland

G. L. WRENN
Mullard Space Science Laboratory, Dorking, Surrey, England

D. T. YOUNG
University of Bern, Bern, Switzerland

Abstract. An attempt has been made to combine data from five instruments on GEOS-1 in order to determine the characteristics of the ambient cold plasma, assess the effects of spacecraft sheaths on the different techniques and establish cross calibration criteria. In addition to measuring plasma density and temperature it is necessary to consider the influence of satellite potential and motion, ionic composition, ion drifts (electric fields), electrons emitted from spacecraft surfaces and any consequent departures from isotropy or Maxwellian distribution.

9 October 1977 was selected for the study, the orbit covers a range of L-values between 3.5 and 7.5 giving outbound and inbound plasmapause crossings with plasma densities spanning more than two orders of magnitude. Eclipse data from 7 February 1978 is used to determine the influence of photoelectron emission.

Three techniques – active sounding of the plasma frequency resonance (S-301), a mutual impedance measurement (S-304) and DC electric field probe determination of floating potential (S-300) – utilize the long boom sensors in either AC or DC modes. Electrons and ions are measured directly by electrostatic analysers (S-302), mounted on one short radial boom. On the day chosen for study here ions are predominantly protons as determined by the body mounted ion composition experiment (S-303).

The techniques using the 20 m long booms agree well in determining density whereas the short boom and body mounted ion detectors are seriously compromised when satellite potential becomes several volts positive. Measurements of satellite potential and plasma temperature agree within $\approx 20\%$ among the several instruments making these measurements. Data obtained during the transition from eclipse to sunlight conditions show no discontinuity in the long boom instruments caused by the sudden appearance of photoelectrons.

1. Introduction

Historically, the first determinations of plasma density in the 'upper ionosphere' out to $5R_E$ were made indirectly in the mid-1950's by studies of whistler propagation. Several different experimental techniques have since been used to measure *in situ* the density and mean energy of selected magnetospheric plasma populations, e.g. the 'hot' component $\gtrsim 50$ eV (DeForest and McIlwain, 1971) or the 'cold' plasmasphere component ~ 1 eV (Chappell *et al.*, 1970; Taylor *et al.*, 1965). To our knowledge

Space Science Reviews **22** (1978) 633–645. *All Rights Reserved*

TABLE I
GEOS experiments and techniques used in density study

Experiment designation	Measurement	Technique	Parameters measured	
			Directly	Derived
S-300 DC	DC electric field	2 × 20 m radial booms	$V_s - V_p$	V_s, N_e
S-301	Relaxation sounding of plasma resonances (0–77 kHz)	VLF transmitter and receiver	$f_p, f_{uhr}\ nf_e$	$N_e, \|B\|$
S-304	Mutual and self impedance (0–77 kHz)	VLF transmitter and receiver	$Z/Z_0(f)$	N_e, T_e, λ_D
S-302	Electron and ion energy spectra (0.5 – 505 eV)	2 × electrostatic analysers, ⊥ and ∥ to spin axis on 1.8 m boom	J_e, J_+	$N_e, T_e, N_+, V_+, T_+, V_s$
S-303	Ion composition ($M/Q = 1$ to > 138) and energy spectra (0–16 keV/e)	Ion mass spectrometer body mounted ⊥ to spin axis	$J_+(M/Q = 1, 2, 4, \ldots)$	N_+, V_+, T_+, V_s

Abbreviations used in above table and text:

M = ion mass
Q = ion charge
V_s = satellite potential
V_p = probe potential
V_{pl} = plasma potential
f_p = plasma frequency
f_{uhr} = upper hybrid frequency
f_c = electron gyro frequency
Z/Z_0 = normalized plasma impedance
I_e, I_+ = electron/ion differential energy fluxes
N_e, N_+ = electron/ion density
T_e, T_+ = electron/ion temperature
λ_D = Debye length
eV_e = Electron energy
$|B|$ = Magnetic field magnitude

there are no reports of reliable cold plasma density and temperature measurements made outside the plasmasphere. Such measurements are crucial to any study of physical processes, they are also important for establishing the electrostatic environment of the spacecraft which could influence instrument performance. We present in this paper a comparison of measurements of the cold plasma within and outside the plasmasphere made independently by five experiments on-board the GEOS-1 satellite for a magnetically quiet day, 9th October 1977.

Table I lists the plasma parameters determined by each experiment which is identified by a reference number used throughout this paper. The location of the experiments and the detector fields of view are shown in Figure 1. Further details of the techniques employed may be found in Pedersen *et al.* (former issue), Decreau *et al.*, Etcheto *et al.*, Johnson *et al.*, Geiss *et al.* (this issue). This is the first time that these techniques are used simultaneously in the magnetosphere and in a coordinated way.

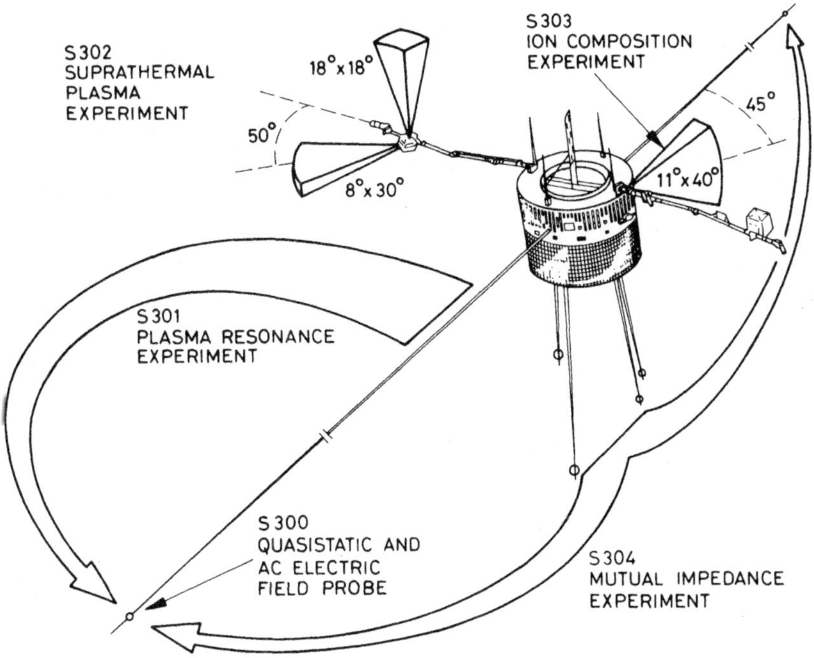

Fig. 1. Schematic diagram of the GEOS satellite showing the location of instruments listed in Table I. The viewing angle of the particle instruments are given.

The DC electric field probes used by S-300 are extension of the well-known Langmuir probe technique specialised for a medium with very long (> 1 m) Debye length. Relaxation sounding (S-301) was first used in space on the highly successful Alouette satellites which obtained density profiles of the topside ionosphere in the early 1960's. The mutual impedance method has been used before in the ionosphere. GEOS is the first high altitude satellite to employ both sounding and mutual impedance techniques. The electrostatic analyser (S-302) and ion mass spectrometer (S-303) are more sophisticated versions of ion traps and spectrometers first used to obtain measurements of magnetospheric plasma in the early and mid-1960's on the IMP and OGO spacecraft.

Basically the techniques may be separated into two groups: those which measure bulk properties of the plasma or some sub-population (e.g. electrons) and those which attempt to measure the plasma velocity distribution function whose moments describe the bulk properties. Thus the two wave propagation experiments respond mainly to the parameters N_e and T_e of ambient electrons interacting with the waves. Electric field probes respond primarily to the total electron flux, with the effect of photoelectrons being taken into account by accurate determination of the probe photoemission characteristics. Although an assumption about V_e is needed in order to obtain N_e, the probes do provide a key parameter, the satellite potential V_s needed by particle experiments to determine density. The two particle experiments measure directly the electron and ion fluxes from which the plasma distribution function can be constructed after the effects of spacecraft potential and motion with respect to the plasma are unfolded. This unfolding can be difficult, depending on details of the satellite potential, whereas wave experiments are not at all affected.

Although the measurement of plasma frequency by the active wave experiments can in principle determine the plasma density very accurately ($\sim 1\%$), inter-comparison of the several methods is more than an academic exercise because each of the instruments has certain advantages and, at the same time, certain limitations. Wave and probe measurements cannot give details of the particle distribution function while on the other hand, particle measurements, for reasons discussed below, cannot always give total density at very low ($\lesssim 10$ eV) energies with a high degree of accuracy. Thus the GEOS-1 techniques complement one another in the sense that a full understanding of the plasma physics requires an accurate measurement of the particle distribution function of which density is only the first moment.

One feature of GEOS which makes the present study unique and at the same time of practical interest is that considerable cost and effort have gone into reducing problems arising from electrostatic charging by grounding all outer surfaces of the satellite and making $\gtrsim 96\%$ of the external surface (including solar cells) as uniformly conducting as possible.

In Section 2 the five experiments are briefly described but only to the extent necessary for understanding the present study. In Section 3 we present a detailed analysis of plasma density, temperature, Debye length, satellite potential and give a comparison for a typical quiet time GEOS-1 pass. Data for this selected pass with apogee at 11:00 LT show that electrons and ions with temperatures < 10 eV are the dominant contributor to total plasma density in the region surveyed ($3 \leqslant L \leqslant 8$), higher energy components may be neglected. At more disturbed times and other regions (e.g. plasma sheet) this is certainly not the case. Some eclipse measurements are included in order to assess the influence of photoelectrons on the wave and probe data.

2. Instrumentation and Measurement

2.1. DC Electric Field Probes (S-300)

The prime purpose of this experiment is to determine quasi-static and AC electric fields by measuring the difference in voltage between vitreous carbon probes mounted on the tips of two 20 m long cable booms (Figure 1). Provision is also made to measure the voltage between one of these sensors and the spacecraft which can be used for the determination of the spacecraft floating potential, and the ambient plasma density.

A special feature of the probes is the use of a current source of high impedance to force a bias current of selectable magnitude and polarity from the satellite to the probes. An 83 mA negative bias current is permanently injected into the spheres and the resulting potential difference $V_s - V_p$ (spacecraft potential with respect to probe potential) constantly transmitted. This bias current has no influence on the satellite floating potential because of its larger current collection area. The biasing of the probe forces the probe to adopt a potential very near to space potential.

With the knowledge of the photoemission characteristic of the spheres we have been able to establish two relationships using electron density and temperature measurements from other experiments (S-301, S-304):

–the dependence of the spacecraft floating potential $V_s - V_{pl}$ on the measured potential difference $V_s - V_p$;

–the dependence of N_e, the plasma density, on the measured parameter $V_s - V_p$.

The floating potential is used by particle experiments for the evaluation of their low energy ($\leqslant 10$ eV) data. The second relationship is used to provide information on density with high time resolution, which cannot be achieved in practice by particle experiments. Continuous operation of the active experiments could give nearly instantaneous density data, but this would exclude the passive measurement of natural wave phenomena and cause interference to other experiments.

2.2. Resonance Sounder (S-301)

The resonance experiment employs the braiding of the long cable booms as an antenna to emit into the plasma wave pulses of 3 ms duration at frequencies between 0.3 and 77 kHz (Figure 1). A pronounced echo is observed from the plasma, the signal being received by the probes at the tips of the long booms. When one of the characteristic frequencies of the plasma is emitted, a pronounced 'ringing' of the plasma occurs. Identification of the plasma frequency leads to the determination of the density. Density measurements are limited to the range less than 70 cm^{-3}.

2.3. Mutual Impedance Experiment (S-304)

The mutual impedance experiment employs two electrical aerials on the axial booms to emit signals in the frequency range 0.3 – 77 kHz (Figure 1). The current injected into the plasma is measured; the plasma impedance is determined by measuring the voltage received at the long boom sensors. The plasma impedance shows a peak near the plasma frequency. The position and shape of this peak versus frequency are directly related to the plasma frequency and to the Debye length. Curve fitting can be applied to determine these two quantities from the impedance versus frequency plots and thus the plasma density and temperature. For a given geometry of the antennas, this technique is efficient in a given range of Debye lengths, namely from 0.5 to 6 m on GEOS. Moreover, the density measurements are limited to the range ~0.3 to 70 cm^{-3}, as f_p must be higher than the electron gyrofrequency and lower than the maximum frequency of the experiment.

2.4. Suprathermal Plasma Analyser (S-302)

Two hemispherical electrostatic analysers covering the energy range 0.5 to 505 eV are mounted on a short radial boom 1.8 m from the satellite surface (Figure 1). The two analysers are selected for either 'electron' or 'proton' mode operation with an energy resolution $\Delta E/E$ of 0.11 yielding a full spectrum in 2.7 s in the simplest mode of operation. The sensors are boom mounted in an attempt to avoid serious contamination of electron fluxes by photoelectrons and secondary electrons originating at the spacecraft surface. In practice the displacement of about one satellite diameter is insufficient to achieve this objective because the electrons represent an equivalent density at the sensor location of about 100 cm^{-3} reducing to 20 cm^{-3} on the shadow side of the satellite.

Proton counting rates are lower by a factor of 50 but the mode is free of secondary contamination. Although longer accumulation times can be used, it is important to examine spin modulation of fluxes in order to take into account ion drift velocities resulting from satellite motion or electric fields.

Analyser '*A*' views parallel to the spin axis. Analyser '*B*' views at an angle of 50° to the radial and roughly perpendicular to the spin axis. The *B* analyser suffers from the effects of sunlight entering the aperture, photoelectrons produced inside the analyser cause contamination and restrict low energy measurements to half of each spin cycle. The analysers are housed in an isolated unit which can be biased selectively between −28 V and +34 V with respect to the satellite ground. Operation close to the plasma potential yields maximum accuracy for temperature measurement but accelerating potentials can be used to improve sensitivity.

2.5. Ion Composition Experiment (S-303)

The Ion Composition Experiment (ICE) covers the energy range 0 to 16 keV e^{-1} although in this study we concentrate only on the *T* mode range 0–110 eV e^{-1}. The ICE is mounted on the satellite body and views perpendicular to the spin axis. In the range 0–110 eV e^{-1} the instrument operates as a 32 step retarding potential analyser and can perform, in addition to mass and energy analysis, an azimuthal angular scan with a resolution of 11°. A full mass-energy-angle scan requires 6 min to provide 2-dimensional distribution functions for any cold ions present at concentrations above ~10^{-1} cm^{-3} (greater sensitivity and higher time resolution are possible in other programmable modes).

In the present study, the ICE also provides a monitor of the more energetic (0.1–16 keV e^{-1}) component of the plasma density. For the case presented here however, the energetic component contributes at most <10% to the total density outside the plasmasphere and <1% inside.

The method of data analysis for both particle experiments S-302 and S-303 is similar. The logarithm of particle phase space density is plotted as a function of particle energy in the plasma frame of reference (i.e. the measured energy is corrected for satellite potential and relative motion of satellite and plasma). If the electron or ion distributions are Maxwellian, which seems generally to be the case for the cold ambient plasma, then the plot yields directly temperature and number density. Any error in estimating the satellite potential V_s leads to an error ~ $\exp(eV_s/T)$ in estimating the number density. Ion drifts can be estimated rather well inside the plasmasphere by assuming that the satellite orbital velocity in a frame of reference corotating with the Earth dominates over plasma drifts. Pitch angle anisotropies, which may sometimes be present, are not accounted for in this analysis. In addition to difficulties arising from incorrect determination of satellite potential, details of the satellite sheaths may also lead to discrepancies between instruments.

In the case of the boom mounted S-302 package, the effect of collimator bias can to some extent cancel that of satellite potential, although the perturbation of particle trajectories by the combined S-302 and satellite sheaths (e.g. possible focussing effects) have not been assessed. Furthermore, the B analyser of S-302 views non-radially and hence samples ions with non-zero impact parameter whereas S-303 ions have zero impact parameter. Finally, S-302 and S–303 sit at different locations in the satellite sheath. A full treatment of these effects is beyond the aims of the present work (c.f. Parker and Whipple, 1975).

3. Results

In Figures 2, 3, and 4 we show measurements of three parameters: satellite potential, plasma temperature and density for the quiet day 9 October 1977 ($\sum Kp = 4-$).

Figure 2 shows the satellite potential V_s determined by two experiments, the electrostatic probes and suprathermal plasma analysers. The agreement is good, within 1 V, over the entire pass. This is within the limit of error, although some discrepancy outside the plasmapause may in part be due to the long (~5 m) Debye length which leaves S-302 (on the short boom) well within the Debye sheath, whereas S-300 is still well outside.

In Figure 3 we give electron and proton temperatures from experiments S-302, S-303, and S-304. The large jumps in temperature at 05:20 UT and 10:40 UT measured by S-303 occur at plasmapause crossings where the cold plasma component drops abruptly in density. Beyond the plasmapause S-303 and S-302 ion measurements are dominated by a hot ion component with density $\lesssim 1\ \mathrm{cm}^{-3}$ and temperature ≫1 eV. Outside the plasmapause the relatively large positive satellite potential (+5 to +6 V) prevents S-303 from detecting a cold (~1 eV) ion component

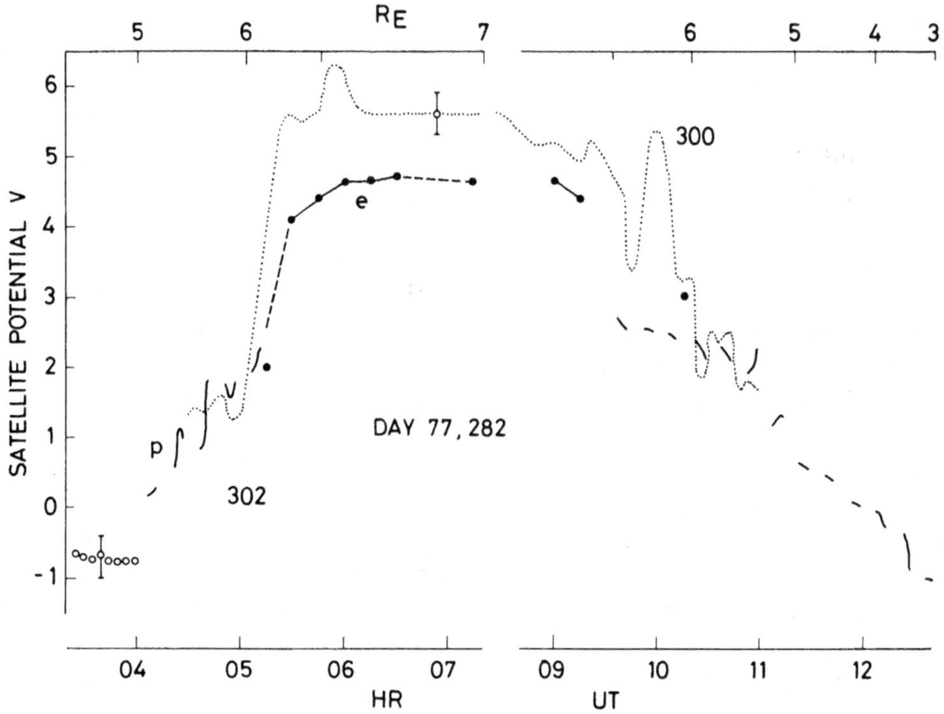

Fig. 2. Satellite potential for the GEOS-1 pass on 9 October 1977 as determined by the S-300 electrostatic probes (dotted line) and S-302 plasma analysers (broken line). S-300 data is obtained continuously whereas S-302 data are collected every 6 min and averaged over 44 s.

at densities below $\sim 50\ \text{cm}^{-3}$. Because it measures electrons as well as ions, S-302 is able to make reliable temperature measurements at large positive potentials and is in good agreement with the impedance measurements of S-304. Before 05:20 UT and after 10:40 UT, the S-302 proton and S-303 ion measurements of temperature are in good agreement. This is as expected since the method of measurement is insensitive to satellite potential so long as it is sufficiently small to allow the ions in question to reach the two detectors.

It is important to note in conjunction with Figure 3 that, so far as we know, GEOS-1 has here provided the first reliable radial profile of temperature in the outer plasmasphere ($L = 3$–6). Clearly the plasma is rather cold, varying only between 3000 K and 10 000 K, which is typical of all plasmasphere data analysed so far.

From S-304 data we find that at about 05:10 UT the plasma Debye length increases by a large amount, reflecting the oppositely directed changes in plasma temperature (Figure 3) and density (Figure 4). Outside the plasmasphere the Debye length is 5 m, larger than all satellite dimensions with the exception of the long radial booms. A sharp but temporary drop in Debye length occurs at 09:35 UT, due to a fluctuation in plasma density, but it is not possible to say whether this is a spatial or

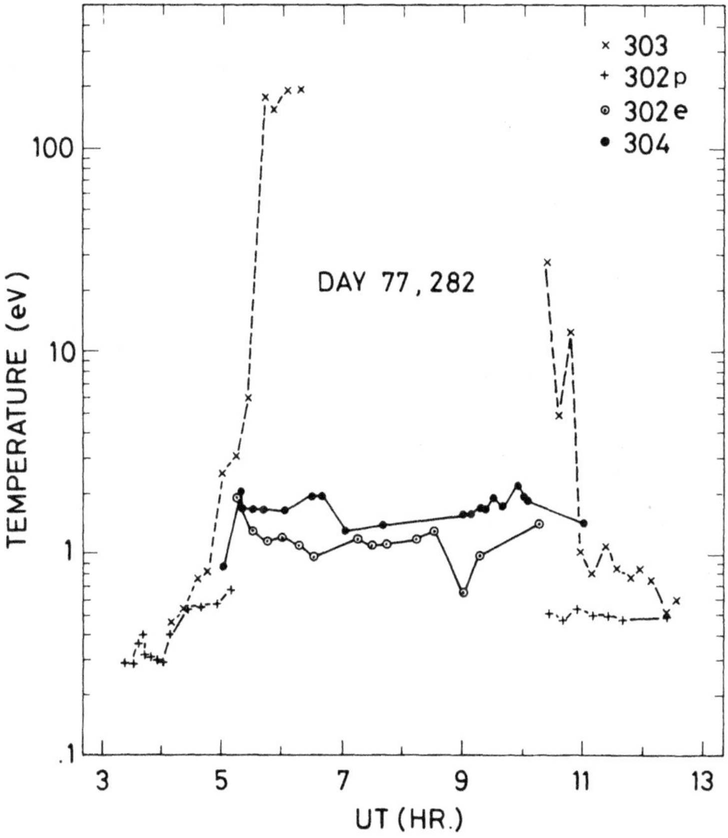

Fig. 3. Comparison of plasma temperatures measured by the S-302, S-303 (ion composition), and S-304 (mutual impedance) experiments. S-303 temperature measurements correspond to the dominant ion component observed by the instrument which, beyond the plasmapause, is composed only of a higher temperature population capable of overcoming the +5 to +6 V satellite potential. S-302e and S-302p curves refer to measurements of electrons and protons respectively.

temporal variation. Between 10:00 UT and 11:00 UT the Debye length decreases as the satellite again crosses into the plasmasphere.

A direct comparison of density measurements by all five instruments is shown in Figure 4. We wish to emphasize that in constructing Figure 4 no arbitrary adjustments were made to make the different instruments agree more closely. The S-300, S-301 and S-304 determinations are plotted directly, whereas the particle measurements of S-302 and S-303 have been corrected for satellite potential (measured by S-300 and S-302) and velocity.

GEOS-1 crossed the plasmapause at about 04:20 UT – 05:00 UT ($L = 5.4 - 6.1$) on the upleg and about 10:00 UT – 11:30 UT ($L = 7.2 - 5.8$) on the downleg of the orbit. During the plasmapause crossings there are differences between various measurements of as much as a factor of ten. The low values of density measured by

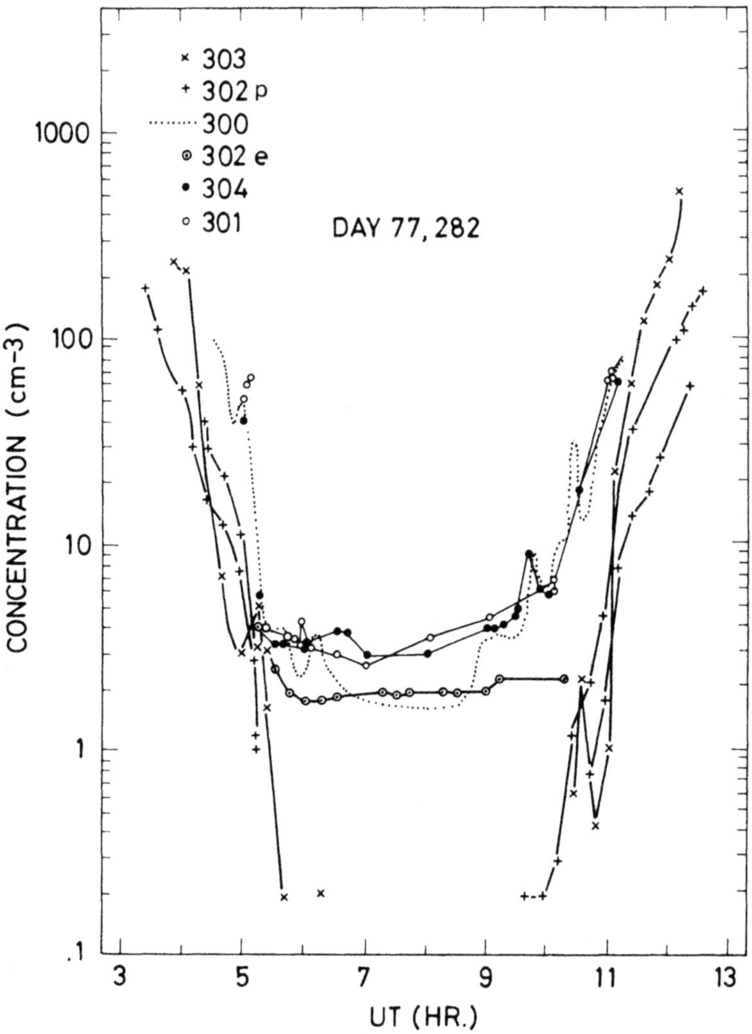

Fig. 4. Cold plasma density as measured by the five instruments in this study. The S-303 measurements correspond to the total ion density which was predominantly H^+ with only 4% – 12% He^+ throughout the pass.

the ion detectors S-302 and S-303 outside the plasmasphere (05:30 UT – 11:00 UT) result from the fact that none of the low density cold (< 1 eV) ions can reach the satellite which is charged to + 5 V at this time (c.f. Figure 2). At the same time the S-302 electron measurements are in fairly good agreement with the S-300 probes and both are ~40% lower than the active experiments S-301 and S-304. At other times the two ion detectors which view radially agree fairly well, although the parallel viewing S-302 ion measurements are somewhat lower. This latter discrepancy may result either from an asymmetry in the satellite sheath or from a genuine anisotropy in the ion distribution. In the two plasmapause transitions there is reasonable agreement among S-300, S-301 and S-304 despite the fact that S-301 and S-304 are clearly undersampled relative to S-300. We generally find that S-300, S-301 and S-304 agree within 5% in regions of higher density.

In this preliminary study we are not prepared to discuss quantitatively the discrepancies among the various instruments. A larger data sample is needed as well as measurements at moderate density ($<10\ cm^{-3}$) in the absence of large temporal variations, e.g. measurements made when the plasmasphere extends out to the apogee region where the satellite moves slowly. For this purpose GEOS-2 measurements at geostationary orbit would also be useful.

Qualitatively, there are several factors contributing to the differences in Figure 4:

– As noted above, there are considerable differences in sampling times and intervals between samples. Shorter term variations may be missed entirely or averaged out e.g. the spike in S-300 and S-303 data at $\approx$10:22 UT. Particle detectors do not measure the full 3-dimensional distribution function which can lead to either under or overestimation of density depending on whether the distribution has a local maximum or minimum within the detector's pitch angle range (e.g. ions may be flowing parallel or perpendicular to B).

– Inaccuracies in determining the satellite potential cause errors proportional to $\exp(eV_s/T)$ in the particle measurements. The accuracy of this determination is affected by several factors including differences in the work functions of materials at the various detector entrances, possible differences in detector grounding for the boom and body mounted packages, sheath asymmetries, the potential distribution within the sheath, and asymmetries in detector viewing directions relative to the sheath (c.f. Figure 1).

– Errors in calibration of the particle detector geometric factors (which are estimated to be good only to $\approx$50% at these low energies).

– Anisotropies or gradients in the ambient particle distribution may affect the sounder and ion experiments (which tend to sample a larger volume of space, e.g. near apogee the 1 eV ion gyroradius is $\sim$1 km) more than the probe, mutual impedance and electron spectrometer measurements.

The effect of photoelectrons on the S-300, S-301, and S-304 experiments is negligible as demonstrated by Figure 5 where we show density and temperatures from the S-300 probes and two active wave experiments as GEOS emerges from eclipse. As noted above, the three methods agree quite well. We draw attention to the smooth transition in the data from eclipse to sunlight conditions, indicating that the photoelectron sheath surrounding the satellite does not interfere with the measurements of ambient plasma.

Considering the above factors, it seems clear that the particle measurements are most susceptible to error, particularly due to possible undersampling in time and space and problems with determining the appropriate correction for satellite potential. One remedy for this is to find an effective geometric factor which takes into account the real potential distribution in sheaths surrounding the respective particle detectors. This could be based on a measured transfer function relating detector counting rates to the ambient density determined by the S-300, S-301, and S-304 experiments.

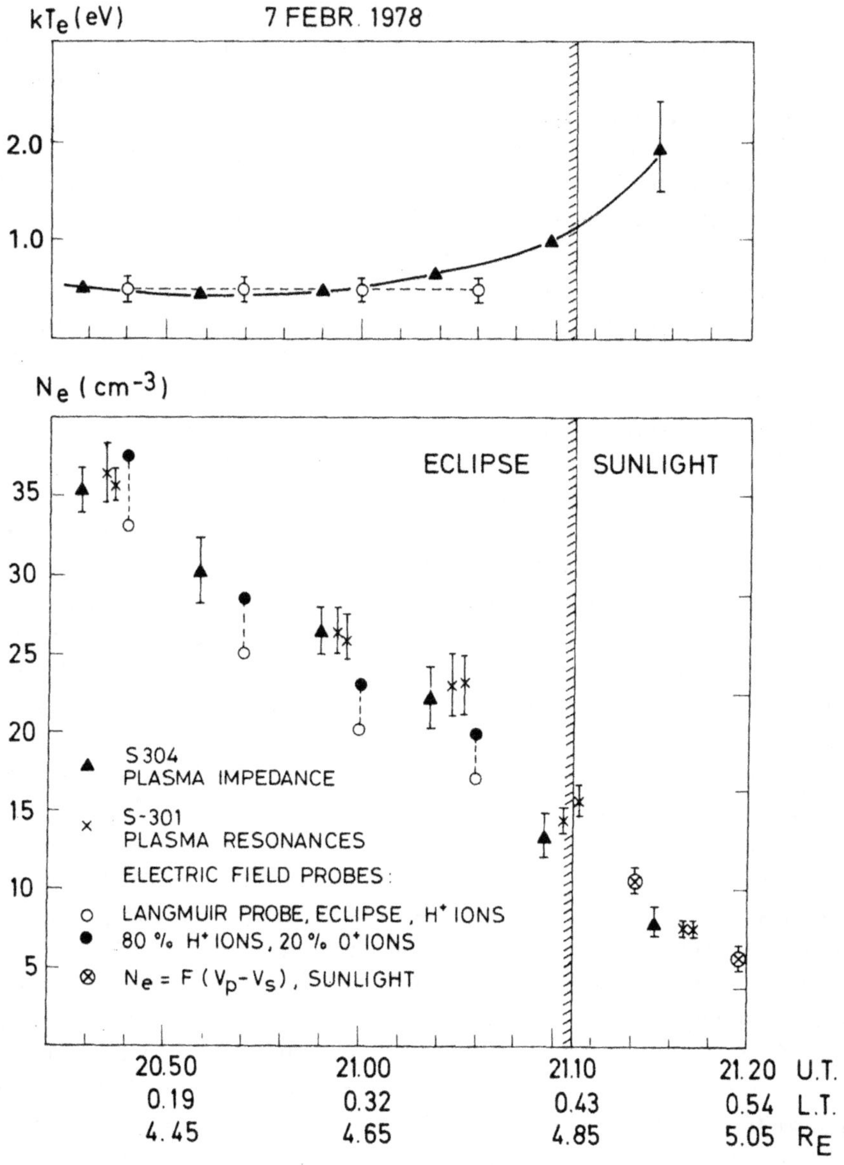

Fig. 5. Density and temperature profiles for an eclipse transition on 7 February 1978. Particle experiment data is not available from this pass.

References

Chappell, C. R., Harris, K. K., and Sharp, G. W.: 1970, *J. Geophys. Res.* **75**, 50.

Decreau, P. M. E., Beghin, C., and Parrot, M.: 1978, *Space Sci. Rev.* **22**, 581.

DeForest, S. E. and McIlwain, C. E.: 1971, *J. Phys. Res.* **76**, 3587.

Etcheto, J. and Bloch, J. J.: 1978, *Space Sci. Rev.* **22**, 597.

Geiss, J., Balsiger, H., Eberhardt, P., Walker, H. P., Weber, G., Young, D. T., and Rosenbauer, H.: 1978, *Space Sci. Rev.* **22**, 537.

Johnson, J. F. E., Sojka, J. J., and Wrenn, G. G.: 1978, *Space Sci. Rev.* **22**, 567.

Parker, L. W. and Whipple, E. C.: 1975, *J. Geophys. Res.* **75**, 4720.

Pedersen, A., Grand, R., Knott, K., Jones, D., Gonfalone, A., and Fahleson, U. V.: 1978, *Space Sci. Rev.* **22**, 333.

Taylor, H. A., Jr., Brinton, H. C., and Smith, R. C.: 1965, *J. Geophys. Res.* **70**, 5769.

SUBSTORM-ASSOCIATED INJECTIONS OF ENERGETIC IONS OBSERVED BY GEOS-1 AND ATS-6 IN AND NEAR SYNCHRONOUS ALTITUDE

B. WILKEN, T. A. FRITZ*, A. KORTH, and G. KREMSER

Max-Planck-Institut für Aeronomie, D-3411 Katlenburg-Lindau 3, F.R.G.

Abstract. Energetic ion measurements of GEOS-1 and ATS-6 are analysed for the period of geomagnetic activity following the arrival of a solar wind shock at 0027 UT on July 29, 1977. GEOS crossed the magnetopause at 6.9 R_E and 0027 UT (1312 LT). Although the difference in local time to ATS at 6.6 R_E is only 2 h ATS seems to remain well inside the magnetopause. During the second orbital pass on this day GEOS crossed the geostationary orbit at the onset time of a major substorm developing at 1120 UT. At this time the local time difference of GEOS and ATS was 12 h. The considerably different energy dispersions are discussed. An azimuthal anisotropy of approximately 20% observed in the GEOS data is interpreted to be the result of a particle density gradient.

1. Instruments

The energetic ion instruments on-board of GEOS-1 and ATS-6 are based on the same detection technique. Magnetic electron/ion separators are combined with solid state detector telescopes for single parameter analysis of ions. Neither of the instruments is capable of identifying ion species, but an additional single heavy ion telescope is included as part of the energetic particle instrument on ATS-6. Details of the GEOS and ATS-6 instruments are described by Wilken *et al.* (1977) and Fritz and Cessna (1975) respectively. The energy passbands of the two instruments are given in Table I and the orientation of the viewing directions is shown in Figure 1. Although the number of telescopes is comparable, the obtainable coverage and resolution in pitch angle are significantly different. The instrument on the spin-stabilised GEOS spacecraft can return very detailed measurements of the ion distribution function even if the local magnetic field vector is parallel to the spin axis. The instrument on ATS, on the other hand, can obtain measurements for a maximum of three different pitch angles since the satellite is three axis stabilized in a coordinate system corotating with the Earth.

2. Introduction

On July 29, 1977 (day 210) a strong solar wind shock arrived at the Earth magnetosphere generating a SSC at 0027 UT which was observed in most of the US groundbased magnetometer. The SSC triggered substantial geomagnetic activities throughout the first part of this day with the last major substorm developing shortly after 1100 UT. At the time of the SSC as well as during the substorm 11 h later GEOS was near the geostationary orbit providing a unique opportunity for correlated measurements with ATS-6.

* NOAA-SEL, Boulder, Colo., U.S.A.

Space Science Reviews **22** (1978) 647–657. *All Rights Reserved*

TABLE I

Energy passbands of the GEOS and ATS energetic ion spectrometer

Energy-channel	GEOS (keV)	ATS (keV)
ΔE1	16.5– 24.0	25.5– 33.4
ΔE2	24.0– 33.0	33.4– 47.8
ΔE3	33.0– 43.0	47.8– 70.8
ΔE4	43.0– 56.0	70.8- 100.0
ΔE5	56.0– 73.0	100.0– 150.5
ΔE6	73.0– 96.0	150.5– 234.0
ΔE7	96.0– 125.0	234.0–2800.0
ΔE8	125.0– 165.0	
ΔE9	165.0– 219.0	
ΔE10	219.0– 294.0	
ΔE11	294.0– 390.0	
ΔE12	390.0–3300.0	

The orbital phases of GEOS and ATS on day 210 are shown in Figure 2. At the time of the SSC GEOS and ATS were almost in conjunction. GEOS was travelling on the outbound leg of its orbit at 1312 LT and $L = 6.8$. The position of ATS was 2 h in local time ahead of GEOS at 1512 LT.

The relative position of the two satellites was maintained for approximately 2 h. At 0530 UT GEOS crossed the geostationary orbit on the inbound trajectory. Twelve hours later GEOS was again outbound and approaching synchronous altitude when a substorm developed at 1120 UT. However, this time the difference in LT to ATS was almost exactly 12 h. The purpose of this paper is to briefly illustrate the different response of the GEOS and ATS instruments to the SSC and to present in more detail the observations during the substorm at 1120 UT.

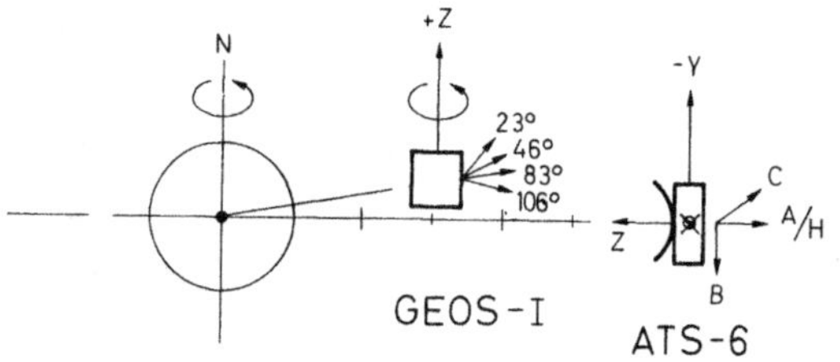

Fig. 1. Viewing directions of the energetic particle telescopes on GEOS-1 and ATS-6 (A, B and C).

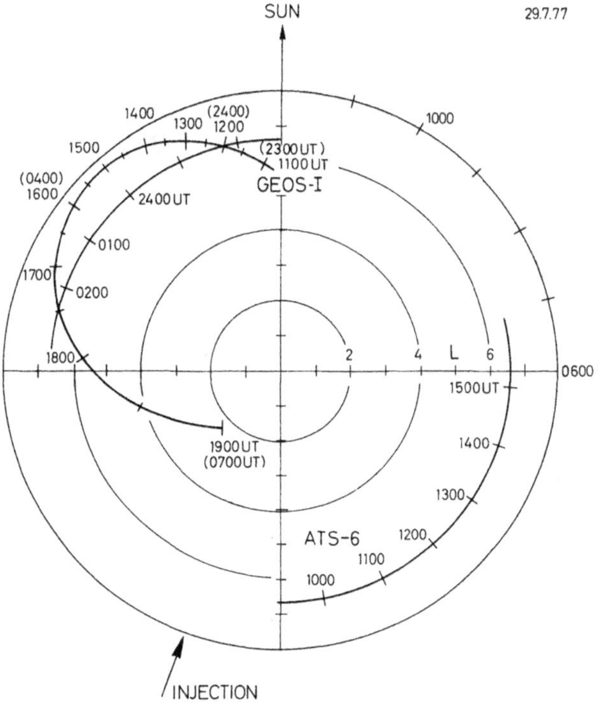

Fig. 2. Orbital phases of GEOS and ATS-6 for July 29, 1977 (d 210) Universal time in brackets refers to the Alaska pass of GEOS.

3. Magnetopause Crossing at 0027 UT

Figure 3 shows a general view of ion intensity variations in selected energy channels observed by ATS on day 210. In response to the solar wind shock (0027 UT) the ion flux started to increase slowly with a characteristic time of approx. 30 min. The shock is followed by an active period with a series of large intensity fluctuations ending with a last strong injection at 1150 UT.

A detailed inspection of the differential spectrum measured at 0027 UT shows an adiabatic increase of the particle flux possibly resulting from a field compression caused by the shock. Unfortunately *in situ* magnetic field measurements are not available since the ATS-magnetometer was no longer providing the 3-component B-vector.

The effect of the SSC at the position of GEOS is demonstrated in Figure 4. The integral flux of protons and electrons above 24 keV and 20 keV respectively shows a well defined sudden flux drop with the electrons starting to decrease at 0026:30 UT (flux level dropped to 50% at 0027 UT). Compared to the electron flux drop-out the ion response seems to be delayed by almost 50 s.

The data indicate that GEOS was passed by the magnetopause at 6.9 R_E. Since the proton and electron flux did not return to the average magnetospheric level for the entire time data coverage was available we are led to the

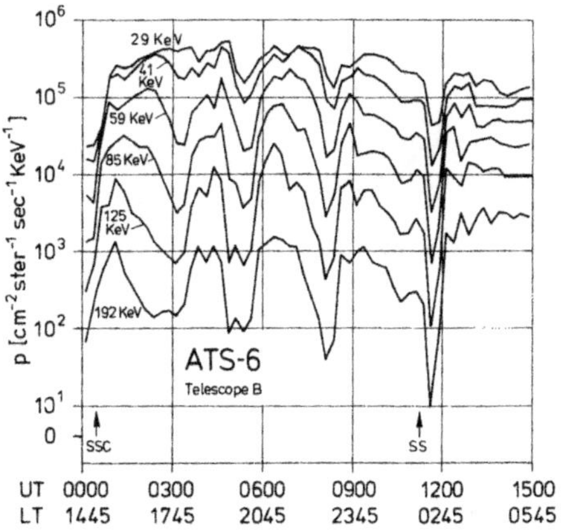

Fig. 3. ATS observation of intensity variations in the energetic ion population following the arrival of the Solar Wind Shock at 0027 UT. SS indicates the onset of a substorm at 1115 UT.

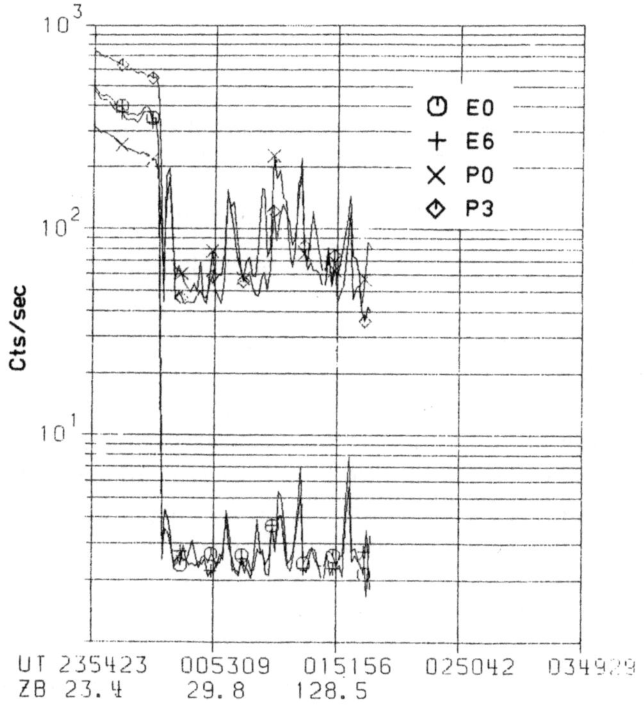

Fig. 4. Integral flux for electrons (detectors E0, E6, energies >20 keV) and protons (detectors P0, P3, energies >24 keV) measured by GEOS during the Alaska pass. Viewing directions of the telescopes are approximately parallel (E0, P0) and perpendicular (E6, P3) to the satellite spin. A steep flux drop out is observed at 0028 UT in association with a SSC which apparently pushed the magnetopause past the satellite.

conclusion that GEOS remained outside the magnetopause for at least 2 h. It is interesting to note that there is no evidence in the ATS measurements indicating that the magnetopause penetrated close enough to the position of ATS to effect the energetic ion spectral distribution by direct interaction with the boundary.

4. Substorm Injected Ions

As shown in Figure 2 GEOS was again outbound and approximately at $L = 6.0$ when a major substorm occurred at about 1100 UT. North American magnetometers in the midnight-sector indicate an onset time of 1115 UT. The combined GEOS/ATS observations of energetic ions at the time of the substorm are shown in Figure 5. At the location of both satellites an injection like flux increase is observed following a slow gradual decrease. However, the injection time and the energy dispersion is significantly different at the two positions in the geostationary orbit as is demonstrated in Figure 6. At the position of ATS (0244 LT) the intensity in the lowest energy channel centered at 28 keV was the first to increase followed by the higher energy channels. A 4 min delay is observed for the arrival of 220 keV ions with respect to the 28 keV channel. A similar inverted energy dispersion has previously been observed by Fritz *et al.* (1974) around $L = 4$ as a result of substorm injected proton fluxes. The inverted dispersion observed by ATS can not easily be explained by a model which assumes that ions are uniformly and simultaneously injected into the nightside magnetosphere and gradient drift them to the position of ATS. Using the concept of an injection boundary as proposed by Konradi *et al.* (1975) and taking $Kp = 6$ for this period of July 29, energetic particles can well be injected down to approximately $L = 5$ in the vicinity of midnight. The relatively small dispersion observed by ATS suggest that the satellite was probably close to the injection region although details of the arrival time remain unexplained.

The corresponding dispersion pattern observed by GEOS at $L = 6.6$ and 1300 LT has an entirely different time dependence. The flux increase at 1204 UT shows neglegible dispersion for all channels above 150 keV. For lower energies down to $E = 50$ keV a distinct energy dispersion is apparent. It seems to be impossible to reproduce the observed dispersion simply by drifting particles in a magnetic and electric field since this would require an unreasonably large retarding convective electric field with a field strength of 18 kV/RE (3 mV m^{-1}). A more likely interpretation is that particles have been injected somewhat before 1204 UT and GEOS intercepted the energetic component of the developing ring current. The fact that the flux increased by more than an order of magnitude in less than 300 km radial distance indicated an extremely sharp boundary. The arrival of low energy particles at GEOS is then interpreted as being essentially a temporal effect. A dispersion curve based on this assumption is shown in Figure 6 as curve 1. This curve implies a particle release at 1141 UT from a position at 2300 LT and an L value between 6.7 and 7.2. This injection time compares

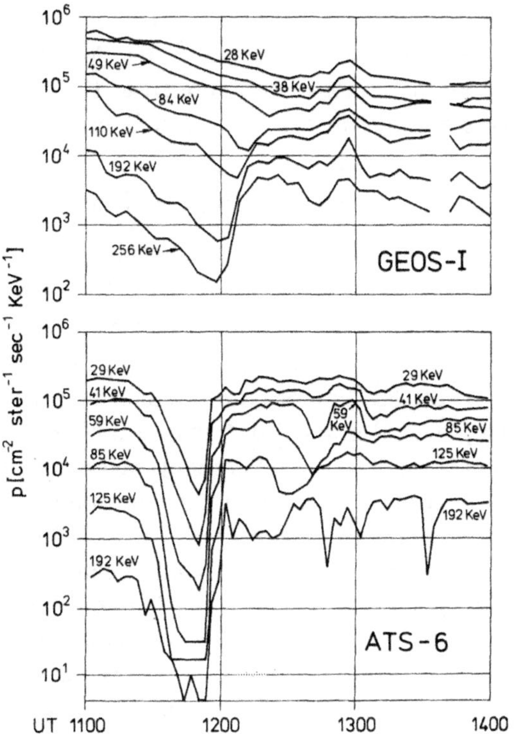

Fig. 5. Combined GEOS and ATS observation at the time of the substorm developing at 1115 UT.

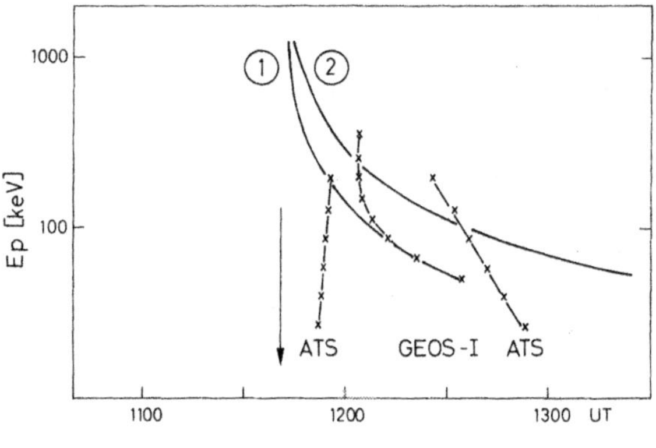

Fig. 6. Energy dispersion of the substorm injected particles observed by GEOS and ATS. Curve 1 is an extrapolation based on the arrival time of low energy particles at GEOS. this curve would imply a particle release at 1141 UT in the vicinity of 2300 hours LT as indicated in Figure 2. Curve 2 gives the expected dispersion at the position of ATS-6 if these particles have drifted around the noontime magnetosphere.

reasonably well with the onset of the flux increase observed by ATS at 1152 UT. Curve 2 in Figure 6 is the energy dispersion predicted for the position of ATS (0320 LT) if the particles are propagated around the dayside magnetosphere. The gross disagreement with the actual dispersion measured by ATS at 1230 UT as shown in Figure 6 indicates that ATS is not responding to particles which passed through the GEOS position.

The steep intensity enhancement observed by GEOS shortly after 1200 UT was associated with a significant change in the pitch angle distribution of the ions. The general observation is that the distribution function broadened when the flux level increased, although this effect exhibited a distinct energy dependence. In fact, the shape of the pitch angle distribution was energy dependent in the sense that the observed pitch-angle anisotropy is increasing with particle energy, i.e., the higher the energy the more sharply is the distribution peaked at $\alpha = 90°$. The changes in the angular distribution associated with the flux increase are illustrated in Figure 7 for two energy channels centered at 28.5 and 145 keV. The angular dependence of 145 keV particles was initially dominated by a $\sin^{12} \alpha$ term but the exponent dropped to $n = 6$ at 1255 UT when the intensity was enhanced. The distribution of the 28.5 keV channel, on the other hand, remained almost uneffected by the flux increase although a minor steepening is noticeable.

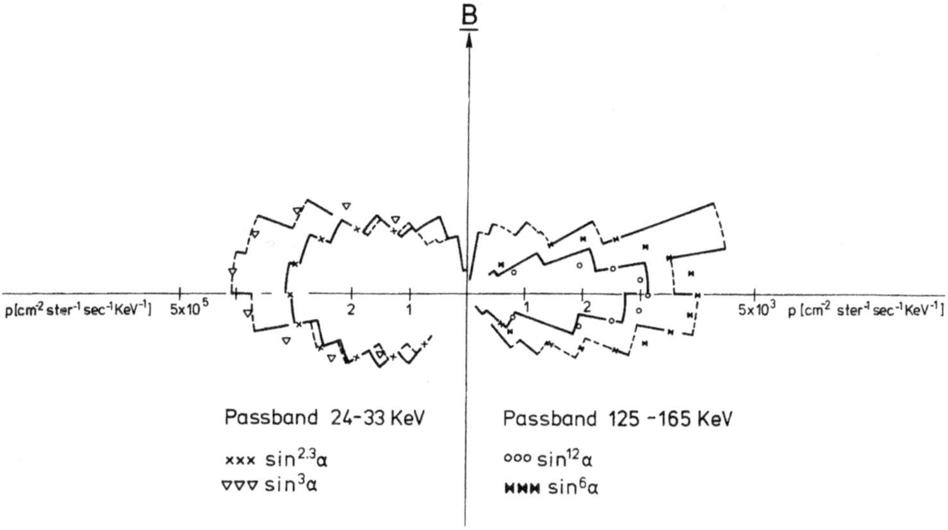

Fig. 7. Pitchangle distributions for two energy passbands observed by the GEOS instrument at 1200 UT (solid line) and at 1255 UT (dashed line). The sudden flux increase shortly after 1200 UT resulted in a considerable broadening of the pitchangle distribution at high energies. The low energy passbands does not show a significant effect.

High time resolution spin-modulated integral data ($E > 24$ keV) from the four proton telescopes of GEOS (P0, P1, P2 and P3) are shown in Figure 8. A detailed analysis showed that the pitch angle interval sampled by P2 and P3 contained the pitch angle $\alpha = 90°$. The phase angles for which the instantaneous pitch angle was

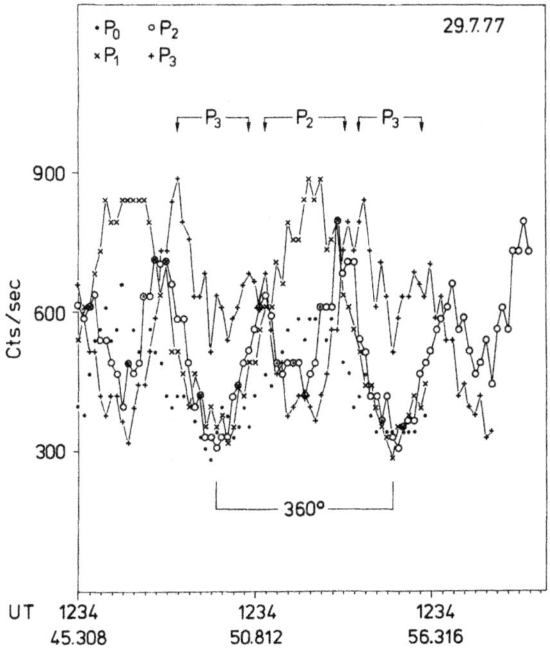

Fig. 8. Spinmodulated variations of the integral counting rate ($Ep > 24$ keV) in the GEOS proton telescopes P0, P1, P2 and P3 measured at 1234 UT shortly after the flux increase was observed. Arrows indicate phaseangle for which the instantaneous look direction of P2 or P3 corresponded to 90° pitch angle. A clear difference in the 90° flux level for a given telescope indicates an azimuthal asymmetry in the ion distribution function.

90° are marked by arrows in Figure 8. It is interesting to note, that for both detectors the measured intensity at $\alpha = 90°$ is not symmetric to a plane containing the satellite spin **S** and the magnetic field vector **B**. This plane can be identified from the particle data by noting that the maximum and minimum counting rate in P0 and P1 occurs when the detectors pass through the plane defined by the vectors **S** and **B**. Specifically the phase angle of the flux minimum corresponds to the phase of the (V, D)-component of the magnetic field vector.

The observed asymmetry in the 90° counting rate implies a significant deviation of the distribution function from rotational symmetry around the magnetic field direction. Figure 9 shows the derived azimuthal distribution of the particle intensity at 90° pitch angle with respect to the sunward direction. The resulting azimuthal anisotropy of the integral counting rate is approximately $A = 11\%$. The corresponding direction of the flow vector **A** and the phase angle of the magnetic field as inferred from the particle data are also shown in Figure 9.

The evolution of the flow vector for the event at 1200 UT is illustrated in Figure 10 together with the observed phase angles of the magnetic field. It appears that the azimuthal anisotropy of the integral ion flux ($E > 24$ keV) did not change appreciably during the flux enhancement. The magnitude of A increased

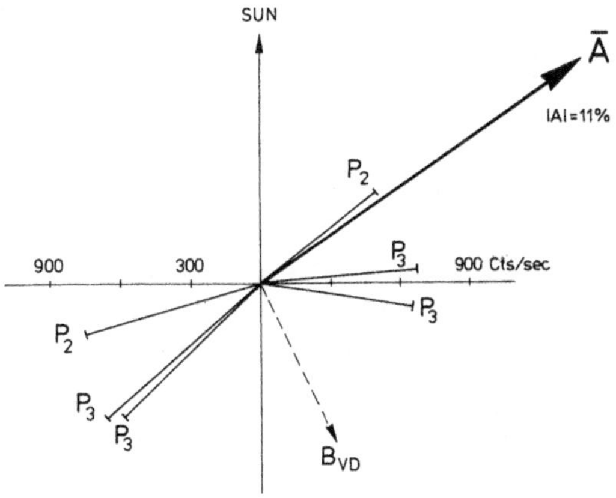

Fig. 9. Azimuthal distribution of the integral ion flux intensities at 90° pitchangles determined from a single spin revolution of GEOS at 1234 UT. The estimated direction of the flow vector is indicated by **A**. the magnitude of **A** is equal to the observed azimuthal anisotropy. The dashed line indicates the phase angle of the magnetic field component B (V, D) as inferred from the particle data.

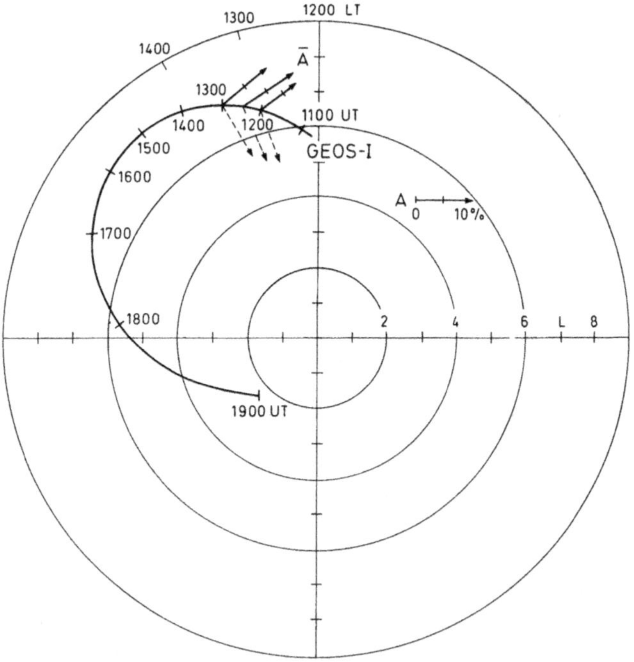

Fig. 10. Orientation and magnitude of the flow vectors **A** along the GEOS orbit determined before (1200 UT) and after the flux increase (1234 and 1255 UT). The inferred phase angles of the (V, D)-component of the local magnetic field (dashed lines) suggest a considerable warping of the field line.

from initially 8% (1200 UT) to a maximum of 11% at 1234 UT shortly after the enhancement and dropped back to 10% at 1255 UT.

The inferred phase angles of the local magnetic field in Figure 10 imply that the (V, D) component of the field vector deviate from a radial direction. This observation can indicate a considerable warping of the magnetic field line in the noontime magnetosphere.

Figure 11 shows the differential energy spectrum for $\alpha = 90°$ particles obtained shortly after the flux increase occurred (1234 UT). The spectra are constructed for the azimuthal directions with maximum and minimum counting rates as shown in Figure 9. Within the statistical limits, the spectral shape appears to be almost identical in both directions. The spectral anisotropy is approximately 20% for all energy channels above 50 keV and reduces to 7% at the low energy limit of the spectrum.

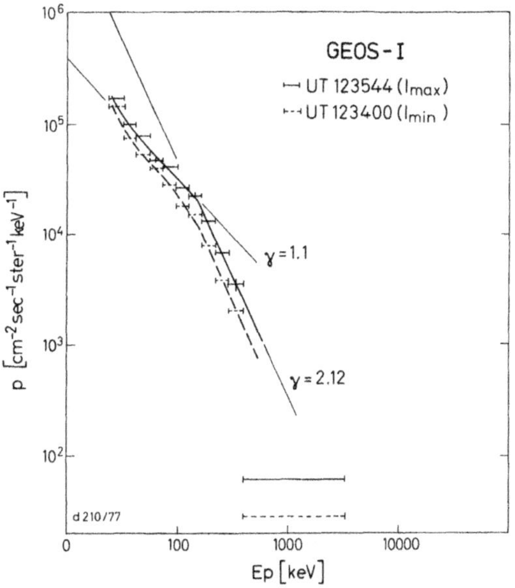

Fig. 11. Spectral distributions of energetic ions measured with P3 at 90° pitchangles. The spectra are constructed for the two azimuthal directions showing a larger and smaller relative maximum in the response of telescope P3 as presented in Figure 7. The two directions are separated in azimuthal phase by ~135° and the difference in the spectra indicate the existance of the azimuthal asymmetry at all energies.

If plasma flow or an electric convection field was to cause azimuthal asymmetries the resulting anisotropy should be larger at low particle energies than at higher energies. It is therefore concluded that GEOS observed anisotropies due to a persistent radial gradient in the particle density. Because of the absence of any noticable energy dependance particularly at higher energies the gradient is unlikely to be the immediate result of an interaction with a magnetospheric boundary.

5. Conclusions

As a result of the SSC at 0027 UT the magnetopause is pushed inside the instantaneous position of GEOS (1312 LT, 6.9 R_E). GEOS seemed to stay outside the magnetopause for at least 2 h. The magnetopause did not reach the position of ATS (1512 LT, 6.6 R_E).

Twelve hours later an injection like ion flux increase associated with a substorm is observed by GEOS-1 and ATS-6 at $L \geqslant 6.6$. The local time difference between the two satellites is almost 12^h.

The energy dispersion observed by GEOS during this substorm is interpreted in terms of spatial (steep radial profile for high energy particles) and temporal (low energy particles) effects. The inferred injection time of 1141 UT compares with the onset of the substorm at 1120 UT and the injection observed by ATS at 1151 UT.

The observed inverse dispersion at ATS remains difficult to explain within the concept of a uniform injection of substorm particles down to an injection boundary at approximately $L = 5$ ($K_p = 6$).

The distribution function of ions at the position of GEOS before and after the flux increase shows a persistent azimuthal asymmetry. The azimuthal anisotropy is approximately 10% for integral rates (≥24 keV) and 20% for all differential energy channels above 50 keV.

Based on the fact that the spectral shape is essentially independent of the azimuth a radial gradient in the particle density is assumed to cause the anisotropy. The gradient is believed to be not the effect of a magnetospheric boundary.

References

Fritz, T. A. and Cessna, J. R.: 1975, *IEEE Trans. Aerospace and Electronic Systems*, Vol. AES-11, No. 6.

Fritz, T. A., Smith, P. H., Williams, D. J., Hoffman, R. A., and Cahill, L. J.: 1974, D. E. Page (ed.), *Correlated Interplanetary and Magnetospheric Observation*, D. Reidel Publ. Co., Dordrecht, Holland, p. 485.

Konradi, A., Semar, C. L., and Fritz, T. A.: 1975, *J. Geophys. Res.* **80,** 543.

Wilken, B., Korth, A., and Keppler, E.: 1977, *Zeitschrift für Flugwissenschaft und Weltraumforschung*, Bd. 1, H. 4, 298.

OBSERVATIONS OF LOW-ENERGY ELECTRONS FROM AE-C IN THE SOUTH POLAR CUSP DURING THE GEOMAGNETIC STORM OF SEPTEMBER 21, 1977

T. A. POTEMRA and C. O. BOSTROM

Applied Physics Laboratory, The Johns Hopkins University, Laurel, Maryland 20810, U.S.A.

and

J. P. DOERING and J. S. LEE

Department of Chemistry, The Johns Hopkins University, Baltimore, Maryland 21218, U.S.A.

Abstract. The AE-C spacecraft skimmed through the southern polar cusp at a 400 km altitude during a large geomagnetic storm on September 21, 1977. This period has been designated as a special IMS period, and the AE-C data were acquired close to the times that data were acquired by the DMSP satellite at nearly the same location over the southern polar cap, and by the GEOS satellite located near the noon-meridian in the northern hemisphere. Low energy electrons (1–500 eV) were measured with the photoelectron spectrometer experiment experiment onboard AE-C. This instrument was operated in the mode which measured precipitating electron fluxes and backscattered electron fluxes in alternating 4 s intervals with two sensors. A region of intense precipitating electron fluxes was observed near 0924 UT on September 21, 1977 extending from 69 degree invariant latitude at 1100 MLT to 72 degree invariant latitude at 1152 MLT. From the spectra of the precipitating electrons, this region is identified as the southern polar cusp. Since the K_p equals 7− during this time, the displacement of the cusp down to these low latitudes is not unreasonable. Particle data obtained from the DMSP satellite on orbits close to AE-C, confirm that the position of the cusp was rapidly changing during this period, and was displaced to latitudes equatorward of the quiet time position. A second region of intense fluxes of precipitating electron was observed by AE-C at approximately 0933 UT from 69 degree invariant latitude near 1700 MLT to 66 degree invariant latitude near 1730 MLT. This region of low energy electron fluxes is characterized by slightly harder energy spectra and is interpreted as being the afternoon auroral zone. The remarkable and fortunate location of the AE-C, DMSP, and GEOS spacecraft during this special IMS period will allow future correlative studies aimed at the determination of the shape of the magnetosphere during very disturbed conditions.

1. Introduction

Studies of particle data acquired with satellite-borne detectors have confirmed that the magnetosheath plasma penetrates down to the ionosphere through the polar cusp (or cleft) (e.g., Heikkila and Winningham, 1971; Frank and Ackerson, 1972; McDiarmid *et al.*, 1975, 1976; Potemra *et al.*, 1977, 1978). The energy spectra of electron fluxes observed in the cusp peak below 100 eV and proton energy spectra peak near 1 keV (Frank, 1971; Winningham, 1972; Doering *et al.*, 1976).

A variety of instruments onboard the Atmosphere Explorer-C (AE-C) spacecraft have been used to measure detailed characteristics of the low-altitude cusp. These instruments were designed principally for aeronomy studies related to the lower atmosphere and ionosphere, but the data they have obtained in the auroral

Space Science Reviews **22** (1978) 659–666. *All Rights Reserved*

and polar regions have been used in studies which have significantly advanced the understanding of magnetospheric processes. For example, direct measurements of ion convection in the polar region have revealed an important convection reversal region near noon (referred to as the 'throat') and the dependence of the location of this region on the orientation of the interplanetary magnetic field (Heelis *et al.*, 1976). Observations of low energy electrons and protons from AE-C have been used to evaluate plasma injection mechanisms at the dayside cusp (Reiff *et al.*, 1977); to distinguish between the dayside auroral region and dayside cusp region (Potemra *et al.*, 1977); to identify important auroral electron acceleration regions (Burch *et al.*, 1976); and to study plasma instabilities in the dayside cusp (Potemra *et al.*, 1978).

The purpose of this paper is to report on the characteristics of low-energy electrons measured over the south polar cusp during the intense geomagnetic storm in progress on September 21, 1977. The electron measurements were obtained with the Photoelectron Spectrometer (PES) experiment on the AE-C spacecraft. The AE-C data were acquired close to the times that particle data were obtained by the DMSP satellite at nearly the same location, and by the GEOS satellite located near the noon meridian in the northern hemisphere. Only the PES/AE-C data will be described here, but future correlations with the DMSP and GEOS spacecraft during this special IMS period will allow the determination of the shape of the magnetosphere during very disturbed conditions.

2. Data

The data were acquired when the Atmosphere Explorer-C (AE-C) satellite was in a circular 68° inclination orbit at approximately 400 km altitude and in the 'despun' mode. In this mode, precipitating and backscattered electron fluxes in the range 1–500 eV were measured by the Photoelectron Spectrometer Experiment (PES) in alternating 4 s intervals with two sensors directed upward and downward, respectively, on the stabilized spacecraft. The PES instrument consists of electrostatic electron energy analyzers with an energy bandpass $\Delta E/E = 2.5\%$ (e.g., Doering *et al.*, 1973, 1975, 1976). The PES data presented here were acquired in the high sweep mode in which one 16 point energy spectrum between approximately 1 and 500 eV is measured every $\frac{1}{4}$ s. In principal, 'precipitating' and 'backscattered' electron fluxes can be measured within a $\frac{1}{4}$ s time separation, but 1 s averages of these fluxes will be presented in the following.

The track of AE-C during orbit 20 627 on September 21, 1977 is shown in Figure 1 superimposed upon a schematic diagram of large-scale field-aligned current regions determined from studies of magnetic data acquired by the TRIAD satellite during the expansion phases of substorms (Iijima and Potemra, 1978). The AE-C orbit was over the southern hemisphere, and the field-aligned current distribution was determined for the northern hemisphere. Preliminary studies indicate that the statistical pattern of field-aligned currents in the south pole are

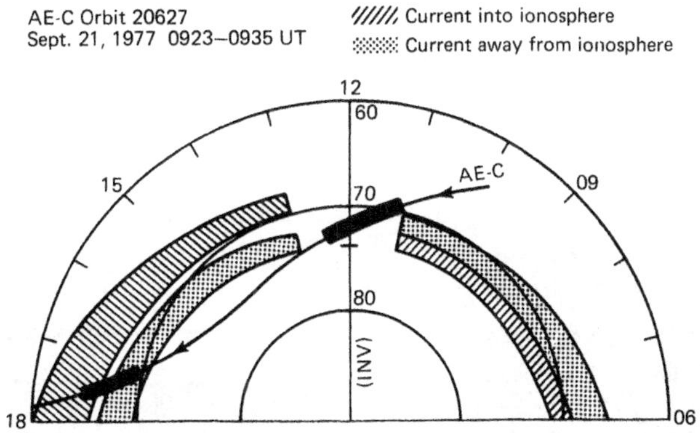

Fig. 1. The track of AE-C orbit 20 627 over the south polar region with the statistical map of large scale field-aligned currents for 'substorm expansion phase' conditions from Iijima and Potemra (1978). The black areas indicate regions of intense electron fluxes observed with the PES instrument.

the same as for the north pole (Potemra, 1977). The distribution of field-aligned currents between 1100 MLT and 1300 MLT has been purposely excluded from Figure 1 because of the complicated nature of the magnetic disturbances observed there during geomagnetic storms.

Regions of intense electron fluxes observed with the PES instrument are shown as the thick black areas along the orbital track in Figure 1. One region extends from 69° invariant latitude at 1100 MLT to 72° invariant latitude at 1200 MLT, and is identified as the cusp for reasons to be discussed later. The second region of intense electron fluxes extends from approximately 69° invariant latitude near 1700 MLT to 66° invariant latitude near 1730 MLT, and is identified as being a portion of the afternoon auroral region. This region of intense precipitating electron fluxes is close to the statistical location of upward field-aligned currents determined by Iijima and Potemra (1978).

Figure 2 is the energy-time spectrogram for the electron fluxes measured by the two sensors of the PES instrument during a portion of AE-C orbit 20 627. The top panel in Figure 2 is the spectrogram for the electron fluxes measured with the PES sensor directed radially upward from the earth and is labelled 'precipitating electrons'. The bottom panel is the spectrogram for the electron fluxes measured with the sensor directed downward toward the Earth, and is labelled 'backscattered electrons'. Common time (universal time in hours and minutes) is listed across the bottom panel with magnetic local time (MLT in hours and minutes) and invariant latitude (Λ in degrees). Electron energy, from 0 to 525 eV, is indicated on the vertical scale of each panel, and the relative electron flux is displayed by the shading. The 'upward' and 'downward' directed sensors are operated in alternate 4 s intervals.

Regions of intense precipitating and backscattered electron fluxes are evident from approximately 0924:30 UT to 0926 UT (near 1200 MLT) and from 0932:30 UT to 0933:30 UT, with less intense fluxes extending to the end of

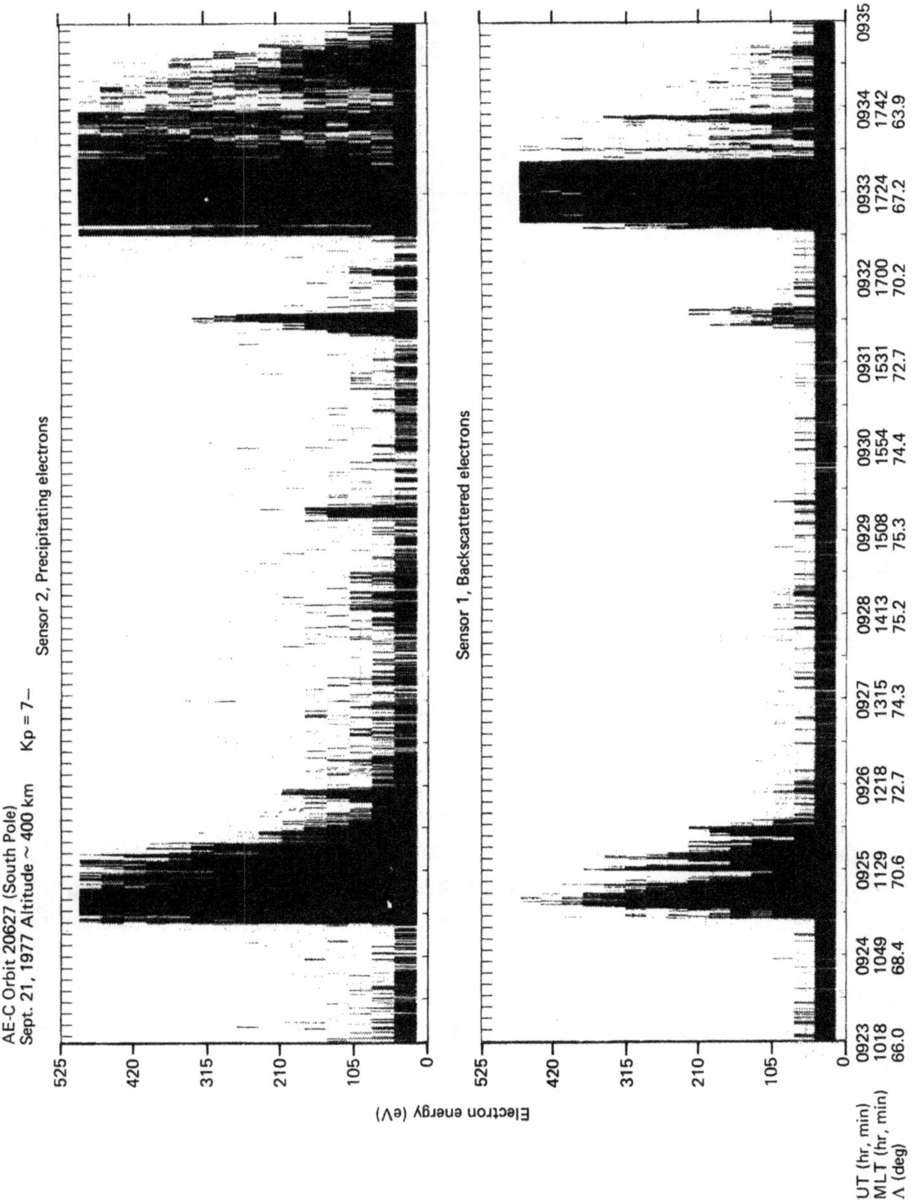

Fig. 2. The energy–time spectrogram for electrons measured with the PES instrument during orbit 20 627.

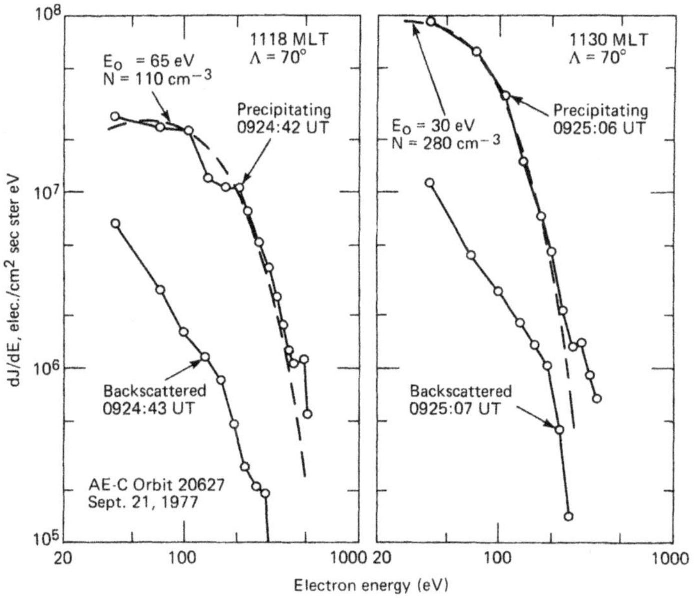

Fig. 3. Differential energy spectra of precipitating and backscattered electrons measured near noon. The dashed curves are model Maxwellian spectra with the important parameters listed in each case.

AC-E's coverage at 0935 UT. The latter region is located close to the afternoon auroral zone associated with upward field-aligned currents (as discussed in reference to Figure 1). The electron fluxes exhibit very 'soft' energy spectra especially near noon and the precipitating fluxes are more intense than the backscattered fluxes.

Figures 3 and 4 show examples of differential electron fluxes, plotted against energy in a more conventional format, observed near noon and in the afternoon auroral region, respectively. These are 1 s averaged energy spectra for precipitating and backscattered electron fluxes in each case. These examples shown some of the most intense electron fluxes observed in each region. The energy spectra of the precipitating electrons measured near noon are very close to Maxwellian, $dJ/dE \sim E/E_0 e^{-E/E_0}$, with $E_0 = 63$ eV and $N = 110$ cm^{-3} in one case and $E_0 =$ 30 eV and $N = 230$ cm^{-3} (as indicated by the dashed lines in Figure 3). The electron fluxes measured near 1700 MLT are also characterized by very soft energy spectra, but are closer to a power-law variation with $dJ\,dE \sim E^{-2}$ as shown by the dashed lines in Figure 4.

The backscattered electron fluxes are 5 to 10 times weaker than the precipitating fluxes in Figure 3, and the former are 10 times weaker than the latter for the spectra in Figure 4. This means that the net flux of precipitating electron fluxes will carry a significant field-aligned current in each region.

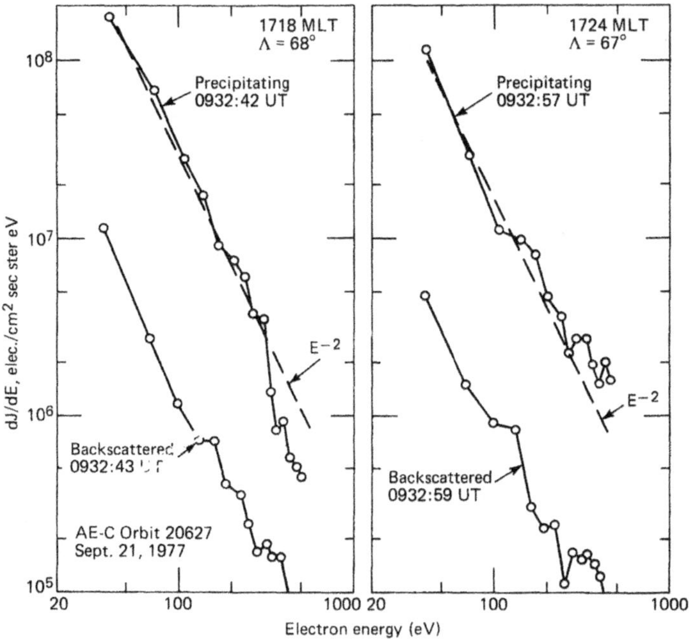

Fig. 4. Differential energy spectra of precipitating and backscattered electrons measured in the afternoon auroral zone. Power-law spectra, $dJ/dE \sim E^{-2}$ are shown as the dashed lines.

3. Discussion

The region near noon characterized by the precipitating electron fluxes with soft Maxwellian energy spectra is identified as the cusp and the region near 1700 MLT with precipitating electron fluxes with soft power-law energy spectra is identified as a portion of the dayside auroral zone. This means that the cusp has been displaced down to 69°–72° invariant latitude during this extremely disturbed period. The cusp was identified at 71° invariant latitude on May 16, 1975 when $K_p = 6-$ by the same instrument on AE-C (Potemra *et al.*, 1978).

The fact that precipitating and backscattered electron fluxes were measured permits an estimate of the corresponding net field-aligned current to be made. For example the density of the current due to the net downward flow of electrons with energies greater than 40 eV and energy spectrum shown in the left panel of Figure 3 is approximately 6 $\mu A\, m^{-2}$, and approximately 7 $\mu A\, m^{-2}$ for the example in the right panel of the same figure. The current density for the net electron flux described by the spectrum in the left panel of Figure 4 (and for electrons > 40 eV) is 30 $\mu A\, m^{-2}$ and is 18 $\mu A\, m^{-2}$ for the example in the right panel of the same figure. The values of 6–7 $\mu A\, m^{-2}$ are within the range of field-aligned currents observed in the cusp region (Iijima and Potemra, 1976), but the values of 18–30 $\mu A\, m^{-2}$ are unusually large in the afternoon auroral region. Proton fluxes can be measured with other instruments on the AE-C spacecraft, but during the period studied here, this detector was directed downward toward the Earth and measured only backscattered protons. Therefore it was not possible to compute the contribution of protons to the field-aligned current density.

4. Summary

Observations of low energy electron fluxes acquired with the Photoelectron Spectrometer experiment onboard the AE-C satellite indicate that the dayside polar cusp was displaced down to 69°–72° invariant latitude during a major geomagnetic storm in progress on September 21, 1977. A region of intense fluxes of precipitating electrons was observed in a region near 1700 MLT, between 66° and 69° invariant latitude, which statistically coincides with the region of upward flowing field-aligned currents in the disturbed afternoon auroral region. The densities of field-aligned currents carried by the net precipitating electron fluxes are very large (6–7 $\mu A\ m^{-2}$ for two examples in the cusp, and 18–30 $\mu A\ m^{-2}$ for two examples in the auroral region).

Preliminary studies indicate that the magnetopause was displaced Earthward of the GEOS spacecraft (located in the northern hemisphere). And preliminary analysis of the particle data acquired by the DMSP satellite, close to the position and time of the AE-C observation reported here, indicate that the position of the cusp was rapidly changing (H. Kroehl and M. Candidi, personal communication, 1978). The fortunate location of the AE-C, DMSP, and GEOS spacecraft during this special IMS period will permit future correlative studies of the detailed and dynamic shape of the magnetosphere during disturbed periods.

Acknowledgements

This work was supported by the National Aeronautics and Space Administration under Contract NASS-11415.

The authors gratefully acknowledge the assistance of S. Favin (APL) in data processing.

References

Burch, J. L., Fields, S. A., Hanson, W. B., Heelis, R. A., Hoffman, R. A. and Janetzke, R. W.: 1976, *J. Geophys. Res.* **81,** 2223.

Doering, J. P., Bostrom, C. O. and Armstrong, J. C.: 1973, *Radio Sci.* **8,** 387

Doering, J. P., Peterson, W. K., Bostrom, C. O., and Armstrong, J. C.: 1975, *J. Geophys. Res.* **80,** 3934.

Doering, J. P., Potemra, T. A., Peterson, W. K., and Bostrom, C. O.: 1976, *J. Geophys. Res.* **81,** 5507.

Frank, L. A.: 1971, *J. Geophys. Res.* **76,** 5202.

Frank, L. A. and Ackerson, K. L.: 1972, *J. Geophys. Res.* **77,** 4116.

Heelis, R. A., Hanson, W. B., and Burch, J. L.: 1976, *J. Geophys. Res.* **81,** 3803.

Heikkila, W. J. and Winningham, J. D.: 1971, *J. Geophys. Res.* **76,** 883.

Iijima, T. and Potemra, T. A.: 1976, *J. Geophys. Res.* **81,** 5971.

Iijima, T. and Potemra, T. A.: 1978, *J. Geophys. Res.* **83,** 599.
McDiarmid, I. B., Burrows, J. R., and Budzinski, E. E.: 1975 *J. Geophys. Res.* **80,** 73.
McDiarmid, I. B., Burrows, J. R., and Budzinski, E. E.: 1976, *J. Geophys. Res.* **81,** 221.
Potemra, T. A.: 1977, in B. Grandal and J. A. Holtet (eds.), *Dynamical and Chemical Coupling*, D. Reidel Publ. Co., Dordrecht, Holland, p. 337.
Potemra, T. A., Peterson, W. K., Doering, J. P., Bostrom, C. O., McEntire, R. W., and Hoffman, R. A.: 1977, *J. Geophys. Res.* **82,** 4765.
Potemra, T. A., Doering, J. P., Peterson, W. K., Bostrom, C. O., Hoffman, R. A., and Brace, L. H.: 1978, *J. Geophys. Res.* **83,** 3877.
Reiff, P. H., Hill, T. W., and Burch, J. L.: 1977, *J. Geophys. Res.* **82,** 479.
Winningham, J. D.: 1972, in B. M. McCormac (ed.), *Earth's Magnetospheric Processes*, D. Reidel Publ. Co., Dordrecht, Holland, p. 68.

DMSP-F2 OBSERVATIONS OF CUSPLIKE ELECTRON PRECIPITATION REGIONS DURING THE SEPTEMBER 19–21, 1977 EVENT

M. CANDIDI, H. W. KROEHL, G. KOSINSKI, and R. W. BUHMANN
NOAA NGSDC D64, Boulder, Colo. 80303, *U.S.A.*

Abstract. The differential particle spectrum of electrons between 50 eV and 20 keV, as measured by the electrostatic analyser on DMSP-F2, is used to identify regions of precipitation in the earth's polar regions above 800 km altitude. A typical cusplike spectrum is defined by analysis of the DMSP data, compared with published data, and used to locate regions of cusplike precipitation. A comparison between the September 19–21, 1977 stormtime location of such regions and the quiet-time location is discussed. In general, the region of cusplike precipitation is significantly displaced equatorward from its quiet-time location.

1. Introduction

DMSP-F2 was launched in July 1977, to a 99-degrees inclined, Sun-synchronous orbit at about 800 km altitude. The Sun-synchronous orbit keeps the satellite in a dawn to dusk plane, inclined 9 degrees from the southern geographic pole toward the Sun, and crosses each polar region every 100 min. When plotted on a magnetic latitude (MLAT) vs. magnetic local time (MLT) polar projection, the 100 km magnetic field footprint of the satellite computed using a Jensen–Cain magnetic field model traverses the dayside southern polar region as shown in Figures 1 and 2, for December 7, 1977. Latitude and time are in corrected geomagnetic coordinates computed according to the Gustaffson model [1].

The data were recorded by the Air Force Geophysics Laboratory low-energy electron analyzer. The instrument (shown in Figure 3) consists of two electrostatic analyzers that sample two energy ranges once every second, between 50 eV and 1 keV and between 1 keV and 20 keV. The accumulation time is about 0.1 s for each of the 8 energy channels of each analyzer, and the view angle is 7 degrees × 10 degrees and 3 degrees × 16 degrees respectively, toward the local zenith. In the latitude range below −60 degrees, the angle between the ambient magnetic field and the local vertical never exceeds 33 degrees. The instrument is therefore looking into the loss cone at these high latitudes.

2. Discussion

December 7, 1977 is a magnetically quiet day with Kp less than or equal to 1+ and the sum of $Kp = 7-$. The DMSP-F2 southern hemisphere pass between 1000 and 1019 UT crosses the −80 degrees MLAT contour on the dayside in the MLT region where the quiet-time southern cusp region ought to be (Figure 1). The differential electron spectra for this pass are shown in Figure 4, as a three

Space Science Reviews **22** (1978) 667–675. *All Rights Reserved*

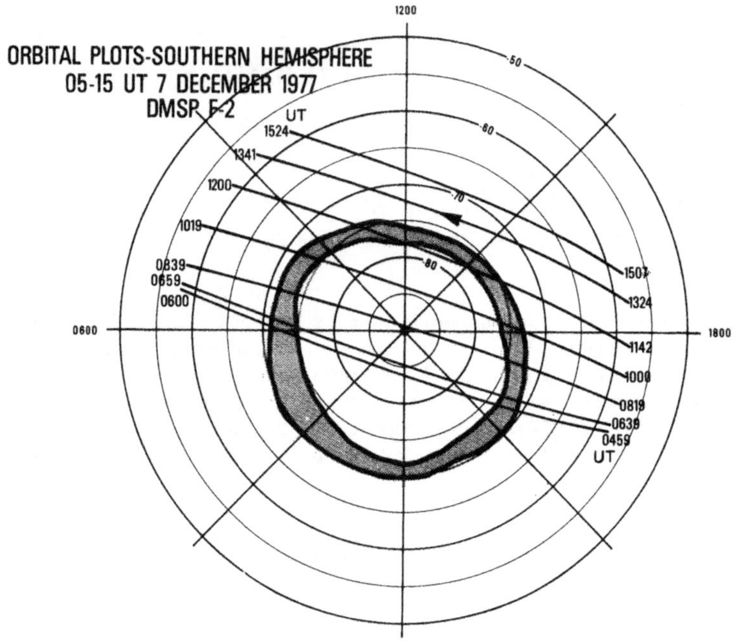

Fig. 1.

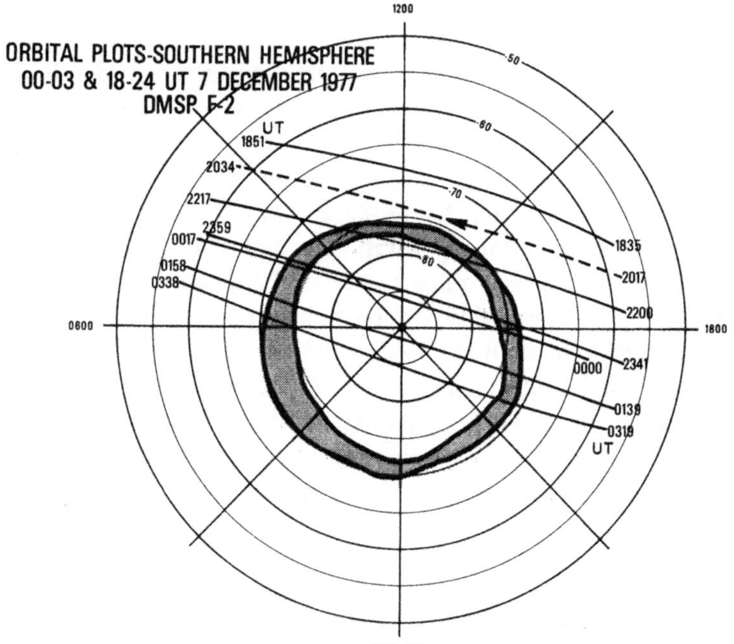

Fig. 2.

Figs. 1 and 2. Plot of DMSP-F2 southern hemisphere orbits for December 7, 1977, when projected from 800 km to 100 km, using a Jensen–Cain field model. Coordinates are corrected geomagnetic latitude and corrected geomagnetic local time. The quiet-time Feldstein auroral oval has been superimposed for reference.

Fig. 3. DMSP-F2 electron energy analyzer from Huber *et al.* [5].

dimensional plot. Three different electron precipitation regimes are identifiable:

(1) 1010–1013 UT: Soft precipitation exists with the peak electron flux around 100 eV in a region between −84 degrees MLAT at 1200 MLT and −78 degrees MLAT at 0900 MLT.

(2) 1013–1015 UT: Hard precipitation exists with the electron flux peak around 1 keV, between −78 degrees MLAT at 0920 MLT and −73 degrees MLAT at 0830 MLT.

(3) After 1015 UT: Instrumental background exists in all energy channels equatorward of −73 degrees MLAT.

Figure 5 shows a different picture for the pass on September 21, 1977, between 0957 and 1015 UT. It roughly corresponds in corrected geomagnetic coordinates to the one previously examined. On September 21, 1977, *Kp* was greater than or equal to 3−. The soft spectrum flux region, labeled (1) in the previous discussion, is now hardly identifiable and, if it does exist, is restricted to a narrow region around −68 degrees MLAT at 0800 LMT. However, the next pass on September 21, 1977 between 1139 and 1156 UT, which explores a lower latitude region on the noon meridian, Figure 6, shows a picture distinctly similar to that in Figure 4.

(1) 1144–1151 UT: Soft precipitation exists between −72 degrees and −77 degrees MLAT in the 1000–1500 MLT sector.

(2) Hard precipitation exists equatorward of −72 degrees MLAT.

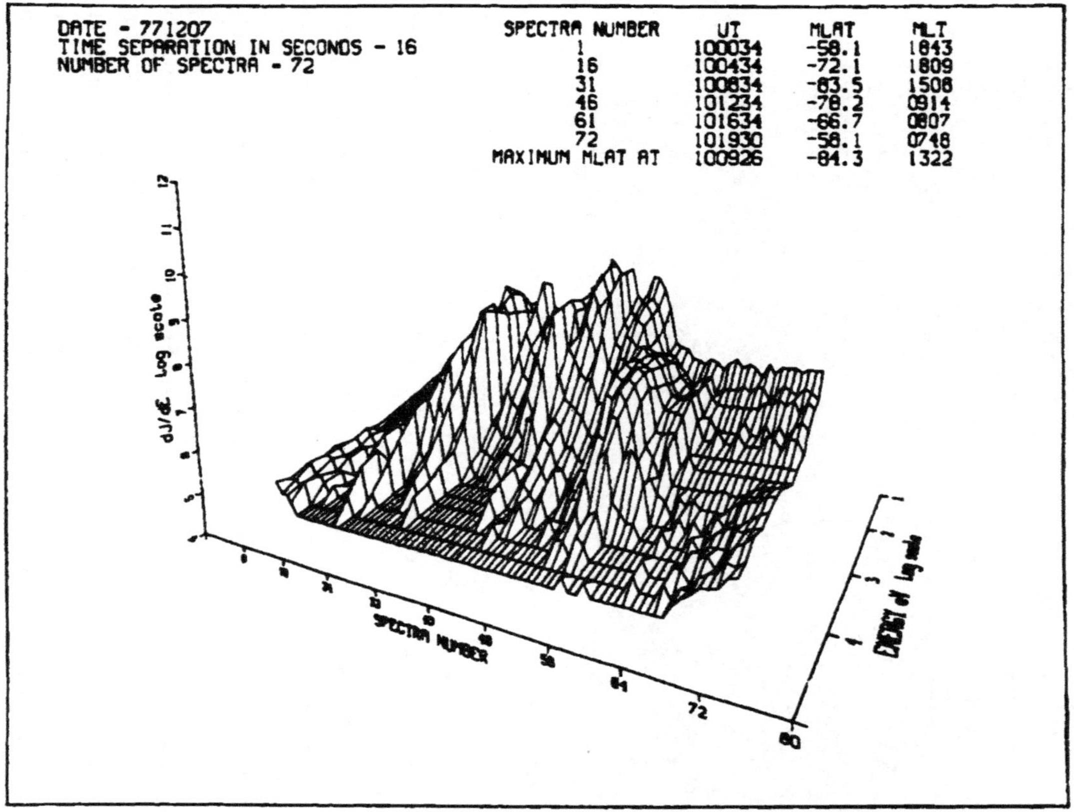

Fig. 4. Three dimensional plot of DMSP-F2 differential electron particle flux for December 7, 1977, 1000–1019 UT. Vertical axis is particle flux in electrons/(cm^2 s ster keV) plotted on a log scale, front axis is time from left to right, and horizontal right axis is energy, on a log scale, from 50 eV on the back to 20 keV at the front of the figure. One spectrum every 16 seconds is plotted. Additional ephemeris information is given at upper right.

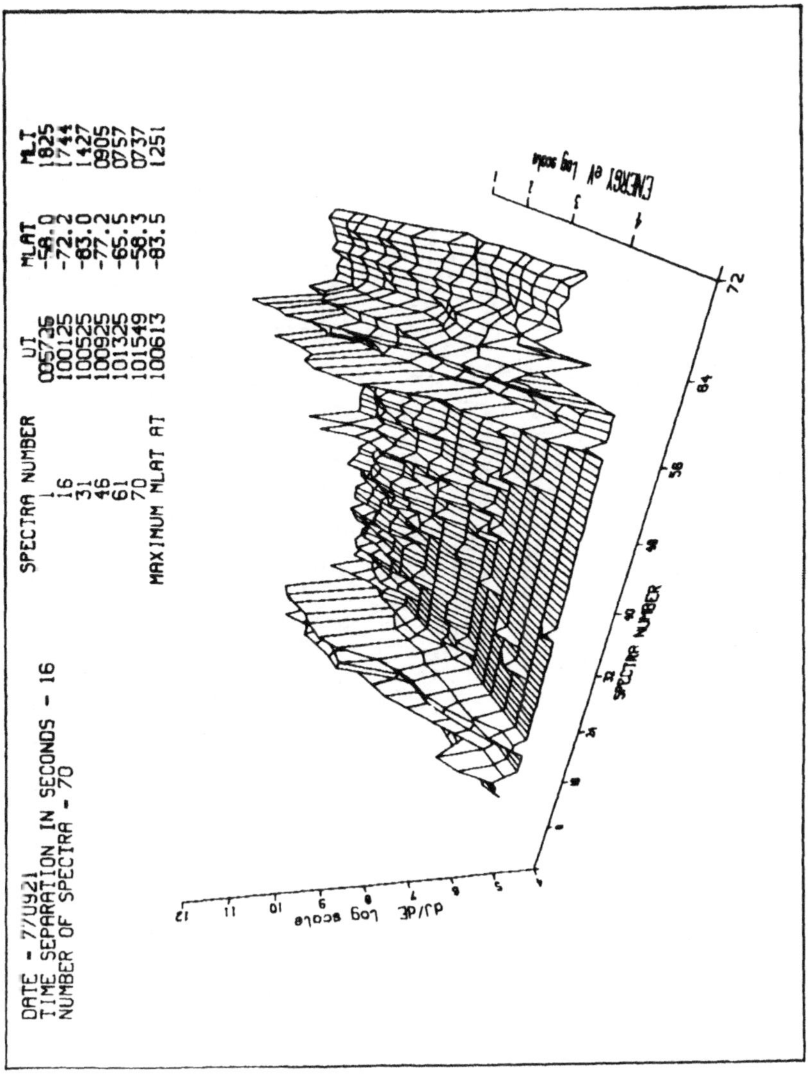

Fig. 5. DMSP-F2 electron differential particle flux for September 21, 1977, 0957–1015 UT.

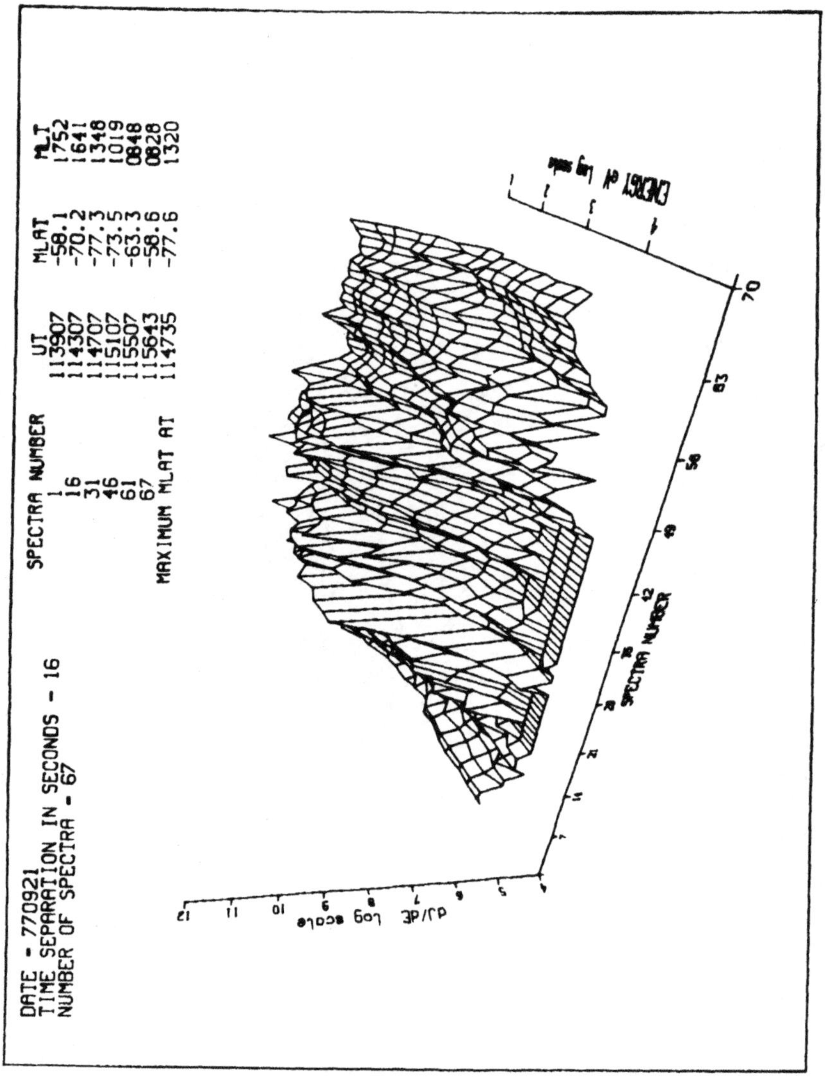

Fig. 6. DMSP-F2 electron differential particle flux for September 21, 1977, 1139–1156 UT.

These differences are explained by an expansion of the polar cap and equatorward movement of the polar cusp and auroral zones during the magnetic storm of September 19–22, 1977.

In order to carry out a more quantitative analysis of our data, the soft spectrum region (1), a spectrum that is typical of the cusp region, was extensively studied. By examining single spectra recorded during magnetically quiet periods and by comparing these results with the data published by Winningham and Heikkila [2] and Craven and Frank [3], the following selection criteria for the polar cusp were defined:

(1) The differential particle flux of at least one of the low energy channels (50, 77, 110 eV) be greater than 10^9 electrons/(cm^2 s ster keV).

(2) The flux at the 1 keV channel be lower than 6×10^6 electrons/(cm^2 s ster keV).

Spectra which satisfy these criteria are not limited to the cusp region. The criteria can be met by spectra detected also on the nightside and other spectra can be observed in the cusp region. Confidence in them increased after a statistical

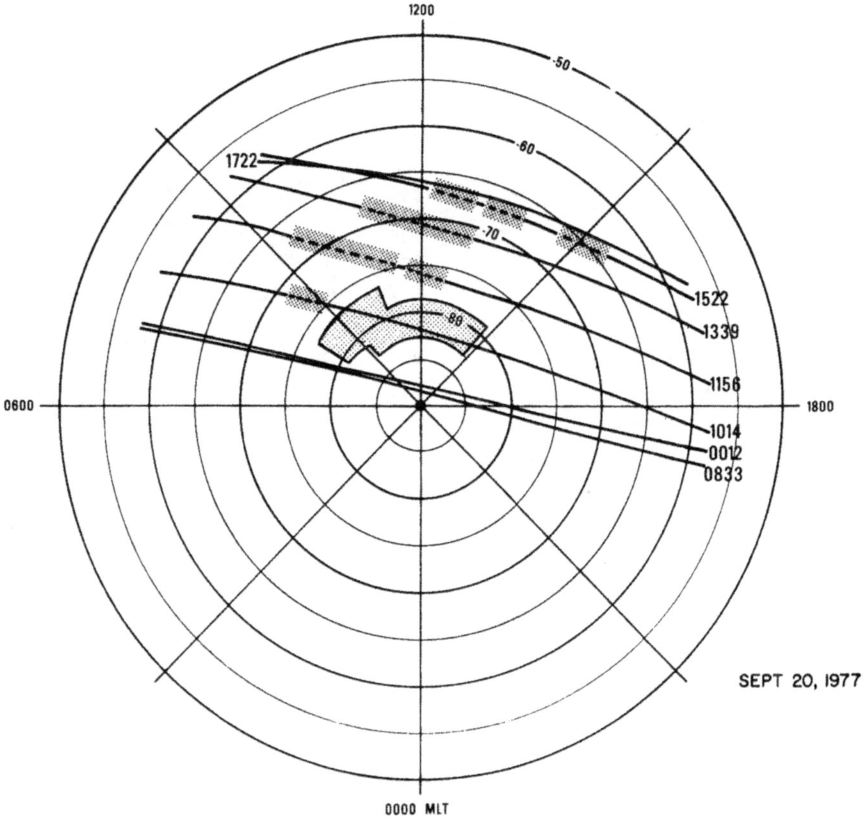

Fig. 7. Identified regions of cusplike precipitation for DMSP-F2 passes on September 20, 1977 (shaded broken foot-tracks), as compared to quiet-time cusplike precipitation region (shaded area) for December 1, 6, 7, and 8, 1977 when *Kp* was less than 2−.

analysis of occurrence frequency during quiet-time conditions placed the cusp between −77 degrees and −83 degrees MLAT.

Figure 7 shows the quiet-time region as defined by our analysis for times on December 1, 6, 7, 8, 1977 when *Kp* was less than 2−, together with the locations where spectra meeting the stated criteria were detected on September 20, 1977. The cusplike precipitation region is significantly displaced equatorward, with its equatorward boundary as low as −68 degrees MLAT, between 0800 and 1500 MLT. The latitudinal extent of this region is at times wider than expected. Dynamic behaviour from one pass to the other is evident as well. Figure 8 shows a similar comparison for September 21. A displacement of the equatorward boundary of the cusplike precipitation region to −72 degrees MLAT is evident, in the sector 1000 to 1500 MLT.

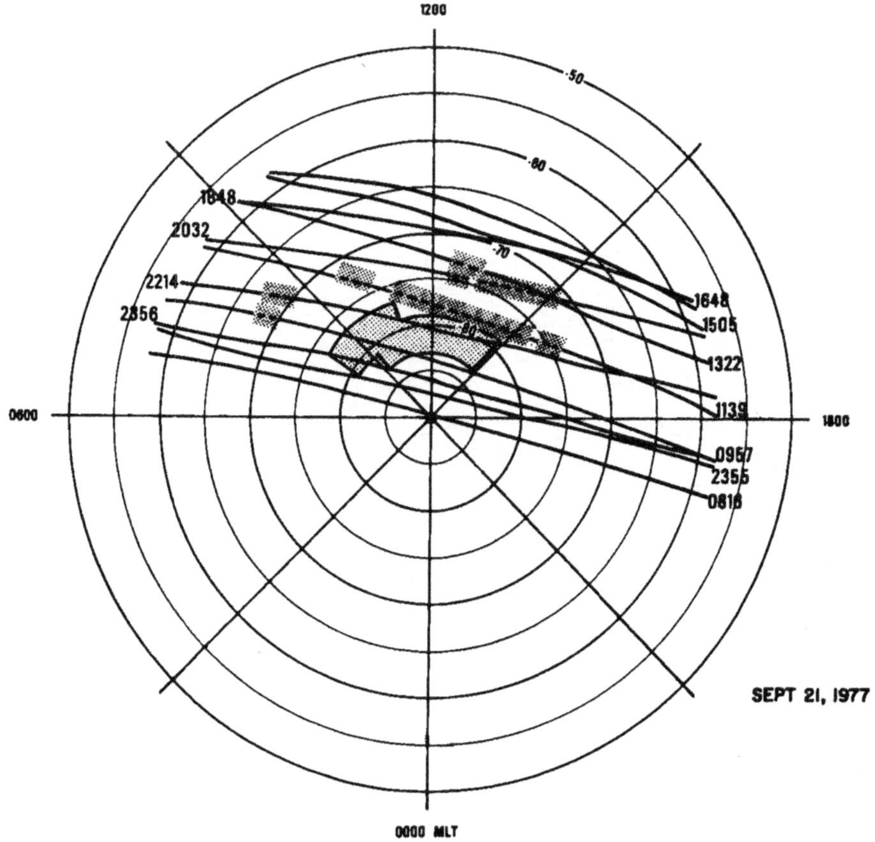

Fig. 8. Same as Figure 7 for September 21, 1977.

3. Conclusions

Having defined a typical energy spectrum for the cusplike electron flux, we observe an equatorward shift to as low as −68 degrees MLAT, for the boundary between soft and hard spectra, at times of enhanced magnetic activity. This has already been observed in correlation with negative Bz [4]. The Goddard Space Flight Center IMP-8 magnetometer data reveal extended periods of negative Bz for September 19–21, 1977.

A comparison of these observations with the GEOS-1 low-energy electron data could be quite interesting, since a depression of the magnetic field to the equatorial plane indicates the transition of GEOS-1 into the magnetosheath around 1000 UT on September 21, when it was at apogee near the noon meridian. This comparison could further clarify the disputed subject of the presence or absence of cusplike flux on closed field lines neighboring the magnetopause.

An extended statistical study is needed to better define the selection criteria for cusplike flux, and to study the dependence of cusp location on magnetic activity and seasons of the year.

Acknowledgements

One of the authors (M.C.) is indebted to the Space Science Department of the European Space Agency for Sponsoring his research fellowship at NOAA NGSDC.

References

[1] Gustaffson, G.: 1969, *A Revised Corrected Geomagnetic Coordinate System*, KGO Report No. 694, Kiruna Geophysical Observatory, Kiruna, Sweden.
[2] Winningham, J. D. and Heikkila, W. J.: 1974, *J. Geophys. Res.* **79**, 949.
[3] Craven, J. D. and Frank, L. A.: 1976, *J. Geophys. Res.* **81**, 1695.
[4] Kamide, Y., Burch, J. L., Winningham, J. D., and Akasofu, S.-I.: 1976, *J. Geophys. Res.* **81**, 698.
[5] Huber, A., Pantazis, J., Besse, A. L., and Rothwell, P. L.: 1977, *Calibration of the SSJ/3 Sensor on the DMSP Satellites*, AFGL-TR-77-0202 Scientific Report No. 2, September 1977.

GEOS dedicated S-band antenna near ESOC, Darmstadt, Germany.

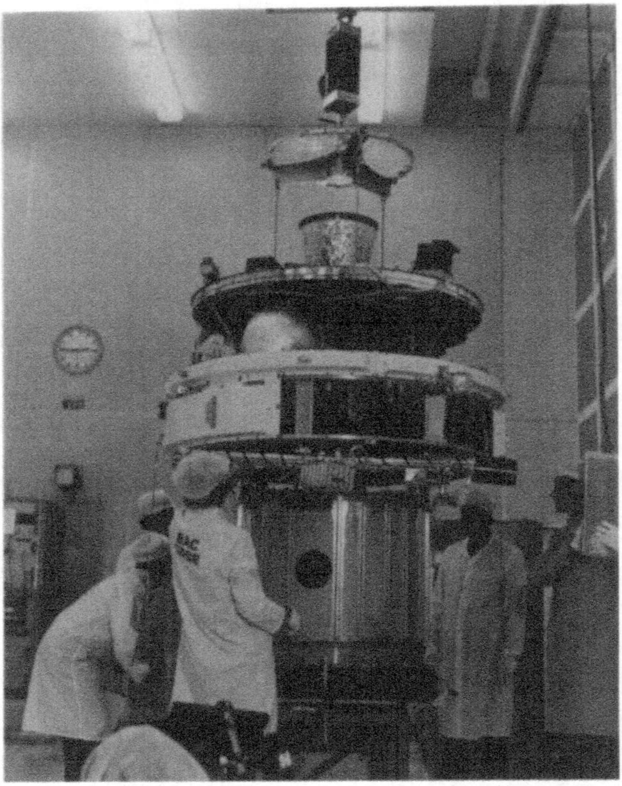

GEOS satellite during mechanical allignment test at ESTEC, Noordwijk, Holland.

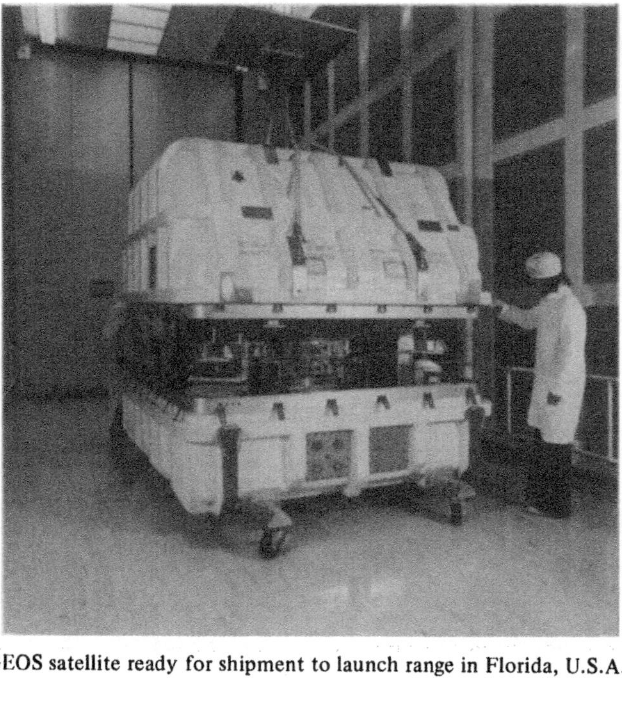

GEOS satellite ready for shipment to launch range in Florida, U.S.A.

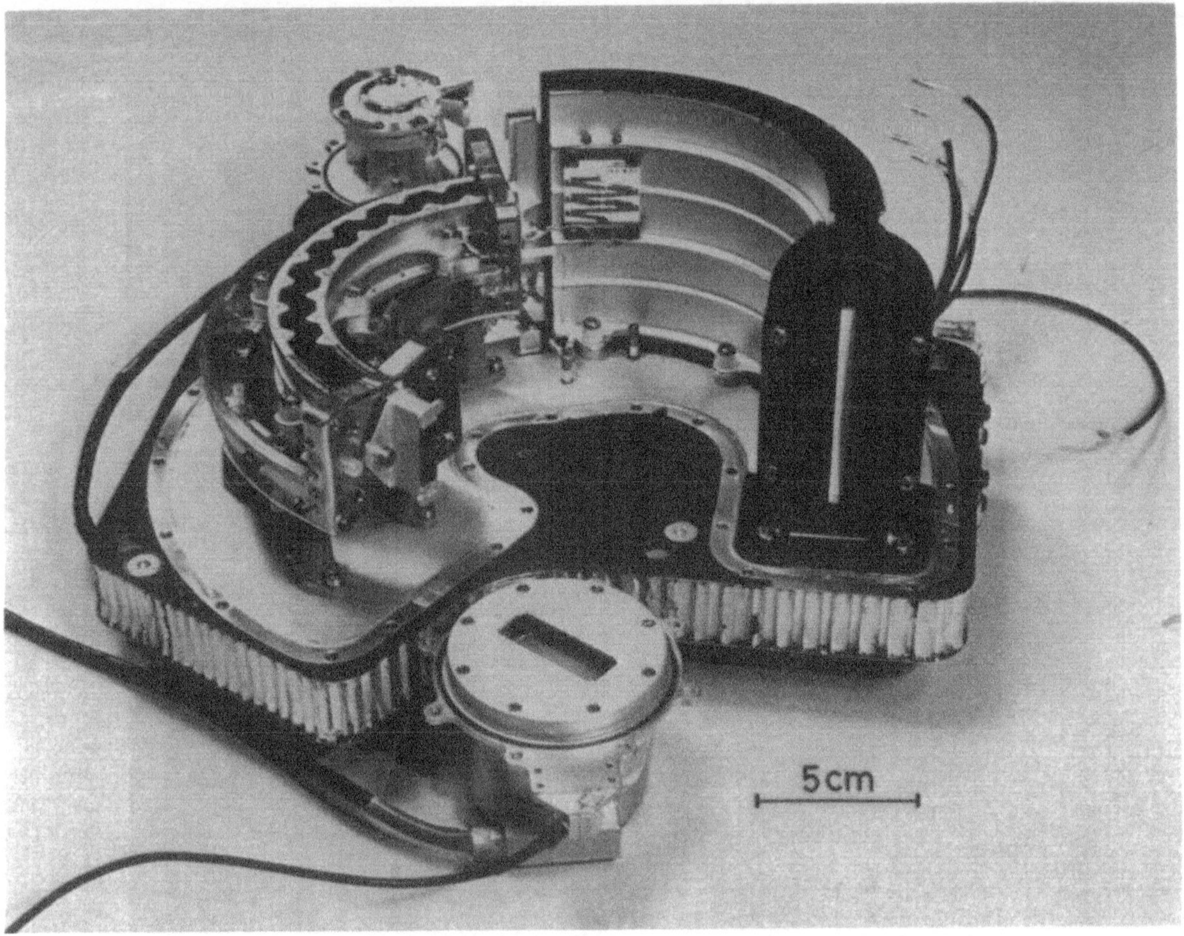

Photograph of GEOS mass spectrometer (experiment S-303), partly dismantled.

ISEE-1 and 2 spacecraft mated on top of each other ready for double-launch.

INTRODUCTION TO THE ISEE MISSION

The novel feature of the ISEE programme is that it uses a pair of spacecraft with carefully matched payloads to travel through the magnetosphere in nearly identical orbits a known and controllable distance apart. These two spacecraft can break the space-time ambiguity inevitably associated with measurements by single spacecraft on thin boundaries which may be in motion, such as the bow shock and the magnetopause. These spacecraft are named ISEE-1 and ISEE-2, they were launched as a stacked pair on a single Thor Delta 2914 from the Eastern Test Range on October 22, 1977; the apogee is 22.6 earth radii (R_E) to produce the maximum number of bow shock crossings, the perigee is about 700 km, the inclination is about 30 degrees and the orbital period approximately 57 hr. Each orbit the spacecraft penetrate the interplanetary medium for up to $\frac{3}{4}$ of the orbital period depending on the time of the year. A third spacecraft, named ISEE-3, was launched on August 12 and will be inserted into a halo orbit around the libration point situated $235 R_E$ upstream between the Earth and the Sun [1]. Plasma passing this point arrives at the Earth about 1 hr later where it may cause changes which can be observed by ISEE-1 and ISEE-2.

ISEE-1 and ISEE-3 spacecraft are based on the IMP design pattern and were built by NASA as their main contribution to the IMS. ISEE-2 was built by ESA; it is smaller, was specially designed for this mission and carries the thrusters which control the separation distance. A more detailed account of the ISEE mission is given in [2].

ISEE-3 had not been launched at the time of the Innsbruck conference so the articles presented here concern only the measurements of ISEE-1 and -2. The apogee of the first orbit was at about 11:30 LT, deliberately planned so that the magnetopause and bow shock would be crossed every orbit for the first two months of the mission. Unfortunately, complete data records for this period only became accessible to the investigators in the spring of 1978, so some of the analyses in this volume, despite intense work, are necessarily rather preliminary.

However, even in their undeveloped state, the data from the sophisticated instruments of this mission are sufficiently rich and significant to give a new impetus to magnetospheric and interplanetary research and deserve publication.

Brief descriptions of all ISEE instruments and their objectives will be found in [3].

References

1. Farquhar, R. W., Muhonen, D. P., and Richardson, D. L.: 1977, *J. Spacecr. Rockets* **14,** 170.
2. Ogilvie, K. W., von Rosenvinge, T., and Durney, A. C.: 1977, *Science* **198,** 131.
3. Contents of *IEEE Trans. Geoscience Electronics* **GE-16,** 1978.

September, 1978

A. C. Durney and K. W. Ogilvie

INITIAL ISEE MAGNETOMETER RESULTS: MAGNETOPAUSE OBSERVATIONS

C. T. RUSSELL and R. C. ELPHIC

Institute of Geophysics and Planetary Physics, University of California, Los Angeles, Calif. 90024, U.S.A.

Abstract. The magnetic field profiles across the magnetopause obtained by the ISEE-1 and -2 spacecraft separated by only a few hundred kilometers are examined for four passes. During one of these passes the magnetosheath field was northward, during one it was slightly southward, and in two it was strongly southward. The velocity of the magnetopause is found to be highly irregular ranging from 4 to over 40 km s^{-1} and varying in less time than it takes for a spacecraft to cross the boundary. Thicknesses ranged from 500 to over 1000 km.

Clear evidence for reconnection is found in the data when the magnetosheath field is southward. However, this evidence is not in the form of classic rotational discontinuity signatures. Rather, it is in the form of flux transfer events, in which reconnection starts and stops in a matter of minutes or less, resulting in the ripping off of flux tubes from the magnetosphere. Evidence for flux transfer events can be found both in the magnetosheath and the outer magnetosphere due to their alteration of the boundary normal. In particular, their presence at the time of magnetopause crossings invalidates the usual 2-dimensional analysis of magnetopause structure. Not only are these flux transfer events probably the dominant means of reconnection on the magnetopause, but they may also serve as an important source of magnetopause oscillations, and hence of pulsations in the outer magnetosphere. On two days the flux transfer rate was estimated to be of the order of 2×10^{12} Maxwells per second by the flux transfer events detected at ISEE. Events not detectable at ISEE and continued reconnection after passage of an FTE past ISEE could have resulted in an even greater reconnection rate at these times.

1. Introduction

The magnetopause has now been studied with in-situ spacecraft measurements for over a decade and a half. Much work has been performed on the average position and shape of the magnetopause (Cahill and Amazeen, 1963; Frank and Van Allen, 1964; Freeman, 1964; Ness *et al.*, 1964; Bridge *et al.*, 1965; Holzer *et al.*, 1966; Wolfe *et al.*, 1966; Cahill and Patel, 1967; Heppner *et al.*, 1967; Ness *et al.*, 1967; Binsack, 1968; Fairfield, 1971). As a result we now know the average position and shape of the forward position of the magnetopause quite well. The magnetopause position also varies with geomagnetic activity (Holzer *et al.*, 1966; Patel and Dessler, 1966; Cahill and Patel, 1967; Gosling *et al.*, 1967; Heppner *et al.*, 1967; Feldstein, 1970; Meng, 1970). It is now known that the cause of such variation is not simply variations in the external (solar wind) and internal (ring current) plasma pressure but also due to the erosion of the dayside magnetospheric flux under the influence of a southward directed interplanetary, and hence magnetosheath magnetic field (Aubry *et al.*, 1970, 1971; Russell *et al.*, 1974; Holzer and Slavin, 1978).

The magnetopause is constantly in motion with velocities greater than that of the spacecraft. This velocity is often periodic leading to multiple crossings. These multiple crossings and, in a few cases, energetic particle shadowing have been

Space Science Reviews **22** (1978) 681–715.

used to deduce velocities along the boundary normal of up to many tens of kilometers per second (Holzer *et al.*, 1966; Heppner *et al.*, 1967; Cummings and Coleman, 1968; Kaufmann and Konradi, 1969, 1973; Aubry *et al.*, 1970, 1971; Ledley, 1971). The structure or magnetic signature of the boundary has also been studied revealing a bewildering variety of signatures (Sonnerup and Cahill, 1967, 1968; Sonnerup, 1971; Aubry *et al.*, 1971; Ogilvie *et al.*, 1971; Sonnerup and Ledley, 1974; Neugebauer *et al.*, 1974). Comparison of these structures with theory has been difficult, although occasional 'classic' signatures are found (Sonnerup and Ledley, 1978). This difficulty is due in part to the complexity of the signatures, in part to the unknown relative velocity and in part to the variability of the orientation of the boundary in the course of a boundary crossing.

The ISEE mission has as one of its prime objectives the unfolding of the mystery of the structure of the magnetopause. To this end it includes high time resolution measurements of the magnetic and electric fields, low energy plasma and energetic particles. Furthermore many of the plasma and particle measurements are three-dimensional so that flows and anisotropies are not missed, nor misidentified. In addition many of the measurements are duplicated, or nearly so on both the closely spaced spacecraft. Thus time of flight measurements can be used to determine boundary velocities and hence thickness for quantitative comparisons with theory, and successive profiles can be compared to determine the time-stationarity of the boundary.

Recently two important papers on reconnection at the magnetopause have appeared. The first by Haerendel *et al.* (1978) proposes on the basis of HEOS-2 magnetic field and plasma data that reconnection occurs at the polar cusps and is impulsive. The second by Mozer *et al.* (1978) proposes on the basis of ISEE-1 magnetic and electric field data that at least one example of quasi-steady state reconnection has been found. In this paper we examine only four magnetopause crossings, and although we attempt to analyze these exhaustively, we cannot claim to have made a sufficient survey to say where reconnection takes place or to say if situations such as discovered by Mozer *et al.* (1978) are common or rare. In fact at this writing, we cannot even subject the set of crossings discussed by Mozer *et al.* to the analysis given in this paper because the requisite ISEE-2 data have not been received. Nevertheless, as will become evident in the sections to follow, we can show that reconnection does occur when the magnetosheath field is southward, very much in the impulsive manner envisioned by Haerendel *et al.* (1978). We cannot, however, pinpoint the location of initiation of the reconnection event from the limited amount of data examined so far.

2. Analysis and Display

Since we are attempting to present data from many different magnetopause crossings in this paper we have adopted certain conventions for presenting the data which will remain the same for all intervals and simplify intercomparisions. Before proceeding to examine the magnetopause crossings we wish to discuss these conventions.

In our data analysis we use three separate temporal resolutions: low resolution, 64-s averages; medium resolution, 12-s averages; and the full resolution of the instrument which ranges from $\frac{1}{4}$ to $\frac{1}{32}$ of a second per sample. The 64-s data is generally used for surveying the data, and calculating average field strengths and directions, e.g., when characterizing a crossing. The 12-s averages are overlapped by two thirds and calculated every 4 s. These averages are used to restrict the bandwidth of the analysis and reduce the number of data points, for example, when high frequency oscillations may obscure the gross structure of a boundary. Finally the full resolution of the instrument is used, whenever necessary, to define accurately the rapid variations often encountered at magnetospheric boundaries. These samples are averaged internal to the instrument and overlapped to avoid aliasing problems. A fuller description of the instrument can be found in Russell (1978) and in Russell and Greenstadt (1979).

For each of the crossings we will use a boundary normal coordinate system to display the medium resolution data (12-s overlapped averaged) across the entire interval of interest. The boundary normal system has its N-direction outwards along the boundary normal, its L-direction along the projection of the solar magnetospheric Z-direction perpendicular to the normal. The M-direction completes the right-hand orthogonal set. The purpose of choosing a local coordinate system rather than a global one is to reduce the variations in the data to two dimensions and thus make them easier to understand. We shall avoid the often confusing practice of displaying data in terms of field magnitude and angles. We feel that the common practice of displaying field directions in terms of ecliptic longitude and latitude has previously obscured certain obvious features, examples of which are conspicuous in data we will be discussing.

We use three techniques to determine normals. If the boundary is a tangential discontinuity, then there is no field along the boundary normal and the vector cross product of the fields on either side of the boundary gives the direction of the normal. Second, if the boundary contains a clear rotation of the field which is relatively noise free and the normal does not appear to rotate during the crossing, a minimum variance technique (Sonnerup and Cahill, 1967) can be used. As we shall see, for the crossings discussed herein, this technique is often far from satisfactory. Thirdly, we can compute the normal from a model. The magnetopause model we use is a conic section with eccentricity 0.4 with focus at the Earth. The normal we use will depend on the particular circumstances of the crossing and will be explicitly stated.

For selected intervals we will also display the full resolution data and do principal axis or minimum variance analysis. The results of this analysis will be displayed in an **ijk** coordinate system with **i** along the eigenvector associated with the maximum eigenvalue and **k** along the direction of minimum variance. The

j-direction completes the right-handed set. These data will be displayed as hodograms of the tip of the field vector in the i-j and i-k planes.

Tables giving the location of the satellites, their separation, field directions normals, etc., will use solar magnetospheric (GSM) coordinates throughout. The GSM system has its X-direction pointing toward the Sun, its Y-direction perpendicular to the Earth's dipole and the Sun direction towards dusk and its Z-axis in the Sun-dipole plane northward.

The magnetometer zero levels are checked periodically by flipping the spin axis aligned sensor into the spin plane. Spacecraft associated fields were determined during a period of crossed spin axis operation by intercomparing data in the spin plane of one spacecraft with that along the spin axis of the other. Other than this one cross-check the data processing of both spacecraft has been completely independent. No adjustments have been made to either the phase in the spin plane, the offsets or the gain of the instruments to bring the data into alignment. When the satellites enter regions of essentially homogeneous fields, as we will see, the data from the two spacecraft coincide to within the resolution of the plots. All differences between spacecraft visible on these plots are due to real differences in the ambient field. None are due to spacecraft or instrument related sources.

3. Northward Magnetosheath Field, Orbit 6 In

One expects the least stable magnetopause to arise when the magnetosheath and magnetospheric magnetic fields are antiparallel; conversely, the most stable configuration should be obtained when the fields are parallel. Thus, we choose to begin our examination of the ISEE magnetopause data with what should be a comparatively simple crossing, in which magnetosheath and magnetospheric field directions differ by less than 25 deg. During this interval the ISEE spacecraft were at a solar magnetospheric position of (10.35, -1.13, 5.17) R_e with a separation of (259, -174, 215) km or 360 km along the model normal.

3.1. The macrostructure of the boundary

Figure 1 shows medium resolution, 12 s averages, of ISEE-1 and -2 measurements across the magnetopause from 16:48 to 17:24 on November 5, 1977. ISEE-1 data are plotted with a heavy line, ISEE-2 with a light line. The coordinate system is the boundary normal system with the normal chosen to be that of a tangential discontinuity. Table I lists many of the parameters relevant to this crossing including the direction of the model normal and the tangential discontinuity (TD) normal. We note that these two normals differ by over 14°, whereas when the field is southward as we will see in later sections the normals agree much more closely.

There are three distinct magnetic field regimes in Figure 1 which can be characterized by the behavior of the field magnitude. First, the period before 16:59 is typified by oscillating field strength which maximizes at 40γ. These

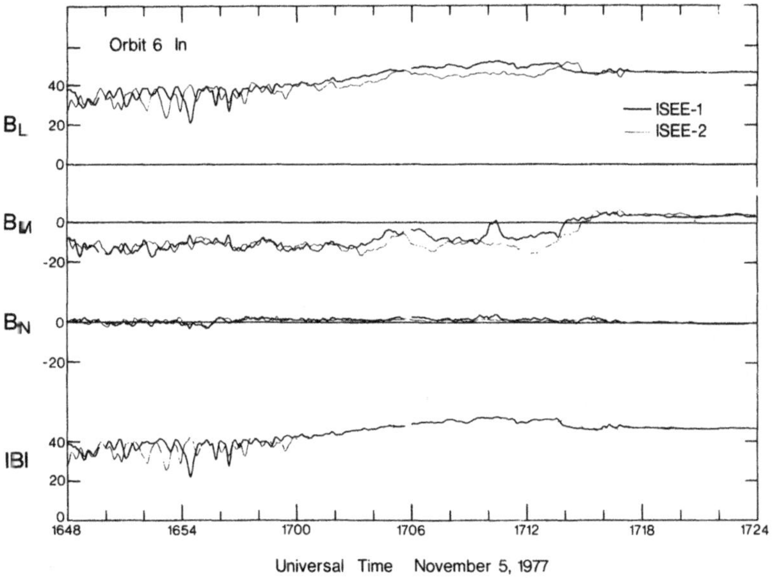

Fig. 1. Magnetic field measured by ISEE-1 (heavy line) and ISEE-2 (light line) across magnetopause on orbit 6 inbound, November 5, 1977. The coordinate system is oriented so that B_N is along the normal of the boundary assuming the boundary is a tangential discontinuity. B_L is along the projection of the solar magnetospheric Z-direction perpendicular to the boundary normal. The data shown here are 12 s averages calculated every 4 s, called medium resolution data. The scales on this and all following plots are in units of gammas ($1\gamma = 1$ nT). The ISEE spacecraft were at $(10.35, -1.13, 5.17)R_e$ in solar magnetospheric coordinates and separated by 360 km along the magnetopause normal.

oscillations in field strength show little coherence between the two spacecraft, even though the separation is only 380 km. We note that the transverse oscillations in B_N and to a lesser extent in B_M show high coherence. After 16:59 the oscillations decrease, and the magnitude rises to a peak of about 62γ before dropping suddenly at 17:14 on ISEE-1, followed by ISEE-2 roughly a minute later. Thereafter the field strength remains steady at about 47γ.

TABLE I
Magnetopause parameters

Time	Nov. 5, 1977 17:15 UT	Nov. 3, 1977 07:46 UT
Orbit	6 inbound	5 inbound
Location $\mathbf{R}_{GSM}$ (R_e)	(10.35, −1.13, 5.17)	(10.53, 0.46, 5.20)
Separation $\mathbf{S}_{GSM}$ (km)	(259, −174, 215)	(261, −106, 236)
$\mathbf{B}^{sh}_{GSM}(\gamma)$	(−13.8, 15.7, 28.1)	(2.0, 13.4, −4.5)
$\mathbf{B}^{ms}_{GSM}(\gamma)$	(−24.3, 1.6, 39.9)	(−15.4, −0.1, 42.3)
$\sphericalangle \mathbf{B}^{sh}, \mathbf{B}^{ms}$	24.8°	110.3°
${}^{m}\mathbf{n}_{GSM}$	(0.943, −0.072, 0.325)	(0.946, 0.031, 0.322)*
${}^{TD}\mathbf{n}_{GSM}$	(0.835, −0.190, 0.516)*	(0.939, −0.025, 0.342)
$\sphericalangle {}^{m}\mathbf{n}, {}^{TD}\mathbf{n}$	14.4°	3.9°
$\mathbf{S} \cdot \mathbf{n}$ (km)	360	320

* Normal used in calculations and plots.

The region before 16 : 59 UT is clearly the magnetosheath. The LASL/MPI fast plasma instrument shows that the plasma in this region is flowing with a velocity of from 80 to 100 km s^{-1} with a density of about 25 cm^{-3}. An interesting feature is that the plasma thermal pressure is anti-correlated with the field pressure here. Thus, the dips in the field strength appear as spikes in the plasma pressure (G. Paschmann, personal communication, 1978).

It is tempting to call the entire region from 16:59 to 17: 14, the magnetopause. The two satellites have obviously entered a field structure, for ISEE-1 and -2 fields begin to differ significantly at this point, with ISEE-2 lagging ISEE-1 in field strength and direction. The time lag and the difference in field values are a complicated function of time and it is difficult to trace the motion of the boundary over the spacecraft until about 17: 13 when the field changes at the two spacecraft are both unidirectional.

The variations in the plasma parameters during this crossing are initially as subtle as the field data (see Paschmann *et al.*, 1978). At about 16:59 the density slowly begins to decrease as does the plasma thermal pressure. At 17:12 on ISEE-1 and 17: 13 on ISEE-2, the density drops more quickly. At about 17:14 on ISEE-1 and 17:15 on ISEE-2 the first proton temperature rise occurs followed by a second rise at 17:16 and 17: 17 on the two spacecraft. This latter behavior is characteristic of the magnetopause boundary layer (Haerendel *et al.*, 1978). Thus the boundary layer straddles our final magnetopause current layer.

3.2. The microstructure of the boundary

Figure 2 shows the full resolution of the instrument in spacecraft coordinates for the ISEE-1 crossing of the main current layer. The bandwidth of the data here is 24 times that of Figure 1. We see an additional change upon crossing into the magnetosphere, namely the high frequency wave amplitude drops. The start of the boundary layer, BL, has been chosen to be the time of the onset of the rise in proton temperature as measured by the LASL/MPI fast plasma analyzer and the end to be the time of the last temperature rise accompanied by a density drop. Internal to the boundary layer a reversed flow was seen, directed towards the Sun with a velocity of greater than 60 km s^{-1}. There is no obvious signature of this flow in the magnetic field.

The correlation with the Berkeley/Toulouse energetic particles is also quite interesting. Enhanced fluxes of 2 keV electrons are first seen on ISEE-1 at 17:12:20 and on ISEE-2 at 17:13:55, roughly at the same location within the boundary crossing as judged by the phase of the field variation. Although the initial electrons were seen on ISEE-1 at 17:12:20, the final and strongest increase in electrons occurred at 17:13:30 coincident with the entry into the boundary layer as defined by the fast plasma instrument. The ISEE-2 data shows only a single electron increase and it is also coincident with the boundary layer defined by the fast plasma instrument. Enhanced fluxes of 2 keV protons are not

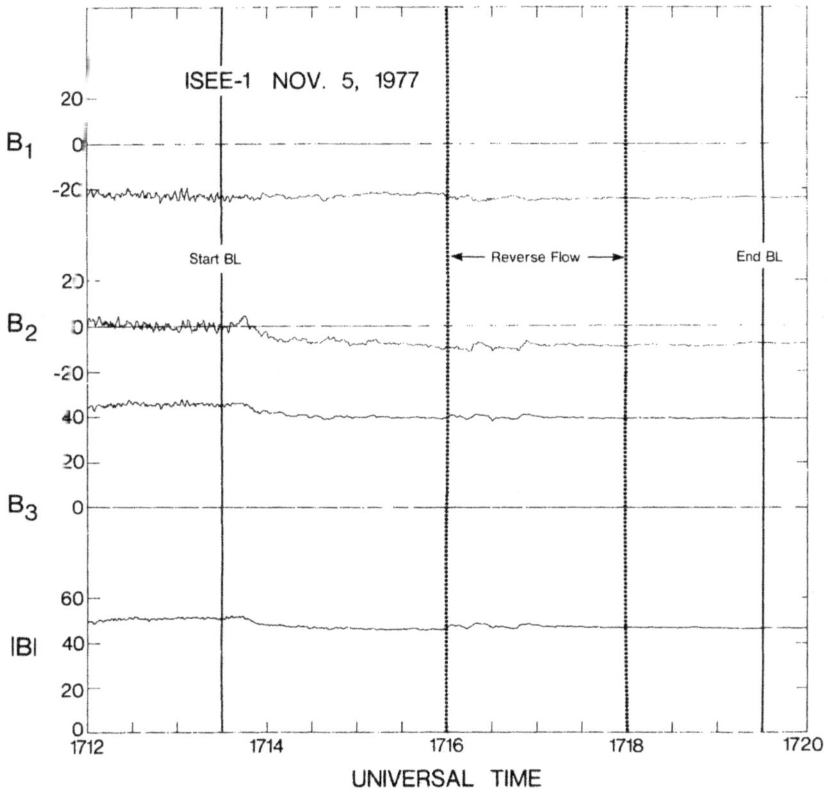

Fig. 2. High resolution ISEE-1 measurements (4 samples every second) through the magnetopause on November 5, 1977. Measurements are in spacecraft coordinates.

seen until somewhat later 17:13:45 on ISEE-1 and 17:14:40 on ISEE-2, essentially when ISEE had completely traversed the final current layer.

Figure 3 shows hodograms of the variation of the field through the magnetopause in boundary normal components. The $B_L - B_M$ plane is the plane of the boundary. We see roughly the same general variation on both ISEE-1 and -2. Initially on the magnetosheath side of the boundary there were strong fluctuations. Eventually the field rotates through about 30° and the fluctuations settle down to much lower magnetospheric levels.

3.3. Velocity and thickness of the magnetopause

If the magnetopause were stationary during this crossing, it would have a thickness of 2200 km, the distance travelled by the spacecraft along the normal from 16:59 to 17:15. However, the data suggest that the magnetopause is moving, sometimes irregularly during this crossing. In order to measure a velocity we have to find the same point on the boundary measured at different times by the two spacecraft. The only location in the magnetopause traversal discussed

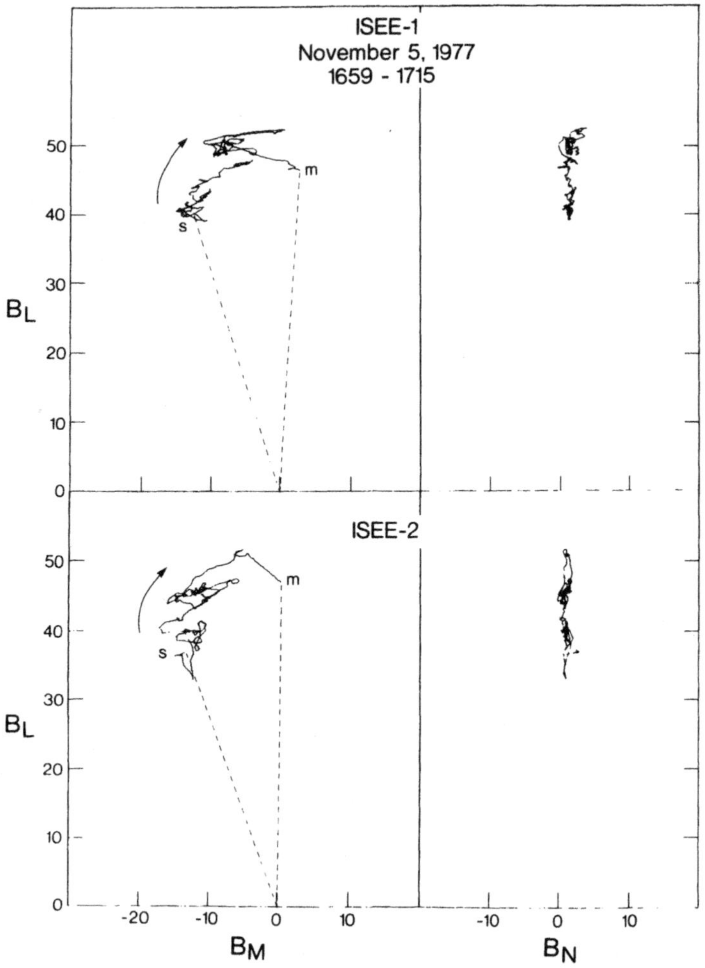

Fig. 3. Hodogram of the variation of tip of the magnetic field vector through the magnetopause on November 5, 1977. The data are 12 s averages rotated into boundary normal coordinates.

here for which this is unambiguously possible, assuming invariant structure, is in the crossing of the final current sheet at about 17:14.

Figure 4 shows the B_M component of the field for ISEE-1 and -2 from 17:13-17:18. At the line labelled '*a*' ISEE-1 and -2 are 360 km apart along the magnetopause normal. In 20 s, ISEE-2 reaches the same field value. Interpreting this as motion of the magnetopause relative to the two spacecraft, the resultant velocity is 18 km s^{-1}.

At this time, as indicated by the line labelled '*b*', ISEE-1, which is 360 km closer to the Earth, measures a new field value. ISEE-2 reaches this value 14 s later for an average velocity of 26 km s^{-1}. However, ISEE-1 has now moved further along the boundary to a new field value at '*c*', which ISEE-2 reaches 8 s later at an average velocity of 45 km s^{-1}.

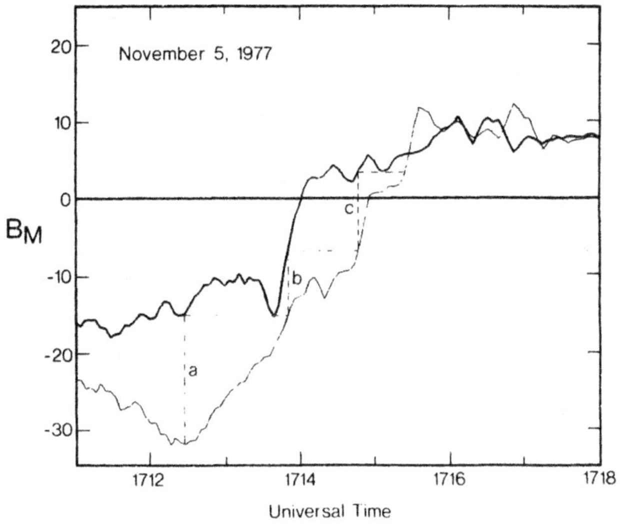

Fig. 4. Medium resolution averages of the B_M component of the field measured on ISEE-1 (heavy line) and ISEE-2 (light line) on November 5, 1977. At the time of the vertical lines labelled '*a*', '*b*', and '*c*', the two spacecraft were separated by 360 km along the magnetopause normal. Thus the current layer causing the B_M variation is over 1180 km thick.

Since it took three steps for us to traverse the boundary the current layer must have been about 1000 km thick. The velocity is quite irregular, varying from 18 to 45 km s^{-1} according to our calculations. However, if the gradients in the field are a measure of velocity, the instantaneous velocity must at times be even greater since we averaged over varying slopes.

3.4. Discussion

In a situation with a strong northward magnetic field, as we have here, we would expect that the magnetopause would be relatively simple. Perhaps it is, in comparison with the crossings we will be examining below. Nevertheless it has its complications. First, the velocity is quite variable and temporal variations in the structure confuse the determination of velocity and therefore thickness. Perhaps surprisingly, the magnetic field strength within the boundary is larger than the field in the magnetosheath or in the magnetosphere. This effect cannot be dismissed as due to a temporal change in external conditions during the crossing. Both ISEE-1 and -2 see the effect at different times as they pass through the boundary. Finally, again perhaps surprisingly, there is a boundary layer of plasma in the magnetosphere adjacent to the magnetopause, and even in the apparent absence of any reconnection between the magnetosphere and magnetosheath reverse flow of plasma to the Sun is seen in the boundary layer.

4. Horizontal Magnetosheath Field, Orbit 5 In

When thinking of the typical magnetopause, one usually considers the case in which the interplanetary and magnetosheath fields are in the ecliptic or equatorial plane and therefore at right angles to the magnetospheric field. The magnetopause crossing inbound on orbit 5 around 07:50 UT, November 3, 1977 occurred under nearly such conditions. The location of the ISEE spacecraft was (10.53, 0.46, 5.20) R_e in solar magnetospheric coordinates and their separation vector was (261, −106, 236) km. ISEE-1 was the lead spacecraft. The magnetosheath field was about (2.0, 13.4, −4.5)γ and the field just inside the magnetosphere was (−15.4, −0.1, 42.3)γ. Thus the angle between the magnetosheath and magnetospheric fields was locally about 110°.

4.1. The macrostructure of the boundary

Figure 5 shows the medium resolution data from 07:34 to 08:10 UT. ISEE-1 data are plotted with a heavy line, ISEE-2 data with a light line. The coordinate system is the boundary normal system with the normal chosen to be that of the model normal. As shown in Table I, the angle between the model normal and the

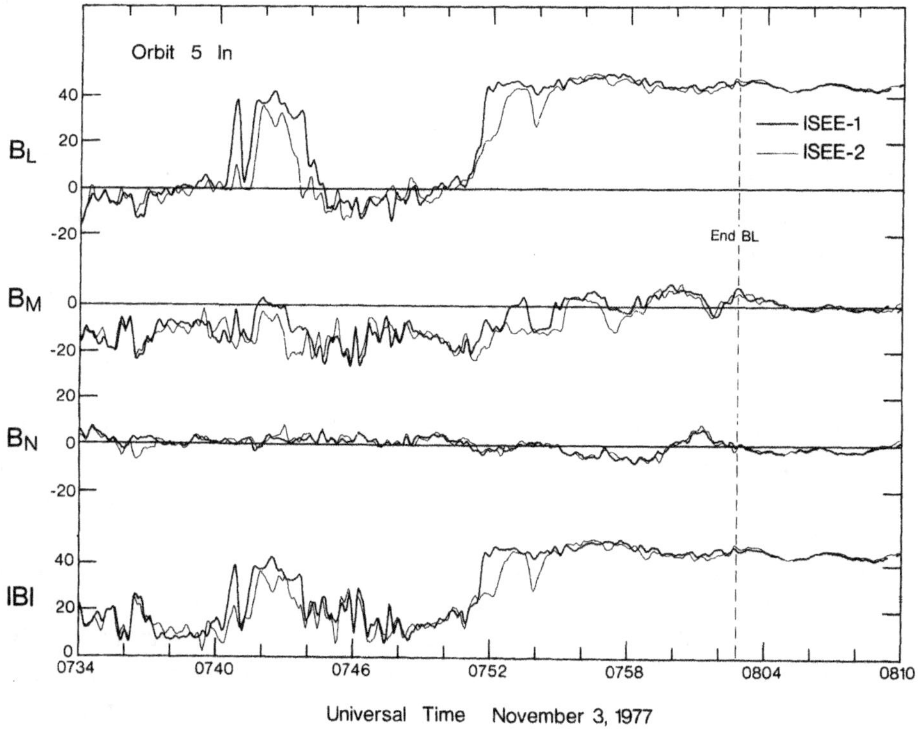

Fig. 5. The medium resolution magnetic field measured by ISEE-1 (heavy line) and ISEE-2 (light line) on November 3, 1977, expressed in boundary normal coordinates. ISEE-1 and -2 were at (10.53, 0.46, 5.20)R_e in solar magnetospheric coordinates and separated by 320 km along the normal at this time.

normal obtained from the assumption that the boundary is a tangential discontinuity is 3.9° deg, which is much closer agreement than the 14° obtained at a very similar position or orbit 6 inbound when the magnetosheath field was strongly northward.

As is most evident in the BL component and the total field there are 7 crossings or partial crossings seen by the two spacecraft, the first being at 07:41 and the last at 07:54. Since B_L is essentially along the Z-GSM direction, the field directions on the far right of this figure are those expected in the magnetosphere. We note how well the spacecraft fields track each other on the far right and far left of this figure; the deviations occur principally at the boundary crossings.

The field along the model normal, B_N, oscillates about zero and is everywhere less than 9γ in magnitude. The oscillations in this component again are highly coherent between the two spacecraft. This suggests that the source of the oscillations is the rocking back and forth of the true boundary normal. The oscillation of the B_N component of the field would be simply the local field magnitude times the sine of the angle between the model normal and the actual normal projected in the field-normal plane. Deviations of B_N of 1.5γ in the magnetosheath and 6γ in the magnetosphere correspond to fluctuations in the normal direction of about 7° but in nearly orthogonal directions.

The fluctuations in B_M are not as immediately interpretable. However the B_M trace does lead to some conclusions not readily evident in the other component. For example, since the B_M component does not reach magnetospheric levels during the 07:51-07:53 crossing on ISEE-2 we know the crossing was not complete. The fact that neither ISEE-1 nor -2 B_M components settle down to their final magnetospheric level until 08:05 suggests that something is occurring in this region that is not evident in the other components.

Perhaps the most useful component, however, is B_L. This component contains most of the field change across the boundary and hence is a good tracer of the distance travelled through the boundary. When ISEE-2 reaches a field value just previously attained by ISEE-1, we know the boundary relative to the satellite has moved along its normal a distance of 320 km, that is the separation distance between ISEE-1 and -2 measured along the model normal. Thus, the time delay between ISEE-1 and -2 can be converted to a velocity. We will defer discussion of the velocities derived from the behavior of B_L until later.

The boundary layer as determined by the LASL/MPI fast plasma instrument begins immediately inside the magnetopause and ceases at 08:03 UT. The field is somewhat more irregular in the boundary layer region than later in the magnetosphere. The variations in the B_M component suggest the presence of field aligned currents within the boundary layer.

4.2. The Microstructure of the Boundary

High resolution plots of the time series of the magnetic field data do not reveal appreciably different detail than is evident in the medium resolution plot shown in

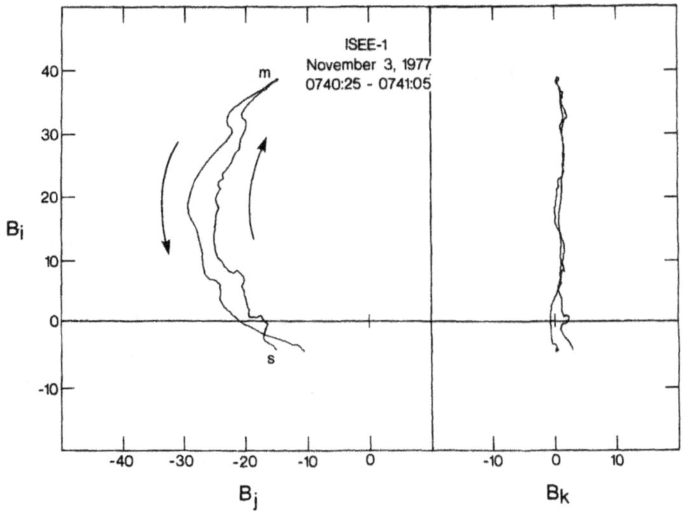

Fig. 6. Hodogram of the top of high resolution measurements of the magnetic field vector on ISEE-1 in the principal axis system for the first pair of in and out crossings on November 3 at 07:41.

Figure 5. Therefore we will next examine the hodograms of these boundary crossings. The first crossing of the magnetopause is centered about 07:40:30. There is almost a complete traversal by ISEE-1 and only a very partial crossing by ISEE-2. Minimum variance analysis of high resolution ISEE-1 data through this double crossing is shown in Figure 6. This pair of crossings is extremely close to the ideal compared to those that follow; first, the statistically determined minimum variance direction is only 3° different from the model normal, and second, the average field along the minimum variance eigenvector is not significantly different from zero, suggesting that this boundary is essentially a tangential discontinuity.

It is first apparent that ISEE-1 enters the boundary layer at 07:40:30, concurrent with the approximate midpoint of the first magnetopause crossing. While ISEE-2 makes a partial penetration of the magnetopause, no sign of the boundary layer is seen, and these two facts point out that the higher energy particles do not appear until part way into the main current sheet. A weak (<50 km s^{-1}) reversed flow was seen by the fast plasma analyzer at 07:40:50.

The third magnetopause crossing at 07:41:20 by ISEE-1 and 07:41:40 by ISEE-2 is accompanied by small variations in the B_N component, suggesting that the boundary normal direction is changing at this time. Since the principal axis directions were not sufficiently well determined to perform meaningful comparisons between ISEE-1 and -2, nor between adjacent crossings we have chosen simply to display the hodograms in boundary normal coordinates, using the high resolution data. These hodograms appear for ISEE-1 and -2 in Figure 7a; it is clear that differences between the two spacecraft traces reflect the variability of

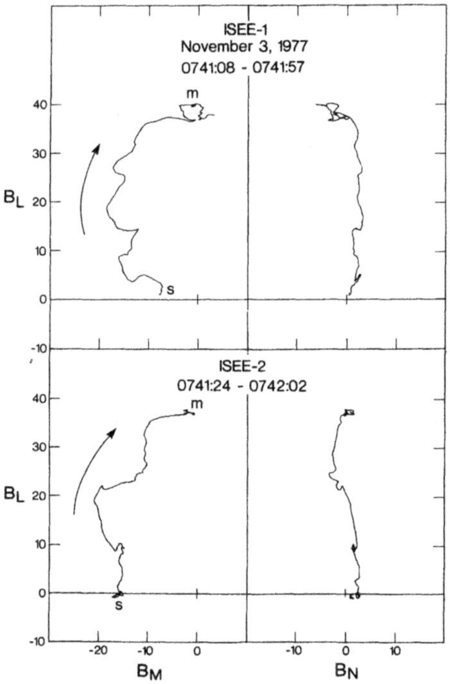

Fig. 7a. Hodogram of high resolution magnetic field data in boundary normal coordinates across 07:42 magnetopause.

the boundary orientation and structure. If, for example, differences in the B_N component in Figure 7a are due principally to changes in the boundary attitude, the maximum variation corresponds to changes in true normal direction of $\sim \pm 5°$. The differences in the ISEE-1 and -2 B_L vs B_M traces, while in broad agreement with Figure 6, point out the variability of the boundary micro-structure on temporal and spatial scales comparable to the crossing time and spacecraft separation distance.

At roughly 07:42:40 ISEE-2 begins the fourth magnetopause traversal, followed by ISEE-1 some 24 s later. This crossing is characterized by extreme departures from the relatively simple signatures of the initial crossings. High resolution hodograms in boundary normal coordinates are displayed in Figure 7b, in which are evident significant differences between the traces of the two spacecraft, as well as variations from the 'ideal' rotation. Most notably, a large drop in field magnitude occurs on ISEE-1 at the same point in the field rotation as a magnitude enhancement occurs on ISEE-2. Similar features appear more prominently in crossings analyzed in later sections of this paper, and discussion of these features is deferred until then. It is noteworthy that prior to this crossing at 07:42:30 on ISEE-2 and 07:42:40 on ISEE-1, results from the fast plasma analyzer indicate a change from boundary layer to a plasma with nearly magnetosheath densities, and flowing weakly reverse to the magnetosheath flow

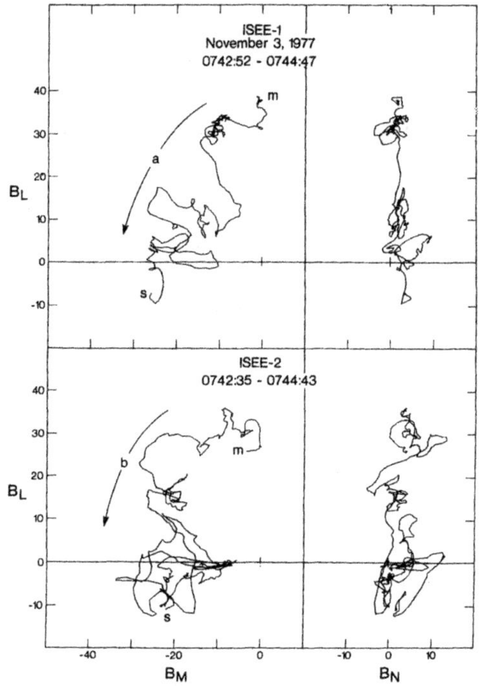

Fig. 7b. Hodogram of high resolution magnetic field data in boundary normal coordinates across 07:43 magnetopause.

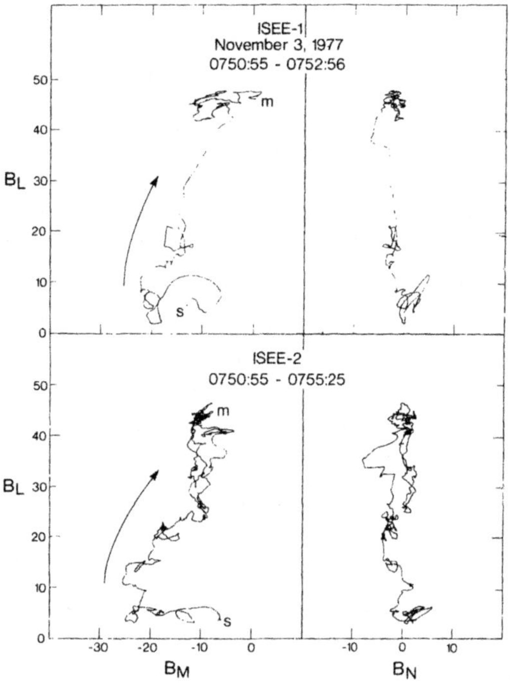

Fig. 7c. Hodograms of high resolution magnetic field data in boundary normal coordinates across 07:52 magnetopause.

direction. A true return to the magnetosheath as evidenced by the field and plasma experiments does not take place until 07:44.

The fifth and final full magnetopause crossing occurred at roughly 07:52 on ISEE-1 and 07:53 on ISEE-2. The boundary traversal is accompanied by entry into the boundary layer, which is seen as a strong reversed flow ($\sim 100\ \mathrm{km\ s^{-1}}$) on ISEE-1, and somewhat weaker ($\sim 50\ \mathrm{km\ s^{-1}}$) on ISEE-2, according to the LASL/MPI fast plasma analyzer. This reverse flow is seen on ISEE-1 until 07:52:30 when the bulk velocity direction begins to swing through anti-sunward flow to final downward flow at roughly 07:52:50. During this flow direction variation the B_M component on ISEE-1 changes from $\sim -10\gamma$ to $\sim 0\gamma$. The signatures of the high resolution hodograms for this crossing, seen in Figure 7c, are yet another variant of the basic rotation seen on the earliest crossings. Not only is the rotation less circular, but the final magnetospheric field strength is greater than in previous crossings. Changes in the B_N component, if due to boundary attitude variations, lead to angular oscillations of roughly $\pm 5°$ or less. The flow direction change seen at 07:52:40 on ISEE-1 corresponds to the topmost rotation, in which B_M goes from $\sim -10\gamma$ to 0γ.

4.3. Velocity and Thickness

To obtain boundary velocities relative to the spacecraft on this pass, we have determined the time delay required for the two spacecraft to consecutively attain the same position in the B_L field gradient. These time delays are shown as the horizontal lines in Figure 8, and have been (wherever possible) spaced uniformly in time, not in pseudo-distance through the boundary. The lines range from 7 to 71 s duration, and given the spacecraft separation along the normal (320 km), these times yield the relative boundary velocity plotted below the field trace. We note that these are average velocities over the time span denoted by the horizontal lines in the B_L trace; negative velocities indicate motion of the magnetopause towards the Earth, positive away. The 07:52 crossing is interesting in that an initial very rapid motion is present, followed by an almost exponential deceleration. The average boundary speed and its standard deviation from the 32 determinations here are $15.9 \pm 9.8\ \mathrm{km\ s^{-1}}$. The median speed was $15.1\ \mathrm{km\ s^{-1}}$.

Measuring the total thickness of the boundary is difficult for several reasons. First, the velocity is quite variable during these crossings. Second, no crossing from unambiguous magnetosheath to unambiguous magnetosphere seems to be made. Third, the relation between B_M field value and distance through the boundary is non-linear. Nevertheless, we can get an approximate value for the thickness of the boundary using the staircase approach of the previous section.

The first magnetopause traversals are partial crossings of the boundary, so only a fractional thickness may be obtained. ISEE-1, however, appears to achieve nearly the magnetospheric configuration, and a stepwise calculation of the entry and exit thicknesses of the magnetopause yield ~ 520 and ~ 640 km, respectively. For the third crossing, in which ISEE-1 appears to traverse the boundary

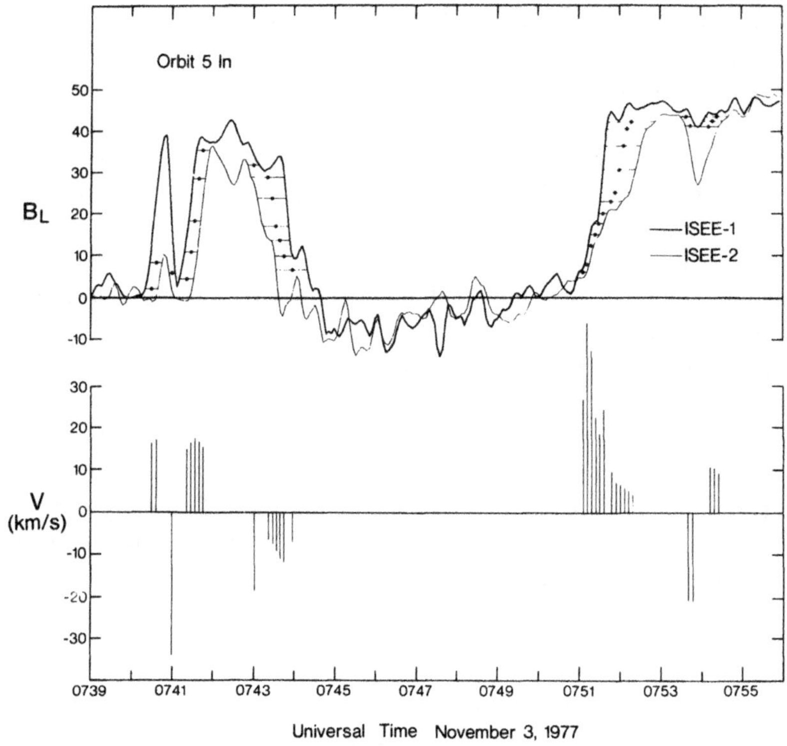

Fig. 8. Medium resolution averages of the B_L component of the ISEE-1 (heavy line) and ISEE-2 (light line) measurements through the magnetopause on November 3, 1977. The short horizontal line segments joining the ISEE-1 and ISEE-2 traces at common values of B_L provide measures of the time delay between ISEE-1 and -2. At the center time of each of these intervals we have calculated the average velocity over the time interval of the line segment, and have plotted these in the bottom panel of the figure. Velocity estimates are equi-spaced in time whenever single-valued velocities could be deduced.

completely, between two and three steps are required, giving a thickness of ~640 to ~960 km. In the fourth magnetopause encounter, boundary speed and structure variations complicate the thickness determination, but it seems clear that at least two and possibly three steps are needed to enter the magnetosheath, giving a thickness once again of approximately 640–960 km. The final full crossing likewise suffers from boundary structure and velocity changes, but appears to require at least three steps for the full crossing. Thus, this magnetopause is evidently the thickest traversed on this pass, being at least ~960 km thick.

5. Southward Magnetosheath Field, Orbits 7 In

The occurrence of geomagnetic storms and substorms is controlled by the southward component of the interplanetary magnetic field (Arnoldy, 1971; Russell and McPherron, 1973; Burton *et al.*, 1975). Further, the magnetopause appears to be eroded when the interplanetary field is southward (Aubry *et al.*, 1971). Thus, we would expect that increasingly southward fields would lead to an increasingly stressed magnetopause. In the next two sections we will examine two such magnetopause crossings. The first crossing occurred on November 8, 1977 at about 02:50 UT. The ISEE spacecraft were at (10.16, −1.77, 5.07) R_e in solar magnetospheric coordinates and were separated by (231, −181, 192) km again in GSM. The separation along the model normal was 299 km. The angle between the average magnetosheath field and the magnetospheric field locally ranged from 115° to 145°.

5.1. The Macrostructure of the Boundary

Figure 9 shows the medium resolution data from 02:00-03:06 UT. The ISEE-1 data are plotted with a heavy line; the ISEE-2 data with a light line. The coordinate system is the boundary normal system using the model normal. As shown in Table II, the angle between the model normal and the tangential discontinuity normal is 0.6°, using the 02:02 sheath field and the 03:04 magnetospheric field and 2.5° using the 02:43 sheath field.

If one were to examine solely the B_L and $|B|$ traces, one would interpret this section of data to contain two partial entries into the magnetosphere: one at 02:12 and one at 02:36 together with a full entry, albeit with much back and forth motion, from 02:44 to 02:52. The B_L traces are nested as expected for an

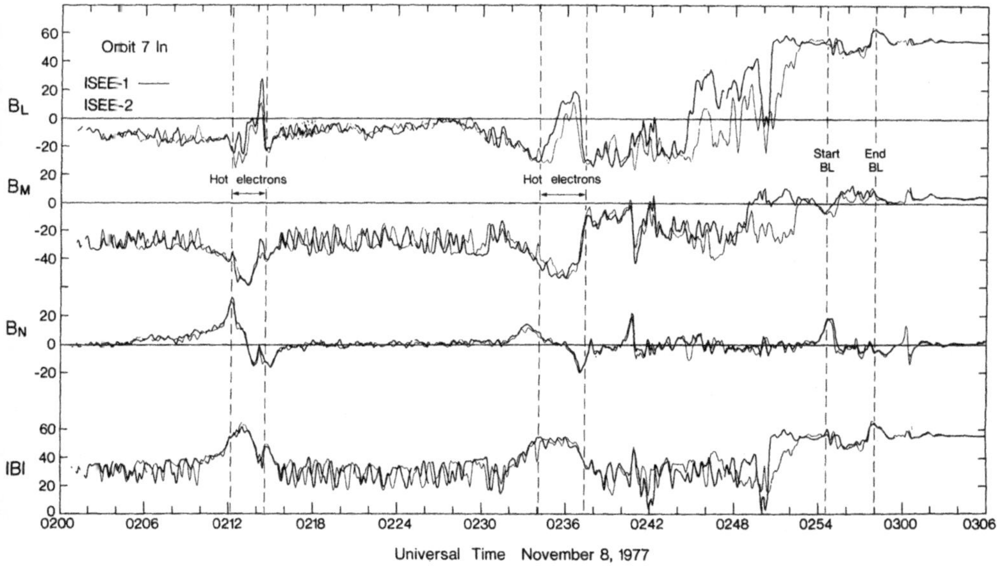

Fig. 9. Medium resolution measurements in boundary normal coordinates across the magnetopause on orbit 7 inbound, November 8, 1977. ISEE-1 measurements are shown with a heavy line, ISEE-2 with a light line. The boundary normal has been taken to be the model normal here. The ISEE spacecraft were at (10.16, −1.77, 5.07)R_e in solar magnetospheric coordinates and separated by 299 km along the normal.

TABLE II
Magnetopause parameters

Time	Nov. 8, 1977 02:45 UT	Nov. 10, 1977 14:50 UT
Orbit	7 inbound	8 inbound
Location $\mathbf{R}_{GSM}$ (R_e)	(10.16, −1.77, 5.07)	(7.09, 0.34, 2.86)
Separation $\mathbf{S}_{GSM}$ (R_e)	(231, −181, 192)	(196, −79, 142)
$\mathbf{B}^{sh}_{GSM}(\gamma)$	(9.2, 18.2, −24.3)	(13.5, 74.0, −48.4)
$\mathbf{B}^{ms}_{GSM}(\gamma)$	(−17.9, −2.0, 52.2)	(−49.5, −2.0, 127.9)
$\measuredangle \mathbf{B}^{sh}, \mathbf{B}^{ms}$	147.0°	124.8°
${}^{m}\mathbf{n}_{GSM}$	(0.940, −0.108, 0.322)*	(0.962, 0.033, 0.271)
${}^{TD}\mathbf{n}_{GSM}$	(0.945, 0.048, 0.323)	(0.930, 0.066, 0.361)*
$\measuredangle {}^{n}\mathbf{n}, {}^{TD}\mathbf{n}$	4.1°	5.9°
$\mathbf{S} \cdot \mathbf{n}$	299	228

* Normal used in calculations and plots.

in and out boundary crossing at 02:12 and 02:36 and the field strength reaches magnetospheric levels. However, the B_M and B_N traces exhibit strange behavior at this time. The B_M component becomes much stronger during these events, whereas it is near zero in the magnetosphere. The B_N component has a very peculiar pattern of an increase to large positive values and then a decrease to rather large negative values and then a return to zero. This behavior of B_N is not restricted solely to these apparent magnetopause crossings. There is a similar event at 02:41 and two in the magnetosphere itself: one at 02:57 and one at 03:00. Before examining the magnetopause let us examine these events in detail.

5.1.1. *Flux Transfer Events*

When the magnetic field has a strong component along the magnetopause normal, the immediate conclusion one is tempted to draw is that there is reconnection. However, if reconnection is occurring what is the explanation for the characteristic behavior of B_N and B_M? The first clue is the coherence of the B_N variations at the two spacecraft. The coherence is simply telling us that the variation is associated with a variation in the orientation of the boundary, and the spacecraft are much closer together than the dimensions of the bump in the boundary. The magnetosheath magnetic field has to wrap around that bump in the boundary. The characteristic increase and then decrease through a negative maximum is simply what one would expect if the magnetopause bulged out and this bulge travelled from the nose regions past the spacecraft. We recall that in the magnetosheath B_N is sensitive to oscillations in the $M-N$ plane mainly since the field is mostly along the M-direction whereas in the magnetosphere it is sensitive to oscillations of the boundary orientation in the $L-N$ plane.

The B_L component variation at 02:12 and 02:36 is therefore the signature of a magnetopause entry as expected. The motion of the boundary outward carries the satellites part way into the magnetosphere. We note that B_M increases at these times. In other words the tension in the field lines increases, i.e., the field is

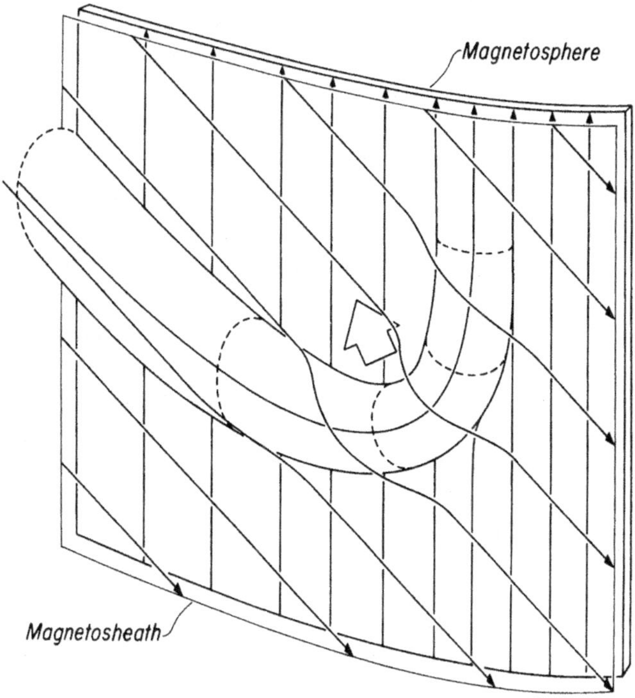

Fig. 10. Qualitative sketch of a flux transfer event. Magnetosheath field lines, slanted arrows, have connected with magnetospheric field lines, vertical arrows, possibly off the lower edge of the figure. As the connected flux tube is carried in the direction of the large arrow by the magnetosheath flow, the stressed field condition at the 'bend' tends to relax, effectively shortening the flux tube and straightening the bend. Magnetosheath field lines not connected to the magnetosphere drape over the connected flux tube and are swept up by its motion relative to the magnetosheath flow.

pulling harder. We note in fact that the B_M component is much stronger in the 'magnetopause' portion of these two events than in the magnetopause crossing at 02:50 Therefore the 'magnetospheric' flux tubes during these times must be bent in the direction of the magnetosheath field.

Our explanation of these events is straightforward. The magnetosphere is undergoing patchy reconnection, and we are seeing flux transfer events. Somewhere upstream of ISEE a magnetospheric flux tube reconnects with the magnetosheath field. Instead of convecting along with the other field lines in the magnetosheath, this reconnected flux tube attempts to shorten itself. This shortening process pulls the magnetospheric field line out of the boundary and twists it out of the vertical direction into the sheath field direction. It also lifts unconnected sheath field lines that overlay the connected lines. Thus we see a B_N increase well before we ever encounter a reconnected tube. These features are depicted schematically in Figure 10.

Patchy reconnection has been postulated before, in connection with high latitude observations (Haerendel *et al.*, 1978). This is the first observation of patchy reconnection at low latitudes. We do not feel that the reason that patchy reconnection has not been reported before has anything to do with its frequency of occurrence. It is a common feature in magnetic field data. We attribute our successful identification in part to having two satellites and in part to the use of boundary normal coordinates and the use of cartesian components.

Before leaving these events we should remark on the behavior of other instruments during these periods. The LASL/MPI fast plasma instrument measured flowing magnetosheath plasma throughout the 02:12 and 02:36 events. There was a slight drop in the thermal pressure when the field strength increased. We presume that the net effect was to keep the total plasma pressure constant. The direction of the flow also changed characteristically during the event but we have not yet correlated individual flow vectors and field vectors.

The Berkeley/Toulouse electron and proton data show clear 'bursts' of particles at the entries to the 'magnetopause' flux tubes. These bursts have almost identical spectra to the spectra seen later in the outer magnetosphere. In short then, the plasma data say the satellites are in the sheath while the energetic particle data (>2 keV) say the satellites are connected to the magnetosphere.

5.1.2. *The Magnetopause and Boundary Layer*

The magnetopause takes 8 min to cross, because it is oscillating back and forth. These oscillations make the calculation of thickness and velocity very difficult. Fortunately, since the B_N component is fairly steady during this time we can assume that no major changes in boundary orientation occurred and we can use the model normal.

In general the main difference between ISEE-1 and -2 is a time lag with profiles appearing somewhat different because of the variability of the velocity. However, right in the middle of the ISEE-1 crossing there is a feature not duplicated on ISEE-2. At 02:50 the field strength suddenly drops to 2γ. This is not part of the regular boundary structure as defined by the trend in the earlier part of the crossing, or the full crossing on ISEE-2. We will discuss the 'hole' in more detail when we treat the hodograms through the boundary. The only feature in the particle data coincident with the 'hole' that we have found thus far is an increase of close to a factor of 2 in the proton thermal pressure as measured by the LASL/MPI instrument on ISEE-1. Thus the total plasma pressure seems to be conserved through the 'hole' as one would expect.

As indicated on Figure 9 the boundary layer is detached from the magnetopause on this orbit. The boundary layer starts at 02:54:30 and ends at 02:58. We note that the boundary layer begins with a B_N event, which we would interpret as signaling the occurrence of a flux transfer event. The field magnitude is also somewhat depressed within the boundary layer. Further, there is some variation of the field in the M-direction, possibly indicating field aligned currents associated with the boundary layer.

5.2. The microstructure of the boundary

As noted above, the magnetopause crossing on this pass was several minutes in duration, owing to the fact that the boundary oscillated about the mean position of the spacecraft, ISEE-1 making the deeper penetration. This is also apparent in the high resolution data for this crossing, shown in Figure 11a and b, in spacecraft coordinates (approximately GSE); in addition, many high frequency field variations become evident in this figure. ISEE-1 begins to enter the magnetopause at roughly 02:44:30, followed by ISEE-2 at 02:45:00. About 10 s before the ISEE-1 entry, the LASL/MPI instrument indicates that directed bulk flow drops below 40 km s^{-1}, and the same occurs on ISEE-2 at 02:44:30. Flows then reappear at ~02:45:20 on ISEE-1, and persist with variable magnitude until roughly 02:48:15, when a very large bulk flow peak appears. This same feature appears 15 s later on ISEE-2, and reaches approximately 200 km s^{-1} peak flow speed at both spacecraft. In the center of this high speed flow, the fast plasma instrument indicates an enhanced plasma temperature, and the proton energy spectra are peaked at 600–800 eV (see Paschmann *et al.*, 1978). Upon entry into the magnetosphere, at 02:50:30 on ISEE-1 and 02:53:00 on ISEE-2, the

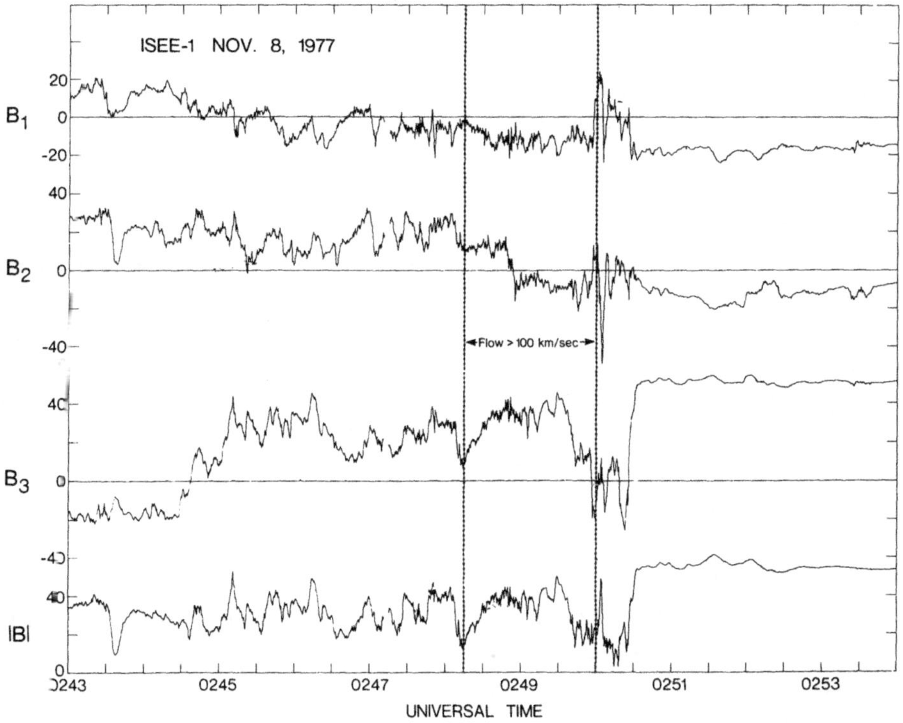

Fig. 11a. High resolution ISEE-1 measurements (16 samples per second) across the magnetopause on November 8, 1977 in spacecraft coordinates. Spacecraft coordinates are within 1° of solar ecliptic coordinates at this time for both satellites.

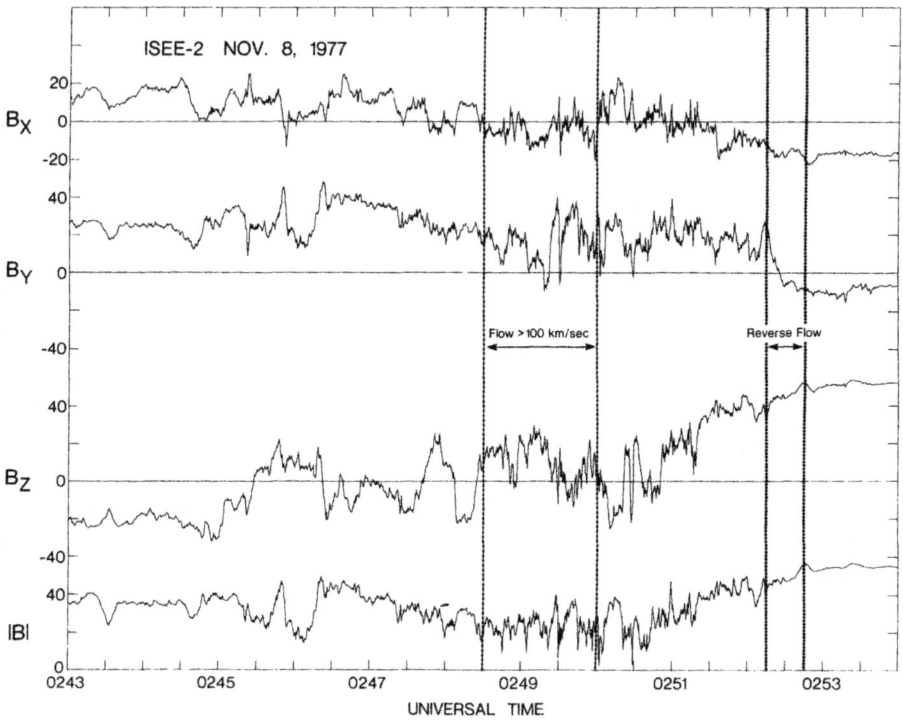

Fig. 11b. High resolution ISEE-2 measurements (16 samples per second) across the magnetopause on November 8, 1977 in spacecraft coordinates.

spacecraft immediately encounter the high temperatures and low densities of outer magnetospheric plasma – no boundary layer is evident until 02:54:30. Also, just before the entry of ISEE-2, a strong ($\sim 100\ \mathrm{km\ s^{-1}}$) reverse flow was seen on that spacecraft.

One of the curious aspects of this crossing is the appearance of regions of very low field strength; brief excursions to 3γ or less are not uncommon. As mentioned above, these 'drop-outs' are anticorrelated with proton thermal pressure enhancements, suggesting overall pressure balance. It is noteworthy that such a magnitude drop follows an apparent flux transfer event at 02:50:05 on ISEE-1, which in turn follows the region of high bulk flow speed.

5.3. The Morphology of Flux Transfer Events

The flux transfer events at 02:12 and 02:36 UT cannot be transformed into a coordinate system that reduces the field variations to two dimensions. Therefore we will use the boundary normal coordinate system to display the hodograms of the field variations through the events. Then we will segment one of the crossings for minimum variance analysis. However, only one of these intervals has a well determined normal.

As shown in Figure 12a, the first flux transfer event (FTE) begins in a

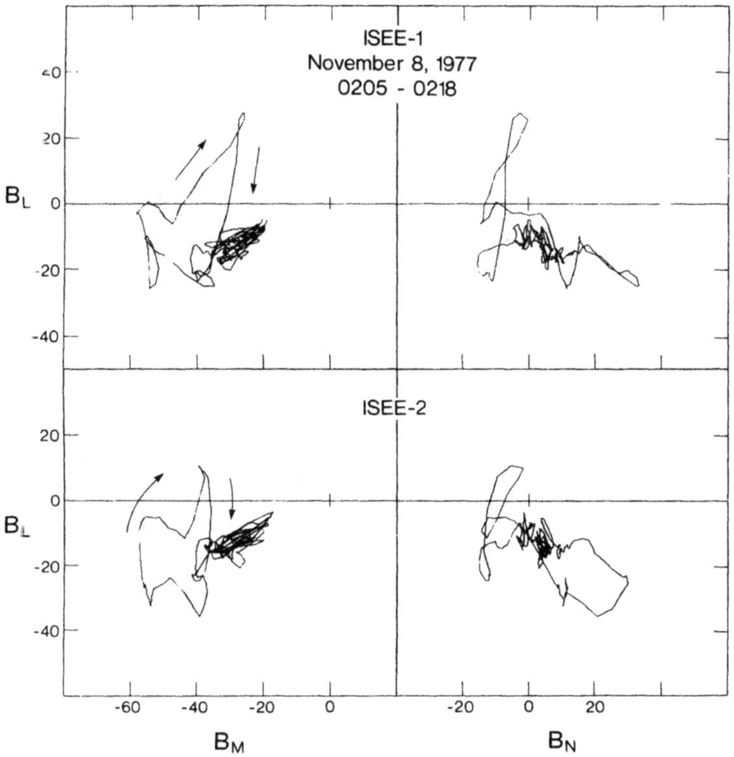

Fig. 12a. Hodogram of medium resolution magnetic field variation on ISEE-1 and -2 during flux transfer event at 02:12 in boundary normal coordinates.

magnetosheath field of roughly $(-10, -25, 0)\gamma$ in the boundary normal coordinates. The B_N component then grows positive, presaging the oncoming 'bump' of the reconnected flux tube, followed by a magnitude enhancement and slight southward turning of the field. Then the B_N component decreases as B_M becomes increasingly negative, corresponding to the stressed magnetosheath field as the 'bump' begins to pass beneath the spacecraft. As the B_M component approaches its extremum, and B_N approaches 10γ, B_L begins to become northward, defining the entry into the flux tube. The excursion reverses itself once as B_M becomes less negative, then continues northward with B_N negative as the spacecraft pass more deeply into the structure. ISEE-1 penetrates the structure more than ISEE-2 as is evident from the higher maximum B_L value. It is during this period that the energetic particles are seen. Exit from the flux tube is then seen with a sudden large decrease in B_L, as B_N remains at negative values, followed by a return to the unperturbed magnetosheath as B_N gradually approaches zero.

The next such occurrence near 02:35, follows the same general sequence of events but with slightly different detail, as shown in Figure 12b. The initial increase in B_N, for example, is much smaller, and entry into the flux tube, as seen

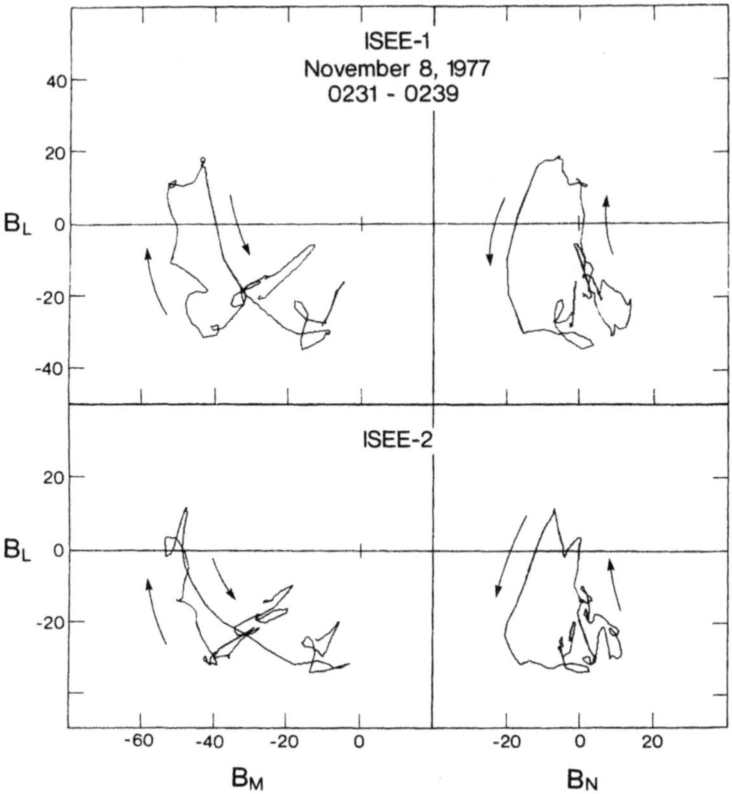

Fig. 12b. Hodogram of medium resolution field variation through flux transfer event at 02:36 UT in boundary normal coordinates.

in the excursion of B_L, occurs from a field somewhat more strongly southward than on the previous occasion. In addition, the final configuration of the magnetosheath field is different from the initial configuration. It should be mentioned that the plasma bulk flow direction between 02:33 and 02:34 is much more nearly antisunward than the unperturbed magnetosheath flow, in accord with the notion that the positive B_N component at this time reflects magnetosheath flow about an obstacle upstream of the spacecraft. Likewise, the flow direction between 02:37:30 and 02:38:00 from the LASL/MPI instrument is more nearly dawnward than unperturbed magnetosheath flow, as would be expected if an obstacle downstream on the boundary were deflecting the magnetosheath around it; this is also suggested by the negative value of B_N at this time. For both events, the final exit through the current layer defining the magnetosheath/flux tube interface at the upstream side is more discrete than initial entry through the downstream interface.

To contrast the behavior of the magnetopause crossing with the two FTE's we show in Figure 12c the magnetopause crossing near 02:50 in the same type of

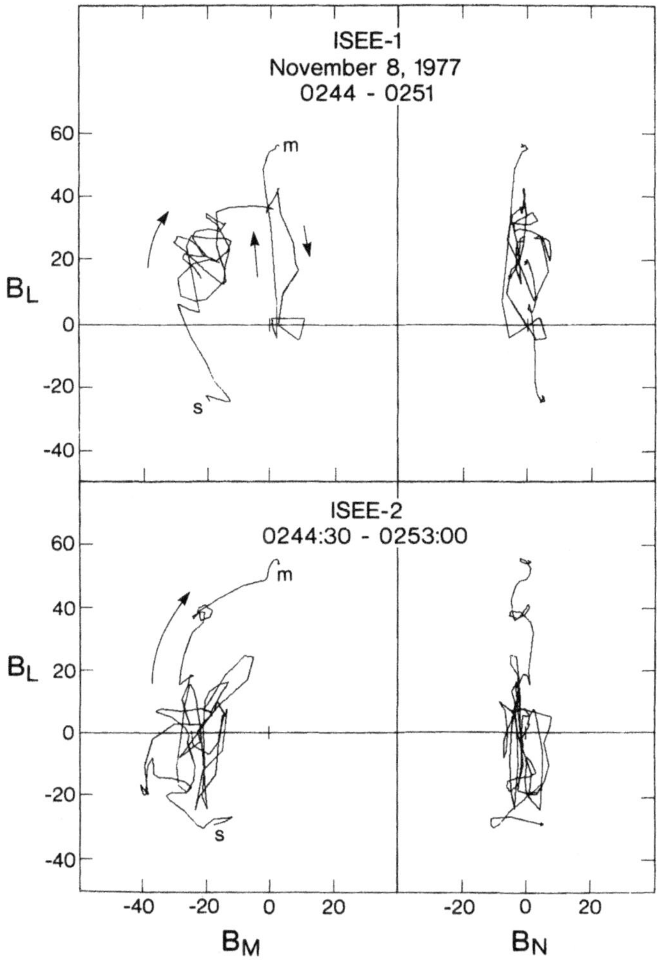

Fig. 12c Hodogram of medium resolution field variation through magnetopause at 02:50 UT in boundary normal coordinates.

display. The initial rotation from magnetosheath to magnetosphere is interrupted by the sudden change in motion of the boundary, which oscillates about the spacecraft. ISEE-1, having penetrated deeper into the magnetopause before the motion change, shows excursions about a different average phase of the rotation than ISEE-2. ISEE-1 then passes into a region of very small field strength, before finally attaining the magnetospheric configuration. ISEE-2, meanwhile, shows no hint of the low field structure. Thus, the scale size of this 'hole' appears to be less than the separation of the spacecraft.

To see if some portions of an FTE might be reduced to a two-dimensional variation we have performed an eigenvalue analysis on the entry and exit from the flux tube. These analyses are shown in Figure 13. The eigenvalues for the entry

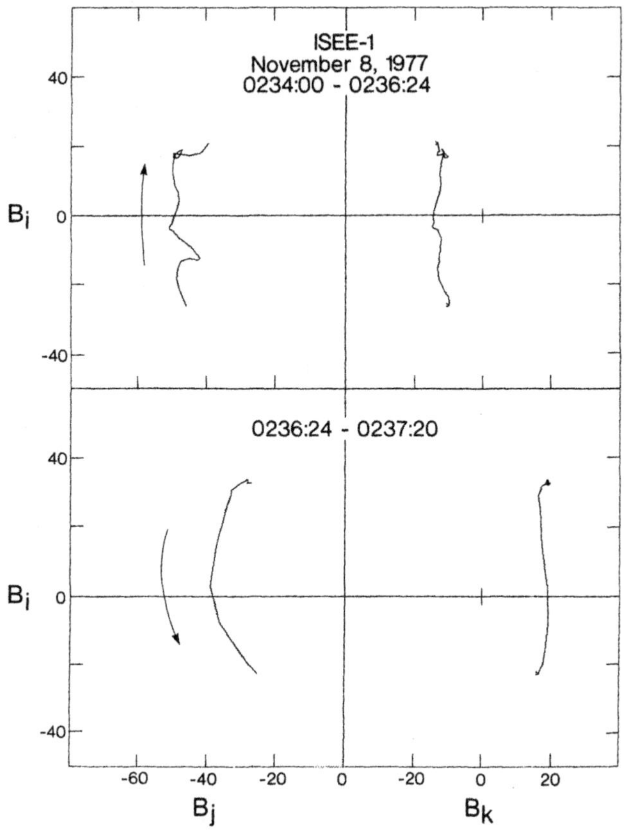

Fig. 13. Hodogram of medium resolution field variation through entry and exit of flux transfer event in principal axis system.

are $(233, 8, 2)\gamma^2$ with eigenvectors associated with the maximum eigenvalue of (0.976, −0.134, −0.174) and with the minimum eigenvalue of (0.205, 0.268, 0.941) in boundary normal coordinates. This direction is 20° from the undisturbed normal. It is also 12° up out of the $M-N$ plane and tilted back 16° from the $L-N$ plane, with a probable error of 14°. These angles are quite consistent with the picture of a flux tube being lifted up and back by reconnection with the magnetosheath field.

A similar analysis for the exit yields eigenvalues of $(451, 16, 1)\gamma^2$, and a minimum variance eigenvector of (−0.435, −0.692, 0.576) in boundary normal coordinates. This vector is 55° from the undisturbed normal, tilted down 26° from the $M-N$ plane, and 44° forward (sunward) from the $L-N$ plane. The probable error in direction of this eigenvector is only 7°, so that the orientation is consistent with passage through the trailing edge of the flux tube. As can be seen in Figure 13, there is a well-defined component of $18 \pm 4\gamma$ normal to the flux tube/magnetosheath interface, suggesting that reconnection may still be occurring at this point.

To approximate the amount of flux eroded during a flux transfer event we must estimate the cross-section of the reconnected flux tube and the flux passing through this cross-section. The duration of the flux tube crossing gives us an estimate of one dimension assuming it is convecting by with the speed of the plasma. This dimension is $4R_e$ for both FTE's, the longer duration of the second FTE being due to a slower flow velocity. Assuming that the orthogonal dimension is similar since the normal appears to have been rotated by about 45° during the passage of the flux tube, and we obtain a flux of 3×10^{15} Mx for the first FTE and 2×10^{15} for the second. At a rate of 2–3 per hour, these events represent an erosion rate of about 2×10^{12} Mx s^{-1}. This rate is similar to that found by Aubry *et al.* (1970) when the magnetosheath field had a southward component of about 30γ, compared with the approximately 10γ southward field before the flux transfer events at 02:12 and 02:36 UT. Thus, these FTE's are not merely curiosities, but in fact large enough to be significant if not dominant in the erosion of the dayside magnetopause.

6. Southward Magnetosheath Field, Orbit 8 In

On orbit 8 inbound the magnetopause was encountered far inward of its usual position at $(7.09, 0.34, 2.86)R_e$ in solar magnetospheric coordinates. This unusual location is manifested in Figure 14 by the large field values both in the magnetosheath and in the magnetosphere. The angle between the model normal and the normal obtained assuming that the boundary was a tangential discontinuity during the quiet intervals on this day (see Table II) was 5.9° while this is a moderately small difference, the data are better ordered by the tangential discontinuity normal and we will use this normal for our discussion and data display. The separation along the normal was 228 km.

6.1. Macrostructure of the Boundary

The medium resolution data in magnetopause normal coordinates in Figure 14 show there are three main magnetopause crossings. The B_N component shows that these crossings are accompanied by many (at least 8) large deviations of the boundary normal. These have all the characteristics of the flux transfer events of the orbit 7 inbound crossing except that the duration of the events is much shorter. The events show a positive enhancement of B_N followed by a negative depression. Variance analysis shows they are three dimensional. They do not have a single principal plane of field change. Further, they contain hot magnetospheric-like electrons (and protons too) when they occur in the magnetosheath. We will not discuss the presence of hot particles at B_N fluctuations in the magnetopause or magnetosphere, because we expect these particles in these regions at all times. Perhaps, the most important new facet of this series of crossings is the fact that the flux transfer events are occurring almost continually. They are in fact

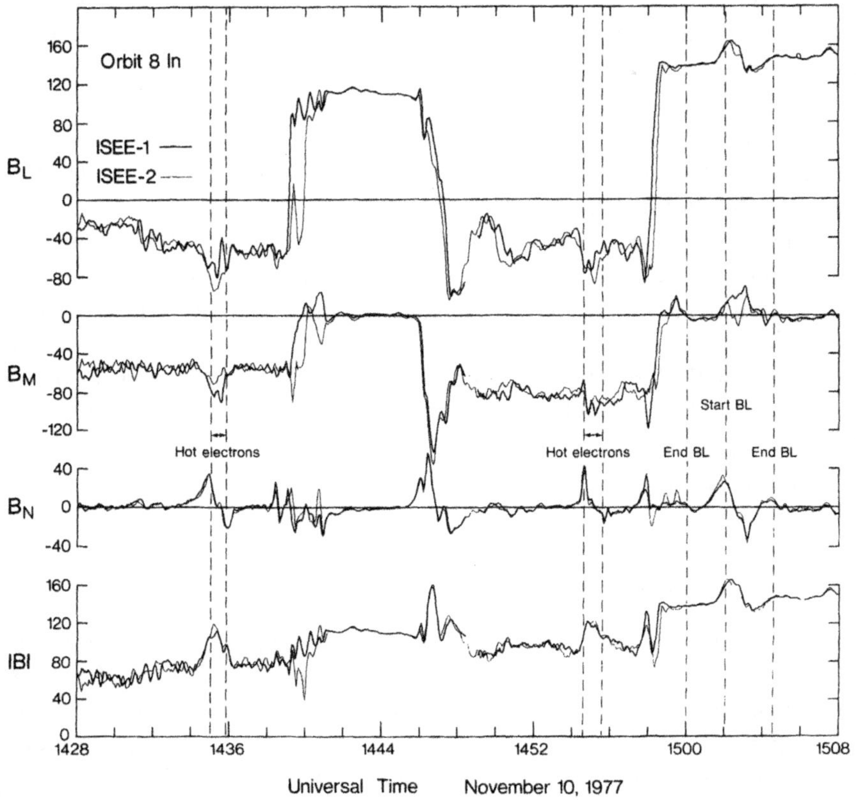

Fig. 14. Medium resolution measurements in boundary normal coordinates across magnetopause on orbit 8 inbound, November 10, 1977. ISEE-1 measurements are shown with a heavy line; ISEE-2 with a light line. The boundary normal has been taken to be the tangential discontinuity normal here. The ISEE spacecraft were at (7.09, 0.34, 2.86)R_e in solar magnetospheric coordinates and were separated by 228 km along the boundary normal.

occurring as the ISEE spacecraft cross the magnetopause on all three occasions. We note that within the magnetosphere the boundary layer ceases on both spacecraft at 15:00 UT and reappears at 15:02, only to cease again at 15:04:30. The behavior suggests that the appearance may be temporal rather than spatial. We also note that the B_M fluctuations in the boundary layer suggest the presence of field-aligned currents.

6.2. Microstructure of the Boundary

The magnetopause crossings observed on this day are all complicated by the presence of flux transfer events in or near the boundary. The first crossing, near 14:40, is shown in high resolution in Figure 15. ISEE-1, 228 km inward along the normal from ISEE-2, appears to enter the magnetopause at 14:39, as evidenced by the strong northward excursion of the B_L component. The variations in the sign of B_N, as well as the enhancement in B_M, also suggest a flux transfer event.

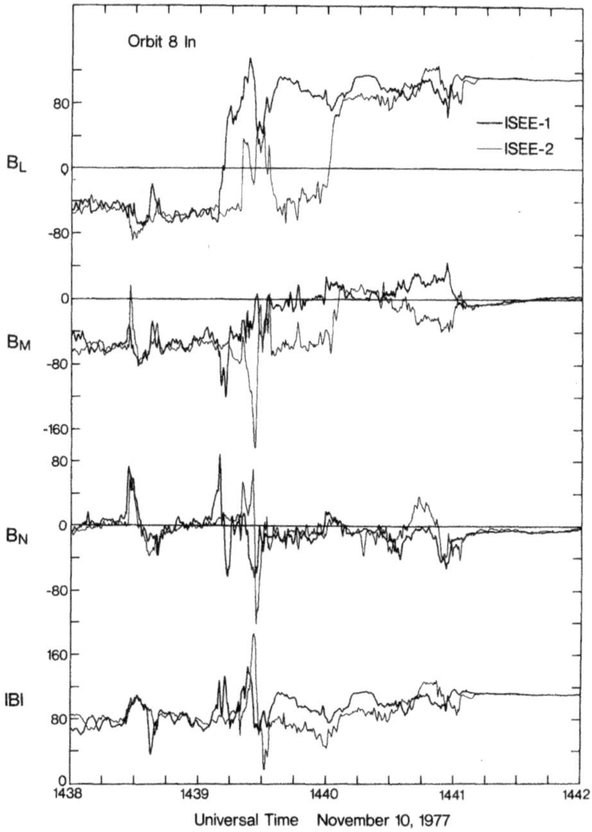

Fig. 15. High resolution measurements (four samples per second) across the first magnetopause crossing on November 10, 1977 expressed in boundary normal coordinates.

At 14:40, however, the field configuration is nearly magnetospheric, though considerable noise is present. Finally, entry into the quiet magnetosphere takes place at 14:41. Meanwhile, ISEE-2 also enters the flux tube near 14:39:20, but the field is more nearly horizontal and of higher magnitude than at the ISEE-1 entry. Then ISEE-2 reenters the magnetosheath soon after 14:39:30, while ISEE-1 is in a magnetospheric configuration; thus, the current layer thickness must be less than the total spacecraft separation distance, 255 km. Entry into the true magnetosphere shortly after 14:41 on ISEE-2 is accompanied by the sudden drop in density and increase in temperature which signals the disappearance of magnetosheath plasma. High speed ($> 100\ \mathrm{km\,s^{-1}}$) flows are seen at 14:39:30 and 14:40:30 on ISEE-2. The first has a dawnward horizontal component, and coincides with the spacecraft's flight through the flux tube while the second flow has a horizontal component directed inward from the boundary, and occurs after ISEE-2's entry into magnetosphere-like field. The plasma and particle instruments on ISEE-1 were turned off at this time.

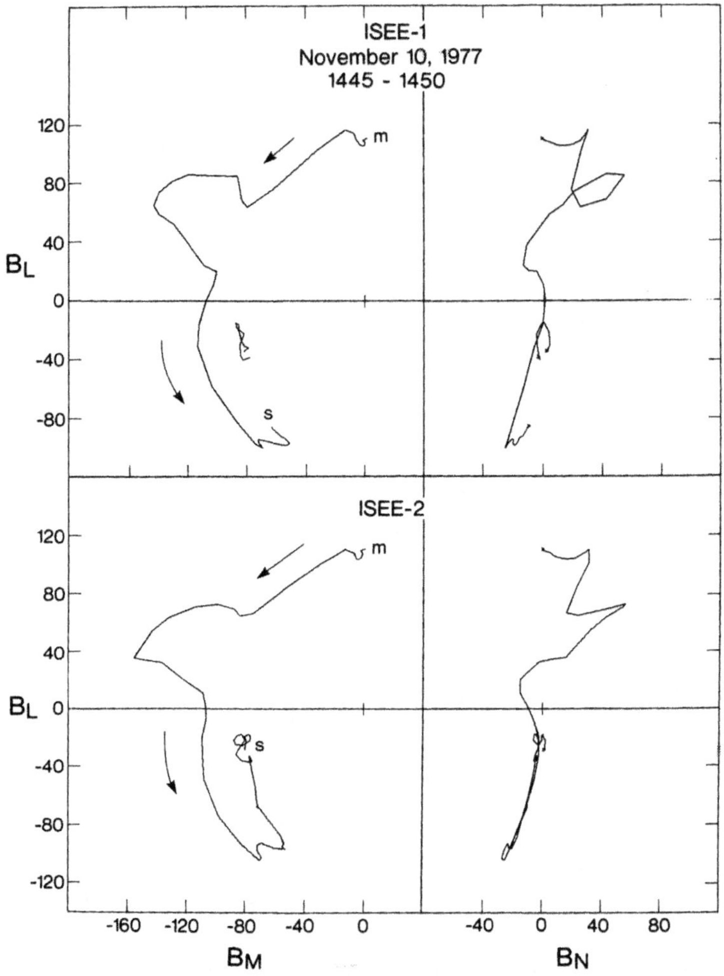

Fig. 16. Hodogram of medium resolution field measurements through the second magnetopause crossing on November 10, 1977 in boundary normal coordinates.

The second crossing of the magnetopause at 14:47 includes another FTE. Here again, the transition from hot, tenuous magnetospheric plasma to magnetosheath plasma occurs before any appreciable field direction change, at 14:46:10. Hodograms of medium resolution data in boundary normal coordinates appear in Figure 16. First, the B_N component increases, followed by the initial phase of the field rotation. Abruptly, the magnitude increases, as does B_N, and the rotation continues. B_N then becomes slightly negative as B_M returns from its excursion. The B_L component continues to turn southward, and a mild enhancement of B_M occurs again, followed by rotation to a southward extremum before returning to unperturbed magnetosheath. A bulk flow maximum of ~ 150 km s^{-1} occurs at the first B_M enhancement, and speeds of $\sim 100\ \mathrm{km\ s^{-1}}$ prevail throughout the rest of the structure.

Another FTE is encountered just before the third and final magnetopause crossing at about 14:58:30. The plasma flow speed seen at the time of this event is $\sim 45\ \mathrm{km\ s^{-1}}$, which increases to $\sim 100\ \mathrm{km\ s^{-1}}$ at the magnetopause. This entry, unlike the previous two, is accompanied by the appearance of a boundary layer.

6.3. Flux transfer events

Because of the many flux transfer events on this day, it is of interest to determine whether a correspondingly large amount of flux is being transported from the dayside. The first event in which significant penetration of the flux tube occurs is also the initial 'magnetopause' crossing. Using, as before, the LASL/MPI bulk flow speed, the peak B value and the duration of the event, an approximate total area of $4 \times 10^7\ \mathrm{km^2}$ results, giving a net flux of 6×10^{14} Mx. The second useable event is during the second crossing, with a duration of about 36 s, giving an area of $\sim 2 \times 10^7\ \mathrm{km^2}$ (linear dimension of about $0.7 R_e$), yielding a transported flux of about 2.2×10^{14} Mx. The last useful event, just before the third crossing, gives a flux of 3×10^{14} Mx. The flux transferred per event here is smaller than on November 8, but if in the period of a half hour eight intermediate (4×10^{14} Mx) events take place, the rate of the flux removal is about $2 \times 10^{12}\ \mathrm{Mx\ s^{-1}}$, a number very similar to that calculated for the previous orbit in Section 5.3. We note that these numbers should be lower limits to the true rate since depending on satellite location events could be missed by ISEE and further reconnection could take place after the FTE has passed by ISEE.

6.4. Velocity and thickness

Since the first two magnetopause crossings are complicated by FTE's, the direction of motion of the boundary is poorly determined. The third crossing, however, has no such problem, and the thickness calculated by stepping through the boundary turns out to be between 680 and 912 km. Speeds range from 11.5 to $20.9\ \mathrm{km\ s^{-1}}$ on this crossing. As discussed in Section 6.2, the current sheet thickness for the crossing near 14:40 must be less than 255 km, but the velocity of the structure is nearly impossible to determine given its three dimensional nature and temporal evolution. On the other hand, the structure seen on the second crossing at 14:47 is apparently much larger than the separation of the two spacecraft, since at any time the two spacecraft see fields which are but slightly different (see Figure 14). Such apparent near simultaneity can be caused by motion largely transverse to the spacecraft separation vector, such as could occur when a fluxtube containing both spacecraft becomes reconnected and is pulled tangential to the boundary, leaving the spacecraft to pass almost simultaneously into the magnetosheath. Thus, on this day, boundary thickness from less than 255 km to 912 km and possibly larger may have occurred.

7. Conclusions

The studies presented here represent only the beginnings of our investigation of the magnetopause. In particular we plan to undertake a survey of the behavior of the magnetopause over a wider range of longitudes including both morning and afternoon local times and under a wider range of magnetosheath field directions. Until such a survey is performed we will not be able to tell how typical are the signatures observed here (near local noon) and exactly where the flux transfer events are initiated. With the present set of data we cannot tell whether the flux transfer events begin at the polar cusp (the south polar cusp for the events we saw) or in the equatorial regions. A study of where they are observed as a function of B_M (or B_Y GSM) should be able to resolve this question.

It is clear from the data that the structure of the magnetopause is complex. This complexity arises in part from the impulsive and three-dimensional nature of the flux transfer event. The boundary is often not time stationary and its normal varies in direction during a crossing. Even when the magnetosheath field is northward and flux transfer events are not seen, the motion of the boundary is irregular, changing markedly in the length of time it takes to cross a boundary, thus complicating our attempts to place dimensions on the thickness of the magnetopause. On the one boundary crossing studied under northward magnetosheath conditions, the major current layer was over 1000 km thick but the total boundary thickness was probably over 2000 km. A further interesting aspect was that the field strength in the magnetopause itself was greater than in either the magnetosheath or the magnetosphere.

When the interplanetary field was southward, the magnetopause appeared to be thinner, usually in the range 500–1000 km thick. However, this determination is complicated by the irregularity in motion, the changes in boundary orientation, and the time variability of the structure. In particular the occasional occurrence of drop-outs in the field strength, to very low values during crossings on one of the spacecraft and not on the other is very intriguing.

Flux transfer events naturally lead to oscillations in the position of the magnetopause as they propagate from the subsolar regions down the flanks of the magnetosphere much in the same way the Kelvin-Helmholtz instability should lead to boundary oscillations. Furthermore, flux transfer events probably, although we have not proven this, occur more readily at high solar wind velocities, just as the Kelvin-Helmholtz instability should be more effective at high solar wind velocities (Boller and Stolov, 1970; Boller and Stolov, 1973). The difference between the two mechanisms is that the Kelvin-Helmholtz mechanism is independent of whether the magnetosheath field is northward or southward. Until investigations of the dependence of boundary motions and pulsation amplitudes on the direction of the north-south component of the interplanetary field are undertaken we must reserve judgment on the efficacy of the Kelvin-Helmholtz mechanism in producing magnetic pulsations in the outer magnetosphere.

A complete study of the magnetopause requires a detailed examination of the joint plasma and field behavior including the three-dimensional flows present. Such work has just begun. However, the excellence of the ISEE data, together with the spirit of cooperation shown by the various experiment groups promises returns limited only by the resources available to support such studies.

Acknowledgements

We are grateful for the spirit of free data exchange and cooperation exhibited by the various investigator groups on the ISEE spacecraft. In particular in this paper we benefited much from the data from the LASL/MPI fast plasma instrument and the Berkeley/Toulouse energetic particle experiment. Discussions of these data with G. Paschmann (LASL/MPI), G. Parks, K. Anderson, and R. Lin (Berkeley/Toulouse) were most welcome. We are grateful to K. Anderson for sponsoring two of the workshops at which much of this exchange took place.

We should mention some of the many people who contributed to the success of the spacecraft and helped maintain a magnetically clean vehicle so that these studies could be performed: J. Madden and M. Davis of GSFC and D. Eaton and R. Gruen of ESTEC. The instrument itself was ably built, tested and calibrated by B. Plitt and his staff of Westinghouse and R. Snare, F. George, and J. Means of UCLA. We are also indebted to R. Walker, H. Singer, and S. Kaye for assisting us with the initial data processing. The flipper design was kindly provided by C. P. Sonett.

This research was conducted for the National Aeronautics and Space Administration under contract NAS 5-20064.

References

Arnoldy, R. L.: 1971, 'Signature in the Interplanetary Medium for Substorms', *J. Geophys. Res.* **76,** 5189.

Aubry, M. P., Russell, C. T., and Kivelson, M. G.: 1970, 'Inward Motion of the Magnetopause Before a Substorm', *J. Geophys. Res.* **75,** 7018.

Aubry, M. P., Kivelson, M. G., and Russell, C. T.: 1971, 'Motion and Structure of the Magnetopause', *J. Geophys. Res.* **76,** 1673.

Binsack, J. H.: 1968, 'Shock and Magnetopause Boundary Observations with IMP-2', in R. L. Carovillano, J. F. McClay, and H. R. Radoski (eds.), *Physics of the Magnetosphere*, D. Reidel Publ. Co., Dordrecht, p. 605.

Boller, B. R. and Stolov, H. L.: 1970, 'Kelvin-Helmholtz Instability and the Semi-Annual Variation of Geomagnetic Activity', *J. Geophys. Res.* **75,** 6073.

Boller, B. R. and Stolov, H. L.: 1973, 'Explorer-18 Study of the Stability of the Magnetopause Using a Kelvin-Helmholtz Instability Criterion', *J. Geophys. Res.* **78,** 8078.

Bridge, H., Egidi, A., Lazarus, A., Lyon, E., and Jacobson, L: 1965, 'Preliminary Results of Plasma Measurements on IMP-A', *Space Research* **V,** North-Holland Publ. Co., Amsterdam, p. 969.

Burton, R. K., McPherron, R. C., and Russell, C. T.: 1975, 'An Interplanetary Relationship Between Interplanetary Conditions and Dst', *J. Geophys. Res.* **80,** 4204.

Cahill, L. J. and Amazeen, P. G.: 1963, 'The Boundary of the Geomagnetic Field', *J. Geophys. Res.* **68,** 1835.

Cahill, L. J. and Patel, V. L.: 1967, 'The Boundary of the Geomagnetic Field, August to November, 1961', *Planetary Space Sci.* **15,** 997.

Cummings, W. D. and Coleman, P. J., Jr.: 1968, 'Magnetic Fields in the Magnetopause and Vicinity at Synchronous Altitude', *J. Geophys. Res.* **73,** 5699.

Fairfield, D. H.: 1971, 'Average and Unusual Locations of the Earth's Magnetopause and Bow Shock', *J. Geophys. Res.* **76,** 6700.

Feldstein, Y. I.: 1970, 'Auroras and Associated Phenomena', in E. R. Dyer (ed.), *Solar Terrestrial Physics*, Part III, D. Reidel Publ. Co., Dordrecht, p. 152.

Frank, L. A. and Van Allen, J. A.: 1964, 'Intensity of Electrons in the Earth's Inner Radiation Zone', *J. Geophys. Res.* **69,** 4923.

Freeman, J. W., Jr.: 1964, 'Detection of an Intense Flux of Low-Energy Protons or Ions Trapped in the Inner Radiation Zone', *J. Geophys. Res.* **69,** 1691.

Gosling, J. T., Asbridge, J. R., Bame, S. J., and Strong, I. B.: 1967, 'Vela-2 Measurements of the Magnetopause and Bow Shock Positions', *J. Geophys. Res.* **72,** 101.

Haerendel, G., Paschmann, G., Sckopke, N., Rosenbauer, H., and Hedgecock, P. C.: 1978, 'The Frontside Boundary Layer of the Magnetosphere and the Problem of Reconnection', *J. Geophys. Res.* **83,** 3195.

Heppner, J. P., Sugiura, M., Skillman, T. L., Ledley, B. G., and Campbell, M.: 1967, 'OGO-A Magnetic Field Observations', *J. Geophys. Res.* **72,** 5417.

Holzer, R. E. and Slavin, J. A.: 1978, 'Magnetic Flux Transfer Associated with Expansions and Contractions of the Dayside Magnetosphere', *J. Geophys. Res.* **83,** 3831.

Holzer, R. E., McLeod, M. G., and Smith, E. J.: 1966, 'Preliminary Results from the OGO-1 Search Coil Magnetometer: Boundary Positions and Magnetic Noise Spectra', *J. Geophys. Res.* **71,** 1481.

Kaufmann, R. L. and Konradi, A.: 1969, 'Explorer-12 Magnetopause Observations: Large-Scale Non-Uniform Motion', *J. Geophys. Res.* **74,** 3609.

Kaufmann, R. L. and Konradi, A.: 1973, 'Speed and Thickness of the Magnetopause', *J. Geophys. Res.* **78,** 6549.

Ledley, B. G.: 1971, 'Magnetopause Attitudes During OGO-5 Crossings', *J. Geophys. Res.* **76,** 6736.

Meng, G.-I.: 1970, 'Variation of the Magnetopause Position with Substorm Activity', *J. Geophys. Res.* **75,** 3252.

Mozer, F. S., Torbert, R. B., Fahleson, U. V., Fälthammar, C.-G., Gonfalone, A., Pedersen, A., and Russell, C. T.: 1978, *Geophys. Res. Letters*, submitted.

Ness, N. F., Sceorce, C. S., and Seek: 1964, 'Initial Results of the IMP-1 Magnetic Field Experiment', *J. Geophys. Res.* **69,** 3531.

Ness, N. F.: 1967, in J. W. King and W. S. Newman (eds.), *Solar Terrestrial Physics,* Academic Press, London, p. 57.

Neugebauer, M., Russell, C. T., and Smith, E. J.: 1974, 'Observations of the Internal Structure of the Magnetopause', *J. Geophys. Res.* **79,** 499.

Ogilvie, K. W., Scudder, J. D., and Sugiura, M.: 1971, 'Magnetic Field and Electron Observations Near the Dawn Magnetopause', *J. Geophys. Res.* **76,** 3574.

Paschmann, G., Sckopke, N., Haerendel, G., Papamastorakis, J., Bame, S. J., Asbridge, J. R., Gosling, J. T., Hones, E. W., Jr., and Tech, E. R.: 1978, 'ISEE Plasma Observations Near the Subsolar Magnetopause', *Space Sci. Rev.* **22,** 717.

Patel, V. L. and Dessler, A. J.: 1966, 'Geomagnetic Activity and Size of Magnetospheric Cavity', *J. Geophys. Res.* **71,** 1940.

Russell, C. T.: 1978, 'The ISEE-1 and -2 Fluxgate Magnetometers', *IEEE Trans. Geosci. Electr.* **GE-16,** 239.

Russell, C. T. and Greenstadt, E. W.: 1979, 'Initial ISEE Magnetometer Results: Shock Observations,' *Space Sci. Rev.* **23,** in press.

Russell, C. T. and McPherron, R. L.: 1973, 'The Magnetotail and Substorms', *Space Sci. Rev.* **15,** 205.

Russell, C. T., Neugebauer, M., and Kivelson, M. G.: 1974, 'OGO-5 Observations of the Magnetopause', in D. E. Page (ed.), *Correlated Interplanetary and Magnetospheric Observations*, D. Reidel Publ. Co., Dordrecht, p. 139.

Sonnerup, B. U. O.: 1971, 'Magnetopause Structure During the Magnetic Storm of September 24, 1961', *J. Geophys. Res.* **76,** 6717.

Sonnerup, B. U. O. and Cahill, L. J., Jr.: 1967, 'Magnetopause Structure and Attitude from Explorer-12 Observations', *J. Geophys. Res.* **72,** 171.

Sonnerup, B. U. O. and Cahill, L. J., Jr.: 1968, 'Explorer-12 Observations of the Magnetopause Current Layer', *J. Geophys. Res.* **73,** 1757.

Sonnerup, B. U. O. and Ledley, B. G.: 1974, 'Magnetopause Rotational Forms', *J. Geophys. Res.* **79,** 4309.

Sonnerup, B. U. O. and Ledley, B. G.: 1978, 'OGO-5 Magnetopause Structure and Classical Reconnection', *J. Geophys. Res.*, submitted.

Wolfe, J. H., Silva, R. W., and Meyers, M. A.: 1966, 'Observations of the Solar Wind During the Flight of IMP-1', *J. Geophys. Res.* **71,** 1319.

ISEE PLASMA OBSERVATIONS NEAR THE SUBSOLAR MAGNETOPAUSE

G. PASCHMANN, N. SCKOPKE, G. HAERENDEL, and J. PAPAMASTORAKIS

Max-Planck-Institut für Physik and Astrophysik, Institut für Extraterrestrische Physik, 8046 Garching, F.R.G.

and

S. J. BAME, J. R. ASBRIDGE, J. T. GOSLING, E. W. HONES, Jr., and E. R. TECH

University of California, Los Alamos Scientific Laboratory, Los Alamos, N.M. 87545, U.S.A.

Abstract. The early ISEE orbits provided the opportunity to study the magnetopause and its environs only a few Earth radii above the subsolar point. Measurements of complete two-dimensional ion and electron distributions every 3 or 12 s, and of three-dimensional distributions every 12 or 48 s by the LASL/MPI instrumentation on both spacecraft allow a detailed study of the plasma properties with unprecedented temporal resolution. This paper presents observations obtained during four successive inbound orbits in November 1977, containing a total of 9 magnetopause crossings, which occurred under widely differing orientations of the external magnetic field. The main findings are: (1) The magnetosheath flow near the magnetopause is characterized by large fluctuations, which often appear to be temporal in nature. (2) Between ~ 0.1 and $\sim 0.3 R_E$ outside the magnetopause, the plasma density and pressure often start to gradually decrease as the magnetopause is approached, in conjunction with an increase in magnetic field strength. These observations are in accordance with the formation of a depletion layer due to the compression of magnetic flux tubes. (3) In cases where the magnetopause can be well resolved, it exhibits fluctuations in density, and especially pressure and bulk velocity around average magnetosheath values. The pressure fluctuations are anticorrelated with simultaneous magnetic field pressure changes. (4) In one case the magnetopause is characterized by substantially displaced electron and proton boundaries and a proton flow direction change from upwards along the magnetopause to a direction tranverse to the geomagnetic field. These features are in agreement with a model of the magnetopause described by Parker. (5) The character of the magnetopause sometimes varies strongly between ISEE-1 and -2 crossings which occur ~ 1 min apart. At times this is clearly the result of highly non-uniform motions. There are also cases where there is very good agreement between the structures observed by the two satellites. (6) In three of the nine crossings no boundary layer was present adjacent to the magnetopause. More remarkably, two of the three occurred while the external magnetic field had a substantial southward component, in clear contradiction to expectations from current reconnection models. (7) The only thick (low-latitude) boundary layer (LLBL) observed was characterized by sharp changes at its inner and outer edges. This profile is difficult to reconcile with local plasma entry by either direct influx or diffusion. (8) During the crossings which showed no boundary layer adjacent to the magnetopause, magnetosheath-like plasma was encountered sometime later. Possible explanations include the sudden formation of a boundary layer at this location right at the time of the encounter, and a crossing of an 'inclusion' of magnetosheath plasma within the magnetosphere. (9) The flow in the LLBL is highly variable. observed directions include flow towards and away from the subsolar point, along the geomagnetic field and across it, tangential and normal to the magnetopause. Some of these features clearly are nonstationary. The scale size over which the flow directions change exceeds the separation distance (several hundred km) of the two spacecraft.

Space Science Reviews **22** (1978) 717–737. *All Rights Reserved*

1. Introduction

The early orbits of the ISEE-1 and -2 satellites crossed the Earth's magnetopause near local noon at ecliptic (and solar magnetospheric) latitudes of approximately 25°, i.e. only a few Earth radii away from the subsolar point. This region is of crucial importance for the understanding of magnetospheric dynamics since it is preferred site of day-side reconnection (and consequently plasma entry) in all theoretical work ever since the original proposal by Dungey (1961). The interest in the subsolar region also stems from the subsonic, stagnated nature of the magnetosheath flow in the vicinity of the subsolar point. Until now, no comprehensive study of plasma properties in the subsolar region has been reported. Crooker and Siscoe (1975), using low time resolution data from Explorer-33, studied a single crossing near the subsolar point. Their emphasis, however, was on magnetosheath subsonic flow characteristics. At medium and high latitudes a number of studies of the plasma near the noon magnetopause have been reported (Haerendel and Paschmann, 1975; Paschman *et al.*, 1976; Eastman *et al.*, 1976; Crooker, 1977; Formisano *et al.*, 1978; Haerendel *et al.*, 1978; Eastman and Hones, 1978). These studies have established that a boundary layer of magnetosheath-like plasma, originally discovered by Hones *et al.* (1972) near the flanks of the magnetosphere and by Rosenbauer *et al.* (1975) in the polar region, exists inside the entire dayside magnetopause. Some interesting questions with regard to the plasma entry processes were raised by these studies. In particular, Haerendel *et al.* (1978) have argued that the lack of a correlation of the existence or thickness of the (dayside) boundary layer with the southward component of the external magnetic field, and the absence of any accelerated tangential flow inside the magnetopause cast serious doubt on the validity of previous reconnection concepts. The ISEE mission now provides the unique opportunity to test ideas of solar wind-magnetosphere interaction mechanisms by studying, in high time resolution, the structure of the magnetopause and low-latitude boundary layer in the subsolar region.

In this paper we present high resolution plasma observations obtained in November 1977. The selected crossings occurred under different external magnetic field orientations and are well suited to demonstrate the wide range of conditions encountered in the subsolar region. It is clear that such a limited data set cannot reveal the entire situation, but it does provide directions for further studies. We have chosen to present the observations in considerable detail, in order that the reader may obtain an impression of the nature of the data (Section 3). This seems important in an area of research where guidance by sound theoretical predictions has not yet been achieved. The essential results are then summarized and discussed in Section 4.

2. Instrumentation and Method of Data Analysis

The measurements discussed in this paper were made with the identical LASL/MPI 'Fast Plasma Experiments' (FPE) on ISEE-1 and -2. The instrumentation is discussed in detail elsewhere (Bame *et al.*, 1978). Briefly, the FPE consists of three quadrispherical electrostatic analyzers, each with large ($0.18\,cm^2$) entrance aperture and fan shaped angular acceptance, defined by a $\pm 55°$ elevation angle, centered about the spacecraft equatorial (i.e. ecliptic) plane, and an azimuthal angle of $\pm 3°$ FWHM at 0° elevation angle. Of the three analyzers, two (denoted '2D') integrate over the 110° elevation angle through the use of a single collector plate at the exit of the analyzers. Secondary electrons produced at the collector are counted by an open electron multiplier. In contrast, the third ('3D') analyzer has a collector/emitter which is split into four equal segments. Each of these segments is sampled separately through the use of appropriately switched retarding potentials, thus resolving the 110° elevation angle range into 4 contiguous channels of 27.5° width. During each spacecraft revolution (~ 3 s) each 2D analyzer measures eight electron and eight ion spectra. For the cases to be discussed here, the energy per charge range, from 11 V to 20 kV for electrons, and from 70 V to 40 kV for ions, is covered in 16 contiguous channels using exponentially decaying analyzer voltages. Combining the measurements of both 2D instruments yields the two-dimensional distribution function of electrons and ions at sixteen energies and sixteen (nearly equidistant) azimuthal angles, integrated over 110° of elevation angle, in only 3 s. The extremely high analyzer sensitivity ensures that adequate counting statistics were obtained in each energy/angle channel. During high bit rate transmission, complete 2D distributions are repeated every 3 s, whereas during low bit rate transmission they were repeated every 12 s. Because the four elevation channels of the 3D analyzer are sampled sequentially, completion of the basic 3D cycle takes four times longer, i.e. 12 s, and repetition rates are correspondingly slower. Two 3D cycles are necessary if an azimuthal resolution equivalent to the 2D distributions is required. Because of the much longer snapshot and repetition times for the 3D distributions, the 2D data are of great value for the study of rapidly varying phenomena.

The data in this paper are presented in terms of moments of the measured two-dimensional and three-dimensional distribution functions. Because of instrumental limitations, the moment calculations are based on a number of assumptions and approximations. A brief outline of the employed method is given in the Appendix.

3. Observations

In this section we will present observations obtained during four successive crossings of the magnetopause region by ISEE-1 and -2 in early November of 1977. All crossings occurred close to local noon (11:20 to 12:15 hr) and ~ 3 to $\sim 5 R_E$ northward of the subsolar point. In three of the four crossings the magnetopause was encountered at fairly nominal geocentric distances (about $11.5 R_E$), whereas the fourth orbit had multiple magnetopause crossings down to a distance of $\sim 7.5 R_E$. External magnetic field conditions ranged from strongly positive B_z through nearly zero B_z to highly negative B_z. We will present the cases in order of decreasing B_z. The magnetic field data for these cases may be found in the paper by Russell and Elphic (1978). In order to facilitate a comparison, the field magnitude is shown together with our data (courtesy C. T. Russell).

Crossing of 5 November 1977. This crossing is characterized by nearly parallel and equally strong internal and external magnetic fields at the magnetopause, thus being well suited for a study of the magnetopause structure for conditions of strongly northward external field and very small magnetopause currents.

Figures 1a and 1b show the plasma data from ISEE-1 and -2 in terms of the two-dimensional moments of the distribution function (see Appendix). A curve denoting the magnetic field pressure, $B^2/8\pi$, has been added to the plasma pressure panel. Beginning with the ISEE-2 data (Figure 1a), plasma density and temperature profiles show a transition from magnetosheath to magnetospheric values over a time span of almost 6 min (17:13 to 17:19 UT). At 17:13 (first dashed line) the density begins to decrease gradually until reaching a plateau near 17:15 UT (solid line). Throughout most of this interval there is still magnetosheath-like flow. The period between 17:13 and 17:15 is characterized also by a depression in an otherwise fairly constant plasma pressure. This depression is compensated by a simultaneous enhancement in magnetic field pressure, which also shows no net change throughout the period. Thus, there is a net pressure balance, although, contrary to the common expectation, it is not a balance between predominantly magnetic field pressure on the inside and plasma pressure on the outside. Instead both pressures are essentially balanced separately in this case.

Nearly simultaneous with the magnetic field enhancement the B_y component shows a gradual change between a fairly steady magnetosheath value of $+10\gamma$ to its final magnetospheric level of -10γ (see Russell and Elphic, 1978). It is tempting, therefore, to associate the period between 17:13 and 17:15 with the magnetopause (see, however, the discussion further below).

Between 17:15 and $\sim$17:20 density and temperature finally show a transition to magnetospheric values in several fairly sudden changes. From the energy spectra (not shown) it appears that the plasma is composed of a mixture of magnetosheath and magnetospheric populations. This is a feature found to be typical of the low-latitude boundary layer by Eastman *et al.* (1976) and Haerendel *et al.* (1978). There is a brief occurrence of plasma flow ($\sim$60 km s^{-1}) directed towards local evening (i.e. opposite to the magnetosheath flow) after 17:17.

We now turn to a discussion of data from ISEE-1, separated at this time by about 475 km from ISEE-2. Comparing Figures 1a and 1b it is clear that the plasma parameter profiles at the magnetopause transition are remarkably similar. In particular, the sharp transitions in density and temperature noted above for ISEE-2 are also clearly visible in the ISEE-1 data. The same description applies also for the plasma flow. From a comparison of the times for the clearly identifiable common features (marked by the solid and dashed lines), one can

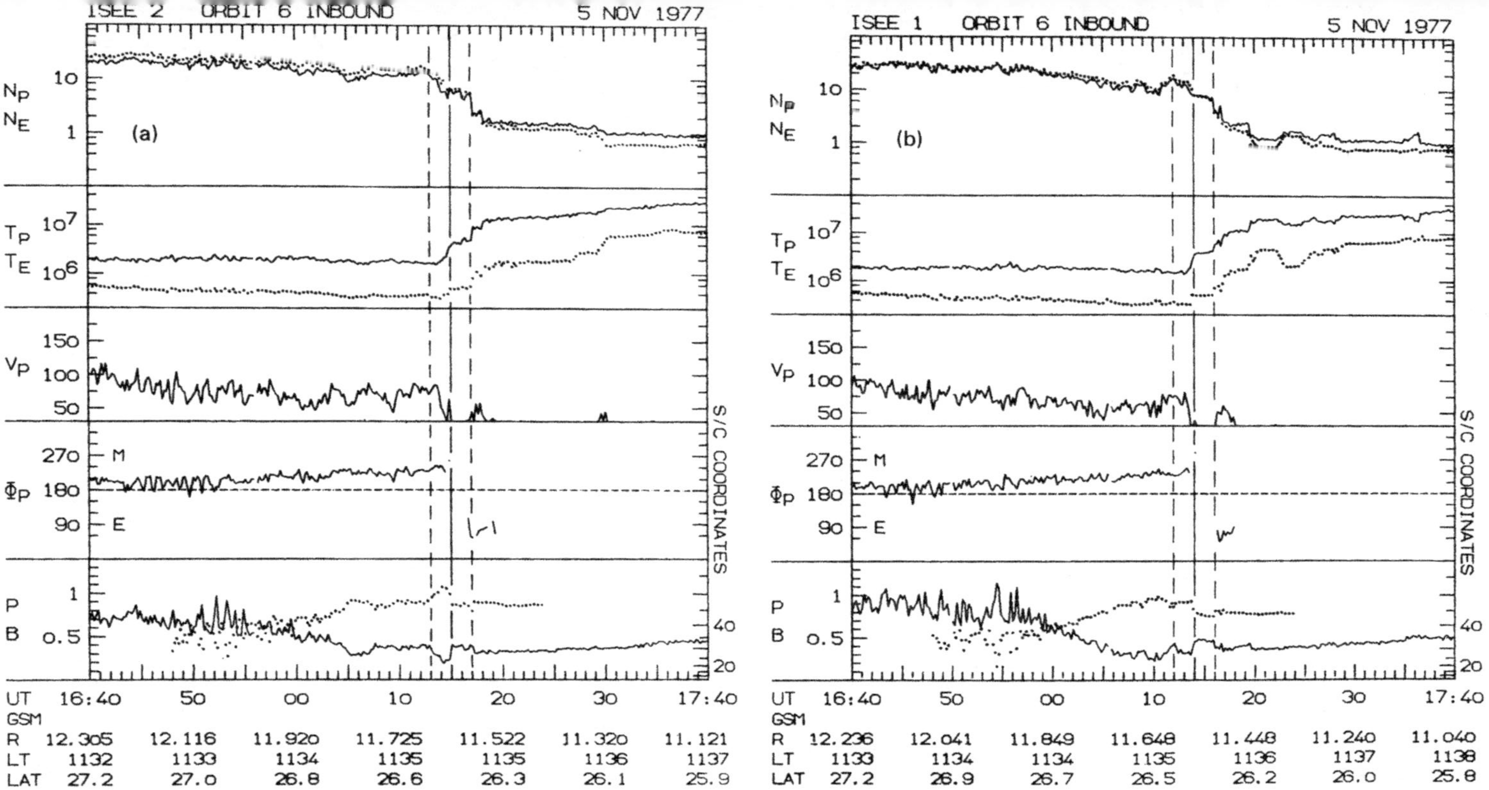

Fig. 1. (a) Two-dimensional proton and electron fluid parameters and magnetic field pressure for a 1 hr interval during an inbound pass of ISEE-2 on 5 November 1977. Spacecraft positions are given as geocentric radial distance, R (in R_E), local time, LT (in hours), and latitude, LAT (in degrees) in solar magnetospheric (GSM) coordinates. Densities, N, are given in cm^{-3}, temperatures in Kelvin, with solid lines for protons and dotted lines for electrons Magnitude (v_p, in $km\,s^{-1}$) and direction (ϕ_p) of the proton bulk velocity correspond to a projection onto the spacecraft equatorial plasma which is nearly identical to the ecliptic plane. $\phi_p = 180°$ indicates an antisunward flow; $\phi_p = 90°$ (E) and $\phi_p = 270°$ (M) indicate flows towards evening and morning respectively. The total plasma pressure (protons plus electrons, solid line) and the magnetic field pressure (dotted line, courtesy of C. T. Russell) are given in units of 10^{-8} dynes cm^{-2} (left scale). The dotted curve can also be interpreted as the magnetic field strength (in gamma), when using the right hand scale (quadratic). The vertical line indicates the time of the magnetopause encounter. (b) Same as Figure 1a, except the data are from ISEE-1.

conclude that ISEE-1, the leading satellite, encountered these features 1 min prior to ISEE-2. It might be pointed out that the observation of the same well defined features a minute apart demonstrates two things; first, these features must be spatial in nature, and second, even sharp gradients in plasma properties can remain stable over a minute. For a satellite separation distance of ~475 km one derives a relative speed between these structures and the ISEE spacecraft of 7.9 km s^{-1}, assuming a steady motion. Considering the spacecraft's inward velocity of ~2.3 km s^{-1}, the magnetopause thus is moving outward at 5.6 km s^{-1}.

Based on this velocity one derives a thickness of ~950 km and ~240 km for the magnetopause as seen by ISEE-2 and ISEE-1, respectively. The large difference could be real, caused by a change in the magnetopause thickness between the two crossings, or it could reflect nonuniformity of motion. There is also the possibility that our identification of the magnetopause with the slow B_y change in the case of ISEE-1 is incorrect and that the magnetopause is marked by the sudden drop in field pressure near 17:15 UT, thus also supporting a thickness in the vicinity of 200 km, which corresponds to slightly more than three proton gyro-radii. If the latter magnetopause identification is correct the enhanced field pressure and reduced plasma pressure and density between 17:13 and 17:15 (17:12 to 17:14 for ISEE-1) are located immediately outside the magnetopause and become part of the 'depletion' region referred to below.

After 17:14 UT, the plasma observed by ISEE-1 has the characteristics of the low-latitude boundary layer. Judging from the more gradual density and temperature profiles (and confirmed by the particle spectra), the inner edge of the layer is less well defined than in the case of ISEE-2.

Finally, we will briefly turn to a discussion of the magnetosheath data. Both Figure 1a and 1b show smooth profiles for density and temperature, but substantial fluctuations in flow velocity. (For the sake of completeness, we just mention here that the 3D analysis indicates an elevation angle of the flow between 40 to 50° throughout the magnetosheath period shown.) Comparison of the measurements from ISEE-1 and -2 indicate a rather poor correlation of the velocity profiles, indicating that a mixture of spatial and temporal variations probably caused these fluctuations.

A feature common to both spacecraft observations is the gradual decrease of plasma density and pressure after ~17:00, which is accompanied by a corresponding rise in field strength. This magnetosheath 'depletion' region is discussed in Section 4.3.

Crossing of 3 November 1977. In Figure 2 we show the data from the 2D analysis in the same format as in Figure 1, but have added the 3D flow velocity (see Appendix). As can be seen from Figures 2a and b, there is excellent agreement between the 2D and 3D azimuth angles of the flow. As expected for the mostly large ($\geqslant 45°$) elevation angles Λ, the 2D flow speed is significantly lower than the 3D speed a large fraction of the time.

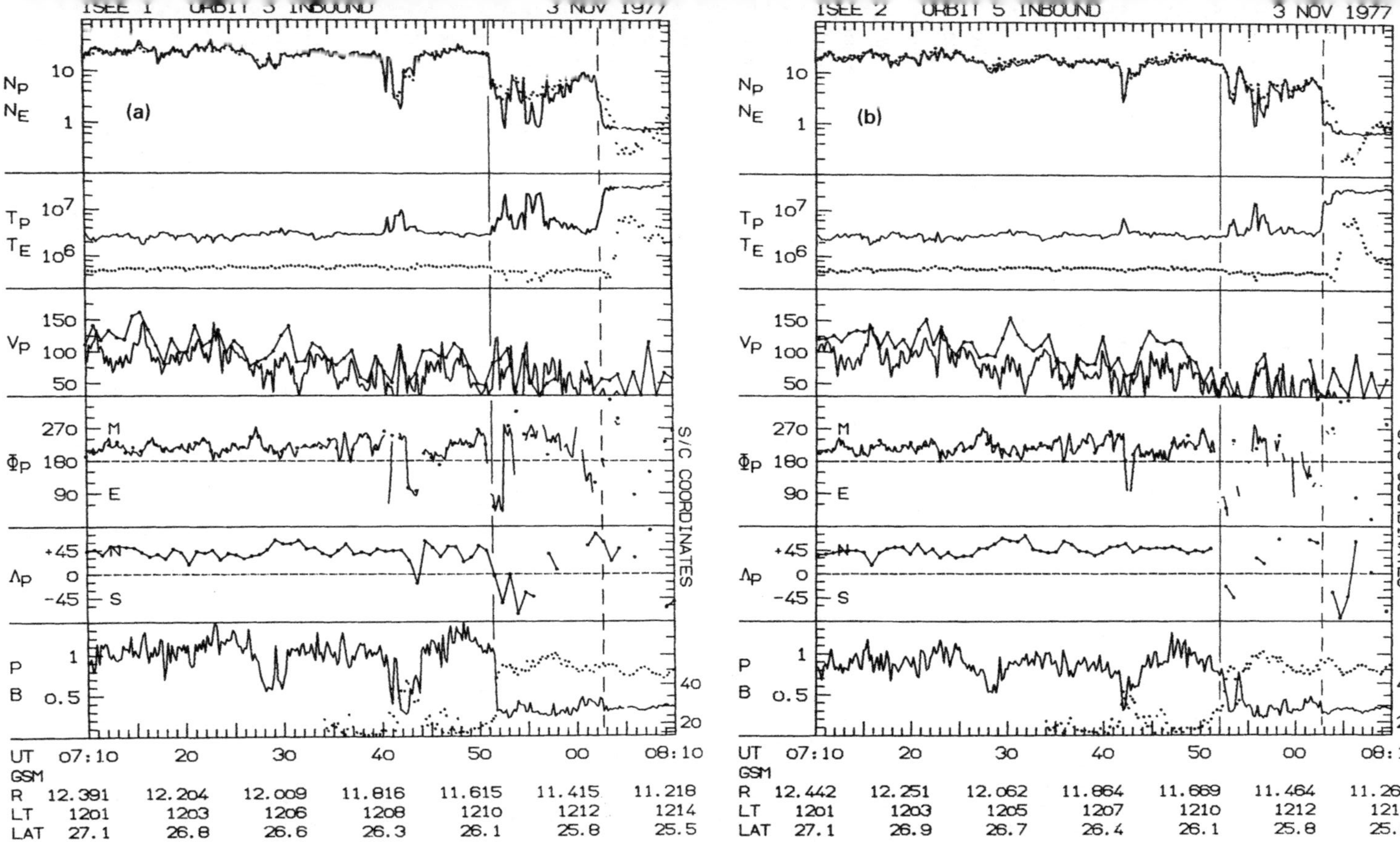

Fig. 2. (a) Two- and three-dimensional plasma parameters and magnetic field pressure for the pass of ISEE-1 on 3 November 1977. The format is the same as for Figure 1a, except that the three-dimensional bulk flow velocity (in terms of speed v_p, azimuth ϕ_p and latitude or elevation, Λ_p, in the spacecraft, i.e. ecliptic, system) has been added. $\Lambda_p = 0°$ indicates a flow in the ecliptic plane and $\Lambda_p = 90°$ indicates a northward flow. The vertical solid line indicates the magnetopause crossing and the dashed line marks the inner edge of the low-latitude boundary layer (LLBL). (b) Same as Figure 2a, except the data are from ISEE-2.

As opposed to the previous example, the magnetopause encountered on 3 November is characterized by large and abrupt changes in the plasma and magnetic field quantities. This is particularly evident in the ISEE-1 crossing (Figure 2a): after a brief entry into the magnetosphere from ~07:41 to ~07:44, the magnetopause is finally crossed at 07:51:30 (solid line), as demonstrated by the onset of an abrupt drop in plasma density and pressure (by about a factor of four), coincident with a likewise sharp increase in magnetic field pressure from magnetosheath to magnetospheric levels resulting in a change in β (the ratio of plasma to magnetic field pressure) from more than 10 to about one third. These changes are completed by about 07:51:40. The plasma properties at this time have not yet reached magnetospheric levels, but remain essentially at the intermediate levels characteristic of the low-latitude boundary layer.

Prior to these abrupt changes there is a short (~3 min) period of gradual decrease in plasma pressure, accompanied by an increase in field pressure, beginning around 07:48:30. At the same time there is a notable decrease in flow speed. This, again, could be a magnetosheath depletion layer. Earlier, the magnetosheath flow exhibits large fluctuations in speed and direction. Right after the onset of the sharp density drop (and slight temperature increase) at 07:51:30, the satellite encounters magnetosheath-like plasma, which, noting from the ϕ and Λ panels, flows in the direction opposite to the plasma in the magnetosheath, and also approximately antiparallel to the local **B** vector (see complete field data in Russell and Elphic (1978). Note in particular that the flow is directed downwards, i.e. towards the equatorial region. A similar reversal occurred at the time of the earlier brief magnetopause encounter near 07:43. Such a flow signature right at the magnetopause has been observed previously in a few of the Heos-2 crossings of the LLBL (Haerendel *et al.*, 1978).

Following 07:51:40, and lasting until 08:02:45 (dashed line), there is a boundary layer of plasma with properties intermediate between magnetosheath and magnetosphere, except for a few brief periods where the plasma characteristics (large temperature, low density) are similar to those observed later in the magnetosphere, which is finally entered at 08:02:45. The sharp drop in density at the magnetopause, followed by intermediate densities until another sharp drop is encountered at the inner edge of the layer, is evidence against the layer being populated by diffusive entry through the nearby magnetopause, as pointed out by Haerendel *et al.* (1978). The sudden density jump (by about a factor of four) at the magnetopause suggests that the magnetopause at this time provides a rather effective barrier against direct particle entry from the magnetosheath along interconnected field lines. At the inner edge, the density change is even larger (by a factor of 10) and is also rather sharp. Note that, in spite of this large plasma density change, no significant changes in plasma and magnetic field pressures are observed. The flow in the LLBL is characterized by large fluctuations in magnitude ($dv/dt \sim 10 \text{ km s}^{-2}$), and directions varying from essentially tangential to the magnetopause and antiparallel to **B** (in the reversal region), to directions more perpendicular to **B**. Note the repeated occurrence of equatorward directed flow components.

The ISEE-2 data displayed in Figure 2b, are characterized by a slower magnetopause transition, commencing at about 07:52 (solid line), and completed about 1 min later. The magnetic field components also shows a much slower change, simultaneous with the plasma changes (Russell and Elphic, 1978). Note again the occurrence of a flow reversal immediately after the onset of the plasma transition but before its steepest gradient. The situation immediately afterward is complicated by a brief return into the magnetosheath or magnetopause layer near 07:54, not observed by ISEE-1. Together with the noted difference in the magnetopause duration, this feature indicates that, even over distances of only a few hundred kilometers, the boundary motion appears as nonuniform and oscillatory. A comparison of the flow characteristics in the low-latitude boundary layer shows that the flow speeds often disagree, whereas the flow directions do agree rather well whenever the speeds are sufficiently high in both cases.

Another important observation resulting from a comparison of the ISEE-1 and -2 data concerns the inner edge of the low-latitude boundary layer. The feature to note here is the complex time sequence: ISEE-1, although encountering the edge first, completes the transition after ISEE-2, with the midpoints occurring simultaneously. This temporal order clearly contradicts the expectations from a simple motion of the two satellites through a spatial boundary. For a spatial model to explain the data, one would have to invoke a boundary with rapidly varying steepness along its front, moving obliquely with respect to the spacecraft separation vector. Although such a model cannot be discarded a priori, a temporal explanation appears to be more likely (see Section 4.2). Under these circumstances it is, of course, impossible to derive a meaningful boundary velocity and boundary layer thickness.

When comparing the magnetosheath data of Figures 2a and 2b, it is evident, that in contrast to the crossing on 5 November, the present case is characterized by a good correlation of flow fluctuations, with no clear temporal displacement. This similarity of conditions over the separation distance of the two spacecraft suggests that the observed fluctuations are temporal in nature.

Crossing of 8 November 77. This crossing has been chosen as an example demonstrating a different character of the magnetopause as seen by the two spacecraft. Another reason was the high data rate which permitted the transmission of complete distribution functions every three seconds. This crossing is therefore particularly well suited for a high-resolution study of the magnetopause.

Figure 3a shows 24 min of data from ISEE-2, with data points spaced 6 s apart. As the spacecraft approaches the magnetopause, one notices first the extremely

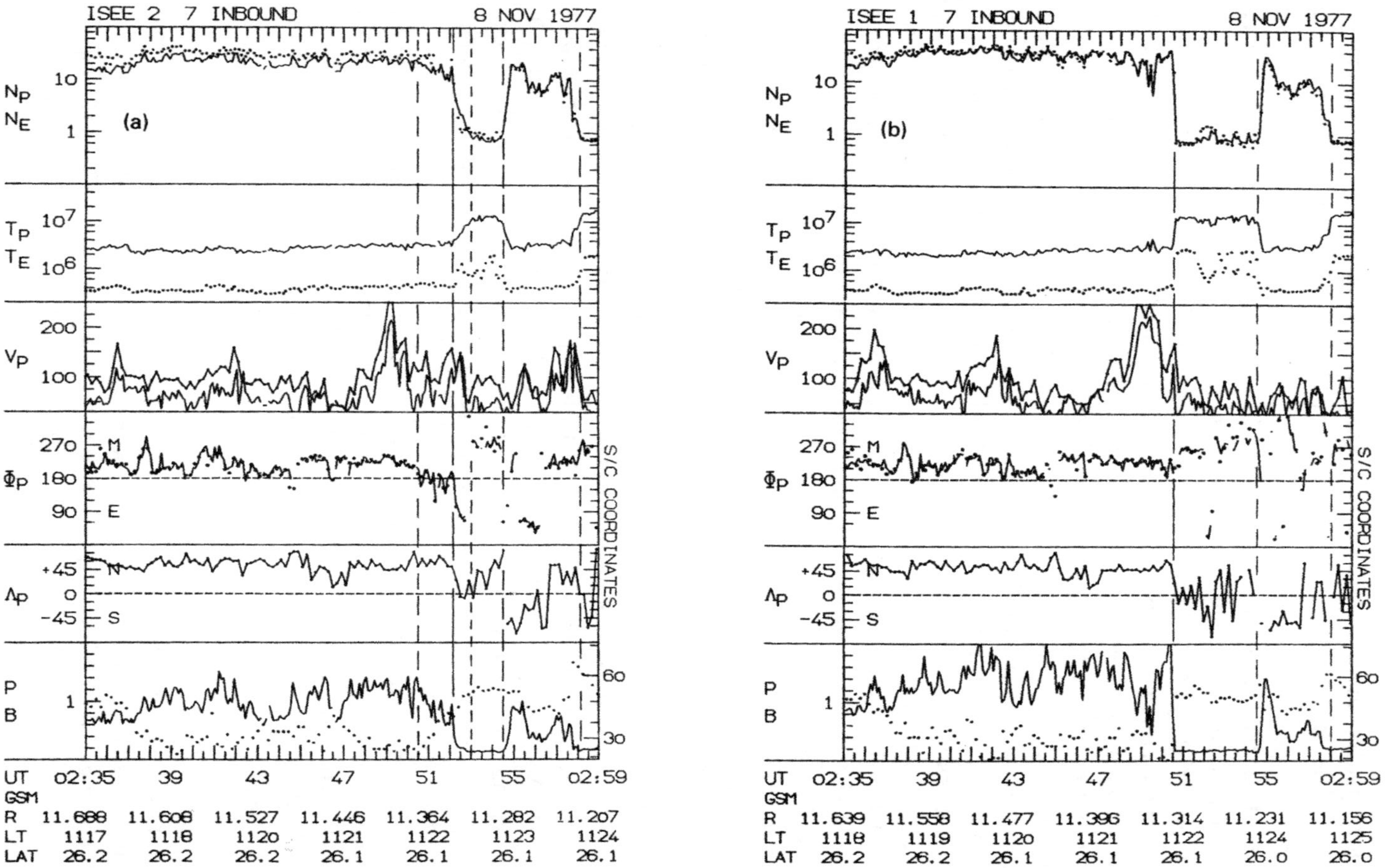

Fig. 3. (a) Two- and three-dimensional plasma parameters and magnetic field pressure for a 24 min interval of the ISEE-2 crossing on 8 November 1977. The format is the same as in Figure 2. The vertical solid line indicates the time of magnetopause crossing. (The dashed lines are described in the text.) (b) Same as Figure 3a, except the data are from ISEE-1.

large fluctuations in magnetosheath flow speed ranging from $\sim 40 \text{ km s}^{-1}$ to about 200 km s^{-1}. Near 02:50:30 a slow decrease in plasma density and pressure is noted, accompanied by a rise in field pressure (as well as changes in B_x and B_z). Note that the plasma flow has changed from a direction upwards and towards morning (as expected for a pre-noon, low-latitude crossing) to a precise noon-signature, namely anti-sunward and upward. At 02:52:10 there is a sharp density drop, the temperature increases (especially for the electrons) and the proton flow becomes suddenly directed towards evening.

Coincident with the plasma density change, there is a sharp $\sim 30\ \gamma$ change in B_y. One would tend to identify the magnetopause with these sudden changes in N and B_y and associate the preceding gradual change in N and $|B|$ with a magnetosheath depletion layer. However, the latter interval could also be part of the magnetopause layer. (For still another identification of the magnetopause see Russell and Elphic (1978).) Inspection of the sharp magnetopause transition reveals that the electron density drops much more rapidly than the proton density. The proton energy spectra indicate that the higher energy protons penetrate deeper into the magnetosphere than the lower energy protons. Together with the proton flow direction change occurring at this time, these features are reminiscent of the magnetopause model originally described by Parker (1967a, b), as discussed in Section 4.1.

After several minutes of magnetospheric environment, magnetosheath-like plasma is encountered, although there is no evidence that a magnetopause crossing has occurred: The flow directions observed between 02:55 and 02:56 are nearly opposite in direction to the magnetosheath flow. Furthermore, the magnetic field, although showing a noticeable change at this time, clearly does not return to magnetosheath levels, with the consequence of a much reduced β. It thus appears that this layer is located within the magnetosphere, either as an 'inclusion' of magnetosheath-like plasma, separated from the magnetopause or as a normal boundary layer (i.e. attached to the magnetopause), but appearing isolated due to the formation of a boundary layer right at this time.

We now turn to the discussion of the ISEE-1 data (Figure 3b). Comparison with Figure 3a immediately reveals the striking dissimilarity of the magnetopause transition. Note in particular the extremely abrupt density and temperature change near 02:50:30 (solid line) in the ISEE-1 data and the great differences in flow directions. The period immediately preceding this transition also differs strongly: whereas in the case of ISEE-2 density and pressure showed just a gradual decrease a few minutes prior to the main density transition, there is considerable structure in the density and pressure profile in the case of ISEE-1, with a pronounced minimum near 02:49:30. Similarly strong differences are also reflected in the magnetic field data. Note in particular the steep minimum in field pressure near 02:50 which opposes the simultaneous plasma pressure increase, without quite balancing it. Consequently, there are substantial variations in total pressure. If this pressure and density structure were to be part of the magnetopause, so would be (at least partially) the large (~200 km s^{-1}) peak in velocity, since it extends up to nearly 02:50. On the other hand, this peak is well correlated with a remarkably similar velocity structure in the ISEE-2 data, and there we interpreted the peak to be clearly located well outside the magnetopause. It appears then that the magnetopause in the ISEE-1 data is probably just the sharp transition of about 10 s duration, centered around 02:50:30. However, even if this identification were correct, the different plasma and field profiles at the magnetopause suggest a sufficiently complicated motion of the boundary to make the determination of a speed probably useless.

Other features in the ISEE-1 and -2 data do, however, agree rather well: for example, there is good correlation between the large variations of magnetosheath flow speeds, which emphasizes the importance of temporal fluctuations in the magnetosheath flow.

Finally, when comparing the observations of the isolated region of magnetosheath-like plasma after 02:54, it appears that the temporal sequence is hard to reconcile with a simple motion of a spatial structure across the two satellites: ISEE-2, although encountering the plasma first, reaches the peak later than ISEE-1, due to a slower rise time, with midpoints recorded simultaneously. Since ISEE-2 is the trailing satellite, the first noted fact would be consistent with a plasma boundary overtaking the satellites. The remaining feature would, however, require a very special type of motion and a spatially varying steepness of the boundary. The described situation is almost identical to that observed at the inner edge of the boundary layer on 3 November (see above), and, as pointed out there, might well be caused by an essentially temporal process. The lack of any significant flow when encountering the region, also excludes an explanation in terms of a passage of a front of injected particles. However, the observation might be explainable by impulsive penetration of filaments of magnetosheath plasma, or by an eddy convection cell from a nearby boundary layer passing the spacecraft (see Section 4.2).

Crossing of 10 November 1977. This pass has been chosen mainly because of its strongly southward external B_z, whereas the other examples had at least slightly northward external B_z. Unfortunately, the ISEE-1 instrument was turned off during part of this orbit, therefore we can only present the data from a single spacecraft, ISEE-2.

In Figure 4 we show the three magnetopause crossings which occurred over a time interval of 20 min, all at distances well below nominal ($<8R_E$). The magnetic field showed very pronounced and smooth changes in all components which were very similar for the three crossings (Russell and Elphic, 1978). In Figure 4 we have indicated (by grey shading) the width of each magnetopause as defined by the time it took the magnetic field to essentially complete its change from magnetosheath to magnetosphere values. We refer to this as the 'magnetopause layer' (MPL), to be distinguished from a boundary layer, which would be located

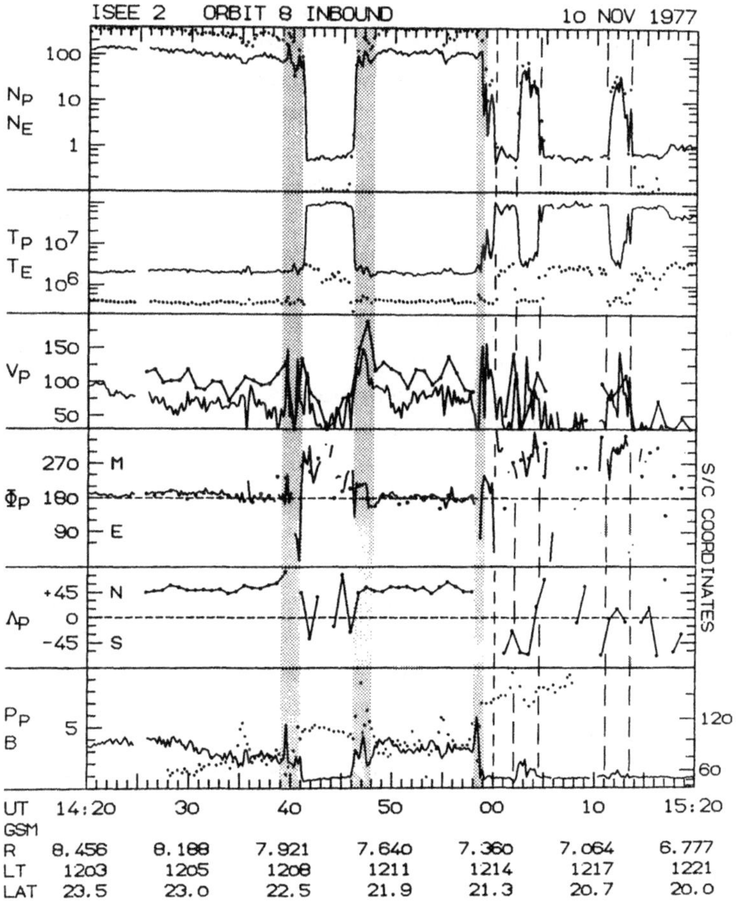

Fig. 4. Observations for the three ISEE-2 magnetopause crossings on 10 November 1977. The format is the same as on the previous figures. The grey-shaded areas indicate the width of the magnetopause as defined by the duration of the change in magnetic field quantities (see Russell and Elphic, 1978). The significance of the dashed lines is explained in the text.

earthward of the MPL. For the first two crossings it is evident that the plasma density and temperature changed very abruptly (within 10 to 20 s) over several orders of magnitude right at the inner (i.e. magnetospheric) edge of the MPL. (The substantial difference between proton and electron densities on either side of the transition is probably caused by particle spectra extending outside the range of instrument energy coverage.) Within the MPL the plasma density and temperature fluctuated, but on average agreed with the nearby magnetosheath values. Note the enhanced flow speeds which occurred in the MPL. The fact that the plasma reached magnetospheric levels precisely at the inner edge of the MPL means that no boundary layer was present during these crossings, similar to the case on 8 November. In other words, at these times and locations the magnetopause represented a very effective barrier against plasma entry (see discussion in Section 4.2).

The third magnetopause crossing exhibits a sharp drop in density near the magnetospheric side of the magnetopause layer as in the previous two crossings. Here, however, a thin (~1 min) boundary layer follows, characterized again by values of density and temperature intermediate between magnetosheath and magnetosphere levels.

A noticeable feature in all three crossings is the pronounced structure of plasma and field pressure and the occurrence of enhanced flow speeds within the magnetopause layers. In the first and third crossing there are sharp maxima in plasma pressure, with simultaneous minima in field pressure. In the second crossing the structure is more complicated, but still the field and plasma pressure variations tend to oppose each other.

In addition to the three clear magnetopause crossings, there appear to be two close approaches near 14:35 and again near 14:55. The plasma parameters at these intervals are very similar to those observed in the magnetopause layers. The large spikes in B_x observed at these times (Russell and Elphic, 1978), together with deflections in the plasma flow direction, ϕ, are suggestive of the presence of folds in the magnetopause at these times.

Following the third magnetopause crossing there are two isolated periods (15:02 to 15:04:30 and 15:10 to 15:13:30) of magnetosheath-like plasma which do not result from still more magnetopause crossings, but appear to be 'inclusions' of magnetosheath-like plasma within the magnetosphere, or the result of a sudden formation of a regular boundary layer, as already discussed in the subsection on the crossing of 8 November. This interpretation again is based on the observed plasma flow directions within these regions, which differ by 60° or more from those seen previously in the magnetosheath, and which are, at times, pointed equatorward. Also, the magnetic field orientation clearly did not return to the direction and magnitude seen in the magnetosheath. Most clearly, however, the different nature of these regions is reflected in the very small value (~0.1) of β, compared to values >1 at the magnetopause and further out.

Finally we note that signatures of a magnetosheath depletion layer are seen prior to the first clear magnetopause crossing, somewhat obscured by the brief magnetopause encounter near 14:35.

4. Summary and Discussion

In the previous section we have presented from four inbound passes of the ISEE-1 and -2 spacecraft. A total of nine clear magnetopause crossings were observed during these orbits counting ISEE-1 and -2 crossings separately whenever data from both experiments were available. In the following the main findings are summarized and discussed.

4.1. Magnetopause structure and motion

Although ISEE-1 and -2 were separated by only a few hundred km during the times discussed, and hence boundaries crossed the two spacecraft typically within a minute or less, there often is a large difference in the character of the two crossings. At times this is clearly due to substantial velocity changes that occurred between crossings, i.e. over short times or distances, but there is also evidence for rapid structural changes. In other examples the structure and width of the profiles were remarkably similar, indicating that both the spatial structures and their velocities can remain constant for periods of a minute or so. Only in these cases is a determination of the speeds and approximate thickness of the structures possible. In order to investigate the question of the characteristic time constant for magnetopause evolution, a wider range of separations between the two spacecraft is needed.

In the crossings of 10 November, where the magnetopause was exceptionally broad (~ 2 min), and thus could be resolved well, the magnetopause layer (MPL) was characterized by a pronounced structure in plasma and magnetic field pressures, with (anti-correlated) excursions of these quantities considerably exceeding the values in the neighboring regions. On the average, however, plasma density and temperature within the MPL agreed with those observed in the magnetosheath. During the same MPL crossings, substantial enhancements of proton flow speed were observed.

In one magnetopause crossing (8 November, ISEE-2) the magnetopause showed an interesting structure, in that the density decrease at the magnetopause occurred at substantially different times for protons and electrons, with the electron density transition completed significantly earlier. The protons, moreover, showed deeper penetration with increasing energy. Simultaneous with the onset of the density decrease, the protons flow turned $\sim 90°$, from an antisolar and upward direction to a duskward direction in the ecliptic. This behaviour is strongly reminiscent of the situation expected for a model of the magnetopause with neutralization of the boundary charge, as described by Parker (1967a, b). In this model the magnetosheath electrons penetrate one electron-gyroradius, R_{ge}, and the protons penetrate one proton gyroradius, R_{gp}, thus leading to different locations of the boundary of magnetosheath electrons and protons. For the condition of this crossing, R_{ge} would be ~ 0.5 km and $R_{gp} \sim 50$ km. In Parker's model, charge neutrality is preserved by ionospheric electrons. Being cold, they might escape detection by our instrument, thus explaining the discrepancy between the measured proton and electron densities. Parker's model also predicts the observed proton flow behavior. First, the Lorentz force of the upward directed geomagnetic field will deflect the protons such that they will turn towards the evening side, explaining the observed change of the flow azimuth from 0° to +90°. Second, the upward motion of the ions, once they have penetrated past the electrons, represents a current which produces a field transverse to the geomagnetic field. In one limiting solution (Parker, 1967a), the proton velocity parallel to the geomagnetic field is reduced to zero by the Lorentz force in the transverse field region. This would explain why we observe the deflected particle flow to be perpendicular to the field, although there was a large upward directed flow outside the magnetopause. Before drawing too many conclusions from this successful model comparison, it should be noted that this case is unique. In fact, the ISEE-1 crossing, which occurred slightly earlier, did not show any of the features described by the model.

Large changes in proton flow directions ('flow reversals') at the magnetopause like the one just discussed have been observed during a number of ISEE crossings. The explanation given above in terms of finite gyroradius effects in the presence of a large density gradient, does, however, not always work. During the pass on 3 November, for example, the reversed flow was directed essentially along the field rather than across it, thus requiring a different interpretation.

4.2. Properties of the Low-Latitude Boundary Layer

Of the nine magnetopause crossings considered, three were characterized by the complete absence of a boundary layer, i.e. there was no magnetosheath-like plasma on magnetospherically oriented field-lines immediately earthward of the magnetopause. At least one more case had a boundary layer whose thickness did not exceed the thickness of the magnetopause layer. Only during the crossing of 3 November did we find a boundary layer with a duration exceeding several minutes. This result should be contrasted with that of Haerendel *et al.* (1978) who studied Heos-2 observations at typically 45° solar magnetospheric latitude. There boundary layer durations of more than 30 min were frequently observed, although at times no boundary layer could be detected (within the constraints imposed by the very coarse temporal resolution of ~4 min). Larger distances from the subsolar region also may account for the fact that, in contrast to our ISEE results, a boundary layer was observed in all IMP-6 crossings (Eastman and Hones, 1978).

The absence of a boundary layer implies that no or little mass is transferred through the magnetopause at the time and location of the crossing. It therefore is particularly interesting that two such crossings (10 November) occurred while the external field had negative (i.e. southward) Bz. Because the crossings were located about $3R_E$ north of the subsolar point, a very pronounced (~500 km) boundary layer should have been expected if magnetic reconnection were taking place (Sonnerup, 1976; Haerendel *et al.*, 1978).

Only one thick boundary layer was observed during the 1977 ISEE magnetopause crossings between ~10 and ~12 hr magnetic local time, if we exclude for the moment the isolated occurrences of magnetosheath-like plasma. It was characterized by very abrupt and large density changes at the outer and inner edges, and thus belongs to the category of low-latitude boundary layers discussed by Haerendel *et al.* (1978). The other, much briefer boundary layers, also often show such a density change, at least at the magnetopause edge. A sharp density change at the magnetopause followed by a density plateau is an important feature, because it puts severe constraints on the formation of the LLBL. Local entry by flow along magnetic field lines having a component normal to the magnetopause would tend to keep the density constant; entry by diffusion would produce a continuous gradual change, such as often observed in the IMP-6 data (Eastman and Hones, 1978). Noting again the larger distance of the IMP-6 crossings from the subsolar point, the different character of the typical IMP-6 density profiles appears to indicate a dependence of the boundary layer profile on latitude and/or local time. This, in turn, would imply different processes governing the entry or subsequent transport of the plasma.

Another class of observations, related to the boundary layer phenomena are the encounters with high-density, magnetosheath-like plasma well inside of and (at least temporally) isolated from the magnetopause. Superficially they appear like 'inclusions' of magnetosheath-origin plasma within an entirely magnetospheric environment. As pointed out in Section 3, these events could also be simply the result of a temporal effect. The strongest evidence for temporal changes is the noted occasional absence of a boundary layer. A temporal change was also invoked to explain the inner edge of the only thick boundary layer observed in the ISEE-data. It thus appears natural to include the isolated encounters in a temporal model of the boundary layer, in which plasma entry and/or transport along the boundary are 'switched' on and off. Under these circumstances it follows immediately that occasionally no boundary layer should be observed at the time of the magnetopause crossings. A later formation of a boundary layer (by local entry or by transport from an already existing boundary layer nearby) then would appear as an isolated encounter, as long as the satellites are still within reach of the boundary layer thickness. In this picture, the duration of a boundary layer encounter might not be related to its thickness at all, but just reflect a temporally limited plasma entry, or the passage of a convection cell (Haerendel, 1978).

Assuming (1) this entirely temporal explanation of the duration of all boundary layer observations, and (2) localized plasma entry, one can estimate the spatial extent of the boundary layer resulting from the subsequent spreading of the plasma, possibly by eddy convection. Taking a typical boundary layer duration of several minutes, and a plasma bulk speed of $\sim 100 \text{ km s}^{-1}$, one derives an upper limit for the size of individual boundary layer elements of several Earth radii.

The sporadic nature and perhaps small spatial scale of the boundary layer has, of course, important implications for the physical processes of plasma entry and transport, as well as the total amount of mass, momentum and energy transferred. The same applies to the existence of 'inclusions'. A qualitative model for an impulsive penetration of filaments of magnetosheath plasma, leading to such inclusions, has recently been proposed by Lemaire and Roth (1978), although the validity of the model for a realistic situation has yet to be demonstrated. In this model the filaments should retain their original magnetosheath magnetic field until the magnetospheric field has had time to diffuse into it. They should also move inward at a substantial speed, giving rise to an inward directed plasma flow, until their energy is dissipated via connection to the ionosphere. Neither of these effects is observed in our data, although it is possible that we observed these filaments only at their final, slowed-down state, already permeated by the magnetospheric field.

Finally, we turn to a discussion of the proton flow properties of the boundary layer (including its isolated variation). Flow speeds are characterized by large changes on short time scales. Speed changes from essentially zero to 100 km s^{-1} within 10 s are common. The flow directions vary, although generally much less rapidly, over the entire possible range. (Time aliasing of the measured distribution function by rapid density changes can simulate speed variations. Therefore it should be pointed out that the noted variations occurred while the density stayed fairly constant. Strong pressure gradients in the plasma will, however, affect the flow calculation, and indeed have been used to explain some of the flow 'reversals' we observed at the magnetopause.) Observed directions range from parallel to perpendicular to the geomagnetic field, from equatorward to poleward, from eastward to westward, from tangential to normal to the magnetopause, even on a given crossing. Some of these flows, notably those directed normal to the magnetopause, cannot exist in a stationary situation; the inward flow would move the boundary layer too far in, and any stationary outward directed flow would, of course have no lasting supply. All these flow directions are, however, consistent with eddy convection as the main source of plasma transport in the boundary layer (Haerendel, 1978).

There often are differences in flow speed of up to 100 km s^{-1} between the ISEE-1 and -2 measurements. Whenever substantial flow speeds are observed simultaneously on both spacecraft, the agreement in flow direction is very good. This implies that, although the flow may be sheared, its direction is coherent over distances larger than the separation distance between the spacecraft (i.e. a few hundred km). To investigate the question of the spatial scale of the convection pattern, it appears that much larger separation distances (>1000 km) will be needed.

4.3. Magnetosheath characteristics

The magnetosheath near the magnetopause in the noon region shows several features characteristic of a stagnation region; namely large fluctuations in direction and magnitude of the flow velocity. On occasions, the flow speed varied from less than 30 km s^{-1} to about 200 km s^{-1} within less than a minute. In a number of cases (e.g. 3 and 8 November) the correlation between variations seen by ISEE-1 and -2 is very good with no significant time displacement. This indicates that these fluctuations are temporal rather than spatial in nature and are coherent over distances of several hundred km. In other cases (e.g. 5 November) the correlation is very poor, indicating a mixture of temporal and spatial changes.

Close to the magnetopause we found evidence for a decrease of plasma density and pressure, in conjunction with an increase in magnetic field strength. Such a plasma depletion was predicted by Midgley and Davis (1963), Lees (1964), and by Zwan and Wolf (1976) as the result of flux-tube squeezing as the magnetic field becomes draped around the magnetopause. The observed thickness of the layer assuming no boundary motion ranged from less than $0.1R_E$ to about $0.3R_E$. The density and pressure decreased by a factor between 1.5 and 2. The total pressure (plasma plus field) stayed approximately constant. These results are in good agreement with the predictions of Zwan and Wolf (1976).

Appendix

This paper uses moments computed from the two-dimensional and three-dimensional distribution functions measured by the LASL/MPE 2D and 3D instruments on ISEE-1 and -2. It is the purpose of this appendix to outline the derivation of the moments from the measurements.

2D-Analysis. The measured 2D distribution consists of an array of count-rates, C_{ij}, at 16 velocities, v_i, and 16 azimuth angles, ϕ_j. (To avoid photo-electron contamination, electron measurements for energies below about 12 eV have been discarded in the present analysis.) The C_{ij} are proportional to $v_i^4 \int_{-55°}^{+55°} f \cos\theta \, d\theta$, where f is the distribution function (in $s^3 \, cm^{-6}$) and θ is the elevation angle (see Montgomery *et al.*, 1970). Except for a difference in integration limits, this is already the integration of f over θ needed for the density computation. The density, n, then is obtained without knowledge of the distribution function. As to the difference in integration limits, its correct treatment must depend on the type of distribution analyzed. For isotropic distributions (more specifically, those with $\partial f/\partial\theta = 0$), the remaining integration between 55° and 90° would be accounted for by applying a correction factor of 1.22 to the computed integral, whereas in the other extreme, that of a distribution fully contained between ±55° (e.g. solar wind ions), no such correction would be needed. In the present analysis we have assumed the former case, since we are mainly dealing with hot, subsonic distributions. Simulation calculations with convected Maxwellian distributions have shown that the computed densities are correct within 20% for typical temperatures and flow speeds up to flow elevation of ~60°.

For the computation of the higher moments, the distribution function must be known explicitly. As the θ dependence of f is not measured by the 2D analyzers, we have computed f under the assumption that it is independent of θ. From the array of $256 f_{ij}$ thus obtained we have computed the bulk velocity, pressure and temperature. The 2D bulk velocity essentially represents the projection of the true velocity onto the symmetry plane of the analyzers (i.e. approximately the ecliptic plane). The simulation has shown that the computed 2D bulk speeds for protons agree with the true projected speed to within 20% even at flow elevations as high as 70° for total flow speeds of $100 \, km \, s^{-1}$.

The pressure was defined as $P = \frac{1}{2}(P_{xx} + P_{yy})$ where P_{xx} and P_{yy} are the diagonal elements of the pressure tensor. The temperature, T, then followed from the equation of state, $P = nkT$, where k is the Boltzmann constant.

3D-Analysis. Analysis of the 3D-data first requires the subtraction of the background contribution to the count-rate of the active collector segment caused by the 'leakage' of secondary electrons from the other three, nominally 'off', segments (Bame *et al.*, 1978). A fraction, α, of the particles hitting any 'off', segment will thus be counted. The magnitude of this background has been determined from laboratory and in-flight calibrations. For electrons, α varied monotonically between 0.09 at 12 eV, 0.15 at 300 eV, and 0.33 at 10 keV. For protons, α ranged from 0.12 at 80 eV to 0.18 at 3 keV and above. Knowledge of the count-rates for each emitter segment at any given energy and azimuth angle then allows the background contributions to be approximately determined.

After subtraction of the above described background, the three-dimensional matrix of 384 countrates C_{ijk} (16 energy channels, 4 elevation channels, 8 and 4 azimuth channels for the inner and outer elevation channels, respectively), obtained in one basic 3D cycle, provides the basis for the three-dimensional moment computations. The distribution function f is calculated from the C_{ijk} under the assumption that it is constant over the width (27.5°) of the elevation channels. For the moment computations f needs to be extrapolated beyond the measured range in elevation angle. For the present analysis we have assumed f to be constant beyond ±55° elevation. Although other moments are computed, only the 3D bulk velocity is presented in this paper. Simulations with Maxwellian distributions have shown that the computed 3D proton bulk speeds are accurate within ~25% up to flow elevations of 83° at temperatures in the vicinity of 10^6 K and bulk speeds up to 300 km s^{-1}.

Acknowledgments

The authors wish to thank J. P. Glore and H. E. Felthauser at LASL, and K. Lehmann and P. Hemmerich at MPI for the dedicated support they provided in the development and preparation of the instrumentation. The MPI contribution would not have been possible without the efforts of the personnel at Dornier System, Friedrichshafen, especially of F. Pittermann. Special thanks are due for the outstanding support from the Project Offices at GSFC and ESTEC. The computer support by M. Halbig at LASL and E. Sckopke at MPI is gratefully acknowledged. T. E. Eastman is thanked for many valuable discussions and his careful review of the manuscript. Los Alamos portions of this project were conducted under the auspices of the US Department of Energy with NASA support under S-50864A, Max-Planck-Institut portions were supported by the Bundesministerium fur forschung und Technologie under grants RV14-B6/74 and 01-OI-02 7-ZA/WF/WRD 275:4.

References

Bame, S. J., Asbridge, J. R., Felthauser, H. E., Glore, J. P., Paschmann, G., Hemmerich, P., Lehmann, K., and Rosenbauer, H.: 1978, 'ISEE-1 and ISEE-2 Fast Plasma Experiment and the ISEE-1 Solar Wind Experiment', *IEEE Trans Geosci. Electr.* **GE-16,** 216.

Crooker, N. V.: 1977, 'Explorer-33 Entry Layer Observations', *J. Geophys. Res.*, **82,** 515.

Crooker, N. V. and Siscoe, G. L.: 1975, 'Subsonic Magnetosheath Observations from Explorer-33', *J. Geophys. Res.* **80,** 4368.

Dungey, J. W.: 1961, 'The Interplanetary Magnetic Field and the Auroral Zones', *Phys. Rev. Letters* **6,** 47.

Eastman, T. E., Hones, E. W., Jr., Bame, S. J., and Asbridge, J. R.: 1976, 'The Magnetospheric Boundary Layer; Site of Plasma, Momentum and Energy Transfer from the Magnetosheath into the Magnetosphere', *Geophys. Res. Letters* **3,** 685.

Eastman, T. E. and Hones, E. W., Jr.: 1978, 'Characteristics of the Low Latitude Boundary Layer and Magnetopause Layer at High Time Resolution', *J. Geophys. Res.*, submitted.

Formisano, V., Domingo, V., and Wenzel, K.-P.: 1978, 'On the Mechanism of Plasma Penetration and Energetic Electron Escape Across the Dayside Magnetopause', *J. Atmospheric Terrest. Phys.* **40,** 293.

Haerendel, G.: 1978, 'Microscopic Plasma Processes Related to Reconnection', *J. Atmospheric Terrest. Phys.* **40,** 343.

Haerendel, G. and Paschmann, G.: 1975, 'Entry of Solar Wind plasma into the Magnetosphere', in B. Hultqvist and L. Stenflo (eds.), *Physics of the Hot Plasma in the Magnetosphere*, p. 23.

Haerendel, G., Paschmann, G., Sckopke, N., Rosenbauer, H., and Hedgecock, P. C.: 'The Frontside Boundary Layer of the Magnetosphere and the Problem of Reconnection', *J. Geophys. Res.* **83,** 3195.

Hones, E. W., Jr., Asbridge, J. R., Bame, S. J., Gilbert, H. E., and Strong, I.: 1972, 'Measurements of Magnetotail Plasma Flow Made with Vela-4B', *J. Geophys. Res.* **77,** 5503.

Lees, L.: 1964, 'Interaction Between the Solar Plasma Wind and the Geomagnetic Cavity, *AIAA J.* **2,** 1576.

Lemaire, J. and Roth, M.: 1978, 'Penetration of Solar Wind Plasma Elements into the Magnetosphere', *J. Atmospheric Terrest. Phys.* **40,** 331.

Midgley, J. E. and Davis, L., Jr.: 'Calculation by a Moment Technique of the Perturbation of the Geomagnetic Field by the Solar Wind', *J. Geophys. Res.* **68,** 5111.

Montgomery, M. D., Asbridge, J. R., and Bame, S. J.: 1970, 'Vela-4 Plasma Observations near the Earth's Bow Shock', *J. Geophys. Res.* **75,** 1217.

Parker, E. N.: 1967a, 'Confinement of a Magnetic Field by a Beam of Ions', *J. Geophys. Res.* **72,** 2315.

Parker, E. N.: 1967b, 'Small-Scale Nonequilibrium of the Magnetopause and Its Consequences', *J. Geophys. Res.* **72,** 4365.

Paschmann, G., Haerendel, G., Sckopke, N., Rosenbauer, H., and Hedgecock, P. C.: 1976, 'Plasma and Magnetic Field Characteristics of the Distant Polar Cusp Near Local Noon: The Entry Layer', *J. Geophys. Res.* **81,** 2883.

Rosenbauer, H., Grünwaldt, H., Montgomery, M. D., Paschmann, G., and Sckopke, N.: 1975, 'HEOS-2 Plasma Observations in the Distant Polar Magnetosphere: The Plasma Mantle', *J. Geophys. Res.* **80,** 2723.

Russell, C. T. and Elphic, R. C.: 1978, 'Initial ISEE Magnetometer Results: Magnetopause Observations', *Space Sci. Rev.* **22,** 681.

Sonnerup, B. U. O.: 1976, 'Magnetopause and Boundary Layer', in D. J. Williams (ed.), *Physics of Solar Planetary Environments*, AGU, p. 541.

Zwan, J. J. and Wolf, R. A.: 1976, 'Depletion of Solar Wind Plasma Near a Planetary Boundary', *J. Geophys. Res.* **81,** 1636.

THREE-DIMENSIONAL PLASMA MEASUREMENTS WITHIN THE EARTH'S MAGNETOSPHERE

L. A. FRANK, K. L. ACKERSON,
R. J. DeCOSTER, and B. G. BUREK

Department of Physics and Astronomy, The University of Iowa, Iowa City, Iowa 52242, U.S.A.

Abstract. First magnetospheric measurements of the three-dimensional velocity distributions for positive ions and electrons within the energy range $1\ \text{eV} \leqslant E/Q \leqslant 45\ \text{keV}$ are reported. These velocity distributions are gained with quadrispherical Lepedeas on board the spacecraft ISEE-1 and -2. Three-dimensional bulk flows of protons in the vicinity of the magnetopause and within the dayside magnetosphere and dawn sector of the magnetotail are presented. Proton drift velocities within the magnetosphere and magnetotail can be directly determined and employed to calculate the corresponding quasi-static perpendicular electric fields and to provide quantitative analyses of kinematical models for plasma motions. Nonmonotonic features in the electron velocity distributions are found simultaneously with the presence of electron cyclotron harmonic electrostatic waves in the dayside magnetosphere. The relationship of the observed electron velocity distributions to expectations for resonant pitch-angle and energy diffusion is discussed, as well as the possibility of the existence of proton cyclotron harmonic instabilities. Examples of the signature of field-aligned acceleration of protons into the magnetosphere and the presence of low-energy ionospheric ions in the near-earth magnetotail are also presented. Perpendicular electrostatic fields can be calculated from the observed three-dimensional velocity distributions and are found to have typical magnitudes of $\sim 1\ \text{mV m}^{-1}$.

1. Introduction

The recently acquired capability of gaining observations of the three-dimensional velocity distributions of electrons and protons within the Earth's magnetosphere with the quadrispherical Lepedeas on ISEE-1 and -2 provides the opportunity to investigate numerous heretofore inaccessible areas of magnetospheric plasma physics. Observational goals encompassing perpendicular electric fields, *in situ* kinematical motions of plasmas, field-aligned currents, plasma instabilities and field-aligned acceleration are possible when such measurements are combined with simultaneous determinations of the magnetic fields, plasma waves, plasma composition and energetic particle distributions. Here we report initial three-dimensional velocity distributions and their relationship to several of these research areas. In Section 2 we summarize briefly the instrumentation and the extension of the commonly used energy-time spectrograms for displaying two-dimensional plasma measurements to three-dimensional summaries via the use of ancillary energy-spin phase spectrograms for multiple analyzers. Macroscopic plasma parameters for two crossings of the dayside magnetopause are presented in Section 3. A preliminary evaluation of the kinematical motions of low-energy charged particles within the terrestrial ring current is given in Section 4. Nonmonotonic electron velocity distributions presumably related to the generation of electron cyclotron harmonic instabilities are discussed in Section 5, as well as the possible signature of diffusion driven by the electrostatic waves. As a final topic

Space Science Reviews **22** (1978) 739–763.

velocity distributions of positive ions in the dawn magnetotail are examined in Section 6 and the perpendicular electric fields are deduced by a simple transformation to gyrotropic coordinates in velocity space.

2. Instrumentation and Spectrograms

The nearly identical quadrispherical Lepedeas on board ISEE-1 and -2 each employ three concentric plates which are 70° lunes. The center plate is biased with a programmed series of voltages to achieve an E/Q range of 1 eV to 45 keV in 64 contiguous passbands for positive ions and electrons. Measurements at a given E/Q are thus gained separately and simultaneously for positive ion and electron intensities with a minimum of hardware requirements and each of the two fan-shaped fields-of-view can be divided into contiguous segments with appropriate sizing of the entrance apertures and placement of the sensors at the exit apertures. For the ISEE instruments seven pairs of sensors are used to segment the overall field of view of $6 \times 162°$ into seven contiguous fields-of-view along the major axis. The plasma analyzers are positioned on the spacecraft such that the major axis (162°) of the field-of-view lies in a plane parallel to the spacecraft spin axis. Thus angular distributions for positive ion and electron intensities are possible over all but $\sim 2\%$ of the unit sphere for particle velocity vectors at the spacecraft position. This coverage of the angular distributions is shown in Figure 1 for the two sampling modes which are Sun-sectored. On the left-hand side of this figure is shown the angular coverage for 16 Sun sectors. The number of solid-angle segments is 112 each for positive ion and electron intensities. This sampling mode is employed for most of our observations at the low telemetry rate for the spacecraft. The lesser solid-angle coverage shown on the right-hand side of Figure 1 can be used when more rapid Sun-sectored scans of the energy range are desired. At high telemetry rate for the spacecraft, the sectoring is slaved to the spacecraft clock and the spin motion is subdivided into approximately 12 contiguous and equal segments. The corresponding number of solid angle elements over the unit sphere is ~ 84. The only significant disadvantage of the ISEE measurements is the temporal resolution for gaining complete velocity distributions, which is limited by the telemetry rate available to the instruments. Simultaneous three-dimensional electron and positive ion velocity distributions can be gained with a repetition rate of 128 s at the high telemetry rate. If telemetry capability were available, this time resolution can be reduced to ~ 10 s. Such improvement will await the next generation of magnetospheric research spacecraft. A more complete description of the ISEE quadrispherical Lepedeas is given by Frank *et al.* (1978).

The quadrispherical Lepedea is essentially an ensemble of seven individual electrostatic analyzers with axes of fields-of-view inclined at seven polar angles referenced to the direction of the spacecraft spin axis: 1(13°), 2(28°), 3(55°),

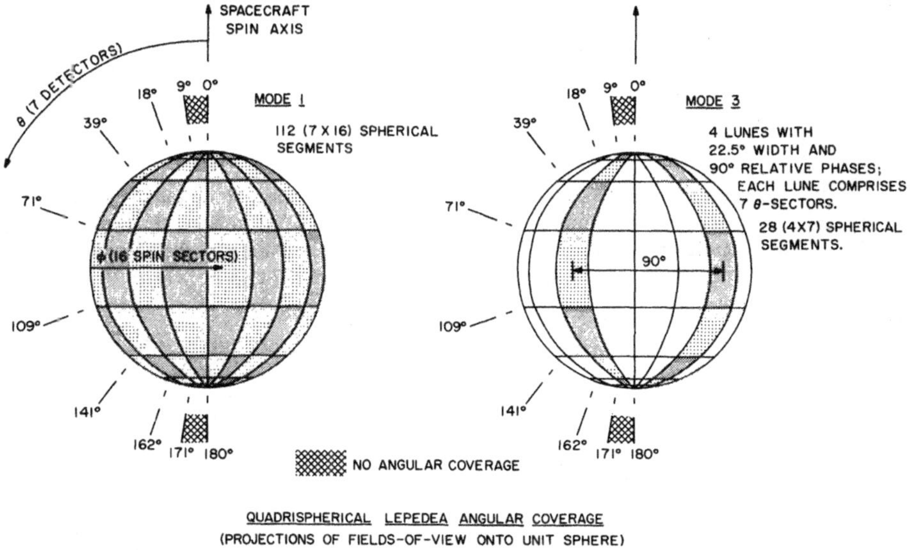

Fig. 1. Coverage of the angular distributions of positive ion and electron intensities in the Sun-sectored modes of operation for the quadrispherical Lepedeas on board ISEE-1 and -2 (after Frank *et al.*, 1978).

4(90°), 5(125°), 6(152°), and 7(167°). The spin axis is aligned nearly perpendicular to the ecliptic plane. Sensor pair 4 then views positive ions and electrons with velocity vectors lying nearly parallel to the ecliptic plane, the viewing geometry of most previous plasma instruments flown in the outer magnetosphere and within the magnetotail. Our energy-time ($E-t$) spectrograms display the responses of this sensor pair 4 for survey purposes. An example of such a spectrogram for the magnetopause crossing with ISEE-2 on November 8, 1977 is given in Plate 1. The upper four spectrograms show the color-coded responses for positive ions in four spin quadrants (beginning from top) centered toward the Sun, toward local evening, away from the Sun, and toward local morning, respectively. The instrument responses are logarithmically encoded according to the color bar at the right-hand side of the plate. This color bar is not labeled since average counts per accumulation interval are used for the display and the same display is used for low and high telemetry rates which have different accumulation times. The color bar has a full range of $\sim 10^4$ and the logarithm of the responses increases linearly from blue to red. The responses are plotted in the spectrograms as functions of Universal Time (hours, abscissa) and logarithm base 10 of the energy (eV, ordinate). Values for the spacecraft position in a coordinate system comprising geocentric radial distance (R_E, earth radii), and solar ecliptic latitude (θ_{SE}, degrees) and longitude (ϕ_{SE}, degrees), are also shown along the abscissa of the bottom positive ion spectrogram. The spin-averaged responses for the electron sensor of pair 4 are shown in the bottom spectrogram of Plate 1 and are color coded and plotted in the same manner as those for positive ions above. For the

Plate 1. Energy-time ($E-t$) spectrogram for proton and electron intensities as ISEE-2 passes from the magnetosheath into the dayside magnetosphere at ~02:50 UT on November 8, 1977. The format for these spectrograms is discussed in the text. (See page 743.)

Plate 2. Energy-spin phase ($E-\phi$) spectrograms for the entire set of quadrispherical Lepedea responses to the three-dimensional velocity distributions of protons and electrons just outside the magnetopause at 02:45 UT on November 8 (cf. Plate 1). See text for details of the spectrogram format. (See page 743.)

Plate 3. Continuation of Plate 2 for 02:48 UT. (See page 743.)

Plate 4. Continuation of Plate 2 for 02:52 UT as the magnetopause is crossed and the presence of an intense beam of protons convecting northward is detected in sensors 6*P* and 7*P*. (See page 743.)

Plate 5. $E-t$ spectrograms for proton and electron intensities measured with the quadrispherical Lepedea on board ISEE-1 as the spacecraft traverses the late morning sector of the magnetosphere and intercepts the magnetopause at ~05:40 UT on November 6, 1977. (See page 750.)

Plate 6. $E-t$ spectrograms for plasma observations in the morning sector of the magnetotail during 06:00–12:00 UT on January 24, 1978. The energy range for these spectrograms is $1\ \mathrm{eV} \leqslant E \leqslant 45\ \mathrm{keV}$. (See page 750.)

Plate 7. Continuation of Plate 6 for the interval 12:00–18:00 UT. The spacecraft enters the magnetosheath at ~15:15 UT. (See page 750.)

Plate 8. $E-\phi$ spectrograms for three-dimensional determinations of proton and electron angular distributions in the magnetotail at 09:28 UT on January 24 (cf. Plate 6). The proton bulk flows, e.g., sensors 4*P* and 5*P*, are generally directed sunward. (See page 751.)

Plate 9. $E-\phi$ spectrograms of ionospheric ions convecting slowly into the magnetotail just outside the energetic particle 'trapping boundary' at 08:05 UT on January 19, 1978. (See page 751.)

Plate 1.

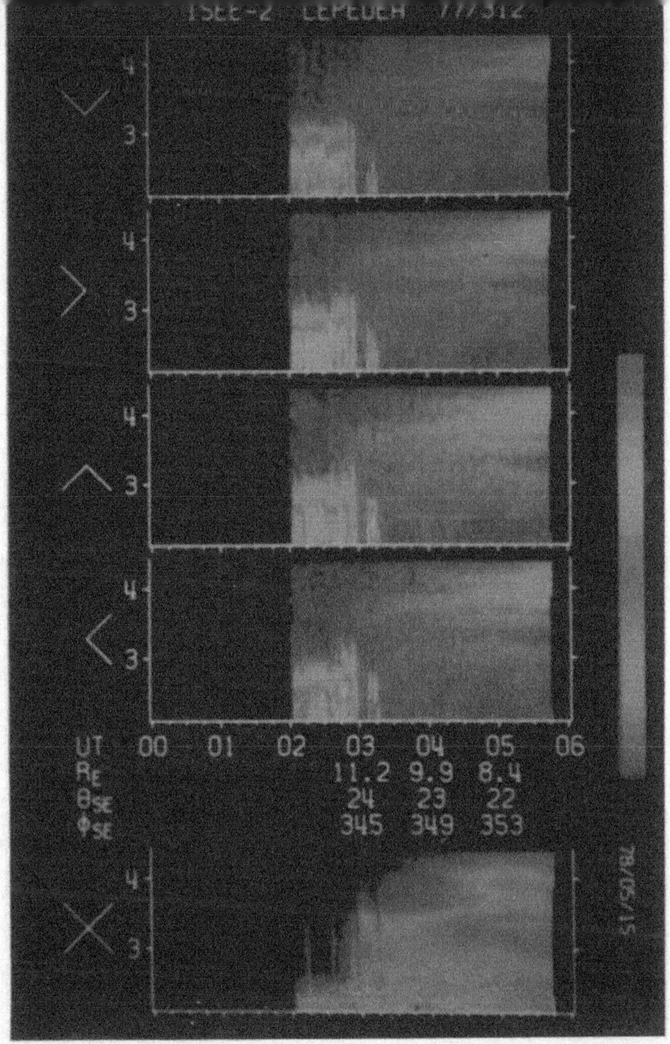

Plate 2.

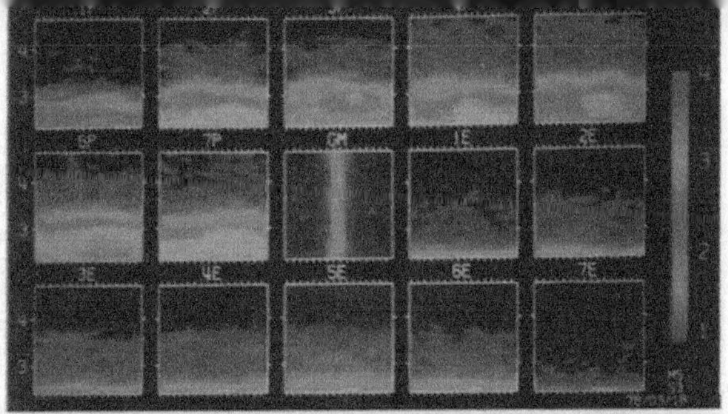

Plate 3.

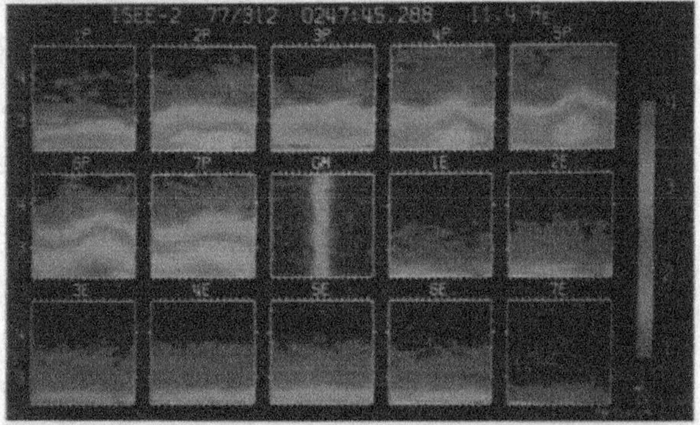

Plate 4.

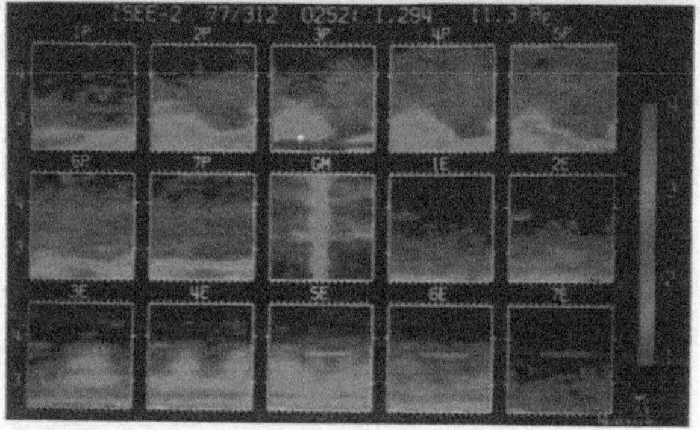

early series of measurements the quadrispherical Lepedeas were limited to the energy range 220 eV to 45 keV by ground command to insure a suitable time for instrument outgassing in order to avoid gain degradation of the electron sensors at the relatively high count rates at lower energies. Examples of measurements over the full energy range are presented later in our present initial survey.

The magnetopause crossing at ~02:50 UT on November 8 is clearly seen in the $E-t$ spectrograms of Plate 1. However, these spectrograms display only the instrument responses for a single plane through the three-dimensional positive ion and electron angular distributions with coarse angular resolution. For each energy scan during an instrument cycle time there is a complete determination of the velocity distributions over the unit sphere of solid angle. In order to display the entire series of plasma measurements we employ energy-spin phase ($E-\phi$) spectrograms such as those shown in Plates 2, 3, and 4 which are a sequence of snapshots of the entire instrument's responses during three chronologically ordered energy scans as the spacecraft passes through the magnetopause on November 8 (cf. Plate 1). Universal Time and geocentric radial distance (R_E) for each set of $E-\phi$ spectrograms are given at the top of each plate. The individual responses of the seven positive ion sensors (1P through 7P) and of the seven electron sensors (1E through 7E) have been color coded according to the color bar at the right-hand side of the plate and plotted as functions of $\log_{10}$ of the energy in eV (ordinate) and spacecraft spin phase (abscissa). The units for the color bar are $\log_{10}$ of the sensor count rate. The instrument's field-of-view crosses the solar direction in the middle of the abscissa scale. For reference the responses of sensors 4P and 4E are used to construct the $E-t$ spectrograms shown in Plate 1. The responses of a thin-windowed Geiger-Mueller tube are shown in the center spectrogram of Plate 2. Here the abscissa is again the spacecraft spin phase but the ordinate is Universal Time increasing from the bottom of the spectrogram. The GM tube responses to solar X-rays are evident in the center of this spectrogram. With inspection of the sequence of $E-\phi$ spectrograms shown in Plates 2 through 4 it becomes readily apparent that the velocity distributions for positive ions near the magnetopause are complex. These velocity distributions are not described adequately as simple convecting Maxwellian or bi-Maxwellian distributions. Further, the bulk flow direction lies far from parallel to the ecliptic plane and is directed dominantly northward. In the immediate vicinity of the magnetopause (Plate 4) only residuals of the magnetosheath proton distributions remain (sensors 1P through 5P) with an intense low-energy positive ion beam directed northward (sensors 6P and 7P). Thus it is not generally possible to perform a thorough analysis of the dynamics of plasmas at the magnetopause without the capability of determining the entire three-dimensional velocity distributions. The two types of spectrograms discussed above will be utilized as 'road maps' of the instrument responses as a first step toward detailed analyses of magnetospheric plasma phenomena.

3. Three-Dimensional Bulk Flows Within the Magnetosheath and Magnetosphere

The results from calculations of the first three moments of the proton three-dimensional velocity distributions for the magnetopause crossing with ISEE-1 on November 8, 1977 are shown as functions of Universal Time in Figure 2 (cf. Plate 1 for ISEE-2). The corresponding proton energy range for this series of measurements is $216\,\text{eV} \leqslant E \leqslant 44.6\,\text{keV}$. The average electron energies ($221\,\text{eV} \leqslant E \leqslant 45.6\,\text{keV}$) are also given. Geocentric radial distance (R, earth radii), solar magnetospheric longitude (ϕ_{SM}) and magnetic latitude (λ_m) are indicated at the top of the figure. The crossing of the magnetopause at ~02:50 UT occurs within ~25° of the subsolar point. Proton bulk flows are shown in the bottom two panels of Figure 2. The upper panel displays the projection of the three-dimensional flow onto the ecliptic plane (V_x, V_y) and the bottom panel gives the V_z component. It is noteworthy that the magnetosheath northward flows (~$150\,\text{km s}^{-1}$) in front of the magnetopause exceed those parallel to the ecliptic, i.e., a plasma instrument viewing only in the (V_x, V_y) plane will be incapable of determining the correct speed and distribution function along the flow vector. We have analyzed several of the proton velocity distributions in the vicinity of the magnetopause and found that such distributions are quite complex (cf. Plates 2, 3, and 4). We shall report these results in a later publication after more examples have been analyzed when the quadrispherical Lepedea scanned the entire energy range $1\,\text{eV} \leqslant E \leqslant 45\,\text{keV}$. In particular evidences of currents in the vicinity of the magnetosphere will be searched for. The increases of average electron and proton energies with the spacecraft crossing of the magnetopause into the outer magnetosphere are readily evident in Figure 2. These average energies are center of mass energies, i.e., with energy associated with the bulk flow subtracted from the second moment. Three-dimensional bulk flows within the outer magnetosphere are relatively small, ~tens of km s^{-1}, and reflect the drift velocities due to electric fields, magnetic field gradients, etc.

A comparison of the simultaneous measurements of three-dimensional bulk flows with the quadrispherical Lepedeas on ISEE-1 and -2 for the magnetopause crossings on November 8 is shown in Figure 3. These bulk flows are very similar both for the magnetosheath and outer magnetosphere, although for the time interval ~02:50–03:15 UT observations just inside the magnetopause differ substantially in detail, indicating a turbulent boundary layer. Statistics for flow determinations within the magnetosphere are at least as good as $\Delta V(V)^{-1} \approx 0.3$. However, asymmetric shielding for the quadrispherical Lepedea on ISEE-2 may be responsible for a systematic error of ~$10\,\text{km s}^{-1}$ in determination of the V_z component due to small sensor background responses to penetrating energetic electrons. This source of error will be eliminated with further analyses of the analyzer responses. A comparison of the simultaneous measurements of three-

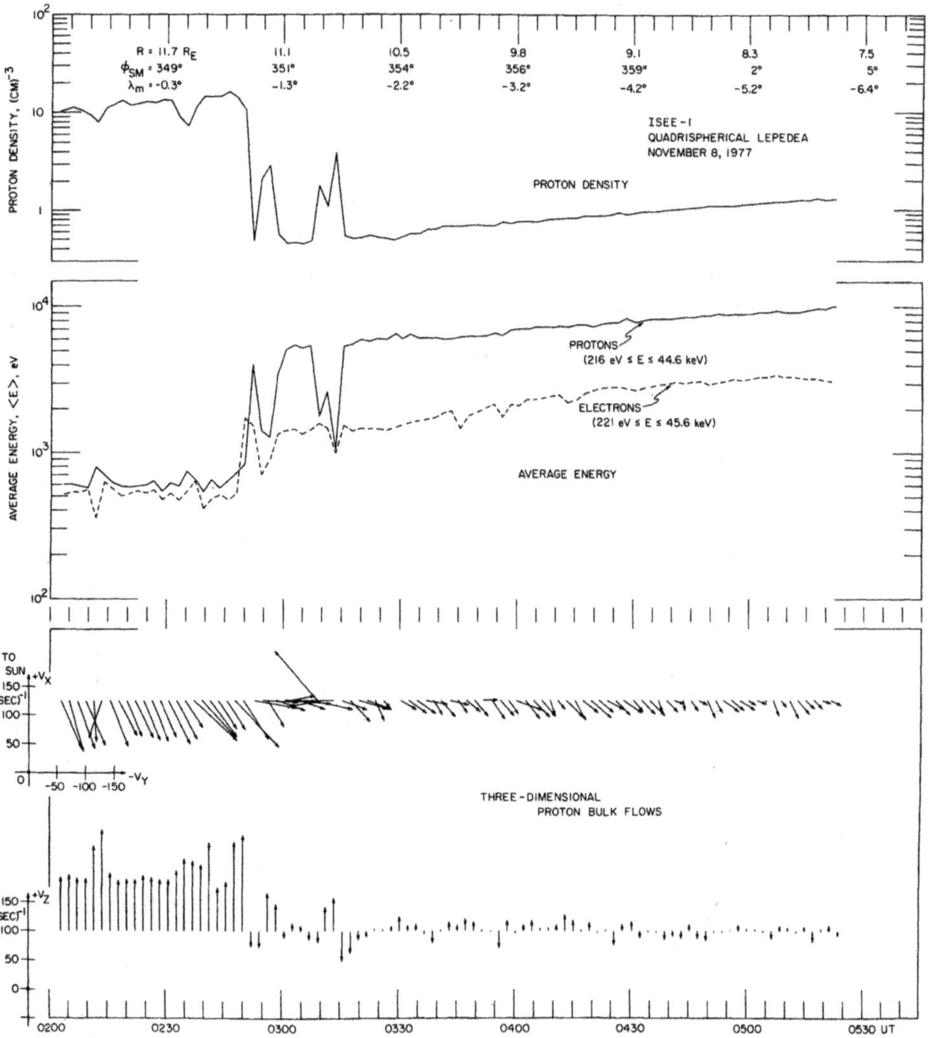

Fig. 2. Plasma parameters for the inbound pass of ISEE-1 through the magnetopause on November 8, 1977. The magnetopause is encountered at ~02:50 UT. Three-dimensional proton bulk flows are summarized in the bottom two panels.

dimensional bulk flows and magnetic fields with ISEE-1 for this series of observations is given in Figure 4. Magnetic field measurements have been provided by C. Russell via the ISEE Data Pool Tape. The vector coordinates for the magnetic field are identical to those used for the display of the bulk flows. In the magnetosheath the magnetic field vector is oriented nearly parallel to the ecliptic plane and the proton bulk flow is dominantly northward. Within the outer magnetosphere the magnetic field is directed northward and the plasma is mainly flowing perpendicular to the geomagnetic field.

Our second series of observations of the features of three-dimensional proton bulk flows in the vicinity of the magnetopause is summarized in Plate 5 and

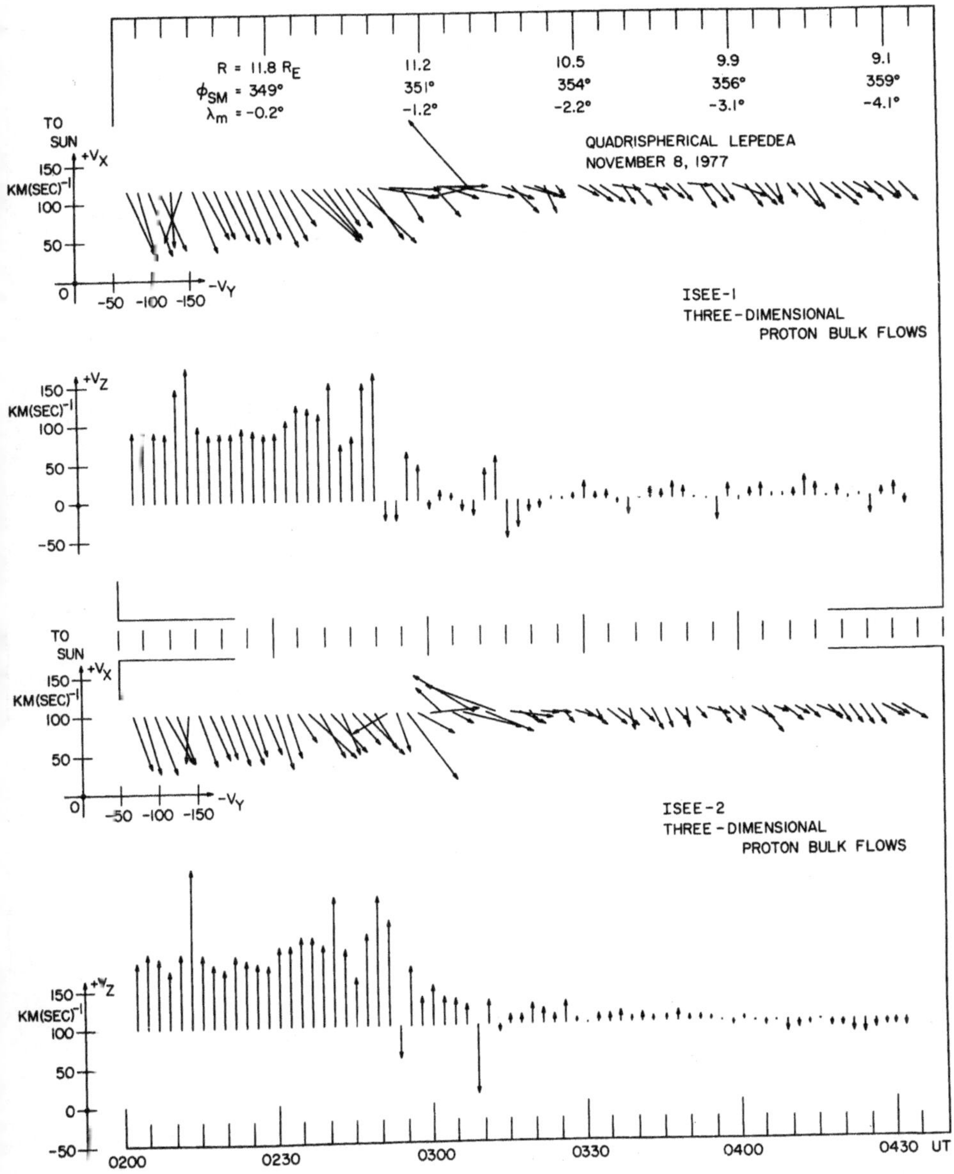

Fig. 3. Comparison of simultaneous determinations of three-dimensional proton bulk flows with the quadrispherical Lepedeas on ISEE-1 and -2 for the inbound pass through the magnetosheath and dayside magnetosphere on November 8, 1977.

Figure 5 for the outbound pass of ISEE-1 on November 6, 1977. These observations are taken at moderate magnetic latitudes ($\lambda_m \sim 25°$) and large angular displacements ($\phi_{SM} \simeq 290°$) from the subsolar point relative to the previous series nearer the subsolar point. The spacecraft is within hot magnetospheric plasmas until ~03:30 UT (cf. Plate 5). After this time, periods of low-energy electron and proton intensities are encountered, e.g., the interval of 15 min

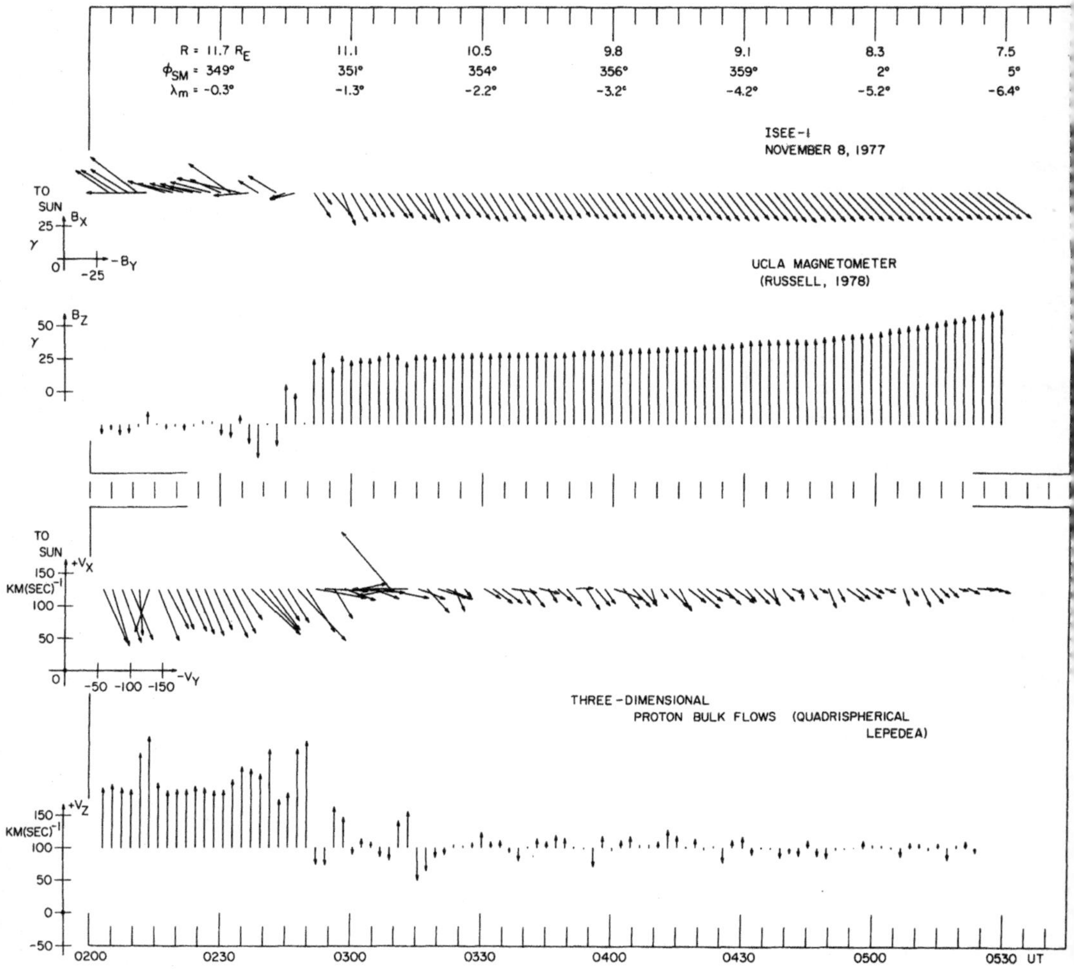

Fig. 4. Comparison of simultaneous measurements of the vector magnetic fields (courtesy of C. Russell) and three-dimensional proton bulk flows for the inbound pass of ISEE-1 on November 8, 1977.

centered at 04:08 UT, intermittently within the hotter plasmas until the spacecraft passes clearly into the magnetosheath at ~05:40 UT. As shown in Figure 5, plasma bulk flows for the interval 03:30–05:40 UT fluctuate greatly in magnitude and direction; ordered flow within the magnetosheath beginning at ~05:40 UT is dominantly in the ecliptic plane, although a substantial northward component of ~50–100 km s^{-1} is present. The increasing proton densities and decreasing average energies during the interval ~05:00–05:40 UT indicate that a magnetospheric boundary layer, or a magnetosheath-to-magnetosphere entry layer, with a thickness of ~1 R_E is present. The overall character of the macroscopic plasma parameters is similar to that reported by Rosenbauer *et al.* (1975) for the plasma mantle in the polar magnetosphere. Our measurements summarized in

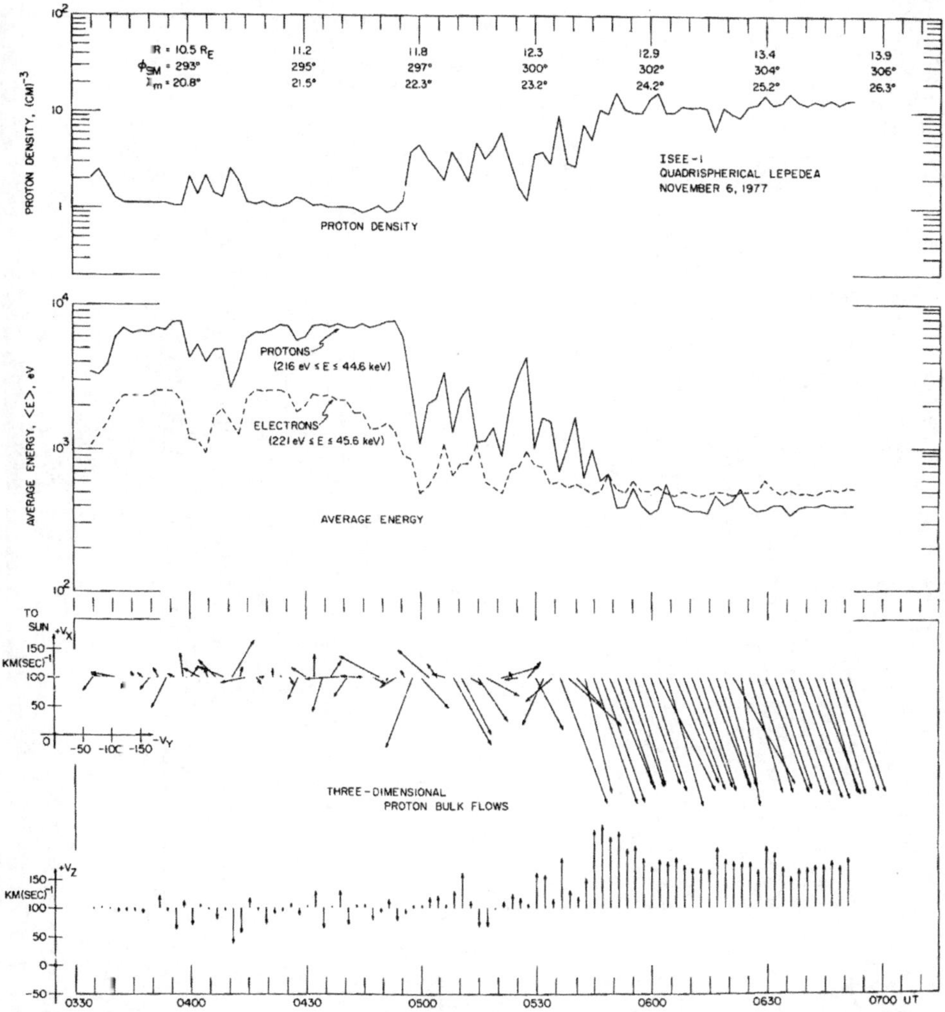

Fig. 5. Plasma parameters for the outbound pass of ISEE-1 through the late morning sector of the magnetosphere on November 6, 1977. Crossing of the magnetopause occurs at ~05:40 UT and is preceded by the presence of boundary layer plasmas during the interval ~05:00–05:40 UT.

Figure 5 show that the northward flow component is in the range of ~25–50 km s^{-1}; bulk flow parallel to the ecliptic plane is generally anti-sunward at ~50–100 km s^{-1}. Hence the overall flow observed with ISEE-1 in this boundary layer is consistent with the interpretation that these low-latitude plasmas are flowing poleward and tailward from the dayside magnetopause into the plasma mantle.

4. Kinematical Motions of Low-Energy Protons in the Outer Magnetosphere

The complex structures to be seen in the $E-t$ spectrograms for proton and electron intensities within the ring current region during ~03:30 to 05:45 UT on

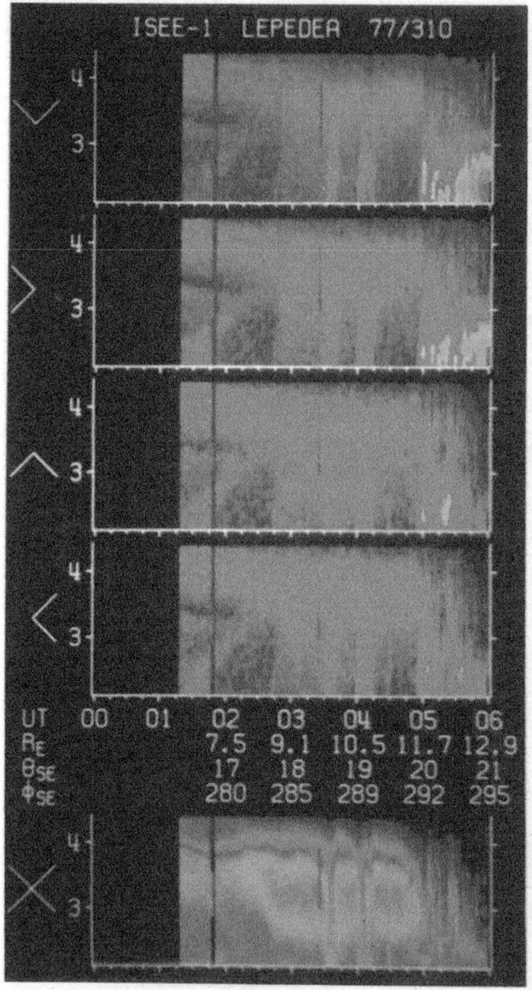

Plate 5.

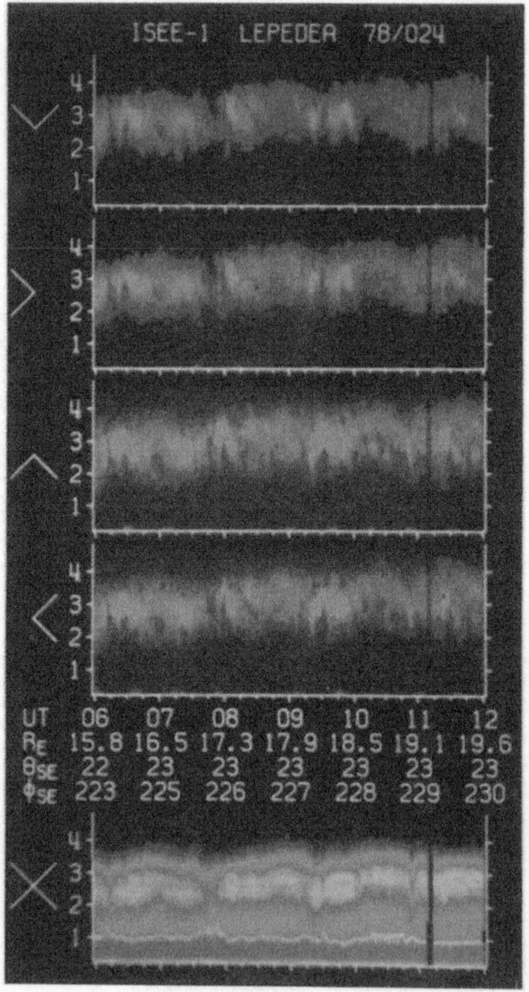

Plate 6.

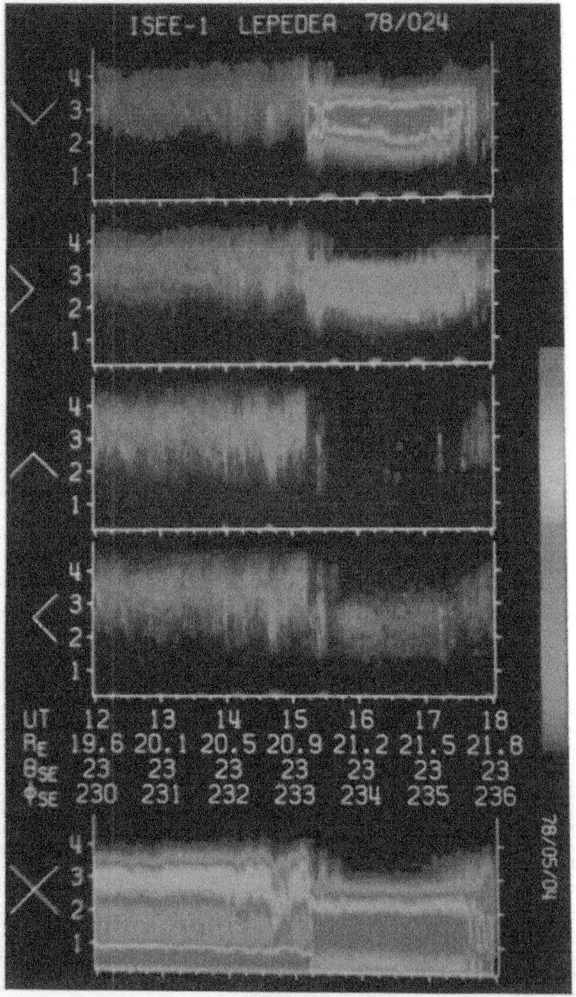

Plate 7.

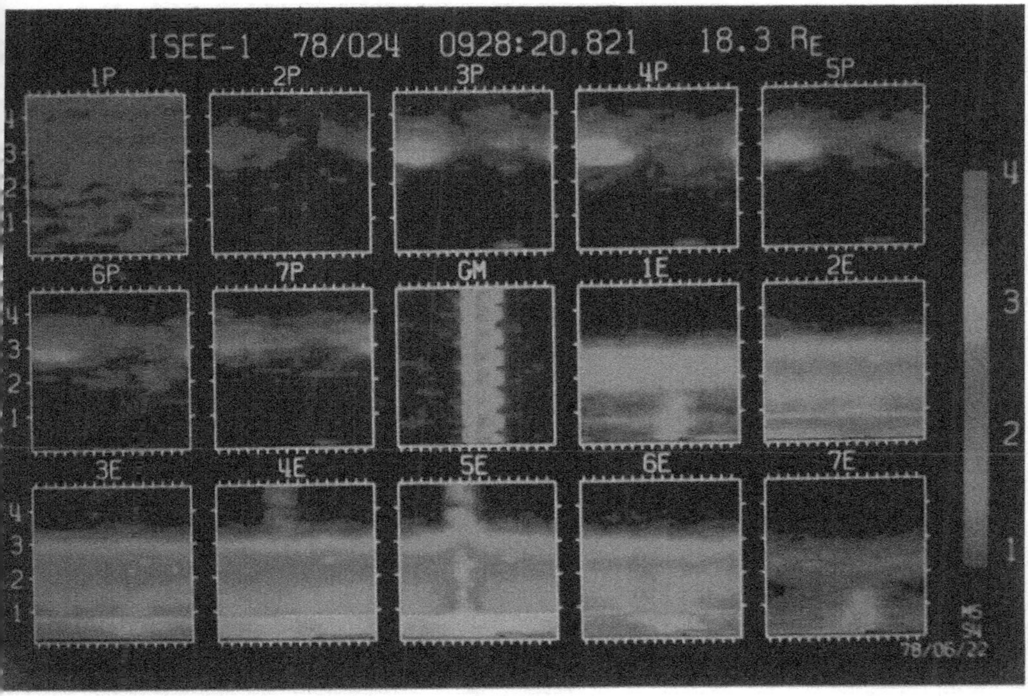

Plate 8.

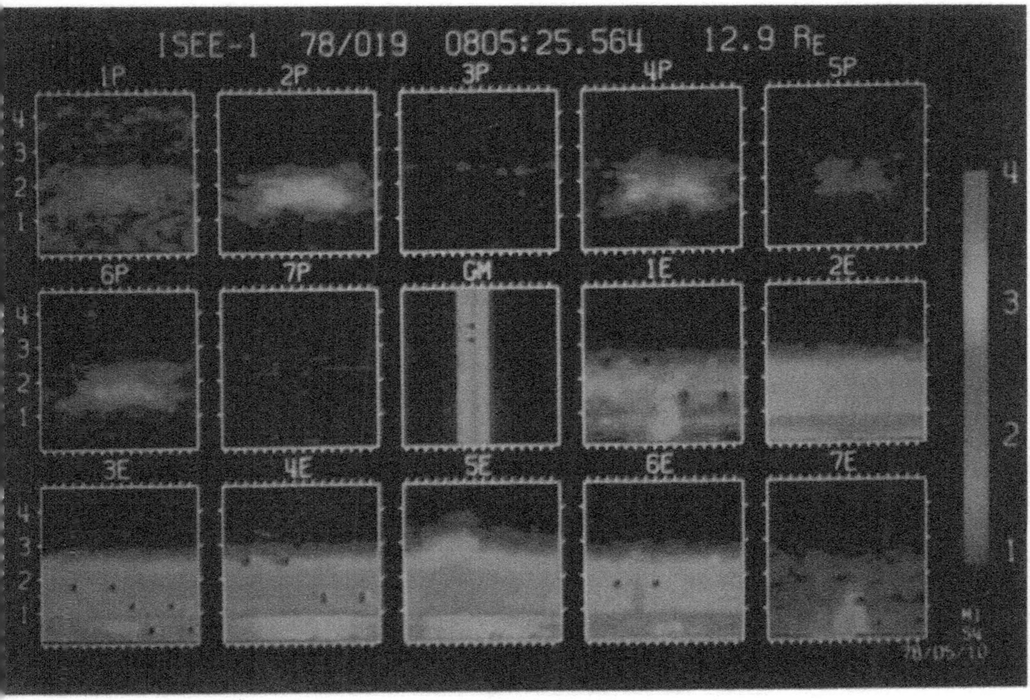

Plate 9.

November 8 (Plate 1) and ~01:20 to 02:50 UT on November 6 (Plate 5) are the signatures of the sources, drifts and loss mechanisms for these plasmas. McIlwain (1972, 1974) has reported extensive analyses of measurements of such proton and electron energy spectra for equatorial pitch angles at the geosynchronous orbit ($6.6 R_E$). His interpretation of these plasma signatures yields a model of the electric fields in the vicinity of the geosynchronous orbit, which is quite successful in reproducing the major features of the measurements. The detailed features of the spectrograms of Plates 1 and 5 for these regions will depend upon the position and time of injection (more than one injection event is possible), the strength and geometry of the geoelectric fields, and species and energy of the individual charged particles. McIlwain has introduced a (U, B) coordinate system that renders the interpretation of the complex structures in Plates 1 and 5 considerably more direct than a straightforward kinematical computation of the trajectory of each charged particle. U is the electric potential and B the magnetic field intensity. The ISEE measurements such as those shown here can be employed to extend McIlwain's model from the geosynchronous orbit to the magnetopause. However, our present effort is directed toward taking advantage of the quadrispherical Lepedea's capabilities for directly determining the *in situ* proton bulk flows. For example, during the interval 04:02–04:17 UT on November 8 the proton three-dimensional bulk flow (V_x, V_y, V_z) is $(-14, -10, 13 \text{ km s}^{-1})$. The average magnetic moment μ for these protons is $\sim 0.1 \text{ keV } \gamma^{-1}$. For a reference value, the gradient drift velocity for a 10 keV proton at $10 R_E$ in a dipole magnetic field is 17 km s^{-1}. The magnetically perpendicular and parallel flows observed with ISEE-2 during this interval are $V_\perp = 19 \text{ km s}^{-1}$ and $V_\| = 11 \text{ km s}^{-1}$, respectively, with a statistical uncertainty of $\sim \pm 5 \text{ km s}^{-1}$. Although the magnetic field is not accurately dipolar in this region, the agreement between the magnitudes of observed and calculated dipolar gradient drifts are encouraging for the topic of direct determination of kinematical motions in the outer magnetosphere. The parallel flow $V_\|$ may reflect an actual field-aligned positive ion flow or, more probably, a vertical motion of the dayside magnetosphere (note the fluctuations of the sign of V_z in Figure 4). The measured perpendicular drift velocity $V_\perp$ given above will also include a substantial component due to an electric field drift. In order to separate the gradient V_g and electric V_E drifts it is most useful to examine the motions of positive ions with low $\mu \simeq 0.002 \text{ keV } \gamma^{-1}$ and high $\mu \simeq 0.2 \text{ keV } \gamma^{-1}$, separately and simultaneously. The positive-ion distributions during the interval ~01:20–01:35 UT on November 6 (Plate 5) are suitable for such analyses and exhibit no notable spatial or temporal fluctuations. As a slight digression, we find this 'nose' event ~01:20–02:30 UT differs in signature from those reported at somewhat lesser radial distances inside the plasmapause (Smith and Hoffman, 1974; Ejiri *et al.*, 1977) in that the ion energies are generally lesser and the 'nose' is pointed outward from the Earth for the present example. However, the same kinetical model should describe both observational results. We assume in the following computations that the magnetic field **B** is directed perpendicular to the ecliptic plane in order to estimate the magnitude and sense of the motion of the plasmas. Later analyses will account for the actual orientation of **B** which departs from the assumed direction substantially for this example. The bulk flows are also computed with the assumption that the positive ions are protons. If the positive ions are indeed heavier ions such as He^+ and O^+, the computed flows are correspondingly reduced. Mass analysis with other instrumentation on board ISEE-1 will resolve this uncertainty. The measured $V_\perp$ for protons in the energy interval $3.9 \leqslant E \leqslant 44.6$ keV during 01:15–01:35 UT is 25 km s^{-1} with components $V_x = -25$ and $V_y = -1$ km s^{-1}. The average proton energy is 20 keV and the corresponding dipolar V_g is 16 km s^{-1}. Since the spacecraft is located near the 09:00 LT meridional plane the components of this calculated drift are approximately $V_x \simeq -16$ and $V_y \simeq 0$ km s^{-1}. Again we note that the measured drift velocity of high-μ protons will include a substantial component due to V_E. In order to evaluate the electric drifts the drift velocity is determined for protons within the energy range $216 \leqslant E \leqslant 538$ eV for which gradient drift will be insignificant. The components for this bulk flow are $V_x = 13$ and $V_y = -24$ km s^{-1} and are in qualitative agreement with the pattern of cold plasma flow predicted by McIlwain's (1972) model. An estimate of the corresponding electric field strength is ~ 2 mV m^{-1}. In summary it is noted that the higher energy protons are observed to be drifting westward under the dominant influence of the magnetic field gradient and the lower ions eastward and radially outward toward the dawn magnetopause under the influence of the geoelectric field. These results for this example are quite encouraging since it appears possible not only to develop a kinematical description of the trajectories of these charged particles via their signatures in the $E-t$ spectrograms but also to determine *in situ* motions of the positive ions. Since the quadrispherical Lepedea provides three-dimensional measurements of the plasma distributions it will be possible to extend the kinematical description to include the modified longitudinal invariant K as described by Whipple (1978).

5. Electron Velocity Distributions Associated with VLF Electrostatic Emissions

It is generally surmised that the intense electrostatic emissions that are observed commonly within the Earth's magnetosphere at frequencies $\sim (n+\frac{1}{2})\Omega_e$, where Ω_e is the electron gyrofrequency, are generated by an instability associated with a nonmonotonic feature of the perpendicular electron velocity distribution $n(V_\perp)$ (Fredricks, 1971; Young *et al.*, 1973; Ashour-Abdalla and Kennel, 1978; Hubbard and Birmingham, 1978). The electron velocity distributions that are employed in the theoretical studies of electron cyclotron harmonic instabilities also require the presence of a cold electron plasma within the hot electron distribution which has presumably been convected into outer magnetosphere from the magnetotail. The cold component is presumably fed into this hot plasma via the

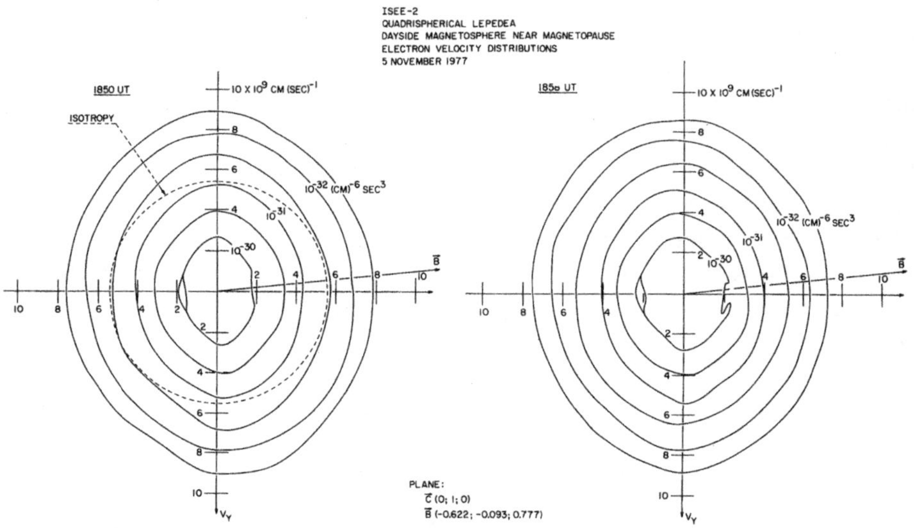

Fig. 6. Two consecutive determinations of the electron velocity distributions in the dayside magnetosphere with ISEE-2 on November 5, 1977. The direction of the magnetic field vector **B** is shown for each distribution. The planes for these two-dimensional cross-sections of the observed three-dimensional velocity distributions are defined by the vectors **C** and **B** (direction cosines).

ionosphere. Heating of this ionospheric electron plasma is expected in the vicinity of the instability. The nonmonotonic feature of the hot plasma basically provides the free energy to drive the instability and the cold electron component greatly decreases the requirement of a severe slope in the loss cone distribution. The capability of obtaining three-dimensional velocity distributions with the quadrispherical Lepedeas on the ISEE spacecraft allows a comprehensive determination of the electron plasmas associated with these half-harmonic emissions. Previously, most measurements of electron intensities have been gained in a viewing plane generally parallel to the ecliptic, thus obviating most possibilities of measuring pitch angle distributions near the equatorial plane since **B** is directed mainly northward. Our first examples of these electron distribution functions are given for a series of observations with ISEE-2 at 18:50 and 18:58 UT on November 5, 1977. These velocity distributions are shown in Figure 6. Intense half-harmonic emissions have been reported during this interval for ISEE-1 by Gurnett *et al.* (1979). Electron distributions gained with the plasma instruments on ISEE-1 and -2 are nearly identical. The isodensity contours shown in Figure 6 are generated by computer software that allows a choice of a plane through the observed three-dimensional distribution function and subsequently generates a plot of the contours in the chosen plane. The specification of the plane is accomplished with two vectors, in this case **C** an arbitrarily chosen vector through the origin and **B** the magnetic field direction. The resulting velocity distributions for electrons at two time intervals beginning at 18:50 and 18:58 UT are

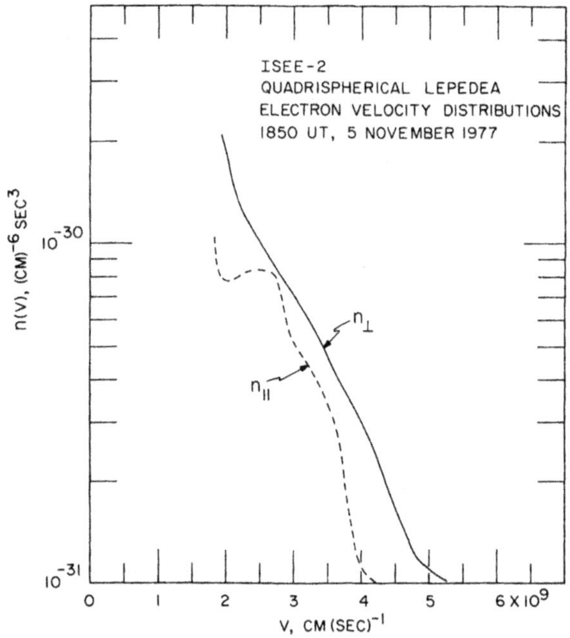

Fig. 7. Perpendicular and parallel velocity distributions corresponding to the two-dimensional electron velocity distributions shown for 18:50 UT in Figure 6.

displayed in Figure 6. The reproducibility of the observed distributions indicate the lack of temporal effects upon the gross features. A strong anisotropy of a factor of ~ 3 and in favor of pitch angle $\alpha = 90°$ is easily seen in the distribution functions. The atmospheric loss cone is only $\sim 5°$ at this position in the equatorial magnetosphere and cannot be resolved with the angular resolution of the present instrumentation. The parallel and perpendicular distributions, $n(v_\parallel)$ and $n(v_\perp)$, respectively, are shown in Figure 7 for the measurements at 18:50 UT. These electron distributions are not Maxwellian over the full velocity range. The characteristic thermal velocity for the hot component at $\sim 2.3 \times 10^9$ to 4.5×10^9 cm s^{-1} is $\sim 2.8 \times 10^9$ cm s^{-1}. The increase of slope of the velocity distribution at $\lesssim 2.3 \times 10^9$ cm s^{-1} is the signature of a cooler electron plasma with a characteristic thermal velocity $\sim 8 \times 10^8$ cm s^{-1}. Thus the ratio of temperatures for the hot and cold components is $T_e^C/T_e^H \simeq 0.1$. The ratio of the densities can be estimated by direct determination of the proton (n_p) and hot electron (n_e^H) number densities and subsequent subtraction to compute the cold electron density (n_e^C). The result of this calculation is $n_e^C = n_p - n_e^H \simeq 0.9–0.3 \simeq 0.6$ electrons cm^{-3}. Hence $n_e^H/n_e^C \simeq 0.5$. An examination of the velocity distributions in Figure 7 shows that the perpendicular and parallel temperatures for the hot electron plasma are similar, with the exception of the nonmonotonic feature in $n(v_\parallel)$. All of the above parameters are coarsely in agreement with the previous theoretical assumptions for generation of the cyclotron harmonic instabilities (cf. Ashour-Abdalla and Kennel, 1978; Hubbard and Birmingham, 1978). However, the nonmonotonic feature in the perpendicular velocity distribution $n(v_\perp)$ is not seen at $v_\parallel = 0$ but at $v_\parallel \sim 2\times 10^9$ cm s^{-1} or near the characteristic thermal velocity for the hot electron plasma (Figures 6 and 7). It will be of immediate interest to know the theoretical expectations for the generation of half-harmonic electrostatic waves, given the electron plasma parameters above.

Lyons (1974) has calculated the pitch-angle and energy diffusion rates for electrons from resonant interactions with intense electrostatic waves at the first half-harmonic of the electron gyrofrequency. One of the results of these computations is the finding that strong pitch-angle and energy diffusion is limited to equatorial pitch angles $\alpha \lesssim 45°$ for electrons near the characteristic thermal velocity of the hot electron plasma. With increasing velocity this pitch angle increases. To examine our measurements for this effect we show another example of electron velocity distributions in the right-hand side of Figure 8. The corresponding position of the ISEE-1 spacecraft is magnetic shell $L = 9.8$ and magnetic latitude $\lambda_m = 20°$. The characteristic electron thermal velocity for the hot component is $\sim 2.3\times 10^9$ cm s^{-1}. For $v \gtrsim 2\times 10^9$ cm s^{-1} the angular distributions for electrons display a prominent anisotropy in favor of $\alpha = 90°$ which decreases to approximate isotropy at $\sim 6\times 10^9$ cm s^{-1} in general qualitative agreement with Lyon's conclusions. In fact a detailed comparison of the isodensity contour at $n(v) = 3\times 10^{-31}$ cm^{-6} s^{-3} and the isotropy circle shows approximate isotropy over

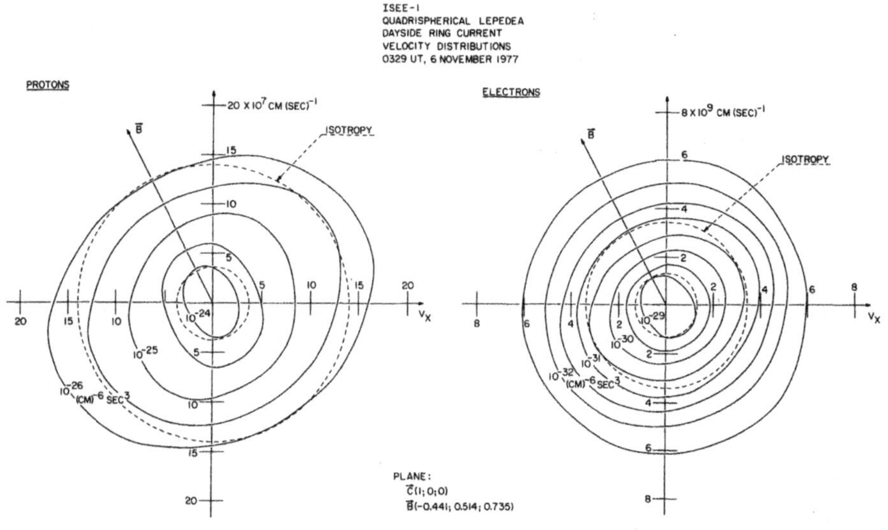

Fig. 8. Simultaneous measurements of the proton and electron velocity distributions with ISEE-1 in the dayside ring current at 03:29 UT on November 6, 1977. The pitch angle anisotropy evolves from field-aligned to perpendicular and then to isotropy with increasing electron velocity. Proton anisotropies are field-aligned at low velocities and are strongly perpendicular at the higher velocities.

a pitch angle cone with half-angle ~30–40° that may be the signature of strong pitch-angle and energy diffusion driven by resonance with the half-harmonic electrostatic waves. This rapid pitch-angle and energy diffusion is probably the source of electron precipitation over diffuse auroras (Lyons, 1974). The presence of a colder electron plasma presumably from the ionosphere and subsequently heated is noted in the field-aligned isodensity contour at $v \simeq 1 \times 10^9 \text{ cm s}^{-1}$. Electrons with these and lesser velocities should not resonantly interact with the half-harmonic cyclotron waves. On the left-hand side of Figure 8 are shown the proton velocity distributions for the same time interval. These proton distributions display the same qualitative features as those for electrons, i.e., a highly anisotropic pitch-angle distribution with a maxima at $\alpha = 90°$ for the higher velocities and a field-aligned cold proton plasma. However, a nonmonotonic feature in the proton velocity distributions has not been found yet in the few examples that we have examined to date. The similitude of the proton and electron velocity distributions would seem to suggest the existence of a concurrent proton cyclotron harmonic instability at lower frequencies. A future comparison of simultaneous plasma wave measurements with those of the proton and electron velocity distributions will yield substantial information concerning these resonant diffusion processes.

6. Velocity Distributions in the Dawn Sector of the Magnetotail

Our first observations of the plasma three-dimensional velocity distributions within the magnetotail are as exciting as those discussed above for the dayside magnetosphere. A series of measurements during the ISEE-1 traversal of the dawn sector of the magnetotail on January 24, 1978 is summarized in the $E-t$ spectrograms of Plates 6 and 7. The energy range for these spectrograms is $1 \text{ eV} \leq E \leq 45 \text{ keV}$. The spacecraft encounters hot plasmas within the magnetotail until ~15:15 UT when passage through the magnetopause occurs. Positive ion intensities within the magnetotail are observed over the energy range $500 \text{ eV} \lesssim E \lesssim 5 \text{ keV}$ and substantial anisotropies in the angular distributions are evident with intercomparison of the spectrograms. No measurable intensities of positive ions with $E \lesssim 100$ eV are found. Peak magnetotail electron intensities occur at energies ~500 eV (bottom spectrograms); the intense fluxes at energies $\lesssim 15$ eV are secondary electrons and photoelectrons. There is a clear instrumental separation of these two electron distributions. Fluctuations of the spacecraft potential of several volts can be seen for the brief periods of plasma sheet disappearance centered at 06:15 and 07:45 UT, for examples. The 'comb-like' structure occurring during 10:15–10:45 UT is the signature of variations of the spacecraft potential relative to that of the ambient plasma due to operation of one of the other science instruments. The corresponding macroscopic plasma parameters for the period 10:00–16:00 UT are shown in Figure 9. Electron average energies

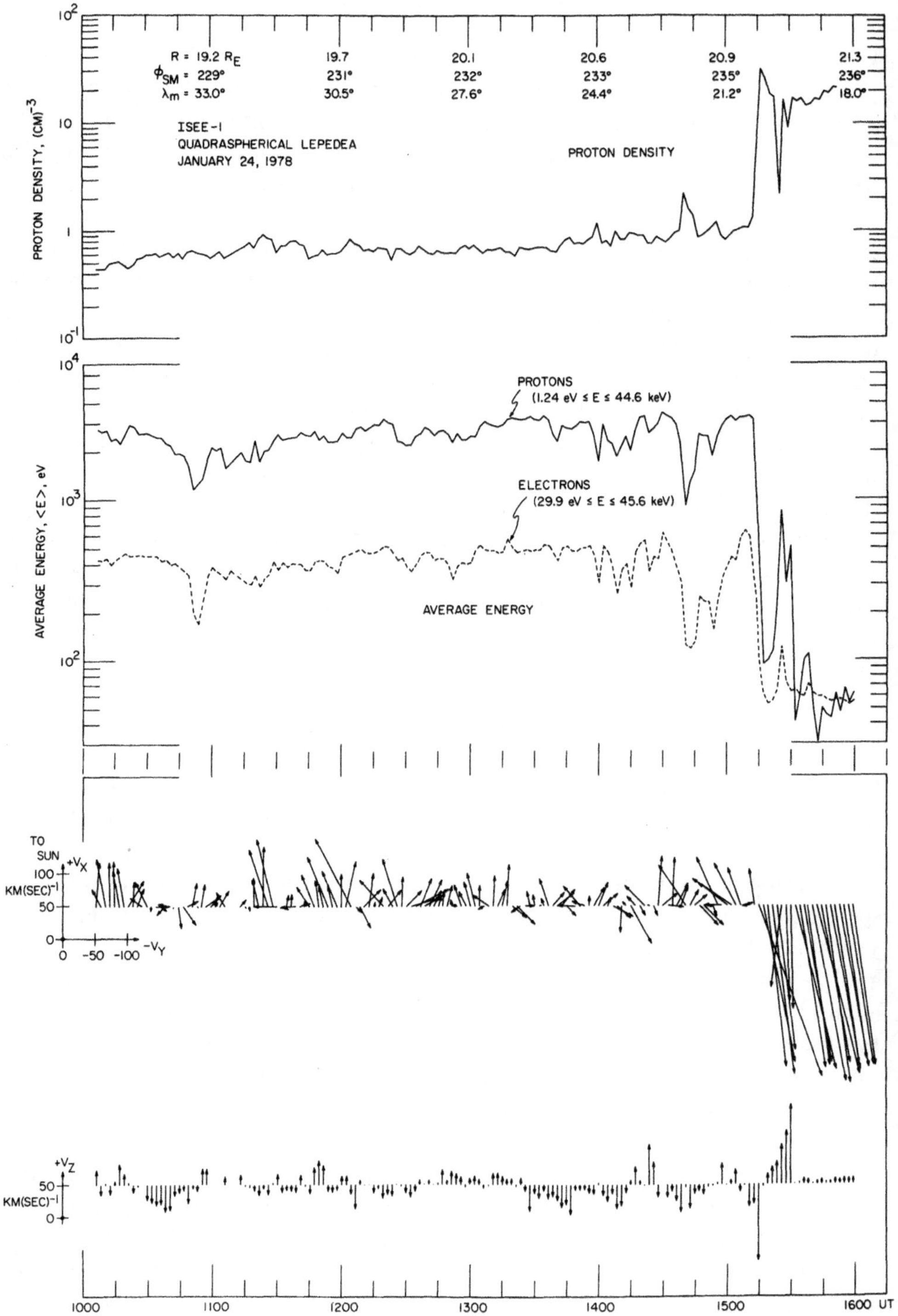

Fig. 9. Plasma parameters for the outbound pass of ISEE-1 through the morning sector of the magnetotail on January 24, 1978. Proton bulk flows are dominantly sunward with fluctuating V_z components in the magnetotail during ~10:00–15:15 UT. Passage through the magnetopause occurs at ~15:15 UT (cf. Plates 6 and 7).

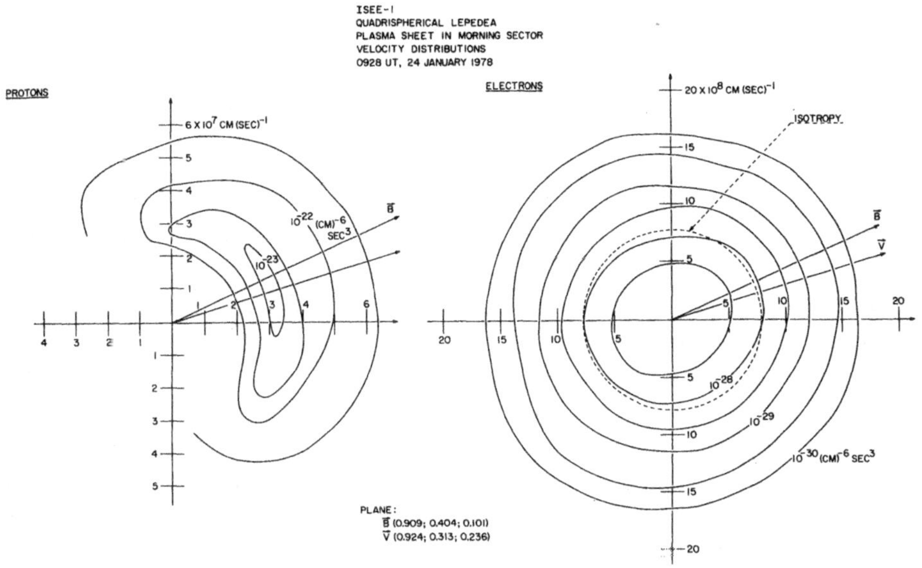

Fig. 10. Proton and electron velocity distributions in the plane containing the magnetic field **B** and proton bulk flow **V** vectors at 09:28 UT on January 24 in the Earth's magnetotail (cf. Plates 7 and 8).

have been computed for the energy interval $29.9\,\mathrm{eV} \leqslant E \leqslant 45.6\,\mathrm{keV}$ to exclude the spacecraft-associated electron fluxes. Proton densities slowly rise with increasing radial distance from $\sim 0.5\,\mathrm{cm}^{-3}$ at 10:00 UT to $1.0\,\mathrm{cm}^{-3}$ just inside the magnetopause. The bulk flows for protons are persistently directed sunward in the range of 25 to $75\,\mathrm{km\,s}^{-1}$ with few exceptions, and there are sustained periods of either southward or northward flows at $\sim$ tens of $\mathrm{km\,s}^{-1}$ (see bottom two panels).

The proton velocity distributions for the series of observations in the magnetotail shown in Plates 6 and 7 often are greatly dissimilar to those for convecting Maxwellians. An $E-\phi$ spectrogram for 09:28 UT is shown in Plate 8. The proton intensities are highly collimated and are generally directed toward the solar direction. On the other hand, the spectrogram for sensor 4E shows that the electron intensities exhibit two maxima. The vertical line of responses for sensor 5E just off the center of the abscissa is the signature of asymmetric instrument responses due to secondary electrons and photoelectrons. The proton and electron velocity distributions in a plane containing the magnetic field vector **B** and proton bulk flow vector **V** are given in Figure 10. The electron distributions are magnetically field-aligned at low velocities and approach pitch-angle isotropy at the higher velocities. The proton velocity distribution is of the type previously identified by DeCoster and Frank (1978) that is the signature of field-aligned acceleration of protons in the magnetotail. For this example, the down-tail field-aligned potential difference is $\sim 550\,\mathrm{V}$ and the source temperature is $\sim 6 \times 10^6\,\mathrm{K}$. Hence the spacecraft is embedded in plasmas recently accelerated into the magnetosphere. The proton bulk velocity is dominantly, but not wholly aligned along the magnetic field vector **B**. Thus there is a measurable drift of the velocity distribution perpendicular to **B**. Since the pressure and magnetic field gradients are small and the centrifugal drifts at these energies are also relatively insignificant, this perpendicular velocity is attributed to the presence of a perpendicular electric field $\mathbf{E}_\perp$. Both the magnitude and direction of $\mathbf{E}_\perp$ can be determined. For this example, the magnitude of $\mathbf{E}_\perp$ is 0.9 mV m^{-1}, which is computed from the observed perpendicular drift velocity of 45 km s^{-1}. Hence the three-dimensional proton velocity distributions in many regions of the magnetosphere can be employed as probes for evaluating $\mathbf{E}_\perp$.

The previous example of magnetotail velocity distributions concerns the field-aligned acceleration of protons in the magnetotail towards the earth; alternatively positive ions are also observed, which are streaming from the ionosphere into the magnetotail. Such ionospheric ions were encountered at latitudes $\lambda_m \sim 40°$ just outside the 'trapping boundary' for energetic charged particles during the outbound pass of ISEE-1 on January 19, 1978. The macroscopic plasma parameters for this series of measurements are shown in Figure 11. The traversal of the trapping boundary occurs at ~08:00 UT. Just beyond this trapping boundary the proton average energy decreases precipitously and bulk flows are directed away from Earth and southward toward the plasma sheet at speeds ~ tens of km s^{-1}. (ISEE-1 is located at northern magnetic latitudes.) The reader is reminded that the positive ions are assumed to be H^+ in the computations of the moments given in Figure 11. The corresponding $E-\phi$ spectrograms for 08:05 UT are shown in Plate 9. The 'patches' in these spectrograms are to be identified as missing telemetry. The electron temperatures are low, $\sim 4 \times 10^5$ K. At first glance, the spectrograms for the positive ion sensors look peculiar, i.e., substantial responses for sensors 2*P*, 4*P*, and 6*P* and barely discernible or no responses for the remaining sensors. The isodensity contours for the spectrograms of Plate 9 are shown in Figure 12 and are computed with the assumption that the ion species is H^+. The plane for this two-dimensional velocity distribution is defined by the bulk flow **V** from the moment calculation and magnetic field **B** determined with the onboard magnetometer. The velocity distribution appears to be the sum of two velocity distributions. A transformation to gyrotropic coordinates (dashed grid) and determinations of the centroids of the two distributions by graphical methods finds that the perpendicular drift of both distributions is ~19 km s^{-1}. The two parallel bulk velocities are ~8 km s^{-1} and 82 km s^{-1}. The accuracy of these determinations of the parallel speeds by such coarse methods is probably not better than ±10 km s^{-1}. Better techniques can improve the accuracy. Frank *et al.* (1977) have reported qualitatively similar velocity distributions for the more distant magnetotail and interpret them in terms of the simultaneous presence of H^+ (maximum at lower velocities) and O^+ (at higher velocities). Within the accuracy of the bulk velocities given in Figure 12, this interpretation is consistent with our present observations. The field-aligned flow of both ion species, directed out of the ionosphere, is thus only ~15 (±10) km s^{-1}, or $\sim 8 R_E$ h^{-1}. The temperature of the proton component of this plasma is $\sim 5 \times 10^5$ K and thus

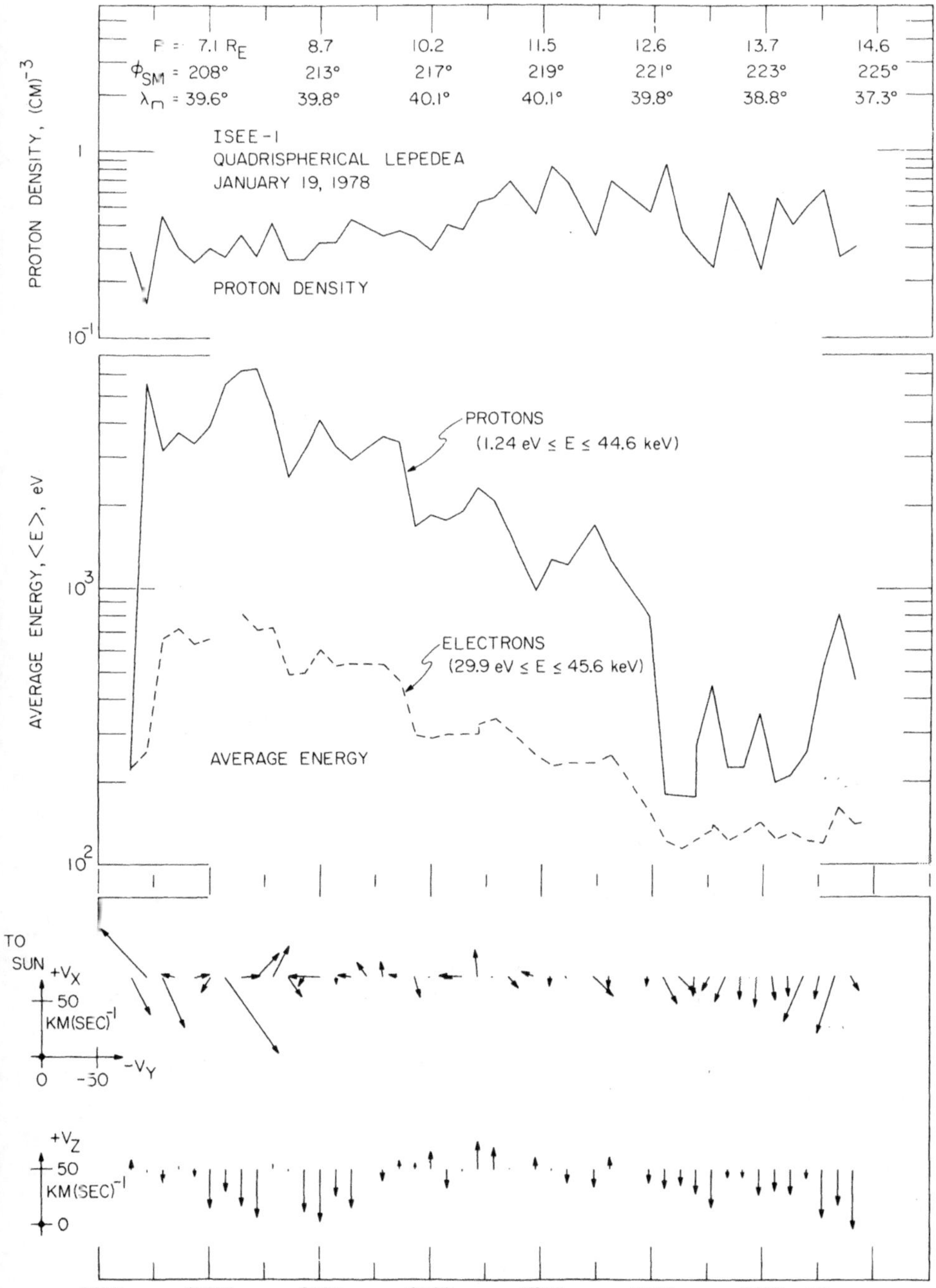

Fig. 11. Plasma parameters for the outbound pass of ISEE-1 on January 19, 1978 through the magnetosphere in the early local morning sector. The crossing of the trapping boundary of energetic charged particles occurs at ~08:00 UT. Ionospheric ions streaming tailward into the magnetotail are observed just outside the trapping boundary. Bulk plasma flows in this region are in the range of tens of km s^{-1}.

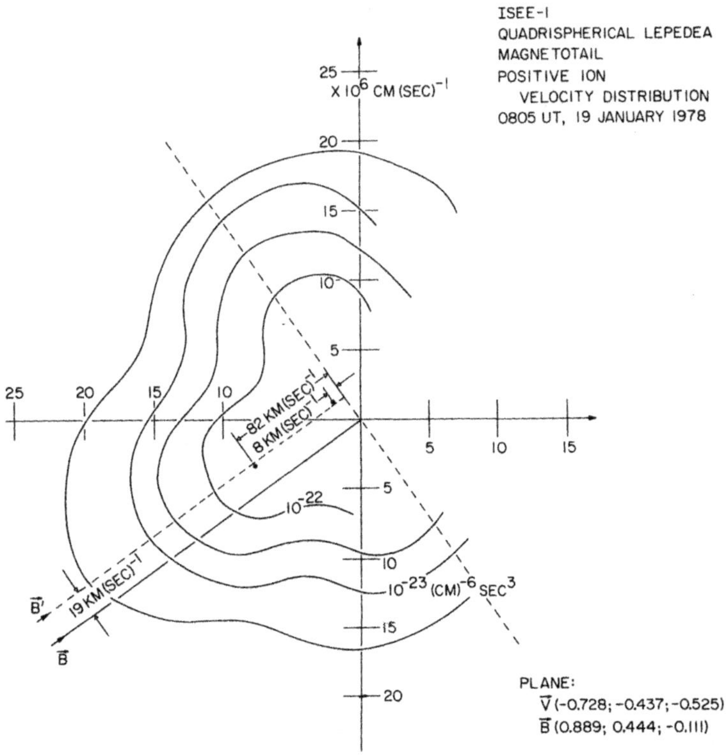

Fig. 12. Positive ion velocity distributions in the (**V**, **B**) plane at 08:05 UT on January 19 (cf. Figure 11). These velocity distributions are interpreted as those for ionospheric H^+ and O^+ streaming into the magnetotail. The gyrotropic reference frame is shown as the dashed axes.

similar to the electron temperature. Other instrumentation on board the spacecraft will be capable of determining the composition along a single line through the origin for the distribution function shown in Figure 12. The computation of the magnitude of $\mathbf{E}_\perp$ corresponding to the perpendicular drift is independent of species. This electric field strength is $0.8\,\mathrm{mV\,m^{-1}}$.

The quadrispherical Lepedeas on board ISEE-1 and -2 offer a first opportunity to measure the three-dimensional plasma velocity distributions in the Earth's magnetosphere. Our examples as discussed above are directed toward stressing the impact that such capability will have in our present studies of several magnetospheric plasma phenomena.

Acknowledgements

We wish to express our appreciation to our many colleagues who participated in the implementation of this project. In particular we acknowledge the efforts of J. Madden and M. Davis of the Goddard Space Flight Center (NASA) and of D. Eaton and R. Grün of the European Space Research and Technology Center

(ESA). R. Beatty and J. Silvers provided valuable support during post-launch operations. Co-investigators for this investigation are Dr C. F. Kennel of UCLA and Dr V. M. Vasyliunas of the Max-Planck Institute at Lindau. Dr C. T. Russell of UCLA has kindly allowed us the use of his magnetometer data from ISEE-1. This research was supported in part by the National Aeronautics and Space Administration under contract NAS5-20094 and grant NGL-16-001-002.

References

Ashour-Abdalla, M. and Kennel, C. F.: 1978, 'Nonconvective and Convective Electron Cyclotron Harmonic Instabilities', *J. Geophys. Res.* **83,** 1531.

DeCoster, R. J. and Frank, L. A.: 1978, 'Observations Pertaining to the Dynamics of the Plasma Sheet', *J. Geophys. Res.* (submitted).

Ejiri, M., Hoffman, R. A., and Smith, P. H.: 1977, 'Energetic Particle Penetrations Into the Inner Magnetosphere', *J. Geophys. Res.* (submitted).

Frank, L. A., Ackerson, K. L., and Yeager, D. M.: 1977, 'Observations of Atomic Oxygen (O^+) in the Earth's Magnetotail', *J. Geophys. Res.* **82,** 129.

Frank, L. A., Yeager, D. M., Owens, H. D., Ackerson, K. L., and English, M. R.: 1978, 'Quadrispherical Lepedeas for ISEE-1 and -2 Plasma Measurements', *IEEE Trans. Geosci. Electr.* **GE-16,** 221.

Fredericks, R. W.: 1971, 'Plasma Instability at $(n+\frac{1}{2})f_c$ and Its Relationship to Some Satellite Observations', *J. Geophys. Res.* **76,** 5344.

Gurnett, D. A., Anderson, R. R., Scarf, F. L., Fredericks, R. W., and Smith, E. J.: 1979, 'Initial Results from the ISEE-1 and -2 Plasma Wave Investigation', *Space Sci. Rev.* **23,** in press.

Hubbard, R. F. and Birmingham, T. J.: 1977, 'Electrostatic Emissions between Electron Gyroharmonics in the Outer Magnetosphere', GSFC Preprint X-602-77-112, Greenbelt, Maryland.

Lyons, L. R.: 1974, 'Electron Diffusion Driven by Magnetospheric Electrostatic Waves', *J. Geophys. Res.* **79,** 575.

McIlwain, C. E.: 1972, 'Plasma Convection in the Vicinity of the Geosynchronous Orbit', in B. M. McCormac, *Earth's Magnetospheric Processes,* D. Reidel Publ. Co., Dordrecht, Holland, p. 268.

McIlwain, C. E.: 1974, 'Substorm Injection Boundaries', in B. M. McCormac. *Magnetospheric Physics,* D. Reidel Publ. Co., Dordrecht, Holland, p. 143.

Rosenbauer, H., Grunwaldt, H., Montgomery, M. D., Paschmann, G., and Schopke, N.: 1975, 'HEOS-2 Plasma Observations in the Distant Polar Magnetosphere: The Plasma Mantle', *J. Geophys. Res.* **80,** 2723.

Smith, P. H. and Hoffman, R. A.: 1974, 'Direct Observations in the Dusk Hours of the Characteristics of the Storm Time Ring Current Particles During the Beginning of Magnetic Storms', *J. Geophys. Res.* **79,** 966.

Whipple, E. C., Jr.: 1978, '(U, B, K) Coordinates: A Natural System for Studying Magnetospheric Convection', *J. Geophys. Res.* **83,** 4318.

Young, T. S. T., Callen, J. D., and McCune, J. E.: 1973, 'High-Frequency Electrostatic Waves in the Magnetosphere', *J. Geophys. Res.* **78,** 1082.

DUAL SPACECRAFT OBSERVATIONS OF ENERGETIC PARTICLES IN THE VICINITY OF THE MAGNETOPAUSE, BOW SHOCK, AND THE INTERPLANETARY MEDIUM*

G. K. PARKS, C. GURGIOLO, and C. S. LIN
Geophysics Program, University of Washington, Seattle, Wash. 98195, U.S.A.

K. A. ANDERSON and R. P. LIN
Physics Department and Space Sciences Laboratory, University of California, Berkeley, Calif. 94720, U.S.A.

F. MARTEL and H. RÉME
Université Paul Sabatier, Toulouse, France

Abstract. This article presents some of the new and important particle features that have been detected in the energy range ~1 keV to ≥290 keV by the ISEE-1 and -2 spacecraft near the magnetopause, bow shock, and the interplanetary space. Only examples of data from the first few orbits, when the spacecraft were on the front side, are shown.

1. Introduction

A particle experiment has been designed to take advantage of the dual ISEE-1 and -2 spacecraft system to study the spatial and temporal effects of energetic particle phenomena in the earth's magnetosphere, magnetopause, magnetosheath, plasmasheet, bow shock, and the interplanetary medium. An identical pair of particle detector assemblies consists of fixed voltage analyzers (FVA's) that detect electrons and ions with center energies about 1.5 keV and 6 keV, and solid state telescopes (SST's) that detect electrons and protons with energies ≳16 keV. These detectors have large geometric factors and a wide dynamic range for particle counting, and together with several modes of operation that include changes of geometric factor for the FVA's and threshold energy changes for the SST's, the experiment permits observations of a wide variety of particle phenomena. During normal operational mode, data from each of the detectors are sampled four times a second, and this is increased to 32 times a second during fast telemetry mode. Table I summarizes the essential characteristics of the detectors. More detailed information on this experiment can be found in Anderson *et al.* (1978).

This paper presents particle data obtained by our experiment during the first few orbits when the spacecraft were on the front side of the magnetosphere. In Section 2 we summarize observations near the magnetopause and the bow shock

* Paper presented at 13th ESLAB Symposium, Innsbruck, Austria (June 5, 1978).

Space Science Reviews **22** (1978) 765–776. *All Rights Reserved*

TABLE I
ISEE-1 and -2 Instrument characteristics

Detector	Geometric factor* (cm² sr)		Aperture (FWHM)		Look direction (angle to spin axis)		Particle type	Energy range (keV)	
	1	2	1	2	1	2		1	2
Fixed voltage analyzers (FVA)									
2e	2.8×10^{-2}	3.3×10^{-2}	6°×14°	6°×16°	0°	0°	e^-	1.3–1.5	1.4–1.6
2p	2.1×10^{-2}	1.8×10^{-2}	7.5°×14°	6.5°×13°	0°	0°	ions	1.3–1.5	1.3–1.7
6e	2.3×10^{-2}	3.2×10^{-2}	5.6°×14°	6.5°×16°	0°	0°	e^-	4.8–6.6	4.5–6.0
6p	1.5×10^{-2}	2.1×10^{-2}	6.5°×11°	7°×13°	0°	0°	ions	5.5–6.5	5.4–6.6
Solid state telescopes (SST)									
Foil	— 0.35 —		— 38° cone —		15°	0°	e^-	19–290	16–280
								42†–290	42†–280
								20–290	
								(16 ch PHA)	
Open	"		"		15°	0°	e^- + ions	19–290	16–280
								42†–290	42†–280
								290–460	280–450
								Coincidence rate	
								22–40 000	
								(48 ch PHA)	

Notes:
* The FVA's also have a high flux mode where the geometric factor is decreased by a factor of $\sim5\times10^2$.
† Can be set to 22, 42, or 80 keV by command.

regions using 32-s averaged particle data. A more detailed picture of particle features near these boundaries is presented in Sections 2.1 and 2.2, where the high resolution particle data ($\frac{1}{4}$-s averages) are used. Section 2.3 discusses the observations of low-energy particles in the upstream interplanetary medium.

The emphasis of this first ISEE paper is on reporting and discussing several new particle features that have been observed in the vicinity of the magnetopause, the bow shock, and the upstream interplanetary medium. A more extensive study of these particle observations is being made in conjunction with magnetic, electric field, and plasma distribution function measurements in order to learn about the fundamental plasma processes that take place in these regimes.

2. Results

The overview of 32-s-averaged particle fluxes detected by the ISEE-2 experiment during its passage through the magnetosphere and the interplanetary medium is shown in Figures 1 and 2 for days 309 (November 5, 1977) and 314 (November 10, 1977). The solar wind was moderately active during these days, as can be inferred from the K_p indices. The three-hour K_p values were 4^+, 3^-, 2, 2^-, 1, 2, 1, 2^- for day 309 and 0^+, 0^+, 2^-, 1^+, 3, 3^+, 2, 3 for day 314 (Lincoln, 1978). Particle

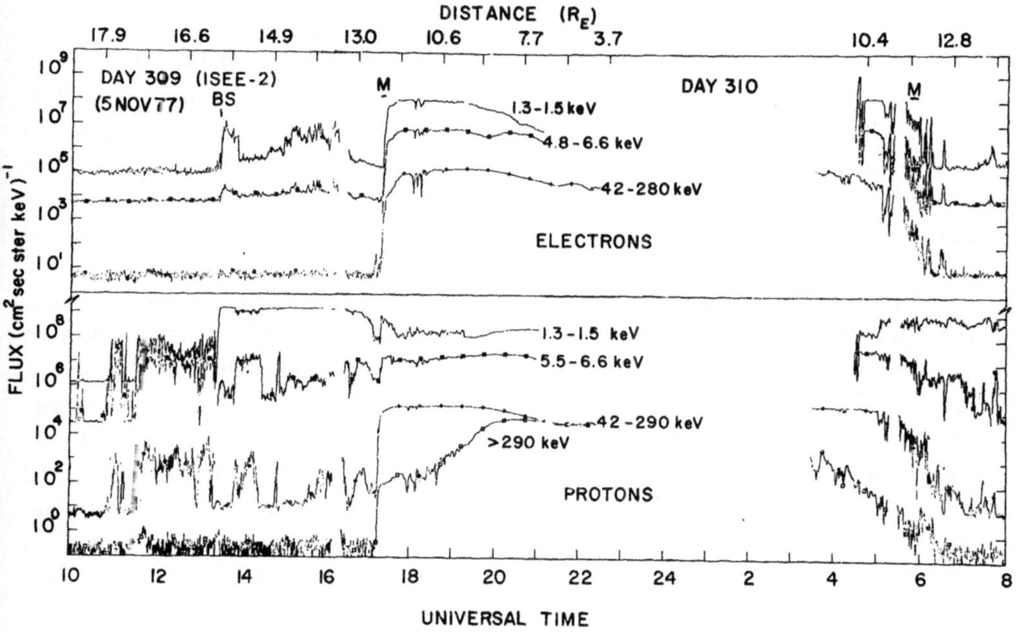

Fig. 1. 32-s averages of particle fluxes as a function of the Universal Time for day 309 (November 5, 1977). The letters 'BS' and 'M' designate bow stock and magnetopause, respectively.

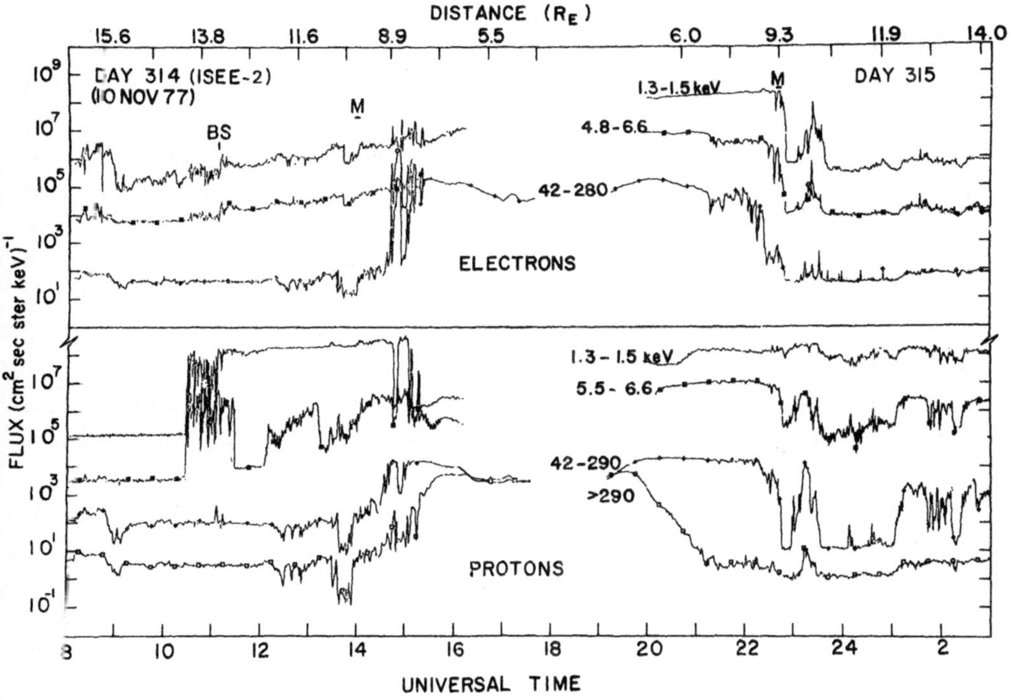

Fig. 2. Same as Figure 1 for day 314 (November 10, 1977).

data of this time scale show some of the well-known particle features that are associated with the magnetopause (marked by *M*) and the bow shock (marked by BS). (For previous studies of the particles near the magnetopause and bow shock, see Gosling *et al.* (1967) and Scarf *et al.* (1977).) For both the inbound and outbound passes, the magnetopause was detected closer to Earth on day 314 than on 309. The inbound magnetopause was much sharper on day 309 than on day 314, while day 314 exhibited the sharper boundary during the outbound pass.

Intense particle fluxes are observed ahead and behind the bow shock. For both the upstream and downstream particle events, the behavior of electrons and ions (protons) is quite different and energy dependent. Downstream from the shock, ion fluxes in the 1.3–1.5 keV interval become $>4\times10^9$. Particles of this energy are no doubt associated with the thermalization of the solar wind particles in the formation of the collisionless bow shock. An important feature to note is that the particle flux here remains high from the magnetosheath to the magnetopause and beyond. Upstream from the shock, particle fluxes are also enhanced, but the ion fluxes undergo rapid spatial and/or temporal variations as indicated by the large and rapid variations in the flux-time profile. The upstream proton events were studied earlier by Asbridge *et al.* (1968) and Lin *et al.* (1974). Particle fluxes in the neighbourhood of the magnetopause and the bow shock are now examined using the high-resolution data, $\frac{1}{4}$-second averages.

2.1. Magnetopause

According to the magnetometer data (courtesy of UCLA), there were multiple crossings of the magnetopause by spacecraft-1 and a single crossing by -2 on day 309. Spacecraft-1's final entry into the magnetosphere was around 17:13:30 Universal Time (UT) and -2's was about 30 s later. Several magnetopause crossings were again encountered on day 314 during the outbound pass, in this case on both spacecraft. This day's final entry into the magnetosphere occurred around 22:43 UT. Figures 3 and 4 show about 10 min of particle data during these magnetopause crossings. The magnetopause positions are indicated by *M* in these figures.

The high resolution data show a number of interesting features concerning the behavior of particle fluxes in the neighborhood of the magnetopause. The features particularly worthy of note are:

(a) Particle discontinuities do not always coincide with the magnetopause position, although the electron fluxes usually change strikingly at the magnetopause.

(b) On day 309 the ion boundary was found earthward from the electron boundary. The two are separated by ~100–150 km in space. At this time the magnetic field just inside the magnetopause was $\sim 50\gamma$ (private communication, UCLA). In this field strength the gyro-radius of 1.5 keV protons is nearly equal to the observed particle separation distance.

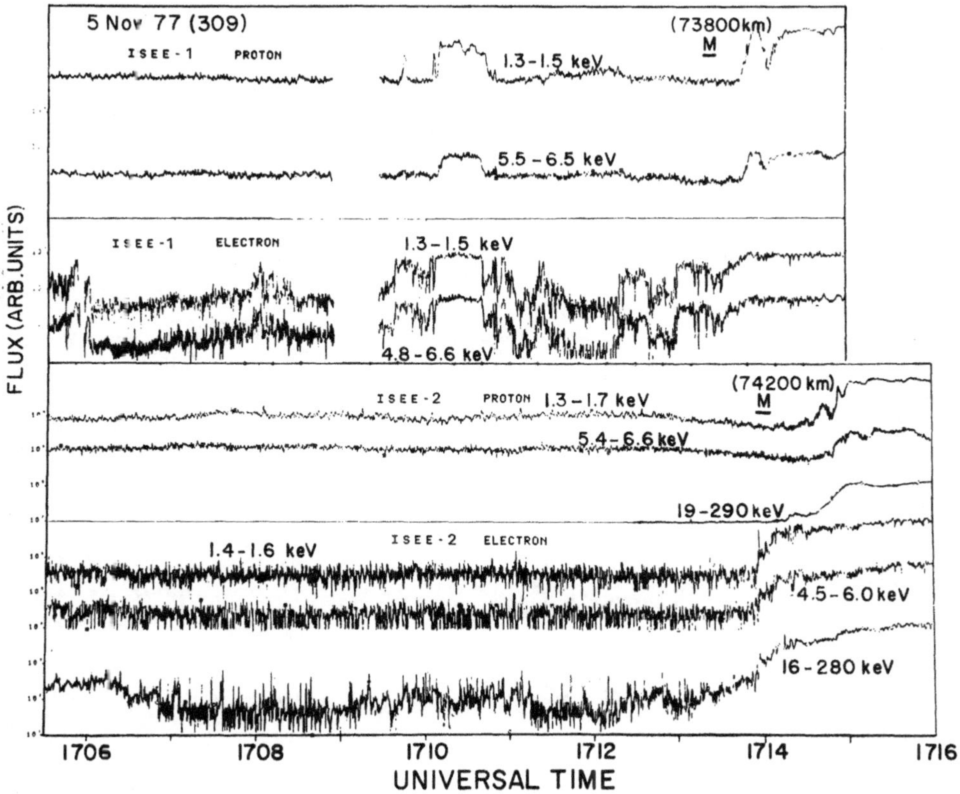

Fig. 3. High-resolution ($\frac{1}{4}$-s averages) of particle data obtained in the neighborhood of the magnetopause. The position of the magnetopause (M) is deduced from the UCLA magnetometer experiment. The geocentric distances of the magnetopause crossings on ISEE-1 and -2 are indicated.

(c) On day 314 the electron boundary was very sharp. For the ions, 1.3–1.5 keV ion fluxes on both ISEE-1 and -2 remain high and do not show a boundary at or near the magnetopause.

(d) The particle fluxes inside the neighborhood of the magnetopause are apparently undergoing temporal changes as well. For example, the ion fluxes just inside the magnetopause are smaller on -2 than on -1 (Figure 3).

Spacecraft orbit information shows that -1 and -2 detected the magnetopause at different geocentric distances. For instance, on day 309, -2 found the magnetopause closer to Earth than did -1. A particular feature detected with a time delay Δt on two spacecraft separated by a distance **d** can be determined to be in motion or stationary by the relation

$$\mathbf{d} \cdot \hat{n} = (\mathbf{V}_f - \mathbf{V}_s) \cdot \hat{n}\, \Delta t. \tag{1}$$

The unit normal vector of a particle front is represented by $\hat{n}$, $\mathbf{V}_f$ is the velocity of a particle front, and $\mathbf{V}_s$ is the spacecraft velocity

On day 309 spacecraft-1 was leading -2 at the time of the magnetopause

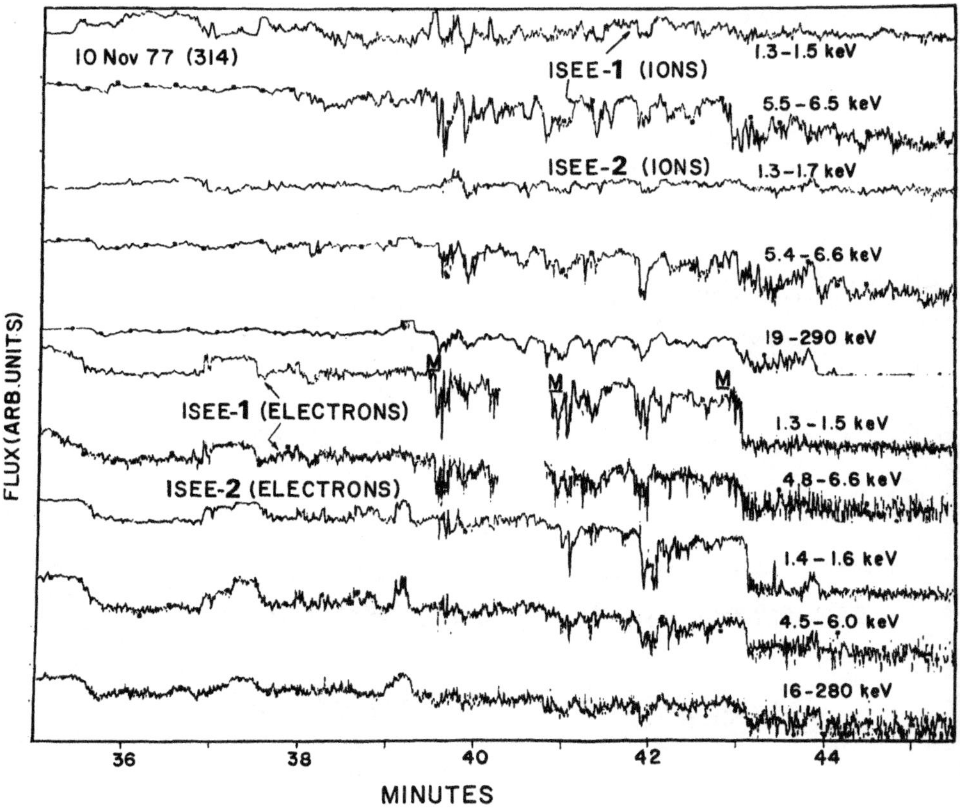

Fig. 4. Same as Figure 3 for the magnetopause crossing of day 314.

crossing by about 408 km. The observed time delay for the ion flux increases that appeared similar on the two spacecraft is about 46 s, and the spacecraft velocity during this portion of the orbit was about 2.4 km s^{-1}. Assuming that the separation distance and velocities are parallel to $\hat{n}$, a propagation speed of 6.5 km s^{-1} is obtained.

Very short time delay was recorded for the day 314 magnetopause crossing (Figure 4). The time difference in the electron data received was $\lesssim 4$ s. Noting that the two spacecraft were separated by about 265 km and that the spacecraft speed was 2.62 km s^{-1}, we observed the magnetopause speed to be $\gtrsim 90$ km s^{-1} toward the Earth. This high magnetopause speed is extremely interesting since it approaches the local Alfvén speed. For several other magnetopause crossings the speed ranges from 10 to 80 km s^{-1}.

The observation that the ion boundary is Earthward from the electron boundary is consistent with the idea that the solar wind plasma there is neutralized by the cold ionospheric particles. A detailed comparison of the particle data with the magnetic and electric field data should give further clues about such interaction processes including the information on the depth of penetration by particles of different energies across the magnetopause.

2.2. Bow shock

Bow shock crossings on ISEE-1 and -2 have been reported by the UCLA magnetometer group (private communication) on day 309 at 13:22:26 UT and 13:23 UT. The high-resolution particle data in the vicinity of the bow shock are shown in Figure 5. Here, BS denotes the position of the bow shock as inferred from the magnetic data. Important features to note are the:

(a) Large particle fluxes, especially ions in the energy range 1.3–1.5 keV found upstream and downstream from the indicated shock position. These intense ion fluxes are probably thermalized solar wind particles created in the process of formation of the collisionless bow shock.

(b) Large variations of the ion fluxes (up to 3 orders of magnitude), while the

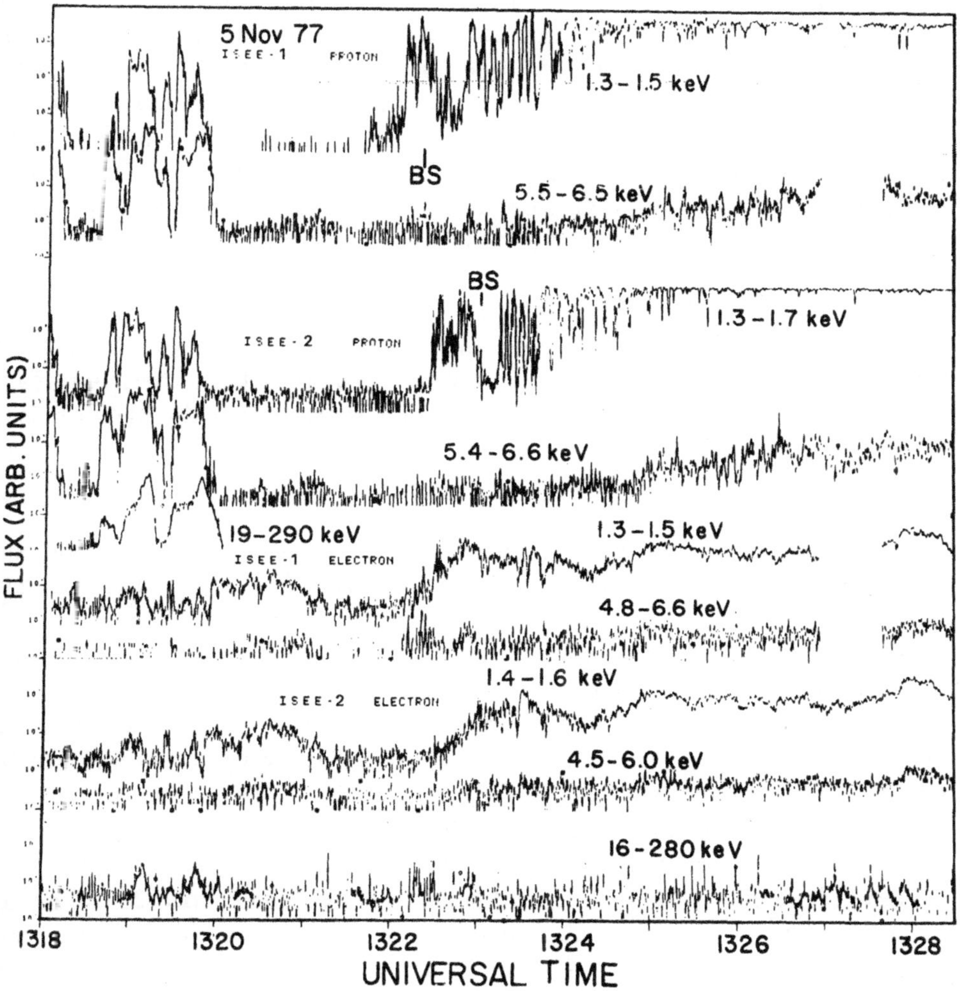

Fig. 5. High-resolution ($\frac{1}{4}$-s averages) of particle data in the vicinity of the bow stock. 'BS' denotes the bow stock position deduced from the UCLA magnetometer data.

electrons are comparatively constant. There is evidence that the ion flux variations are spatial. These structures appear extremely stable since they persist even as the shock moves.

(c) The 6 keV ion fluxes increase a few minutes after the shock crossing. These particles show periodic variations that are close to the spin period of the spacecraft. This indicates that these particles are highly anisotropic.

(d) Observations of the well-known upstream proton events ahead of the shock (to be discussed more below). These upstream particle events have hard energy spectra compared to the shock-associated thermalized particles.

The speed of the bow shock-associated particle structures using the ion features and the observed time delay of 40 s is 6.5 km s^{-1}. This speed is comparable to the magnetopause motion detected on this day ~4 hr later. For the inbound shock of day 314 (not shown), very little time delay was observed between similar particle features on the two spacecraft. The speed for this shock motion is estimated to be ~90 km s^{-1}, very close to the speed determined for the magnetopause 12 hr later.

It is apparent that energetic particle phenomena associated with the bow shock are complex and highly variable from day to day. No doubt these effects will be ordered, at least partially, when correlated with the Mach number, the direction of the bow shock normal, and bêta (β) of the solar wind.

2.3. Upstream energetic particles

Both the electron and ion fluxes are found upstream from the bow shock. Data from the -2 spacecraft will be shown to illustrate the properties of these upstream particle events, since the -1 spacecraft detector shows essentially identical features.

2.3.1. *Protons*

Examples of ion (mainly proton) fluxes detected by ISEE-2 in the upstream region are shown in Figures 6 and 7. Characteristic features of these particle fluxes are:

(a) Large and rapid time variations from a few seconds to about one minute. These events characterize the ones detected in the neighborhood of the bow shock, around $14R_e$. The upstream particle events at large distances, for example at geocentric distances of ~$22R_e$, have durations lasting up to several tens of minutes and hours (not shown). This agrees with the earlier observations of the upstream events detected at the lower orbit (Lin *et al.*, 1974).

(b) Variable energy spectrum. Some bursts are only detected by the 5.5–6.5 keV ion detector; others have proton energies extending above 280 keV (see Figure 7). The energy spectrum within a burst can change. For example, the energy spectrum of the day 309 event (Figure 6) is much harder at the trailing edges of the events.

(c) These flux variations are strongly spin-modulated, showing that the fluxes are highly anisotropic.

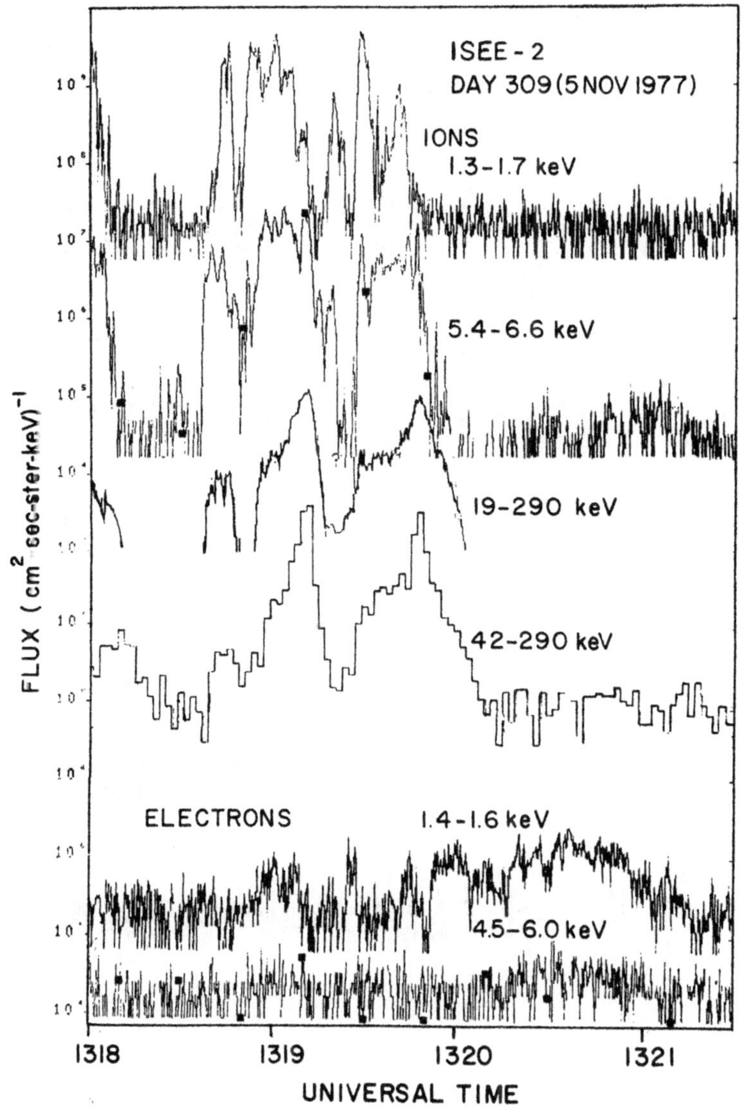

Fig. 6. Example of upstream particle events detected ahead of the bow shock on day 309 in expanded time scale.

(d) The protons are usually accompanied by low fluxes of 1.3–1.5 keV electrons. The electron fluxes on day 314 (Figure 7) also show spin-modulation.

2.3.2. *Electrons*

A new particle feature has been observed in the upstream region. On many occasions, we have detected abundant fluxes of low energy electrons, predominantly in the interval 1.3–1.6 keV. Figure 8 shows these electron events detected by the spacecraft near the apogee ($\sim 22R_e$ geocentric distance). This figure shows

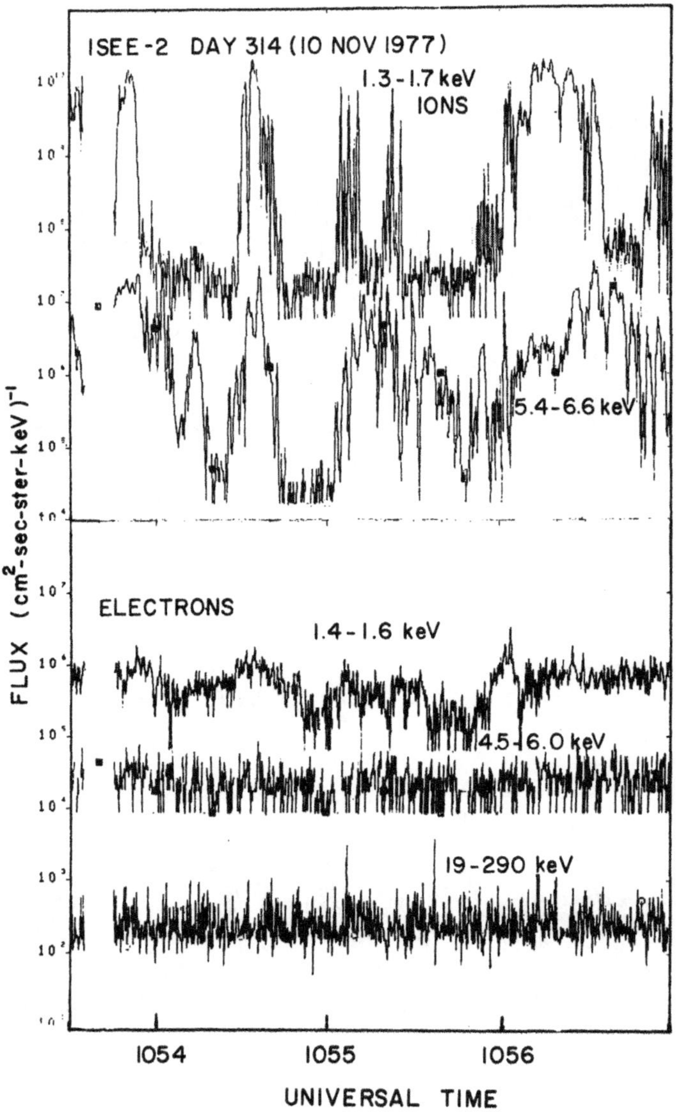

Fig. 7. Same as Figure 6 for the events detected on day 314.

that:

(a) The clusters of low energy electrons are detected in regions near the upstream proton events. Note that these electrons can be detected as well without the upstream proton events.

(b) The upstream electron events are detected primarily in the energy interval 1.4–1.6 keV. The flux increase in this energy interval is typically a factor of 2–10.

(c) The duration of the upstream electron events is from several minutes to about one hour. The onset of the events is gradual (unlike the sharp onset of upstream proton events).

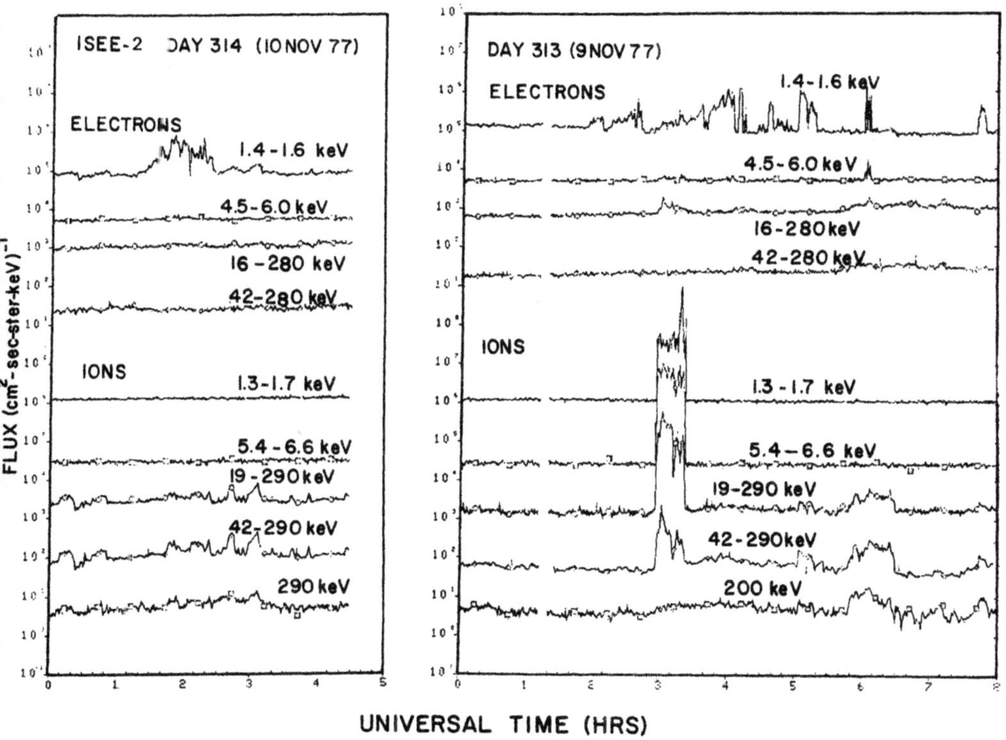

Fig. 8. An example of interplanetary low energy electron fluxes. These data were obtained at $21R_e$ geocentric distances.

A preliminary survey of the data that we have acquired to date indicates that the flux of these upstream electron events shows little or no dependence on the radial distances where they are detected. This feature is similar to that found in the upstream proton events (Lin *et al.*, 1974) and raises an important question concerning the origin of these upstream electrons. A thorough investigation of the upstream particle events, together with the direction of the magnetic field and the observation of plasma and Alfvén waves, should provide additional information on the acceleration of these particles.

Acknowledgement

The ISEE-1 and -2 research at Berkeley and Seattle is conducted under NASA Contract NAS 5-20079 and at Toulouse under CNES grant number 212.

References

Anderson, K. A., Lin, R. P., Paoli, R. J., Parks, G. K., Lin, C. S., Réme, H., Bosqued, J. M., Martel, F., Cotin, F., and Cros, A.: 1978, 'An Experiment to Study Energetic Particle Fluxes in and Beyond the Earth's Outer Magnetosphere', *IEEE Trans. Geosci. Electr.* **GE-16,** 213.

Asbridge, J. R., Bame, S. J., and Strong, I. B.: 1968, 'Outward Flow of Protons from the Earth's Bow Shock', *J. Geophys. Res.* **73,** 5777.

Gosling, J. T., Asbridge, J. R., Bame, S. J., and Strong, I. B.: 1967, 'Vela-2 Measurements of the Magnetopause and Bow Shock Positions', *J. Geophys. Res.* **72,** 101.

Lin, R. P., Meng, C., and Anderson, K. A.: 1974, '30 to 100 keV Protons Upstream from the Earth's Bow Shock', *J. Geophys. Res.* **79,** 489.

Lincoln, J. V.: 1978, 'Geomagnetic and Solar Data', *J. Geophys. Res.* **83,** 1191.

Scarf, F. L., Frank, L. A., and Lepping, R. P.: 'Magnetosphere Boundary Observations along the Imp-7 Orbit, 1, Boundary Locations and Wave Level Variations', *J. Geophys. Res.* **82,** 5171.

EARLY RESULTS FROM ISEE-1 ELECTRIC FIELD MEASUREMENTS

J. P. HEPPNER, N. C. MAYNARD, and T. L. AGGSON

Electrodynamics Branch, Laboratory for Planetary Atmospheres, Goddard Space Flight Center, Greenbelt, Md. 20771, U.S.A.

Abstract. The double probe, floating potential instrumentation on ISEE-1 is producing reliable direct measurements of the ambient DC electric field at the bow shock, at the magnetopause, and throughout the magnetosheath, tail plasma sheet and plasmasphere. In the solar wind and in middle latitude regions of the magnetosphere spacecraft sheath fields obscure the ambient field under low plasma flux conditions such that valid measurements are confined to periods of moderately intense flux. Initial results show: (a) that the DC electric field is enhanced by roughly a factor of two in a narrow region at the front, increasing B, edge of the bow shock, (b) that scale lengths for large changes in E at the sub-solar magnetopause are considerably shorter than scale lengths associated with the magnetic structure of the magnetopause, and (c) that the transverse distribution of B-aligned E-fields between the outer magnetosphere and ionospheric levels must be highly complex to account for the random turbulent appearance of the magnetospheric fields and the lack of corresponding time-space variations at ionospheric levels. Spike-like, non-oscillatory, fields lasting <0.2 s are occasionally seen at the bow shock and at the magnetopause and also intermittently appear in magnetosheath and plasma sheet regions under highly variable field conditions. These suggest the existence of field phenomena occurring over dimensions comparable to the probe separation and $\leq c/\omega_{pe}$ (the 'characteristic' electron cyclotron radius) where ω_{pe} is the electron plasma frequency.

1. Introduction

The symmetric double probe, floating potential technique for measuring DC and low frequency electric fields which was initially described in a 1964 proposal (Aggson and Heppner, 1964) has since been employed on a number of rockets and satellites. In dense plasmas (e.g., >100 el cm^{-3}) the technique has been highly successful with relatively modest probe separations. In weak plasmas (e.g., <10 el cm^{-3}) results from OGO, IMP, S^3, and Hawkeye satellites have shown that acquisition of valid measurements of the ambient DC field is a sensitive function of the probe separation and the ambient plasma flux and that each increase in the probe separation has extended the space-time applicability of the technique (see e.g., Aggson and Heppner, 1977; Heppner *et al.*, 1978). Thus with the substantial increase in probe separation on ISEE-1 (see Figure 1) it was anticipated that valid measurements would be obtained over more extensive spatial regions and for greater periods of time in regions of marginal plasma flux. Preliminary scans of data taken in various plasma regimes confirm this anticipation.

The following statements briefly characterize the status of the measurements.

(1) At the bow shock, at the magnetopause, throughout the magnetosheath, and near and inside the plasmapause, the measurements examined to date have been valid and accurate.

Space Science Reviews **22** (1978) 777–789.

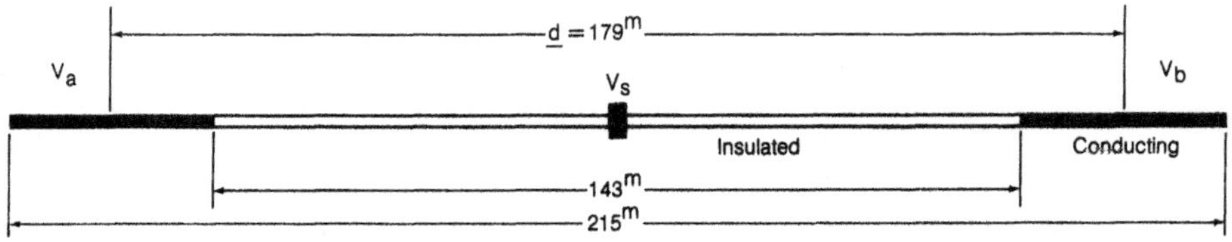

"COMMON MODE" VOLTAGE = $V_a - V_s$

DIFFERENTIAL VOLTAGE = $V_a - V_b$

ELECTRIC FIELD = $E_{xy} = (V_a - V_b)/\underline{d}$

Fig. 1. Double probe, wire antenna dimensions and measured DC parameters.

(2) In the solar wind and in middle latitude regions of the outer magnetosphere, the measurements during periods of low activity are most commonly dominated by strong solar oriented sheath fields associated with the spacecraft and its conducting appendages. Valid measurements are obtained under moderately disturbed conditions when the plasma density increases.

(3) From smaller samplings of data in the geomagnetic tail it appears that measurements in the plasma sheet will be valid most of the time.

(4) Data taken under eclipse conditions in the near earth tail are free of sheath effects and thus demonstrate the photoelectron character of the sheath in sunlight in low density plasmas. From eclipse data it is apparent that DC fields as weak as 20 μ V m^{-1} can be accurately measured. This accuracy also appears evident, but is less definitively proven, in the magnetosheath when the component in the spin plane is small.

(5) Attention has been directed toward examining the signal behavior when the electron gun is active and/or currents are drawn by the spheres or shields of the electric field probes (F. Mozer experiment) extended perpendicular to our (Figure 1) double probe axis to discover whether or not these bias functions, which drive the spacecraft positive relative to the probes, will favorably alter the sheath parameters. In all cases and modes examined the bias functions have introduced effects which do not help the measurement of ambient fields with the Figure 1 probes. The effects involve both large amplitude sheath voltages and asymmetries and these remove any possibility of obtaining valid measurements in weak plasmas when the bias functions are operative.

(6) High voltage spacecraft charging events have not been encountered in the data examined sunward from the Earth. Several charging events have been detected in near tail regions but their statistical prevalence and coincidence with geophysical events has not been evaluated.

The present paper is oriented toward illustrating typical electric field measurements in raw data form. This illustrates that the floating potential measurements are direct measurements, without correction or manipulation. The only computation for the instantaneous electric field in the spin plane (E_{xy}) is the division by **d**, the probe separation shown in Figure 1. The 'Common Mode' voltage, $V_a - V_s$, is a highly useful diagnostic of plasma conditions surrounding the spacecraft. It is also sampled frequently using the same 14 bit A/D converter that gives the differential voltage, $V_a - V_b$. At low bit rates, $V_a - V_b$ and $V_a - V_s$ are sampled 24 and 12 times per spin cycle, respectively.

Additional experiment outputs include: a number of sub-commutated measurements at selected angles indexed to the Sun and the magnetic field directions, measurements of AC fields in 8 filter channels between 0.19 and 1900 Hz, and measurements of the plasma impedance. These functions and the various options, such as blanking-out or leaving-in the Sun pulse that occurs in weak plasmas when the probes are aligned with the Sun, are described in Heppner *et al.* (1978).

Presentation in the form of raw data also enables one: (a) to illustrate invalid data (see discussions related to Figures 2, 3, and 6), and (b) to illustrate the existence and prevalence of short duration (e.g., 0.1 to 0.2 s) spikes and the sharpness of electric field changes. These latter features, which will be more completely studied using high bit rate data in the future, are of particular interest in that they suggest short-lived or small scale electrostatic departures from the hydromagnetic $\mathbf{E}+\mathbf{v}\times\mathbf{B}=0$, relationship.

In general, whether examining shock, magnetopause, magnetosheath, or plasma sheet data the short period or small scale variability of the measured (i.e., as opposed to 'model') electric field is its most evident characteristic. Thus, raw data presentation is appropriate from this standpoint as well as those noted above. It is envisaged that various modes of time-space data averaging will have to be tried to arrive at the most meaningful representations of large scale distributions. Particularly within the magnetosphere finding an averaging process that yields the large scale features evident at ionospheric levels represents a major challenge. In turn this implies that many of the small scale magnetospheric field structures, or equivalent time variations, do not directly map along equipotential magnetic field lines to high latitude ionospheric levels.

2. Coordinates and Solar Wind Measurements

Figure 2 is a sample of the field E_{xy} in the spin plane in the solar wind at a time when the ambient field magnitude was variable. It also serves to illustrate the coordinates. $\phi = 0°$ marks the time when the **a** probe is pointing at the Sun which is also the direction of the **x** axis in solar-ecliptic coordinates. As shown the maximum $+E_{xy}$ field is directed toward $\phi = 90°$ which is also the **y** axis in solar-ecliptic coordinates. Subject to the assumption that the bulk velocity, **v**, of the solar wind is radially away from the Sun, the validity of the field measured is evident from the fact that the ambient field is oriented normal to **v**. The signal distortion, or spike, which occurs at $\phi = 0°$ and $\phi = 180°$ is a consequence of the sudden decrease in photoemission from the probes when they align with the Sun. This example illustrates changes in the appearance of the $\phi = 0°$ and 180° Sun

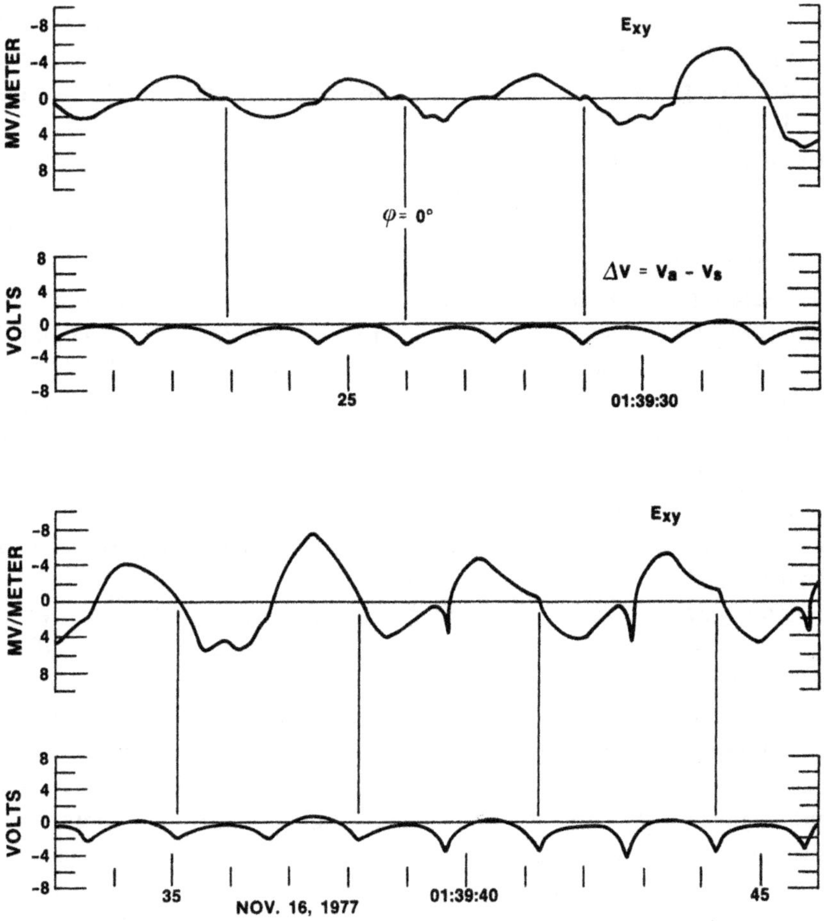

Fig. 2. Measurements of a variable field in the solar wind (raw data).

pulse under variable conditions in the ambient plasma, and also that the measurements near $\phi = 90°$ and 270° are not affected by either the pulses or the spacecraft sheath.

In contrast to the valid ambient field measurements in Figure 2, invalid measurements appear in the solar wind in Figure 3 in the regions labeled SW. In these cases one notes that the signal maximum occurs coincident with the Sun aligned pulse-like distortions at $\phi = 0°$. The electric field that is measured in these cases is simply a Sun-aligned field stemming from the spacecraft. In the presence of extremely low densities in some regions of the magnetosphere the rather broad pulses at $\phi = 0°$ and 180° in Figure 3 become much narrower and it becomes more appropriate to call them 'Sun pulses' or 'Sun spikes'. The pulses located at peak and trough values of the sheath field evident between 22:59:04 and 22:59:10 in Figure 6 roughly illustrate this characteristic.

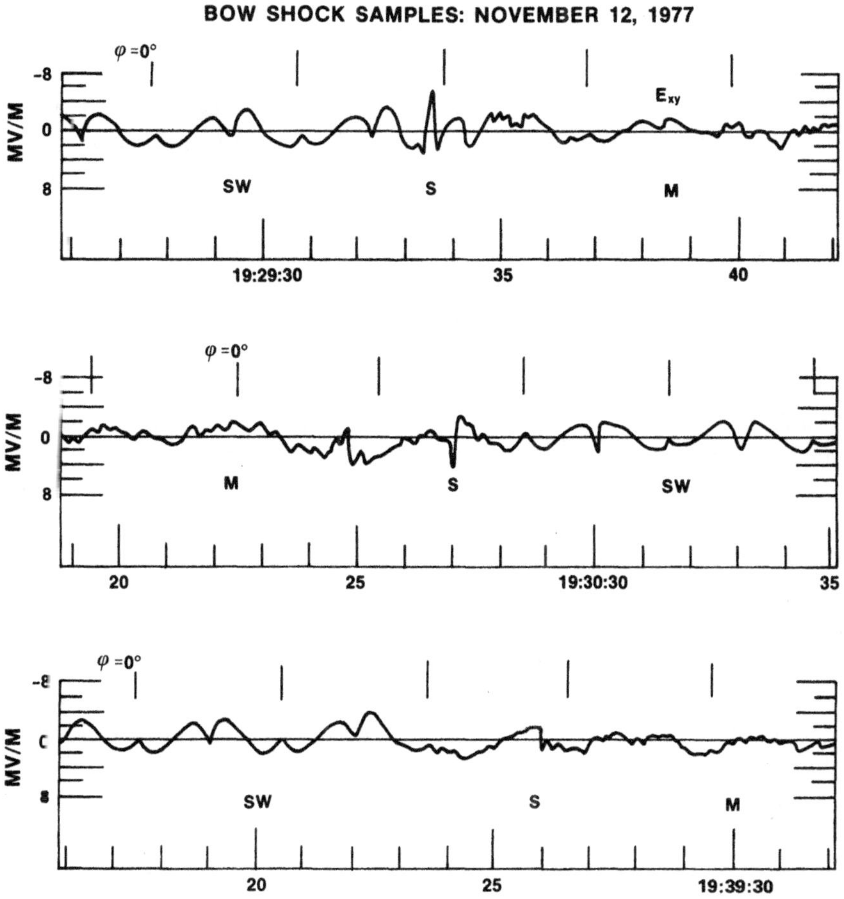

Fig. 3. Sequence of bow shock crossings at locations labeled (S). (M) designates magnetosheath. Measurements in the solar wind (SW) are dominated by the spacecraft sheath.

3. Bow Shock Fields

The three shock encounters, all within ten minutes, shown in Figure 3 illustrate the abruptness of the electric field change at the shock. This abruptness must also be present in the plasma flux to produce the transition from sheath measurements in the solar wind to ambient field measurements at the shock and on the magnetosheath side of the shock. In the first example (top trace) the transition occurs within a fraction of a second, between 19:29:33 and 19:29:34 and coincides with the spike response having twice the magnitude of the adjacent magnetosheath field. In the next encounter (middle trace) a small spike again appears. The third (bottom) trace suggests a less abrupt transition without a spike field; however, it must be kept in mind that for a feature lasting only a fraction of a second the probes can, at that instant, be in an orientation transverse to the field and thus miss its occurrence.

Relative to the many past studies of the magnetic field structure in the bow shock it is desirable to know the relative location of the abrupt electric field transition and the associated small (i.e., roughly a factor of two) enhancement in the electric field magnitude which is frequently, but not always, present at the shock. Figure 4 provides an example relative to the magnetic field. In this case the transition from sheath field to ambient field measurements, near 13:22:23 is separated about 2 s from the largest spike at 13:22:25 and there is a total period of about 5 s (to 13:22:30) when the average field is more intense than at subsequent times. When related to the magnetic field structure it is apparent that the enhanced DC E-field is coincident with the leading edge and with increasing **B** at the shock front. Moving further into this shock, which has a magnetic profile very much like the model shock presented by Heppner *et al.* (1967), the enhanced **B** section which lasts about 30 s coincides with a region of highly irregular electric fields with amplitudes of several V km^{-1}. As shown this is evident in the 6–19 Hz band as well as in the DC measurements.

From observations in the VLF band it has previously been argued (Fredricks *et al.*, 1970): (a) that the scale length for electrostatic turbulence in the bow shock is defined by c/ω_{pe}, the 'electron inertial length' or 'characteristic electron cyclotron radius' where ω_{pe} is the electron plasma frequency, and (b) that the bow shock electrostatic turbulence is associated in time with regions of steep slope in B. These arguments appear to also fit the DC observations and in fact appear to be even more definitive at DC and low frequencies in placing the peak DC field at the leading edge of the shock.

4. Magnetosheath Fields

As anticipated from the well known variability of plasmas and magnetic fields in the magnetosheath, and from previous Imp-6 and -8 electric field measurements, an endless variety of electric field behaviors is encountered. Figure 5 provides several examples from a single magnetosheath crossing. The measurements in this region are free of sheath effects and one does not see the Sun pulses that were apparent in the solar wind. This particular pass near 11 hr local time was chosen for illustration to provide continuity with Figure 6 and to show an interval of exceptional strong fields. The strong field (bottom trace) occurred about 7 min after an interval when the spacecraft was inside the magnetosphere for several minutes. Between the magnetopause crossing preceding the Figure 5 bottom trace and the next magnetopause crossing there was an interval of about 75 min. Thus the strong field occurred after the magnetopause had moved much closer to the Earth. When compared with data from the on-board electron spectrometer (Ogilvie experiment) it was found that the occurrence of this strong field (12 V km^{-1}) coincides exactly with a brief interval when the bulk flow jumped to 450 km s^{-1} with essentially no change in plasma density.

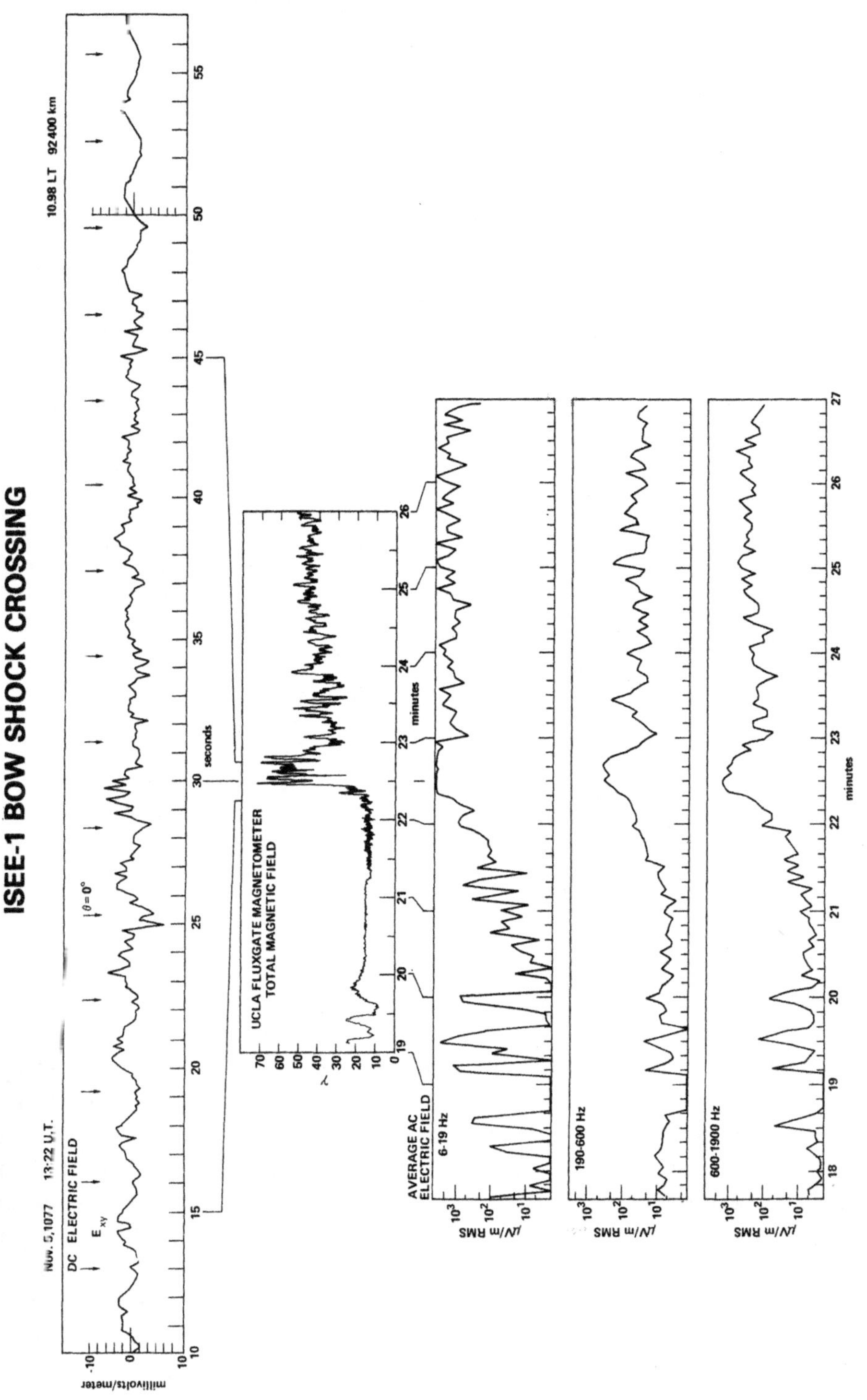

Fig. 4. Bow shock crossing (see text).

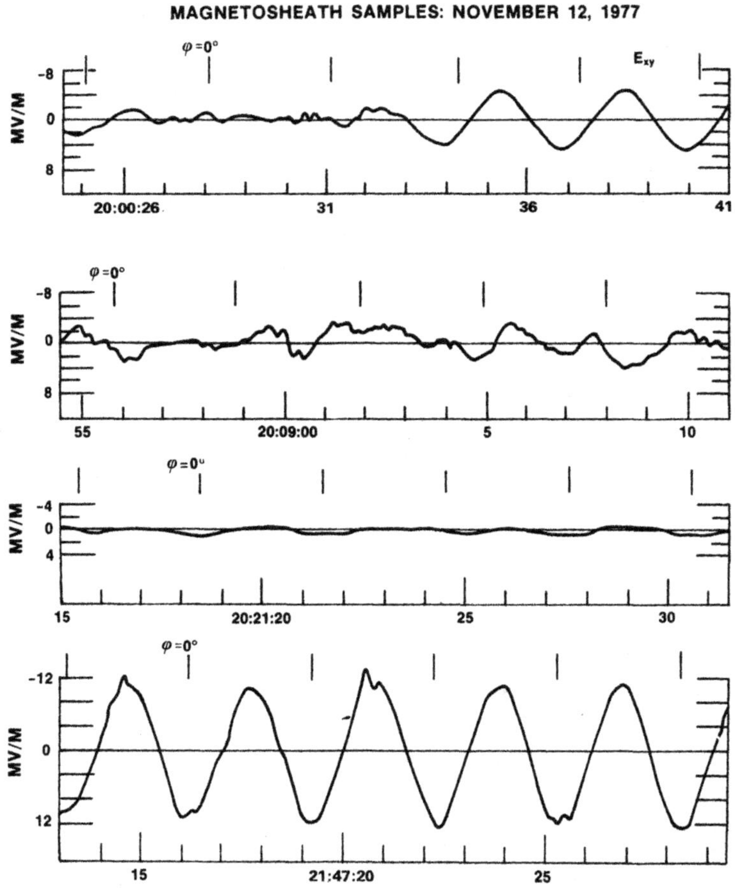

Fig. 5. Sampling of field variations in the magnetosheath.

5. Magnetopause Fields

The magnetopause crossing shown in Figure 6 near 22:58:55 is one of several that occurred more than an hour after the last Figure 5 example. This particular crossing is presented because it shows the strongest electric field observed near the sub-solar point. As shown, the field reached a magnitude of 23 V km^{-1} for about 0.1 s when the probes were at a ϕ angle of 298°. This, of course, does not reveal the maximum field during this 0.1 s interval as it would be fortuitous for the antenna-probe orientation to be parallel to the total electric field vector. Similar spikes of lesser magnitude also appear in this crossing at other angles; all, however, have components toward the Sun. In total, these imply the existence of sharp transients or, alternatively, narrow (e.g., 0.1 to 1.0 km) strong field shells or filaments.

Other interesting aspects of the Figure 6 example are: (a) the apparent stagnation in the magnetosheath between 22:58:34 and 22:58:50 (i.e., a 16 s interval), (b) the total duration of strong E between roughly 22:58:54 and 22:59:02 (i.e., an interval of <8 s), and (c) the abrupt transition to low plasma flux and spacecraft sheath fields between 22:59:01 and 22:59:03 (i.e., an interval of <2 s). Prior to the study of numerous crossings and examination of the simultaneous plasma and magnetic field signatures detailed interpretation would be premature. It does, however, appear clear from this example and others that near the sub-solar nose of the magnetosphere scale lengths for large changes in the electric field are considerably less than the scale lengths associated with the magnetic field structure of the magnetopause. In terms of the classical pictures of the microstructure of the magnetopause (e.g., see review by Willis, 1975) this implies an association between strong DC E-fields and the electron penetration, c/ω_{pe}, as opposed to associations with dimensions as large as the ion cyclotron radius. In viewing the large amplitude spikes of <0.2 s duration one must,

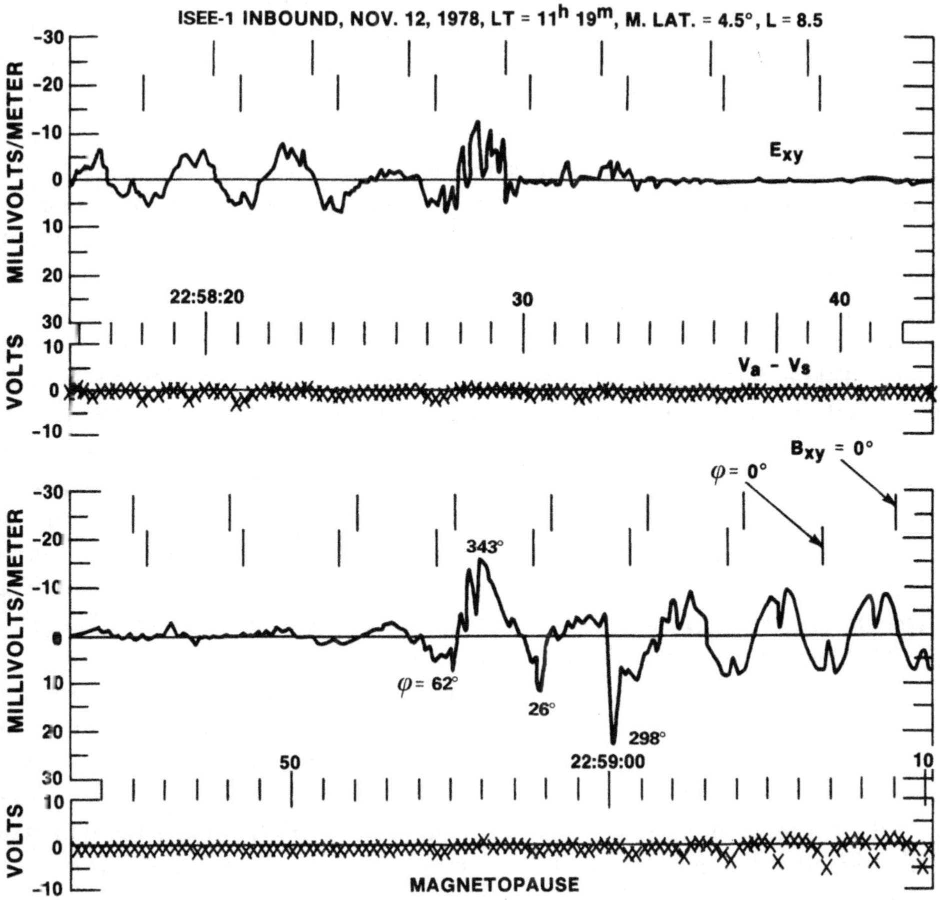

Fig. 6. Example of strong fields at the magnetopause (22:58:54 to 22:59:02) (see text).

however, keep two reservations in mind: (a) these features could be a unique form of time transient at the boundary that is not correctly interpreted in terms of spatial scale, and (b) if these spikes represent a spatial structure it is also possible that they are dimensionally thinner than c/ω_{pe}. These reservations may also be applicable to the spikes observed at the bow shock.

6. Mid-Night, Near-Tail Fields

In advance of routine tape processing, strip chart records were taken in real time on a number of passes in the midnight sector to examine signal characteristics while passing into and out of solar eclipse. The details which reveal the photo-electron character of the sheath will be presented elsewhere. The item of immediate interest is that measurements on some passes showed a greater diversity of ambient field behaviors than one would expect from extrapolation of previous low altitude (e.g., <2000 km) observations in the midnight auroral belt. Figure 7 provides a 'quiet' field reference near the magnetic equator. In this case the field was relatively stable and weak with $E_{xy} \doteq 0.1$ to $0.15\ \mathrm{V\ km^{-1}}$ but may have gradually changed direction as indicated by the phase relative to the $\phi = 0°$ direction (note: because the solar direction, $\phi = 0°$, is extrapolated from sunlit regions the phase angle could be in error). The small fluctuations of 0.01 to $0.02\ \mathrm{V\ km^{-1}}$ superimposed on the spin signal are believed to be ambient field fluctuations. In total there is no reason to suspect that extrapolation of this field to ionospheric levels would conflict with an assumption of equipotential mapping along magnetic lines.

Figure 8, in contrast to Figure 7, shows a diversity of field behaviors, all occurring within several minutes of time, during eclipse near a magnetic latitude of 22°. Taking into account reasonable values for the convergence of the flux tube at ionospheric levels the magnitudes, ranging from 0.1 to $4.0\ \mathrm{V\ km^{-1}}$, are not out of line with magnitudes that might be present at ionospheric levels. However, in terms of existing observations an ionospheric counterpart for the 'turbulent' or random appearance of this magnetospheric field is highly unlikely. From the scanning of data taken in the tail plasma sheet it is also apparent that the Figure 8 example is not unique in that similar randomness appears frequently. This diversity of field behavior is also somewhat analogous to the frequent changes in field behavior encountered in the magnetosheath.

At this writing, dominant or characteristic time and/or space scale factors for these field changes have not been identified. The middle trace of Figure 8, for example, is dominated by spatial scale factors comparable to the probe dimensions as indicated both by the lack of maximum amplitude at the spin period (3 s) and the existence of several spike type field changes. Rotations of the dominant DC field at rates comparable to the spin period are also present (top trace between 16:19:20 and 16:19:23) in the presence of strong fields. In total it appears safe to say that large field changes can be found over any time scale one

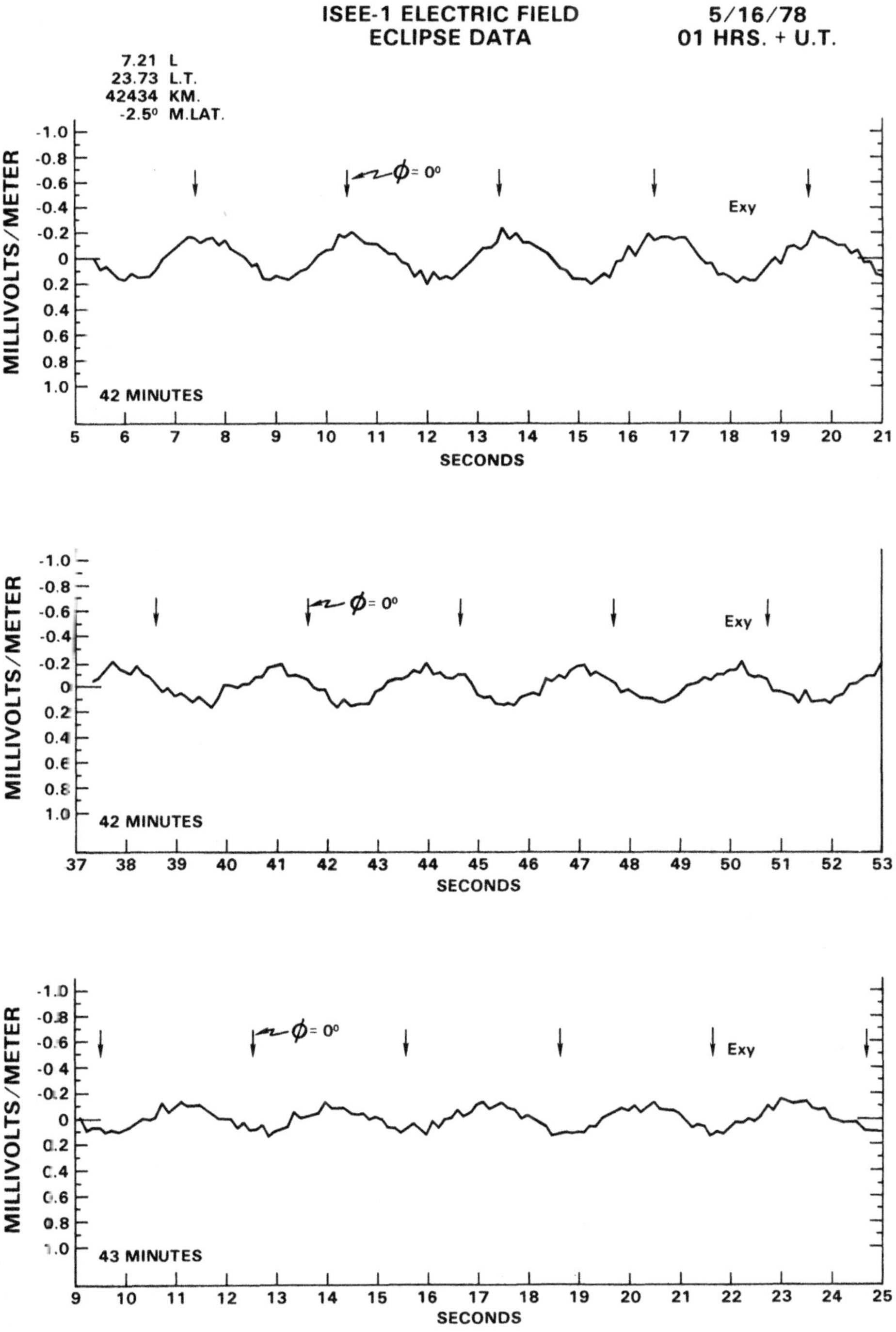

Fig. 7. Samples of a weak field in the midnight magnetosphere.

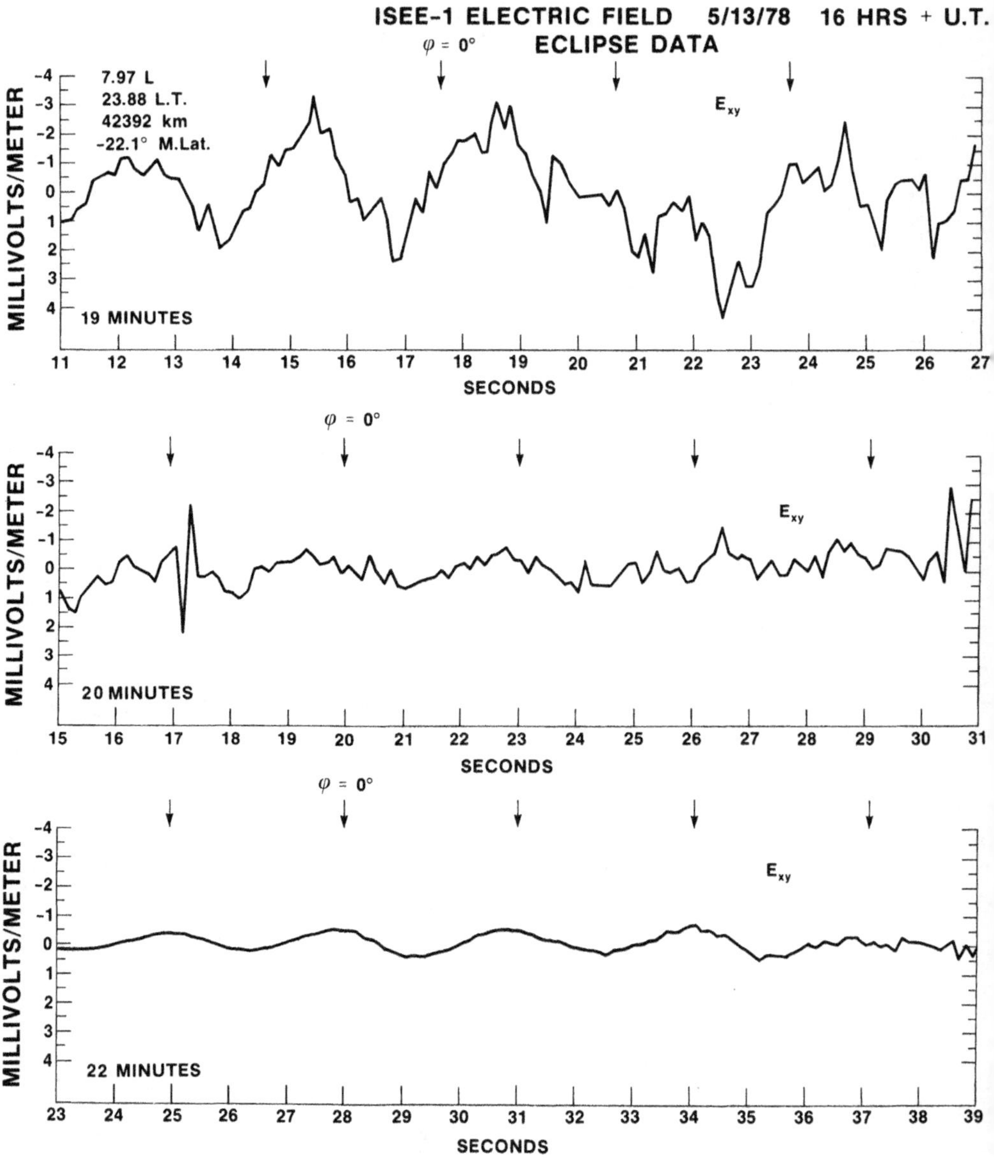

Fig. 8. Samples illustrating a diversity of field behaviors in the midnight magnetosphere.

chooses. Thus analysis techniques that will reveal the outer magnetosphere counterpart of the comparatively orderly patterns of the global ionospheric electric field are not presently apparent. This also implies that between the outer magnetosphere and the ionosphere the distribution of B-parallel components of the electric field, as viewed through surfaces transverse to B, must be highly complex.

References

Aggson, T. L. and Heppner, J. P.: 1964, 'A Proposal for Electric Field Measurements on ATS-1', NASA/GSFC, (preprint).

Aggson, T. L. and Heppner, J. P.: 1977, 'Observations of Large Transient Magnetospheric Electric Fields', *J. Geophys. Res.* **82,** 5155.

Fredricks, R. W., Crook, G. M., Kennel, C. F., Green, I. M., Scarf, F. L., Coleman, P. J., and Russell, C. T.: 1970, 'OGO-5 Observations of Electrostatic Turbulence in Bow Shock Magnetic Structures', *J. Geophys. Res.* **75,** 3751.

Heppner, J. P., Bielecki, E. A., Aggson, T. L., and Maynard, N. C.: 1978, 'Instrumentation for DC and Low Frequency Electric Field Measurements on ISEE-1', *IEEE Trans. Geosci. Electr.* **GE-16,** 253.

Heppner, J. P., Sugiura, M., Skillman, T. L., Ledley, B. G., and Campbell, M.: 1967, 'OGO-A Magnetic Field Observations', *J. Geophys. Res.* **72,** 5417.

Willis, D. M.: 1975, 'The Microstructure of the Magnetopause', *Geophys. J. Roy. Astron. Soc.* **41,** 355.

ELECTRIC FIELD MEASUREMENTS IN THE SOLAR WIND, BOW SHOCK, MAGNETOSHEATH, MAGNETOPAUSE, AND MAGNETOSPHERE

F. S. MOZER and R. B. TORBERT

Physics Department and Space Sciences Laboratory, University of California, Berkeley, Calif., U.S.A.

U. V. FAHLESON and C.-G. FÄLTHAMMAR

Department of Plasma Physics, The Royal Institute of Technology, Stockholm, Sweden

A. GONFALONE and A. PEDERSEN

Space Science Department, European Space Agency, Noordwijk, The Netherlands

Abstract. Electric field measurements are reported at 11 magnetopause crossings that occurred during a single in-bound ISEE-1 satellite pass near a local time of 10:30. In combination with magnetic field data, these measurements show the existence of electric field components tangential to the actual magnetopause in the frame of rest of the magnetopause on every crossing of the current carrying layers associated with the 11 magnetopause traversals. These tangential electric field components were oriented with respect to the magnetopause sheet currents such that there was an electrical power dissipation of between 30 and 110 W km^{-2} on 10 of the 11 crossings. These results are in agreement with requirements of reconnection theories. Histograms of the normal electric field components and of the orientation, velocity, and thickness of the current carrying layer are presented. Suggestions of the existence of a parallel electric field in the magnetosheath near the magnetopause and of propagation of large amplitude waves along the magnetopause are also made.

1. Introduction

The quasi-static electric field experiment on the ISEE-1 satellite has successfully measured electric fields in the solar wind, bow shock, magnetosheath, magnetopause, and magnetosphere. The instrument that performs these measurements consists of a pair of vitreous carbon spheres which are mounted on the ends of wire booms and which are separated by 73.5 m in the satellite spin plane by the centrifugal force associated with the satellite spin at a period of about 3 s. The potential difference between the pair of spheres is measured with a high input impedance voltmeter that is controlled by an 8080 type microprocessor. Further information on the instrument design and operation is given elsewhere (Mozer *et al.*, 1978a).

Observations of the electric field over a radial distance of 16 to 7 R_E on a single inbound satellite pass have been reported (Mozer *et al.*, 1978b). A significant feature of these data is the presence of a 1–2 mV m^{-1} dawn-to-dusk electric field over the entire region, which suggests that a tangential electric field component exists at the magnetopause. A detailed analysis of one magnetopause crossing confirmed the existence of significant tangential and normal electric fields in the

Space Science Reviews **22** (1978) 791–804. *All Rights Reserved*

frame of rest of the magnetopause and also produced estimates of the magnetopause orientation, velocity, thickness, and $\bar{I} \cdot \bar{E}$ power dissipation. It is the purpose of the present paper to expand on several features of this analysis, to apply it to all 11 magnetopause crossings on the orbit of interest, and to present histograms obtained from the 11 crossings of the various magnetopause parameters. This further analysis confirms the presence of a significant tangential electric field at the magnetopause and of power dissipation due to magnetopause currents which flow in the direction of the tangential electric field. These observations provide direct experimental evidence for the existence of processes crucial to magnetospheric dynamics and that have been described in reconnection (Vasyliunas, 1976) and other (Alfvén, 1968; Alfvén and Fälthammar, 1971) models.

2. Verification of Instrument Operation

Before discussing the magnetopause data, solar wind electric field measurements made earlier on the pass of interest will be presented in order to verify that electric field measurements have been made with an accuracy of $\sim 0.2 \text{ mV m}^{-1}$. In the collisionless solar wind it is expected that

$$\bar{E} + \bar{v} \times \bar{B} = 0, \quad (1)$$

where $\bar{v}$, the solar wind velocity, is largely in the $-X$ direction in the solar ecliptic frame of reference. Under the assumption that $\bar{v} = (v_X, 0, 0)$, Equation (1) requires that $E_Y = v_X B_Z$. Figure 1 presents plots of E_Y and $-B_Z$ during a 4-min interval in the solar wind that was randomly selected with the sole criterion being that B_Z varied during the period of interest. As expected from Equation (1), E_Y was proportional to $-B_Z$. This proportionality suggests that E_Y was measured

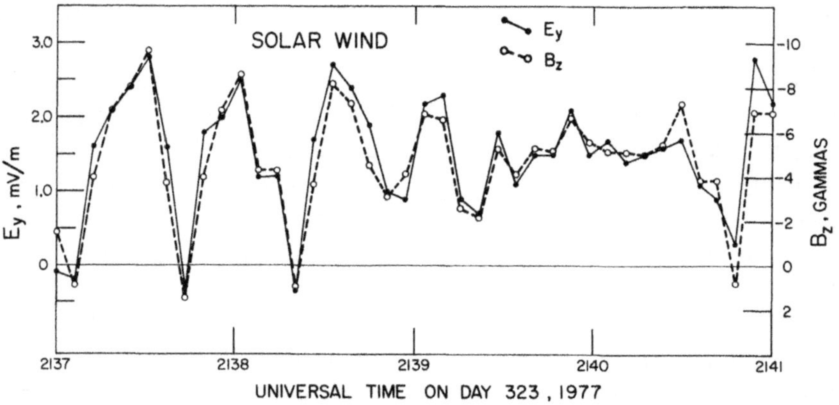

Fig. 1. Measurements of the solar ecliptic Y-component of electric field and Z-component of magnetic field in the solar wind during a four minute interval selected randomly around the sole criterion that the magnetic field exhibited amplitude variations during the selected time interval.

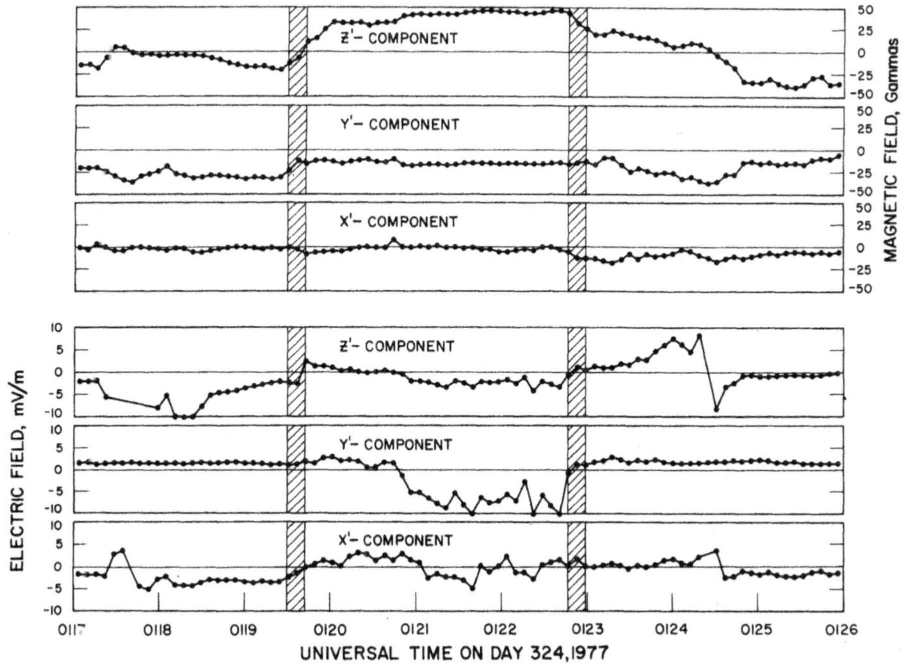

Fig. 2. Nine minutes of electric and magnetic field data plotted in a frame of reference oriented along the average magnetopause, during the passage of the magnetopause back and forth over the spacecraft.

with a sensitivity of $\sim 0.2\ \mathrm{mV\ m^{-1}}$. The ratio of the ordinate scales of E_Y and B_Z yields the solar wind velocity estimate, $v_X = -300\ \mathrm{km\ s^{-1}}$. This value compares favorably with the measured velocity, which varied between 260 and 320 $\mathrm{km\ s^{-1}}$ during the data interval (Paschmann, private communication, 1978). Thus, the magnitude of E_Y was probably measured to an accuracy of $\sim 5\%$.

3. Magnetopause Data and Analyses

A detailed discussion of the first magnetopause crossing back and forth over the spacecraft on day 324 of 1977 was presented in the earlier paper (Mozer *et al.*, 1978b). To illustrate and expand on certain features of the earlier analysis, and as a prelude to presenting results on all 11 magnetopause crossings during the orbit of interest, the second magnetopause crossing back and forth over the spacecraft will be discussed. In Figure 2, magnetic and electric field data during a 9-min time interval including crossings of major current carrying layers at the vertical shaded bars (as determined by changes in the components of the magnetic field) are presented. The prime coordinate system, in which the data of Figure 2 are presented, is rotated from the solar ecliptic frame such that the X'-axis is normal to the average magnetopause orientation at the satellite location (10:30 local time and 23° solar ecliptic latitude). The Y'-axis is in the plane of the average magnetopause and oriented in the dawn-to-dusk sense. As seen in the top curve of Figure 2, B_Z changed from negative to positive values at the current carrying layer near 01:19.6 UT. Because B_Z had its magnetospheric direction and magnitude while B_X and B_Y were small for about three minutes after 01:19.6 UT, the satellite must have been in the magnetosphere during the time interval bounded by the vertical shaded bars in Figure 2. Hence, the current carrying layer at 01:19.6 UT must have been the magnetopause and the magnetic field lines in the magnetosphere must have been oriented close to their average configuration.

The current carrying layer that existed for about 15 s before 01:23 UT is not so clearly identified as the outbound magnetopause, since it might be argued that the satellite was in the magnetopause for nearly two minutes before entering the magnetosheath somewhat before 01:25 UT. Such arguments are unimportant to the following discussions since the direction of the electric field at the major current carrying layers determines whether processes that are important to reconnection theories are occurring. Because the satellite exit from the magnetosphere near 01:23 UT has properties that differ from the other 10 events, its discussion will be postponed in favor of examining the more normal magnetospheric entry near 01:19.6 UT.

Electric field measurements at and near the magnetopause crossings are given in the bottom curves of Figure 2. The X' and Y' components result essentially from direct measurements while the Z' component results essentially from a calculation, because of the following reasons. The electric field booms rotate in the solar ecliptic X–Y plane within about 2°. Thus, the solar ecliptic X and Y components of the electric field have been obtained from sine wave least squares fits of 2 spin periods (about 6 s) of data. In the absence of electric field sensors aligned along the spin axis, the solar ecliptic Z-component of the electric field was computed from knowledge of the other electric and magnetic field components by requiring that the parallel electric field be zero. Since the $X'Y'Z'$ average magnetopause system differs from the XYZ solar ecliptic frame by relatively small rotations, it is essentially correct that the X' and Y' components of the electric field result from measurements, while the Z' component is the result of a calculation.

From the second curve from the bottom of Figure 2, it is seen that the dawn-to-dusk electric field component tangential to the average magnetopause orientation was significant and essentially constant in the magnetosheath, magnetopause, and magnetosphere during the entry near 01:19.6 UT. The change of $E_{Y'}$ to large negative values near 01:21 UT occurred well after the magnetopause passage when the satellite was inside the magnetosphere. This important electric field variation will be discussed in connection with the atypical magnetospheric exit occurring near 01:23 UT. Since it was not associated with the magnetospheric entry under discussion, it may be concluded from Figure 2 that the electric field tangential to the average magnetopause was significant and essentially constant just before, during, and just after the magnetosphere entry.

The above evidence for the existence of a tangential electric field at the magnetopause is not rigorously conclusive, both because the electric field component tangential to the average, and not the actual, magnetopuase was plotted, and because the measurement was made in the satellite frame of reference and not in a frame fixed to the magnetopause. To show the presence of an electric field tangential to the actual magnetopause in a frame fixed to the magnetopause, the measured field vectors must be transformed from the satellite to the magnetopause frame. The measured vectors in the satellite frame of reference are:

(a) the magnetosheath electric field, designated as $\bar{E}_{1S}$, which is the average of 30 s of data centered at 01:19.1 UT;

(b) the magnetosheath magnetic field, designated as $\bar{B}_{1S}$, which is the average of 30 s of data centered at 01:19.1 UT;

(c) the magnetospheric electric field, designated as $\bar{E}_{2S}$, which is the average of 30 s of data centered at 01:20.2 UT;

(d) the magnetospheric magnetic field, designated as $\bar{B}_{2S}$, which is the average of 30 s of data centered at 01:20.2 UT.

Transformation of these four vectors to the magnetopause frame of reference introduces three arbitrary parameters, which are the magnetopause speed normal to itself and relative to the spacecraft, and two angles that give the orientation of the unit vector normal to the actual magnetopause. As discussed earlier (Mozer *et al.*, 1978b), the requirements of Maxwell's equations in a planar, homogeneous, time stationary magnetopause model reduce the number of arbitrary parameters from 3 to 1 and, in addition, require that

$$(\bar{E}_{1S} - \bar{E}_{2S}) \cdot (\bar{B}_{1S} - \bar{B}_{2S}) = 0. \qquad (2)$$

Hence, the magnetopause orientation and velocity and the electric and magnetic field components in the frame of the magnetopause may be plotted as a function of the one arbitrary parameter, which is chosen to be the angle between the normal to the magnetopause and the vector $\bar{E}_{1S} - \bar{E}_{2S} \equiv \overline{\Delta E}$, in Figure 3. The effects of variations of this angle can be interpreted geometrically by recognizing that the requirement of a continuous normal component of $\bar{B}$ forces the vector $(\bar{B}_{1S} - \bar{B}_{2S})$ to be tangential to the magnetopause. Hence, given the magnetic field data, the only possible variations in the magnetopause orientation consist of rotations about an axis parallel to $(\bar{B}_{1S} - \bar{B}_{2S})$. Such rotations are represented by the abscissa of Figure 3. Since the major variation of $\bar{B}$ during the magnetopause crossing near 01:19.6 UT occurred in the Z'-component, variations of the abscissa correspond roughly to rotations of the magnetopause around the Z'-axis. For an abscissa value of either 0° or 180°, the vector $(\bar{E}_{1S} - \bar{E}_{2S})$ is normal to the magnetopause. Hence, for such angles, the tangential components of $\bar{E}$ are continuous across the magnetopause, so the satellite and magnetopause frames of reference must be one and the same. For other angles the magnetopause frame must move with respect to the satellite in order that the $\bar{v} \times \bar{B}$ term in the electric field transformation from the satellite to the magnetopause frame of reference can

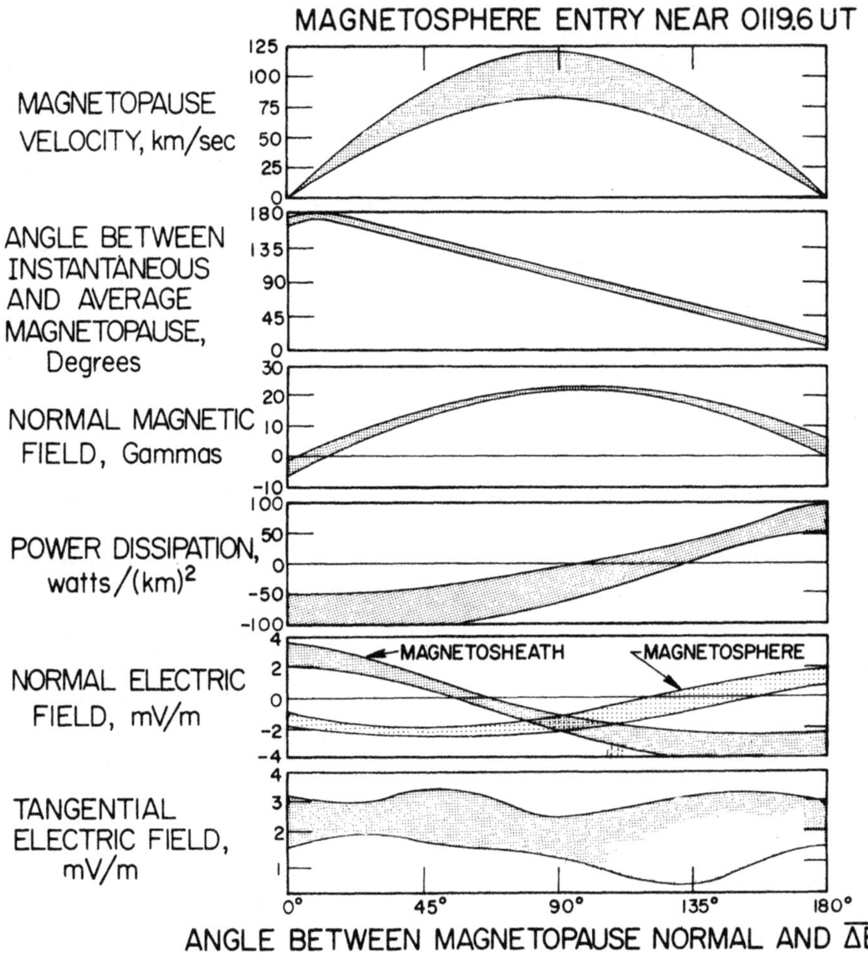

Fig. 3. Magnetopause parameters obtained from electric and magnetic field measurements on opposite sides of the magnetopause in a aplanar, homogeneous, time stationary model of the magnetopause. Positive normal components of all the fields are directed from the magnetosphere toward the magnetosheath.

cause the tangential components of $\bar{E}$ to be continuous in the magnetopause frame. Hence, the general characteristics of the top curve of Figure 3 can be understood.

Equation (2) is a requirement imposed on the measured fields by the model assumptions that the magnetopause is planar, homogeneous, and a time stationary frame. In fact, the fields measured at the magnetosphere entry near 01:19.6 UT did not satisfy this equation if E_Z in the solar ecliptic frame was obtained from the requirement that the parallel electric field in both the magnetosheath and the magnetosphere was zero. An interpretation of the failure to satisfy Equation (2) will be given in a later section. For the present, it is important to assess uncertainties in the curves of Figure 3 resulting from the fact that Equation (2) was not satisfied. The width of each curve of Figure 3 represents the range of values obtained by varying the measured fields within their uncertainties, by selecting E_Z in either the magnetosphere or magnetosheath to satisfy Equation (2), or by allowing Equation (2) to remain unsatisfied.

From the bottom curve of Figure 3 it is seen that no combination of allowed fields, magnetopause orientations, and magnetopause velocities produces a zero tangential electric field in the magnetopause frame of reference near 01:19.6 UT. Hence, the electric field in a frame of reference fixed to the magnetopause must have had a component tangential to the actual magnetopause. The magnitude of this component can be obtained by constraining the abscissa of Figure 3 via physical arguments. Since abscissa values greater than 90° correspond to an average $\bar{E} \times \bar{B}$ drift towards dawn on both sides of the magnetopause, and since the magnetopause passage occurred before noon (which is the region of expected flow towards dawn), the most likely value for the abscissa of Figure 3 is in the range greater than 90°. This parameter can be further constrained by assuming that the normal component of $\bar{B}$ at the magnetopause was small. This assumption combines with the third curve from the top of Figure 3 to suggest that the appropriate value of the abscissa was near 180°. Alternatively, one may arrive at the same conclusion by utilizing the second curve from the top of Figure 3 and the evidence from Figure 2 that the magnetopause may have had an orientation close to its average value, since the magnetic field lines in the magnetosphere were undistorted.

For an abscissa value near 180°, the bottom curve of Figure 3 suggests that the electric field component tangential to the actual magnetopause and in a frame of reference fixed to the magnetopause was 2.2 ± 0.7 mV m^{-1}. This tangential electric field had a component parallel to the magnetopause current (deduced from the change of $\bar{B}$) such that the $\bar{I} \cdot \bar{E}$ power dissipation (or magnetic field annihilation rate) at the magnetopause was 75 ± 25 W km^{-2} (see the third curve from the bottom of Figure 3). When integrated over the magnetopause (a circle of radius $10 R_e$), this rate of power dissipation produces a total power of 10^{12} W that must be deposited in magnetopause particles if the local heating rate is present over the front of the magnetopause.

The pair of curves in the second panel from the bottom of Figure 3 show the electric field component normal to the magnetopause pointed outward and had a magnitude of 1–2 mV m^{-1} in the magnetosphere and pointed inward and had a magnitude of 2–4 mV m^{-1} in the magnetosheath. These field directions correspond to $\bar{E} \times \bar{B}$ flows towards dawn on both sides of the magnetopause, as is expected for a pre-noon crossing.

The top panel of Figure 3 shows that the speed of the magnetopause relative to the satellite was less than about 25 km s^{-1} for an abscissa value near 180°. Since the magnetic field change took about 30 s (see Figure 2), the magnetopause thickness must have been less than about 750 km, or about 7 gyroradii of a 1 keV proton. It is emphasized that the entity whose thickness is estimated in this way is the current carrying layer, which may or may not correspond to the magnetopause, depending on definitions and the detailed nature of the crossing.

4. Histograms of Magnetopause Parameters

Eleven crossings of current carrying layers identified with the magnetopause were made during a 45 min period of the ISEE-1 satellite's in-bound orbit 12 (see Figure 1 of Mozer *et al.*, 1978b). These crossings have been analyzed as in the preceding section to prove histograms of the various magnetopause parameters which are presented below. To obtain these parameters, the angle between the magnetopause normal and the vector $(\bar{E}_{1S} - \bar{E}_{2S})$ (which is the abscissa of Figure 3) was chosen to minimize the normal component of the magnetic field at the magnetopause.

In Figure 4 a histogram of the angle between the normals to the actual and the average magnetopauses is presented. In most cases the magnetopause was found to be oriented within 30° of its average orientation, while in 1 atypical case, it was apparently tilted by more than 90°.

In Figure 5, the histogram of the 11 values of the tangential electric field in the frame of rest of the magnetopause is presented. Except for the atypical case of Figure 4 which had an apparent tangential field of nearly 5 mV m^{-1}, the many tangential field estimates cluster about a mean value of 1.6 mV m^{-1}. Estimates of the $\bar{I} \cdot \bar{E}$ power dissipation associated with these tangential fields, and which are presented in Figure 6, also cluster about an average value of 63 W km^{-2}, with the exception of the same atypical single case. Thus, one may conclude that at 10 of the 11 magnetopause crossings, similar and important tangential electric fields and electric power dissipations were observed.

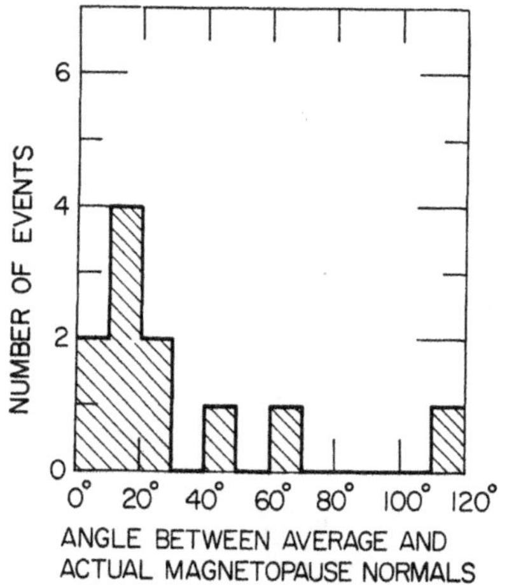

Fig. 4. Histogram of the angles between the average and actual magnetopause normals for the 11 magnetopause crossings.

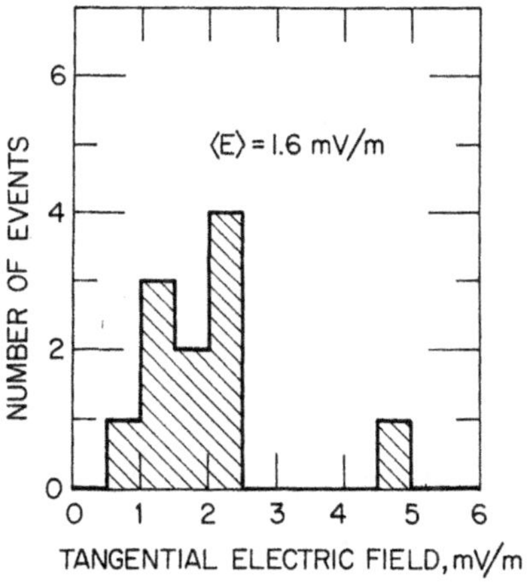

Fig. 5. Histogram of the electric field components tangential to the actual magnetopause in the frame of rest of the magnetopause determined from the 11 magnetopause crossings.

The magnitudes and directions of the normal components of the electric field on the magnetospheric and magnetosheath sides of the magnetopause are presented in Figures 7 and 8. Again with the exception of the single atypical case, the normal electric field on the magnetospheric side of the magnetopause pointed outward in every case, corresponding to an $\bar{E} \times \bar{B}$ flow towards dawn. While some of the normal electric field components on the magnetosheath side of the magnetopause were positive, the average normal field in the magnetosheath was negative, which corresponds to an average $\bar{E} \times \bar{B}$ flow towards dawn in the magnetosheath also.

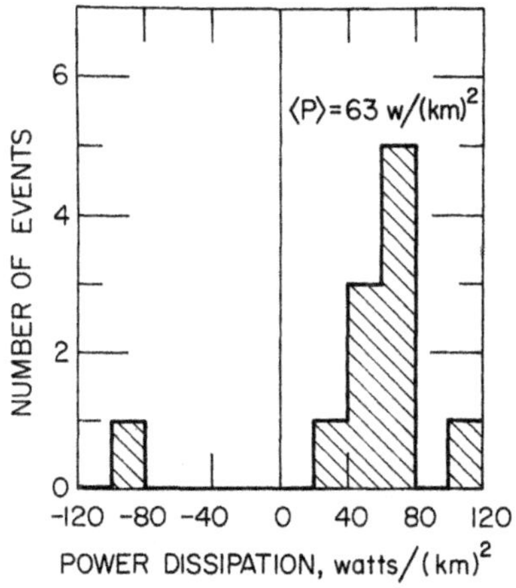

Fig. 6 Histogram of the $\bar{I} \cdot \bar{E}$ power dissipations at the magnetopause determined from the 11 magnetopause crossings.

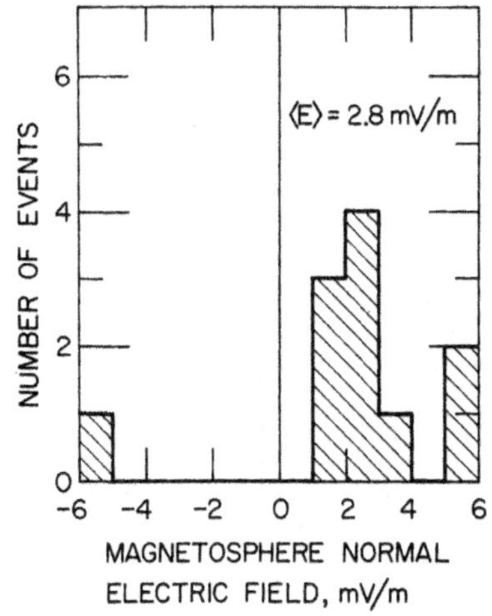

Fig. 7. Histogram of the electric field components normal to the magnetopause on the magnetospheric side of the boundary as determined from the 11 magnetopause crossings.

A histogram of the magnetopause velocity estimates is given in Figure 9. With the exception of the one atypical case, the velocities were all less than 30 km s^{-1}. Because the condition of minimum normal magnetic field used to determine the abscissa of Figure 3 resulted in values of the abscissa near 180° for most cases (as it was in Figure 3), the estimate of the magnetopause velocity at most crossings was uncertain by a factor of about two.

The thicknesses of the current carrying layers obtained from the appropriate time of crossing and magnetopause velocity are presented in Figure 10. Neglecting the one large estimate, the average thickness of the current carrying layer was

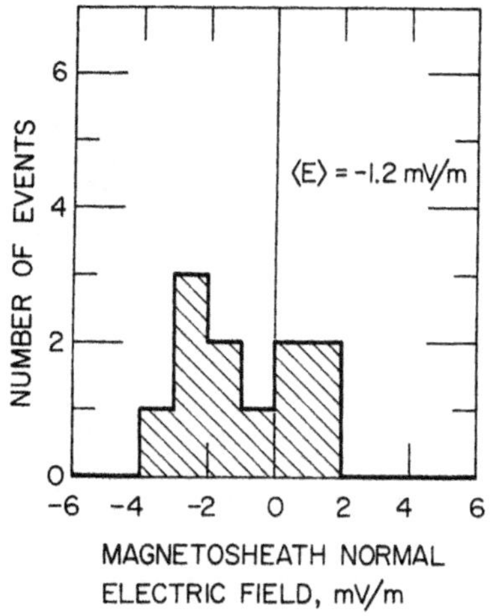

Fig. 8. Histogram of the electric field components normal to the magnetopause on the magnetosheath side of the boundary as determined from the 11 magnetopause crossings.

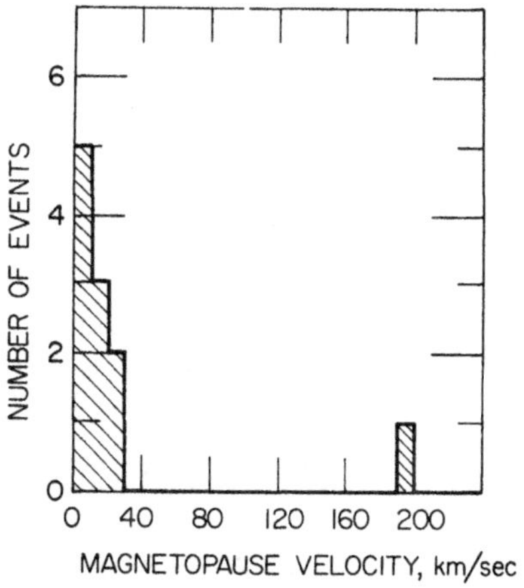

Fig. 9. Histogram of the magnetopause speeds for the 11 magnetopause crossings.

found to be 270 km, with an uncertainty of about a factor of 2 due to the above discussed uncertainty in the magnetopause velocity. Hence, the typical thickness of the current carrying layer was the order of several gyroradii of a 1 keV proton (which is about 100 km in the observed magnetospheric magnetic field).

A histogram of the angle between $\Delta E = (\bar{E}_{1S} - \bar{E}_{2S})$ and $\Delta B = (\bar{B}_{1S} - \bar{B}_{2S})$ is given in Figure 11. The requirement from Equation (2) that the angle between these vectors be 90°, is not satisfied by the experimental data. This discrepancy is not easily explained by relaxing the assumptions of the model that resulted in Equation (2). For example, explaining the angles of Figure 11 by relaxing the assumption that the magnetopause frame was a time stationary frame requires that the magnetic field in the magnetopause frame changed by about $10\ \gamma\ s^{-1}$,

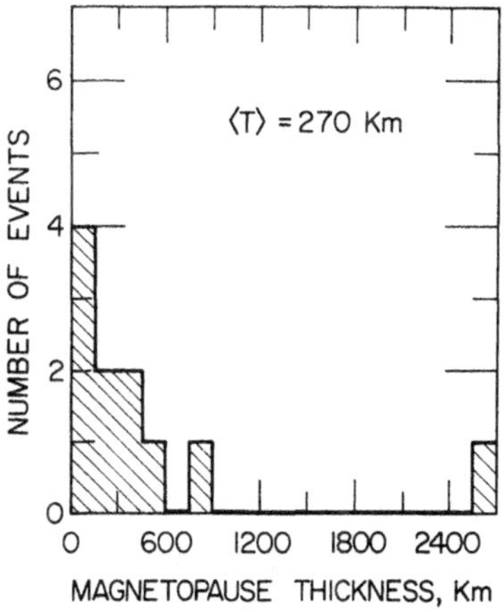

Fig. 10. Histogram of the thickness of the current carrying layers associated with the 11 magnetopause crossings.

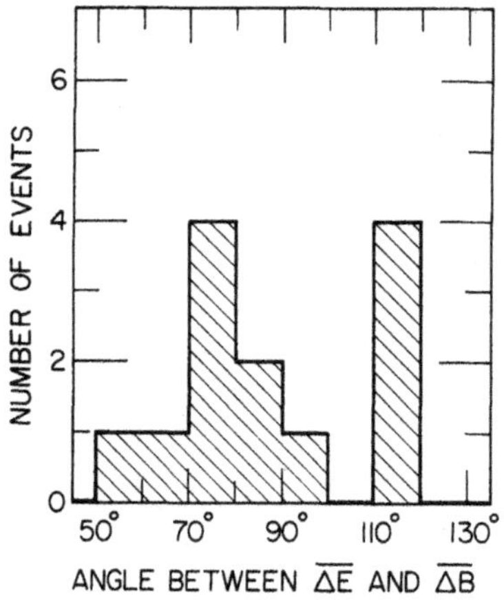

Fig. 11. Histogram of the angles between the vectors $\overline{\Delta E} = (\bar{E}_{1S} - \bar{E}_{2S})$ and $\overline{\Delta B} = (\bar{B}_{1S} - \bar{B}_{2S})$ determined from the electric and magnetic field data on both sides of the magnetopause during the 11 crossings.

which is unreasonably large. Relaxing the assumption that the fields did not vary with distance along the magnetopause allows explanation of Figure 11 if the normal electric field changed by an amount comparable to its magnitude in a distance along the magnetopause the order of the thickness of the current carrying layer. Since such large variations are also unlikely, the remaining possibility is that the unmeasured Z-component of electric field in the solar ecliptic frame of reference was estimated incorrectly from the assumption that the parallel electric field was everywhere zero. Figure 12 presents a histogram of the parallel electric fields in the magnetosphere that are required to satisfy Equation (2) under the

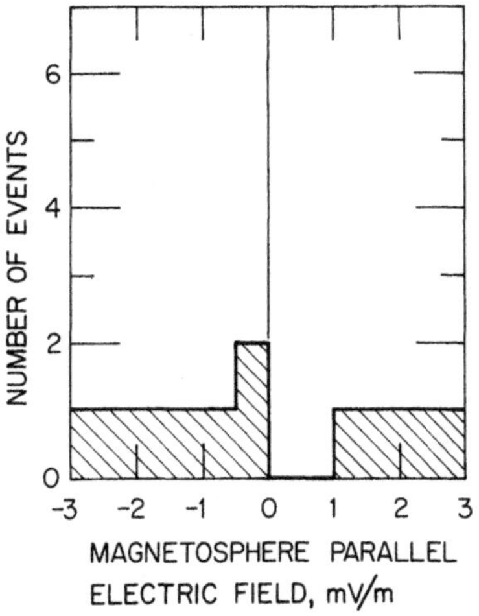

Fig. 12. Histogram of the values of the parallel electric field required in the magnetosphere in order to make the vector change of electric field perpendicular to the vector change of magnetic field at the 11 magnetopause crossings.

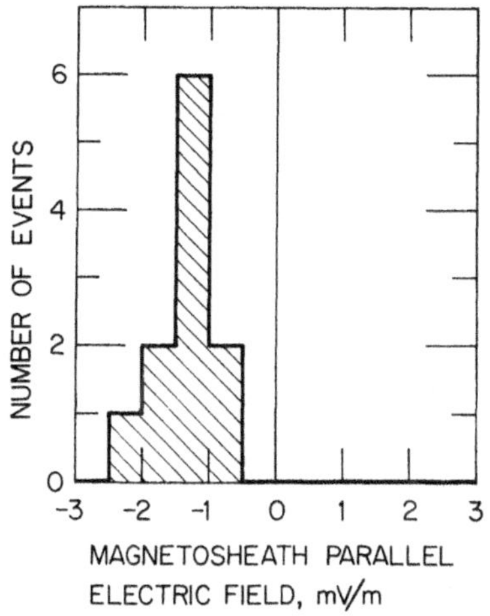

Fig. 13. Histogram of the values of the parallel electric field required in the magnetosheath in order to make the vector change of electric field perpendicular to the vector change of magnetic field at the 11 magnetopause crossings.

hypotheses that the model assumptions are valid and that the parallel electric field in the magnetosheath was zero. The nearly uniform distribution of resultant parallel field values provides little credence to this explanation.

Figure 13 presents a histogram of the parallel electric fields in the magnetosheath that are required to satisfy Equation (2) under the hypotheses that the model assumptions are valid and that the parallel electric field in the magnetosphere was zero. In this case, the data points cluster tightly about an average value of about $-1\ \text{mV}\ \text{m}^{-1}$. Such a clustering is unlikely to occur by chance. For example, the probability that a random selection of 11 numbers would all be negative is 1/2048. That the numbers would all be negative and also cluster about a single value as they do in Figure 13 might occur by chance with an even lower probability. Hence, there is statistically significant experimental evidence that a parallel electric field of about $-1\ \text{mV}\ \text{m}^{-1}$ may have existed near the magnetopause in the magnetosheath on the day of interest.

5. The Atypical Case

Most of the histograms in Figures 4 through 10 exhibit a single anomalous value that has been referred to as the atypical case. The boundary crossing associated with this case is the magnetospheric exit that occurred near 01:23 UT in Figure 2. The major atypical feature of the field data is the existence of a large negative Y'-component of electric field before the crossing and the large change of this component to its nominal value at the crossing. Such a large field change requires that the normal to the current carrying layer was qualitatively oriented largely in the Y'-direction in order that the tangential component of the electric field was continuous. This result explains the $>90°$ angle between the average and actual magnetopauses illustrated in Figure 4.

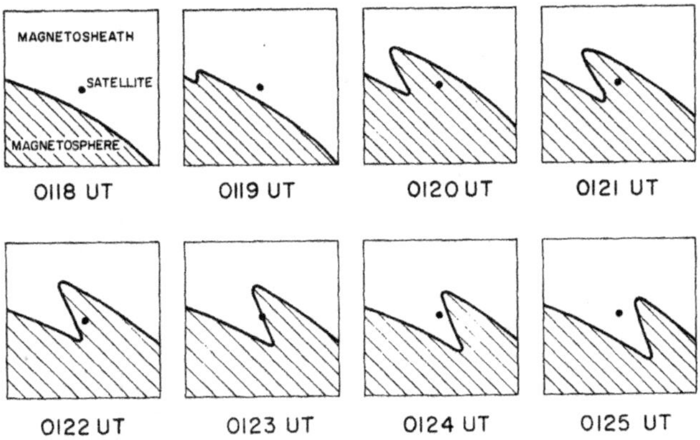

Fig. 14. Model of a possible magnetopause configuration and motion that explains the field data obtained at the magnetopause exit which occurred near 01:23 UT.

One possible interpretation of the atypical orientation and velocity of this current carrying layer and of the fields measured near it is given in Figure 14, which is a sequence of sketches illustrating the local magnetopause orientation near the spacecraft at one minute intervals. In these sketches the outbound passage of the magnetopause before 01:20 UT was a normal expansion which was followed by the propagation of an irregularity along the magnetopause at a speed of a few hundred km s^{-1}. The orientation and propagation of this irregularity were such that the fields and flows near its boundary were largely inverted and the current carrying layer encountered at the satellite exit deviated by more than 90° from its average orientation.

Acknowledgements

This work was supported under NASA contract NAS 5-20080 and by the European Space Agency and the Swedish Board of Space Activities.

References

Alfvén, H.: 1968, 'Some Properties of Magnetospheric Neutral Surfaces', *J. Geophys. Res.* **73,** 4379.

Alfvén, H. and Fälthammar, C.-G.: 1971, 'A New Approach to the Theory of the Magnetosphere', *Cosmic Electrodynamics* **2,** 78.

Mozer, F. S., Torbert, R. B., Fahleson, U. V., Fälthammar, C.-G., Gonfalone, A., and Pedersen, A.: 1978a, 'Measurements of Quasi-Static and Low Frequency Electric Fields with Spherical Double Probes on the ISEE-1 Spacecraft', *IEEE, Trans. Geosci.* **G E -16,** 258.

Mozer, F. S., Torbert, R. B., Fahleson, U. V., Fälthammar, C.-G., Gonfalone, A., and Pedersen, A.: 1978b, 'Direct Observation of a Tangential Electric Field Component at the Magnetopause', submitted to *Geophys. Res. Letters.*

Vasyliunas, V. M.: 'Theoretical Models of Magnetic Field Line merging, 1', *Rev. Geophys. Space Phys.* **14,** 411.

THREE-DIMENSIONAL ENERGETIC ION OBSERVATIONS AND THE TANGENTIAL MAGNETOPAUSE ELECTRIC FIELD

D. J. WILLIAMS and T. A. FRITZ
NOAA Space Environment Laboratory, Boulder, Colo, U.S.A.

E. KEPPLER and B. WILKEN
Max Planck Institute for Aeronomy, Lindau, F.R.G.

G. WIBBERENZ
University of Kiel, Kiel, F.R.G.

Abstract. Three-dimensional distributions for 24.0–44.5 keV protons (ions) are presented from the ISEE-1 medium energy particles instrument during a magnetopause traversal at ~01:10 UT on 20 November 1977. Local time of the traversal was ~10:30. Ion fluxes were observed coming generally from the subsolar region, but over a wide range of latitudes. Enhanced fluxes were observed at the magnetopause crossing with strong components from the subsolar region and from the $+Z_{SE}$ direction. These observations are compared with the simultaneous electric field observations presented by Mozer *et al.* (1978). Ion streaming in a direction consistent with the Y-component of the $\bar{E}\times\bar{B}/_{B^2}$ drift velocity was observed whereas streaming along the X and Z-components is not seen. Based on energy arguments we conclude that in this case, $\geqslant 24$ keV ions are not the major energy carrier of the locally measured $\bar{I}\cdot\bar{E}$ dissipation.

1. Introduction

In this report we present data from the ISEE-1 portion of the medium energy particles experiment flown on ISEE-1 and -2 (Williams *et al.*, 1978). Briefly, the ISEE-1 instrument uses a focusing magnet and solid state detectors to separate and measure, in sixteen energy channels, protons (ions) from 24 keV to 2000 keV and electrons from 24 keV to 1000 keV.

The detection head is mounted on a scan platform which continuously scans in polar angle relative to the satellite spin axis which, in turn, is oriented normal to the ecliptic plane. This angular scan is synchronized to the satellite spin rate and takes 12 spins to scan an angular range from 10° to 170° at which point the scan is reversed. During the scan, 8 samples per spin (in low bit rate) are made in azimuth angle around the spin axis, thus giving a three-dimensional sampling (96 points) of the unit sphere every 12 spins or ~36 s. The geometry is shown in Figure 1.

2. Data

ISEE-1 encountered several magnetopause crossings at ~10:30 LT on November 20, 1977, from 01:00–02:00 UT (Mozer *et al.*, 1978). Particle and magnetic field signatures of these crossings are shown in Figure 2. The particle

Space Science Reviews **22** (1978) 805–811. *All Rights Reserved*

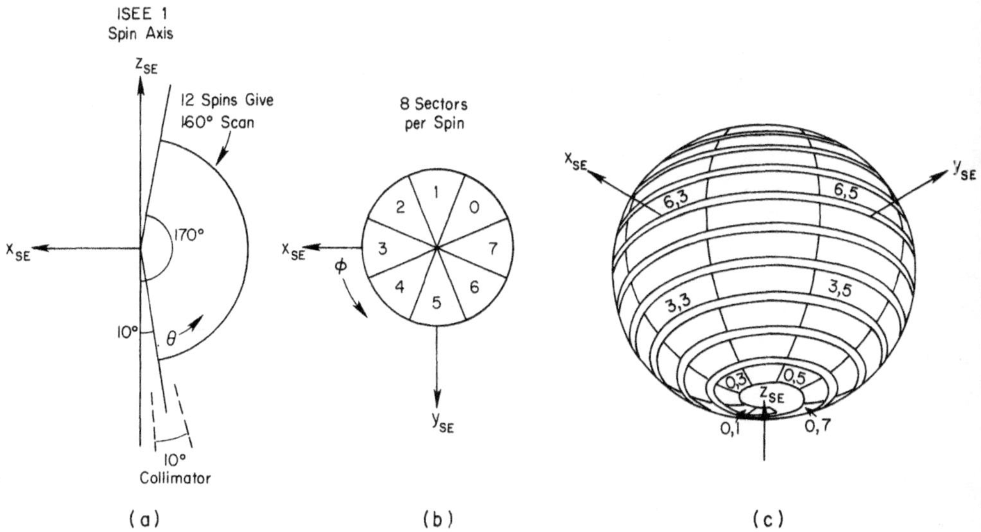

Fig. 1. Scanning geometry for the ISEE-1 medium energy particle instrument. (a) Scan geometry in the X_{SE}, Z_{SE} plane. (b) Low bit rate sectoring in the X_{SE}, Y_{SE} plane. Sixteen sectors are obtained in the high bit rate. (c) Perspective of unit sphere sampling resulting from the geometry of (a) and (b). Each full box represents the 10° detector cone angle plus the 45° azimuthal sample. Eight energy channels in low bit rate and sixteen in high bit rate are obtained in each sample for ions (24–200 keV) and electrons (24–1000 keV). Numbers (M, N) indicate the position of the sample in terms of spin number (M) and sector number (N).

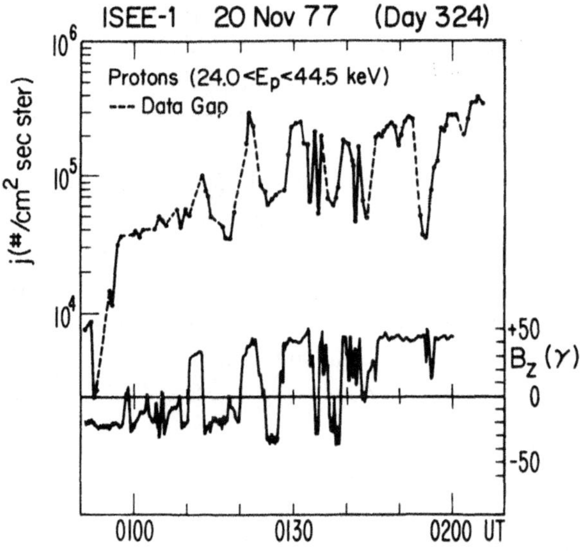

Fig. 2. Peak fluxes of 24–44.5 keV ions and B_z magnitude versus time for ISEE-1 magnetopause encounters on 20 November 1977. Peak ion fluxes are shown at the centre time of the instrument's ~36 s scan cycle.

responses for each of these crossings have been analysed individually as significant differences are observed. The crossing at ~01:45 has been used to infer magnetopause velocities and the existence of boundary waves (Williams, 1978a). In this note we show data from the crossing at ~01:10 for which Mozer *et al.* (1978) have reported simultaneous electric field data.

Figure 3 shows the magnetopause crossing at ~01:10 in more detail. The components of the electric field, the magnetic field, and the $\bar{E} \times \bar{B}/_{B^2}$ velocity (from Mozer *et al.*, 1978) are plotted in a coordinate frame such that the X component is along $\bar{n}$, the magnetopause normal at the location of the spacecraft, and Y is tangential to the average magnetopause and in the dawn-to-dusk direction. Particle fluxes are shown in the 'grey code' diagrams at the top of the figure. The data are shown as individual ~36 s scans of the unit sphere. Each scan shows 24.0–44.5 keV ion fluxes in a polar angle, θ, versus azimuth angle, ϕ, format. These unit sphere displays are referenced to a solar ecliptic system such that $\phi = 0$, $\theta = 90°$ corresponds to X_{SE}, $\phi = 90°$, $\theta = 90°$ corresponds to Y_{SE} and $\theta = 180°$ corresponds to Z_{SE}. The solid and dashed lines, obtained from the magnetic field vector measured at the centre time of the 36-s scan give, respectively, the locus of 90° and ±60° local pitch angles.

Unfortunately, three samples are missing due to electron density transmitter interference at the time ISEE-1 entered and remained in the magnetosphere (from ~01:10 to 01:12). However it is clear that as ISEE-1 left the magnetosphere (particle scans 01:12:09, 01:12:47 and 01:13:22) enhanced ion fluxes were observed to come from the $+Y_{SE}$ and $+Z_{SE}$ directions. The arrival of a steady flux of 24.0–44.5 keV ions from the $+Y_{SE}$ direction is consistent with the generally negative Y component of the $\bar{E} \times \bar{B}/B^2$ velocity shown in the figure. However the enhanced fluxes near ~90° local pitch angle are also consistent with a particle density gradient at the magnetopause for the satellite local time of ~10:30 Williams (1978a).

3. Results

The ion spectrum observed by ISEE-1 in the direction of maximum fluxes seen in the 01:12:09 panel at the top of Figure 3 is shown in Figure 4. Integrating this spectrum over energy and angle yields an energy flux into our unit sphere of ~15 W km^{-2}. The source of this energy flux is located at some unknown distance to the west and north of the satellite. Mozer *et al.* (1978) report a measured local $\bar{I} \cdot \bar{E}$ dissipation of about 65 W km^{-2}. This is presented schematically in Figure 5.

It is not clear how (and it may not be possible) to relate local power dissipation to energy flow from a distant source. Mozer *et al.* (1978) obtain a total dissipation of 10^{12} W by using a magnetopause source surface represented by a circle of $10 R_E$ radius. As ISEE-1 is about $5 R_E$ from the subsolar point and we do not see significant ion fluxes from the $-Y_{SE}$ hemisphere, the source surface probably should have a radius of $\leqslant 5 R_E$ and Mozer's figure of 10^{12} W should be considered an upper limit.

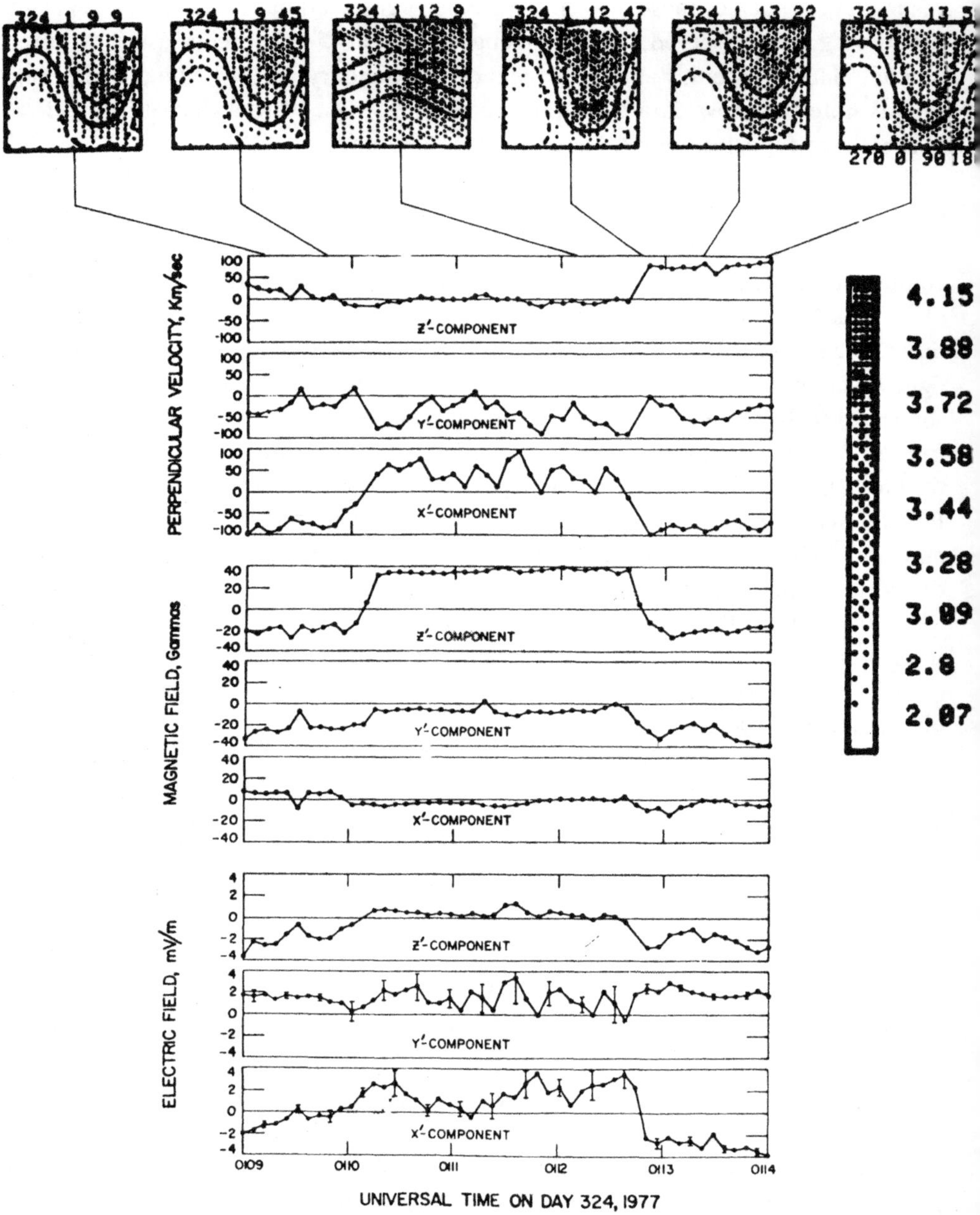

Fig. 3. Details of the magnetopause crossing at about 01:10 on 20 November 1977. Electric field, magnetic field and velocity data are from Mozer *et al.* (1978). Data showing energetic ion fluxes in 'grey code' format are shown at the top of the figure. Intensities of 24–44.5 keV protons (ions) are shown in plots with coordinates of polar angle (θ) as the vertical axis and azimuthal angle (ϕ) as the horizontal. The solid and dashed lines show, respectively, the locus of 90° and ±60° local pitch angles as determined from the measured magnetic field vector obtained at the centre time of the ~36 s scan. The centre time of each scan is shown at the top of each panel in day of the year, hour, minute and second. The grey code used in the top panels gives the log of the differential flux as shown in the scale.

If we assume a shell with thickness $\sim 0.5 R_E$ and radius of $5 R_E$ and integrate the ion flux through it we obtain a total energy flow of $\sim 10^{10}$ W. Thus it does not appear in this case that ions of energy $\geqslant 24$ ke V are major carriers of the power dissipation implied by the measurement of Mozer *et al.* (1978).

We have shown in Figure 4 ion fluxes observed in the magnetosheath at ~17:00 LT observed by Helios-1 on 10 December 1974 and Helios-2 on 15 January 1976 (Richter *et al.*, 1978). It is interesting to notice that all three spacecraft observe similar spectra and that ion fluxes are observed streaming roughly parallel to the magnetopause and away from the subsolar region. The ISEE observations (~10:30 LT) also show a strong streaming component from above the spacecraft, i.e., from the $+Z_{SE}$ direction. The total magnetosheath energy flow obtained from the ISEE measurements is ~ two orders of magnitude smaller than that obtained from the Helios measurements. This may be due to possible time variations and the smaller integrated magnetosheath area available for ISEE.

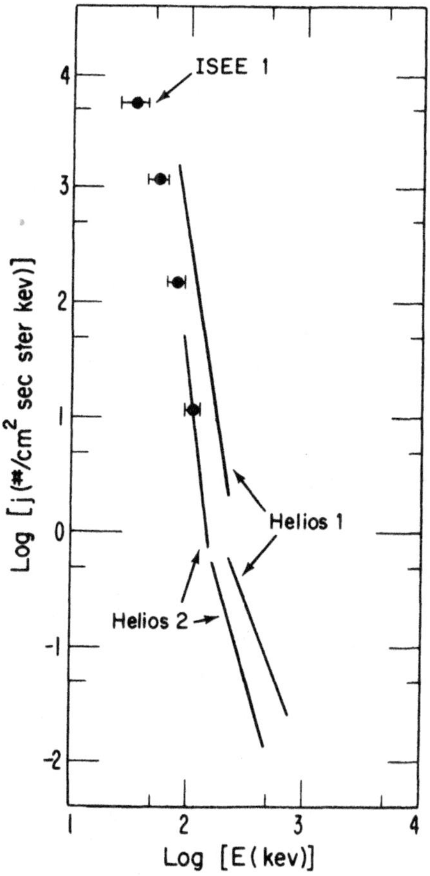

Fig. 4. Energetic ion spectrum observed during peak fluxes encountered in the ISEE-1 magnetopause crossing of about 01:10 on 20 November 1977. The local time was about 10:30. Data from Helios-1 and -2 from 17:00 LT are shown for comparison (Richter *et al.*, 1978).

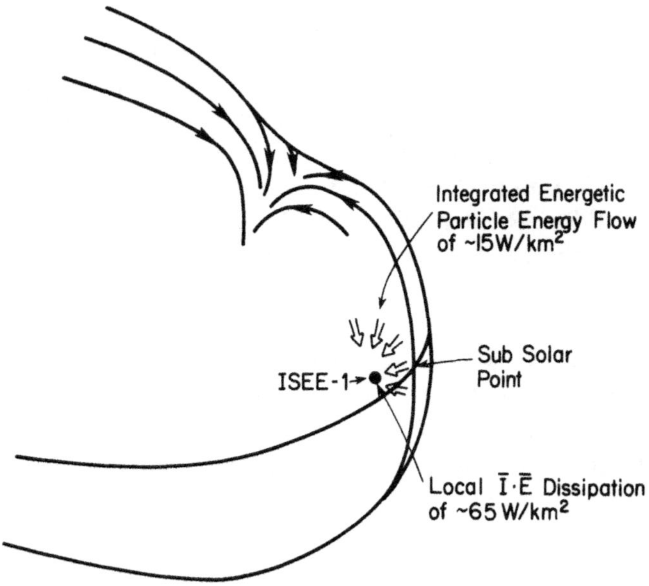

Fig. 5. Schematic showing the energetic ion energy flow and local $\bar{I}\cdot\bar{E}$ dissipation observed by ISEE-1 during the magnetopause crossing at about 01:10 on 20 November 1977.

The spectra shown in Figure 4 are also similar to the two point spectra obtained from IMP-7 for $\geqslant 50$ keV magnetosheath ions streaming in an antisolar direction and parallel to the magnetopause at $35 R_E$ along the flanks of the magnetopause (Williams, 1978b). Again, due to possible time variations and the smaller integrated magnetosheath area, the total energy flow obtained from the ISEE observations is two orders of magnitude smaller than that obtained from the IMP-7 observations.

4. Summary and Conclusions

Three-dimensional energetic ($\geqslant 24$ keV) ion observations were obtained from the ISEE-1 energetic particles instrument during a magnetopause crossing at ~10:30 LT on 20 November 1977. These observations show:

(1) A steady flux of 24–44.5 keV ions arriving at the spacecraft from the direction of the subsolar region as the magnetopause is approached. The arrival is independent of magnetic field orientation over a wide range ($\geqslant 120°$) of latitude (see the top panels of Figure 3).

(2) As the magnetopause is crossed, enhanced fluxes of 24–44.5 keV ions are observed arriving from the general direction of the subsolar region. During these enhanced fluxes, a strong component is observed arriving from high northern latitudes, i.e., from the $+Z_{SE}$ direction.

(3) The flux component arriving from low latitudes has a direction consistent with that implied by the simultaneously observed y-component of the $\bar{E}\times\bar{B}/_{B^2}$

drift velocity (Mozer *et al.*, 1978). Streaming consistent with the X- and Z-components of the drift velocity is not observed.

(4) The total energetic ion energy flow into a unit sphere at the satellite position is $\sim 15\ \mathrm{W\,km^{-2}}$ ($0.015\ \mathrm{ergs\,cm^{-2}}$).

(5) ISEE-1 was $\sim 5R_E$ from the subsolar point. As no significant energetic ion fluxes were observed from the $-Y$ hemisphere, the source region was likely to be $\leqslant 5R_E$ from the subsolar point in the dawn direction.

(6) Mozer *et al.* (1978) measure a local $\bar{I} \cdot \bar{E}$ dissipation of about $65\ \mathrm{W\,km^{-2}}$. It is not clear how (and it may not be possible) to relate this local dissipation to the energy flow from a remote source such as was simultaneously observed by our energetic particle instrument.

(7) The ion spectra observed by ISEE are similar to spectra observed by the Helios (Richter *et al.*, 1978) and IMP-7 (Williams, 1978b) spacecraft in other magnetosheath locations. The streaming observed by all three spacecraft is also similar in that the streaming direction is parallel to the magnetopause and away from the subsolar region.

(8) Integrating over a reasonable but arbitrary magnetosheath area, a total energy flow of $\sim 10^{10}$ W is obtained from the ISEE-1 energetic particle results. This is about 2 orders of magnitude smaller than the IMP-7 results and may be due to time variations and/or simply to the larger (and also arbitrary) magnetosheath area of integration used for Helios and IMP-7. The total magnetopause dissipation of about 10^{12} W obtained by Mozer *et al.* (1978) possibly may be reconciled with the $\sim 10^{10}$ W simultaneously observed in our particle fluxes by using more accurate source region geometries, the creation of narrow, jetting particle beams (which were not observed) and consideration of other energy sinks such as plasma heating, wave generation, magnetopause compression, etc. However in this case it does seem unlikely that energetic ($\geqslant$24 keV) ions were the major energy carrier of the $65\ \mathrm{W\,km^{-2}}$ $\bar{I} \cdot \bar{E}$ dissipation observed by Mozer *et al.* (1978).

References

Mozer, F. S., Torbert, R. B., Fahleson, U. V., Fälthammar, C.-G., Gonfalone, A., and Pedersen, A.: 1978, 'Direct Observation of a Tangential Electric Field Component at the Magnetopause', submitted to *Geophys. Res. Letters* .

Richter, A. K., Keppler, E., and Axford, W. I.: 1978, 'Dynamics of Low-Energy Electrons ($\geqslant$17 keV) and Ions ($\geqslant$80 keV) in the Vicinity of the Low-Latitude, Dusk-Side Magnetopause: Helios-1 and -2 Observations', Max-Planck Institute for Aeronomy, Preprint MPAE-W-60-78-23.

Williams, D. J., Keppler, E., Fritz, T. A., Wilken, B., and Wibberenz, G.: 1978, 'The ISEE-1 and -2 Medium Energy Particle Experiment', *IEEE Trans. Geosci. Electr.* **GE-16,** 270.

Williams, D. J.: 1978a, 'Magnetopause Characteristics Inferred from Three-Dimensional Energetic Particle Distributions', accepted by *J. Geophys. Res.*

Williams, D. J.: 1978b, 'Observations of Significant Magnetosheath Antisolar Energy Flow', submitted to *J. Geophys. Res.*

ISEE-2 spacecraft with experiment booms stowed in dynamic test chamber at ESTEC, Noordwijk, Holland.

The ISEE-1 and 2 spacecraft ready to be mated at Eastern Test Range, Florida, U.S.A.

The launch of Delta rocket number 135 carrying the two ISEE-1 and 2 spacecraft into orbit.

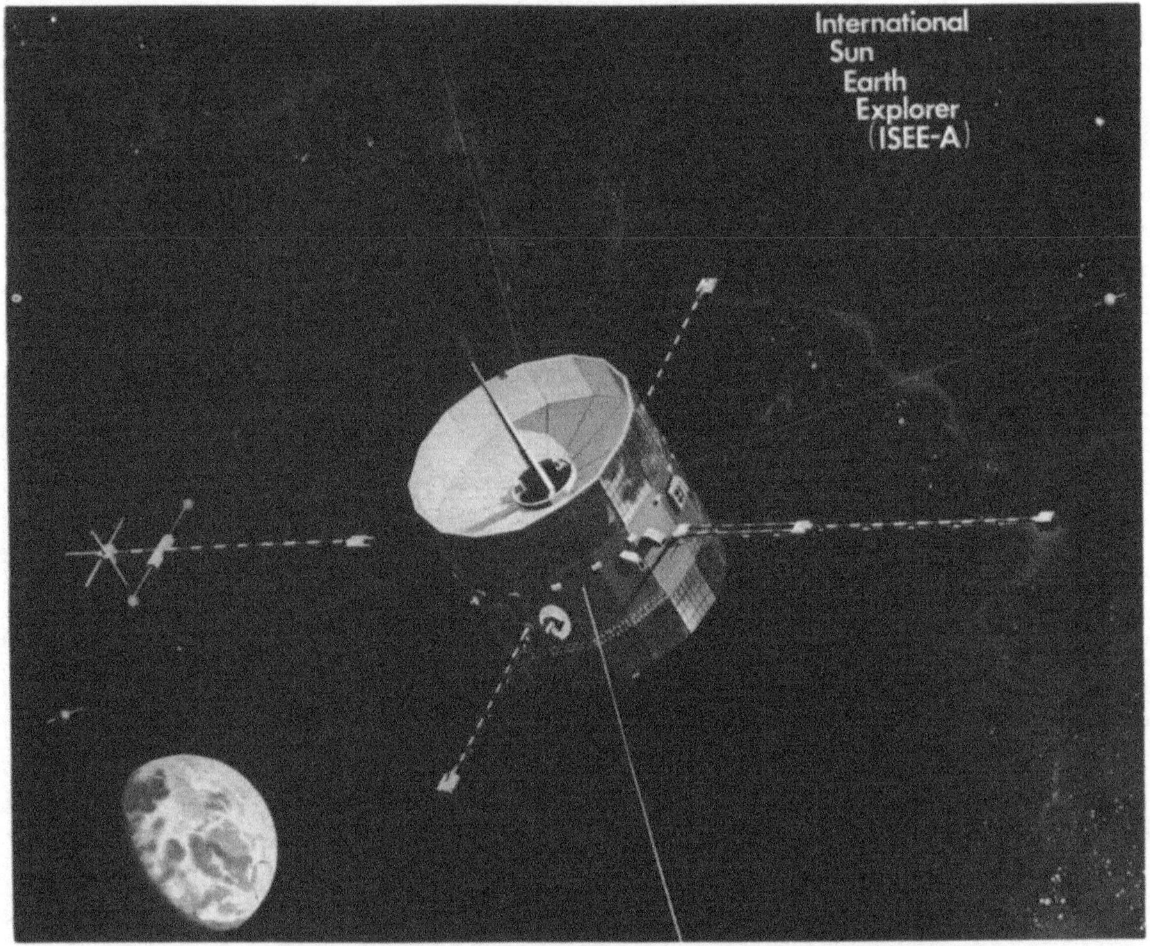

The ISEE-1 spacecraft in orbital configuration with all booms deployed.

INITIAL ISEE MAGNETOMETER RESULTS: SHOCK OBSERVATION

C. T. RUSSELL

Institute of Geophysics and Planetary Physics, University of California, Los Angeles, Calif. 90024, U.S.A.

and

E. W. GREENSTADT

TRW Defense and Space Systems Group, Redondo Beach, Calif. 90278, U.S.A.

Abstract. ISEE-1 and -2 magnetic field profiles across 6 terrestrial bow shock and one interplanetary shock are examined. The interplanetary shock illustrates the behavior of a low Mach number shock. It had an upstream whistler wave precursor with an apparent wavelength of 180 km. The shock thickness was about 90 km for the thickness of the final field jump or 270 km for the exponential growth of the precursor wave packet. The ion inertial length was 50 km, upstream of the shock.

Three examples of low or moderate β, high Mach number, quasi-perpendicular shocks are examined. These did not have upstream waves, but rather had waves growing in the field gradient. The growth length for these waves and the shock profile was of the order of the ion inertial length.

Two examples of high β shocks showed little coherence in field variation even though the two vehicles were only a few hundred kilometers apart. Thus, we cannot gauge their velocity and turn the time profiles into distances. The final crossing examined shows clearly the effect of changing the orientation of the interplanetary magnetic field. Initially the upstream magnetic field made an angle of about 80° to the shock normal and the shock position remained fairly steady. Then the field rotated to 45° to the normal and the field profiles became very irregular and the shock position very unstable. Discrete wave packets appeared.

Finally, we present the joint behavior of wave, particle and field data across some of these shocks to show some of the myriad of shock features whose behavior we are now beginning to investigate.

1. Introduction

The dynamics of collisionless plasmas have come to preoccupy specialists in a wide assortment of subjects, including the natural behavior of the extraterrestrial environment, the technology of energy production, the evolution of astrophysical bodies, and the basic study of ionized gases as an exercise of extreme complexity in classical physics. One dynamic phenomenon of great interest is the collisionless plasma shock wave, which occurs both in nature and in the laboratory. Indeed, natural shocks provide an excellent resource for the study of many fundamental plasma processes.

Unsettled theoretical questions about collisionless shocks abound: what heats the ions? What micro-instabilities dominate, or participate in, the heating process? How, if at all, does the array of instabilities vary with the parameter set in the preshocked plasma? Do stable parallel shocks really exist? How are suprathermal ions energized?

As in any complicated, real physical system, theory must be verified by observation before being accepted as valid, while in natural systems of geophysical or astrophysical scale and complexity, theory serves at best only as a guide to

Space Science Reviews **23** (1979) 3–37. 0038-6308/79/0231-0003 $05.25.

likely ranges of measurable quantities in terms of the ranges of controlling parameters furnished by nature.

Satellite-borne instruments offer opportunities to test theoretical analyses of shock phenomena with unprecedented certainty because the observational apparatus is small, if not minute, in comparison with the dimensions of the shocks to be explored, namely the planetary bow shocks and the interplanetary shocks of solar origin. A number of studies have been useful in illuminating shock phenomenology. Those of Fredricks *et al.* (1968), Montgomery *et al.* (1970), Neugebauer *et al.* (1971), Greenstadt *et al.* (1975, 1977), Formisano *et al.* (1975), Fairfield and Feldman (1975), Fairfield and Behannon (1976), and Morse and Greenstadt (1976) are examples. Further examples can be found in the texts and references of reviews by Greenstadt (1976), Formisano (1977), and Greenstadt and Fredricks (1978).

Unfortunately, the observational side of shock phenomenology has its own limitations and unsettled questions: what are the actual dimensions of structural features? What particle effects accompany magnetic gradients? What plasmawave spectrum corresponds to precursor ion distributions? What upstream parameters prevail for particular structural profiles?

Most of these questions remain open because in a dynamic system governed by the constant input of fresh plasma (the solar wind) with continually changing distribution parameters, variations in observed quantities cannot be assigned a temporal or spatial origin on the basis of measurements at a single location and independent parameters cannot be assumed to be known or constant but must be measured all together, all the time.

The paired ISEE-1 and -2 spacecraft contribute to the investigation of shocks in space in three ways. First, they can resolve most space/time ambiguity by making simultaneous two-point measurements; second, they are instrumented to make comprehensive field and particle observations covering all the important quantities; third, they are capable of recording data at high time resolution so as to take advantage of the small size of satellites compared to the shifting plasma features in which they reside. The UCLA magnetometers in particular are on both spacecraft, are identical, and are designed to make measurements at a rate of up to 32 samples per second.

In the following report, we illustrate with a selection of early measurements the three advantages brought by ISEE to the study of natural collisionless shocks. We display the magnetic field records from shocks of various types: laminar, supercritical, quasi-perpendicular, high β, and quasi-parallel. We use the spacing of the two spacecraft and the resolution of the system to develop numerous kinematic descriptions of the shocks and the waves that compose and surround them. Then, finally, we correlate data from a single particle experiment with the field data for three cases to show the important role comprehensive instrumentation will play in unravelling the mysteries of this intricate subject.

2. Instrumentation

The fluxgate magnetometers have been described in detail elsewhere (Russell, 1978). However, knowledge of a few details of instrument characteristics will be useful in understanding the data to be presented in later sections. Each of the ISEE-1 and -2 spacecraft carries an essentially identical triaxial fluxgate magnetometer. These instruments have two ranges, selectable by command, of $\pm 8192\gamma$ and $\pm 256\gamma (1\gamma = 1$ nT). The data is sampled by a 12-bit analog to digital converter internal to the instrument, accurate to $\frac{1}{4}$ of the least significant bit, at a rate of 512 samples per second. These samples are then averaged in six averaging registers, two for each axis, to provide overlapped 16-bit averages at rates from 4 to 32 samples per second. The resulting transfer function of the instrument shown in Figure 1 has an exact zero at the Nyquist frequency and is identical for all three sensors and both spacecraft. Any contiguous eight of the 16 bits can be transmitted to Earth at twice the data rate of 16-bit samples but the normal operating mode is to transmit all 16 bits at either 4 or 16 samples per second, depending on the telemetry rate of the spacecraft. The resolution of the magnetometer is $\pm 4m\gamma$ in the 256γ range and $\pm 125m\gamma$ in the 8192γ range.

Both magnetometers are equipped with flippers which interchange a sensor along the spin axis with one in the spin plane. With these flippers and the initial

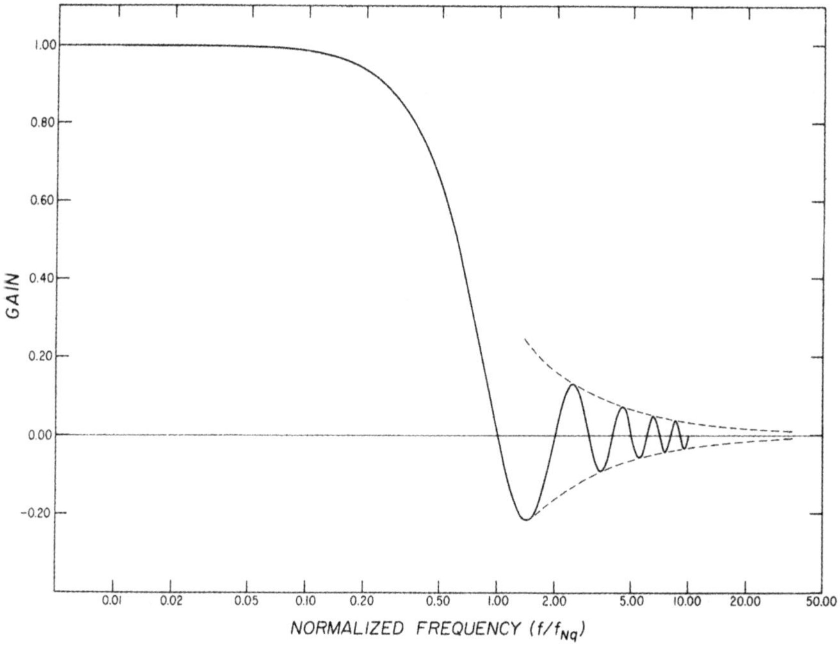

Fig. 1. Transfer function of ISEE magnetometers. Dashed lines show asymptotic behavior of high frequency gain. Gain expressed a function of Nyquist frequency, half the sample frequency, is identical for the six sensors on the two spacecraft for all modes.

operation of the spacecraft with perpendicular spin axis, we estimate we know the absolute field in the 256γ range regions to better than 0.1γ. The direction of the magnetic field has been checked using field models in high field regions near perigee and found to be accurate to ±0.1°.

3. An Interplanetary Shock

Shortly after the launch of ISEE-1 and -2, during a test of one of the IMS magnetometers* at the San Gabriel Canyon Test site just outside of Los Angeles, a geomagnetic storm began, preceded by a sudden commencement at about 23:30 UT on October 26, 1977. The ISEE spacecraft were in the solar wind at a position (14.08, −3.27, 6.74) Re GSE at this time, so we examined our data to find the interplanetary event responsible for this sudden commencement. Figure 2 shows 5 min of magnetic field records from both spacecraft surrounding the interplanetary shock associated with the sudden commencement. The shock caused an increase in the interplanetary field from 8.5γ to 15.3γ, a small change compared to the increases encountered at the Earth's bow shock. ISEE-1 was transmitting data at its highest rate so that the magnetometer was providing 16 samples per second. However, ISEE-2 was providing only 4 samples per second. Since the frequency of the upstream wavetrain, so evident in the ISEE-1 record,

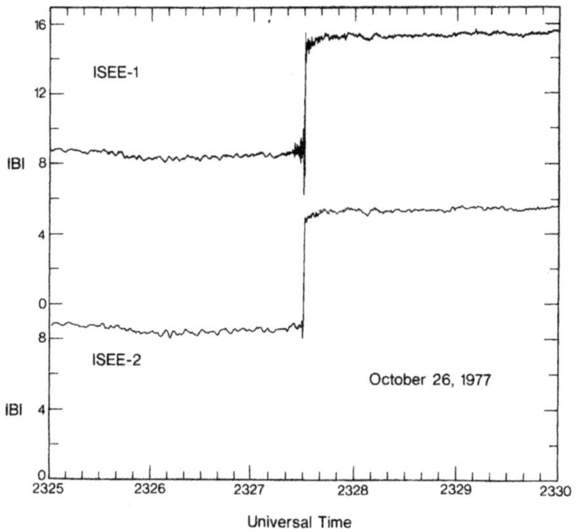

Fig. 2. Field strength of interplanetary magnetic field on ISEE-1 and -2 for 5 min surrounding passage of shock on October 26, 1977. Field strength in this plot and all following plots is in gammas. (1γ = ln T).

* The International Magnetospheric Study, or IMS, is a coordinated program of ground and space data collection and analysis. The IMS magnetometers, which we were testing, were destined for emplacement at mid-latitudes for the detection of substorm-associated currents in near real-time via a satellite communication link.

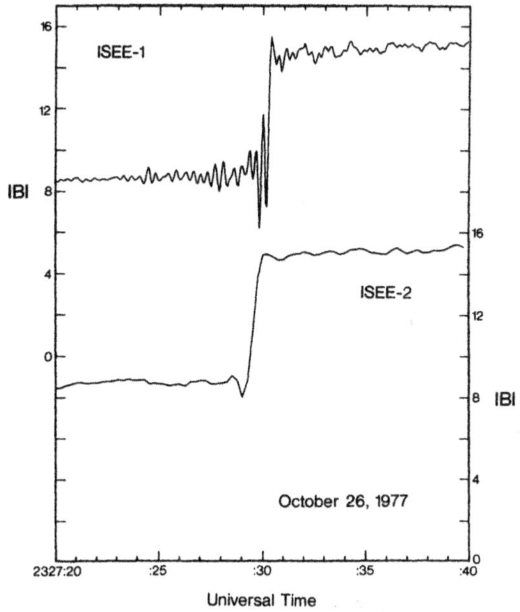

Fig. 3 Field strength for 20 s surround interplanetary shock of October 26.

was between 0.4 and 0.5 Hz, just above the ISEE-2 Nyquist frequency, it was effectively attenuated in the ISEE-2 records. This example shows clearly how well the Nyquist filtering in this instrument works. This characteristic is extremely desirable, for it provides confidence in the reality of signals which do appear. If these signals had not been attenuated before ISEE-2 sampled their waveform, their amplitude envelope would have been irregular and their apparent frequency would have been in error.

Since the magnetometer samples the magnetic field at a rate of 512 Hertz before averaging and the ground stations keep time to the nearest millisecond, we should theoretically be able to time the motion of the shock from ISEE-2 to ISEE-1 to 3 ms. We have attempted this by properly* averaging the ISEE-1 data to the ISEE-2 sample rate, and comparing the two profiles across the shock. The resulting time delay was 0.65 s, which we feel is only accurate to 10 ms, because temporal fluctuations in the boundary prevent the theoretical limit from being achieved.

Figure 3 shows the magnetic field magnitude at ISEE-1 and -2 for the 20 s surrounding the shock. The upstream wavetrain is clearly seen in the ISEE-1 record. The period of this wave ranges from 0.3 s to 0.5 s with an average of 0.39 s. Since it is clear from this picture that the shock front grew out of this wave, one might take the half-period of this wave, i.e., 0.195 s to be the thickness of the

* Since each instrument uses two internal averaging registers to create overlapped averages, a proper comparison requires computing running averages of four even-numbered ISEE-1 data points and of four odd-numbered points and interleaving the averages.

shock. In fact, if one measures the thickness to be the duration of the rise between the last minimum before the shock and the first maximum in field strength after the shock, the thickness was 0.19 ± 0.06 s.

3.1. The shock normal

The separation between ISEE-1 and -2 at this time is 503 km.* However, we cannot simply use this separation to change the above wave periods into wavelengths and crossing duration into thickness without knowing the direction of propagation of the shock. In the case of the terrestrial bow shock we know the average shock location and hence its shape. Further assuming that the shock simply expands as it moves, rather than developing significant ripples in its shape, we can calculate a model normal to the shock surface, i.e., a direction of propagation. The best assumptions we could make about an interplanetary shock is that its normal was radial from the Sun if we thought that it was flare associated, or perpendicular to the Archimedean spiral direction of the average magnetic field, if we thought the shock was caused by the interaction of co-rotating streams.

Another technique is to use the coplanarity theorem which states that $\mathbf{B}_u$, $\mathbf{B}_d$ and $\mathbf{n}$ are co-planar where the subscripts u and d refer to upstream and downstream of the shock surface respectively. If $\mathbf{B}_u$, $\mathbf{B}_d$ and $\mathbf{n}$ are coplanar then $\mathbf{B}_u \times \mathbf{B}_d$ is a vector in the plane of the shock. Another vector in the plane of the shock is $\mathbf{B}_u - \mathbf{B}_d$, since $\mathbf{B} \cdot \mathbf{n}$ is continuous across the shock. If we take the cross product of these two vectors in the shock plane we obtain a vector parallel to the shock normal. Therefore,

$$\mathbf{n} = \frac{(\mathbf{B}_u \times \mathbf{B}_d) \times (\mathbf{B}_u - \mathbf{B}_d)}{|(\mathbf{B}_u \times \mathbf{B}_d) \times (\mathbf{B}_u - \mathbf{B}_d)|}.$$

This formula obviously breaks down when $\mathbf{B}_u$ and $\mathbf{B}_d$ are parallel, as is the case for shock normals exactly perpendicular to the field. The formula also breaks down when $\mathbf{B}_u - \mathbf{B}_d$ is zero which could happen when the magnetic field lies along the shock normal. One might question what is the nature of a shock when there is no change in magnitude or direction. As we shall see, even the plasma has difficulty answering this question.

There are other difficulties with the coplanarity theorem, both practical and theoretical (Holzer *et al.*, 1972). First, there must be a well-defined magnetic field, both upstream and down. This criterion is certainly satisfied here, but is not satisfied for most terrestrial shock encounters. The theoretical difficulties lie in the assumptions used in deriving the coplanarity theorem: the electric field $\mathbf{E}$ on either side of the shock must equal $-(\mathbf{V} \times \mathbf{B})/c$ where $\mathbf{V}$ is the solar wind velocity and there must be no downstream electric field along the shock normal in a moving frame in which there is no upstream electric field along the shock normal.

* The accuracy of the separation vector is nominally 1% or 1 km whichever is greater. However, the true accuracy is still under discussion.

TABLE I
Interplanetary shock parameters

Date/Time	10/26/77 23:27:30	
Location, $\mathbf{R}_{GSE}$	$(14.08, -3.27, 6.74)R_e$	
Separation, $\mathbf{S}_{GSE}$	$(-340, 309, -205)$ km	
$\mathbf{B}^u_{GSE}$	$(3.7, -7.3, -2.0)\gamma$	
$\mathbf{B}^d_{GSE}$	$(3.6, -14.4, -3.5)\gamma$	
${}^1\mathbf{n}_{GSE}$	$(-0.994, -0.014, +0.109)$	
${}^2\mathbf{n}_{GSE}$	$(-0.988, -0.005, +0.156)$	
$\measuredangle {}^1\mathbf{n}, {}^2\mathbf{n}$	2.8°	
$\measuredangle {}^1\mathbf{B}, {}^1\mathbf{n}$	117.4° (62.6°)*	
$\measuredangle {}^2\mathbf{B}, {}^2\mathbf{n}$	116.9° (63.1°)*	
$\mathbf{S} \cdot \mathbf{n}$	307 km	
Preliminary V km s^{-1}	up: 290	down: 340
N cm^{-3}	20	35
T_e K	1.5×10^5	1.9×10^5

* Supplement of angle.
N.B.: Left superscripts refer to satellite; right superscripts to upstream of the shock or downstream; subscripts to coordinate system.

In view of these difficulties we use the coplanarity shock normals with some caution, always checking them against our expectations and intercomparing the two spacecraft.

Table I lists various parameters for the shock crossing. As can be seen in Figure 2 the field has both fluctuations and long-term trends during this interval so the value of the field depends somewhat on the averaging interval. We have used a variety of different intervals and obtained shock normals varying by several degrees. The values quoted here are for one minute averages on either side of the shock. The coplanarity normals derived from ISEE-1 and -2 separately differ by 2.8°. The average normal pointed up 7.5° from the ecliptic plane and its ecliptic plane projections was within 0.5° of the Earth Sun line. Since the two derivations agree and give a physically reasonable result we will use this normal in our future calculations.

The angle between the upstream field measured on each of the satellites and the derived normals was 63°. Thus we would classify this shock as a quasi-perpendicular shock. Finally we note that the field strength increased only 81% across the shock. Thus, it was much weaker than the terrestrial shock and probably has a Mach number of less than 2. If we use the preliminary solar wind parameters provided by G. Paschmann (personal communication, 1978) the magnetosonic Mach number was 2.5, and the plasma beta was 3.

3.2. Shock thickness

The separation of the two spacecraft was 307 km along the shock normal and 399 km projected into the shock plane. Thus, the 0.65 s separation in time implies

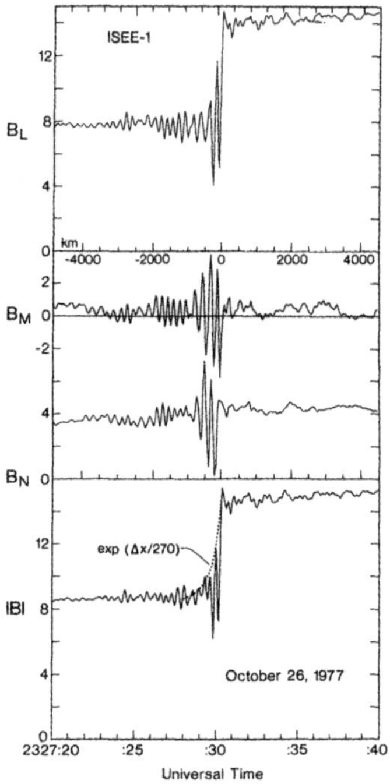

Fig. 4. ISEE-1 magnetic field measurements in shock normal coordinates across interplanetary shock. B_N is along coplanarity shock normal. B_L is along projection of upstream field into shock plane.

a shock velocity along the normal of 472 km s^{-1}. Using either the half period of the upstream waves or the rise of the field magnitude from minimum to maximum we obtain a shock thickness of 90 km. Again, using the preliminary solar wind parameters for the upstream region we obtain for c/ω_{pe}, 1.2 km, and for c/ω_{pi}, 50 km. Thus, the shock thickness was close to twice an ion inertial length.

Figure 4 shows the three components of the magnetic field and the total field magnitude in shock normal coordinates where N is along the shock normal and L is along the projection of the upstream field on the shock plane. Superimposed on the field magnitude is an exponential curve illustrating the final growth of the wave packet in front of the shock. The e-folding length for this curve is 270 km.

Ideally in this display the M-component of the field should average to zero, the N-component should be constant across the shock and the change in field across the shock should all take place in the L-component. The M-component is in fact non-zero since the magnetic field was slowly changing during this interval and we used one minute averages on either side of the shock to calculate the shock normal coordinate system.

3.3. The upstream wave train

We can calculate the direction of propagation of the phase velocity, **k**, of the upstream wavetrain by creating the variance ellipsoid of the field fluctuations and finding its eigenvalues and eigenvectors. The eigenvector associated with the minimum eigenvalue we take to be the direction of propagation. The wavetrain has been divided into three intervals for analysis. Figure 5 shows hodograms of two orthogonal views of the wave. On the left is the $i-j$ variation, the plane of maximum variation. The direction of propagation is into the page. The right hand hodograms show this variation edge-on i.e. the $i-k$ plane. The variation in the $i-j$ plane is counterclockwise, and since the field points out of the page, the waves are right-handed. This is to be expected if the waves were whistler-mode waves propagating upstream.

The direction of the wave normal changes during the passage of the wavetrain. This can be seen in the left-hand side of Figure 5 as the migration of the centroid of the hodogram and on the right-side as a rotation of the major axis of variations. More quantitatively, Table II gives the results of our analysis of the three intervals. As the shock was approached the wave normal tended to be closer to the shock normal. This can be seen in the direction of **k** expressed in GSE coordinates and in θ_{Bk}, the angle between the upstream field and the coplanarity shock normal. The average upstream field expressed in principal axis coordinates, B^u_{PA}, is principally along the wave normal and the major axis of the perturbation ellipse. This behavior is characteristic of whistler mode waves. Further we note that $\mathbf{B}^u$, **k**, and **N**, the upstream field, the wave normal and the shock normal are coplanar.

3.4. Discussion

This interplanetary shock provides a good example of the behavior of a subcritical shock propagating into the upstream plasma slow enough and at such an angle to

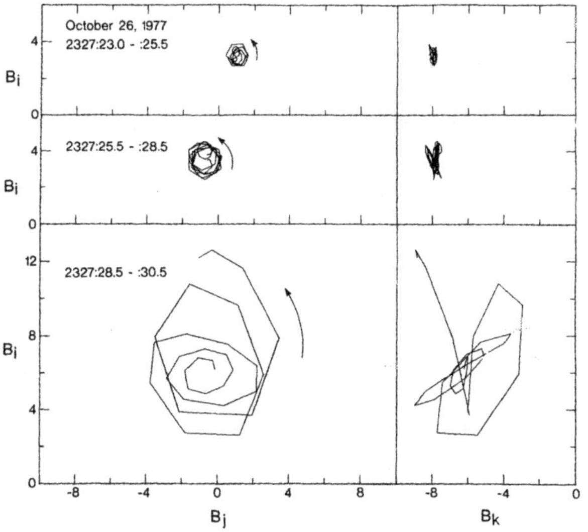

Fig. 5. Hodogram of tip of magnetic field vector in two orthogonal principal planes of variance ellipsoid for three times during upstream wave train.

TABLE II
Eigenvalue analysis of upstream wave

Time interval:	23:27:23.0–25.5
Eigenvalues:	$0.107/0.089/0.013\gamma^2$
$\mathbf{k}_{GSE}$	(−0.724, 0.669, 0.170)
$\mathbf{B}^u_{pA}$	$(-3.2, -1.0, -8.0)\gamma$
θ_{Bk}*	157.3° (22.7°)
θ_{BN}**	140.5° (39.5°)
θ_{kN}	43.2°
$\mathbf{k}_N$	(0.612, −0.055, 0.729)
Time interval	23:27:25.5–28.5
Eigenvalues	$0.354/0.295/0.044\gamma^2$
$\mathbf{k}_{GSE}$	(−0.770, 0.627, 0.115)
$\mathbf{B}^u_{pA}$	$(-3.7, +0.8, -7.8)\gamma$
θ_{Bk}	154.4° (25.6°)
θ_{BN}	144.3° (35.7°)
θ_{kN}	39.7°
$\mathbf{k}_N$	(0.582, −0.105, +0.769)
Time interval	23:27:28.5–30.5
Eigenvalues	$5.297/3.444/2.165\gamma^2$
$\mathbf{k}_{GSE}$	(−0.846, 0.407, 0.345)
$\mathbf{B}^u_{pA}$	(−5.5, +0.4, −7.1)
θ_{Bk}	142.4° (37.6°)
θ_{BN}	152.5° (27.5°)
θ_{kN}	29.2°
$\mathbf{k}_N$	(0.318, 0.157, 0.873)

* The supplement of the angle is given for reference in brackets.
** For reference $\mathbf{B}^u_N$ is (6.68, 0.00, −3.79).

the IMF that whistler mode waves can propagate upstream. Some plasma waves were seen in association with this shock (D. A. Gurnett and F. L. Scarf, personal communication, 1978), but there were no energetic particles seen (D. J. Williams and K. A. Anderson, personal communication, 1978). The only clear energetic particle effect to be seen was acessation of upstream bow shock electrons after the passage of the shock. We expect to study these correlated events in later publications. However, some insight into this shock can be obtained by comparing it with earlier observations of Earth's bow shock under similar conditions.

Examples of quasi-perpendicular, laminar bow shocks were published by Greenstadt *et al.* (1975) and Fairfield and Feldman (1975). Curiously, the eight cases of Greenstadt *et al.* and the 34 cases of Fairfield and Feldman failed to furnish a single example in which even two of the three parameters M, β, θ_{Bn} were the same as those of the interplanetary shock we describe here. The principal reason is that in this case M was slightly, and β much, higher than in any of the previously documented cases. Nevertheless, two of the earlier examples are worth comparing with this new case; one from each paper.

All the previous examples had $\beta \leqslant 0.1$ and $M_{MS} \leqslant 2.5$, at a wide range of θ_{Bn}. One of Fairfield and Feldman's cases, at 02:13 on 17 May, 1972, had $\theta_{Bn} \simeq 65°$, close to our estimate here of 63°; their shock had $M \simeq 1.5$ and $\beta \simeq 0.02$, and a sharp profile with no significant waves upstream. One of Greenstadt *et al*'s cases at 00:49, 12 February 69, had $\theta_{Bn} \simeq 65°$; their shock had $M < 2.1$ and $\beta < 0.1$, and a sharp shock profile with small standing whistlers in the foot of the shock. The major contrast, then, between the parameter sets of the three shocks appears to be a progression in β, while the major contrast in profiles involved a corresponding progression in the amplitude of the upstream waves (the interplanetary shock had $\beta \simeq 3$). Low β in ΔB gradients encourages drift instability which tends to reduce the gradients, hence dampen the waves, while high β should discourage instability, leaving the waves undamped, except by dispersion (Greenstadt *et al.*, 1975, discussion). This reasoning may explain the possible inverse relationship between β and upstream whistler wave amplitude among the three low M shocks. If so, we see at once the benefit of using observations of interplanetary shocks, which are usually subcritical, to fill in gaps in our inventory of laminar structures. Preliminary examination of the first three months of ISEE bow shock crossings failed to uncover any very low M, quasi-perpendicular cases.

4. Quasi-Perpendicular Bow Shocks

The structure of bow shock must be quasi-steady if we are to use time-of-flight measurements between the two spacecraft to calculate the shock velocity and the shock thickness. When the interplanetary field is locally within about 30° of being perpendicular to the shock normal, this condition occurs with fair reliability; within 40° the steadiness of the structure is less secure, but in either case we have what we call a quasi-perpendicular shock. The interplanetary shock discussed above was a steady shock, but at low Mach number. In this section, we will examine the structure of some high Mach number quasi-perpendicular shocks.

4.1. October 27, 1977, 18:44–19:07 UT

Figure 6 shows the magnetic field strength measured at ISEE-1 and -2 from 18:44–19:07 UT on October 27. The two spacecraft were separated by 565 km at this time, and situated $8.82R_e$ in front of the Earth and $12.29R_e$ from the Earth Sun line. The interplanetary field was strong, 12.8γ, and rather steady at this time. Table III lists relevant parameters of this shock. The model normal ${}^m\mathbf{n}$ is derived from a simple unaberrated conic section with its focus at the center of the Earth and an eccentricity of 0.7 (Holzer *et al.*, 1972). The four normals derived from coplanarity all point more toward the Sun than the model. This suggests that, at least under the solar wind conditions present at this time, the shock front was flatter than the Holzer formula predicts. The angle that the normals made with the Sun form an interesting progression in time. The first normal from ISEE-2 at 18:45 was 29.0° from the Sun; the second from ISEE-1 at 18:47 was 23.9° from

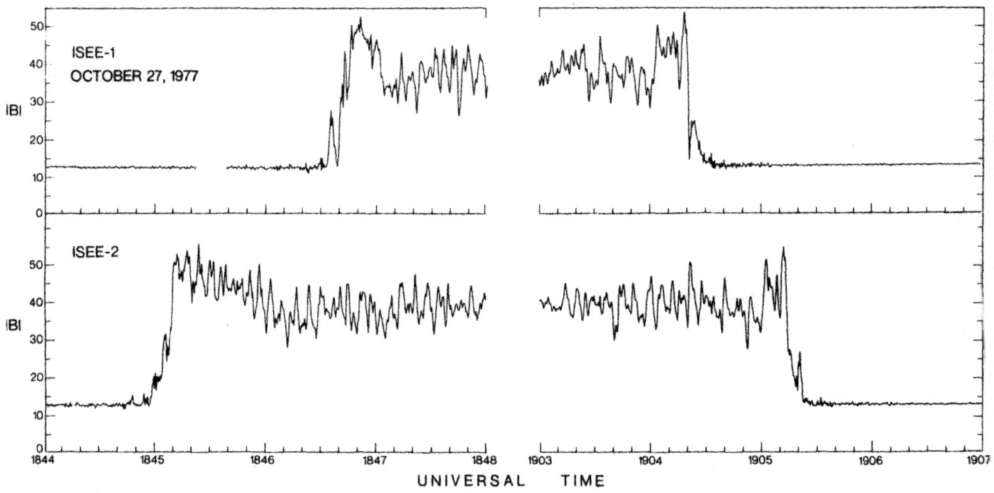

Fig. 6 Magnetic field strength across two shocks on October 27, 1977 as measured on ISEE-1 and -2.

TABLE III
Bow shock parameters: orbit 3 out

Location $\mathbf{R}_{GSE}$	(8.82, −10.95, 5.57) R_e
Separation $\mathbf{S}_{GSE}$	(470, −207, 236) km
Time	18:45 UT 10/27/77
$\mathbf{B}^u_{GSE}$	(−1.7, 2.9, −12.2)γ
$\mathbf{B}^d_{GSE}$	(4.2, 5.0, −37.5)γ
${}^1\mathbf{n}_{GSE}$	(0.914, −0.375, 0.155)
${}^2\mathbf{n}_{GSE}$	(0.875, −0.445, 0.192)
${}^m n_{GSE}$	(0.845, −0.477, 0.242)
$\sphericalangle\, {}^1\mathbf{n}, {}^2\mathbf{n}$	5.1°
$\sphericalangle\, n, {}^m n$	6.2°
$\sphericalangle\, \mathbf{B}^u, \mathbf{n}$	112.4° (67.6°)
$\mathbf{S} \cdot \mathbf{n}$	547 km
Time	19:05 UT
$\mathbf{B}^u_{GSE}$	(−1.6, 3.5, −12.4)
$\mathbf{B}^d_{GSE}$	(3.9, 7.8, −37.1)
${}^1\mathbf{n}_{GSE}$	(0.926, −0.351, 0.140)
${}^2\mathbf{n}_{GSE}$	(0.933, −0.322, 0.159)
$\sphericalangle\, {}^1\mathbf{n}, {}^2\mathbf{n}$	2.0°
$\sphericalangle\, \mathbf{n}, {}^m\mathbf{n}$,	9.4°
$\sphericalangle\, \mathbf{B}^u, n$	110.7° (69.3°)
$\mathbf{S} \cdot \mathbf{n}$	549 km
Preliminary plasma parameters	
V	320 km s^{-1}
N	15 cm^{-3}
T_e	1.1×10^5

TABLE IV
Quasi-perpendicular shock thickness

	Orbit 3 out		Orbit 6 in
	18:46	19:05	13:23
$\mathbf{S} \cdot \mathbf{n}$	547	549	228 km
ΔT	94	53	33 s
V	5.8	10.4	6.9 km s^{-1}
Δ^t	11	9	8 s
$\Delta^2 t$	12	9	11 s
^1D	64	94	55 km
^2D	70	94	76 km
θ_{Bn}	68°	69°	63°
Preliminary M_A	4.5		6.0
M_S	3.6		4.0
β	0.7		1.6
c/ω_{pi}	60 km		80 km

the Sun; the third from ISEE-1 at 19:04 was 22.2° from the Sun; and the last from ISEE-2 at 19:05 was 21.1° from the Sun. This is the behavior expected if the ISEE spacecraft were beyond the average position of the bow shock and an oscillation in the bow shock position, convecting down from the subsolar region, engulfed the two spacecraft and then receded past them. This seems to be the situation since the crossing at 19:05 is the last one in a series beginning at 16:48 UT. Even these two most distant crossings were 10% closer to the Earth than the average shock location of Fairfield (1971).

The separation between the two spacecraft along the shock normal at 18:46 and 19:05 was essentially constant, 547 km in the first case and 549 km in the second. The angle between the field and shock normal also were similar, about 68°. The time delay between the crossings is quite different, however, 94 s for the first set and 53 s for the second set giving shock velocities of 5.8 and 10.4 km s^{-1}. Using these velocities to obtain a thickness is not as clear cut as determining the velocity. There were clearly non-stationary waves, beginning at the base of the shock, whose amplitude was a significant fraction of the shock jump. These waves had periods of about 5 s, or about half the apparent thickness of the shock. Table IV lists the four crossing times based on a linear ramp approximation to the shock profile. The average crossing time was 10 s and the average thickness was 80 km. The ion inertial length at this time, corresponding to a density of 15 cm^{-3}, was 60 km. Thus the average shock thickness is close to one ion inertial length.

Figure 7 shows the ISEE-1 vector field components and the total field in shock normal coordinates. The components orthogonal to the main field jump clearly show the rapid build-up of the 5-s wave. In the bottom panel we have plotted an exponential decay on the total field trace at each of the shocks. The decay times of 30 and 40 km are slightly less than an ion inertial length.

Eigenvalue analysis of these shocks has been performed for each of the four crossings. Figure 8 shows one of these four crossings. The behavior of the other three crossings was similar but not identical. The eigenvalues of the variance ellipsoid for this crossing were 234, 12.2, and $5.4\gamma^2$. Thus the direction of the maximum variation was well determined but the direction of the wave normal was not. As is obvious from Figure 8, the variation through the shock was principally in one component.

A rather noticeable difference between these shocks and the weaker interplanetary shock discussed in the previous section is the downstream behavior. First we see an overshoot in the field in which field strengths were achieved which were much higher than magnetosheath fields further from the shock. The overshoot region seems to be followed by a slight depression in the field magnitude. However, this feature was not as pronounced as the overshoot. Preliminary examination of particle data suggest that the electron density and field strength have a high positive correlation in overshoots, but that the protons may not be so correlated (J. Gosling, S. Bame, and G. Paschmann; K. Ogilvie and J. Scudder; K. Anderson, R. Lin, and G. Parks; personal communications, 1978). Thus the overshoot may be fundamental in the physics of at least some types of collisionless

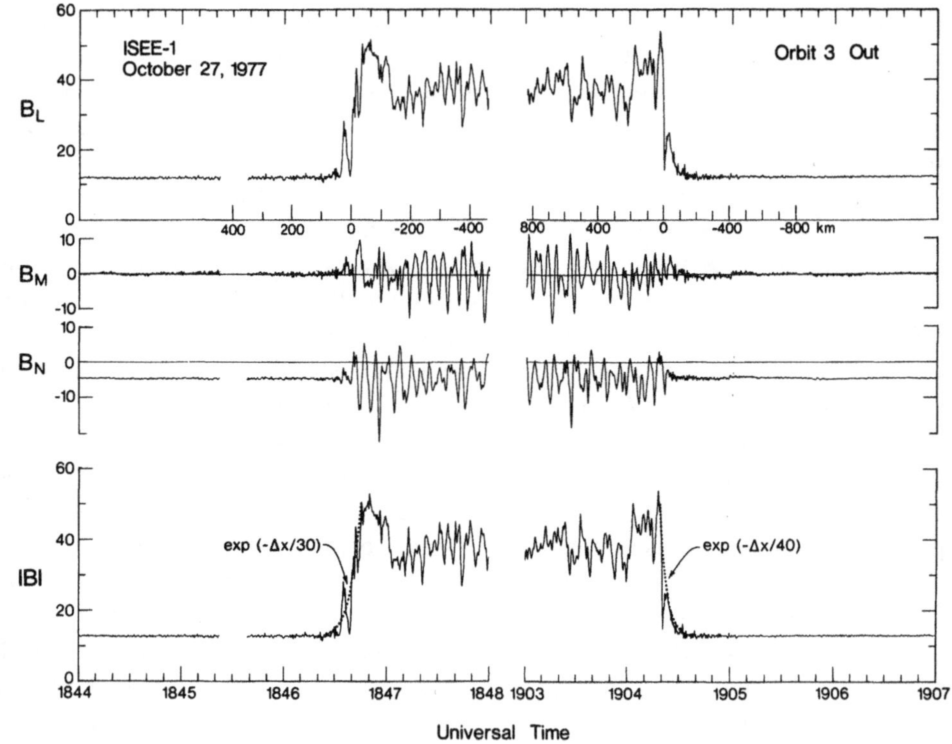

Fig. 7. ISEE-1 magnetic field measurements in shock normal coordinates across two shocks on October 27, 1977.

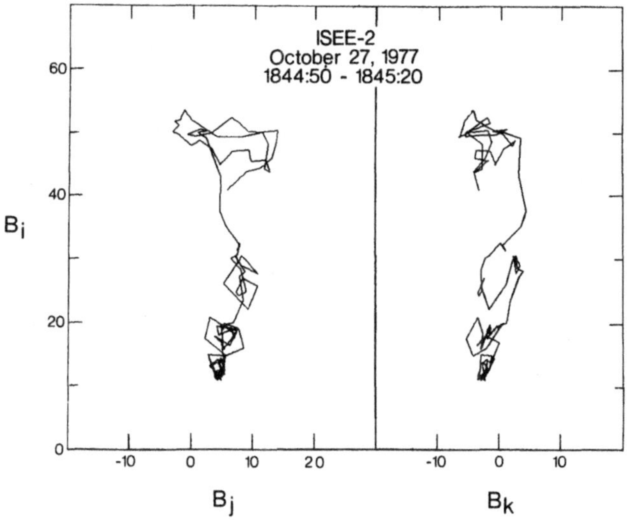

Fig. 8. Hodogram of tip of magnetic field vector in two orthogonal planes of variance ellipsoid for the 18:45 UT shock crossing on October 27, 1977.

shocks. The average thickness of the overshoot region was 160 km or slightly less than three ion inertial lengths.

The wave levels are also more intense behind the shock. However these waves are much reduced in amplitude in the downstream magnetosheath. For example, while the waves in the total field strength are 25% of the field strength just behind the shock they are only 5% of the field strength 10 min later (not shown).

4.2. November 5, 1977, 13:23 UT

Figure 9 shows the magnetic field strength measured at ISEE-1 and -2 from 13:19–13:26 UT on November 5, 1977. The two spacecraft were separated by 277 km at this time and at a position $13.18 R_e$ in front of the Earth and $8.30 R_e$ from the Earth Sun-line. This crossing was within 2% of the average shock location for this local time and altitude. As shown in Table V, the interplanetary field strength for this shock was only slightly greater than average. The normal was somewhat variable. The ISEE-1 and -2 normals differed by 3.8°. The average normal differed 6.2° from the model normal. The field immediately upstream from the shock made an angle of 60.9° with the shock normal. We note three distinct regions of upstream behavior in Figure 9. First, prior to 13:20, there were low frequency, 20 s period, waves present. From 13:20 to 13:21:15 there were essentially no waves. Finally beginning at 13:21:15 and continuing to the shock there were high frequency waves with about a 1 s period. Each of these different regions was associated with a different direction of the interplanetary magnetic field relative to the shock normal. The low frequency waves were associated with a θ_{BN} of 46.5°; the high frequency low amplitude waves with a θ_{BN} of 60.9°; and the quiet field with a θ_{BN} of 68.6°.

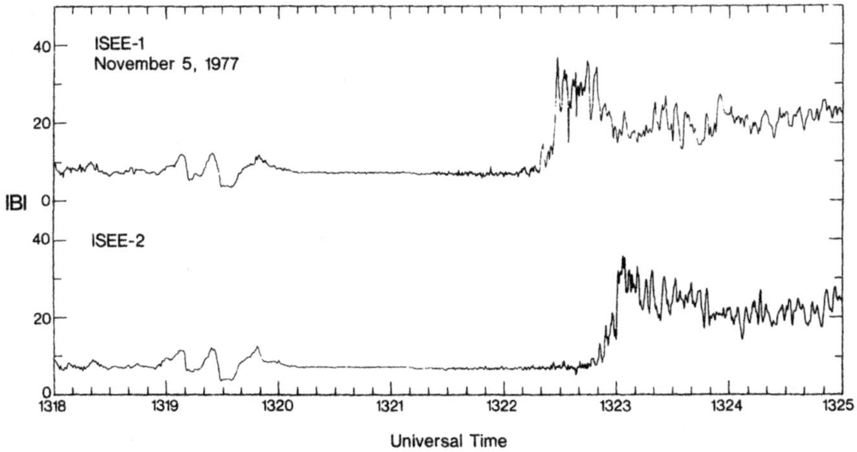

Fig. 9. Magnetic field strength at ISEE-1 and -2 across bow shock on orbit 6 in, November 5, 1977.

The separation in time between the two crossings is 33 s and the separation resolved along the shock normal is 228 km. Hence the shock velocity is 6.9 km s^{-1}. The thickness of the shock again is a more difficult parameter to measure because of the non-stationary waves on the shock front. Using a linear ramp approximation we obtain crossing times of 8 and 11 s for thicknesses of 55 and 76 km or a little less than one ion inertial length. Figure 10 shows the ISEE-2 data in shock normal coordinates. The exponential fit to the magnitude gives an e-folding length of 40 km. The apparent wavelength of the waves on the shock transition is 24 km. However, these waves do not appear to be stationary in the shock frame.

TABLE V
Bow shock parameters: orbit 6 in

Location $\mathbf{R}_{GSE}$	(13.18, −5.18, 6.49) R_e
Separation $\mathbf{S}_{GSE}$	(−143, 202, −123) km
Time	11/5/77 13:23 UT
$\mathbf{B}^u_{GSE}$	(−1.8, 6.6, 1.1)
$\mathbf{B}^d_{GSE}$	(1.6, 21.4, 6.5)
${}^1n_{GSE}$	(0.924, −0.303, 0.234)
${}^2n_{GSE}$	(0.904, −0.308, 0.297)
${}^mn_{GSE}$	(0.945, −0.203, 0.255)
$\sphericalangle\, {}^1\mathbf{n}, {}^2\mathbf{n}$	3.8°
$\sphericalangle\, n, {}^m\mathbf{n}$	6.2°
$\sphericalangle\, \mathbf{B}^u, \mathbf{n}$ 13:19	133.5 (46.5°)
13:21	111.4 (68.6°)
13:22	119.1 (60.9°)
$\mathbf{S}\cdot\mathbf{n}$	228 km
Preliminary plasma parameters	
V	330 km s^{-1}
N	7.5 cm^{-3}
T_e	1.5×10^5 K

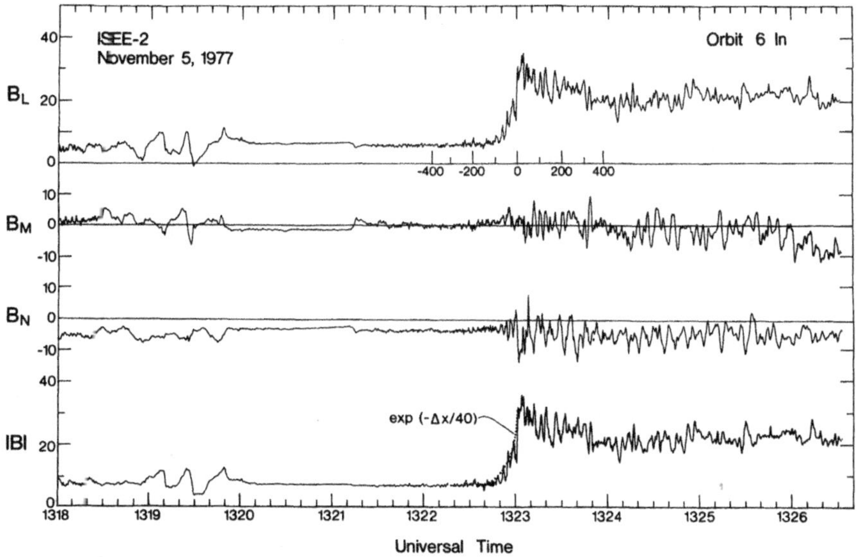

Fig. 10. Magnetic field components in shock normal coordinates across bow shock on November 5, 1977.

Again there is an overshoot region of high field strength immediately behind the shock. However, the duration and shape of the overshoot appears quite different in the two records. The ISEE-1 overshoot lasted 12 s (83 km) and was followed by a sudden decrease in field strength. The ISEE-2 overshoot decreases in strength more gradually. We note that both ISEE-1 and -2 overshoot regions contain large amplitude waves which make precise determinations of thickness impossible. Also again there was a slight depression of the field value below later magnetosheath values immediately downstream of the overshoot region.

Figure 11 shows the results of our eigenvalue analysis of the ISEE-2 data across the shock. The eigenvalues for this crossing were 93.6, 10.4, and $4.1\gamma^2$. The same analysis for ISEE-1 did not result in a good separation of eigenvalues (66.0, 7.4, 5.0) and thus the k-vector was poorly defined for ISEE-1. The ISEE-2 k-vector was (−0.195, 0.182, −0.964) in GSE coordinates making an angle of 91° with the upstream magnetic field and 80° with the shock normal. Thus the direction of minimum variance is essentially in the shock plane perpendicular to the upstream (and downstream) magnetic field.

Finally we should say a few words abut the upstream waves that occurred before 13:20 when the interplanetary magnetic field made an angle of 46.5° with the shock normal. Figure 12 shows a hodogram of the ISEE-2 data for the period from 13:18–13:20 UT in the principal axis system. The predominant motion here is counterclockwise in the $i-j$ plane and, since the average field is into the page i.e. along $-\mathbf{k}$, the waves are left-handed. The direction of the k-vector is (0.925, −0.287, 0.250) and is moderately well determined. The eigenvalues are 5.9, 3.3, and $1.7\gamma^2$. The k-vector makes an angle of only 3° with the shock normal.

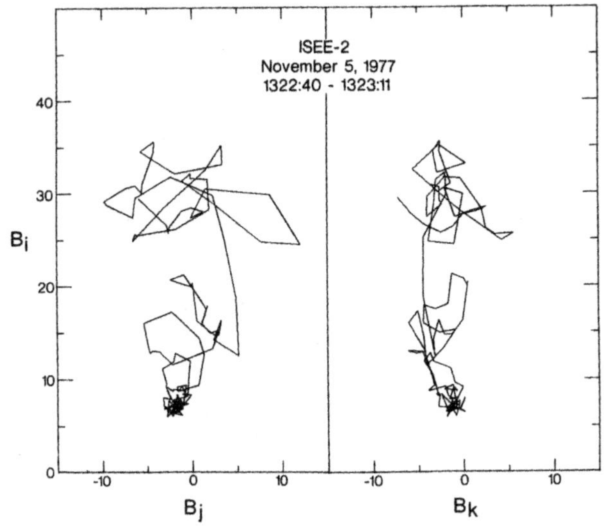

Fig. 11. Hodogram of tip of magnetic field vector in principal axis system through bow shock on November 5, 1977.

4.3. Discussion

Quasi-perpendicular shocks make up one of the two major, instantly-recognizable categories of shock structure (Greenstadt, 1976; Formisano, 1977; Greenstadt and Fredricks, 1978). Indeed, it has not yet been established observationally that a stationary, quasi-parallel structure even shocks the plasma flow in the traditional sense. The clearly defined *q*-perpendicular category is certainly the first choice for early study.

The major subjects of interest in the *q*-perpendicular, collisionless shock

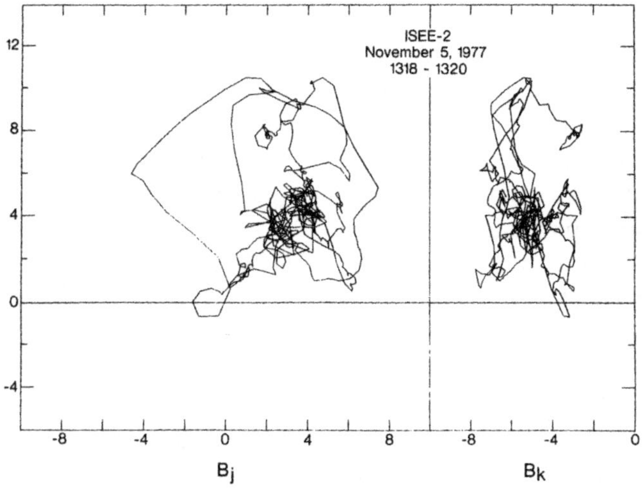

Fig. 12. Hodogram of upstream waves from 13:18 to 13:20 on November 5, 1977.

revolve around the microscopic processes whereby electrons and, more interestingly protons (and ions generally) are deflected and ultimately thermalized or at least partially thermalized. This issue must be taken up with high resolution measurements from all the relevant instruments on ISEE, of course, so we have confined ourselves at this date to kinematic measurements based on the magnetometers. Nevertheless, several important results stand out, not so much because they are all new, but because they now rest on the authority of double satellite observations.

First, like low-M shocks, moderately supercritical, quasi-perpendicular shocks have reproducible features indicating structural stability over a time scale of a minute. These features include an exponentially rising foot containing small waves, a principal, sharp magnetic gradient, a post-gradient overshoot giving the appearance of a square wave on which are superposed a wavetrain of several cycles, a post-overshoot relaxation in which the field returns to a level about half the peak of the overshoot, and a post shock region containing visually identifiable oscillations of somewhat larger amplitude and longer period than those further downstream. Additional features will be described in a later section on Field-Particle Correlation.

Second, the features just described can vary in the spacecraft time frame in a scale of a minute possibly because shock motion is so variable that even between the closely-spaced ISEE satellites the shock changes velocity by a factor of two or three.

Third, the thickness scales of the principal shock features are on the order of c/ω_{pi} or a small multiple of c/ω_{pi}. This is not to say that the proton inertial length has been shown to be the appropriate scale length for the bow shock, but only that under the solar wind parameter conditions that have prevailed, shock thicknesses seen with ISEE have been consistent with those deduced from earlier measurements. There are versions of the proton cyclotron radius, depending on the particle velocity chosen, that can also serve as consistent measures of shock thickness within the uncertainties still governing thickness estimates.

The thicknesses of shock features are closely related to the dissipation mechanisms governing them (Morse and Greenstadt, 1976; Lemons and Gary, 1978; Greenstadt *et al.*, 1978) and we foresee a fruitful, early attack on the outstanding problems of quasi-perpendicular collisionless processes with the ISEE instruments.

5. High β Shocks

The bow shocks we have examined thus far have had relatively low ratios of plasma pressure to magnetic field pressure, 0.7 and 1.6. When this ratio increases the nature of the bow shock changes drastically. The magnetic field downstream of the shock becomes very irregular and the magnetic field strength reaches very high values compared to its upstream value (Formisano *et al.*, 1975). In this section we examine two such intervals in the ISEE data.

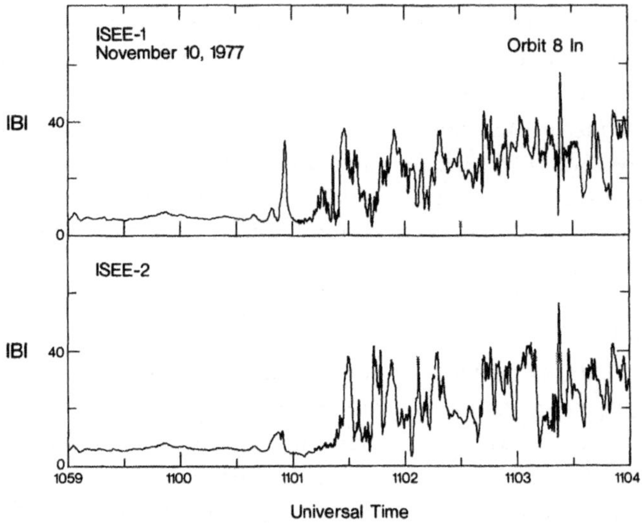

Fig. 13. Magnetic field strength at ISEE-1 and -2 across bow shock on orbit 8 in, November 10, 1977.

5.1. November 10, 1977, 11:01 UT

Figure 13 shows the magnetic field strength at ISEE-1 and -2 across the bow shock on orbit 8 inbound. As shown in Table VI, the magnetosonic Mach number for this shock is 3.3 and the plasma beta is 3.1. Several characteristics are immediately obvious. First, the magnetic field is highly irregular downstream from the shock. Second, the field strength peaks are very much larger than the upstream field. One peak seen on both ISEE-1 and -2 reaches 57γ, a factor of 9 greater than the average upstream field strength. Third, the field variations are

TABLE VI
Bow shock parameters

	Orbit 8 out	Orbit 8 in
Time	18:53 UT 11/8/77	11:01 11/10/77
Location	(7.25, −13.05, 6.09)	(10.99, −4.05, 5.15) R_e
Separation	(288, −94, 90)	(−86, 114, −73) km
B^u_{GSE}	(−0.1, 2.2, 0.6)	(−3.7, 4.9, 2.1)γ
${}^m n_{\text{GSE}}$	(0.790, −0.556, 0.260)	(0.950, −0.193, 0.245)
$\measuredangle \mathbf{B}^u, {}^m\mathbf{n}$	120.1 (59.9)	127.5 (52.5°)
$\mathbf{S} \cdot \mathbf{n}$	303	122 km
Preliminary plasma parameters		
V	300	254
$\mathbf{V} \cdot \mathbf{n}$	237	241
N	11	12.5
T_e	1.3×10^5	1.5×10^5
M_A	15.8	6.0
M_S	4.0	3.3
β	19.1	3.1

quite different at the two spacecraft even though they are separated by 160 km, only a few ion inertial lengths. The behavior of the field is, of course, qualitatively similar at ISEE-1 and -2, and there are some features which appear to be the same at the two spacecraft. This brings us to our fourth characteristic, the near simultaneity of the shock crossing at the two vehicles. The shock appears to cross from ISEE-1 to -2 in no more than 3 s for a velocity of over 40 km s^{-1}. However, the clear non-time stationary character of the boundary makes such an interpretation ambiguous.

Since the field is so highly irregular we have not attempted to calculate an average downstream field value, nor a coplanarity normal. Thus, we restrict our calculations to use of the model normal. The separation quoted above is along the model normal. Also the angle of the field to the model normal is 52.5°. Thus, the shock should be considered a borderline case, with a clearly defined quase-perpendicular 'front', but with substantial upstream wave activity usually associated with quase-parallel geometry.

Figure 14 shows hodograms of the field variation in the $i-j$ and $i-k$ planes of the variance ellopsoid. The eigenvalues are 180.0, 52.2, and 15.8γ^2, so the k-direction is well determined. It makes an angle of 138° with the upstream field and 13.5° with the model shock normal.

Finally, the impulsive nature of the field data suggests rapid motion of the shock back and forth across the spacecraft. However, this cannot be true for two reasons. First, the profiles on the two spacecraft are not nested as they would be for multiple crossings, back and forth, of a single boundary. If anything the impulses seem to convect by one and then the other spacecraft. Secondly, examination of the Berkeley/Toulouse instrument shows very little modulation of the magnetosheath particles after the shock entry. After 11:02 the 72 keV protons and electrons are quite steady in contrast to the behavior of the field. (K. Anderson, and G. Parks, personal communication, 1978) (see Figure 19).

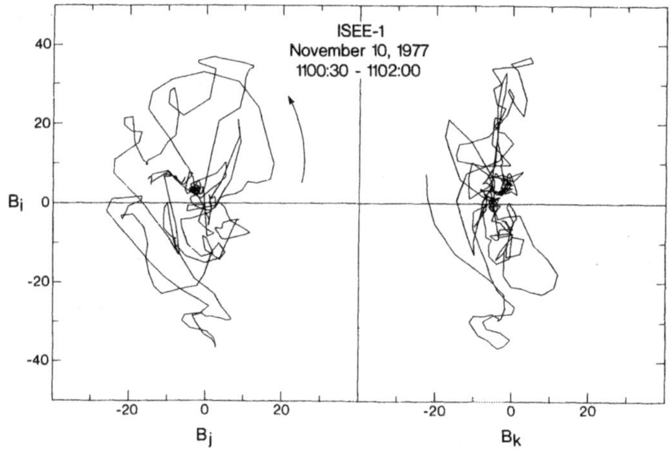

Fig. 14. Hodogram of variation through the shock on November 10, 1977.

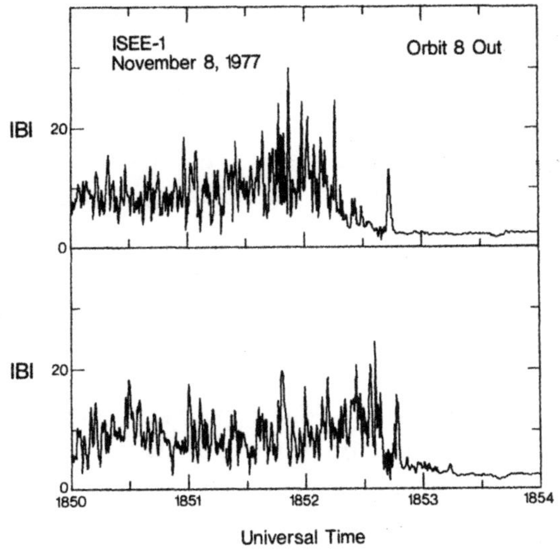

Fig. 15. Magnetic field strength at ISEE-1 and -2 across bow shock on orbit 8 out, November 8, 1977.

5.2. November 8, 1977, 18:53 UT

Figure 15 shows the magnetic field strength at the ISEE spacecraft across the bow shock on orbit 8 outbound. As shown in Table VI the magnetosonic Mach number for this shock is 4.0 and the plasma beta is 19. The upstream field strength here is 2.3γ. Thus the peak downstream field strengths represent an increase of up to a factor of 13 over the upstream value. The two spacecraft are further apart here than in the previous case, 316 km versus 160 km before. There is almost no correspondence of individual features. We have made no attempt to calculate a velocity under such circumstances. The eigenvalue analysis of the variance ellipsoid produced moderately good separation. The eigenvalues were 21.5, 12.7, $4.3\gamma^2$. The k-vector made an angle of 106° with the magnetic field and 62° with the shock normal. Finally we note that the upstream field made an angle of 60° with the shock normal. Thus this shock is close to being classified as quasi-perpendicular based on the field direction alone. We note the relative weakness of the upstream waves compared to the previous example.

5.3. Discussion

Neither of the two high β shocks shown above was as extraordinary as two of the three described by Formisano *et al.* (1975), which had β's of 49 and 170. The main value of these new examples, however, is that the Mach numbers were lower. Thus the shocks of 8 and 10 November serve to isolate the influence of β and to verify that the extreme excursions of the field were indeed associated with the excess of thermal over magnetic energy in the solar wind plasma. One early application of this result suggests itself: it has become traditional, largely for theoretical reasons, to refer to plasma with $\beta \lesssim 1$ as 'high β' plasmas, in contrast to those in which $\beta \ll 1$. The existing observations in the collisionless solar wind support the notion that the term 'high β', insofar as it is intended to imply a distinguishable shock signature, should be reserved for the range $\beta > 2$ or even $\beta \gg 1$. Further investigation may disclose some reproducible threshold above which the 'typical' amplitude extremes appear.

6. Shocks of Orbit 14 Inbound

The shocks encountered on orbit 14 inbound illustrate clearly the effect of changing orientation of the interplanetary magnetic field and hence deserve special mention. We wish to discuss two periods: the period before 16:28 on 11/24/77 when the field was quite perpendicular to the shock normal and the period after 16:28 when it was quasi-parallel.

6.1. The period before 16:28 UT

Figure 16 shows the magnetic field strength at ISEE-1 and -2. The field is quite steady upstream of the bow shock on both satellites except when the foot of the shock is at the satellite. Perhaps surprisingly ISEE-2 enters the shock at 16:22:40 while ISEE-1 did not encounter it until 16:27:30. The satellites moved 330 km along the shock normal during this period and were 278 km apart. Thus the shock was essentially stationary for this period of time. Table VII lists the model normal direction and the angle the upstream field makes with this normal. This is the

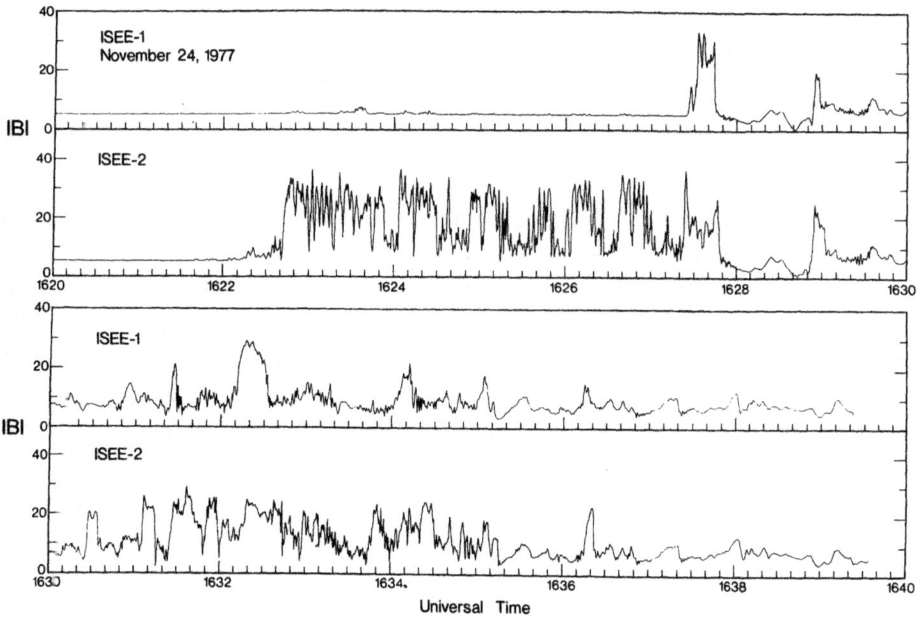

Fig. 16. Magnetic field strength at ISEE-1 and -2 across bow shock on orbit 14 in, November 24, 1977.

TABLE VII
Bow shock parameters, orbit 14 in

Location	(10.84, −8.76, 6.09) R_e	
Separation	(154, −295, 126) km	
Time	<16:28	>16:28 11/24/77
B^u_{GSE}	(−1.4, −4.8, 2.0)	(3.5, 0.8, 4.2)
$^m n_{GSE}$	(0.896, −0.365, 0.254)	
$\sphericalangle \mathbf{B}^u, {}^m\mathbf{n}$	79.2°	45.0°
$\mathbf{S} \cdot \mathbf{n}$	278	278
Preliminary plasma parameters		
V	280	245
$\mathbf{V} \cdot \mathbf{n}$	251	220
N	6	10
T_e	1.75×10^5	2.1×10^5
M_A	5.2	5.7
M_S	3.1	2.7
β	2.5	4.7

most nearly perpendicular shock we have examined to date. It is probably more nearly perpendicular than indicated since the upstream and downstream field directions differ by less than 1°. The use of coplanarity here to derive a shock normal produces a totally strange result as one would expect for almost parallel upstream and downstream fields. Furthermore, this was a high beta shock, $\beta = 2.5$, so that the downstream field was quite irregular. In the Berkeley/Tolouse keV particles this shock was also unusual, in that electrons were heated as usual but protons only weakly. Further analysis is required before the full implications of this observation can be understood.

6.2. The Period After 16:28 UT

At about 16:28 UT the field direction abruptly rotates and the configuration changes from that of a quasi-perpendicular shock to that of a quasi-parallel shock. The plasma parameters also change somewhat, but that might be due to the change in the upstream effects of the shock. The plasma β is high, 4.7, and the separation is large also, 356 km. Thus, there is little correlation between ISEE-1 and -2 in this region. We have taken the region from 16:37 to 16:39 to define the direction of the upstream field. The angle between this direction and the model shock normal is 45°, but instantaneous readings of **B** give angles as low as 33°.

After 16:37 there did appear events that were highly correlated between ISEE-1 and -2. These were the so-called discrete wave packets discovered first in OGO-5 data (Russell *et al.*, 1971). These have the appearance of a damped wavetrain with simultaneously changing period. Hodograms of the event occurring at 16:37:20 are shown in Figure 17. They are very similar on the two spacecraft and extremely two-dimensional. The eigenvalues on ISEE-1 were 31.9,

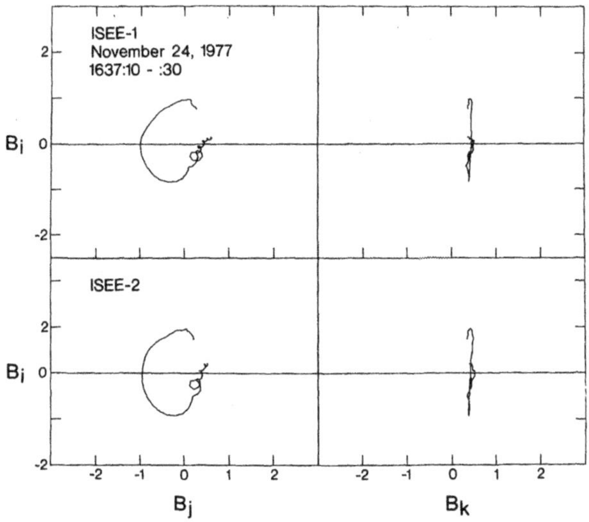

Fig. 17. Hodograms of discrete wave packets seen on November 24, 1977.

15.7, and $0.19\gamma^2$. The k-direction makes an angle of 47° with the ambient magnetic field and is less than 5° from the model normal direction. The ISEE-2 k-vector is within $\frac{1}{4}$ deg of the ISEE-1 k-vector. A second event occurring at 16:38 has a well defined k-direction 25° from this one, making angles of 22° from **B** and 23° from **n**. We believe these fascinating events may bear directly on the process of shock formation and deserve much future attention.

6.3. Discussion

The interval after 16:28 illustrates clearly the difficulty in studying quasi-parallel shock structure. It is very difficult even to decide where the shock is under these conditions. One interesting property of earlier analyses of quasi-parallel shocks was the lack of a significant velocity drop upon entry into the magnetosheath. Preliminary analysis shows that this behavior is also observed here (J. Gosling and S. Bame, personal communication, 1978). This pass has been selected for further study.

The 24 November data also extend somewhat our experience with 'high' β shocks. The crossing of 16:22:40 was at lower β than that of November 10, and the field maxima were an even smaller multiple of the IMF than they had been earlier. In a sense, two sources of structural irregularity were in competition, θ_{Bn} and β, and θ_{Bn} seems to have been the more influential factor. These combinations support the inference expressed earlier (Section 5.3) that 'high' β in shock terminology should mean $\beta > 2$, with an additional provision that the recognized pattern of high β shock structure may require higher and higher β, or may disappear as θ_{Bn} diminishes.

7. Field-Particle Correlations

Thus far our analysis has concentrated almost solely on the magnetic field. A principal objective of the ISEE program, however, is intercomparison of data from many field and particle instruments. Although extensive multidiagnostic analysis has been out of the question at this early stage of investigation, we describe next the correlation of observations from one particle detector with field data covering three bow shock crossings.

The particle data we employ was supplied by the California/Washington/Toulouse group led by Kinsey Anderson of UCB. The instruments measure ions and electrons up to hundreds of keV at high time resolution, but we shall be interested here only in the lowest energy channels at about 1.5 and 6 keV. The detectors look southward and have a highly directional sensitivity which, in the final orientations of ISEE-1 and -2, allows them to sample particles moving northward within roughly ±7.5° of the north ecliptic pole. By examining data from these experiments, we are correlating non-solar wind particles associated with the shock at the specified orientation for the first time. We assume in the following that the ions are not necessarily all protons.

7.1 The 5 November shock

The addition of charged particle to field data exposes the character of the shock much more than either alone, and comparison of data from ISEE-1 and -2 reveals both similarities and differences in the images of the same shock recorded at the two spacecraft. The similarities are most striking and we emphasize those here. Figure 18 displays selected ISEE-1 and -2 field and particle data for three and one-half minutes surrounding the shock crossing of 5 November. In the three panels of the bottom half of the figure, we see that the principal magnetic gradient of the shock wave at ISEE-1 was preceded by a well defined burst of northward directed 1.5 keV protons beginning about 30 s before the shock front and ending exactly at the front. The onset of the burst was not accompanied by any sharp event in the field strength (or in any component) but coincided, to visual approximation, with the forward edge of the exponential envelope of the waves comprising the 'foot' of the shock. The 1.5 keV proton flux then returned to a level about three times what it had been upstream, except for a short pulse, and 20 s later, coincident with the end of the 'plume' or train of 'overshoot oscillations' of the field strength, rose again to the pre-shock burst maximum. The flux then underwent a series of very regular oscillations after which the flux minima gradually increased until they settled a minute and a half later at a high, constant value near the earlier maxima.

At the same time the pre-shock proton burst began, the 1.5 keV electron flux, bottom panel, began a fairly consistent, average increase to a post-shock level about 25 times that in the solar wind.

If we now look at the upper three panels, we see at once the close similarity of

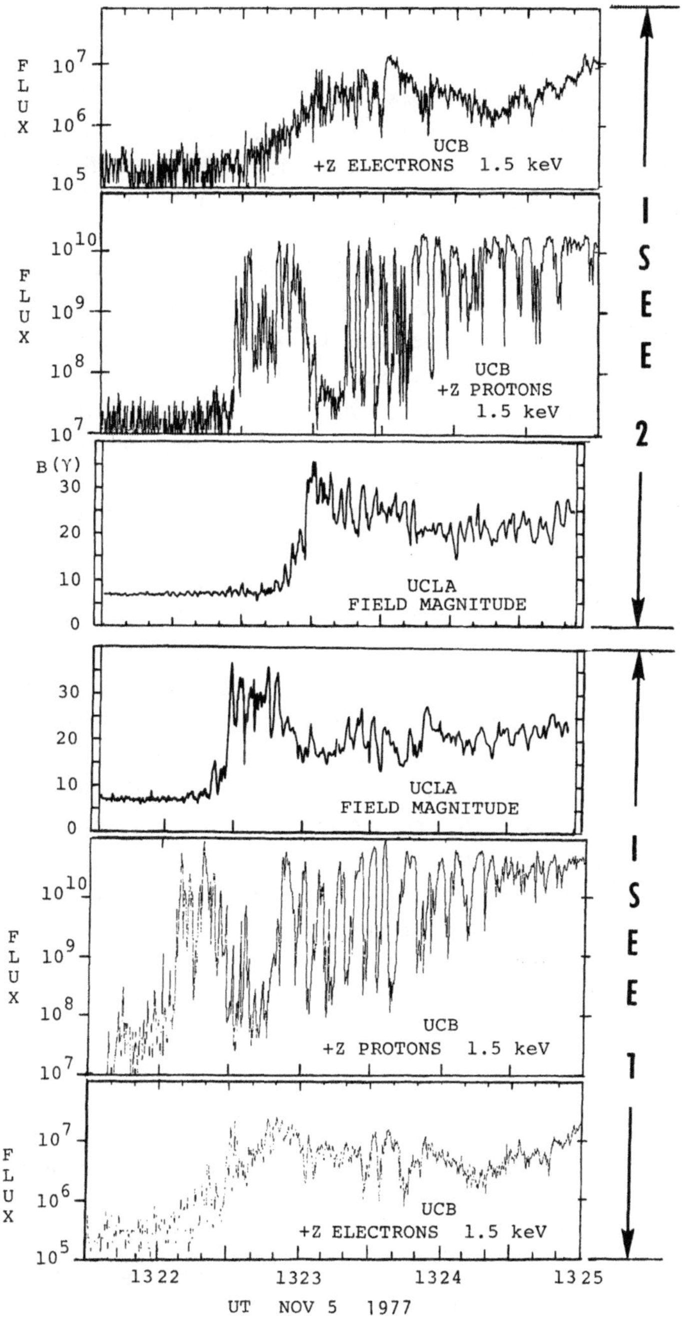

Fig. 18. Simultaneous magnetic field and charged particle data from both ISEE-1 and -2 during bow shock crossings of 5 November, 1977. The charged particles are moving northward ($+Z$) in solar ecliptic coordinates. The two field plots are adjacent at the center, and the respective proton plots are next to them for easy comparison with the field profiles.

structure of the shock at the two locations. The field gradient was preceeded at ISEE-2 by a burst of 1.5 keV proton flux, which dropped at the gradient, then rose and oscillated to settle at peak value a minute and a half or so after the crossing. The 1.5 keV electrons at ISEE-2 executed the same gradual rise as at ISEE-1 in more or less the same relationship to the magnetic profiles.

Certain differences between the crossing signatures tend to emphasize the comparatively stationary character of major shock features in this case. The most notable example seems to be the slightly longer dip in proton flux following the main magnetic gradient at ISEE-1 then at ISEE-2. We notice that the unequal durations of the dips corresponded closely to unequal durations of the overshoot trains in the field at the two locations.

There appeared to be analogous 'zones', or 'isomorphic layers', so to speak, in the two shock profiles. Both profiles included rapid, large amplitude oscillations of ion flux, followed by less rapid oscillations in which the minima were damped upward until the oscillations ended. The two layers of large oscillations were of different absolute duration, but were clearly similar in nature and similarly situated in their respective shock structures. These oscillations stopped simultaneously at the two spacecraft, however, although they began separately, each after its corresponding dip in flux. This coincidence draws attention to the kind of cautious approach analysis of the data required. The reader will observe that beginning about 13:23:26 the general behavior of the northward electron flux was substantially the same at the two spacecraft at the same time, i.e., they no longer showed the 25–30 s delay between ISEE-1 and -2 that applied to the major analogous features before. In fact, the large ion flux oscillations ended at both satellites shortly after the occurrence of *concurrent* large excursions in the electron flux. Thus certain standing features of the shock may have been modified or disrupted by the rapid passage of inhomogeneities in the solar wind, and care must be taken in ascribing every detail to the shock structure itself. The necessity of having simultaneous observations at two points is nowhere more apparent.

7.2. The 10 November shock

Another case exhibiting structural stability is shown in Figure 19, where quite different parameter conditions prevailed (Table VI). The differences in conditions conspired to make the 10 November shock of the figure appreciably looser in structure than that of 5 November. Energy ratio β was significantly higher, and the field-normal angle was lower, in fact, near or below the 45–52° changeover to quasi-parallel configration. Moreover, the actual magnetic character of the shock, as seen at either vehicle was clearly as irregular as would be expected from the upstream parameter set.

Despite these differences, the shock profiles at ISEE-1 and -2 were essentially as similar to one another as in the previous example. The two lowest panels of Figure 19 show that the major magnetic pulses of about 20–30 s period that made

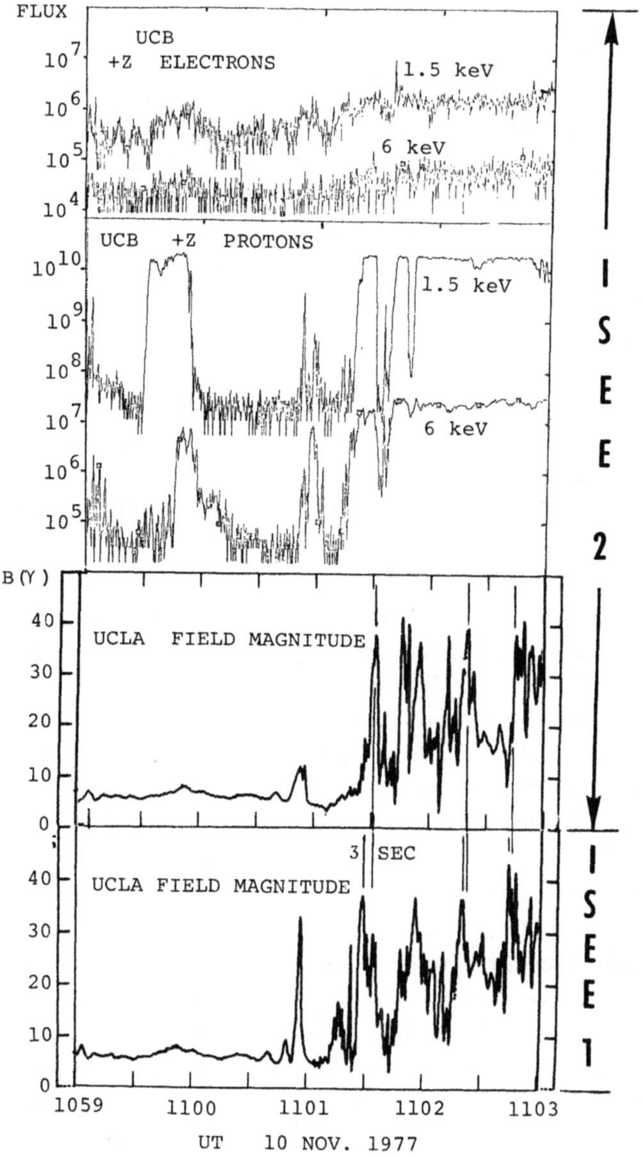

Fig. 19. Comparison of field at ISEE-1 (bottom) with field and particle data from ISEE-2 during bow shock crossing of 10 November, 1977.

up the shock front were repeated at both locations, although they were not identical in detail.

There were several interesting aspects of the 10 November crossing beyond the evident local general uniformity of its internal magnetic structure. In particular, the one-for-one match of the large pulses was made with no time shift in either plot of B in Figure 19. That is, the crossings took place at essentially the same time at both spacecraft even though one was 122 km closer to Earth than the other along the nominal (i.e., model) shock normal. The vertical lines in the bottom panels of the figure mark what appear to be identifiable analogous features of the two magnetic profiles. The delay of these features between the satellites was no more than about 3 s. Either the entire 'front' was moving outward at 40 km s^{-1} or faster, or it had a 'kink' of at least 122 km amplitude. Shock velocities over 40 km s^{-1} are above average but not unusual.

A second aspect of the 10 November crossing was the rapid and permanent rise in northward proton flux coincident with the outermost pulses of the field. Again, the flux rose ahead of the first large field gradient at ISEE-2, then subsided, but ultimately remained high within 30 s of the front crossing. Ongoing oscillations following the initial rise, as seen on 5 November (Figure 18), were absent on 10 November despite the far more prominent field oscillations. At the same time, the northward 1.5 keV electron flux, top panel, reached its post shock level by the end of the first large magnetic pulse, although the level was considerably less than that of 5 November. These particle data confirm the contrast between one type of structure and another.

It will be recalled that the UCB instrument looks southward with a 15° acceptance cone so that field geometry can play a central role in determining what the detector sees. The Z-component of the IMF in both 5 and 10 November crossings was small, with the difference in field-to-shock-normal geometry hinging on a larger X-component in the 10 November than in the 5 November case. Thus the IMF was essentially in the ecliptic, normal to the detector's direction of sensitivity on both occasions and the detection geometry was the same, to first order, at least upstream. The presence of substantial ion flux in the foot ahead of the first large field gradient would therefore seem to be a persistent signature of the shock even under moderately divergent plasma parameter conditions, and even when the total shock profiles differ.

Finally, a third aspect of the 10 November crossing was the sudden appearance and disappearance of strong ion flux in the solar wind ahead of the shock, between 10:59:30 and 11:00. Since we now know, as evidenced in this report, that such ion pulses may accompany the foot of the shock, the incident could have been associated with a close approach of the shock to the spacecraft. However, we understand from the UCB group that comparable pulses are often seen in the solar wind when upstream waves are present, and we interpret this to be an example of a detached particle event.

7.3. The 8 November Shock

The crossing of 8 November, when β was very high in the solar wind (Table VI), demonstrated again the 'layering' of the particle profile of the bow shock and its reproducibility. Figure 20 compares the UCB and UCLA data from the two spacecraft, with the field magnitudes at the center, northward electron fluxes immediately above and below, and northward proton fluxes at top and bottom. In

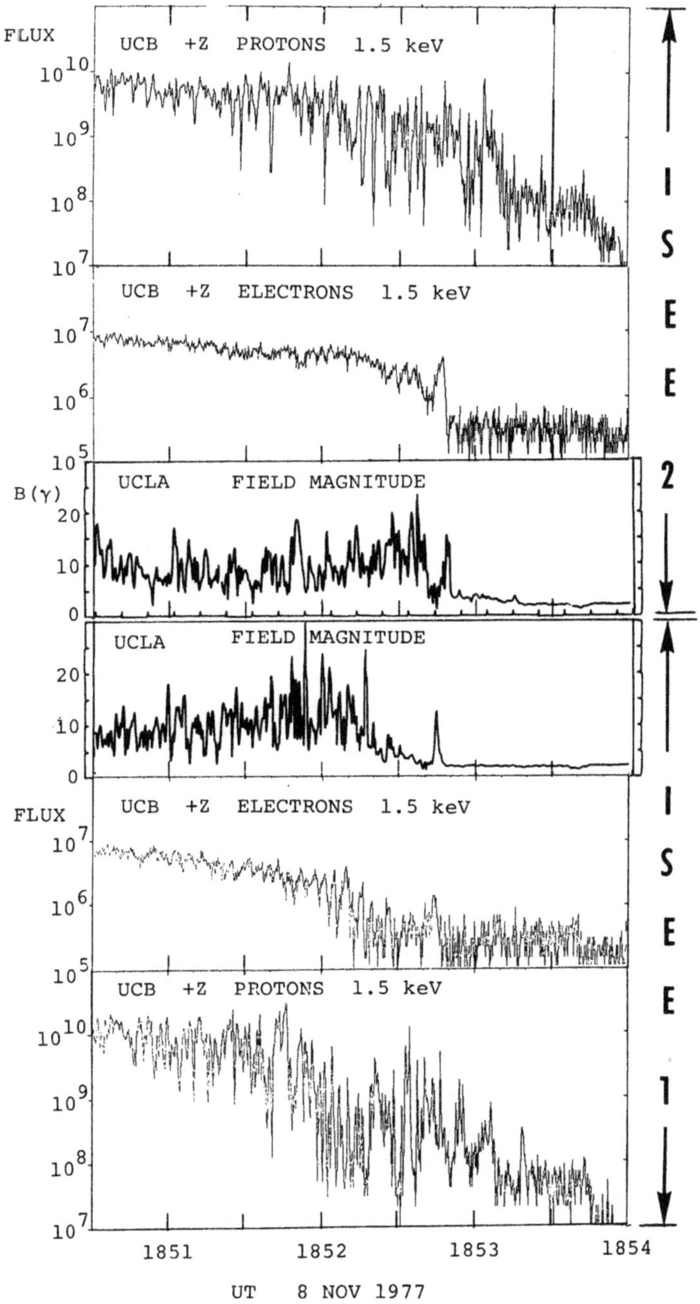

Fig. 20. Simultaneous data from ISEE-1 (lower half) and 2 (upper half) during bow shock crossings of 8 November, 1977. In this figure, the UCB electron observations are adjacent to their corresponding field plots because they give a sharper comparison with B than do the protons.

this case we see that the outermost, isolated pulse at either satellite was accompanied by a sharp rise in electron flux, behind which the electron flux dropped and then gradually rose again to its downstream level, at the left.

The drop in field behind the initial pulse, and the separation between the pulse and the general rise in average field strength, were different at the two spacecraft, but the electron fluxes followed their respective field profiles, showing the same relative spacings in their features. One explanation would be that the shock was stationary or moving very slowly until about 18:52:30, then suddenly contracted, compressing the remainder of the structure in the ISEE-2 record. The main points we wish to make here are that the regularity of the shock's local, general features persisted in the parameter regime of 8 November and that comparison of field and particle data are essential to detecting that regularity.

The protons showed less identifiable similarity of structure than the electrons. They are included in Figure 20, first, to emphasize the importance of examining every diagnostic, since the protons without the electrons would not have confirmed the apparent structural regularity of the shock; and second, to exhibit again the occurrence of large proton fluxes upstream in a particle 'foot' ahead of the outermost significant magnetic feature. The proton feet in this case accompanied very small subtle waves in the interplanetary field. The IMF was again close to the ecliptic, as in the preceding two cases.

7.4. Discussion

We have used the terms proton and ion interchangeably because we cannot be certain at this stage of analysis how many alpha particles contributed to the positive ion fluxes detected. We expect that protons make up the overwhelming majority of flux almost everywhere, so that we equivocate only with a particular phenomenon in mind, namely, the pulses of flux that appeared in the feet of the shock fronts.

A stream of reflected protons is expected just upstream from a supercritical shock, which the crossing of 5 November seems to have been, but the energies could also have been those of alphas as well. Given the limited energy channel sampling capability of the ion instrument, we do not know the energy at which the ion flux was actually maximized. We also note that similar particles were detected on 10 November, when numerous reflected protons were not expected because of the lower Mach number.

It seems improbable that the flux pulses ahead of the shock could have represented solar wind thermalization before the first magnetic gradients were crossed. If the pulses were indeed composed of protons, we would take these to represent a distribution of shock-associated, energized protons separate and distinct from the solar wind bulk distribution. Such a double distribution might satisfy the conditions for generating ion-ion waves which have been suggested as one possible cause of ion thermalization in collisionless shocks.

Clearly, resolution of the kinds of issues raised above will depend on comprehensive study of all outputs from more instruments, and more cases too. What these examples demonstrate is how dependent our comprehension of the shock is on the diagnostics employed. Neither the particle nor the field detector alone could have been used to construct the shock profiles recorded by the other instrument. One way of expressing our aim in further study of the shock is to say that we shall understand this collisionless phenomenon when we can make such constructions reliably.

8. Conclusions

The above results have but scratched the surface of information on collisionless shocks contained in the ISEE data. We have examined but a few bow shock crossings at high Mach number. We wish to examine ultimately bow shocks over a wide range of Mach numbers. Our range of plasma beta during these crossings was also quite limited. The data certainly suggest important differences in the shocks at high and low beta, but examination over a range of β is required to elucidate the effect. Perhaps most obvious in our data is the control of the shock structure by the direction of the magnetic field. When the interplanetary field is perpendicular to the shock normal, the shock is more regular and perhaps more amenable to analysis. However, on the dawn side of the shock where the initial ISEE data were obtained the number of such occasions is small. We await the data on the afternoon side where perpendicular shocks will be much more frequent. The overshoot in field magnitude seems to be a regular feature of at least high Mach number quasi-perpendicular shocks. We expect that the overshoot region must play an important role in the physics of collisionless shocks. The scale length of the leading edge of the quasi-perpendicular bow shock appears to be of the order of an ion inertial length.

The study of interplanetary shocks forms a nice complement to the study of the terrestrial bow shock. These shocks tend to be much weaker than the terrestrial bow shock, but since they are travelling faster than the solar wind speed they tend to appear much sharper in the satellite records and thus benefit from the higher of the ISEE data rates. We must depend on solar largesse to provide us with a sufficient number of interplanetary shocks to study. In the initial ISEE data such shocks were rare.

The dual spacecraft concept of the ISEE mission has proved essential to the success of our investigations thus far. Not only is it critical to the determination of velocities and thickness so that we can compare quantitatively with theory, but it also can tell us what features are time stationary and which are non-time stationary. Even the simple redundancy has proven useful when determining shock normals and thicknesses.

In short ISEE is fulfilling its promise. There is much more data to be taken, much data to be analyzed and many interrelated data sets to compare. However, the nature of the initial results, and the spirit of cooperation evident at the workshops and in the papers of this volume all indicate that the ISEE mission will succeed perhaps even beyond the wildest hopes of its original advocates of over a decade ago.

Acknowledgements

We are extremely pleased and grateful for the spirit of cooperation and free data exchange exhibited during these early stages of the ISEE mission even while software problems were still being sorted out and while the investigators had not even had time to digest their own data. We are particularly grateful to S. J. Bame, G. Paschmann, and J. Gosling for allowing us free access to the LASL/MPI fast plasma data, spending much time explaining it to us, and sharing their interpretation of the data. We are particularly grateful to K. A. Anderson, R. P. Lin, and G. Parks who provided us with literally hundreds of plots of their highest resolution data and for sponsoring two workshops at Berkeley that were instrumental in initiating many of these studies. We also wish to thank K. W. Ogilvie and J. Scudder for freely sharing their electron data and D. A. Gurnett, R. Anderson, and F. L. Scarf for freely sharing their plasma wave data. We expect these data sets to become even more useful in future analyses.

Finally, we would not be able to even begin these analyses if we did not have excellent instruments and clean spacecraft with which to work. The number of persons contributing to this effort was large but special mention must go to B. Plitt and his team at Westinghouse and R. C. Snare, F. R. George, and J. D. Means at UCLA who were responsible for the magnetometers and to M. Davis and J. J. Madden of GSFC and D. Eaton and R. Gruen of ESTEC who were responsible for the ISEE-1 and -2 spacecraft respectively.

This work has been sponsored by the National Aeronautics and Space Administration under contract NAS 5-20064 at UCLA and under NASA Contract NASW-3087 at TRW.

References

Fairfield, D. H.: 1971, 'Average and Unusual Locations of the Earth's Magnetopause and Bow Shock', *J. Geophys. Res.* **76,** 6700.

Fairfield, D. H. and Behannon, K. W.: 1976, 'Bow Shock and Magnetosheath Waves at Mercury', *J. Geophys. Res.* **81,** 3897.

Fairfield, D. H. and Feldman, W. C.: 1975, 'Standing Waves at Low Mach Number Laminar Shocks', *J. Geophys. Res.* **80,** 515.

Formisano, V.: 1977, 'The Physics of the Earth's Collisionless Shock Wave', *J. Phys.* **38,** suppl. 12, C6-65.

Formisano, V, Russell, C. T., Means, J. D., Greenstadt, E. W., Scarf, F. L., and Neugebauer, M.: 1975, 'Collisionless Shock Waves in Space: a Very High β Strucutre', *J. Geophys. Res.* **80,** 2013.

Fredricks, R. W., Kennel, C. F., Scarf, F. L., Crook, G. M., and Green, I. M.: 1968, 'Detection of Electric-Field Turbulence in the Earth's Bow Shock', *Phys. Rev. Letters* **21,** 1761.

Greenstadt, E. W.: 1976, 'Phenomenology of the Earth's Bow Shock System. A Summary Description of Experimental Results', in B. M. McCormac (ed.), *Magnetospheric Particles and Fields*, D. Reidel Publ. Co., Dordrecht, Holland, p. 13.

Greenstadt, E. W. and Fredricks, R. W.,: 1978, in C. F. Kennel, L. J. Lanzerotti, and E. W. Parker (eds.), *A Twentieth Anniversary Review*, North-Holland Publ. Co., in press.
Greenstadt, E. W., Russell, C. T., Scarf, F. L., Formisano, V., and Neugebauer, M.: 1975, 'Structure of the Quasi-Perpendicular Laminar Bow Shock', *J. Geophys. Res.* **80,** 502.
Greenstadt, E. W., Russell, C. T., Formisano, V., Hedgecock, P. C., Scarf, F. L., Neugebauer, M., and Holzer, R. E.: 1977, 'Structure of a Quasi-Parallel, Quasi-Laminar Bow Shock', *J. Geophys. Res.* **82,** 651.
Greenstadt, E. W., Formisano, V., Russell, C. T., Neugebauer, M., and Scarf, F. L.: 1978, 'Ion Acoustic Stability Analysis of the Earth's Bow Shock', *Geophys. Res. Letters* **5,** 399.
Holzer, R. E., Northrop, T. E., Olson, J. V., and Russell, C. T.: 1972, 'Study of Waves in the Earth's Bow Shock', *J. Geophys. Res.* **77,** 2264.
Lemons, D. S. and Gary, S. P.: 1978, 'Current-Driven Instabilities in a Laminar Perpendicular Shock', *J. Geophys. Rev.* **83,** 1625.
Montgomery, M. D., Asbridge, J. R., and Bame, S. J.: 1970, 'Vela-4 Plasma Observation Near the Earth's Bow Shock', *J. Geophys. Res.* **75,** 1217.
Morse, D. L. and Greenstadt, E. W.: 1976, 'Thickness of Magnetic Structures Associated with the Earth's Bow Shock', *J. Geophys. Res.* **81,** 1791.
Neugebauer, M., Russell, C. T., and Olson, J. V.: 1971, 'Correlated Observations of Electrons and Magnetic Fields in the Earth's Bow Shock', *J. Geophys. Res.* **76,** 4366.
Russell, C. T.: 1978, 'The ISEE-1 and -2 Magnetometers', *IEEE Trans. Geosci. Electr.* **GE-16**, 239.
Russell, C. T., Childers, D. D., and Coleman, P. J.: 1971, 'OGO-5 Observations of Upstream Waves in the Interplanetary Medium: Discrete Wave Packets', *J. Geophys. Res.* **76,** 845.

EARLY RESULTS FROM THE ISEE ELECTRON DENSITY EXPERIMENT

C. C. HARVEY, J. ETCHETO,* and A. MANGENEY
Observatoire de Paris, 92190 Meudon, France

Abstract. The ISEE-1 and 2 spacecraft contain two complementary experiments to measure the ambient electron density by radio techniques: a propagation experiment which measures the integrated electron density between ISEE-1 and 2, and a resonance sounder which measures the electron density in the vicinity of ISEE-1, and also provides AC electric field data. These experiments have been described elsewhere (Harvey *et al.*, 1978). Results from these two experiments are presented here for the first time. The propagation experiment permits high time resolution studies of density fluctuations in the solar wind and magnetospheric frontier regions. The sounder experiment has detected for the first time plasma resonances in the solar wind and in the Earth's magnetosheath, as well as in the regions of the magnetosphere where resonances have already been observed by the spacecraft GEOS-1. We present here a preliminary review of the different types of electric field noise observed in the solar wind and magnetosheath, and discuss their relationship to the measured plasma density.

1. Introduction

The electron number density, equal to the sum of the number densities of the different ion species present, is probably the most important single parameter characterising any plasma. Its measurement is essential for understanding both macroscopic and microscopic processes in the plasma, as is illustrated by the following examples:

– The study of the magnetohydrodynamic fluid properties of spatial plasma, such as the pressure balance across the magnetopause or bow shock, or the MHD stability of the magnetotail, requires measurement of the plasma density.

– Accurate plasma density measurements are necessary for studying the micro-instabilities which, in a collisionless plasma, are the only means of creating the dissipation required, for example, in shocks, in regions of magnetic flux tube reconnexion, and in other regions with anomalous transport coefficients. The growth rates, or even the existence of the microinstabilities, is often very sensitive to the ratio of the wave frequency to the local plasma frequency.

– High time resolution density measurements permit the study of MHD waves, for example, to investigate the nature of solar wind density irregularities responsible for the interplanetary scintillation of distant radio sources, and the relation of these density irregularities to dissipation and acceleration processes in the solar wind.

To meet these objectives the French electron density experiment aboard the ISEE spacecraft measures the electron density by two independent and complementary radio techniques. Both techniques determine the plasma frequency via the dispersion relation for waves, which are electromagnetic in one case and

* CRPE/CNET, 92131 Issy-les-Moulineaux, France.

Space Science Reviews **23** (1979) 39–58. 0038-6308/79/0231-00039 $0.3.00.

mainly electrostatic in the other. Therefore the measurements are not affected by spacecraft charging. Furthermore, for both measurements, the waves propagate to large distances from the spacecraft, so that the measurements are influenced negligibly by the sheath surrounding the spacecraft, whether it be caused by photoelectrons or by spacecraft charging. Previous spacecraft have relied upon particle experiments or on natural wave observations to determine the electron density. Particle experiments may count either electron or ion fluxes; both are perturbed by the floating potential of the spacecraft, as has been studied extensively, for instance, by de Forrest (1972) and Whipple *et al.* (1974). In addition, the measured electron fluxes include the contribution from the photoelectrons emitted by the spacecraft. The determination of the plasma frequency from the observation of naturally occurring waves is based upon plausability arguments; and clearly it would be a circular argument to try to use the density derived from natural wave observations for the study of the interactions producing these waves.

One part of the present experiment determines the integrated electron density between the ISEE-1 and 2 spacecraft by measuring the phase advance of a wave transmitted from ISEE-1 and received on ISEE-2. This experiment is supplied by the Department de Recherches Spatiales of the Observatoire de Paris. The other part of the experiment is a resonance sounder, similar to the one on GEOS, which is supplied by the Centre de Recherches en Physique de l'Environnement. A technical description of the instrumentation on board both spacecraft has been given elsewhere (Harvey *et al.*, 1978).

We present here the first results from these two experiments. Sounder data for the first four weeks after launch (1977, October 22) has been processed systematically, and a number of characteristic features have been identified and are described in Sections 5 and 6. Data from the propagation experiment has not yet been processed systematically.

2. The Propagation Experiment

This, the simpler of the two experiments, measures phase velocity of a radio wave of frequency 683 kHz transmitted from ISEE-1 and received on ISEE-2. The choice of frequency has been explained by Harvey *et al.* (1978), and is related to the characteristics of the long wire antenna used for transmission. The wave velocity is measured by comparing the phase of the wave received on board the ISEE-2 with the phase of the same wave modulated on to a carrier of frequency high enough to be unaffected by the intervening plasma. The phase advance is related to the electron density by

$$\phi = \frac{Z_0 e^2}{2m} \frac{1}{\omega} \int_1^2 N_e \, dl \text{ (radians, MKS units)}$$

which may be conveniently expressed

$$\phi = 7.092°/(\text{cm}^{-3} \times 100 \text{ km}).$$

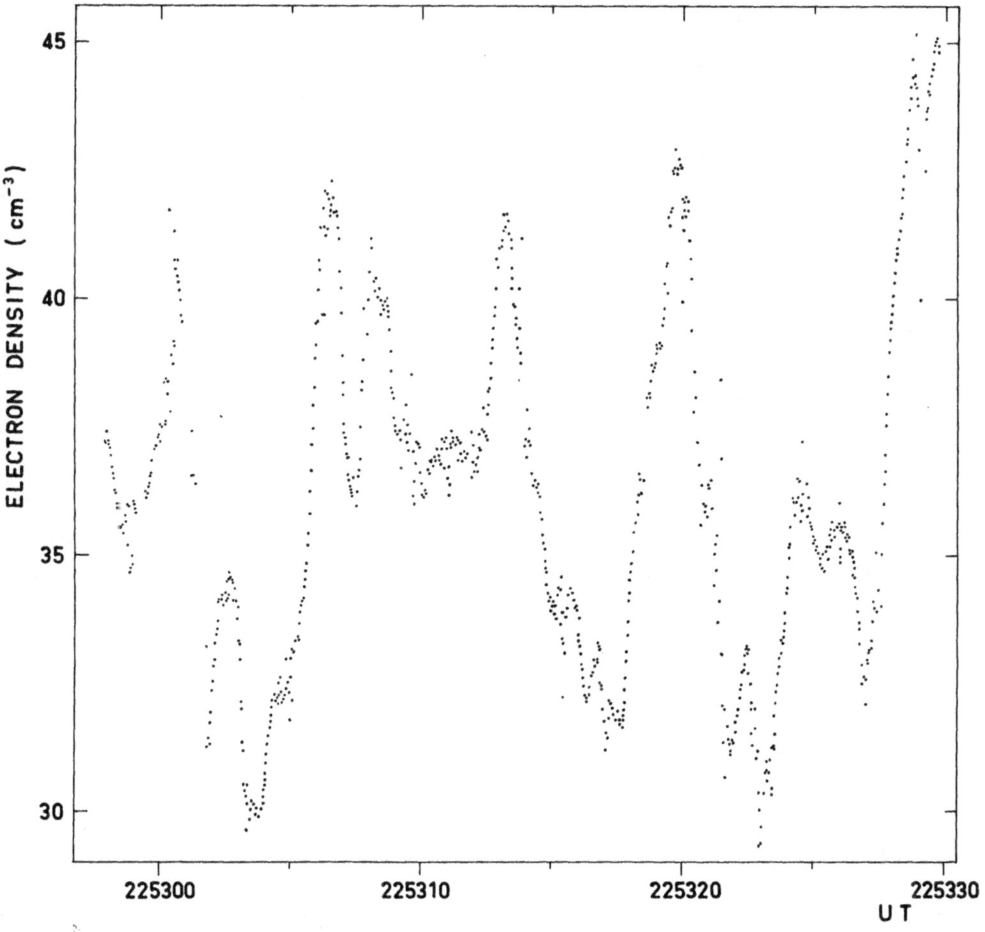

Fig. 1. Densities obtained from the propagation experiment in high bit rate on day 77/11/07 just behind the bow shock. The experiment was working in a 32 s per 17 min 4 s duty cycle, and data from one 32 s interval is shown here. Each point represents one independent determination of the density. As explained in the text, the phase advance has a 360° p ambiguity, where p is an integer for which, in the present example, the value $p = 1$ has been taken.

The random error in the phase determination is of the order of 1°; systematic errors (separation distance, small antenna phase shifts) are presently under study. In high bit rate the experiment makes 32 independent measurements per second of the integrated electron density between the two spacecraft. The maximum separation distance for an adequate signal to noise ratio is of the order of 1000 km. The spin of the two spacecraft produces (in general) four 180° jumps in the phase of the received signal per spin period; these are removed during data processing using spacecraft orientation data. The 683 kHz signal is electromagnetic, and thus the propagation experiment measures the total electron density, independently of the velocity distribution.

Figure 1 shows 32 s of typical data from the propagation experiment, obtained on day 77/11/07 from 22:52:58 to 22:53:30 UT, when the two spacecraft were in the Earth's magnetosheath at a distance of $15.4R_E$ and approximately 10:30 LT. Large phase fluctuations with a typical time scale of 7 s and (peak-to-peak) amplitude 180° can be seen. The separation was about 252 km, so these fluctuations correspond to density fluctuations of the order of 10 cm^{-3}. The mean phase advance (modulo 360°) is 315°: we think that the mean total phase advance was 675°, corresponding to a mean electron density of 38 cm^{-3}, which is above the range of the sounder at that time.

3. The Resonance Sounder

When an electric antenna is excited in a plasma, it is found that for certain frequencies a pronounced 'ringing' of the plasma is observed after the excitation has stopped (Calvert and McAfee, 1969; Muldrew, 1972). This phenomenon occurs at frequencies for which there exists a mode of wave propagation with zero or small group velocity (Fejer and Calvert, 1964; Etcheto and Bloch, 1978). These frequencies are called resonant frequencies, and an experiment to systematically detect and measure these resonances is called a resonance sounder. Such experiments have been constructed by the Centre de Recherches en Physique de l'Environnement for the GEOS and ISEE-1 spacecraft.

The operation of the ISEE-1 sounder is illustrated in Figure 2. A radio receiver with 400 Hz bandwidth is driven by a frequency synthesizer so that it steps in frequency. The frequency step is 395 Hz, and there are 128 steps giving an overall frequency range of 0 to 51 kHz. Each frequency step lasts 125 ms, thus leading to a period of 16 s for a complete frequency sweep. The receiver output is sampled at a frequency of 64 Hz; a buffer memory is used in low bit rate. In fact, this experiment has alternative values for the receiver bandwidth, frequency step, frequency range, step duration and sampling rate, as detailed by Harvey *et al.* (1978); but the above mode of operation is the only one used to obtain the data described in this paper.

Figure 2 also shows the time sequence during each frequency step. It takes about 10 ms for the frequency synthesizer to stabilise at the correct frequency. Approximately 11 ms after the beginning of the frequency step a wave-train of duration 2.5 ms and frequency corresponding to the centre of the receiver passband is transmitted. The duration is chosen so that the spectrum of the transmitted pulse fills the receiver passband; the receiver is off during transmission to avoid saturation. 17 ms after the beginning of the frequency step the receiver is activated, and six values of the receiver output are obtained during the remainder of the frequency step.

In the magnetosphere resonances are seen at the plasma frequency, the upper hybrid frequency, the electron cyclotron frequency and its harmonics, and at the other frequencies above the upper hybrid frequency, the so-called 'f_Q' frequencies, at which the Bernstein mode has zero group velocity (Etcheto and Petit, 1977). Figure 3 shows data obtained from the sounder when ISEE-1 was in the

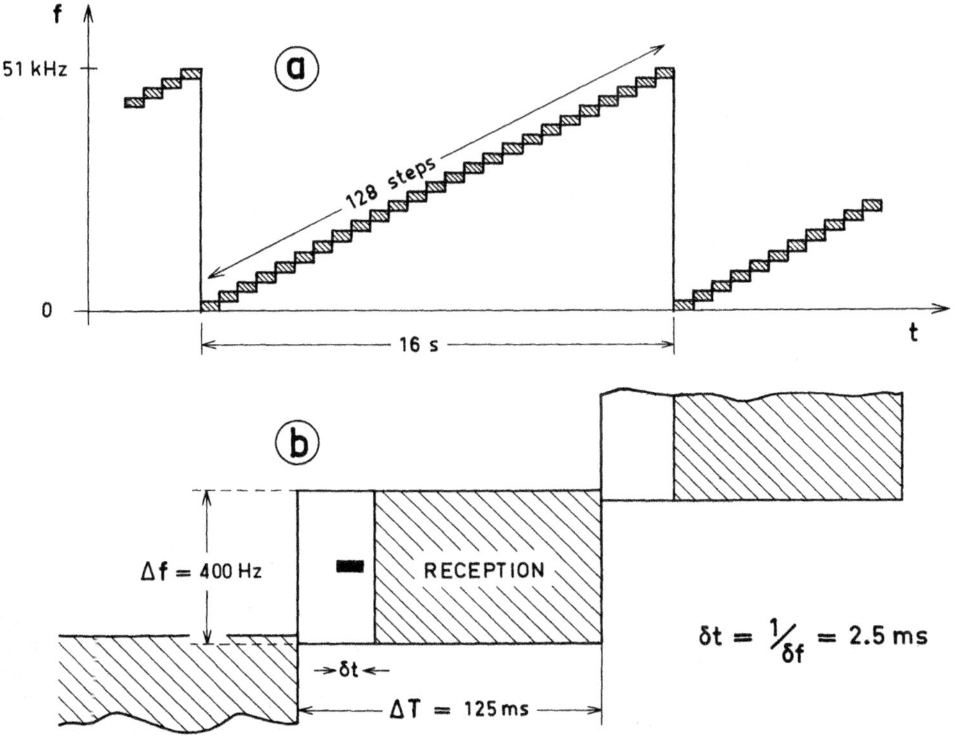

Fig. 2. Schematic diagram of the operation of the resonance sounder at the times considered in this paper. The sounder receiver, with bandwidth 400 Hz, steps in 16 s through 128 frequency steps between 0 and 51 kHz. The lower part of the figure shows the time-sequence inside each frequency step, as described in the text. During pulse transmission unused telemetry words are filled with engineering data.

plasma trough at about 11:00 LT; this type of display has for historical reasons, been called an ionogram, but echogram seems to be a more descriptive and appropriate name. Resonances are clearly seen, and have been identified as indicated. The quasi-exponential decay of the resonances is seen. The characteristic decay time is a function of the type of resonance and its frequency, and is longest for the resonance at the plasma frequency.

How close to the plasma frequency is the plasma resonance? This is a complicated question, and a definitive reply must await a detailed analysis of the analog telemetry data. Close to resonance, the medium is highly dispersive, and hence wave reflexion is produced by only a small density gradient. The observed plasma resonance corresponds to a plasma wave propagating away from the spacecraft, and then being refracted back and intercepted by the spacecraft (McAfee, 1968, 1969). The intercepted wave constitutes the resonance. In fact, there are often two such ray trajectories, corresponding respectively to rays propagating forwards and backwards at their point of reflexion. At any time each echo has a very precise frequency, and the ray tracing computations (McAfee,

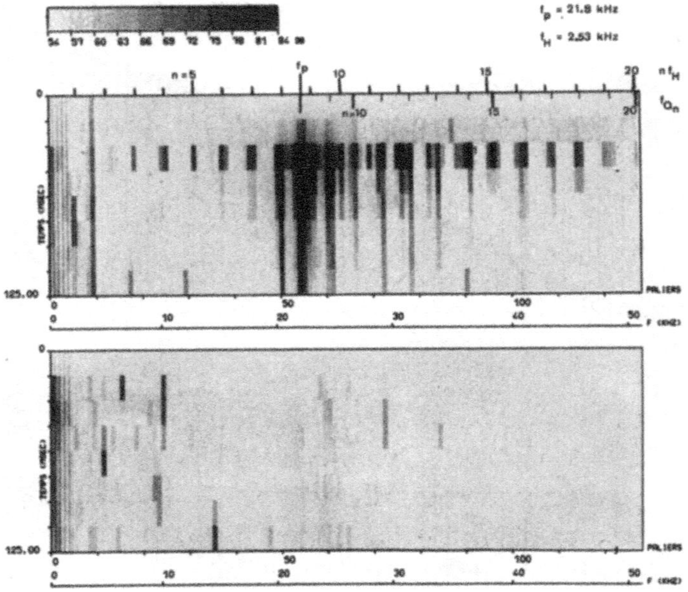

Fig. 3. Pair of echograms of sounder output. The abscissa is frequency step number, the ordinate is time after the beginning of the frequency step, and the intensity (blackness) is proportional to the sounder receiver output, in decibels with respect to an arbitrary level. The first two of the eight data points on each frequency step do not contain useful information. The upper echogram corresponds to a sweep with the transmitter active, the lower one to an adjacent sweep with the transmitter off; the difference is due to the presence of resonance signals, which have been identified as indicated. These echograms were obtained on day 77/11/1 at 01:16 UT when the spacecraft was in the plasma trough.

1969) for ionospheric conditions show that the frequencies of the forward and backward echoes are different, being within a ratio of 10^{-3} of the true plasma frequency. The difference causes modulation of the resonance signal, the beat frequency being a function of the electron temperature. When these calculations were applied to plasma resonance observations, the temperature derived agreed with the expected values (Warnock *et al.*, 1970; Feldstein and Graff, 1972), which suggests that the theory is substantially correct. Using a particular geometry (horizontally stratified ionosphere with the satellite travelling parallel to the horizontal magnetic field), Fejer and Yu (1970) were able to confirm the validity of the ray tracing approach used by McAfee, and also to derive the expression

$$\omega = \omega_p\left[1+\frac{t}{\tau}\pm\frac{v}{V}\left(\frac{3t}{8\tau}\right)^{\frac{1}{2}}\right] \tag{2}$$

for the frequency of the two (forward and backward) plasma frequency echoes at times

$$t \gg t_1 = \surd(\tau/4\omega_p). \tag{3}$$

Here,

$$\tau = 2\omega_p H/VA = 2H/\lambda_D A, \tag{4}$$

$$A = \omega_p \Omega_e/\surd(\omega_p^2 - \Omega_e^2), \tag{5}$$

v is the spacecraft velocity with respect to the medium which has Debye length λ_D, thermal velocity $V=(KT/m)^{\frac{1}{2}}$ and a (linear) density gradient of scale height H. The two waves with frequency described by (2) have equal amplitude and decay as $t^{-\frac{1}{2}}$; the amplitude of the wave accompanying the spacecraft after the emission (i.e., not reflected back) decays as $t^{-\frac{5}{2}}$ (Deering and Fejer, 1965), and the two are equal when $t=t_1$. With the ionospheric values used by McAfee (1969) and Fejer and Yu (1970), namely $H=400$ km, $f_p=1.5$ MHz, $V=1.75\times10^5$ m s^{-1} ($\lambda_D=1.8$ cm), and $A=4.13\times10^6$ rad s^{-1}, we obtain $\tau=10$ s and $t_1=0.5$ ms.

The ISEE resonances are observed to decay during each frequency step, which lasts for about 100 ms after the emission has ceased. In the magnetosphere the second term in Equation (2) is larger than the third, and the observed resonance is within 1% of the plasma frequency if $\tau>10$ s, that is, if $H/\lambda_D A \sim H/\lambda_D \Omega_e \gtrsim 5$ s. This condition is probably satisfied everywhere within the magnetosphere, except possibly near the plasmapause. Even if $t<t_1$ and the accompanying wave is observed, we may note that this wave must have a group velocity equal to the spacecraft velocity v relative to the plasma; using the Bohm–Gross dispersion relation, it may be shown that the Doppler shift simply reduces the magnitude of the dispersion shift, giving

$$\omega=\omega_p[1+\tfrac{1}{2}(v/V)^2] \tag{6}$$

which, in the magnetosphere, is also within 1% of the true plasma frequency.

The situation in the solar wind is more complicated. The particular geometry used by Fejer and Yu could be a reasonable representation of the situation when the magnetic field is radial. Using the solar wind values $H=50$ km (~ thermal proton gyro-radius) $f_p=30$ kHz, $V=1.35\times10^6$ m s^{-1} (temperature 1.2×10^5 K, $\lambda_D=7$ m), $A=2\times10^3$ rad s^{-1}, we obtain $\tau=7$ s, $t_1=3$ ms. At time $t\sim20$ ms condition (3) is just satisfied. For a typical solar wind speed $v=400$ km s^{-1} we have $v/V\sim0.3$; the frequency shift comes mainly from the Doppler (third) term in (2), and is of the order of 1%. It is even smaller if H is larger, but in this case condition (3) indicates that the accompanying wave may be observed. If so, it must satisfy (6), which yields an error of 4% using the above values of v and V. However, this wave is probably not observed at all, because its phase velocity is $v_\phi=V^2/v\sim3.4\,V$; although not appreciably damped on the main part of the electron distribution, it will be damped within a few wave periods by the 'halo' part of the distribution, which has a typical velocity of $3.4\,V$ ($E\sim60$ eV) and normally represents about 5% of the total electron density (Feldman *et al.*, 1975).

The above analytic expressions have been extended to a more general geometry by Graff (1971). The conclusion remains basically unchanged in the magnetosphere, but in the solar wind there is an additional Doppler shift of approximately $vt\cos\alpha/2H$ (~$4\cos\alpha$% using the above values), where α is the angle between the density gradient and the solar wind velocity. This term decreases as H increases; and if the Doppler shift is as large as 4%, the wave is probably completely Landau damped. The analysis of Graff also shows that both Doppler

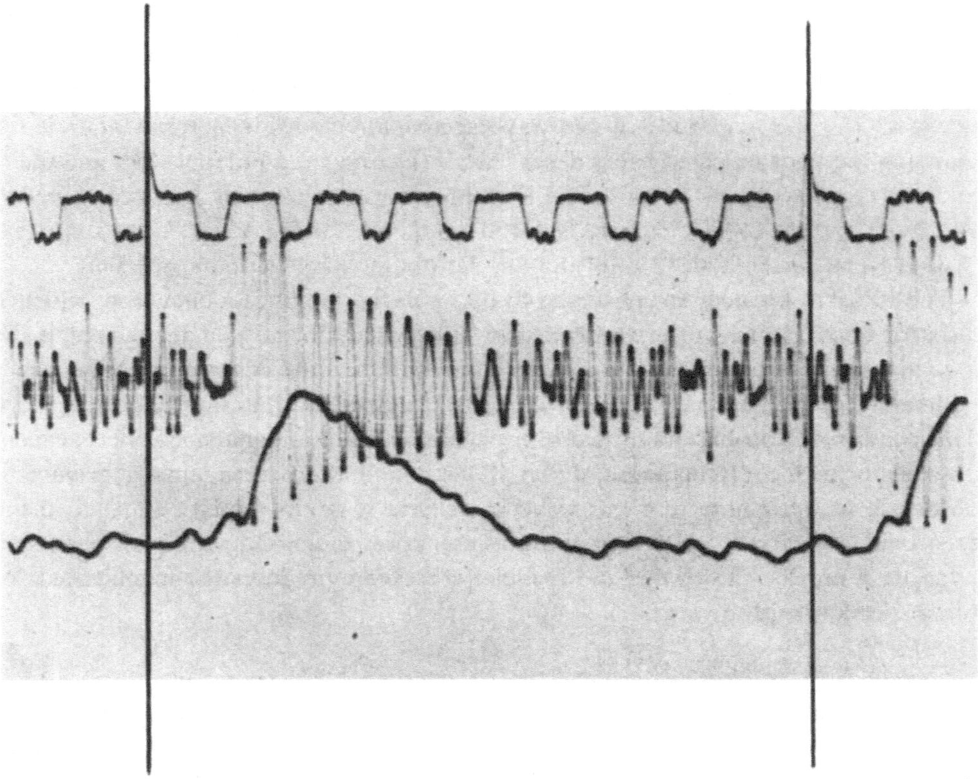

Fig. 4. Typical resonance waveform, as seen using the analog telemetry. The vertical lines delimit the data from one frequency step. The top trace is a digital status and timing signal, and the middle trace is the waveform in the 400 Hz receiver bandwidth, after dynamic compression by the receiver AGC. The bottom trace is the receiver AGC voltage, which indicate the total power in the receiver passband on a quasi-logarithmic scale; this is the same signal as is digitised and sent to the digital telemetry, and is represented by the intensity in Figure 3. The quasi-monochromatic resonance signal stands out clearly from the noise. This data was obtained in the solar wind.

terms increase as the angle between the magnetic field and the density gradient decreases; but at the same time, the amplitude of the resonance signal decreases.

In summary, the observed resonance is within 1% of the ambient plasma frequency in the magnetosphere, and within 4% and probably within 1% in the solar wind. In any case, by analysing the waveform of the receiver output when using the analog telemetry, it will be possible to obtain the precise plasma frequency and to study the different contributions to the frequency shift in both the magnetosphere and the solar wind. The analog telemetry will be used too to study the other resonances observed, to discover whether the sounder may eventually be used to determine other macroscopic parameters of the ambient plasma. An example of analog telemetry data is shown in Figure 4; the quasi-monochromatic resonance signal stands out clearly from the noise.

4. Presentation of the Data

For geophysical studies a more compressed time scale than used for the echogram is required. By averaging the sounder output over each frequency step, a spectrogram is produced. An example, from a typical inbound magnetospheric passage, is shown in Figure 5; this format will be used for all the sounder data presented here.

To make a spectrogram, the data is first converted to power in terms of $V^2\,Hz^{-1}$ at the preamplifier input, and then the data from each frequency step is averaged and this average is plotted as a function of frequency step number ('pas' in French) and time for a period of 3 hr. Everywhere in this paper there are 128 frequency steps with centre frequencies ranging from 0.5 to 50.7 kHz. Owing to the notoriously poor resolution of grey-scales, the spectrogram has been produced twice over, to adapt to the dynamic range of the data; the grey-scale at the left of the spectrograms has ten 3 dB steps, and the value of the top (intensity 1) of the scale used in the top spectrogram and the value of the bottom (intensity 10) of the scale used in the lower spectrogram are marked in the rectangular boxes (MAX

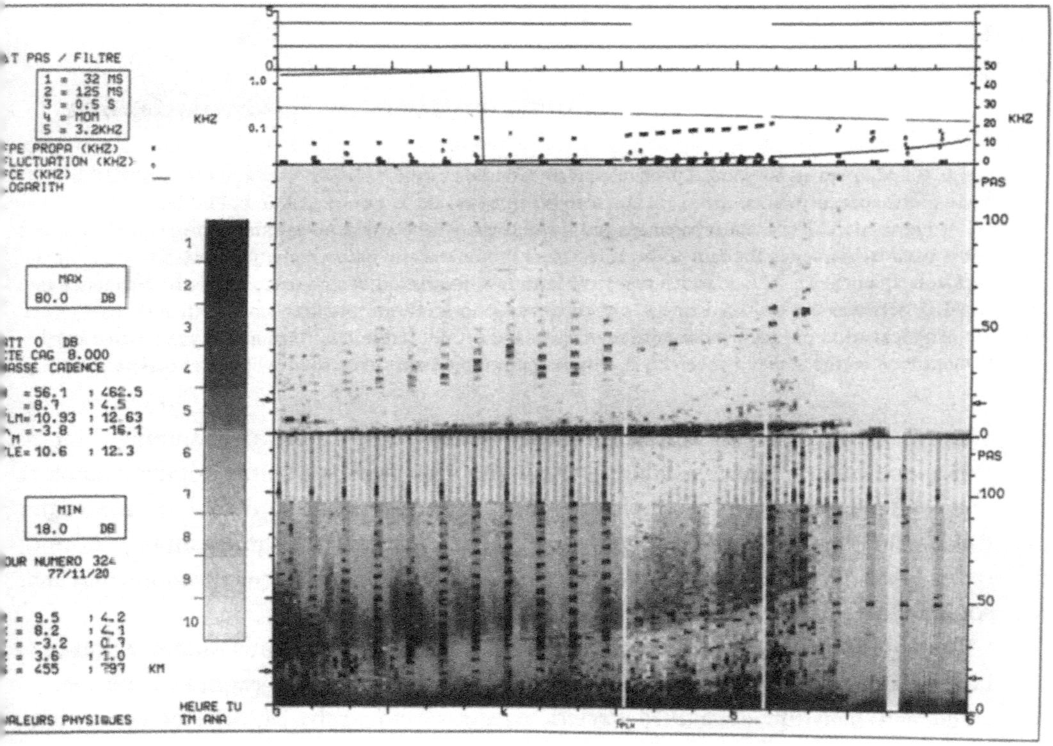

Fig. 5. A typical sounder spectrogram. The data shown here are from an inbound magnetospheric passage, on day 77/11/20 between 03:00 and 06:00 UT, when the satellite was in the altitude range 9.5 to 4.2R_E within 30° of the sub-solar point. At this time the experiment was working in a duty cycle of $\frac{1}{2}$ min every 8 min, and the resonances produced by the sounder are clearly seen. In addition, strong natural continuum radiation can be seen, as well emission at $\frac{1}{2}\Omega_e$. The experiment mode change from 04:32 to 05:08 was associated with use of the analog telemetry transponder.

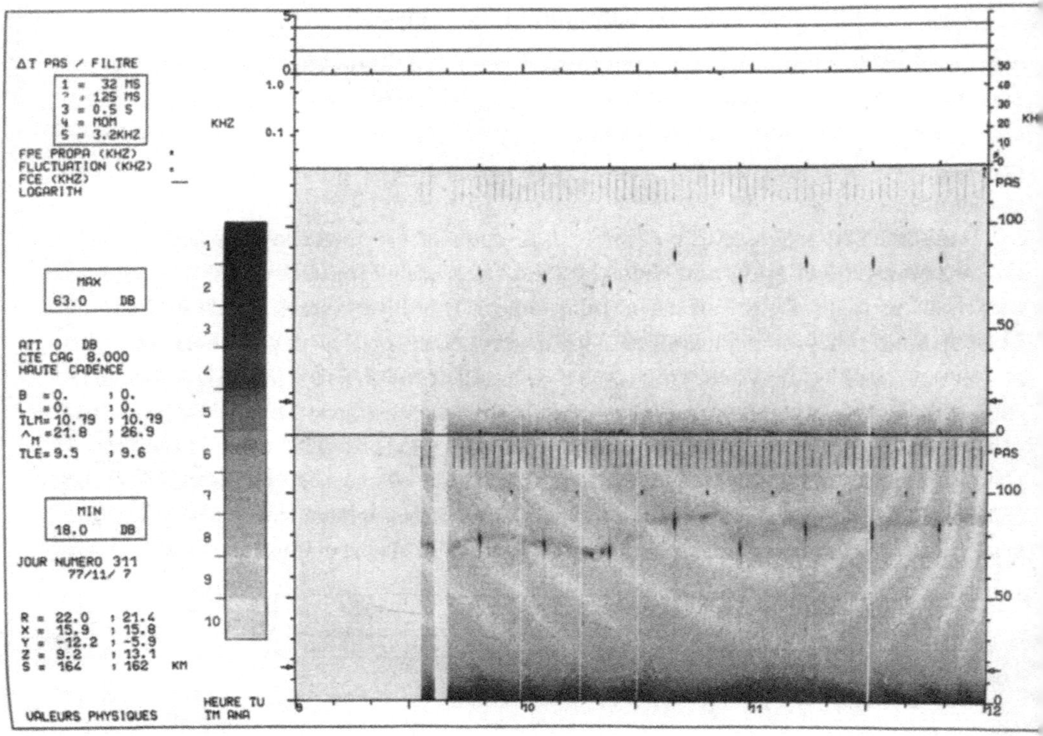

Fig. 6. The 'quiet' solar wind. This data was obtained on day 77/11/07 between 09:00 and 12:00 UT, when the spacecraft was operating in the high bit rate and the experiment internal buffer memory was not being used. The plasma resonances are quite clear. The continuous band of noise slightly above the plasma frequency is thought to be Doppler-shifted naturally occurring noise close to the ambient plasma frequency. The points which resemble faint resonances, but which occur only on frequency step No. 101 between the real resonances, are converter interference at 40 kHz associated with activity of the propagation experiment transmitters. This is the only interference from the propagation experiment seen by the sounder receiver. A common power converter is shared by the two experiments.

and MIN), in dB with respect to 10^{-10} V m^{-1} Hz$^{-\frac{1}{2}}$ at the preamplifier input. When the buffer memory is being used, the 16 s sweeps of the sounder receiver are artificially dilated to completely fill the space available on the spectrogram: this is the case in Figure 5, but in Figure 6, for example, the memory was not being used. The ticks along the upper edge of the upper spectrogram indicate sounder activity.

Immediately above the spectrograms is a panel which gives a summary presentation of data from the propagation experiment. The crosses represent the mean electron density in each 32 s period, expressed in terms of the corresponding electron plasma frequency: and the circles represent the r.m.s. fluctuation δN of the density during the same 32 s period, expressed (curiously, but conveniently) in terms of the frequency $\delta f = f(\delta N/N)^{\frac{1}{2}}$. There is an inevitable uncertainty of 360° p in the phase determination, where p is a small integer; the points are plotted assuming $p = 0$, and the continuous thin line indicates the density which is required to produce a 360° phase shift. A linear frequency scale is used to facilitate comparison with the spectrograms below.

Also represented by a thicker line in the upper panel is the electron gyrofrequency obtained from the data pool tape or, when this is unavailable, from a magnetospheric model: it is plotted on the linear frequency scale when above 2 kHz, and on the logarithmic frequency scale on the left of the panel in the range $20\ \text{Hz} \leq f < 2\ \text{kHz}$.

The top panel in Figure 5 contains status information (sounder step duration, frequency step, antenna used) as indicated in the square box. Ancillary information (B/L coordinates, magnetic local time, magnetic latitude, ecliptic local time, the geocentric radial distance R and GSE coordinates X, Y, and Z expressed in Earth radii, and the separation distance S in kilometres) are written on the left of the Figure; the two sets of values correspond to the beginning and the end of the three hour interval.

In the following two sections we give a rapid review of the different types of characteristic density fluctuations and natural electric field noise which have been identified after examination of only the first four weeks of data. The figures have been carefully selected to illustrate the type of noise under discussion, but it should be remembered that often in the data two or more types of characteristic noise can be present simultaneously. We consider the data in order of decreasing distance from the Earth.

5. Solar Wind Observations

An enormous variety of phenomena are observed in the solar wind. Figure 6 shows, for reference, what we consider to be the 'quiet' solar wind. The experiments were active for $\frac{1}{2}$ mn every 8 mn, alternatively sounder and propagation, and the resonances every 16 mn (actually every $16 \times 64\ \text{s} = 17\ \text{min}\ 4\ \text{s}$) are clearly visible.

5.1. f_p Continuum

Also visible in Figure 6 is a low level, narrow band, continuum noise close to the plasma frequency. This noise, which is a characteristic feature of longer 'quiet' intervals, has not been previously observed; the 'plasma frequency noise' observed by Gurnett and Anderson (1977) is sporadic, and is probably the 'electron burst noise' described below. The noise occurs in a band of width about 2 kHz which is generally situated above the frequency of the plasma resonance. For example, at 10:06 UT the maximum noise intensity ($3 \times 10^{-8}\ \text{V m}^{-1}\ \text{Hz}^{-\frac{1}{2}}$) occurred at 30.2 kHz, while the plasma resonance was at 29.3 kHz. Two possible interpretations are under study:

– This may be simply due to 'thermal fluctuations' of the solar wind plasma, which are enhanced at the plasma frequency owing to the increased antenna impedance; the observed spectrum resembles the theoretical one. However, the antenna impedance required turns out to be an order of magnitude greater than the preamplifier (including stray capacitance) input impedance, so that there is a considerable power mis-match. Furthermore, the intensity of this noise correlates with the intensity of noise at $2f_p$ (see below) when the latter is observed, and this implies some non-linear effects in the plasma.

– More probably, the f_p noise really corresponds to plasma waves. Assuming a reasonable solar wind bulk velocity (400 km s^{-1}) and electron temperature (1.2×10^5 K), by solving the Doppler-shifted Bohm-Gross dispersion relation we find that the noise at (for example) 10:06 UT corresponds to waves with a phase velocity 7 times the electron thermal velocity and negligible Doppler effect. This phase velocity is comparable with the particle velocity at the intersection of the 'core' and 'halo' components of the electron velocity distribution (Feldman *et al.*, 1975), which suggests that the noise is related to the non-Maxwellian tail of the solar wind electron velocity distribution.

5.2. $2f_p$ Continuum

The above narrow band noise close to the plasma frequency is often associated with noise at twice the plasma frequency (e.g., Figure 8); the narrow bandwidth

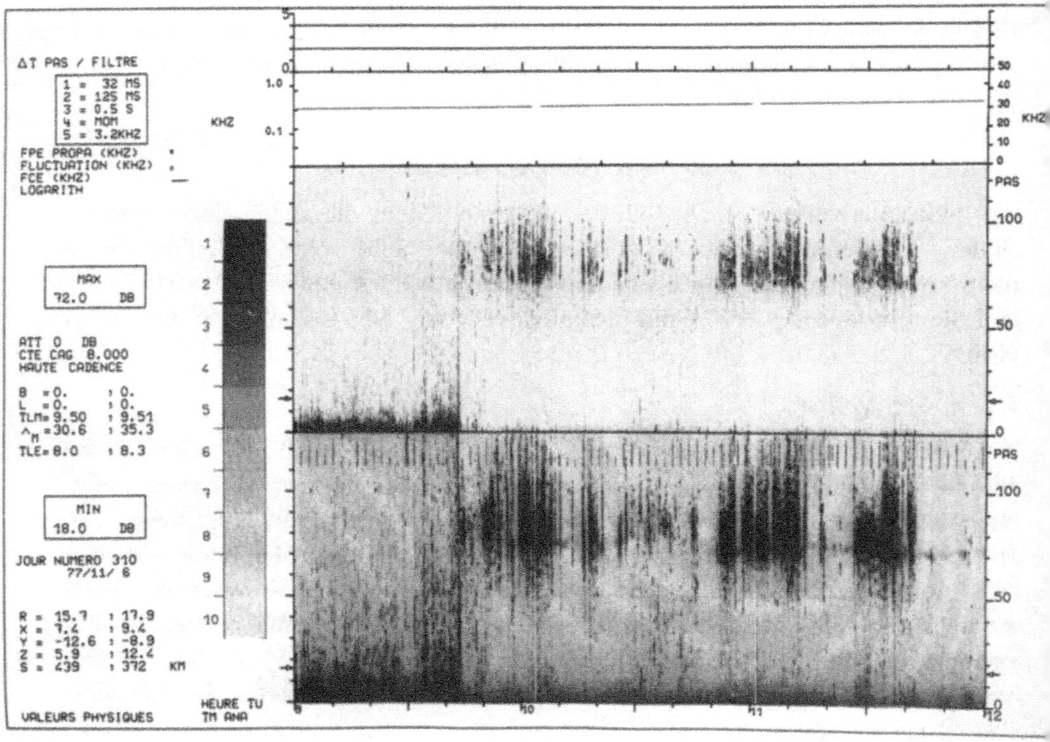

Fig. 7. An outbound crossing of the bow shock, observed on day 77/11/06 at 09:43 in high bit rate. Resonances can be seen inside the shock at 09:13 and 09:30, and outside the shock at 11:47 and 11:55; further resonances can be identified from the digital data. Inside the shock there is characteristic low frequency magnetopause noise, and outside the shock, in addition to the continuum, there are intense electron burst noise events.

and close harmonic relationship, even during periods of density fluctuation, suggests that both emissions are generated locally.

5.3. Electron burst noise

Another type of noise clearly associated with the electrons is shown in Figure 7. After the shock crossing at 09:43 UT, intense noise bursts are seen, each one lasting several minutes. The noise is highly irregular, but appears to have a low-frequency partial cut-off at the local plasma frequency; two plasma resonances can clearly be seen after 11:45 UT. Similar noise observed well away from the shock is shown in Figure 7, and this example has been found to correlate with bursts of electrons of energy about 1.5 keV (Rème, private communication). After examination of several days of data, it appears that whenever electrons are seen this noise is observed, and only occasionally is the noise seen in the absence of electrons (but the electron detector is collimated perpendicularly to the ecliptic plane). In view of its sporadic, bursty character, and its intensity, we believe this to be the noise called 'electron plasma oscillations' by Gurnett and Anderson (1977). Its bursty character and correlation with electron events suggest a local generation mechanism; mode identification will be easier following polarisation studies presently being undertaken.

In contrast to the electron associated turbulence, for which the protons can be considered as a uniform background, this experiment also detects MHD waves, which can either be Doppler shifted into the passband of the radio receiver or, at lower frequencies, appear as large amplitude electron density fluctuations.

5.4. Ion acoustic turbulence

Figure 8 shows, from 09:00 to 09:19, an example of a frequently observed type of noise event with an upper frequency cut-off close to the plasma frequency. This noise correlates with the presence of enhanced fluxes of protons with energy greater than 6 keV (Rème, private communication). This type of turbulence has been extensively studied by Gurnett and Frank (1978) and Kurth *et al.* (1978), and is thought to be Doppler-shifted ion acoustic turbulence. The new factor in the present observations is that the upper frequency limit of the Doppler shifted turbulence seems to be closely associated with the ambient plasma frequency, corresponding to a maximum k vector satisfying $k\lambda_D \sim 8\pi$ (assuming the same solar wind velocity and temperature as in Section 5.1).

5.5. Density fluctuations

Figure 9 shows an abrupt onset of density fluctuations in the solar wind at about 06:33 on day 77/326. From before 06:00 to 06:30 the electron density was fairly constant, at 16 cm^{-3} with fluctuations less than 1 cm^{-3}. At 06:32 there is an unfortunate data gap, but after 06:35 the density fluctuates irregularly between about 8 and 24 cm^{-3}. The data from the propagation experiment for this period have not yet been processed. Simultaneously with the onset of the density

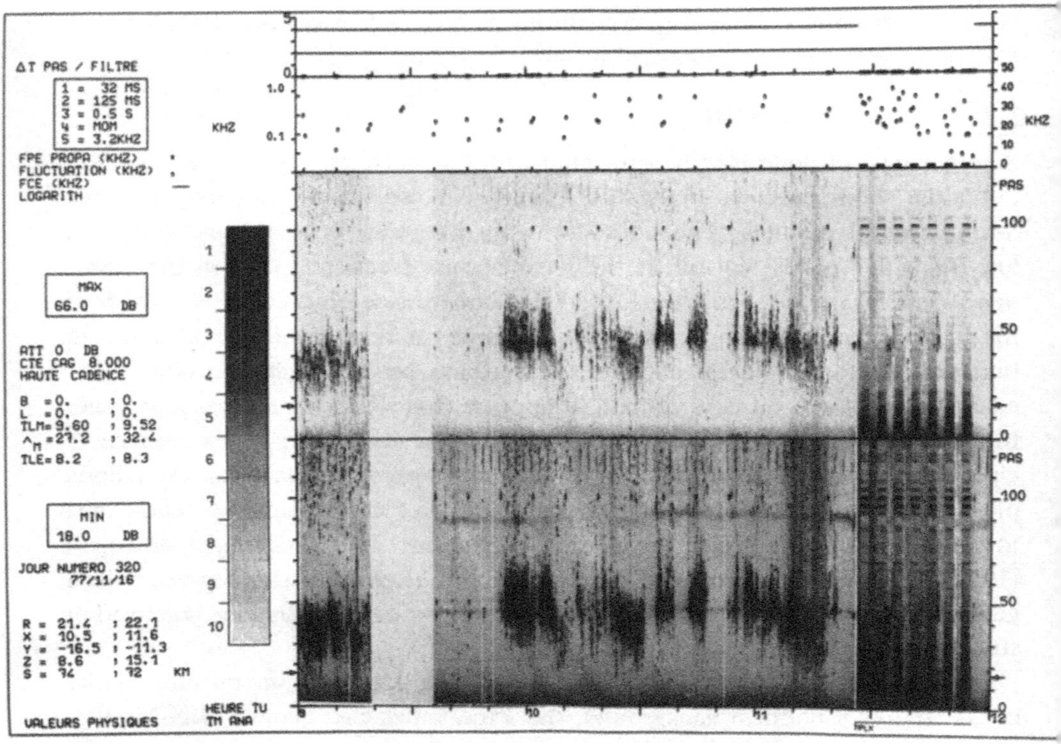

Fig. 8. Solar wind characteristic noise. This three hour period on day 77/11/16 is particularly rich in phenomena. After the data gap from 09:18 to 09:35 UT, both f_p continuum and $2f_p$ continuum can be seen, as well as electron burst noise events. Before the data gap low frequency ion acoustic wave turbulence can be seen. After 11:26 the experiment was in a different mode for the analog telemetry.

fluctuations a large increase in the proton flux between 8 and 50 keV was observed (Lin, private communication). This event probably corresponds to a change in the interplanetary magnetic field direction, which becomes 'connected' to the Earth's bow shock (magnetic field data for this period is not presently available). Note the presence, before 06:30, of noise around the electron plasma frequency: this ceased at 06:32, but a new, strong low frequency noise appeared below 15 kHz, which is probably Doppler shifted ion acoustic waves generated by the protons.

6. Shock and Magnetosheath Observations

These regions are more difficult to study with our present four week data set: the spacecraft spend little time there, and the active time is even less owing to the duty cycle operation. Moreover, the plasma frequency in the magnetosheath sometimes went above the frequency range of the sounder, which then yielded no density determinations. With these reservations, we present the following observations.

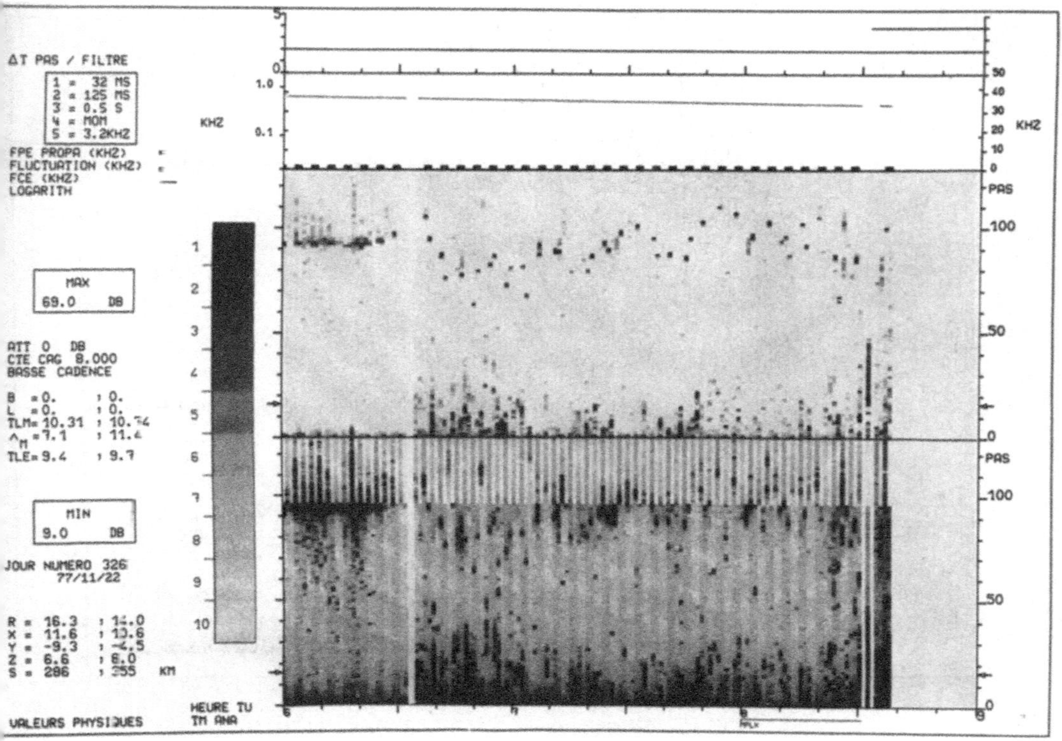

Fig. 9. Density fluctuation in the solar wind, close to the bow shock on day 77/11/22.

6.1. Shock crossings

On passing inbound through the shock we observe a large increase of both the density and its fluctuations. Figure 10 (see also Figure 6) shows an example of an outbound crossing. The density in the magnetosheath varies between about 20 and 30 cm^{-3}. Between 12:41 and 12:43 the spacecraft passed through the shock, as is clearly indicated by the magnetic field, and the density stabilised at 11 cm^{-3}. Such behaviour of the electron density is typical, and can be seen with much higher time resolution using the propagation experiment.

Figure 11 shows propagation experiment data from an inbound pass during which the shock was identified between 22:51 and 22:52 by the magnetometers aboard both ISEE-1 and 2 (Russell, private communication). The propagation experiment was active for 32 s every 16×64 s. The first series of measurements was made in the solar wind: the density was 11.7 cm^{-3} and the fluctuations ± 0.2 cm^{-3}. The sounder was active 8 min earlier and 8 min later, and interpolating with the aid of the natural f_p continuum we estimate the sounder density to be 10.6 ± 0.3 cm^{-3}; the possibility of the difference between the propagation and sounder densities being due to a 10% error in the separation distance (nominally 242 km) is being studied. The next series of measurements, which is detailed in

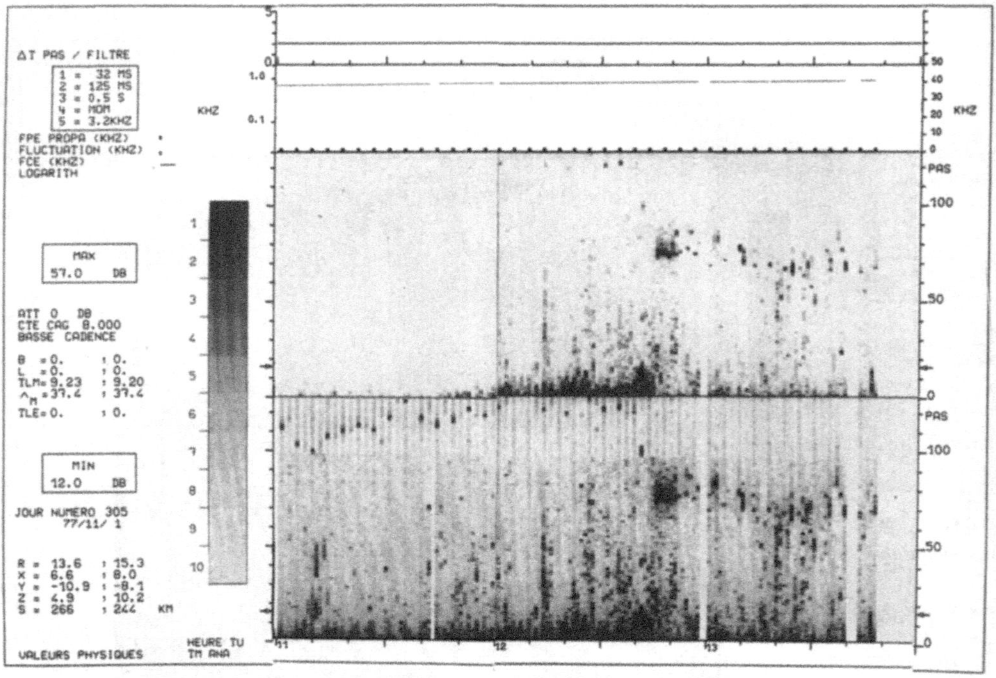

Fig. 10. Inbound shock crossing on day 77/11/01.

Figure 1, was made in the magnetosheath just behind the shock: the density fluctuated between 30 and 45 cm^{-3} (thus the sounder was out of range). Such large density fluctuations are not unusual in the magnetosheath, but in the example they have a periodicity of about 7 s, whereas usually it is somewhat longer. The ISEE-1 and 2 separation distance was 252 km, and these fluctuations of the integrated density can be interpreted either as large scale density irregularities or as a density gradient moving back and forth somewhere between the two spacecraft. The third series of measurements in Figure 11 is made further into the magnetosheath: here the density is slowly varying in the range 35.0 ± 1.5 cm^{-3}.

A shock crossing is seen on the natural waves as a sharp discontinuity between the solar wind noise previously described and a low frequency (<10 kHz) noise in the magnetosheath. An example of an outbound shock crossing is seen in Figure 7 at 09:43 UT. During the shock crossing a strong wide-band noise is observed, the characteristics of which depend upon the example studied. Typically, there is an increase by about an order of magnitude in amplitude in a frequency range between zero and several tens of kH, but as the spectrum is variable from shock to shock, further study, including possible antenna effects, is required. Noise

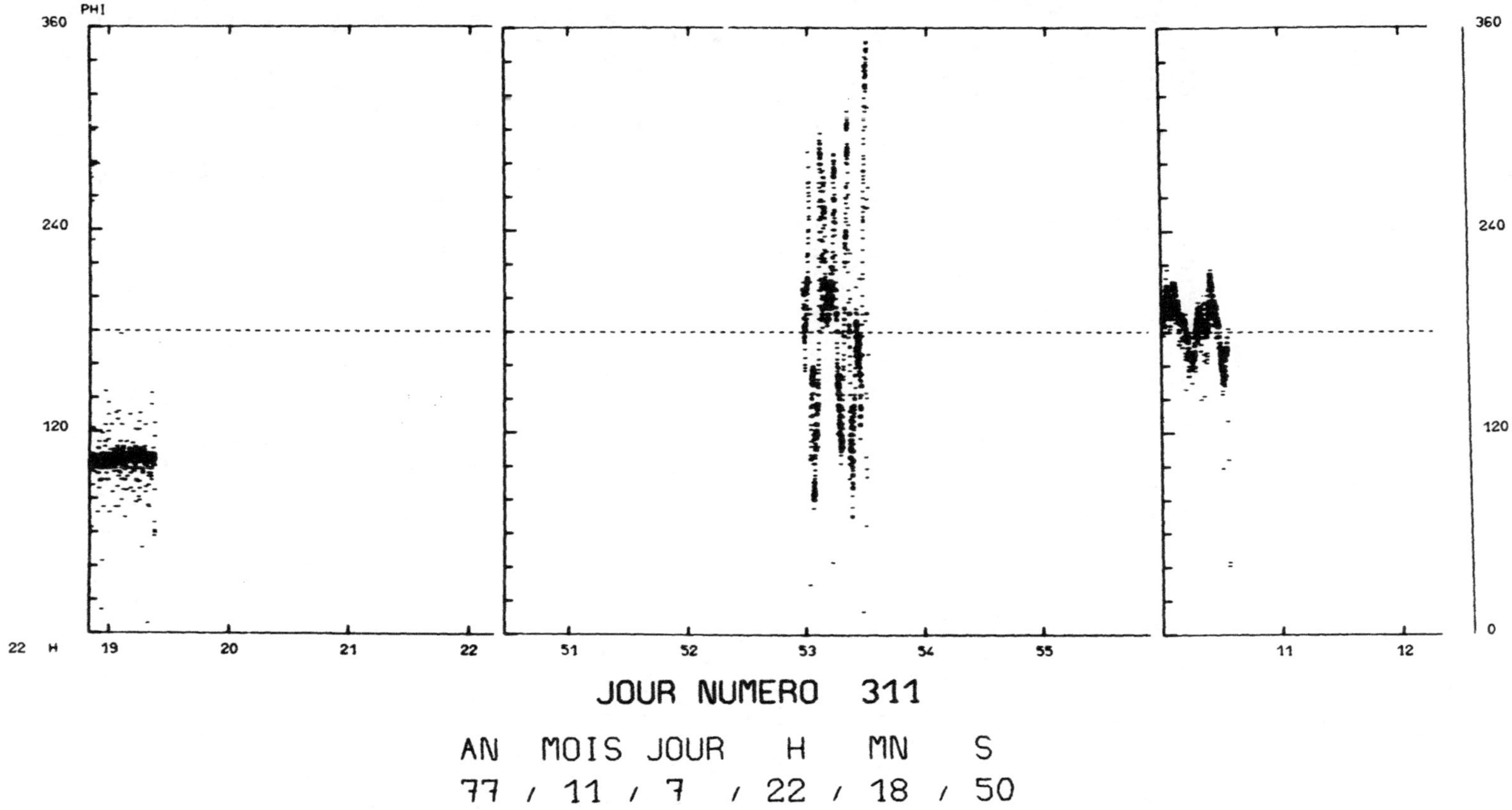

Fig. 11. Propagation experiment data on day 77/11/07 when the experiment was active for 32 s every 17 min 4 s. The phase advance is plotted in the range $0 \leqslant \phi < 360°$. The first 100 s of data was obtained in the solar wind, and the next 100 s of data were lost through interference. The large oscillations seen starting at 22:53 UT were observed just behind the shock, but further into the magnetosheath the oscillations were smaller.

measurements on each side of the shock are similar to those reported by Rodriguez and Gurnett (1975).

At the magnetopause the magnetic field increases and simultaneously the cyclotron harmonic and f_Q resonances become visible as, for example, are seen in Figure 5, albeit at lower altitudes.

7. Discussion and Summary

Both experiments, the propagation and the sounder, measure the plasma density by probing a volume of space sufficiently large for the perturbing effects of the photoelectrons or the spacecraft floating potential to be negligible. Of course, as they both use the properties of the dispersion relation, they really measure the density of that part of the particle population which furnishes the main contribution to the dispersion properties of the medium at the frequency in question; but, for magnetospheric or solar wind plasma, there is no reason to believe that this population differs significantly from the thermal plasma, particularly for the propagation experiment.

After examination of the first four weeks' data, we can draw the following general conclusions:

– The sounder observes resonance signals everywhere, except possibly in very hot tenuous plasma as discussed elsewhere in this volume in connexion with the GEOS sounder (Etcheto and Bloch, 1978).

– This is the first time a sounder experiment has been launched into the Earth's magnetosheath and the solar wind. Plasma frequency resonances are clearly seen in both regions (except at times, in the magnetosheath, when the plasma frequency was observed to creep above the upper limit of the sounder's frequency range). In particular, fears that resonances would never be seen in the solar wind, owing to Landau damping, have been shown to be unfounded: they are always present there.

– Last, but not least, in the admittedly few cases where a direct comparison has been made, the densities obtained from the sounder and propagation experiments agree to within the uncertainty of the separation distance.

In summary, both experiments provide a new precision for the measurement of the electron density, which is particularly important for the study of wave-particle interactions. However, they differ in certain respects.

(1) *Time resolution.* The propagation experiment makes 32 independant measurements per second of the electron density in high bit rate, or 8 per second in low bit rate. The sounder, in the most commonly used mode of operation has a time resolution of only 16 s. This can be improved to 4 s, at the expense of sometimes not being able to distinguish easily between a resonance and an interference; further improvement is possible by limiting the range of the frequency sweep, but this increases the chance of the plasma frequency being outside the instrument range.

(2) *Spatial resolution.* The sounder measures the density within a few kilometres of the ISEE-1 spacecraft, a distance small enough for the density normally to be considered a point measurement, but far enough for the satellite plasma sheath to have negligible effect. The propagation experiment, on the other hand, measures the integrated electron density between the ISEE-1 and 2 spacecraft, so that the above temporal resolution is, in fact, only useful in regions of high plasma convection speed, or where fairly large scale density fluctuations propagate with a large phase velocity.

(3) *Density resolution.* As already mentioned, the sounder is thought to measure the plasma frequency to within about 1%, except possibly when the bulk plasma flow velocity is large. When the experiment does not have use of the analog telemetry, below about 20 kHz the frequency precision is limited to about ±200 Hz owing to the receiver bandwidth. The propagation experiment measures the phase advance, which is proportional to the separation distance, with a precision which also varies with the separation distance (received S/N ratio); it is not presently clear which has the greater precision: the measured phase advance or the computed ISEE-1 and 2 separation distance.

(4) The propagation experiment is easy to interpret, the electron density is obtained from the phase advance via the relation given in Section 2. The identification of the plasma resonance frequency is sometimes more delicate, especially when there is no analog telemetry.

(5) The sounder continuously provides natural wave electric field data with the same frequency and time resolution as the resonance data.

(6) Both experiments interfere with other experiments on board the ISEE-1 spacecraft. It is not surprising that sounder transmission perturbs the passive experiment which observes natural waves in the same frequency range. However, the sounder itself can measure natural waves owing to the synchronisation of the operation of its receiver with that of the transmitter. The sounder must obtain 'passive' data for comparison with 'active' data to detect resonances (Figure 3). Note that, because of modification after pre-integration tests, the Mozer DC electric field experiment can also function with the sounder active, using the same antenna. It is almost inevitable that the propagation 683 kHz transmitters interfere with other experiments using the same antenna to observe natural phenomena; however, the other (Mozer) antenna can still be used for wave observations (see Figures 5 through 8, and especially 6) if the necessary precautions have been taken. The propagation experiment also interferes with some particle experiments, but apparently not by changing the floating potential of the spacecraft, so as to attract one species of particle while repelling the other. Further study of the interference is needed before it is possible to say whether it could have been avoided by taking greater precautions before launch, or whether it is inevitable owing to the very nature of the particle detectors being used. In a collaborative effort to optimise the scientific return of the entire payload, these two active experiments are presently constrained to operate on a 10% duty cycle.

Acknowledgements

The authors would like to thank all who contributed to the success of this experiment, particularly Y. de Javel and R. Manning, the engineers responsible respectively for the sounder and propagation experiments, and also the NASA and ESA engineers who paid such careful attention to the particular needs of this experiment.

The data analysis was realised thanks to the invaluable help of the personnel of the CNES Centre Spatial de Toulouse, where all the spectrograms were realised. Thanks are also due to J.-G. Trotignon and C. Wachtel for their assistance with the data reduction.

The experiment was financed by the Centre National d'Etudes Spatiales under contracts nos. SL.180.01.E and CNES/214.

References

Calvert, W. and McAfee, J. R.: 1969, 'Topside Sounder Resonances', *Proc IEEE* **57,** 1089.

Deering, W. D. and Fejer, J. A.: 1965, 'Excitation of Plasma Resonances by a Small Pulsed Dipole', *Phys. Fluids* **8,** 2066.

De Forrest, S. E.: 1972, 'Spacecraft Charging at Synchronous Orbits', *J. Geophys. Res.* **77,** 651.

Etcheto, J. and Bloch, J.-J.: 1978, 'Plasma Density Measurements from the GEOS-1 Relaxation Sounder', *Space Sci. Rev.* **22,** 597.

Etcheto, J. and Petit, M.: 1977, 'Une nouvelle methode de caracterisation du plasma magnetospherique', *Acad. Sci., serie B* **285,** 329.

Fejer, J. A. and Yu, W. M.: 1970, Excitation of Plasma Resonances by a Small Pulsed Dipole in a Weakly Inhomogeneous Plasma', *J. Geophys. Res.* **75,** 1919.

Fejer, J. A. and Calvert, W.: 1964, 'Resonance Effects of Electrostatic Oscillations in the Ionosphere', *J. Geophys. Res.* **69,** 5049.

Feldman, W. C., Asbridge, J. R., Bame, S. J., Montgomery, M. D., and Gary, S. P.: 1975, 'Solar Wind Electrons', *J. Geophys. Res.* **80,** 4181.

Feldstein, R. and Graff, P.: 1972, 'Day Time and Night Time Electron Temperatures from Topside Resonances', *J. Geophys. Res.* **77,** 1896.

Graff, P.: 1971, 'General Expression of the Frequencies of the Oblique Echoes at the Plasma Resonance', *J. Geophys. Res.* **76,** 1060.

Gurnett, D. A. and Anderson, R. R.: 1977, 'Plasma Wave Electric Fields in the Solar Wind: Initial Results from HELIOS-1', *J. Geophys. Res.* **82,** 632.

Gurnett, D. A. and Frank, L. A.: 1978, 'Ion Acoustic Waves in the Solar Wind', *J. Geophys. Res.* **83,** 58.

Harvey, C. C., Etcheto, J., De Javel, Y., Manning, R., and Petit, M.: 1978, 'The ISEE Electron Density Experiment', *IEEE Trans. Geosci. Electr.* **GE-16,** 231.

Kurth, W. S., Gurnett, D. A., and Scarf, F. L.: 1978, 'High Resolution Spectrograms of Ion-Acoustic Waves in the Solar Wind', University of Iowa, Preprint No. 78-15.

McAfee, J. R.: 1968, 'Ray Trajectories in an Anisotropic Plasma near Plasma Resonance', *J. Geophys. Res.* **73,** 5577.

McAfee, J. R.: 1969, 'Topside Resonances as Oblique Echoes', *J. Geophys. Res.* **74,** 802.

Muldrew, D. B.: 1972, 'Electron Resonances Observed with Topside Sounders', *Radio Science* **7,** 779.

Rodriguez, P. and Gurnett, D. A.: 1975, 'Electrostatic and Electromagnetic Turbulence Associated with the Earth's Bow Shock', *J. Geophys. Res.* **80,** 19.

Warnock, J. M., McAfee, J. R., and Thompson, T. L.: 1970, 'Electron Temperature from Topside Plasma Resonance Observations', *J. Geophys. Res.* **75,** 7272.

Whipple, E. C., Warnock, J. M., and Winckler, P. H.: 1974, 'Effect of Satellite Potential on Direct Ion Density Measurements Through the Plasmapause', *J. Geophys. Res.* **79,** 179.

PRELIMINARY RESULTS FROM THE SOLAR WIND EXPERIMENT ON THE ISEE-2 SATELLITE

M. B. BAVASSANO-CATTANEO, C. BONIFAZI,
P. CERULLI-IRELLI, L. DIODATO, M. DOBROWOLNY,
A. EGIDI, G. MORENO, and S. ORSINI

Laboratorio Plasma Spazio, CNR, Via Galileo Galilei, Frascati, Italy

Abstract. The EGD plasma experiment, launched on board of the ISEE-2 spacecraft, is briefly described. Preliminary results concerning three bow shock crossings occurred in November 1977 are discussed. Our attention is focused on some non-thermal features of the distribution function of proton velocities: in particular, we discuss double peaked structures of solar wind spectra and flows of particles backstreaming from the bow shock.

1. The Experiment

1.1. Instrumentation

A detailed description of the experimental apparatus has been given elsewhere (Bonifazi *et al.*, 1978). Here it will be sufficient to recall the following points.

The instrument (Figure 1) consists of two identical hemispherical electrostatic energy selectors, to which variable voltages are applied in order to perform the simultaneous measurement of positive ions in two different energy windows. The ratio between the voltages applied simultaneously to the two deflectors is kept constant ($V_1/V_2 = 1.5$). During a complete spin revolution of the satellite (i.e. ~3 s) the voltages V_1 and V_2 are held constant and then varied in steps according to a programmed sequence to examine different portions of the ion energy spectrum.

A mechanical collimator at the entrance of the deflector is used to obtain the desired azimuthal response.

The relative energy response is $\Delta E/E \simeq 4\%$, the azimuthal and polar angular acceptance (in satellite coordinates) are respectively about $\pm 1.5°$ and $\pm 40°$ at 1/10 of the maximum. Figure 2 shows the response of one of the deflectors in velocity space on the equatorial plane of the spacecraft.

The positive ion fluxes are detected by two channel electron-multipliers of the Channeltron type, placed, as seen in Figure 1, at the exit of the electrostatic deflectors.

1.2. The measurement routine

The deflecting voltages V_1 and V_2, applied to the electrostatic analysers, are varied according to two different programmed sequences, which allow ion energy spectra to be obtained with increasing temporal resolution and decreasing details.

These two modes of operation will be referred to as 'wide energy spectrum' (WES) and 'high time resolution' (HTR) modes.

Space Science Reviews **23** (1979) 59–74. 0038-6308/79/0231-0059 $02.40.

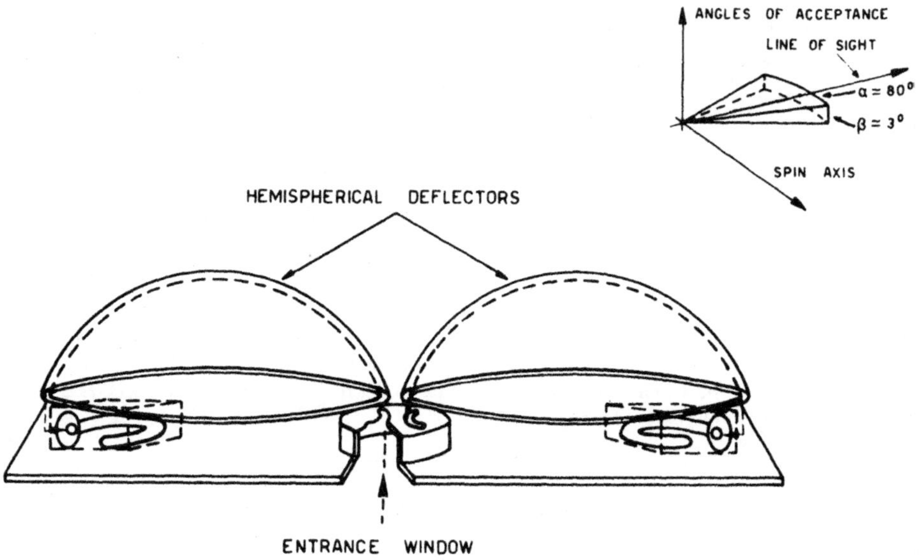

Fig. 1. Schematic representation of the electrostatic deflectors.

The complete cycle of measurements takes 512 spacecraft revolutions. During this time, the measurements are made partly in the WES and partly in the HTR mode. The relative duration of the two modes is controlled by means of telecommands.

The WES mode supplies a detailed energy spectrum every 32 revolutions (i.e. every 96 s). The sequence of the energy steps is illustrated in Figure 3. The range

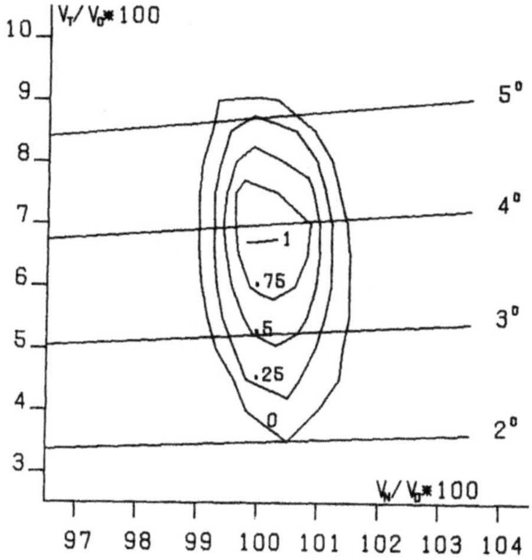

Fig. 2. Cross section in the spacecraft equatorial plane of the deflector response represented in velocity space. The curves are lines of constant normalized response. V_N and V_T indicate respectively the velocity component along the geometrical axis of the instrument and perpendicular to it. V_0 is the value of the velocity corresponding to the center of the accepted window. The figure shows also that the fringing field of the deflector has the effect of displacing the true line of sight about 4° with respect to the geometrical axis.

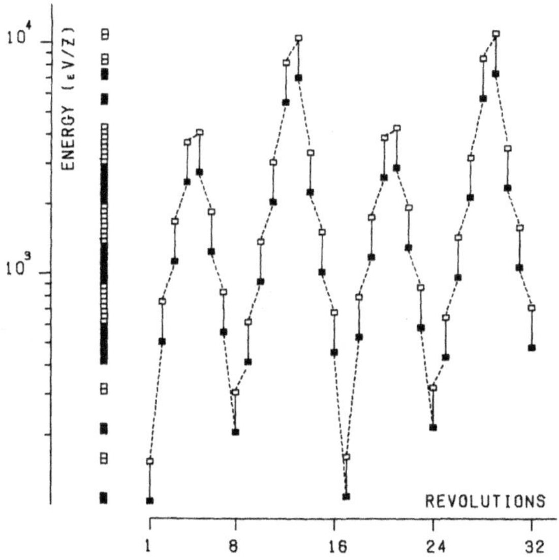

Fig. 3. Energy steps obtained by stepping the voltage across the two deflectors at each spin revolution. The complete measurement cycle is accomplished in 96 s, in the WES mode of operation.

of energy per unit charge, from ~100 eV/Z to ~10 keV/Z is covered with 64 windows (contiguous in the interval ~400–~4000 eV and more spaced at lower and higher energies).

When operating in the HTR mode, only the first 16 voltage pairs, i.e. 32 energy windows (see Figure 3), can be applied to the electrostatic deflectors. The first measurement in the HTR mode is performed at the 'peak' energy, i.e. in the window corresponding to the maximum channeltron counting rate measured during the last energy spectrum taken in the WES mode. A second measurement is then taken at the next higher voltage step. The accumulated counts during these two measurements are then compared and the voltage for the next measurement is shifted in the direction of increasing counting rate. This compare-and-shift process is repeated continuously to allow tracking of the maximum counting rate.

The plasma experiment is mounted on the ISEE-2 satellite in such a way that its line of sight is perpendicular to the spin axis. The 360° field, explored during each spacecraft revolution, is divided in a different number of sectors according to the transmission bit rate of the satellite.

When the satellite is operating at the 'normal bit rate' (NBR), i.e. 2048 bps, 15 angular measurements are performed, as shown in Figure 4a; 12 measurements are taken within 5.6° sectors ('solar sectors') and 3 measurements within 90° sectors ('antisolar sectors'). (During the 22.5° 'blind sector' no measurements are performed). When the satellite is operating at 'high bit rate' (HBR), i.e. 8192 bps, 30 angular measurements are performed (the number of both the solar and the antisolar sectors being doubled, see Figure 4b).

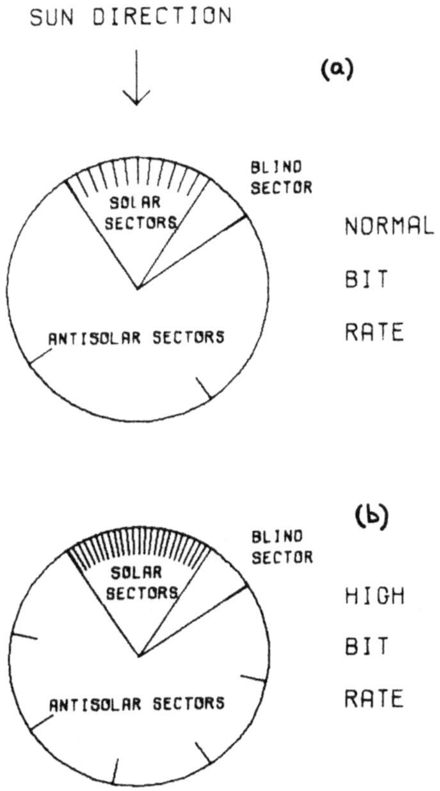

Fig. 4. Position of the angular sectors when the spacecraft operates at normal and high bit rates.

The dynamic range of the instrument depends on the angular and energy spread of the incident ion beam. For a well-collimated and monoenergetic beam, the dynamic range extends from $\sim 10^4$ ion cm^{-2} s^{-1} up to $\sim 10^8$ ion cm^{-2} s^{-1}. For higher fluxes, only a lower limit indication will be attained: however, due to the relatively narrow band of the accepted energies, this situation should occur very rarely.

1.3. The method of analysis

The analysis of the plasma data was done through a method not implying any a priori hypothesis on the analytical form of the distribution of the ion velocities.

Under the hypothesis of a distribution which is locally Maxwellian in ϕ (the azimuthal angle in the equatorial plane of the spacecraft), by fitting the counts in three consecutive angular sectors (in the same energy channel) to a gaussian function, one can unfold the effect of the instrumental response from the counts and convert the count matrix into a bi-dimensional matrix giving the values of the distribution function. The details of the method, which is similar to that sketched by Hundhausen *et al.* (1967), will be given elsewhere.

Through numerical integrations of the matrix of the distribution function, any moment of the distribution can then be computed (density, radial and transversal bulk velocity, temperature, thermal anisotropy, heat flux).

In the results to be presented here, our attention will be confined to protons. However, the same method of analysis can also be applied to derive the parameters of the α-particle component of the solar wind.

2. Evidence of Non Thermal Features in the Solar Wind Upstream of a Bow Shock Transition

The results of analysis we present in the following refer to the vicinity of the bow shock transitions occurring on days 309, 312 and 314, 1977.

In Figure 5 we have plotted the main proton parameters (bulk velocity, flow direction, density and most probable thermal speed) along with the magnetic field longitude, obtained for an hour and a half period around the bow shock crossing on day 309. The crossing, which occurs at ~13:23 UT is indicated by the vertical dashed line in the Figure. Gaps in the plots correspond to the HTR mode of our instrument described in Section 1.2. Although there is a gap across the shock transition, the jumps in the parameters can still be deduced from Figure 5. For a discussion of the magnetic field parameters of this shock we refer to the paper by Russell and Greenstadt (1979).

The main reason for reporting data around this bow shock transition lies in the kinetic state of the upstream solar wind which is indicated by our observations.

The ion distributions, as a function of energy, appear to show a double peaked structure (both for protons and α-particles), for a period of at least half an hour upstream of the transition. Furthermore, the details of the structures appear to be highly variable from one spectrum to the next. An example is given in Figure 6 which shows individual plasma spectra for the period from 12:18 to 12:36 UT upstream of the bow shock.

The reasons for the variability appear clearly from Figure 7 which shows the time history of the counts (integrated over the azimuthal angle) in each energy channel, including a period where the instrument was working in the HTR mode. Looking at the counts in channel 17 (near to the maximum of the distribution), we see strong fluctuations on the HTR time scale (6 s), indicating the existence of a turbulence with variations on temporal scales smaller than or equal to 6 s. The existence of turbulent waves in these periods seem to be confirmed by the plasma density experiment (Harvey, private communication).

As successive measurements in a given channel are separated by 96 s, the presence of a turbulent background, besides being of interest in itself, explains the variability between successive spectra mentioned before.

To eliminate such variability and obtain more or less stationary features of the ion distribution function, we have computed averages over different numbers of consecutive spectra until reproduceable features have been obtained. The result is

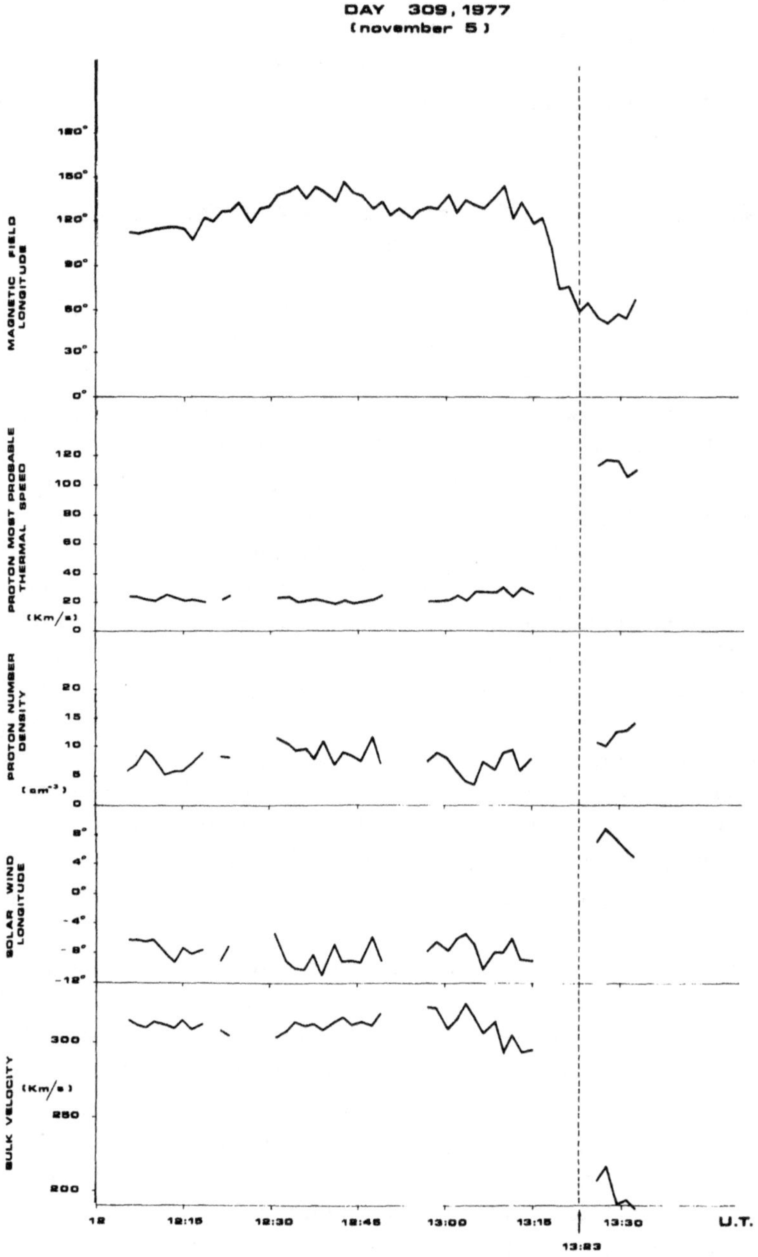

Fig. 5. Time history of the proton parameters during the bow shock crossing occurred at 13:23 UT on day 309, 1977. From the bottom: the bulk velocity, the solar wind longitude (measured from the Sun direction), the number density and the most probable thermal speed are given. At the top, the angle between the direction of the Sun and that of the interplanetary magnetic field, projected onto the $X_{SE}Y_{SE}$ plane, is given for reference. The broken vertical line indicates the shock crossing.

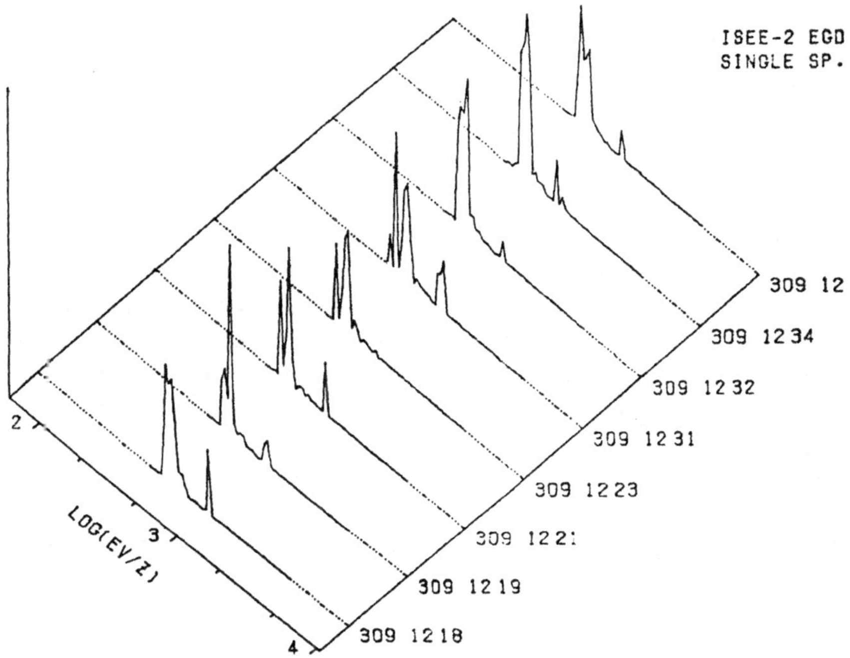

Fig. 6. Sequence of individual ion energy per unit charge spectra recorded on day 309 from 12:18 UT to 12:36 UT. Times refer to the beginning of each spectrum. Counts, along vertical axis, are given in arbitrary units.

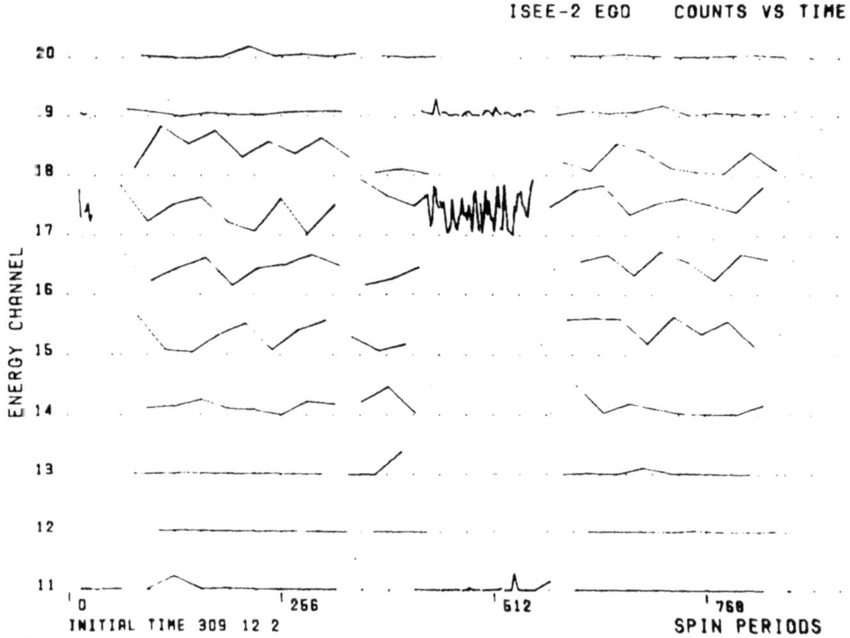

Fig. 7. Counts in the energy channels from 11 to 20, as a function of time, during a period starting from 12:2 UT (day 309, 1977). When the experiment was operating in the HTR mode, only measurements in channels 11, 17, and 19 were available. Note the very rapid fluctuations of the ion flux in channel 17, during HTR periods.

shown in Figure 8. The Figure refers to the last period of time upstream of the bow shock (from 12:57, to 13:06 UT) with the instrument operating in the WES mode (12 spectral measurements). Spectra shown in the figure are running averages made over 6 consecutive spectra.

As appear clear from Figure 8, starting with a normal solar wind spectrum with only one proton and one α-particle peak, a situation develops with a second peak (for both protons and α-particles). The secondary proton peak appears at lower energies with respect to the first one and its relative importance increases towards the shock. The peaks are observed in the same energy channels (corresponding to velocities of about 291 and 321 km s^{-1} respectively), their separation ($\sim$30 km s^{-1}) remaining constant during the time interval under consideration. This number is comparable with the average thermal speed over the same period, as can be deduced from Figure 5 (referring to the whole proton distribution).

Further information (which will be investigated more quantitatively later) is that the two proton peaks appear to come from approximately the same direction.

Non thermal features of this type have been reported earlier (Feldman *et al.*, 1974) in connection with the high velocity streams in the solar wind. However, our observations do not seem to be of the same type of phenomenon as the period considered is one of rather low velocity without any strong gradient (see Figure 5). Going backwards in time, with respect to the period referred to in Figure 8, either the double peaked structure is absent or it is much less evident. Although

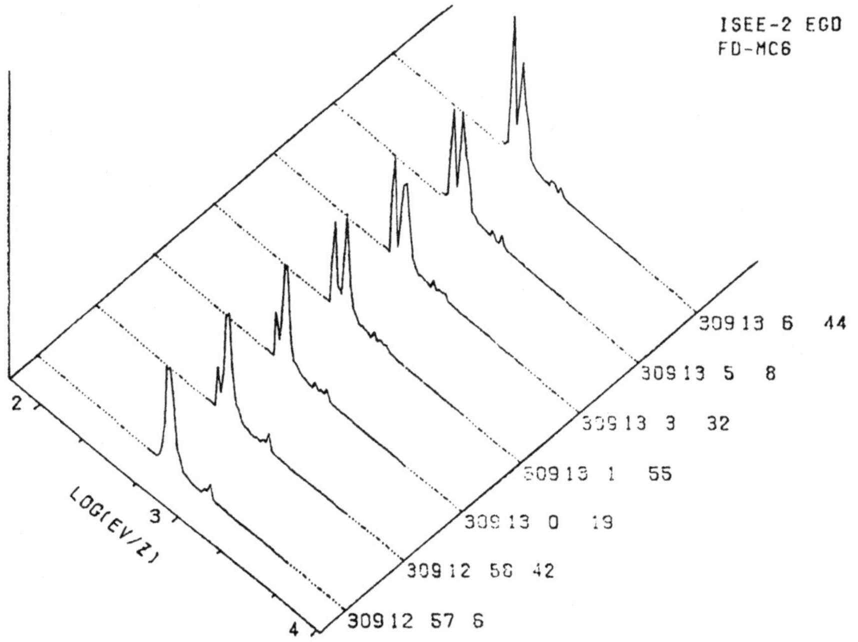

Fig. 8. Sequence of running averages of the ion distribution function, performed over groups of six consecutive spectra, in the period from 12:57 UT to 13:16 UT (day 309, 1977). This time interval is the last, before the bow shock crossing, during which the instrument was operating in the WES mode.

the analysis must be completed, this would seem to indicate that the double peaked feature is somehow connected with the bow shock transition. Note also that, with reference to phenomena connected with the bow shock, non thermal features of the distribution have been found only within, or in the immediate vicinity of, the shock itself (Montgomery *et al.*, 1970).

It is too early to attempt any interpretation of the observed distribution upstream of the shock crossing of day 309. A moment analysis of the separate peaks of the distribution should lead, for example, to conclusions about their stability. Correlation with the plasma data from ISEE-1, and with magnetic field data from both spacecraft as well as information about the features of the turbulence (of which we found evidence from our HTR data) are necessary to advance possible explanations of the observed phenomenon.

3. Outward Flows Upstream of the Bow Shock

Flows of protons backstreaming from the bow shock, in the antisolar direction, have been identified in several previous experiments (Asbridge *et al.*, 1968; Scarf *et al.*, 1970). These particles are thought to be reflected backward and accelerated at the shock front (Sonnerup, 1969). The mechanism of such phenomenon is however not yet completely understood (at least in a quantitative way) and constitutes a problem in the framework of bow shock connected phenomena, where the ISEE-1 and -2 experiments will hopefully lead to some clarification.

We will present here evidence for such backstreaming particles which has been obtained from the EGD Plasma experiment.

A first example refers to the bow shock crossing of day 312. This is a very turbulent shock (see Russell and Greenstadt, 1979) with probable multiple crossings and an angle between magnetic field and shock normal of ~60°. Figure 9 gives the time history of several parameters across the transition which occurs at 18:53 UT (vertical dashed line in the figure). The lower five panels give, as in Figure 5, the bulk velocity, flow direction, density and most probable thermal speed of solar wind protons and the magnetic field longitude (again gaps in the plot correspond to periods when the experiment was operating in the HTR mode). The top panel gives the ratio between the ion flux in the antisolar direction and the solar wind flux. It is seen that the presence of antisolar particles is strongly correlated with changes in the magnetic field direction.

Using the method sketched by Diodato *et al.* (1976), we have checked whether the presence or absence of backstreaming ions was consistent with the observed direction of the interplanetary magnetic field. As shown in Figure 10, the agreement turned out to be very satisfactory. This figure shows the portion of ISEE-1 trajectory corresponding to the time interval considered in Figure 10, along with a bow shock profile. The thick lines refer to periods when backstreaming protons were detected. Three cases (respectively referred to as 1, 2, 3) are

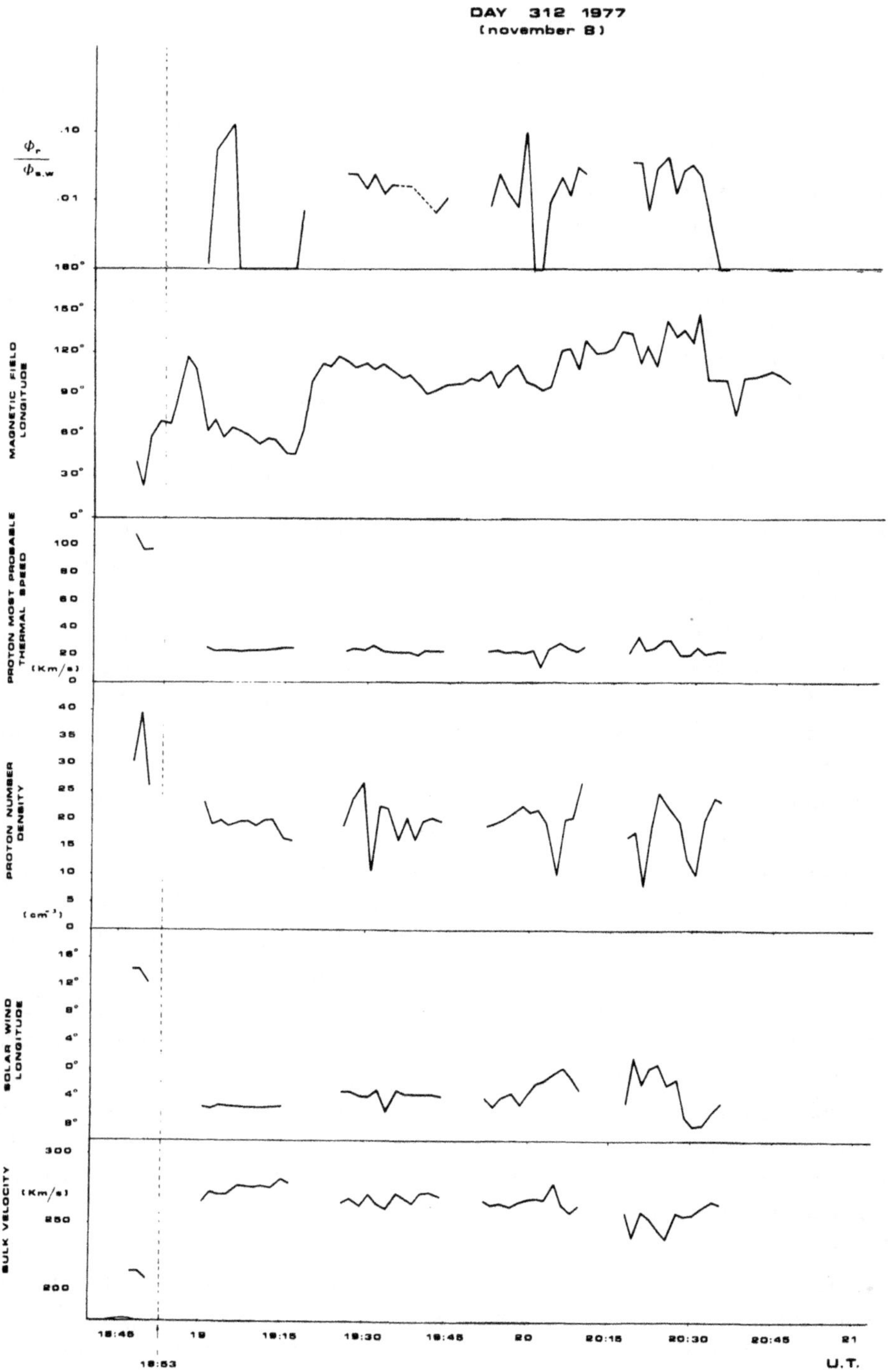

Fig. 9. Time history of the proton parameters during the bow shock crossing occurred at 18:53 UT, on day 312, 1977. The parameters plotted in this figure are the same as in Figure 5, except for the top panel which gives the ratio between the ion flux in the antisolar direction and the solar wind flux.

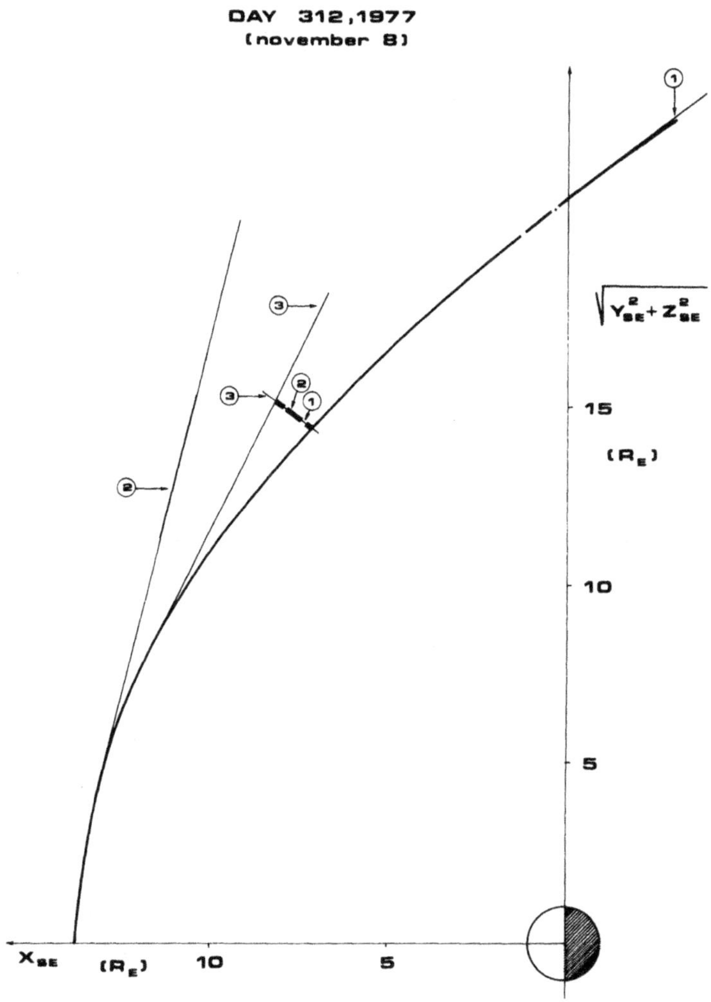

Fig. 10. Sketch of the bow shock and ISEE-2 trajectory on day 312, 1977. The portion of the orbit, shown in the figure, refers to the same period of time considered in Figure 10. Parts of the orbit when backstreaming protons were detected are indicated by thicker lines. The straight lines labelled 1, 2, 3 represent the boundary of the 'upstreaming ion region' when the satellite was at the locations indicated with the corresponding numbers 1, 2, 3 (see text for details).

illustrated. The first one corresponds to the period from 19:07 UT to 19:25 UT when backstreaming particles were not observed. The direction of the interplanetary magnetic field implies that particles, reflected from the bow shock, should reach only the region between the shock front and the straight line labelled 1 in the figure, in agreement with the lack of antisolar particles at the ISEE-1 position. The second case corresponds to the period from 19:25 UT to 20:00 UT when backstreaming particles were always observed: the position of the expected boundary of the 'upstreaming ion region' is again in agreement with observations.

The last case corresponds to the final disappearance of the antisolar ions (20:34 UT): the boundary of the upstreaming ion region (straight line labelled 3 in the figure) crosses the ISEE-1 trajectory exactly at the point where the antisolar ion suddenly disappear.

Such a very good agreement between kinematic predictions and observations strongly support the hypothesis that antisolar ions detected by ISEE-1 are protons reflected by the Earth's bow shock.

Figure 11a and 11b give a sequence of ion energy spectra (integrated over the azimuthal angle ϕ) in the period from 19:27 UT to 19:39 UT (day 312).

Each energy per unit charge spectrum has been split into two spectra: the first was taken in the solar direction (i.e. in the 12 solar angular sectors, see Section 1.2); the second in the antisolar direction (i.e. in the 3 antisolar sectors). The latter spectra are amplified by a factor 40, with respect to solar wind spectra, to make their features more clearly visible.

The following comments can be derived from the observations. The solar wind spectra appear to be rather normal, even though with some variability. By looking at the sequence of antisolar spectra, one notes, on the other hand, a much stronger variability from one spectrum to the next. This is indicative of the presence of turbulent fluctuations at frequencies much higher than our sampling time (96 s).

Apart from this variability, one can distinguish (both in the period of Figure 11 and in other upstream data which we have analyzed) two somewhat different types of antisolar spectra. In some of these spectra, a main peak of particles, at an energy about twice the peak energy of the solar wind particles, is clearly distinguishable against much smaller fluxes observed in the other energy channels (see for example the spectrum of 19:27 UT in Figure 11a or the spectrum of 19:37 in Figure 11b).

In other cases the spectrum of antisolar particles appears to extend more towards lower energies and no prominent peak is exhibited (see for example the sequence of three spectra starting from 19:29 in Figure 11a).

A clear example of an antisolar spectrum with a very dominant peak is given in Figure 12. This spectrum (taken at UT 19:05) corresponds to the very large peak in antisolar counts visible in the top panel of Figure 9. The antisolar particle flux amounts here to as much as 11% of the solar wind flux and its average velocity is estimated to be $\sim$473 km s^{-1} (i.e. about 1.75 times the bulk velocity of solar wind protons).

It is interesting to recall that, in the period of time during which this spectrum was recorded, the high resolution magnetic data (Greenstadt, private communication) show an enhanced magnetic activity, with high amplitude fluctuations ($\partial B/B$ up to 0.5) of both shorter ($\sim$1 s) and longer periods ($\sim$20 s).

Although a more detailed correlated study is needed, such upstream waves appear to be generated by the high intensity beam of reflected protons shown in Figure 12.

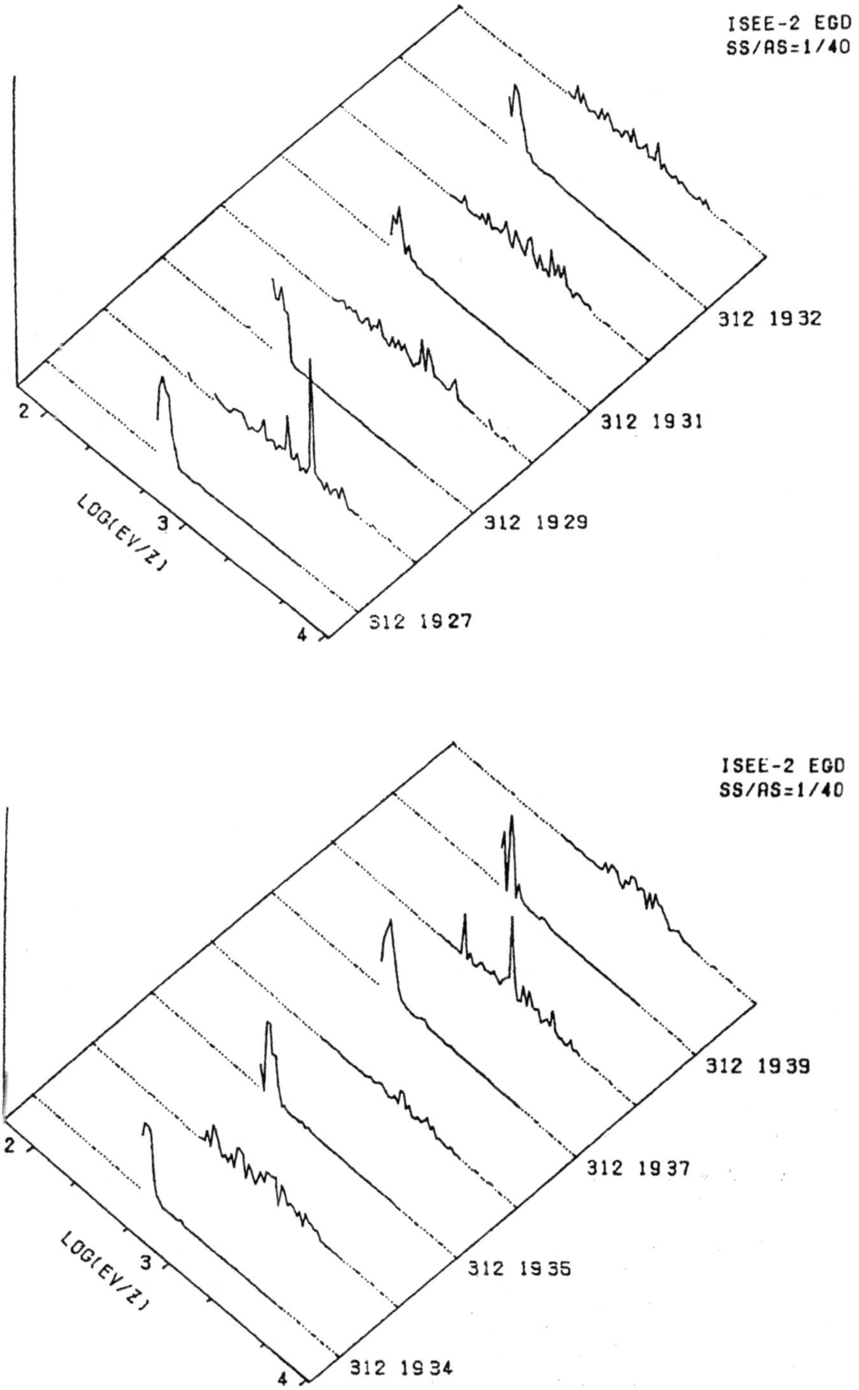

Fig. 11a, b. Sequence of solar and antisolar ion energy spectra recorded on day 312, from 19:27 to 19:39 UT. Note that the antisolar counts were multiplied by a factor 40.

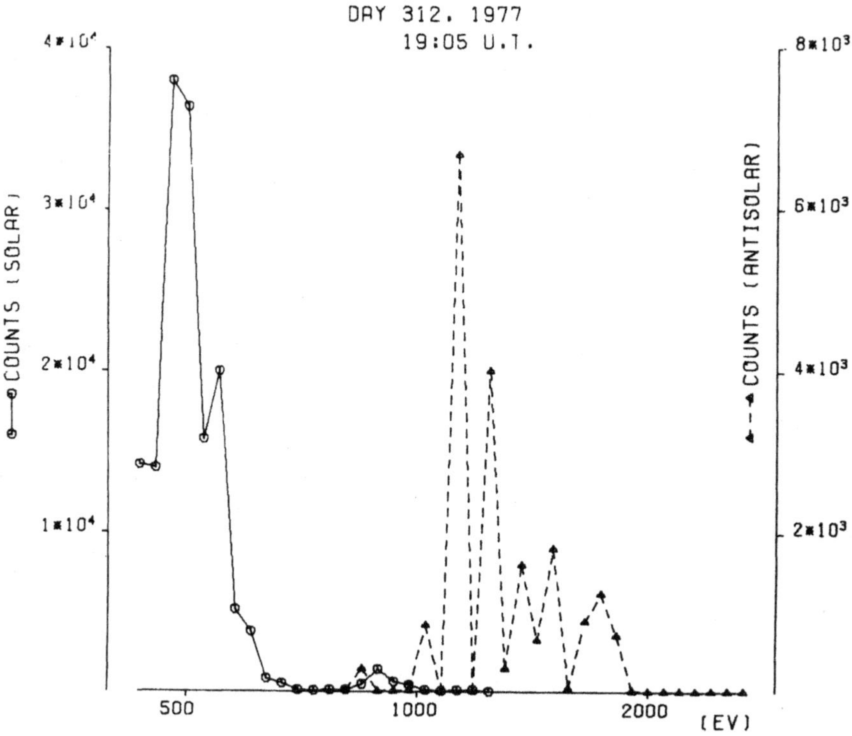

Fig. 12. Ion energy per unit charge spectrum observed at 19:05 UT on day 312, 1977. The continuous line refers to solar fluxes (left scale); the broken line to antisolar fluxes (scale on the right).

As for the more extended energy spectra of antisolar particles (referring, as a typical one, to that of 19:29 UT shown in Figure 11a), we find that the spectrum extends in the range of energies from ~0.5 to ~3.5 keV. The percentage of antisolar particles is here only ~1.5% and their average velocity ~470 km s^{-1} (i.e. a factor 1.7 higher than the bulk velocity of the solar protons). For comparison we recall that previous observations of backstreaming protons (Asbridge *et al.*, 1968; Scarf *et al.*, 1970) referred to the range of energies from 2 to 7 keV. In our case we do not find energies as high: this difference may be related to the rather low bulk velocity of the wind (260–270 km s^{-1}, as can be seen from Figure 9).

Further study is needed before drawing conclusions on the origin of the more extended antisolar spectra of the type shown in Figure 11. As it has been pointed out by Greenstadt (1975), the kinematics of particle reflection from the shock may be consistent with a very wide range of energies for the outward moving particles, especially in cases of complex shock structures such as that of day 312; therefore, the wide spectra observed by our experiment could be still made up essentially of reflected protons.

Further examples of antisolar particle spectra are given in Figure 13, referring

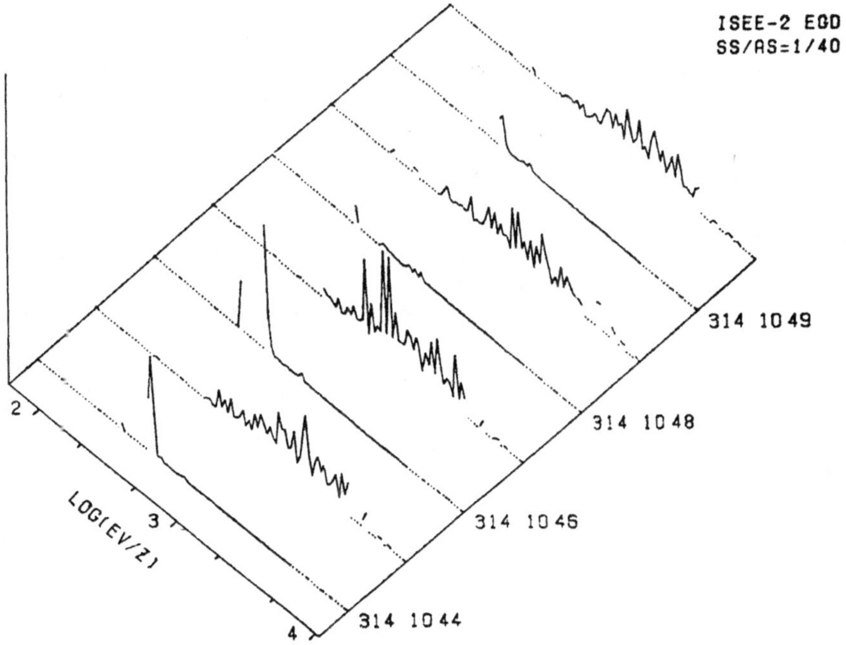

Fig. 13. Sequence of solar and antisolar ion energy spectra recorded on day 314 from 10:44 UT to 10:49 UT. Note that the antisolar counts were multiplied by a factor 40.

to the upstream state of the shock transition which occurred at 11:00 UT of day 314.

This is again a highly turbulent shock characterized by multiple crossings (Russell and Greenstadt, 1979). However, the features of the antisolar ion fluxes are somewhat different from those of day 312. The energy spectra of the upstreaming particles are now even wider than the widest spectra observed on day 312. For example, at 10:48 UT the energy spectrum of the backstreaming ions extends from ~100 eV to ~10 keV (i.e. over the whole energy spectrum explored by the experiment).

The features of the observations seem to suggest for these particles an origin other than the reflection from the bow shock.

Such antisolar spectra should be studied in more detail to derive their average characteristics and plasma parameters and correlate them with data on both magnetic and electrostatic fluctuations. An investigation of this type should throw some more light on the mechanisms of particle reflection at collisionless shock fronts and on the subsequent wave phenomena generated by the backward moving particles.

References

Asbridge, J. R., Bame, S. J., and Strong, I. B.: 1968, 'Outward Flow of Protons from the Earth's Bow Stock', *J. Geophys. Res.* **73,** 5777.

Bonifazi, C., Cerulli-Irelli, P., Egidi, A., Formisano, V., and Moreno, G.: 1978, 'The EGD Positive Ion Experiment on the ISEE-2 Satellite', *IEEE Trans., Geosci. Electr.* **GE-16,** 243.

Diodato, L., Greenstadt, E. W., Moreno, G., and Formisano, V.: 1976, 'A Statistical Study of the Upstream Wave Boundary Outside the Earth's Bow Shock', *J. Geophys. Res.* **81,** 199.

Feldman, W. C., Asbridge, J. R., Bame, S. J., and Montgomery, M. D.: 1974, 'Interpenetrating Solar Wind Streams', *Rev. Geophys. Space Phys.* **12,** 715.

Greenstadt, E. W.: 1975, in V. Formisano (ed.), *The Magnetospheres of the Earth and Jupiter,* D. Reidel, Dordrecht, Holland, p. 3.

Hundhausen, A. J., Asbridge, J. R., Bame, S. J., Gilbert, H. E., and Strong, I. B.: 1967, 'Vela-3 Satellite Observations of Solar Wind Ions: a Preliminary Report', *J. Geophys. Res.* **72,** 87.

Montgomery, M. D., Asbridge, J. R., and Bame, S. J.: 1970, 'Vela-4 Plasma Observations near the Earth's Bow Shock', *J. Geophys. Res.* **75,** 1217.

Russell, C. T. and Greenstadt, E. W.: 1978, 'Initial ISEE Magnetometer Results: Shock Observations', *Space Sci. Rev.* **23,** 3.

Scarf, F. L., Fredricks, R. W., Frank, L. A., Russell, C. T., Coleman, P. J., Jr., and Neugebauer, M.: 1970, 'Direct Correlations of Large Amplitude Waves With Suprathermal Protons in the Upstream Solar Wind', *J. Geophys. Res.* **75,** 7316.

Sonnerup, B. U. O.: 1969, 'Acceleration of Particles Reflected at a Shock Front', *J. Geophys. Res.* **74,** 1301.

HIGH TEMPORAL RESOLUTION OBSERVATIONS OF ELECTRON HEATING AT THE BOW SHOCK*

S. J. BAME, J. R. ASBRIDGE, J. T. GOSLING, and M. HALBIG

University of California, Los Alamos Scientific Laboratory, Los Alamos, N.M. 87545, U.S.A.

G. PASCHMANN and N. SCKOPKE

Max Planck Institut für Physik und Astrophysik, Institut für Extraterrestriche Physik, Garching, F.R.G.

H. ROSENBAUER

Max Planck Institut für Aeronomie, Lindau 3, F.R.G.

Abstract. High temporal resolution measurements of solar wind electrons at the Earth's bow shock on the dawn side have been made with the LASL/MPI fast plasma experiments on ISEE-1 and 2. One dimensional, 1-*d*, temperatures, T_e, and densities, N_e, are obtained every 0.3 s and 2-*d* values are obtained every 3 s. Profiles of T_e and N_e at the shock usually are found to be similar to one another and also to the profile of the magnetic field magnitude. The time scale of electron thermalization varies from about 0.5 s to greater than 1 min, depending importantly on the shock motion and the orientation of the magnetic field. Typical thermalization times from 05:00–12:00 LT are ~10 s, considerably shorter than proton thermalization times at the shock. This time scale corresponds to a distance of ~100 km, comparable to but somewhat larger than the typical ion inertial length. The electron thermalization times are significantly longer than some of the values frequently cited in the past. At the end of the electron thermalization there typically is an overshoot in electron thermal pressure followed by an undershoot which give the pressure profile of the shock the appearance of a damped wave. Ion thermalization is essentially completed by the time the electron pressure wave is 'damped'. The most probable value of the electron temperature ratio across the shock is 1.7, and this value is relatively independent of the Sun–Earth–satellite angle, θ_{SS}, for θ_{SS} between 25° and 100°.

1. Introduction

Studies of physical processes occurring at the Earth's bow shock have, in the past, been hampered by the lack of rapid temporal measurements of plasma distribution functions. Magnetometer studies of the bow shock structure (e.g. Russell and Greenstadt, 1978) have revealed that the major changes in field strength which occur there take place on a distance scale comparable to one ion inertial length (c/ω_{pi}, where c is the speed of light and ω_{pi} is the ion plasma frequency). Typical ion inertial lengths in the solar wind are ~50 km, and since shock motions are typically ~10 km s^{-1} the shock transition in the field as observed by a nearly

* By acceptance of this article, the publisher recognizes that the U.S. Government retains a non-exclusive, royalty-free license to publish or reproduce the published form of this contribution, or to allow others to do so, for U.S. Government purposes.

The Los Alamos Scientific Laboratory requests that the publisher identify this article as work performed under the auspices of the Department of Energy.

Space Science Reviews **23** (1979) 75–92. 0038-6308/79/0231-0075 $02.70.

stationary satellite (e.g. Russell and Greenstadt, 1979) occurs on a time scale of ~5 s. By way of contrast, snapshot times for plasma instruments penetrating the bow shock have generally been several 10's of seconds with repetition times of 1 min or greater. It is clear that with such instrumentation, the structure of the bow shock is poorly resolved, and spectra obtained there have generally suffered from severe time aliasing. It was for this reason, among others, that the LASL/MPI fast plasma experiments (FPE) on ISEE-1 and 2 were designed for very rapid plasma measurements. Complete 2-dimensional ion and electron countrate spectra are obtained every 3 s (1 spin) on each spacecraft. At high data rate transmission (HDR) the repetition time for spectra is also 3 s, whereas at low data rate (LDR) the repetition time between spectra is 12 s. For the electrons, useful 1-dimensional spectra are obtained in 140 ms with a repetition time of 375 ms (HDR). This represents a substantial improvement in temporal resolution compared to previous experiments, particularly when data are transmitted at HDR.

A detailed description of the instruments and their operation can be found elsewhere (Bame *et al.*, 1978), and a brief description of the instruments as well as a discussion of data analysis techniques can be found in the accompanying paper by Paschmann *et al.* (1978b). The reader should bear in mind that, because our analysis is still in its preliminary stages, complete corrections for the effects of photoelectrons and spacecraft charging have not yet been made, but these effects are not expected to influence seriously the results reported here. Solar wind electron temperatures calculated from the ISEE data obtained immediately upstream from the bow shock compare reasonably well with electron temperatures obtained simultaneously far from the shock by LASL instrumentation on IMP-8, but the ISEE estimates tend to be ~15% high. At least a part of this difference is real because heat from the shock flows upstream to preheat the electrons in the immediate vicinity of the shock (Feldman *et al.*, 1973; Scudder *et al.*, 1973). Corrections for photoelectrons and spacecraft charging should be less important within the magnetosheath where temperatures and average particle energies are higher.

Our primary purpose here is to present new results deduced from the highly time-resolved electron plasma profiles of the bow shock obtained with this new instrumentation. Emphasis is placed on those crossings which occurred during periods of high data rate transmission so that the detailed structure of the shock transition for the electrons is revealed. (In several examples the ion inertial length, c/ω_{pi}, (see e.g. Kennel and Sagdeev, 1967) has been indicated as one possible measure of shock thickness. We do not mean to imply by its exclusive use here that it is the only appropriate measure of shock thickness.) The measurements indicate that electron thermalization and density compression generally are synchronized, although exceptions to this rule occur. Further, in the few examples where a direct comparison with magnetic field measurements has been possible, the electron temperature and density profiles at the shock are nearly identical to that of the field intensity. Thus, for example, electron thermalization at the shock generally occurs on time scales ranging from ~0.5 s to greater than 1 min depending importantly on the speed of the shock past the spacecraft and the orientation of the interplanetary field. A new feature of the shock profile is revealed by the measurements, namely an electron pressure overshoot lasting several tens of seconds and generally followed by an undershoot which give the shock profile the appearance of a damped wave.

2. The Observations

2.1. Electron shock profiles at high temporal resolution

The ISEE-1 and 2 satellites were launched into elliptical orbits with apogee at $22.6R_E$ on 22 October, 1977. The initial line of apsides was close to the plane of the noon meridian, and through mid-January 1978 traversals of the dawn bow shock occurred between about 05:00 and 12:00 LT. Because of the inclination of the orbital plane to the ecliptic plane, shock crossings occurred at Sun–Earth–satellite angles (θ_{SS}) varying from about 25 to 105 deg.

Our intent here is to show several shock crossings which are representative of many of the more than 100 shock crossings observed prior to mid-January 1978. For the most part the crossings to be shown occurred during high data rate transmission, and usually data are available from both of the ISEE spacecraft. Ion shock profiles are not included here, primarily because the energy-angle sampling of the ISEE FPE's is insufficient to resolve adequately solar wind ion spectra and also because ion spectra in the vicinity of the shock are often so chaotic that moment calculations inadequately describe the changes which occur there. However, ion thermalization times are always considerably longer than electron thermalization times (see e.g. Montgomery *et al.*, 1970), and in several examples an approximate ion thermalization time has been indicated.

Figure 1 shows electron density, temperature, and pressure variations observed at the bow shock by ISEE-1 on 29 October, 1977 at two temporal resolutions. In the upper panels the parameters are derived from 2-dimensional spectra sampled every 3 s, while in the lower panels the parameters are derived from 1-dimensional spectra (averaged over 2 simultaneous measurements made at 180° separations in spacecraft longitude) sampled approximately every 375 ms. Very small amplitude fluctuations at periods of ~1.5 s in the lower panels (and also in subsequent figures) are artifacts of the analysis caused by imperfect calibration between detectors sampling simultaneously in opposite directions. The units of the ordinate axes for this and subsequent plots are cm^{-3} (N_e), K (T_E), and 10^{-8} dynes cm^{-2} (P_E). Electron density and temperature vary nearly in concert with one another across the shock, the major increases going from the solar wind to the magnetosheath occurring during the 11-s interval from 12:48:52 to 12:49:03 UT. The major shock transition was preceded by a more gradual rise in density and temperature beginning at ~12:48:43 UT. Similar 'foots' to the

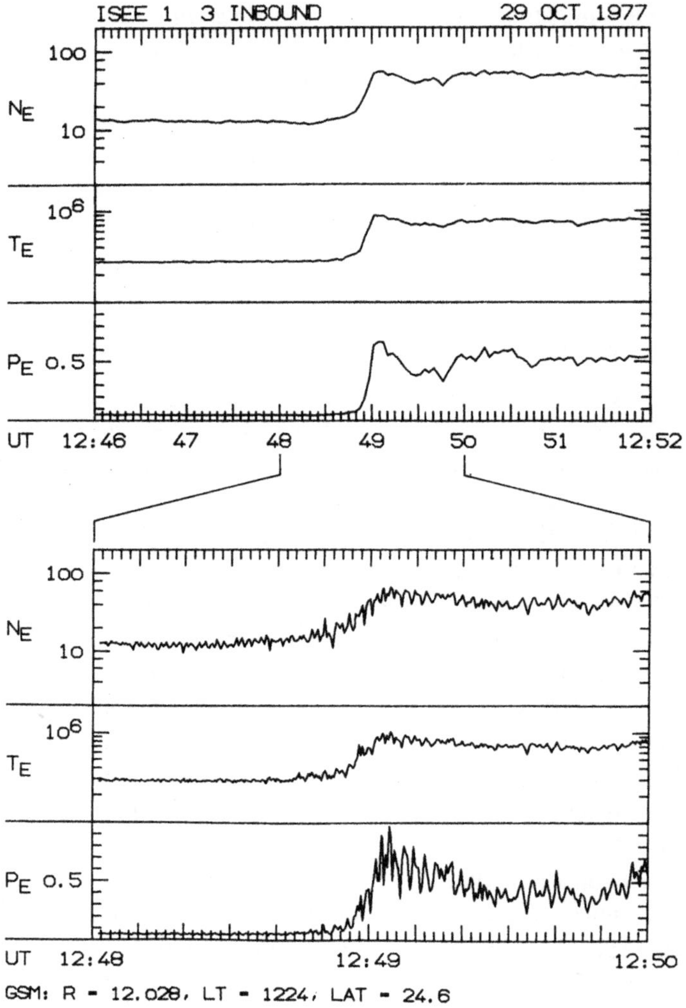

Fig. 1. Two- and one-dimensional (top and bottom panels respectively) calculations of electron density ($\#cm^{-3}$), temperature (K), and pressure (10^{-8} dynes cm^{-2}) for an inbound crossing of the Earth's bow shock on 29 October, 1977. Six minutes worth of data from ISEE-1 are displayed in the upper panels with data sampling every 3 s. Two minutes worth of data are shown in the lower panels with sampling approximately every 375 ms.

electron thermal rise at the shock have been reported previously (e.g. Montgomery *et al.*, 1970) and are probably directly associated with similar 'foots' observed in magnetic field data (e.g. Russell and Greenstadt, 1978). The time scale for the major electron thermalization (11 s) is somewhat longer than has been reported previously using data with poorer temporal resolution. Although both considerably shorter and considerably longer thermalization times have been found in the data, this time scale is representative of a 'typical' crossing as observed with the ISEE instruments.

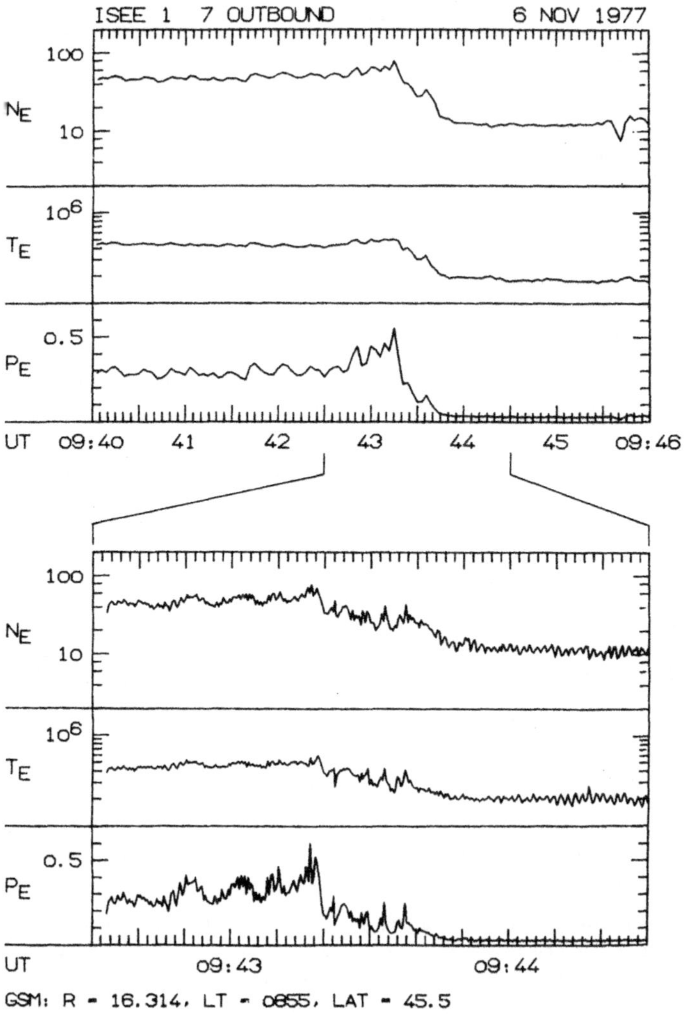

Fig. 2. Electron parameters for an ISEE-1 crossing of the shock on 6 November, 1977. The format is identical to Figure 1.

An additional important feature of this particular crossing is the overshoot in electron pressure which occurs at the end of the electron thermalization. This overshoot of ~15 s duration is followed by a region ~30 s wide in which the pressure is lower than in the sheath which follows. Altogether the overshoot and undershoot give the shock transition the appearance of a wave damped within 2 oscillations.

Electron profiles for an outbound pass of ISEE-1 on 6 November, 1977 are shown in Figure 2 at the same 2 temporal resolutions as used in the previous figure. Again, the density and temperature profiles are similar to one another, although neither is as smoothly varying as in the previous example. In this case the electron thermalization time going from the solar wind to the sheath is ~30 s

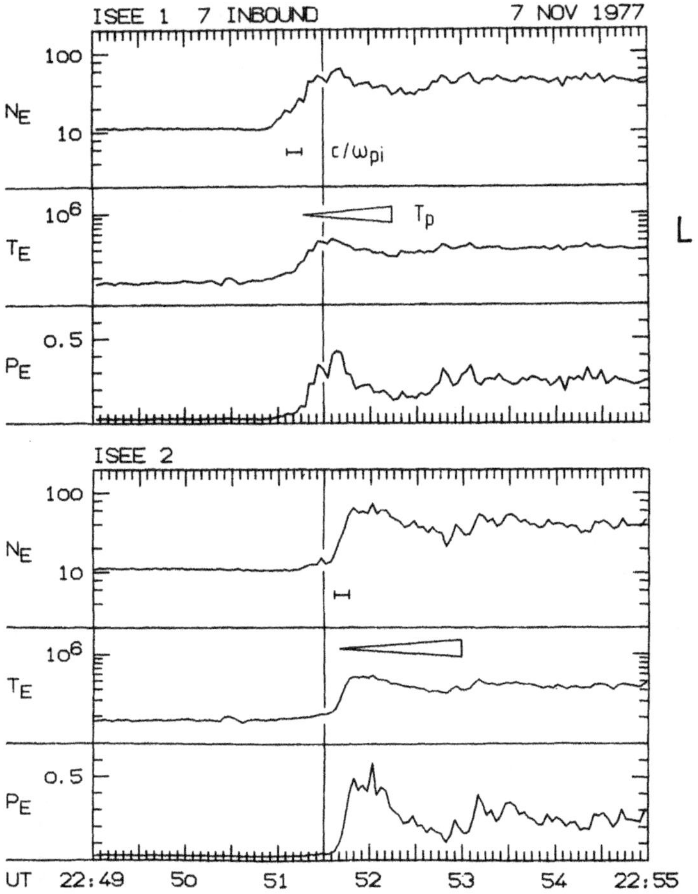

Fig. 3. Two-dimensional electron parameter plots for ISEE-1 (upper panel) and ISEE-2 (lower panel) for an inward crossing of the shock on 7 November, 1977. The wedge above the electron temperature plot in this and in subsequent figures indicates the approximate time required for ion thermalization. The ion inertial length (c/ω_{pi}) is indicated below the density profile.

and appears to occur in several discrete steps. At the end of the electron thermalization (actually observed first because of the outward motion of the satellite) there again is a pronounced overshoot in electron pressure which lasts ~30 s. No corresponding undershoot is apparent in this case although oscillations in pressure with a period of ~10 s occur within the overshoot as well as elsewhere within the sheath.

Figure 3 shows data from both ISEE-1 and ISEE-2 during an inbound shock crossing on 7 November, 1977. The electron parameters from both spacecraft are derived from 2-dimensional spectra sampled every 3 s. ISEE-1 was leading ISEE-2 by 210 km at this time and so encountered the shock first. From the ~24 s time delay between shock encounters at the 2 satellites we calculate a shock speed of 8.5 km s^{-1} parallel to an assumed average shock normal. The shock profiles observed at the 2 satellites are roughly similar, although small differences in detail are evident. For example, the shock transition at ISEE-2 is slightly faster than that at ISEE-1. Both profiles contain small 'feet' of ~20 s duration wherein both the density and temperatures rise slowly prior to the main shock transition which lasts ~10 s. For a shock speed of 8.5 km s^{-1} these time scales correspond to distances of 170 km and 85 km respectively. Both distances are comparable to but somewhat larger than the measured ion inertial lengths (c/ω_{pi}). A pronounced pressure overshoot followed by an undershoot is evident at both spacecraft, again giving the shock transition the appearance of a damped wave. The ion thermalization time for this shock, indicated by the wedges above the electron temperature profiles, is considerably longer than the electron thermalization time, as expected. It is interesting and perhaps significant that ion thermalization is essentially completed by the time the electron pressure undershoot ends.

One of the faster shock transitions observed during high data rate transmission occurred on 27 November, 1977, and is shown in Figure 4. ISEE-2 was leading

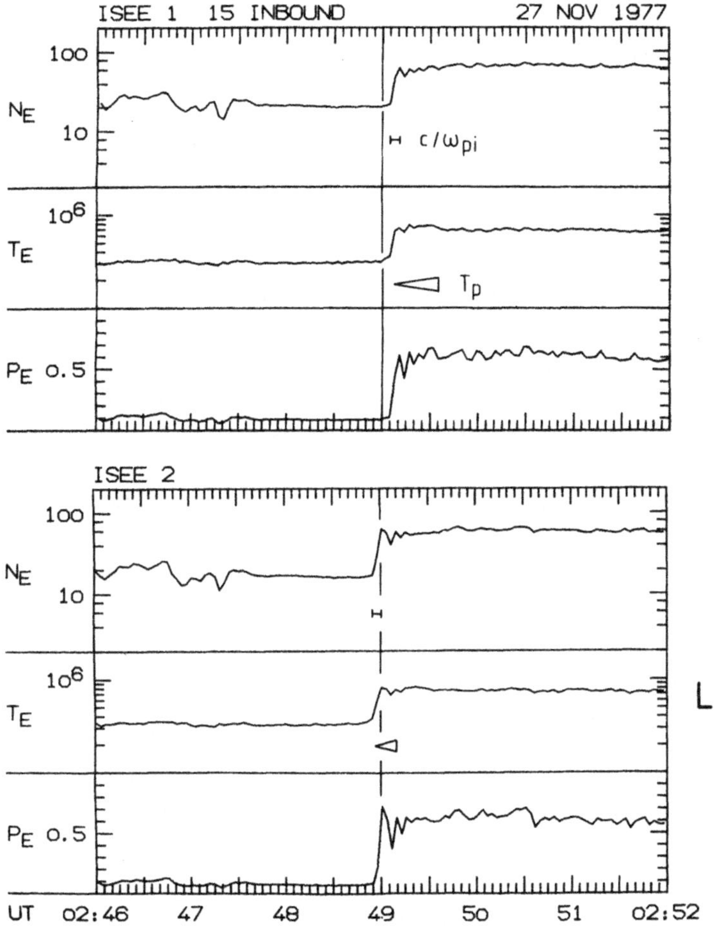

Fig. 4. Same as Figure 3 except that the date is 27 November, 1977.

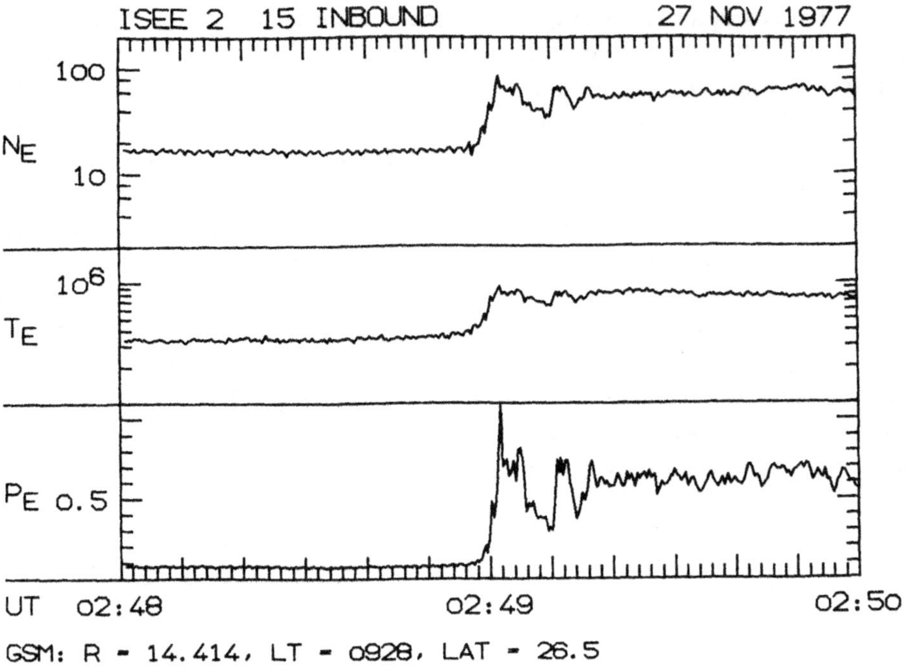

Fig. 5. The ISEE-2 one-dimensional electron parameters for the shock crossing of Figure 4.

ISEE-1 by 401 km at this time and so encountered the shock first. From the ~9 s delay between the encounters at the 2 spacecraft we calculate a shock speed of 43 km s^{-1} parallel to an assumed average shock normal. The electron profiles are nearly identical at the 2 spacecraft; Figure 5 shows the ISEE-2 data at higher temporal resolution using the 1-dimensional data sampled approximately every 375 ms. A small foot of ~3 s duration precedes the major shock transition which also requires ~3 s. For a shock speed of 43 km s^{-1}, 3 s corresponds to a distance of 129 km, not noticeably different from the distance scales obtained for the previous example. The electron pressure overshoot at the end of the electron thermalization is most pronounced at ISEE-2, but both satellites observed several rapid oscillations of pressure as the satellites began to penetrate the sheath. This oscillation pattern again gives the shock transition the appearance of a damped wave. The width of the region of electron pressure oscillation is comparable to the width of the region where ion thermalization occurs.

Before leaving the 27 November example, we would like to discuss briefly the large amplitude fluctuations in solar wind electron density which end abruptly at 02:47:30 UT at both spacecraft (Figure 4). The identical appearance of these fluctuations at both ISEE-1 and 2 demonstrates conclusively that the fluctuations are real and not just statistical perturbations. In fact, the abrupt end of the fluctuations coincides with the sudden disappearance of diffuse upstream ions (see e.g. Gosling *et al.*, 1978) at both spacecraft. A more complete discussion of this and other similar upstream events can be found elsewhere (Paschmann *et al.*, 1978a).

A double shock crossing which occurred on 1 January, 1978 nicely illustrates the various time scales involved in typical crossings of the dawn shock and is shown in Figure 6. At the time of this pair of crossings ISEE-2 was 427 km closer to the earth than was ISEE-1. Thus, as the shock first expanded outward and then moved inward, ISEE-2 sampled the sheath for a longer period than did ISEE-1. Because the shock moved outward considerably slower than it moved inward, the time lag between the shock transitions at the two satellites was longer for the first crossing than it was for the second. For the same reason, the shock transition at both satellites was considerably slower for the first crossing than for the second. Our estimates for the shock speed parallel to an average shock normal are 14 km s^{-1} for the outward motion and 48 km s^{-1} for the inward motion. Estimates

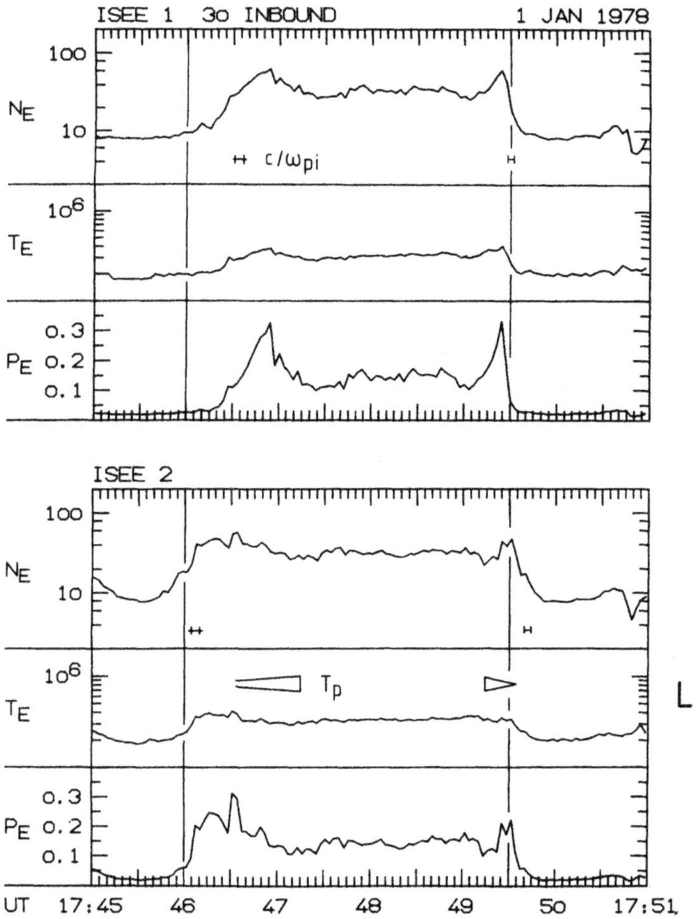

Fig. 6. Two-dimensional electron parameter plots for ISEE-1 (upper panels) and ISEE-2 (lower panels) for a double shock crossing on 1 January, 1978.

of the electron thermalization time for the first crossing are difficult because of the ragged appearance of the temperature profile and because the profiles are not identical at the 2 satellites. For the second crossing we obtain a thermalization time of ~10 s corresponding to a distance of 480 km. This distance is several times larger than the calculated ion inertial length (c/ω_{pi}). Finally, let us note that the electron pressure profile for each crossing has the aforementioned appearance of a damped wave. As in the other examples, the 'damping' occurs on a distance scale comparable to the ion thermalization distance.

All of the shock profiles shown up to this point are examples where the electron density and temperature vary nearly in concert with one another across the shock. This is the usual situation. However, occasionally shock transitions are observed where this rule is violated. Figure 7 shows an example of a shock crossing observed by both ISEE-1 and 2 where the density transition occurs much more rapidly than does the temperature transition, and Figure 8 shows an example

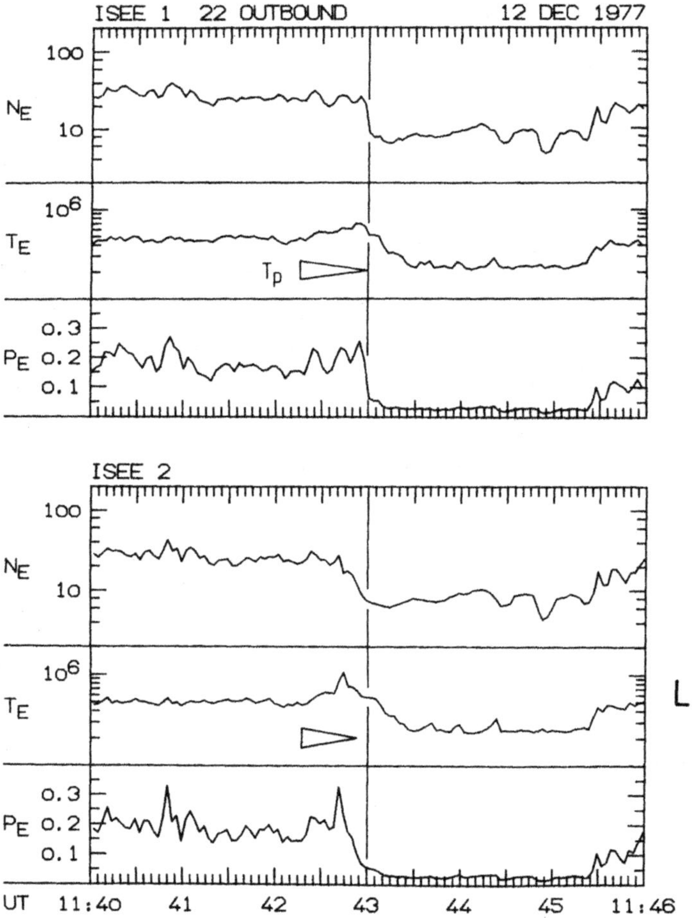

Fig. 7. Same as Figure 6 but for a single shock crossing on 12 December, 1977.

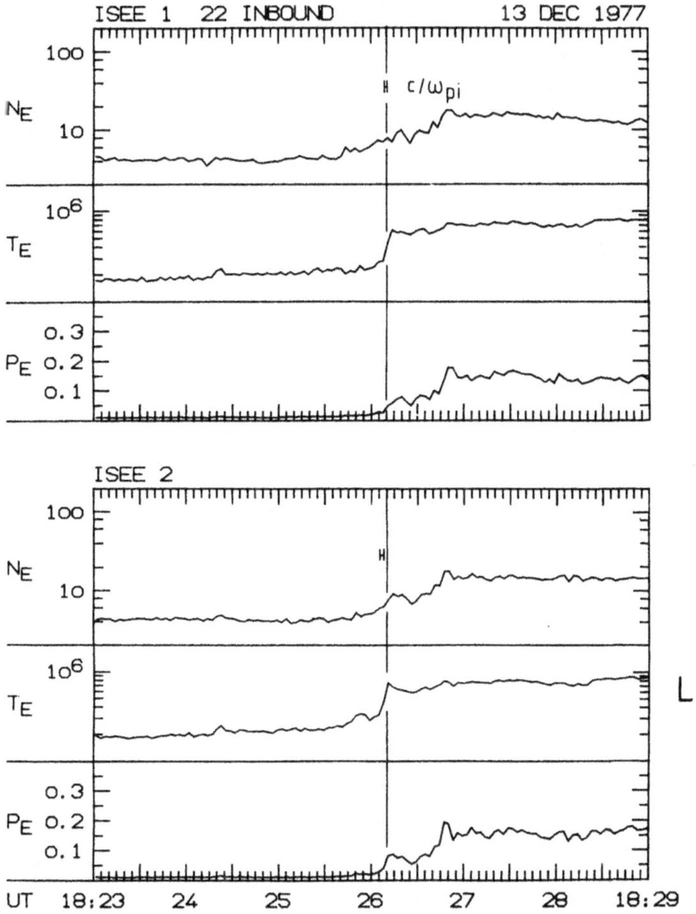

Fig. 8. Same as Figure 6 but for a single shock crossing on 13 December, 1977.

where the temperature transition is more rapid than the density transition. The nearly identical appearance of these shock profiles at both ISEE-1 and 2 assures that these anomalous shock profiles are real.

Thus far we have considered shock crossings where the transition from solar wind to magnetosheath flow is reasonably well defined. Although comparisons with the magnetic field have not yet been made for most of these crossings, our expectation is that most of these are quasi-perpendicular shocks. There exists another type of crossing in the ISEE data where the shock profile is less clearly defined. The December 7, 1977 outward crossing shown in Figure 9 is an example of a shock profile where the transition from solar wind to sheath flow is so ragged that it is difficult to state with certainty just where the shock lies. Yet at 17:02 UT both satellites were clearly in the magnetosheath, and by 17:07 UT both satellites were clearly in the upstream solar wind. Magnetic field data for this crossing (C. Russell, private communication) shows that this is a quasi-parallel shock. The data

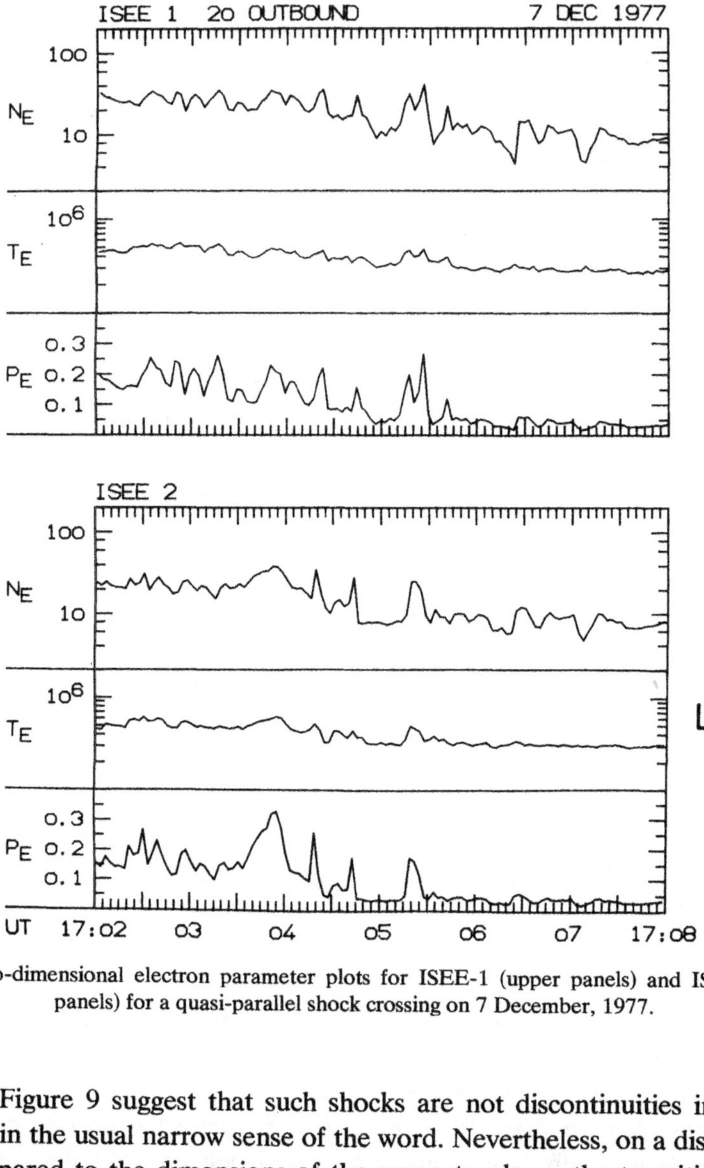

Fig. 9. Two-dimensional electron parameter plots for ISEE-1 (upper panels) and ISEE-2 (lower panels) for a quasi-parallel shock crossing on 7 December, 1977.

shown in Figure 9 suggest that such shocks are not discontinuities in the solar wind flow in the usual narrow sense of the word. Nevertheless, on a distance scale small compared to the dimensions of the magnetosphere, the transition is made from unshocked solar wind to compressed, heated magnetosheath flow.

2.2. A COMPARISON WITH MAGNETIC FIELD PROFILES OF THE SHOCK

Magnetic field data for a limited number of the early ISEE shock crossings have been provided to the ISEE experiment groups (courtesy C. Russell and E. Greenstadt). Unfortunately the ISEE satellites were transmitting at low data rate during these crossings so that the temporal resolution of the LASL/MPI fast plasma experiment is ~12 s at these times. Nevertheless this rate of sampling is

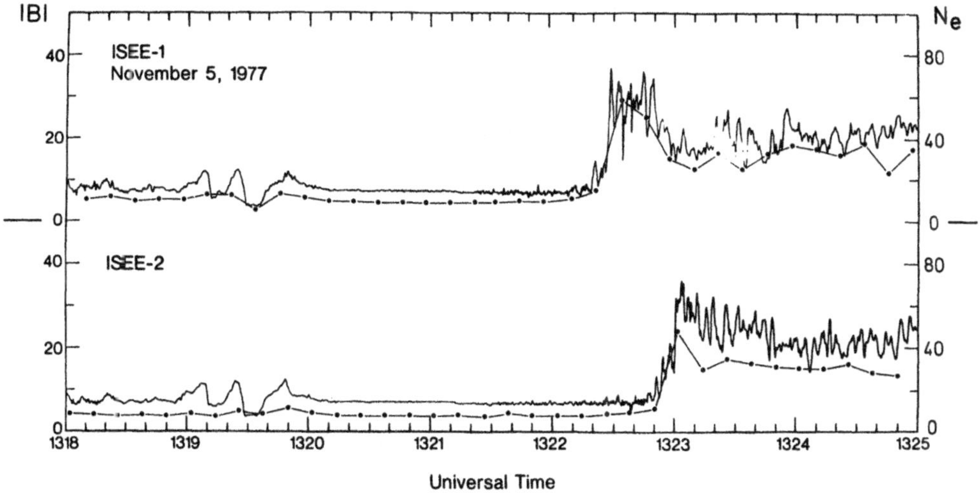

Fig. 10. Superposed magnetic field magnitude (UCLA magnetometer) and electron density (sampled every 12 s) for ISEE-1 and 2 on 5 November, 1977.

sufficient to establish a general similarity between the electron density and temperature profiles of the shock and that of the magnetic field. In Figure 10 the 12-s samples of electron density from ISEE-1 and 2 have been superposed upon the simultaneously measured magnetic field magnitude (UCLA magnetometer) for an inbound crossing on 5 November, 1977. Although not plotted, the electron temperature profile for this crossing is similar in shape to the density profile. Within the limits imposed by the rate of sampling, the shapes of the field and density profiles are nearly identical. This result is the same as reported by Neugebauer *et al.* (1971). A comparison of field and plasma data for an outbound crossing on 8 November, 1977 (Figure 11) yields a similar result.

2.3. The electron temperature ratio across the shock

The increase in electron temperature observed at the shock probably depends jointly upon several factors, including the Mach number of the shock, the plasma β, the direction of the interplanetary field, and the location where the shock is crossed. In this preliminary study we have not attempted to isolate which of these factors are dominant. However, it is a relatively simple matter to determine the electron temperature ratio (T_{MS}/T_{SW}), where T_{MS} is the electron temperature in the sheath after the pressure wave has 'damped', and T_{SW} is the electron temperature immediately upstream from the shock. In Figure 12 we have plotted this ratio as a function of the Sun–Earth-satellite angle, θ_{SS}, of the shock crossing. The temperature ratio is nearly constant for multiple crossings occurring during a single orbit, so in constructing Figure 12 we have taken only single representative

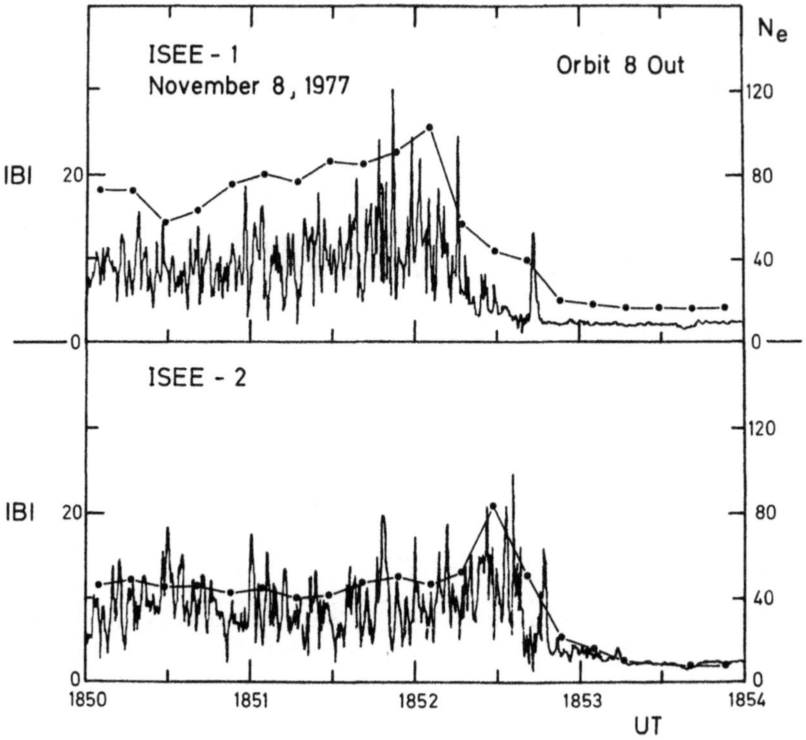

Fig. 11. Same as Figure 10 but for 10 November, 1977.

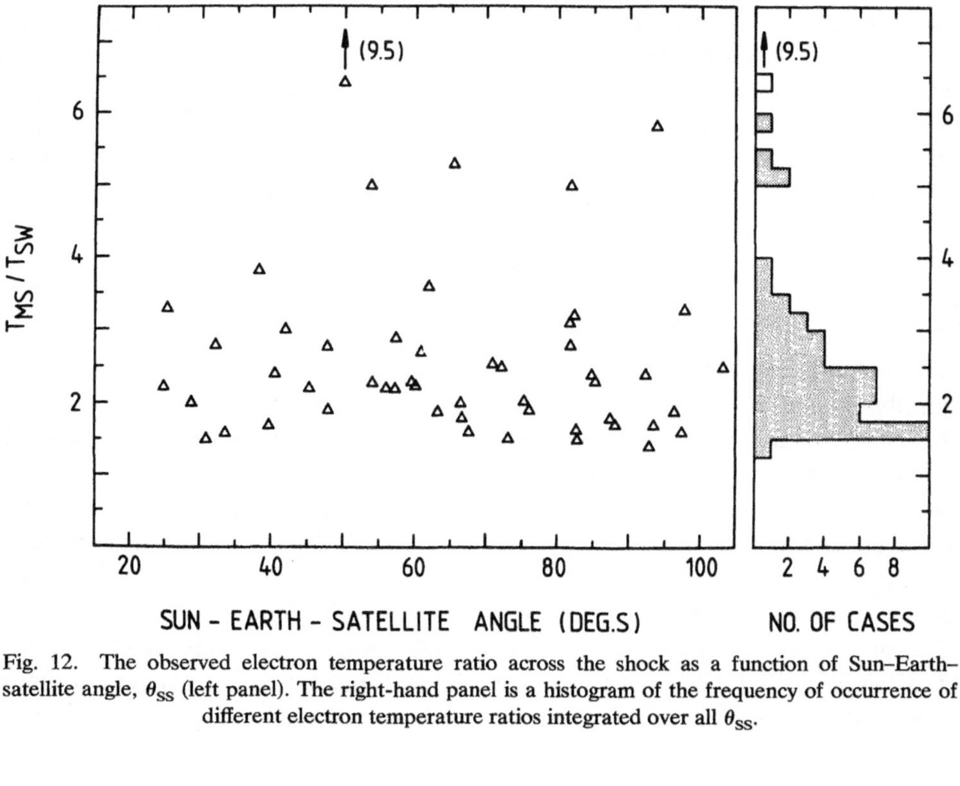

Fig. 12. The observed electron temperature ratio across the shock as a function of Sun–Earth–satellite angle, θ_{SS} (left panel). The right-hand panel is a histogram of the frequency of occurrence of different electron temperature ratios integrated over all θ_{SS}.

crossings for the inward and outward portion of each orbit through 15 January, 1978. It is readily apparent that the temperature ratio is variable but is not a strong function of θ_{SS} for the range of θ_{SS} angles encountered by ISEE-1 and 2. Note that the satellites never crossed the shock closer than about 25° from the subsolar point. The average value of T_{MS}/T_{SW} is 2.7, while the most probable value is 1.7. These values are lower than values reported previously by Montgomery *et al.* (1970) and Scudder *et al.* (1973).

2.4. Double ion beams at and behind the shock

The ion angle and energy sampling by the LASL/MPI fast plasma experiment (FPE) on ISEE-1 and 2 is insufficient to resolve adequately most solar wind ion spectra. It is for this reason that the present paper has concentrated on electron observations at the shock. On the other hand, the FPE is ideally suited for studying the rapid changes in the broader ion distributions which occur within the shock itself and downstream within the sheath. Further, shock profiles comparable to those presented for the electrons will be possible in the future by combining the FPE ion data with the ISEE-1 solar wind experiment data (Bame *et al.*, 1978). Here we wish only to provide the reader with a sample of the science residing in the ion FPE shock data.

Figure 13 displays consecutive ion countrate spectra averaged over the sunward hemisphere, ion angular distributions at four energies (averaged over four energy channels), and electron count rate spectra averaged over all look angles from 22:50:41 to 22:52:08 UT on 7 November, 1977. This interval contains the shock crossing previously presented in Figure 3. At the beginning of the 'foot' of the electron thermal rise at 22:50:53 a second ion peak first appears at $\sim 4E_{SW}$, where E_{SW} is the energy associated with the bulk flow of the solar wind. (Note that countrate spectra emphasize phenomena at the higher energies. The distribution function, f, is proportional to Cv^{-4}, where C is the countrate, and v is the velocity associated with a given energy.) The higher energy ions arrive from ~70° E of the Sun–satellite line. Thus these are probably solar wind ions which have been reflected and accelerated at the shock (e.g. Asbridge *et al.*, 1968; Gosling *et al.*, 1978). In subsequent panels this second group of particles retains its identity as the bulk of ions in the solar wind beam are decelerated and heated. Note that most of the ion deceleration has occurred by 22:51:20, which is near the end of the region of electron thermalization. As the ions in the main beam thermalize (most of this occurs after 22:51:20), it becomes less easy to distinguish the secondary beam. Even so, it is still barely resolved at 22:52:08, well behind the major gradient in field strength and electron pressure. (On some crossings the second beam can be identified all the way across the sheath.) Crossings such as this indicate that double ion beams in and behind the shock may be the result of particle reflection at the shock as first suggested by Montgomery *et al.* (1970). Other possibilities for the origin of the secondary beam have been discussed by Formisano and Hedgecock (1973).

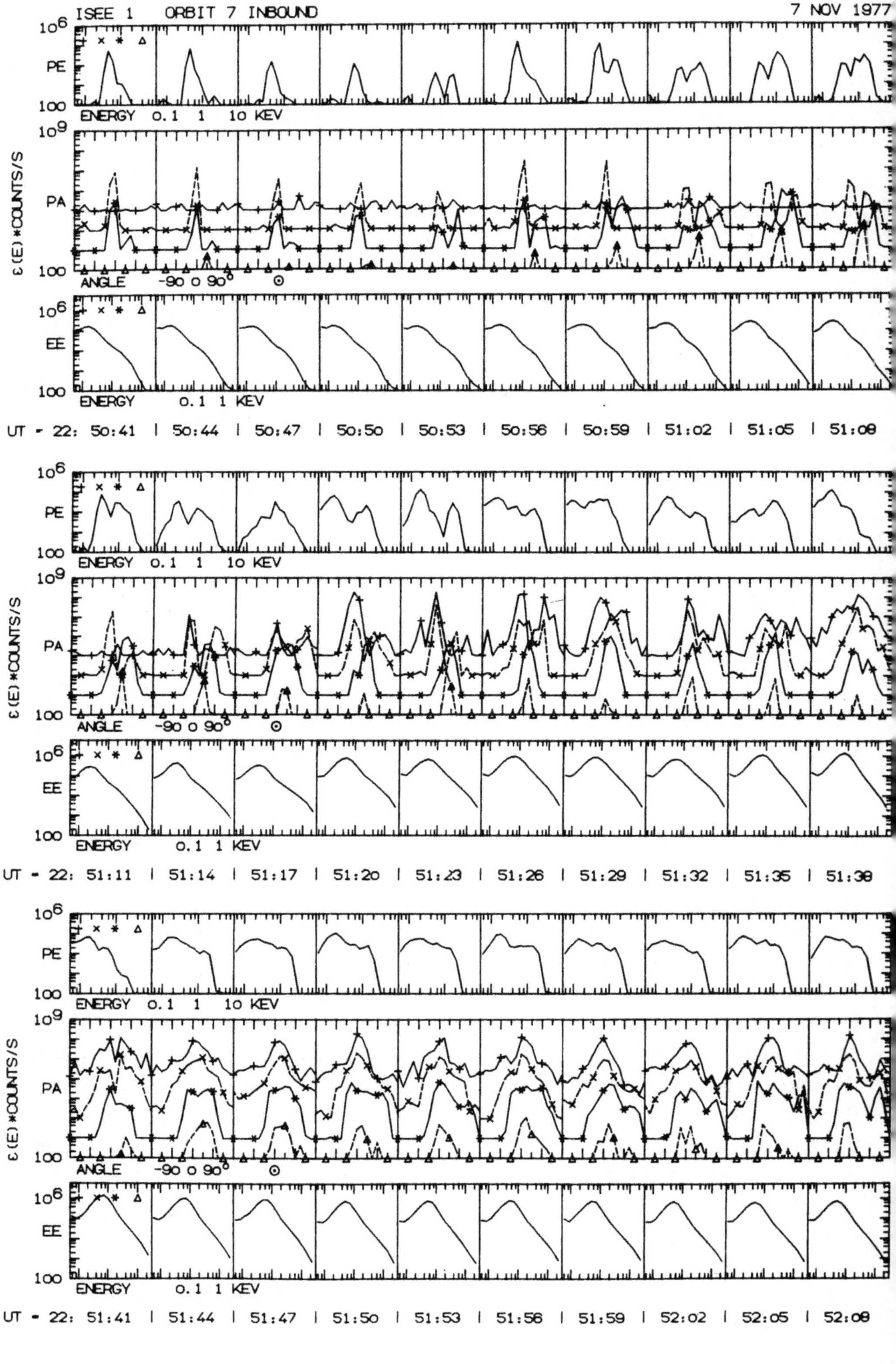
ISEE 1 ORBIT 7 INBOUND
7 NOV 1977
PE
PA
EE
ε(E)*COUNTS/S
ENERGY 0.1 1 10 KEV
ANGLE -90 0 90°
ENERGY 0.1 1 KEV
UT - 22: 50:41 | 50:44 | 50:47 | 50:50 | 50:53 | 50:56 | 50:59 | 51:02 | 51:05 | 51:08
UT - 22: 51:11 | 51:14 | 51:17 | 51:20 | 51:23 | 51:26 | 51:29 | 51:32 | 51:35 | 51:38
UT - 22: 51:41 | 51:44 | 51:47 | 51:50 | 51:53 | 51:56 | 51:59 | 52:02 | 52:05 | 52:08

3. Conclusions

In the preceding pages an attempt has been made to present representative bow shock data from the fast plasma experiments on ISEE-1 and 2. Although this report is preliminary, several important conclusions can be drawn.

(1) Electron density and temperature usually vary in concert with one another and with the magnetic field strength across the bow shock.

(2) The major electron thermalization at the shock typically occurs on a time scale of ~10 s, but can vary from ~0.5 s to greater than a minute depending importantly upon the speed of the shock past the satellite and the orientation of the interplanetary magnetic field vector. A 10-s electron thermalization time scale is considerably longer than that frequently cited in the literature (e.g. Montgomery *et al.*, 1970; Neugebauer *et al.*, 1971), and for typical shock speeds of 10 km s^{-1} corresponds to a thickness of ~100 km.

(3) At the end of the region of electron thermalization at the shock there often exist one or more major oscillations in electron pressure lasting for 20–60 s which give the shock profile the appearance of a damped wave. The width of the 'damping' region is comparable to the distance required for ion thermalization. Some theoretical treatments of shocks in plasmas predict pressure profiles similar to those reported here (see e.g. Sagdeev, 1966; Tidman and Krall, 1971), and such profiles have been observed for magnetic-field-free collisionless ion acoustic shocks in the laboratory (Taylor *et al.*, 1970).

(4) In the case of at least some quasi-parallel shocks, the transition from unshocked solar wind to compressed, heated magnetosheath flow is not discontinuous in the usual sense of the word. Rather, the transition occurs over a broad region perhaps 5 min wide and is ragged in appearance.

(5) The electron temperature jump at the shock is variable but does not vary significantly with θ_{SS} in the range 25°–110°. On the average the temperature increases by a factor of 2.7 at the shock, although the most probable increase is only a factor of 1.7. These values are lower than those previously reported by other authors.

(6) Double ion beams are common both within the shock itself as well as behind the shock in the sheath, and, in fact, have been observed all the way across the sheath. The ion data from at least one shock crossing suggest that the second, higher energy beam may be the result first of reflection and acceleration at the shock, and then convection through it.

Fig. 13. A series of 'snapshots' of the ion countrate spectra and angular distributions and electron countrate spectra around the time of the shock crossing of 7 November, 1977 shown in Figure 3. The ion spectra are averages of the countrate obtained when viewing in the sunward hemisphere. The symbols above the first ion spectra in each row denote the central energies for the ion angular distributions in the middle row. The electron spectra at the bottom are averages of the countrates over all viewing directions.

Acknowledgements

The authors wish to thank J. P. Glore and H. E. Felthauser at LASL, and K. Lehmann and P. Hemmerich at MPI for the dedicated support they provided in the development and preparation of the instrumentation. The MPI contribution would not have been possible without the efforts of the personnel at Dornier System, Friedrichshafen, especially of F. Pittermann. Special thanks are due for the outstanding support from the Project Offices at GSFC and ESTEC. The computer support by E. Tech at LASL and J. Papamastorakis and E. Sckopke at MPI is gratefully acknowledged. We wish to thank C. T. Russell and E. W. Greenstadt for sharing their magnetic field data with us. Los Alamos portions of this project were conducted under the auspices of the U.S. Department of Energy with NASA support under S-50864A, Max-Planck-Institut portions were supported by the Bundesministerium für Forschung und Technologie under grants RV14-B6/74 and 01-OI-02 u-ZA/WF/WRD 275:4.

References

Asbridge, J. R., Bame, S. J., and Strong, I. B.: 1968, 'Outward Flow of Protons from the Earth's Bow Shock', *J. Geophys. Res.* **73,** 5777.

Bame, S. J., Asbridge, J. R., Felthauser, H. E., Glore, J. P., Paschmann, G., Hemmerich, P., Lehmann, K., and Rosenbauer, H.: 1978, 'ISEE-1 and ISEE-2 Fast Plasma Experiment and the ISEE-1 Solar Wind Experiment', *IEEE Trans. Geoscience Electronics* **GE-16,** 216.

Feldman, W. C., Asbridge, J. R., Bame, S. J., and Montgomery, M. D.: 1973, 'Solar Wind Heat Transport in the Vicinity of the Earth's Bow Shock', *J. Geophys. Res.* **78,** 3697.

Formisano, V. and Hedgecock, P. C.: 1973, 'On the Structure of the Turbulent Bow Shock', *J. Geophys. Res.* **78,** 6522.

Gosling, J. T., Asbridge, J. R., Bame, S. J., Paschmann, G., and Sckopke, N.: 1978, 'Observations of Two Distinct Populations of Bow Shock Ions in the Upstream Solar Wind', *Geophys. Res. Letters* **5,** 957.

Kennel, C. F. and Sagdeev, R. Z.: 1967, 'Collisionless Shock Waves in High β Plasmas', *J. Geophys. Res.* **72,** 3303.

Montgomery, M. D., Asbridge, J. R., and Bame, S. J.: 1970, 'Vela-4 Plasma Observations near the Earth's Bow Shock', *J. Geophys. Res.* **75,** 1217.

Neugebauer, M., Russell, C. T., and Olson, J. V.: 1971, 'Correlated Observations of Electrons and Magnetic Fields at the Earth's Bow shock', *J. Geophys. Res.* **76,** 4366.

Paschmann, G., Sckopke, N., Asbridge, J. R., Bame, S. J., Gosling, J. T., and Russell, C. T.: 1978a, *Geophys. Res. Letters,* in press.

Paschmann, G., Sckopke, N., Haerendel, G., Papamastorakis, J., Bame, S. J., Asbridge, J. R., Gosling, J. T., Hones, E. W. Jr., and Tech, E. R.: 1978b, 'ISEE Plasma Observations near the Sub-Solar Magnetopause', *Space Sci. Rev.* **22,** 717.

Russell, C. T. and Greenstadt, E. W.: 1979, 'Initial ISEE Magnetometer Results: Shock Observations', *Space Sci. Rev.* **23,** 3.

Sagdeev, R. Z.: 1966, 'Cooperative Phenomena and Shock Waves in Collisionless Plasmas', in M. A. Leontovich (ed.), *Review of Plasma Physics,* Publ. Consultants Bureau Enterprises, New York, No. 4, p. 23.

Scudder, J. D., Lind, D. L., and Ogilvie, K. W.: 1973, 'Electron Observations in the Solar Wind and Magnetosheath', *J. Geophys. Res.* **78,** 6535.

Taylor, R. J., Baker, D. H., and Ikezi, H.: 1970, 'Observations of Collisionless Electrostatic Shocks', *Phys. Rev. Letters* **24,** 206.

Tidman, D. A. and Krall, N. A.: 1971, *Shock Waves in Collisionless Plasmas,* Wiley-Interscience, New York, p. 106.

INITIAL OBSERVATIONS OF LOW ENERGY CHARGED PARTICLES NEAR THE EARTH'S BOW SHOCK ON ISEE-1

F. M. IPAVICH and G. GLOECKLER
University of Maryland, Md., U.S.A.

C. Y. FAN
University of Arizona, Ariz., U.S.A.

L. A. FISK
University of New Hampshire, N.H., U.S.A.

D. HOVESTADT and B. KLECKER
Max-Planck-Institut, Garching, F.R.G.

J. J. O'GALLAGHER
University of Chicago

M. SCHOLER
Max-Planck-Institut, Garching, F.R.G.

Abstract. We report initial measurements from the ULECA sensor of the Max-Planck-Institut/University of Maryland experiment on ISEE-1. ULECA is an electrostatic deflection – total energy sensor consisting of a collimator, deflection analyzer and an array of solid state detectors. The position of a given detector, which determines the energy per charge of an incident particle, together with the measured energy determine the particle's charge state. We find that a rich variety of phenomena are operative in the transthermal energy regime ($\sim$10 keV/Q to $\sim$100 keV/Q) covered by ULECA. Specifically, we present observations of locally accelerated protons, alpha particles, and heavier ions in the magnetosheath and upstream of the Earth's bow shock. Preliminary analysis indicates that the behavior of these locally accelerated particles is most similar at the same energy per charge.

In this paper we present some preliminary results on charged particles near the Earth's bow shock based on measurements made with the ULECA sensor (Ultra Low Energy Charge Analyzer) of the HOM experiment on ISEE-1.

Figure 1 shows a schematic drawing of the ULECA sensor. It consists of three separate sub-sections, each of which consists of a collimator to focus incoming particles, high voltage plates to deflect the particles, mask slits to define their final deflection (and hence their energy per charge), and solid state detectors to measure their total energy. Discriminators for each detector allow one to separate H, He, and the charge group $Q > 3$ using only the counting rate data. The heavy ions may be further resolved using pulse height analysis techniques.

Space Science Reviews **23** (1979) 93–101. 0038-6308/79/0231-0093 $01.35.

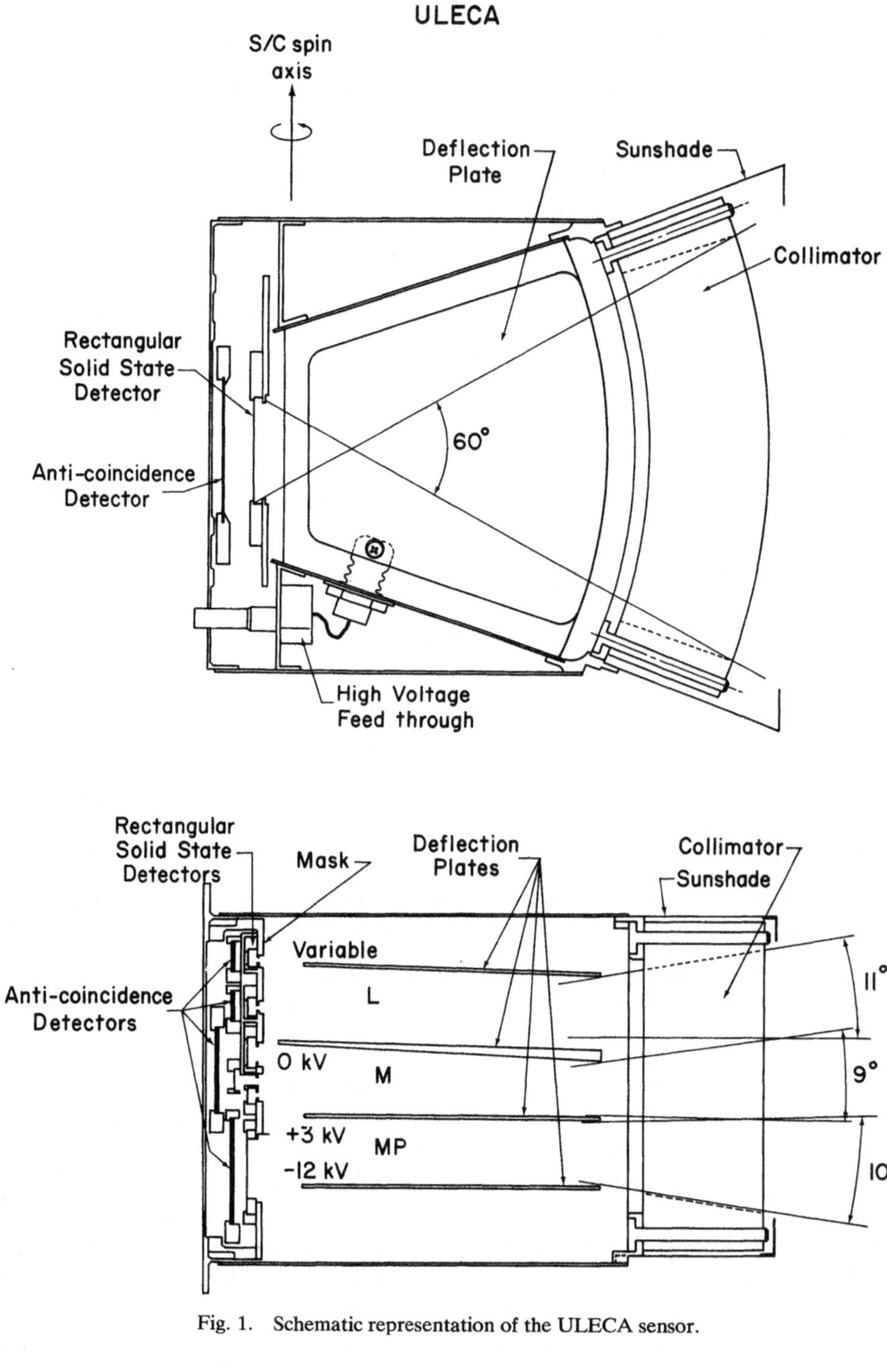

Fig. 1. Schematic representation of the ULECA sensor.

The high voltage in the L-section is stepped through 32 logarithmic steps, one complete cycle taking ~2 or ~8 min, depending on the spacecraft bit rate. In addition, three different energy ranges in the L-section are selectable by command: 6–19, 9–29, and 14–45 keV per charge. The energy bandwidth at any given voltage step is ~16%. The M-section consists of three solid state detectors (M1, M2, and M3), with respective mean energy per charge responses of 30, 60, and 120 keV/charge, and respective bandwidths of 20%, 25%, and 35%. The factor of 2 difference in energy per charge response in the M-detectors facilitates the comparison of protons and alpha particles at the same energy per nucleon and the same rigidity. Thus, for example, alpha particles in M2 detector have the same velocity as protons in the M1 detector, and the same rigidity as protons in the M3 detector.

Figure 2 shows the position of the ISEE-1 during one of its early orbits, along with the nominal position of the Earth's bow shock and magnetopause. Thus, time periods 1 and 3 are normally in the magnetosheath while time period 2 is near apogee, some $8R_E$ from the bow shock. During these times the magnetosphere was reasonably quiet, with the K_p index never exceeding 2.

The energy spectra obtained from the rate data for protons and helium during time period 1 are given in Figure 3. Although ISEE-1 was nominally in the magnetosheath at this time, the plasma data (kindly provided by S. Bame, G. Paschmann, and N. Sckopke) show that the bow shock was actually crossed at ~03:50 UT. Hence the spacecraft was in interplanetary space for ~75% of this time period. However, there was no large change in the energy spectra as the bow shock was crossed. Two points should be noted from Figure 3:

(1) The spectra are not power laws – rather they consistently steepen with energy.

(2) The data are best organized in terms of energy per charge in that the composition is most steady when evaluated at equal energy per charge. At equal

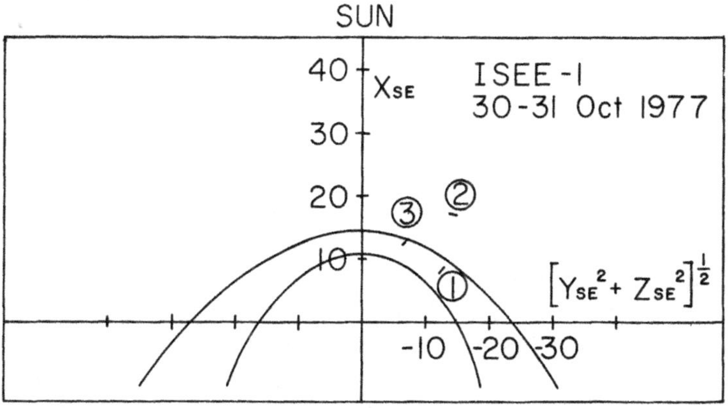

Fig. 2. Position of ISEE-1 during the three time periods discussed in this paper. Also shown are the nominal bowshock and magnetopause positions.

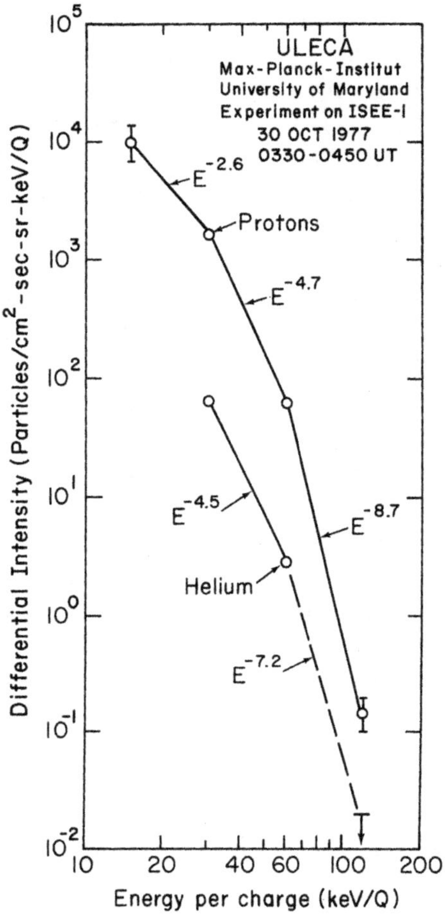

Fig. 3. Differential energy spectra for protons and helium.

energy per nucleon the proton to alpha ratio varies dramatically and also is quite large – for example at 60 keV nucl^{-1} the p/α ratio has a lower limit of 10^3.

Figure 4 contains the same data as Figure 3, but now plotted against energy per charge on a linear scale. (A straight line would correspond to an exponential spectrum in energy per charge.) We note that the protons are fit very well by such an exponential spectrum, and that the helium spectrum is consistent with the same e-folding energy per charge. Thus the alpha to proton ratio is constant, at about 4%, over the measured energy range.

Time period 3 (19:20–20:20 UT) is also in the nominal magnetosheath. However, the plasma data indicates that the spacecraft was actually in the magnetosheath only from ~19:38 to ~20:12 UT. Hence the energy spectra shown in Figure 5 are actually based on data taken during two bow shock crossings. The rates during this time were sufficiently high to allow us to calculate a three-point

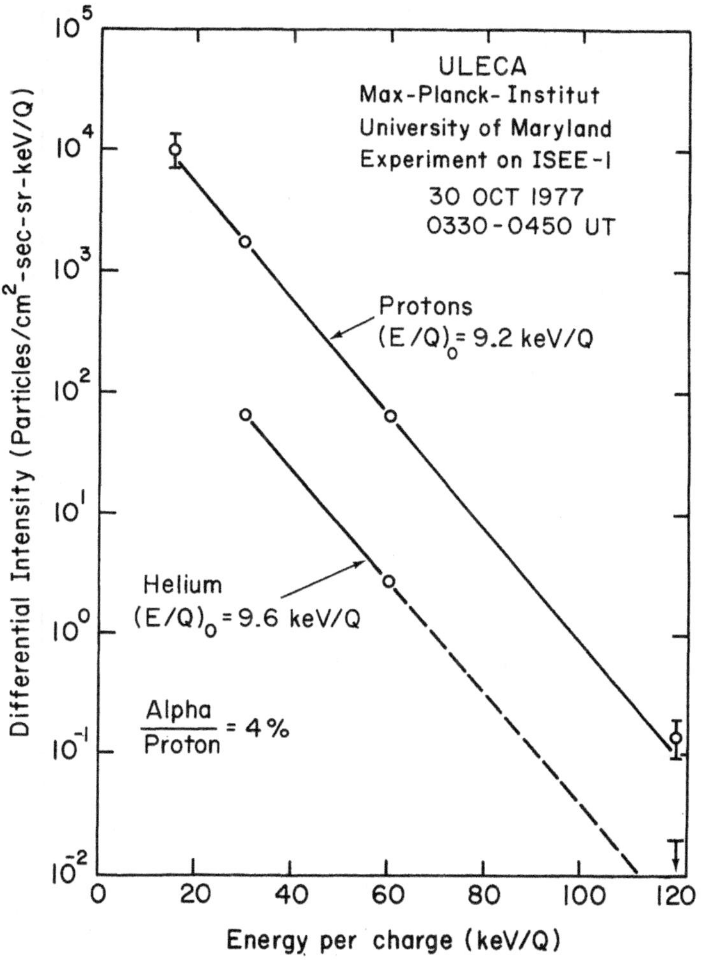

Fig. 4. Differential energy spectra using same data as in Figure 3, but plotted linearly against energy per charge.

spectrum for helium and also for the CNO group. We note that protons are again well described by an exponential in energy per charge, and that the helium and CNO spectra are also well represented by such a spectrum. In addition, the three spectra have essentially the same *e*-folding values, (~17 keV/charge) so that the composition is constant when evaluated at equal energy per charge.

We turn now to time period 2 when the spacecraft was clearly in interplanetary space. The rate profiles for protons at ~30 and ~120 keV are shown in Figure 6. Observations of upstream protons have been described previously by Lin *et al.* (1974). Briefly, they find that the rate-time profiles have square-wave signatures with the protons present when and only when the interplanetary magnetic field connects the spacecraft with the bow shock. They also find that the energy spectrum is generally a power law with a high energy cut-off at ~100 keV.

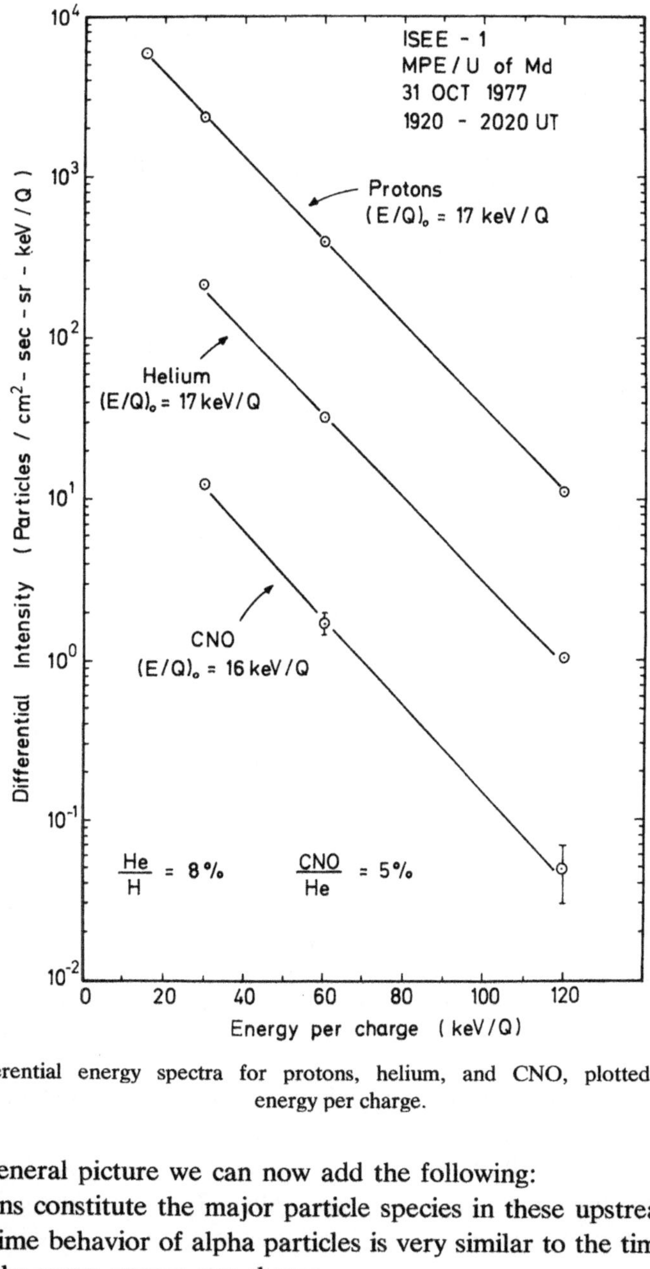

Fig. 5. Differential energy spectra for protons, helium, and CNO, plotted linearly against energy per charge.

To this general picture we can now add the following:

(1) Protons constitute the major particle species in these upstream events.

(2) The time behavior of alpha particles is very similar to the time behavior of protons at the same energy per charge.

(3) In this event there is a large change in the energy spectrum with time (notice from Figure 6 the very different time behavior of the two different proton energies), implying that some important parameter characterizing the acceleration mechanism is changing with time. Although the average interplanetary magnetic field direction (kindly supplied by C. T. Russell) changes substantially near the beginning and end of this particle event, it is quite steady in the time period

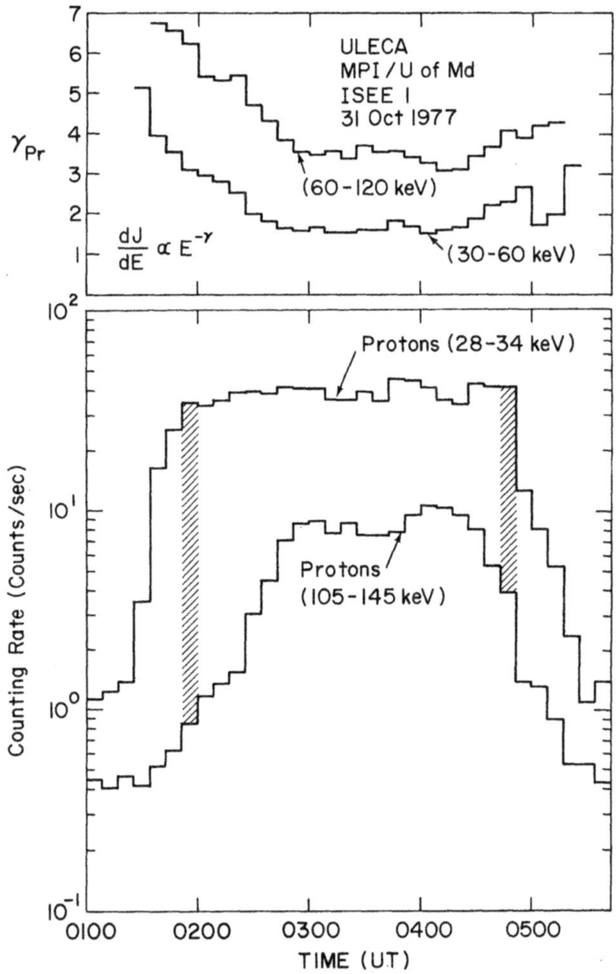

Fig. 6. *Top panel:* time behavior of proton spectral indices during an upstream particle event. *Bottom panel:* time profile of 30 keV and 120 keV proton counting rate during the event.

between the shaded bars shown in Figure 6. (There is considerable wave activity in the field at this time, which is not unexpected since the particle intensity is of the order of 10% of the magnetic field energy density.) Furthermore, the plasma velocity and density are very steady throughout this event. It is as yet not clear what caused the observed changes in the energy spectrum of these low energy particles.

(4) The behavior of the spectral indices in Figure 6 shows that throughout this event the spectrum is steeper at higher energies.

The energy spectra for P, helium, and CNO during the time when the higher energy protons are steady (03:00–04:00 UT) are shown in Figure 7. Here we see that the proton spectrum is quite flat at ~10 keV, but falls off quite sharply at

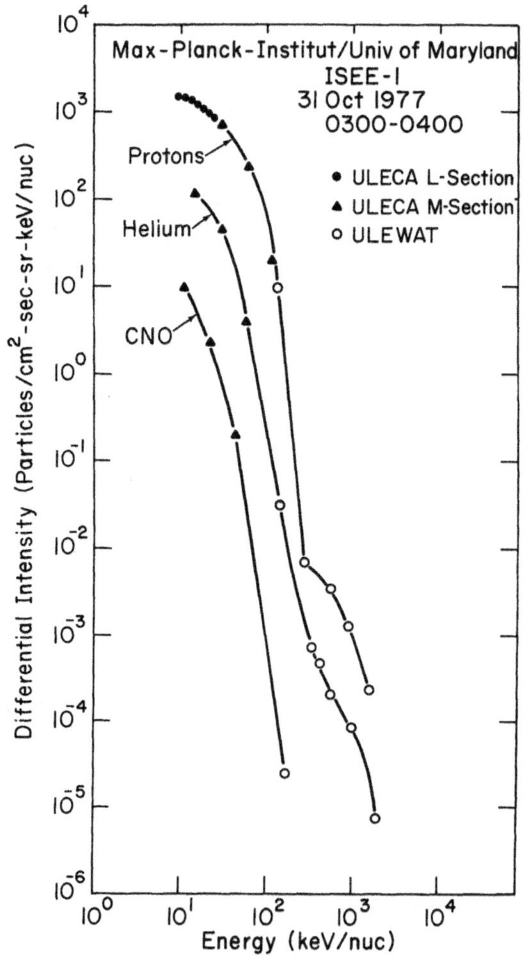

Fig. 7. Differential energy spectra for protons, helium, and CNO obtained during the peak of the upstream particle event shown in Figure 6.

higher energies (at 150 keV the special index is ~−11). The particles with energies above ~150 keV $nucl^{-1}$ constitute an independent population which did not increase in intensity during this event.

If one examines this energy spectrum on an energy per charge basis one finds certain similarities with the spectra shown in Figures 4 and 5 which were obtained very near the bow shock. In particular, the energy spectrum is reasonably well represented as an exponential in energy per charge. Also the composition is similar: the He/H ratio is ~10%, and the CNO/He ratio is ~7%. Thus it is possible that particles observed in the magnetosheath and in upstream events have a common origin. A more systematic study to evaluate this possibility is now underway.

Acknowledgements

This work was supported by NASA Contract NAS5-20062. We are grateful to E. Tums, J. Cain, R. Lundgren, R. Cates, L. Reed, and G. Umberger for the design and fabrication of the ULECA sensor and other portions of the HOM experiment, to A. Galvin and E. Tums for checkout and integration of the experiment, and to K. Hope, A. Galvin, and R. Sciambi for their programming efforts. We would also like to thank S. Bame, G. Paschmann, and N. Sckopke for providing us plasma data, and C. Russell for providing magnetic field data. Finally we acknowledge the willing cooperation of the ISEE project office, in particular of M. A. Davis and J. Madden, throughout the development, launch, and tracking of the instrument.

Reference

Lin, R. P., Meng, C. I., and Anderson, K. A.: 1974, '30- to 100 keV Protons Upstream from the Earth's Bow Shock', *J. Geophys. Res.* **79,** 489.

INITIAL RESULTS FROM THE ISEE-1 AND -2 PLASMA WAVE INVESTIGATION

D. A. GURNETT and R. R. ANDERSON

Department of Physics and Astronomy, The University of Iowa, Iowa City, Iowa 52242, U.S.A.

F. L. SCARF and R. W. FREDRICKS

TRW Defense and Space Systems Group, One Space Park, Redondo Beach, Calif. 90278, U.S.A.

E. J. SMITH

Jet Propulsion Laboratory, 4800 Oak Grove Drive, Pasadena, Calif. 91103, U.S.A.

Abstract. In this paper we present an initial survey of results from the plasma wave experiments on the ISEE-1 and -2 spacecraft which are in nearly identical orbits passing through the Earth's magnetosphere at radial distances out to about $22.5R_e$. Essentially every crossing of the Earth's bow shock can be associated with an intense burst of electrostatic and whistler-mode turbulence at the shock, with substantial wave intensities in both the upstream and downstream regions. Usually the electric and magnetic field spectrum at the shock are quite similar for both spacecraft, although small differences in the detailed structure are sometimes apparent upstream and downstream of the shock, probably due to changes in the motion of the shock or propagation effects. Upstream of the shock emissions are often observed at both the fundamental, f_p^-, and second harmonic, $2f_p^-$, of the electron plasma frequency. In the magnetosphere high resolution spectrograms of the electric field show an extremely complex distribution of plasma and radio emissions, with numerous resonance and cutoff effects. Electron density profiles can be obtained from emissions near the local electron plasma frequency. Comparisons of high resolution spectrograms of whistler-mode emissions such as chorus detected by the two spacecraft usually show a good overall similarity but marked differences in detailed structure on time scales less than one minute. Other types of locally generated waves, such as the $(n+\frac{1}{2})f_g^-$ electron cyclotron waves, show a better correspondence between the two spacecraft. High resolution spectrograms of kilometric radio emissions are also presented which show an extremely complex frequency-time structure with many closely spaced narrow-band emissions.

1. Introduction

In this paper we present a survey of the initial results from the plasma wave investigation on the ISEE-1 and -2 spacecraft, which were launched simultaneously by the same rocket on October 22, 1977. These two spacecraft, which are known as the mother and daughter, are in nearly identical eccentric earth orbits with apogee geocentric radial distances of about $22.5R_e$. Because of the closely matched orbits these two spacecraft provide a unique new capability to resolve spatial and temporal variations in the magnetosphere and solar wind. The separation distance can be controlled by a thruster on ISEE-2 to meet various scientific objectives throughout the mission. For a complete description of the International Sun-Earth Explorer (ISEE) program and the instrumentation on ISEE-1 and -2, see Ogilvie *et al.* (1977).

The plasma wave instruments on ISEE-1 and -2 are designed to provide measurements of the electric and magnetic field of plasma waves over the

Space Science Reviews **23** (1979) 103–122. 0038-6308/79/0231-0103 $03.00.

frequency range from about 5 Hz to 300 kHz. This frequency range includes most of the important characteristic frequencies of the plasma encountered in the magnetosphere and solar wind by these spacecraft. The experimental approach used in the design of these instruments is to provide very complete measurements of plasma wave characteristics (simultaneous electric and magnetic field amplitudes, wave normal and Poynting flux determinations, and high resolution frequency-time spectrograms) on the mother spacecraft (ISEE-1) and somewhat simpler measurements (selected to provide basic space-time correlations) on the daughter spacecraft (ISEE-2). Since the detailed description of the experiment is somewhat involved, the reader is referred to Gurnett *et al.* (1978) for a complete description of the instrumentation. Basically the mother instrument consists of four main elements, (1) a 24-channel high time resolution spectrum analyzer, (2) a narrow-band sweep frequency receiver, (3) a wave normal and Poynting flux analyzer, and (4) a wide-band receiver. These elements can be connected in various combinations to the six plasma wave antennas on ISEE-1, which consist of three electric dipole antennas with tip-to-tip lengths of 215 m, 73.5 m, and 0.61 m, and a set of triaxial search coil antennas. The daughter instrument consists of two main elements, (1) a 16-channel high time resolution spectrum analyzer, and (2) a wide-band receiver, both with bandwidths and characteristics matched to the corresponding elements on the mother spacecraft. The ISEE-2 plasma wave antennas consist of two electric dipole antennas, with tip-to-tip lengths of 30 m and 0.61 m and one search coil magnetic antenna. Signals from only one of these antennas can be analyzed at a time.

The primary purpose of this report is to present some of the initial results from this investigation and to comment on the overall performance and operation of the instrument. Since only a few months of data are currently available for analysis, the depth of analysis and comparisons with other instruments are necessarily limited. However, as will be seen, the plasma wave instruments on the ISEE-1 and -2 spacecraft are providing very exciting new measurements which have the potential of making a significant advance in our understanding of magnetospheric plasma waves.

2. A Survey of Some Representative Passes

To illustrate the variety of observations which are being obtained we have selected several representative passes for detailed analysis and comment. The first example selected is the inbound pass on day 309, November 5, 1977. This pass enters the magnetosphere on the day side of the Earth, near local noon, and has been selected because of the well defined bow shock and magnetopause crossings which occurred on this day. The electric field intensities from the high time resolution spectrum analyzers on ISEE-1 and -2 during this day are shown in Figures 1 and 2, respectively. The field strengths are on a logarithmic scale with a 100 dB dynamic range from the baseline of one channel to the baseline of the

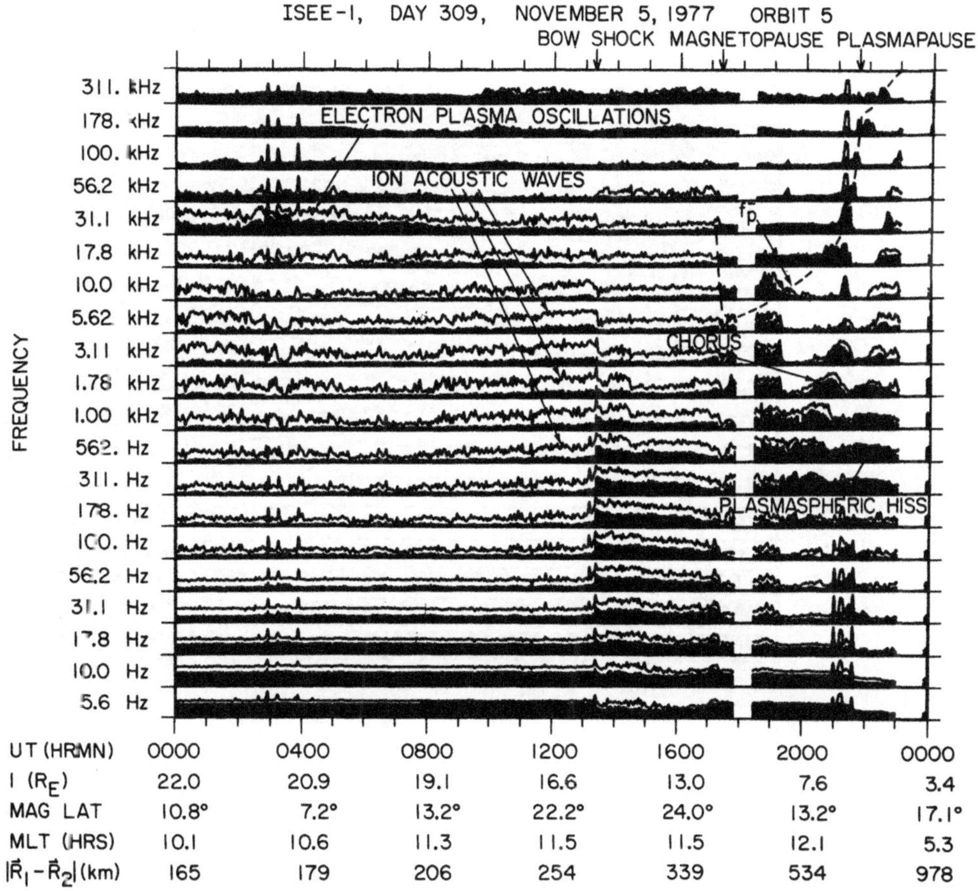

Fig. 1. The electric field amplitudes from the 24-channel high time resolution spectrum analyzer on ISEE-1 for the inbound magnetospheric pass on day 309, 1977. The dynamic range for each channel is 100 dB. The solid line gives the peak field strength and the solid black area indicates the average field strength.

next higher channel. This range extends from about 0.1 μV m^{-1} to 10.0 mV m^{-1} for the ISEE-1 data and from about 0.15 μV m^{-1} to 15.0 mV m^{-1} for the ISEE-2 data. The solid lines gives the maximum electric field strength and the solid black area gives the average electric field strength. The bow shock crossing is evident from the very abrupt increase in the electric field intensities at about 13:23 UT. The magnetopause crossing on this pass occurs at about 17:10 to 17:20 UT, and the plasmapause crossing occurs at about 21:35 UT.

On the coarse time scale shown in Figures 1 and 2 the electric field intensity variations show a remarkably close correspondence between ISEE-1 and -2 in all regions of the magnetosphere and solar wind. In the solar wind the most prominent types of waves are the electron plasma oscillations in the 31.1 kHz channel and the ion-acoustic waves in the channels from about 100 Hz to 10 kHz. Previous studies have shown that the electron plasma oscillations are produced by

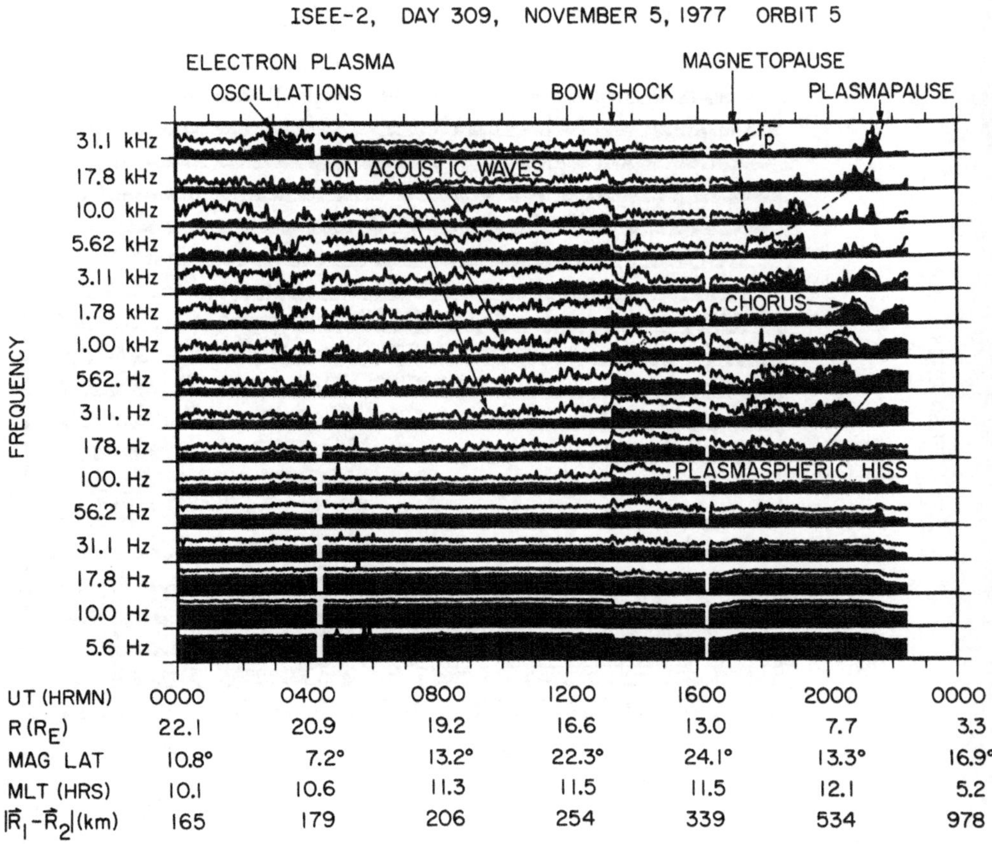

Fig. 2. The electric field amplitudes from the 16-channel spectrum analyzer on ISEE-2 for the same pass as in Figure 1. On this coarse time scale the electric field intensities on ISEE-1 and -2 show a very close correspondence.

electrons streaming into the solar wind from the bow shock (Scarf *et al.*, 1971) and that the ion-acoustic waves, based on the mode identification by Gurnett and Frank (1978), are produced by protons streaming into the solar wind from the bow shock (Scarf *et al.*, 1970). Comparisons of the electric field strength obtained by the different length antennas on the two spacecraft provide an interesting check on the wavelength of these two types of waves. Detailed comparisons show that the electric field strengths of the plasma oscillations, computed using an effective length equal to one-half of the tip-to-tip length of the antennas, are in very close agreement (±10%) between the two spacecraft. This close agreement confirms that the wavelength of the electron plasma oscillations is substantially longer than the length of the longest antenna (215 m). This conclusion is in agreement with the results of Gurnett and Frank (1975), who estimated that the wavelength of the upstream electron plasma oscillations is of the order of several kilometers. In contrast to the electron plasma oscillations, the computed electric field strengths of the upstream ion-acoustic waves near the frequency of maximum intensity (3.11 and 5.62 kHz) are about a factor of 5 to 15 larger for the shorter (30 m) ISEE-2 antenna compared to the longer (215 m) ISEE-1 antenna. This difference in the computed electric field strengths indicates that these waves have wavelengths substantially shorter than the length of the longest antenna ($\lambda \lesssim$ 215 m). This conclusion confirms the earlier results of Gurnett and Frank (1978), who showed that these waves have relatively short wavelength, near $2\pi\lambda_D$ where λ_D is the Debye length, as would be expected for the ion-acoustic mode. For such short wavelengths the frequency is mainly determined by the Doppler shift.

At the bow shock substantial differences in the electric field intensities are evident when the data are examined on a fine time scale. A detailed high time resolution comparison of the electric field intensities near the shock crossing is shown in Figure 3, together with the magnetic field strength obtained from the UCLA magnetometer (data provided by C. Russell). As can be seen in Figure 3 the difference in the times of the shock crossing is readily apparent between the two spacecraft. These time delays have been analyzed by Russell *et al.* (1978) to provide estimates of the shock velocity. Comparisons of the electric field intensity variations through the shock show the same general overall characteristic for both spacecraft, with a broad peak lasting several tens of seconds near the ramp in the magnetic field, and significant differences in the fine structure. The width of the transition region appears to be somewhat larger for ISEE-1 than for ISEE-2.

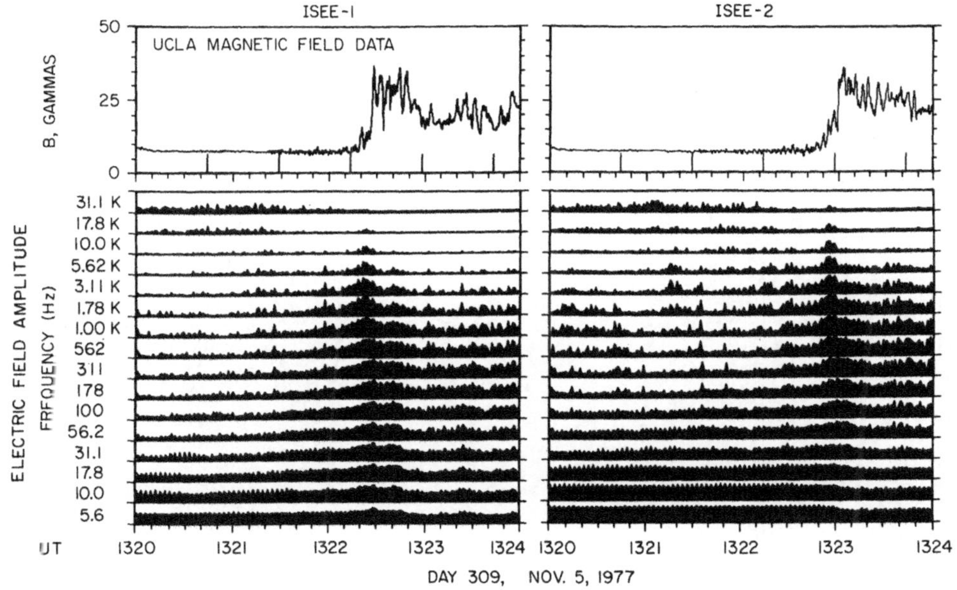

Fig. 3. High time resolution details of the electric field intensity variations at the bow shock crossing on day 309, 1977. The difference in the shock crossing times due to the separation between the two spacecraft is clearly evident. Although the overall structure of the electric field turbulence is similar at the two spacecraft, significant differences can be seen in the fine time scale structure.

Differences in the detailed temporal structure of the electron plasma oscillations are also evident upstream of the shock. Although differences in detail can be seen upstream and downstream of the shock, the electric field spectrums averaged over 16 s around the times of peak intensity are quite similar for the two spacecraft, as shown in Figure 4. Both spectrums show a very broad nearly flat plateau extending up to about 3 kHz and are very similar to the shock spectrums previously published by Rodriguez and Gurnett (1975) from the IMP-6 spacecraft. Above 3 kHz the ISEE-1 and -2 spectrums deviate considerably, with the largest field strengths being detected by ISEE-2. This deviation at high

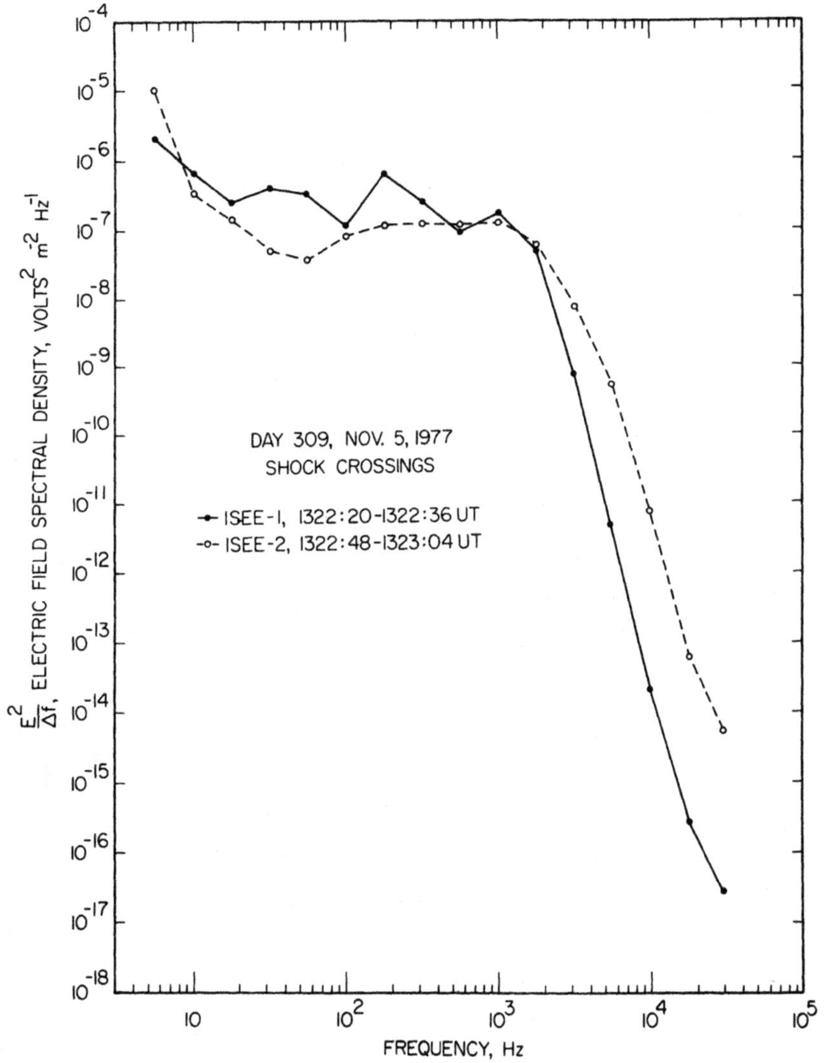

Fig. 4. The electric field spectrums at the times of peak intensity for the two shock crossings shown in Figure 3. The two spectrums are very similar except at high frequencies.

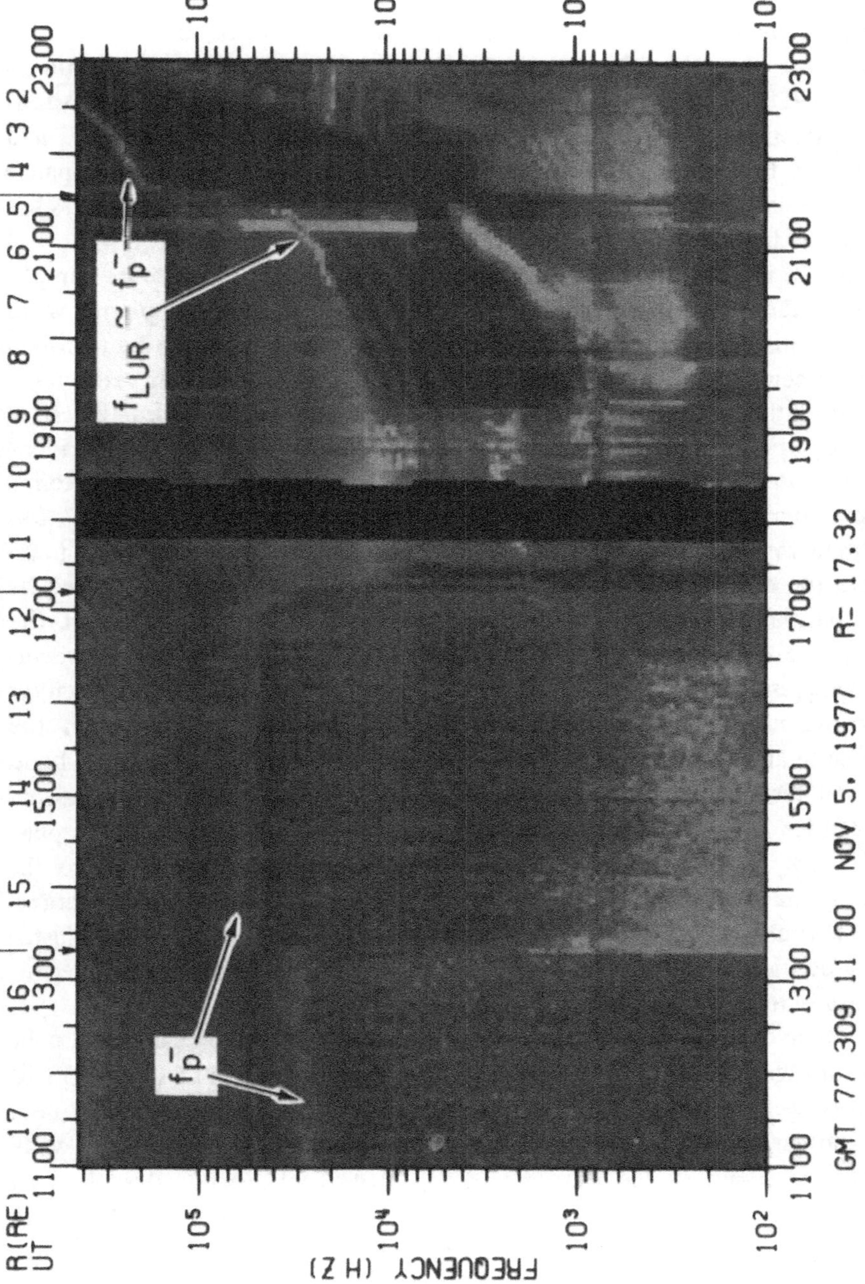

Fig. 5. The narrow-band sweep frequency receiver data for the same inbound pass shown in Figure 1. This receiver provides essentially constant fractional frequency resolution over a very wide frequency range, from 100 Hz to 400 kHz, and dynamic range, 100 dB. Very accurate electron density contours can be obtained from the emissions at the electron plasma frequency and upper hybrid resonance frequency and from the continuum radiation cutoff at the electron plasma frequency.

frequencies could be an indication that wavelengths shorter than 215 m are being detected at frequencies above 3 kHz.

In the magnetosheath behind the shock a broad spectrum of electric field turbulence is evident extending from below 5.62 Hz up to about 100 kHz. This electric field turbulence has been previously described by Rodriguez (1978) and is believed to be due to either ion-acoustic waves or Buneman mode waves (Buneman, 1958) at low frequencies, <10 kHz, and electron plasma oscillations at high frequencies, ~56.2 kHz. A slight enhancement in the electric field intensities can be seen at low frequencies, <100 Hz, near the magnetopause boundary, possibly indicating the generation of plasma wave turbulence in association with this boundary.

Inside of the magnetosphere the plasma wave spectrum is much more complicated than in the solar wind and magnetosheath. The analysis of the plasma wave spectrums in this region is greatly aided by the data from the narrow-band sweep frequency receiver shown in Figure 5. This receiver provides an essentially constant resolution logarithmic frequency scan, with $\Delta f/f \approx 6.5\%$, from 100 Hz to 400 kHz. The bow shock, magnetopause and plasmapause can be readily identified, as indicated at the top of the spectrogram. One of the most striking features of this spectrogram is the extent to which the electron plasma frequency, f_p^-, can be identified through essentially the entire orbit. In the solar wind and magnetosheath the electric field emissions near the plasma frequency are caused by electron plasma oscillations, of the type previously discussed. Inside of the magnetosphere the electron plasma oscillations are no longer present. However, the electron plasma frequency can still be identified from the low frequency cutoff of the continuum radiation trapped in the low density cavity between the magnetopause and plasmapause (Gurnett and Shaw, 1973) and from narrow-band emissions at the upper hybrid resonance frequency, f_{UHR}, inside of the plasmasphere (Mosier *et al.*, 1973). For the high density conditions inside of the plasmapause, $f_g^- \ll f_p^-$, the upper hybrid resonance is approximately equal to the electron plasma frequency ($f_{UHR} = \sqrt{f_p^{-2} + f_g^{-2}}$, where f_g^- is the electron gyrofrequency). This ability to identify the electron plasma frequency from the plasma wave data provides a valuable method of obtaining very accurate sheath independent measurements of the plasma density.

Many other characteristic emission frequencies and cutoff effects can also be identified in Figure 5. Beyond the plasmapause, from about 19:15 to 21:35 UT, several very weak, few μV m^{-1}, narrow-band emissions can be identified near half-integral harmonics of the electron cyclotron frequency, $(n+\frac{1}{2})f_g^-$. These weak emissions correspond to the diffuse $(n+\frac{1}{2})f_g^-$ electrostatic bands described by Shaw and Gurnett (1975). At larger radial distances, from about 17:30 to 19:15 UT, the $(n+\frac{1}{2})f_g^-$ electrostatic emissions at $(\frac{3}{2})f_g^-$ and $(\frac{5}{2})f_g^-$ become much more intense ranging from several hundred μV m^{-1} to about one mV m^{-1}. These more intense $(n+\frac{1}{2})f_g^-$ electrostatic emissions are believed to correspond to the $(\frac{3}{2})f_g^-$ and $(\frac{5}{2})f_g^-$ electrostatic waves first reported by Kennel *et al.* (1970) and discussed by Young *et al.* (1973), Ashour-Abdalla and Kennel (1978) and others. At about 21:05 to 21:10 UT a very intense narrow-band emission occurs near the upper hybrid resonance frequency, $f_{UHR} \simeq 30$ kHz. This emission is so strong that it saturates the ISEE-1 narrow-band receiver, causing the dark vertical band in Figure 5. The field strength of this emission can be determined from the multi-channel spectrum analyzer, which is not saturated, and is about 7.73 mV m^{-1}. Intense narrow-band emissions of this type are frequently observed near the upper hybrid resonance outside of the plasmapause (see Figure 13 of Gurnett, 1975) and are thought to be related to the generation of the continuum radiation which occurs in this frequency range. These narrow-band emissions near the upper hybrid resonance are one of the most intense electrostatic waves encountered anywhere in the magnetosphere. Other effects which can be identified in Figure 5 include (1) signals from ground VLF transmitters, at about 20 kHz inside of the plasmasphere from 22:10 to 23:00 UT, (2) an upper frequency cutoff for the whistler-mode noise inside the plasmasphere at about $0.6 f_g^-$, due to cyclotron damping, (3) plasmaspheric hiss, from about 300 Hz to 3 kHz inside of the plasmasphere, (4) chorus from about 1 to 3 kHz outside of the plasmasphere and (5) several periods of intense auroral kilometric radiation at frequencies above 100 kHz.

To further illustrate the complex structure of the magnetospheric electric field emissions Figures 6 and 7 show two more spectrograms from the narrow-band sweep frequency receiver. These spectrograms show the inbound portions of two successive passes through the magnetosphere approximately two days apart, on December 28 and December 30, 1977. Essentially all of the features remarked on in Figure 5 are again evident on these passes. One noteworthy effect is the abrupt decrease in the lower hybrid resonance frequency, hence plasma density, at about 11:55 UT on December 28 and an almost identical decrease on the same L-shell two days later, at about 21:15 UT on December 30. Evidently this depression in the plasma density represents a relatively long-lived feature of the plasmaspheric density distribution which took several days to decay. Another noteworthy effect is the series of type III solar radio events evident in Figure 6 at frequencies above about 100 kHz. These events can be clearly identified by their characteristic frequency-time dispersion, decreasing in frequency with increasing time. One of these events, from about 11:20 to 12:00 UT, shows a sharp low frequency cutoff in almost exact correspondence with the narrow-band lower hybrid resonance emission. This propagation cutoff provides a particularly convincing identification of the local electron plasma frequency, which for the conditions present ($f_g^- \ll f_p^-$) is essentially identical to the lower hybrid resonance frequency.

3. High Resolution Comparisons of Chorus and $(n+\frac{1}{2})f_g^-$ Noise Bands

For purposes of making very high time resolution comparisons both the ISEE-1 and ISEE-2 plasma wave instruments contain identical wide-band receivers which

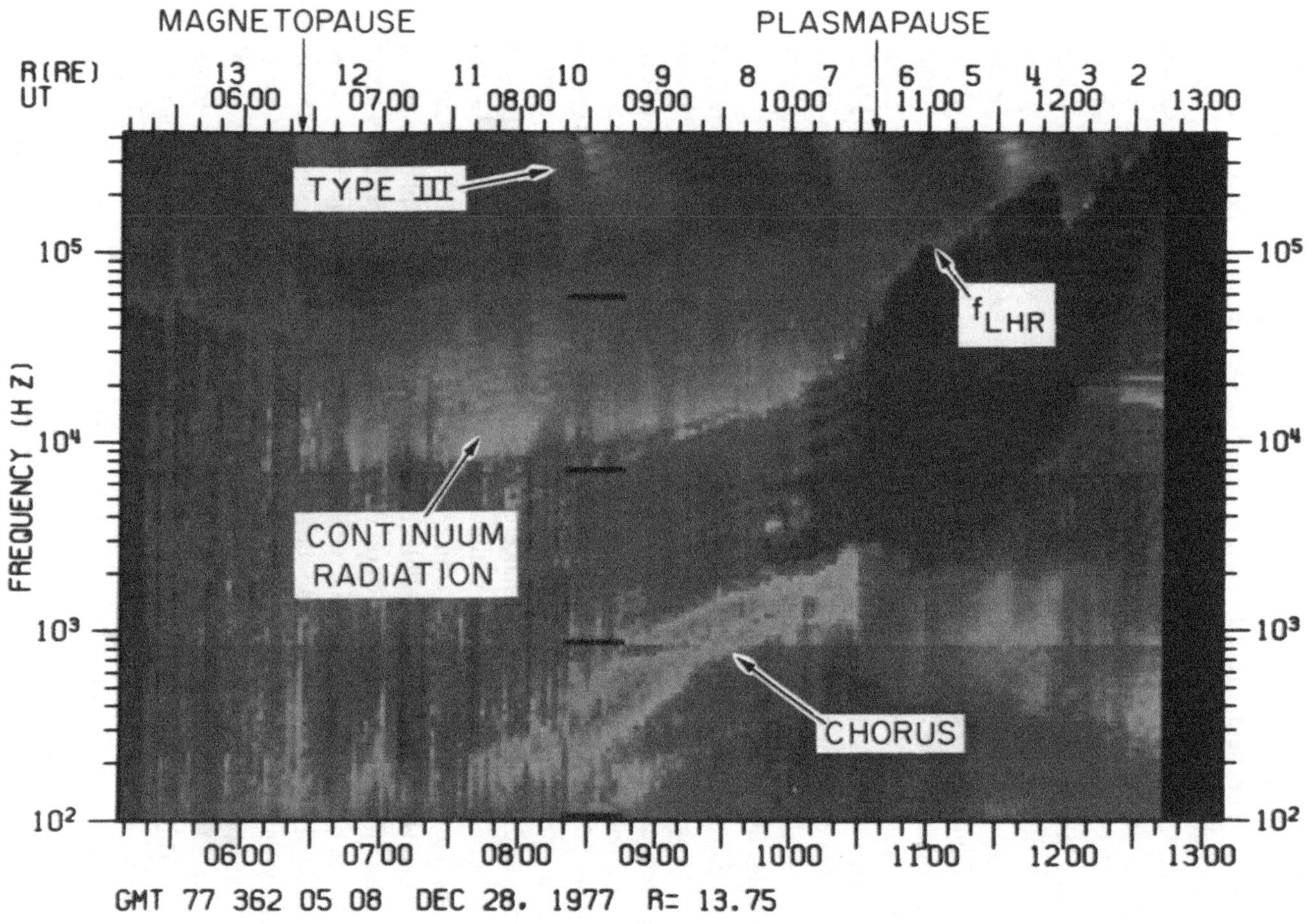

Fig. 6. Another spectrogram from the narrow-band sweep frequency receiver on ISEE-1 illustrating the repeated occurrence of many of the same plasma wave emissions shown in Figure 5.

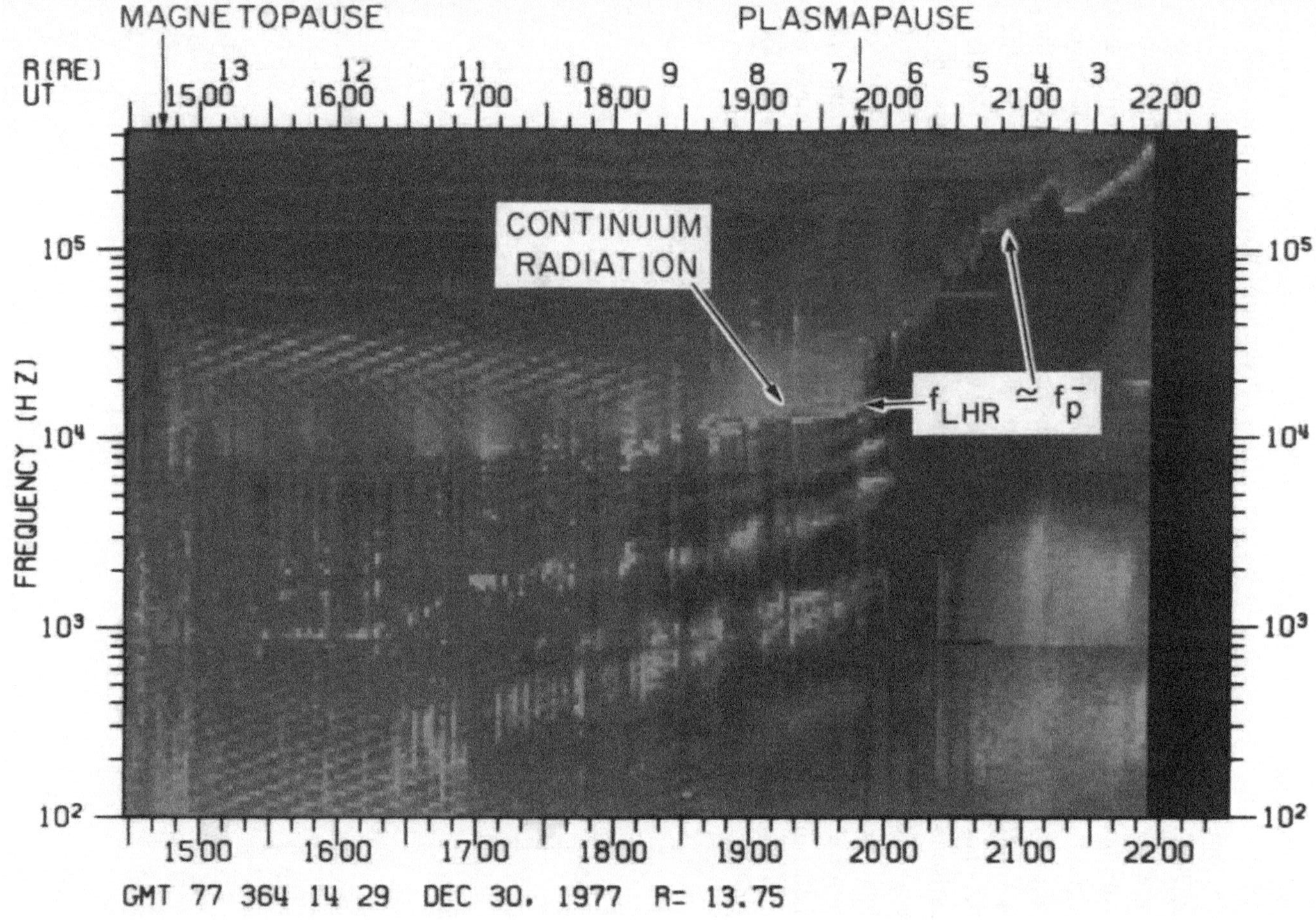

Fig. 7. The sweep frequency receiver spectrogram for the next pass following pass in Figure 6. Of special interest is the small depression in the electron density within the plasmasphere which apparently persists for the two-day interval between these passes.

can be used to transmit electric and magnetic field waveforms to the ground for detailed spectrum analysis. This capability permits measurements to the limit of the uncertainty principle, $\Delta f \Delta t \sim 1$, so that very small propagation delays can be detected and analyzed. A typical example of the wide-band electric field data from ISEE-1 and -2 is shown in Figure 8. This illustration shows frequency-time spectrograms of whistler-mode chorus emissions selected from the inbound pass on day 309, 1977, which was discussed earlier. Although the spectrums at the two spacecraft show good overall similarity on a time scale of a few minutes or more, it is extremely difficult to identify corresponding bursts of noise on time scales of a few seconds or less. During this period the spacecraft are separated by distances on the order of 450 km. For this separation distance the time difference for the two spacecraft to pass through the same region of space is about 160 s, and the whistler-mode time delay is on the order of a few tenths of a second. Detailed comparisons by overlaying the two spectrograms show that it is impossible to arrive at a simple time delay, either due to the spacecraft transit time or due to propagation delays, which can bring the two spectrums into alignment. Thus, for separation distances of 400 km the electric field spectrums do not correlate well on small time scales of a few minutes or less. Chorus spectrograms examined on other orbits show a similar effect. Although a more involved cross-correlation analysis may later reveal a better correlation, simple comparisons of the chorus

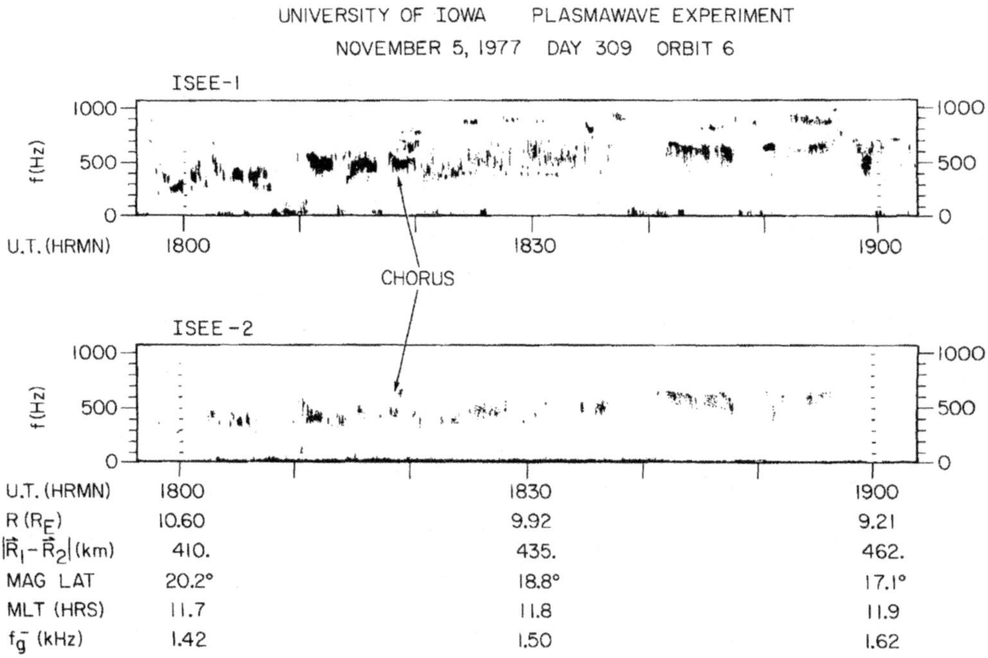

Fig. 8. High time resolution spectrograms of chorus emissions from the wideband receivers on ISEE-1 and -2. Although the chorus emissions are seen to occur in the same general frequency range at the two spacecraft, very marked differences are evident in the detailed temporal structure which cannot be accounted for by simple time delays.

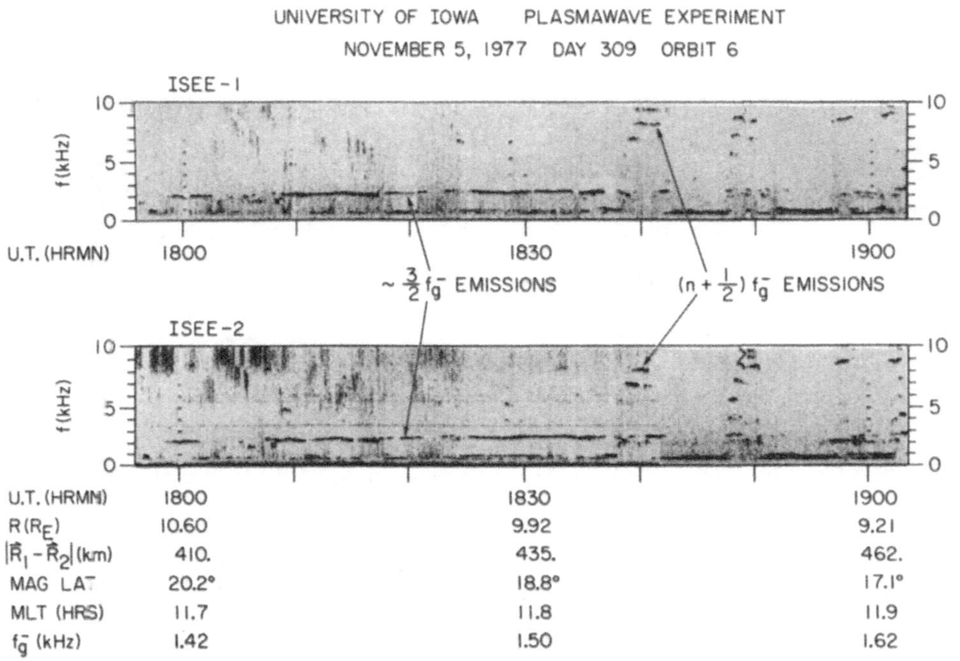

Fig. 9. High time resolution spectrums of $(n+\frac{1}{2})f_g^-$ electrostatic emissions detected by ISEE-1 and -2. In this case a reasonably good correspondence in the overall structure of the emissions can be obtained if the spectrums are shifted to account for the transit time of spacecraft, which in this case is about 160 s. Significant differences in the fine structure are, however, still evident.

spectrograms indicate that the size of the wave packets responsible for the chorus emissions is small compared to a few hundred km and that considerable small scale spatial structure exists in the chorus emissions.

Similar high resolution frequency-time spectrograms of the $(\frac{3}{2})f_g^-$ electrostatic emissions observed on day 309, 1977, are shown in Figure 9 from about 17:30 to 19:15 UT. These spectrograms again show that on a time scale of a few minutes or longer the 'same' emission can be identified on both spacecraft, but that on shorter time scales the detailed correlation tends to be rather poor. In this case careful comparisons show that the correlation between certain features, such as the gaps in the $(\frac{3}{2})f_g^-$ band and the bursts of high order $(n+\frac{1}{2})f_g^-$ emissions, is improved if the ISEE-2 spectrogram is shifted to the left by about 160 s to account for the spacecraft transit time. Many small scale differences, however, still exist. Another example of the $(n+\frac{1}{2})f_g^-$ electrostatic emissions from day 326, November 22, 1977, is shown in Figure 10. In this case the detailed correspondence in the frequency variations caused by changes in the magnetic field strength is well correlated between the two spacecraft, and can be brought into good correspondence by shifting the ISEE-2 spectrogram to the left by about 160 s to account for the spacecraft transit time. Again, even though the large scale features can be brought into good correspondence, many of the features on time scales less

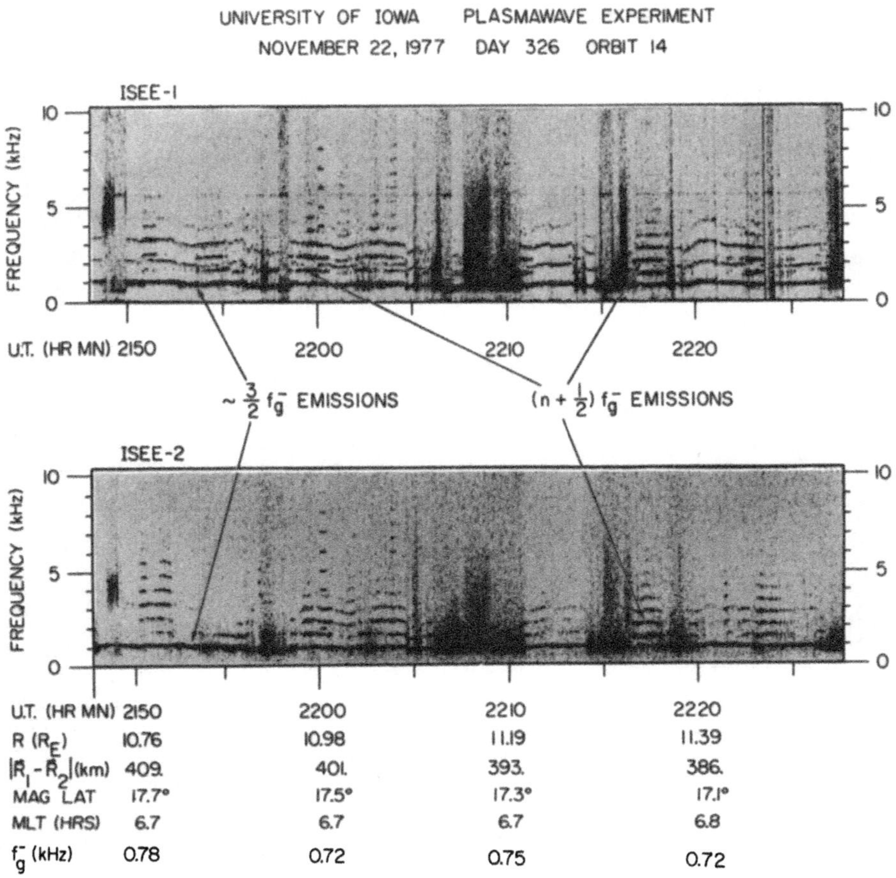

Fig. 10. Another example of the $(n+\frac{1}{2})f_g^-$ electrostatic noise bands illustrating the good overall correspondence in the structure and frequency variations observed by the two spacecraft as well as the differences in the fine structure.

than one minute are not well correlated. The overall conclusion from these comparisons is that the $(n+\frac{1}{2})f_g^-$ electrostatic emissions show a better correlation between the two spacecraft than the chorus emissions. However, neither type of emission shows a good correspondence on small time scales, less than one minute. This result is not considered unexpected since the electrostatic $(n+\frac{1}{2})f_g^-$ emissions have much lower phase velocities than the electromagnetic chorus emissions and should display a spatial structure characteristic of a localized non-propagating wave.

4. High Resolution Spectrums of Auroral Kilometric Radiation

In order to provide high resolution frequency-time spectrograms of high frequency radio emissions from the earth and other solar system radio sources the ISEE-1 and -2 wide-band receivers can be tuned, using frequency conversion techniques, to a variety of frequencies extending up to 2.0 MHz. This frequency conversion mode of operation has produced some particularly interesting results for periods when intense bursts of auroral kilometric radiation were being detected by ISEE-1 and -2. This radiation is generated at relatively low altitudes in the auroral regions and is closely correlated with the occurrence of discrete auroral arcs (see for example, Gurnett, 1974; Kurth *et al.*, 1975; Alexander and Kaiser, 1976; and Green *et al.*, 1977).

An example of a period when intense auroral kilometric radiation is being detected by ISEE-1 and -2 is shown in Figure 11. A long period of intense kilometric radiation activity can be seen in the 100, 178, and 311 kKz channels, extending from about 04:30 UT to the end of the day. During this period the wide-band receiver was operated for a short period, from 04:45 to 06:40 UT, in the 500 kHz frequency conversion mode. In this mode waveform signals in the frequency band from 500 to 540 kHz are converted to 0 to 40 kHz and transmitted back to the ground. The electric field intensity in the 500 to 540 kHz frequency band is shown in the top channel of Figure 11. As can be seen the

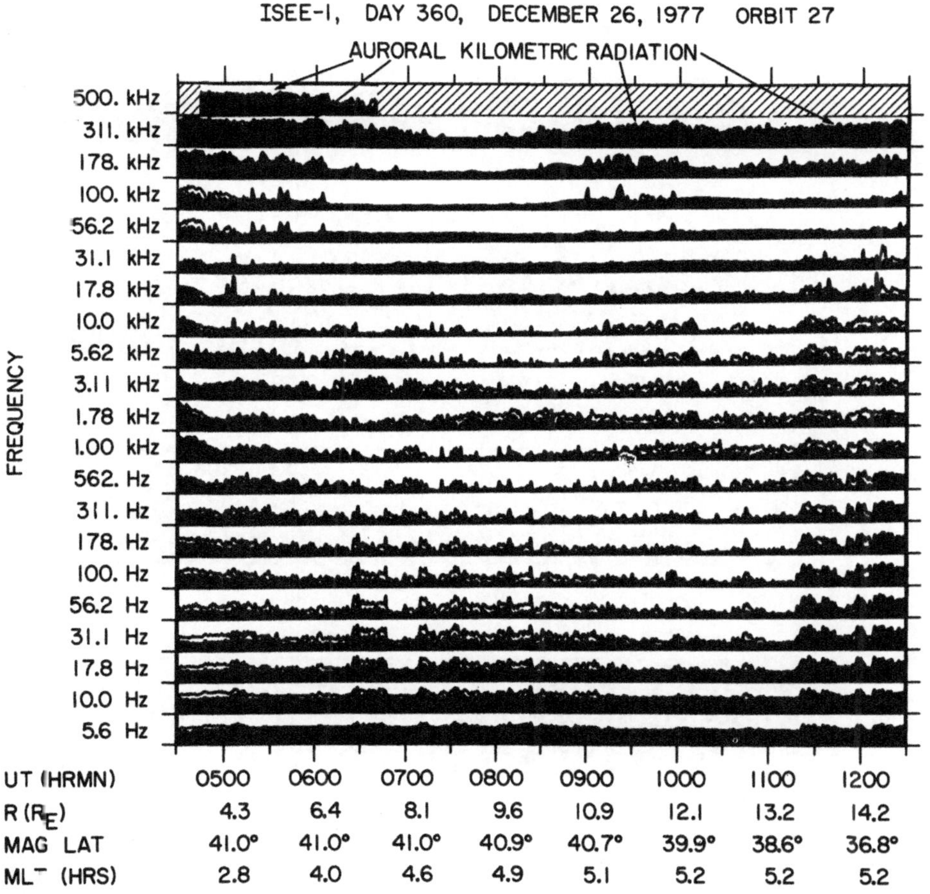

Fig. 11. An example of a period when intense kilometric radio emissions are being detected by ISEE-1 in the frequency range from 100 to 500 kHz.

auroral kilometric radiation spectrum at this time extends from about 100 kHz to well above 500 kHz.

The high resolution frequency-time spectrograms of the auroral kilometric radiation detected during this time in the narrow frequency band from 500 to 540 kHz are shown in Figure 12. As can be seen the spectrum has an astonishing amount of fine structure, consisting of many narrow-band emissions with rapidly

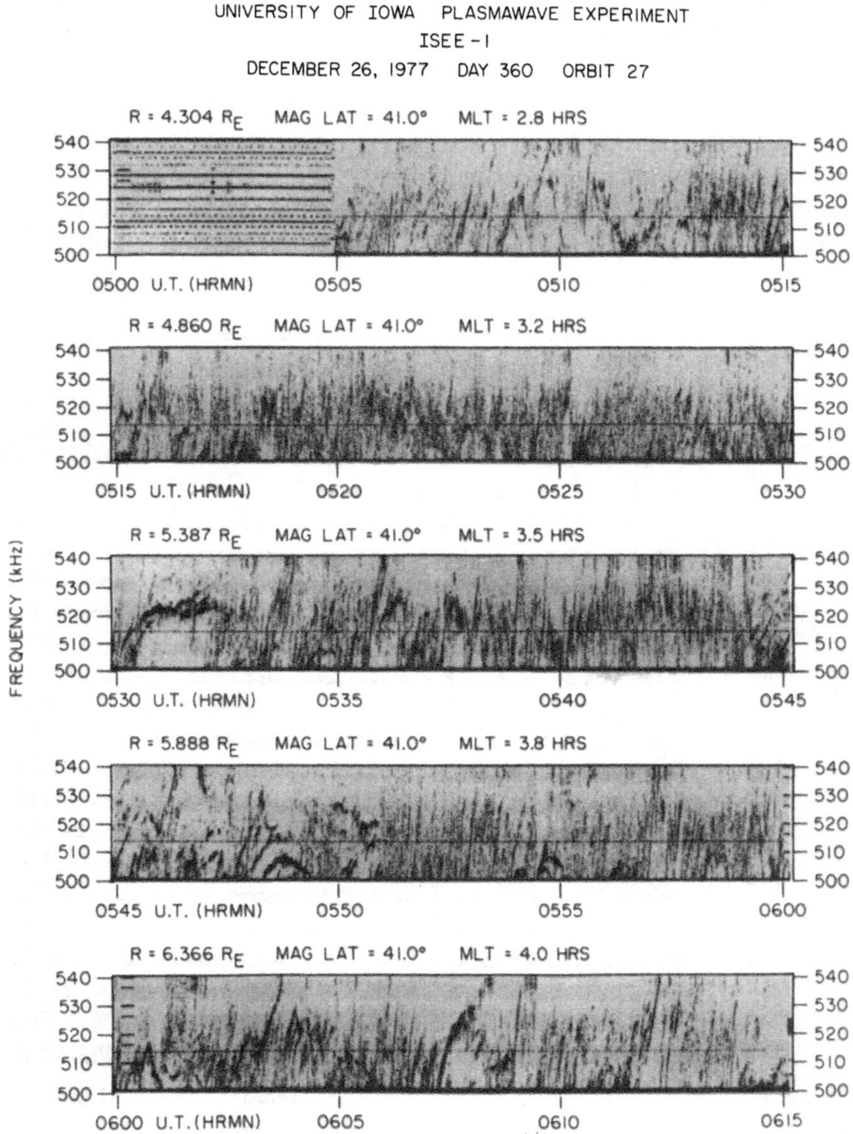

Fig. 12. High resolution frequency time spectrograms of the terrestrial kilometric radiation from the wideband receiver on ISEE-1 for the same event shown in Figure 11. These spectrograms, which cover the narrow frequency range from 500 to 540 kHz, show that the kilometric radiation has a great deal of fine structure, often consisting of narrowband emissions increasing in frequency with increasing time.

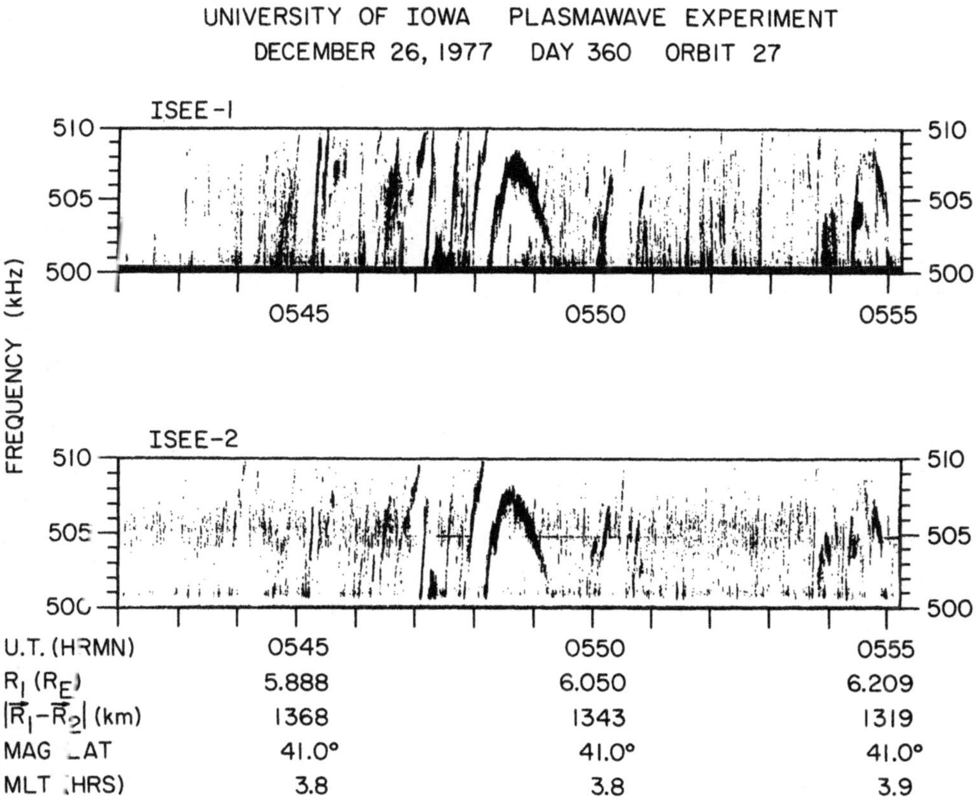

Fig. 13. A comparison of the wideband spectrograms of the kilometric radiation detected simultaneously by ISEE-1 and -2. Because of the small separation distance the time delay between the two spacecraft, ~4 ms, is too small to be resolved on these spectrograms. Later, more detailed analyses using techniques of long baseline interferometry will give very accurate information on the source position and motion from data such as these.

varying center frequencies. In many cases the frequency of these emissions appears to be increasing with increasing time. A few cases are evident for which the frequency rises, reaches a maximum, and then starts to decrease, for example at about 05:44 UT, giving an inverted U-shaped appearance on the frequency-time spectrogram. Comparisons of ISEE-1 and -2 show almost identical spectrums for this radiation, as would be expected since the radiation is propagating at frequencies well above the local electron plasma frequencies, where local plasma effects should not be important. Corresponding ISEE-1 and -2 spectrograms in the frequency range from 500 to 510 kHz are shown in Figure 13. Since the radiation is propagating at the speed of light, the time delay expected between the two spectrums is quite small, only about 4.5 ms. Because the time delay is so small we have not yet attempted to measure it, since a fairly elaborate cross-correlation analysis will be required to provide good accuracy. However, such measurements are possible and can be used to provide information on the source position and motion using standard techniques of long baseline interferometry. Appropriate reference signals are transmitted to the ground from each spacecraft so that phase interferometry measurements can be made between the two spacecraft.

The complex frequency-time structure of the auroral kilometric radiation is a new characteristic which has never been previously observed. Although unexpected, the occurrences of bursts with rapidly drifting center frequencies is not unusual in other astrophysical radio sources, and is a common feature of solar radio bursts (Kundu, 1965) and Jovian decametric radio emissions (Warwick, 1967). Usually such features are interpreted as being due to the motion of the emitting particles, or some other disturbance, through a plasma which has a spatial gradient in the characteristic emission frequencies, as for example in type III and type IV solar radio bursts. If we adopt this explanation for the frequency drift of the auroral kilometric radiation, an estimate can be made for the velocity of the source. From Figure 12 the drift rate of a typical burst can be estimated to be about $df/dt = 2.0\ \mathrm{kHz\ s^{-1}}$. This value is representative, for example, of the bursts from about 05:51 to 05:54 UT. To translate this drift rate into a source velocity some assumptions must be made about the characteristic frequencies of the plasma in the source region. For the terrestrial kilometric radiation it is widely believed that the radiation is generated at relatively low altitudes of about 1 to $2R_e$ along the auroral field lines. Although it is not known whether the emission frequency is more closely associated with the electron gyrofrequency or the electron plasma frequency, at these altitudes the radial gradients of these two characteristic frequencies are roughly comparable. Using the ionospheric model given by Gurnett (1974), the radial gradients in the electron gyrofrequency and plasma frequency are approximately, $df/dR = 0.2\ \mathrm{kHz\ km^{-1}}$. For the observed drift rates the corresponding velocity of the source is $dR/dt = (dR/df)(df/dt) = 10\ \mathrm{km\ s^{-1}}$. Since the drift rate is positive and the characteristic frequencies of the plasma increase with decreasing radial distance the source motion is toward the Earth. Exceptions to this motion do of course occur, for example in the case of the inverted-U shaped bursts. However, most of the bursts in Figure 12 can be accounted for by a source moving down the auroral field lines at a velocity of about $10\ \mathrm{km\ s^{-1}}$.

Having estimated the approximate velocity of the source attention can now be given to the question of whether this source motion corresponds to an actual motion of the emitting particles, or to some other type of disturbance propagating down the auroral field lines. If the frequency drift is caused by the motion of the emitting particles then the corresponding energies for electrons and protons would be 2.8×10^{-4} and 0.5 eV. These energies are so small compared to the thermal energies of the auroral plasma that it is not reasonable to attribute the frequency drift directly to the motion of the particles emitting the radiation. Some type of propagating disturbance is needed to explain the observed frequency drift. One interesting possibility is that the source motion may be related to the transient motion of an electrostatic shock or double layer along the auroral field line since the inferred source velocity of $\sim 10\ \mathrm{km\ s^{-1}}$ is comparable to the ion-acoustic speed in the auroral ionosphere. This mechanism would be somewhat analogous to the generation of a drifting type IV solar radio burst by a shock propagating through the solar corona.

5. Conclusion

In this paper we have presented a survey of the initial results from the ISEE-1 and -2 plasma wave investigation. These results demonstrate the very high quality data which are being obtained from these instruments and illustrate the wide range of magnetospheric plasma physics problems which can be attacked using these spacecraft. The plasma wave spectrum comparisons between the two spacecraft illustrate the great advantages of using two spacecraft in similar orbits to unravel the complex spatial and temporal variations which occur in the magnetosphere. Clearly, the complexities are such that a great deal of additional work is needed to fully understand the spatial and temporal structure of magnetospheric plasma waves and their relationship to the magnetospheric charged particle distribution. These studies will be our primary objective during the next several years.

Acknowledgements

We are very thankful to Dan Odem and his team at the University of Iowa, consisting of Roger D. Anderson, Bill Walker, Keith Baker, and Don Kirchner, who carried out the design and construction of these large and complex instruments. Special thanks are also given to Robert Shaw of the University of Iowa who helped direct this project while the Principal Investigator (D. Gurnett) was on sabbatical leave. We are also indebted to Allan Frandsen of JPL and Paul Virobik of TRW for their contribution to the design and construction of the search coil and short electric antenna. We are also very grateful for the fine support and cooperation of the many members of the ISEE Project at GSFC and ESTEC who made this mission a success. Special thanks is also given to C. Russell of UCLA for providing the magnetic field data and to P. Jepsen of JPL and W. Kurth of the University of Iowa for their efforts in producing the color spectrograms.

This research was supported by NASA under Contract NAS5-20093 from Goddard Space Flight Center and Grant NGL-16-001-043.

References

Alexander, J. K. and Kaiser, M. L.: 1976, 'Terrestrial Kilometric Radiation, 1. Spatial Structure Studies', *J. Geophys. Res.* **81,** 5948.

Ashour-Abdalla, M. and Kennel, C. F.: 1978, 'Nonconvective and Convective Electron Cyclotron Harmonic Instabilities', *J. Geophys. Res.* **83,** 1531.

Buneman, O.: 1958, 'Instability, Turbulence and Conductivity in a Current Carrying Plasma', *Phys. Rev. Letters* **1,** 8.

Green, J. L., Gurnett, D. A., and Shawhan, S. D.: 1977, 'The Angular Distribution of Auroral Kilometric Radiation', *J. Geophys. Res.* **82,** 1825.

Gurnett, D. A.: 1974, 'The Earth as a Radio Source: Terrestrial Kilometric Radiation', *J. Geophys. Res.* **79,** 4227.

Gurnett, D. A.: 1975, 'The Earth as a Radio Source: the Non-Thermal Continuum', *J. Geophys. Res.* **80,** 2751.

Gurnett, D. A. and Frank, L. A.: 1975, 'Electron Plasma Oscillations Associated with Type III Radio Emissions and Solar Electrons', *Solar Phys.* **45,** 477.

Gurnett, D. A. and Frank, L. A.: 1978, 'Ion-Acoustic Waves in the Solar Wind', *J. Geophys. Res.* **83,** 58.

Gurnett, D. A., Scarf, F. L., Fredricks, R. W., and Smith, E. J.: 1978, 'The ISEE-1 and -2 Plasma Wave Investigation', *IEEE Trans. Geosci. Electr.* **GE-16,** 225.

Gurnett, D. A. and Shaw, R. R.: 1973, 'Electromagnetic Radiation Trapped in the Magnetosphere Above the Plasma Frequency', *J. Geophys. Res.* **78,** 8136.

Kennel, C. F., Scarf, F. L., Fredricks, R. W., McGehee, J. H., and Coroniti, F. V.: 1970, 'VLF Electric Field Observations in the Magnetosphere', *J. Geophys. Res.* **75,** 6136.

Kundu, M. R.: 1965, *Solar Radio Astronomy*, Interscience Publishers, New York.

Kurth, W. S., Baumback, M. M., and Gurnett, D. A.: 1975, 'Direction-Finding Measurements of Auroral Kilometric Radiation', *J. Geophys. Res.* **80,** 2764.

Mosier, S. R., Kaiser, M. L., and Brown, L. W.: 1973, 'Observations of Noise Bands Associated with the Upper Hybrid Resonance by the IMP-6 Radio Astronomy Experiment', *J. Geophys. Res.* **78,** 1673.

Ogilvie, K. W., von Rosenvinge, T., and Durney, A. C.: 1977, 'International Sun-Earth Explorer: a Three-Spacecraft Mission', *Science* **198,** 131.

Rodriguez, P.: 1978, 'Magnetosheath Electrostatic Turbulence', *J. Geophys. Res.*, accepted for publication.

Rodriguez, P. and Gurnett, D. A.: 1975, 'Electrostatic and Electromagnetic Turbulence Associated with the Earth's Bow Shock', *J. Geophys. Res.* **80,** 19.

Russell, C. T. and Greenstadt, E. W.: 1979, 'Initial ISEE Magnetometer Results: Shock Observations', *Space Sci. Rev.* **23,** 3.

Scarf, F. L., Fredricks, R. W., Frank, L. A., Russell, C. T., Coleman, P. J., Jr., and Neugebauer, M.: 1970, 'Direction Correlations of Large-Amplitude Waves with Suprathermal Protons in the Upstream Solar Wind', *J. Geophys. Res.* **75,** 7316.

Scarf, F. L., Fredricks, R. W., Frank, L. A., and Neugebauer, M.: 1971, 'Non-Thermal Electrons and High-Frequency Waves in the Upstream Solar Wind, 1. Observations', *J. Geophys. Res.* **76,** 5162.

Shaw, R. R. and Gurnett, D. A.: 1975, 'Electrostatic Noise Bands Associated with the Electron Gyrofrequency and Plasma Frequency in the Outer Magnetosphere', *J. Geophys. Res.* **80,** 4259.

Warwick, J. W.: 1967, 'Radiophysics of Jupiter', *Space Sci. Rev.* **6,** 841.

Young, T. S. T., Callen, J. D., and McCune, J. E.: 1973, 'High-Frequency Electrostatic Waves in the Magnetosphere', *J. Geophys. Res.* **78,** 1082.

FIRST RESULTS FROM THE SIX-AXIS ELECTRON SPECTROMETER ON ISEE-1

K. W. OGILVIE and J. D. SCUDDER

NASA/Goddard Space Flight Center, Laboratory for Extraterrestrial Physics, Greenbelt, Md. 20771, U.S.A.

Abstract. A survey, using results from the first 25 orbits of ISEE-1, has been made of some aspects of electrons in the dawn magnetosheath. There are indications that the flow of plasma is not uniformly turbulent over this region. The electron heat flux is observed to be directed away from the shock and to have an average value of about twice the interplanetary heat flux. Many magnetopause crossings were observed and usually resemble abrupt transitions from one well-defined plasma state to another. The ejection of plasma from flux tubes convected up against the magnetopause is observed for about half the time, and its thickness and dependence on the solar wind mach number agree with theoretical predictions. A full traversal of the whole forward hemisphere of the magnetosheath is required to fully confirm these deductions.

1. Introduction

In this paper the aim is to illustrate the capabilities and operation of this instrument and the data reduction system by conducting a preliminary partial survey of the magnetosheath using appropriate data obtained during the first 25 orbits of ISEE-1. This survey, which is confined to the region of the sheath between the subsolar point and the dawn meridian, will be concerned with observations of the electron distribution functions, and the flow of heat, and also the behavior of the plasma density in the region adjacent to the magnetopause. Examples of magnetopause crossings will be given, and the discussion of this data set indicates that the picture of a magnetosheath unstable to the mirror instability, with large amplitude disturbances, and bounded by a magnetopause whose surface carries wave-like disturbances is generally applicable. The reductions of density adjacent to the magnetopause are found to be consistent with the theory of Zwan and Wolf (1976), and result from the expulsion of plasma from flux tubes as these are convected up to the magnetopause. Such reductions occur in at least half of the traversals and seem to depend weakly if at all upon the angle between the interplanetary magnetic field and the solar wind flow direction.

2. The Instrument

The instrument forms the subject of a separate paper (Ogilvie *et al.*, 1978); we therefore give only a very condensed description here. In order to determine the distribution function of electrons in three dimensions, it measures the electron spectrum along both senses of three perpendicular directions fixed in the coordinate system of the spacecraft. One such 6-axis measurement is taken for every sixty degrees of spacecraft rotation, which takes approximately one-half second. From the 36 spectra taken during each spin, the velocity distribution function is

Space Science Reviews **23** (1979) 123–133. 0038-6308/79/0231-0123 $01.65.

derived and its velocity moments obtained. This process is much simplified because simultaneous measurements are taken in each sense along each of the three axes. There are three energy ranges, 7.6 to 512 eV, 11 to 2062 eV, and 109 to 7285 eV, but only the first two were used in the present work. (The third energy range is designed to be used for the study of solar electron events.) An individual data point always has the same time resolution, and when the spacecraft is operating at high bit-rate (16 Kb s^{-1}) the telemetry system can transmit a complete measurement for every spin. At the low bit-rate (4 Kb s^{-1} for 80% of the time) the instrument obtains a measurement every fifth spin and is inoperative while this is being transmitted. Each of the six cylindrical electrostatic analyzers has two channeltron detectors, and the present measurements were made using both, in order to improve counting statistics at the higher energies. The angular field of view of each analyzer is 8.5×11 deg, and sixteen energy steps are sampled with 5.5% FWHM energy resolution.

3. Observations

Figure 1 shows an example of a complete outbound traversal of the magnetosheath, on October 30, 1977, from the magnetopause at 00:45 UT to the bow shock at 03:50 UT. In this plot we show the density and bulk speed, the azimuth and latitude angles describing the bulk speed direction, the temperature, temperature anisotropy, the heat flux (in the rest frame of the plasma) and two angles

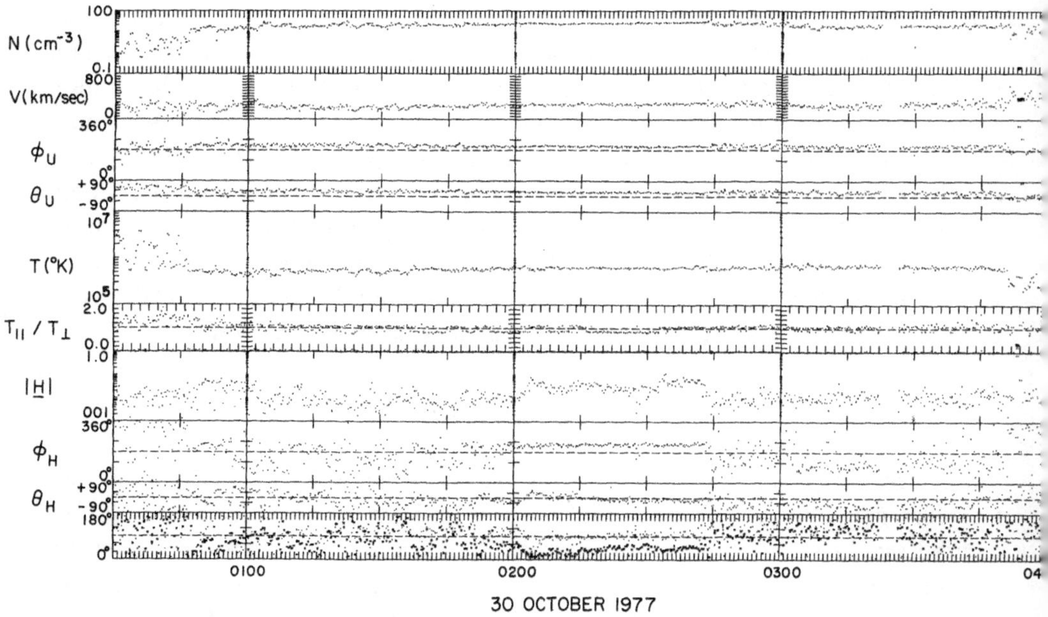

Fig. 1. A complete traversal of the magnetosheath on October 30, 1977, showing the variation of density, bulk speed, bulk speed azimuth and elevation angles, temperature and temperature anisotropy, and the heat flux (in units of ergs cm^{-2} s^{-1}) and its azimuth and elevation angles. Note the especially stable period between 02:00 and 02:44 UT.

defining its direction. We use this figure to illustrate qualitatively some general features of the data. Average conditions ($n = 8\ cm^{-3}$, $U = 425\ km\ s^{-1}$, $T_e = 2.2 \times 10^5$ K) were indicated in the solar wind, across the bow shock these parameters jump to ($n = 22\ cm^{-3}$, $U = 250\ km\ s^{-1}$, $T_e = 5.5 \times 10^5$ K). The directional changes are of the correct magnitude and in the correct direction to agree with the aerodynamic model of the magnetosheath flow. The heat flow is towards the Sun outside the bow shock, and towards the Earth inside the magnetosheath. Upstream heat flows have been previously described by the Los Alamos and GSFC groups (Feldman *et al.*, 1975; Scudder *et al.*, 1973), and are seen as a matter of course when measurements are made at a point connected to the bow shock by the interplanetary magnetic field. The flux of heat in the magnetosheath, larger in magnitude than that in the interplanetary medium, seems not to have been previously evaluated and discussed.

A feature commonly seen in the present magnetosheath traversals is the presence of a region or regions of space where the flow is less turbulent than the flow in the magnetosheath as a whole. In the present example one such region may be seen from about 02:07 UT to about 02:44 UT. The temperature and flow directions are seen to be more constant and the heat flux both larger and more constant there than elsewhere along the trajectory. These regions of less turbulent flow are usually seen in the mid-magnetosheath between the disturbed region immediately downstream of the bow shock, and the region of waves adjacent to the magnetopause. Another region of low turbulence, observed when the spacecraft was in the high bit-rate mode, is shown in Figure 2. A change in the magnetic field and heat flux directions occurred at 03:17 UT and the plasma became less turbulent, remaining so until 03:35 UT, and during this interval both the density and heat flux decreased. The form of the distribution function changed during these intervals corresponding to the change in density and heat flux as shown in Figure 3; the heat is carried by electrons with energies above 50 eV. The magnetosheath heat flux magnitude, which normally lies between 5×10^{-3} and 5×10^{-2} ergs $cm^{-2}\ s^{-1}$, and has a mean value for the present observations of 1.7×10^{-2} ergs $cm^{-2}\ s^{-1}$, is considerably larger than the interplanetary heat flux of $\sim 8 \times 10^{-3}$ ergs $cm^{-2}\ s^{-1}$ (Feldman *et al.*, 1975). Striking a particle energy balance across the bow shock requires extensive detailed simultaneous measurements of solar wind and upstreaming ions and electrons, and of protons and electrons in the magnetosheath, and this cannot be carried out here. However, inserting average values into the equation

$$[\text{K.E. flux} + \text{Enthalpy flux} + Q_p + Q_e] = 0$$

indicates that approximate balance can be obtained to ~20% accuracy, assuming that Q_e dominates over Q_p in the sheath as it does in the solar wind.

The electron temperature anisotropy usually changes from less than unity to greater than unity when the magnetopause is crossed, Figure 1, and a density reduction, present in 25 of the 48 traversals examined to date, is clearly seen as

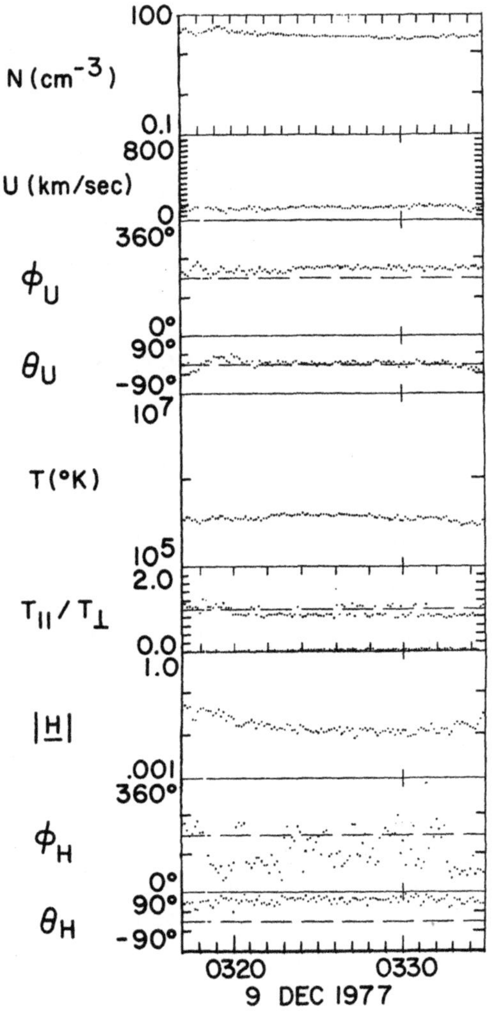

Fig. 2. A period of low turbulence observed on December 9, 1977, between 03:17 and 03:35 UT.

the magnetopause is approached. At the magnetopause crossing, in this example situated 37.5 deg from the subsolar point in the ecliptic plane and 25 deg above it, the flow direction changes in qualitative agreement with the presence of a boundary flow.

The flow speed U has been calculated everywhere from the relation,

$$U = \frac{\int f_e(\bar{v})\bar{v}v^2 \, d\omega \, dv}{\int f_e(\bar{v})v^2 \, d\omega \, dv} \tag{1}$$

and a discussion of factors influencing this determination is in order. The value of U depends upon the absolute differences between the counting rates of the

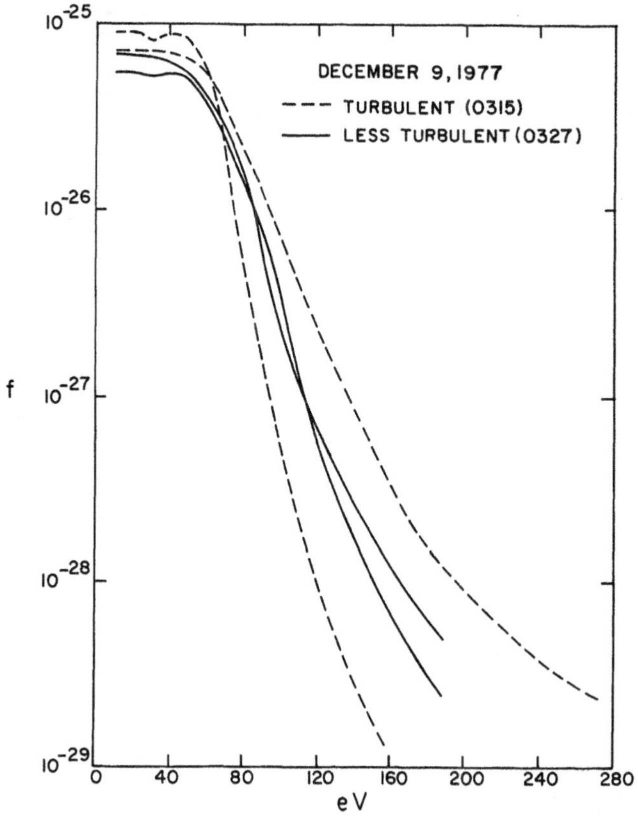

Fig. 3. Distribution functions observed during the period shown in Figure 2, along and perpendicular to the *average* magnetic field direction.

detectors in opposition in the two triads; such differences can arise because of detector area and efficiency differences, but these have been successfully controlled in this experiment. Certain detectors in the triad show a small effect at some deflection potentials due to photoelectrons produced inside the energy analyzer by sunlight scattered from the collimator, although the Sun does not shine directly into any of the analyzers at any time. This is negligible in the solar wind and magnetosheath, but, if uncorrected, would lead to the appearance of spurious bulk speeds in low density plasmas, such as occur in the magnetosphere and magnetotail. A deflection-potential and aspect-angle dependent correction, based upon backgrounds determined deep in the magnetosphere, is applied to remove these effects.

There is also the question of successfully evaluating the density, the denominator in the ratio (1) above, in the presence of spacecraft potential. The energy given to the incoming electrons while falling through the photoelectron sheath around the spacecraft may be comparable to or greater than the energy of bulk motion of the electrons. The ISEE spacecraft were constructed in such a way

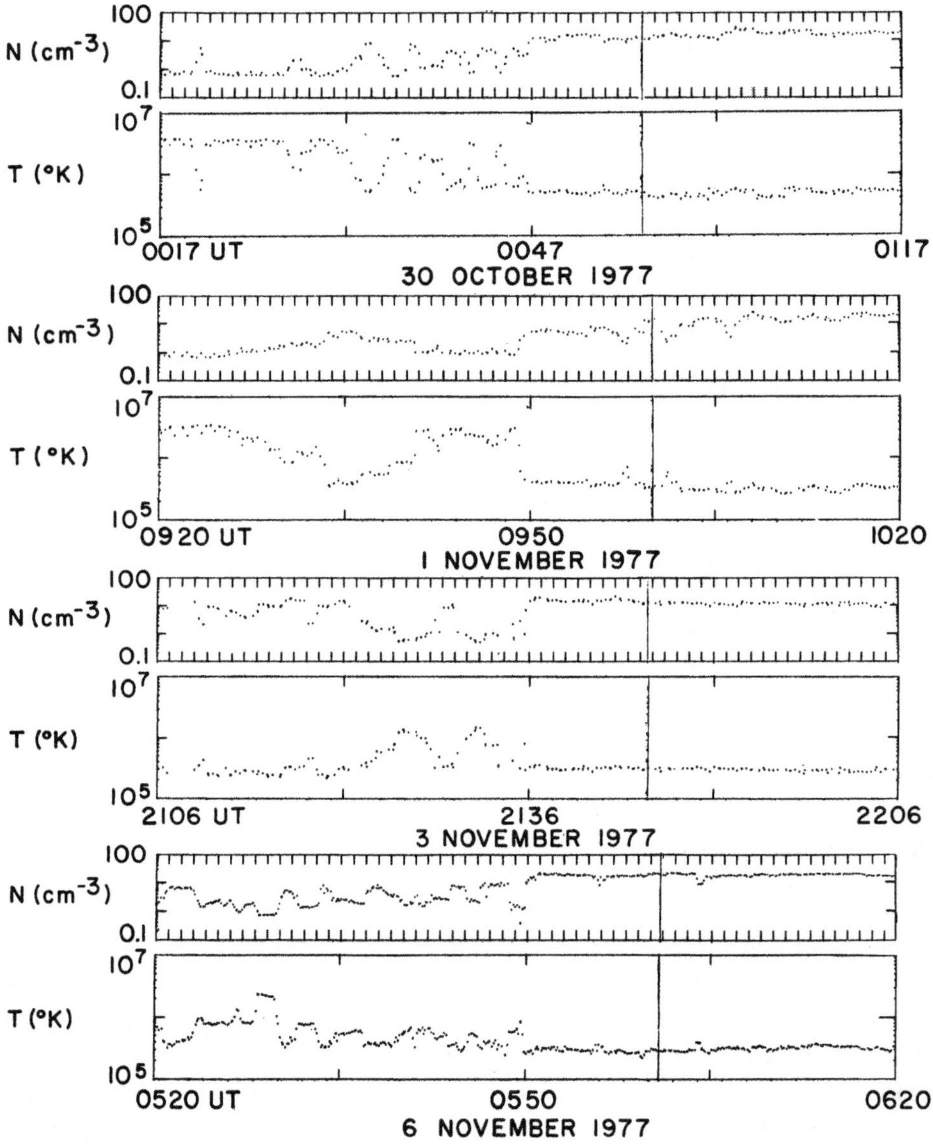

Fig. 4. Four examples of observations of crossings of the magnetopause. The magnetosphere is on the left in the diagram.

as to make their surfaces approximate equipotentials, thus avoiding the appearance of a potential difference between the dark and sunlit sides. Thus, although the spacecraft doubtless rides up and down in potential in response to incident plasma flux (Montgomery *et al.*, 1973), the bulk speed and temperature determinations should not be affected, provided the apparent density can be successfully corrected to the ambient density, since the photoelectron sheath should be symmetric. Due to the low potentials normally present there, the observations in the solar wind and magnetosheath can be readily reduced, but in the magnetosphere, especially at higher lattitudes and in the magnetotail, each period of data must be treated individually.

In Figure 4 we see four further sets of magnetopause crossing observations; the magnetosheath is characterized by densities of 10–30 and temperatures of $3–4\times 10^5$ K and is on the right in the figure. The outer magnetosphere typically has much higher temperatures and lower densities; $kT_{e/n}$ changes from ~ 2 to ~ 200 in crossing the magnetopause. The upper two crossings show density reductions adjacent to the boundary, and significant wave activity (Crooker *et al.*, 1977; Crooker and Siscoe, 1977). Between 10:05 and 10:20 UT on November 1 (third and fourth panels) we can see quasi-sinusoidal oscillations with n and T in phase (but out of phase with B). Multiple magnetopause transitions are very common, and these frequently have, in the electron data, the appearance of transitions between two definite stable states. This is consistent with the motion of a surface carrying waves whose amplitude is large compared to its thickness as suggested for the magnetosphere by (Aubry *et al.*, 1971).

4. Discussion

A prominent feature of the magnetosheath is the density reduction which occurs adjacent to the magnetopause on about half of the traversals. This effect has been studied by Cummings and Coleman (1968) and Freeman *et al.* (1968), and modeled by Zwan and Wolf (1976). Observations by the plasma analyzer and magnetometer on IMP-6, analyzed by Crooker *et al.* (1977), exhibited such a density decrease in eleven of the seventeen crossings studied. It is of interest because it is a hydromagnetic effect (models without a magnetic field predict a density maximum at the magnetopause), and because the process is inconsistent with the occurrence of large-scale magnetic merging. Crooker and Siscoe (1977) have pointed out that this effect is a likely cause for a pressure anisotropy, in the sense of $P_\parallel < P_\perp$, in the dayside magnetosheath. The magnetosheath crossing shown in Figure 1 is a typical sample from the present measurements, and it can be seen that the electron temperature anisotropy $A(=T_\parallel/T_\perp)$ usually lies between 0.8 and 1.2. During the quiet periods it is characteristically less than unity, and in the region of waves adjacent to the magnetopause it is usually greater than unity. In the magnetosheath, the ions make a somewhat greater contribution to the pressure than do the electrons, so there is no contradiction in the present observations to the picture of Crooker and Siscoe (1977) of an inner magnetosheath unstable to the mirror instability over a large portion of the dawn sector.

Density reductions were observed adjacent to the magnetopause in 25 out of 48 magnetosheath traversals and we can compare their magnitude with the model of Zwan and Wolf (1976). Table I shows parameters for 15 cases for which sufficient information has been obtained to make an approximate quantitative test of the

TABLE I

Measured parameters for 15 magnetopause crossings showing density depletions adjacent to the magnetopause. DD is the decimal day of year, HH the hour, Z and Y are specified in R_e, B_T and B_z (the interplanetary magnetic field and its Z component) in γ, and D in km.

DD	HR	Z^2+Y^2	B_T	B_Z	M_A	D	$\cos^{-1} X/R$
302	0047	10.00	9.4	+2.1	6.0	1800	62.5
304	0949	10.35	5.6	0.9	9.6	1920	65.7
306	0804	5.94	3.0	−0.5	15.0	742	26.2
308	1716	5.44	8.3	3.2	5.4	1700	26.3
313	2239	11.08	8.6	4.2	6.3	400	73.6
315	2300	4.13	10.0	–	8.5	650	26.0
316	0825	10.55	6.5	–	7.5	1890	77.6
320	1524	6.99	5.5	–	12	970	34
327	1947	8.15	5.3	−2.0	8.4	1440	40.0
332	1509	6.42	2.5	−2.0	19	340	34.5
337	0820	9.29	4.5	–	13.5	580	47.8
338	0123	15.9	7.5	–	7.9	700	88
339	1844	8.94	5.0	0.5	9.8	850	50
349	0645	12.00	7.5	–	6.0	1400	60
351	1639	11.44	5.5	6.0	8.0	1360	61

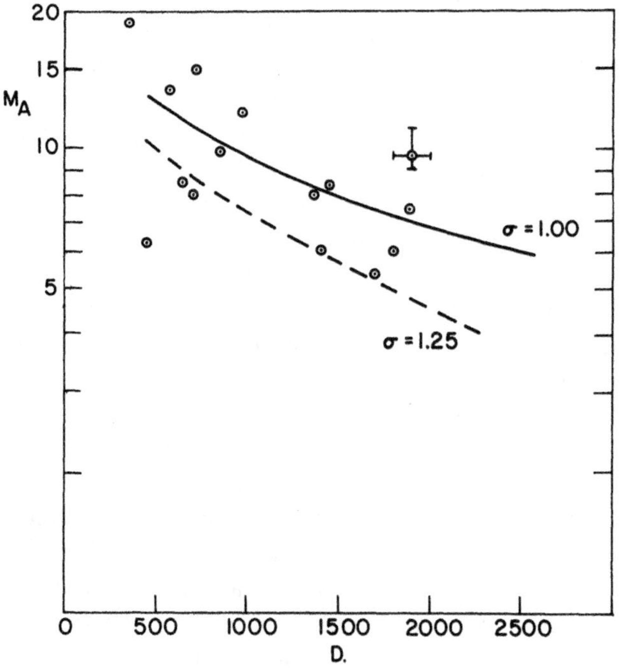

Fig. 5. Measured values of D, the thickness of the depletion layer adjacent to the magnetopause, plotted against the solar wind Alfvén mach number determined at the nearest shock crossing.

relation between D, the effective thickness of the layer of reduced density and the Alfvén mach number of the solar wind. This relation predicts that the higher the mach number characterizing the solar wind flow, the thinner the layer of reduced density will be. D is defined as the distance from the magnetopause to the point where the density is half of the eventual value. The mach number is that obtained from the observations made in the solar wind a short distance upstream of the bow shock. Since the magnetopause is in motion, and all the necessary quantities change with time, we cannot expect to achieve a highly precise test, but approximate agreement and the correct slope of the dependence will increase our confidence in the correctness of the physical mechanism suggested to explain the effect. Figure 5 displays the data from Table I, together with the prediction of Zwan and Wolf. The theoretical curves marked $\sigma = 1.00$ and $\sigma = 1.25$ show the

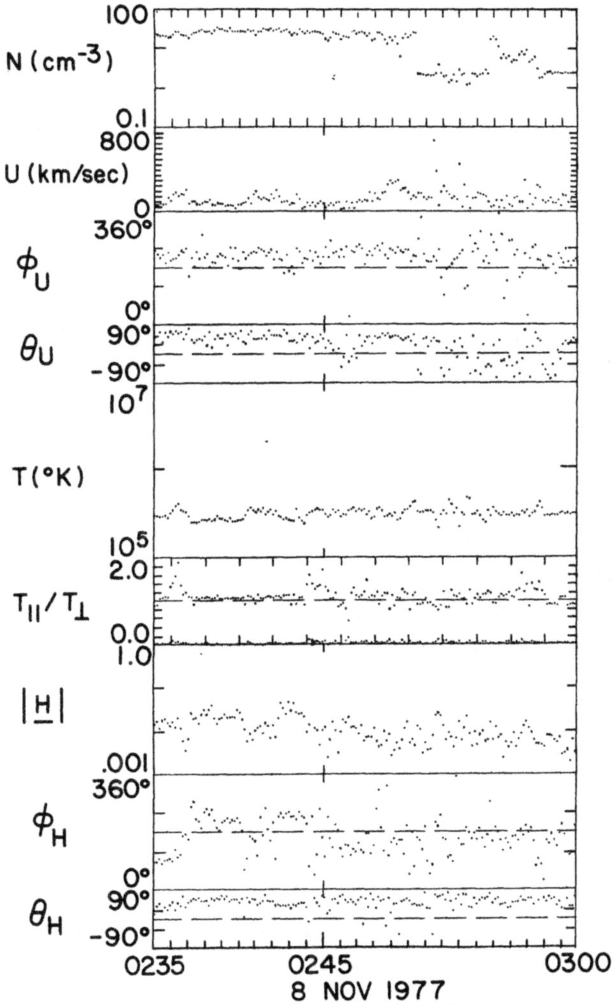

Fig. 6. A magnetopause crossing which did not show a depletion layer.

predicted effect of magnetic tension. The value $\sigma = 1.00$ represents the case in which the magnetic field line approximates a great circle on the magnetopause and the tension has no effect in accelerating the flux tube along the boundary, $\sigma = 1.25$ represents the case where there is a component along the boundary. There is considerable scatter among the points in Figure 5, and this is comparable to the distance between the two theoretical curves, but both the order of magnitude of D and the slope of its variation with the mach number are in agreement with the theory. Since as a result of the average direction of the interplanetary magnetic field, the dawnside shock tends to be parallel and therefore more often unstable or pulsating than the shock on the duskside, data from the dusk magnetosheath may improve the precision of this work when it becomes available, since the same phenomenon should occur on each side of the noon meridian.

Figure 6 shows an example of a magnetopause crossing not showing a density reduction. The crossing takes place abruptly, although there is an indication of wave motion in the three-minute period immediately before the crossing. Examples can be found where there are multiple crossings of the magnetopause during a magnetosheath passage and some of these exhibit density reductions, while the others do not.

5. Conclusions

As a result of this preliminary survey of electron data obtained in the dawn magnetosheath it is hardly appropriate to draw other than tentative conclusions, with the understanding that they must be confirmed by future work. When that is understood, one can put down some features of these observations which appear to be correct and also to be relevant to understanding of the magnetosheath and magnetopause.

(1) The flow is not uniformly turbulent all over the magnetosheath, regions of quiet flow often occurring sandwiched between turbulence associated with the two boundaries, the bow shock and the magnetopause.

(2) The average electron heat flux in the magnetosheath is about twice that in the solar wind and is directed away from the bow shock.

(3) Magnetopause crossings observed by the behavior of the low energy electrons often have the appearance of rather sharp transitions between two definite states. Although this is not always observed, it indicates that the magnetopause is often a thin boundary in motion or carrying large amplitude surface waves.

(4) The density reduction, predicted as a result of plasma flowing out of flux tubes as they are convected up against the magnetopause, is observed to occur for about 50% of magnetosheath traversals. The width of this depletion layer, and its variation with solar wind mach number, are approximately as predicted by theory. Large amplitude waves are observed in the depletion layer. None of these

features of the data necessarily occur in a given crossing, but all are frequently observed. Since it takes approximately four hours for the spacecraft to cross the magnetosheath it is not to be expected that Figure 1, for example, represents anything like a time-stationary situation. The present observations do, however, provide support for the theoretical models of Zwan and Wolf (1976) and Crooker and Siscoe (1977). Since the presence of a depletion layer is inconsistent with large scale reconnection, these observations favor the idea of the predominance of locally confined merging processes, as suggested by Haerendel *et al.* (1978).

Acknowledgements

The authors wish to thank Dr C. T. Russell for his cooperation in supplying magnetic field data in advance of publication. There are many others to whom we are indebted for assistance, principal among them being Mr H. Doong and Mr F. Ottens of GSFC for their work on the instrument and data analysis programs respectively.

References

Aubry, M. P., Kivelson, M. G., and Russell, C. T.: 1971, 'Motion and Structure of the Magnetopause', *J. Geophys. Res.* **76,** 1673.

Crooker, N. U. and Siscoe, G. L.: 1977, 'A Mechanism for Pressure Anistropy and Mirror Instability in the Dayside Magnetosheath', *J. Geophys. Res.* **82,** 185.

Crooker, N., Eastman, T., and Stiles, G. S.: 1979, 'Observations of Plasma Depletion in the Magnetosheath at the Dayside Magnetopause', *J. Geophys. Res.*, in press.

Cummins, W. D. and Coleman, P. J.: 1968, 'Magnetic Fields in the Magnetopause and Vicinity at Synchronous Orbit', *J. Geophys. Res.* **73,** 5699.

Feldman, W. C., Asbridge, J. R., Bame, S. J., and Montgomery, M. D.: 1973, 'Solar Wind Heat Transport in the Vicinity of the Earth's Bow Shock', *J. Geophys. Res.* **78,** 3697.

Feldman, W. C., Asbridge, J. R., Bame, S. J., Montgomery, M. D., and Gary, S. P.: 1975, 'Solar Wind Electrons', *J. Geophys. Res.* **80,** 4181.

Freeman, Jr., J. W., Warren, C. S., and McGuire, J. J.: 1968, 'Plasma Flow Directions at the Magnetopause on January 13 and 14, 1967', *J. Geophys. Res.* **73,** 5719.

Haerendel, G., Paschmann, G., Sckopke, N., Rosenbauer, H., and Hedgecock, P. C.: 1978, 'The Front Side Boundary Layer of the Magnetosphere and the Problem of Reconnection', *J. Geophys. Res.* **83,** 3195.

Montgomery, M. D., Asbridge, J. R., Bame, S. J., and Hones, E. W.: 1973, 'Low Energy Electron Measurements and Spacecraft Potential, Vela-5 and Vela-6', in R. J. L. Grard (ed)., *Proton and Particle Interactions with Surfaces in Space*, D. Reidel Publ. Co., Dordrecht, p. 247.

Ogilvie, K. W., Scudder, J. D., and Doong, H., 1978, 'The Electron Spectrometer Experiment on ISEE-1', *IEEE Trans. Geosci. Electr.* **16,** 261.

Scudder, J. D., Lind, D. L., and Ogilvie, K. W.: 1973, 'Electron Observations in the Solar Wind and Magnetosheath', *J. Geophys. Res.* **78,** 6535.

Zwan, B. T. and Wolf, R. A.: 1976, 'Depletion of Solar Wind Plasma Near a Planetary Boundary', *J. Geophys. Res.* **81,** 1636.

GPSR Compliance

The European Union's (EU) General Product Safety Regulation (GPSR) is a set of rules that requires consumer products to be safe and our obligations to ensure this.

If you have any concerns about our products, you can contact us on ProductSafety@springernature.com

In case Publisher is established outside the EU, the EU authorized representative is:

Springer Nature Customer Service Center GmbH
Europaplatz 3
69115 Heidelberg, Germany

Batch number: 10809436

Printed by Printforce, the Netherlands